I0055503

Plant Nematology

3rd Edition

Dedication

This edition is dedicated to alumni from the Ghent University MSc Nematology course who have gone on to develop Nematology programmes of their own across the world.

Front Cover Images

1. Female of the rice root-knot nematode, *Meloidogyne graminicola*, on rice roots, with egg mass, stained pink with acid fuchsin. Courtesy of Zobaida Lahari. From: Lahari, Z., Ullah, C., Kyndt, T., Gershenzon, J. and Gheysen, J. (2019) Strigolactones enhance root-knot nematode (*Meloidogyne graminicola*) infection in rice by antagonizing the jasmonate pathway. *New Phytologist* 224, 454–465. DOI: 10.1111/nph.15953. OA licence https://creativecommons.org/licenses/by/4.0/
2. Confocal image of a second-stage juvenile of *Globodera pallida* stained with PKH26, migrating through cells of a potato root 6 days post-infection. Courtesy of Dr Vivian Blok and Dr Slawek Janakowski, The James Hutton Institute, UK.

Plant Nematology

3rd Edition

Edited by

Roland N. Perry
School of Life and Medical Sciences, University of Hertfordshire, UK, and Biology Department, Ghent University, Belgium

Maurice Moens
Flanders Research Institute for Agriculture, Fisheries and Food (ILVO), Merelbeke, Belgium, and Laboratory for Agrozoology, Ghent University, Belgium

John T. Jones
Cell and Molecular Sciences Department, James Hutton Institute, UK, and School of Biology, University of St Andrews, UK, and Biology Department, Ghent University, Belgium

CABI is a trading name of CAB International

CABI
Nosworthy Way
Wallingford
Oxfordshire OX10 8DE
UK

Tel: +44 (0)1491 832111
E-mail: info@cabi.org
Website: www.cabi.org

CABI
200 Portland Street
Boston
MA 02114
USA

Tel: +1 (617)682-9015
E-mail: cabi-nao@cabi.org

A catalogue record for this book is available from the British Library, London, UK.

Library of Congress Cataloging-in-Publication Data

Names: Perry, Roland N., editor. | Moens, Maurice, editor. | Jones, John (John Trefor), editor.
Title: Plant nematology / edited by Roland N. Perry, Maurice Moens, John T. Jones.
Description: Third edition | Boston, MA : CABI, [2024] | Includes bibliographical references and index. | Summary: "Provides a broad introduction to plant -parasitic nematodes,, including biological and chemical control. This 3rd edition updated to include molecular aspects, the application of genomics, and nematode ecology"-- Provided by publisher.
Identifiers: LCCN 2024010437 (print) | LCCN 2024010438 (ebook) | ISBN 9781800622425 (hardback) | ISBN 9781800622449 (ebk) | ISBN 9781800622456 (epub)
Subjects: LCSH: Plant nematodes. | Nematode-plant relationships. | Nematode diseases of plants.
Classification: LCC SB998.N4 P56 2024 (print) | LCC SB998.N4 (ebook) | DDC 632/.6257--dc23/eng/20240416
LC record available at https://lccn.loc.gov/2024010437
LC ebook record available at https://lccn.loc.gov/2024010438

ISBN-13: 9781800622425 (hardback)
9781800622449 (ePDF)
9781800622456 (ePub)

DOI: 10.1079 9781800622456.0000

Commissioning Editor: Rebecca Stubbs
Editorial Assistant: Emma McCann
Production Editor: Rosie Hayden

Typeset by Straive, Pondicherry, India

Contents

The Editors

Roland N. Perry

Professor Roland Perry is Senior Visiting Research Scientist at The University of Hertfordshire, UK, with teaching and supervisory links to Ghent University, Belgium, where he is Guest Professor. He graduated with a BSc (Hons) in Zoology from Newcastle University, UK, where he also obtained a PhD in Zoology on physiological aspects of desiccation survival of *Ditylenchus* spp. After a year's postdoctoral research at Newcastle, he moved to Keele University, UK, where he taught Parasitology; after 3 years at Keele, he was appointed to Rothamsted Research. He moved from Rothamsted to the University of Hertfordshire in 2012. His research interests have centred primarily on plant-parasitic nematodes, especially focusing on nematode hatching, sensory perception, behaviour and survival physiology, and several of his past PhD and post-doctoral students are currently involved in nematology research.

He co-edited *The Physiology and Biochemistry of Free-living and Plant-parasitic Nematodes* (1998), *Root-knot Nematodes* (2009), *Molecular and Physiological Basis of Nematode Survival* (2011), *Cyst Nematodes* (2018), *Techniques for Work with Plant and Soil Nematodes* (2021) and the first (2006) and second (2013) editions of this textbook, *Plant Nematology*. He is author or co-author of over 40 book chapters and refereed reviews and over 120 refereed research papers. He is co-Editor-in-Chief of *Nematology* and Chief Editor of the *Russian Journal of Nematology*. He co-edits the book series *Nematology Monographs and Perspectives*. In 2001, he was elected Fellow of the Society of Nematologists (USA) in recognition of his research achievements; in 2008 he was elected Fellow of the European Society of Nematologists for outstanding contribution to the science of Nematology; in 2011 he was elected Honorary Member of the Russian Society of Nematologists; and in 2022 he was elected Honorary Member of the Association of Applied Biologists.

Maurice Moens

Professor Maurice Moens is Honorary Director of Research at the Institute for Agriculture and Fisheries Research (ILVO) at Merelbeke, Belgium and honorary professor at Ghent University, Belgium where he gave a lecture course on Agro-nematology at the Faculty of Bioscience Engineering. He is a past-director of the Post Graduate International Nematology Course (MSc Nematology) and coordinator of the Erasmus Mundus – European Master of Science in Nematology, where he gave five lecture courses on Plant Nematology. The MSc course is organised in the Faculty of Sciences of Ghent University.

He graduated as an agricultural engineer from Ghent University and obtained a PhD at the same University on the spread of plant-parasitic nematodes and their management in hydroponic cropping systems. Within the framework of the Belgian Cooperation, he worked from 1972 to 1985 as a researcher in crop protection, including nematology, at two research stations in Tunisia. Upon his return to Belgium, he was appointed as senior nematologist at the Agricultural Research Centre (now ILVO). There, he expanded the research in plant nematology over various areas covering molecular characterization, biology of host–parasite relationships, biological control, resistance and other forms of non-chemical control. He was appointed head of the Crop Protection Department in 2000 and became Director of Research in 2006. He retired from both ILVO and Ghent University in 2012 but continued to supervise PhD students until 2017. In 2001 he was elected Fellow of the Society of Nematologists (USA) for outstanding contributions to Nematology; in the same year he was elected Fellow of the European Society of Nematologists for outstanding contribution to the science of Nematology; and in 2012 he was elected Honorary Fellow of the Chinese Society for Plant Nematology. He supervised 23 PhD students, who now are active in nematology all over the world. He was president of the European Society of Nematologists (2010–2014). He co-edited *Root-knot Nematodes* (2009), *Cyst Nematodes* (2018) and the first (2006) and second (2013) editions of this textbook, *Plant Nematology*. He is author or co-author of ten book chapters and refereed reviews and over 150 refereed research papers. He is a member of the editorial board of *Russian Journal of Nematology*.

John T. Jones

Professor John Jones is head of the Cell and Molecular Sciences Department at The James Hutton Institute and holds a joint appointment as Professor of Biology at The University of St Andrews. He is also a Guest Professor at Ghent University, Belgium. John graduated with a BSc (Hons) in Zoology from Newcastle University and obtained a PhD from University College of Wales, Aberystwyth, on the structure and function of nematode sense organs. This project was undertaken jointly with Rothamsted Research. Following a 2-year postdoctoral position at the University of South Carolina, John moved to The Scottish Crop Research Institute in 1993 and has remained there since then. The Scottish Crop Research Institute merged with the Macauley Land Use Research Institute in 2011 to form The James Hutton Institute. John's research is focused on understanding the molecular basis of the interactions between plant-parasitic nematodes and their hosts and has included extensive analysis of nematode genomes and transcriptomes. John is author or co-author of over 100 refereed research papers and has co-edited three books. John is a member of the editorial board of *Nematology* and is currently the elected vice president of the International Federation of Nematology Societies.

Contributors

Antonio Archidona-Yuste, Institute for Sustainable Agriculture, Department of Crop Protection, 14004 Córdoba, Spain
e-mail: antonio.archidona@ias.csic.es
ORCID: 0000-0003-4446-0642

Matthew A. Back, Department of Agriculture and Environment, Harper Adams University, Newport, Shropshire TF10 8NB, UK
e-mail: mback@harper-adams.ac.uk
ORCID: 0000-0003-4774-5491

Thomas H. Been, Wageningen University and Research Centre, Plant Research, PB 9101, 6700 HB Wageningen, The Netherlands
e-mail: thomas.been@wur.nl
ORCID: 0000-0001-8706-2017

Stéphane Bellafiore, PHIM Plant Health Institute, Université de Montpellier, IRD, CIRAD, INRAE, Institut Agro, 911 Av. Agropolis, 34394 Montpellier, France
e-mail: stephane.bellafiore@ird.fr
ORCID: 0000-0002-4394-0866

Wim Bert, Department of Biology, Ghent University, Ledeganckstraat 35, 9000 Ghent, Belgium
e-mail: wim.bert@UGent.be
ORCID: 0000-0002-5864-412X

Pablo Castillo, Institute for Sustainable Agriculture, Department of Crop Protection, 14004 Córdoba, Spain
e-mail: pcastillo@csic.es
ORCID: 0000-0003-0256-876X

David J. Chitwood, *formerly* Mycology and Nematology Genetic Diversity and Biology Laboratory, Beltsville, MD 20705, USA
e-mail: chitworm@gmail.com
ORCID: 0000-0002-2440-1365

Abiodun O. Claudius-Cole, Department of Crop Protection and Environmental Biology, University of Ibadan, Ibadan, Nigeria
e-mail: bi_cole@yahoo.com
ORCID: 0000-0003-1206-6325

Sofia R. Costa, CBMA – Centre of Molecular and Environmental Biology, University of Minho, Campus de Gualtar, 4710-057 Braga, Portugal
e-mail: sofia.costa@bio.uminho.pt
ORCID: 0000-0003-0100-1518

Danny L. Coyne, International Institute of Tropical Agriculture (IITA), Grosvenor House, 125 High Street, Croydon CR0 9XP, UK
e-mail: d.coyne@cgiar.org
ORCID: 0000-0002-2030-6328

Keith G. Davies, School of Life and Medical Sciences, University of Hertfordshire, Hatfield AL10 9AB, UK
e-mail: k.davies@herts.ac.uk
ORCID: 0000-0001-6060-2394

Wilfrida Decraemer, Department of Biology, Ghent University, Ledeganckstraat 35, 9000 Ghent, Belgium
e-mail: wilfrida.decraemer@UGent.be
ORCID: 0000-0001-5535-0129

Etienne Geraert, Department of Biology, Ghent University, Ledeganckstraat 35, 9000 Ghent, Belgium
e-mail: etienne.geraert@UGent.be
ORCID: 0000-0001-7798-7307

Godelieve Gheysen, Department of Biotechnology, Ghent University, Coupure links 653, 9000 Ghent, Belgium
e-mail: godelieve.gheysen@UGent.be
ORCID: 0000-0003-1929-5059

Johannes Hallmann, Institute for Epidemiology and Pathogen Diagnostics, Julius Kühn Institute, Federal Research Centre for Cultivated Plants, Messeweg 11/12, 38104 Braunschweig, Germany
e-mail: johannes.hallmann@julius-kuehn.de
ORCID: 0000-0002-3386-0600

Martin C. Hare, Crop and Environment Research Centre, Harper Adams University, Newport, Shropshire TF10 8NB, UK
e-mail: mhare@harper-adams.ac.uk
ORCID: 0000-0002-7269-5054

Sue Hockland, Independent Plant Nematology Consultant, Harrogate, North Yorkshire, UK
e-mail: sue.hockland@btinternet.com

Lindy Holden-Dye, School of Biological Sciences, University of Southampton, University Road, Southampton SO17 1BJ, UK
e-mail: l.m.holden-dye@soton.ac.uk
ORCID: 0000-0002-9704-1217

David J. Hunt, CABI Europe-UK, Bakeham Lane, Egham TW20 9TY, UK
e-mail: tabanids.bite@gmail.com

Renato N. Inserra, Florida Department of Agriculture and Consumer Services, Division of Plant Industry, P.O. Box 147100, Gainesville, FL 32614-7100, USA
e-mail: renato.inserra@fdacs.gov

John T. Jones, Cell and Molecular Sciences Department, James Hutton Institute, Invergowrie, Dundee DD2 5DA, UK
e-mail: john.jones@hutton.ac.uk
ORCID: 0000-0003-2074-0551

Lisa M. Kohl, U.S. Department of Agriculture, Animal and Plant Health Inspection Service, 4700 River Road, Riverdale, MD 20737, USA
e-mail: lisa.m.kohl@usda.gov
ORCID: 0009-0002-0106-966X

Catherine Lilley, Centre for Plant Sciences, University of Leeds, Leeds LS2 9JT, UK
e-mail: c.j.lilley@leeds.ac.uk
ORCID: 0000-0001-8958-0272

Maurice Moens, Flanders Research Institute for Agriculture, Fisheries and Food (ILVO), Burg. Van Gansberghelaan 96, 9820 Merelbeke, Belgium
e-mail: maurice.moens@ilvo.vlaanderen.be
ORCID: 0000-0001-7428-9749

Juan E. Palomares-Rius, Institute for Sustainable Agriculture, Department of Crop Protection, 14004 Córdoba, Spain
e-mail: palomaresje@ias.csic.es
ORCID: 0000-0003-1776-8131

Roland N. Perry, School of Life and Medical Sciences, University of Hertfordshire, Hatfield AL10 9AB, UK
e-mail: r.perry2@herts.ac.uk
ORCID: 0009-0003-9169-6612

James A. Price, Cell and Molecular Sciences Department, James Hutton Institute, Invergowrie, Dundee DD2 5DA, UK
e-mail: james.price@hutton.ac.uk
ORCID: 0000-0001-9134-0892

Thomas Prior, Fera Science Limited, York Biotech Campus, Sand Hutton, York YO41 1LZ, UK
e-mail: thomas.prior@fera.co.uk
ORCID: 0009-0006-8823-9420

Corrie H. Schomaker, Quantitative Solutions, Randwijk, Lipperstpad 5, 6668 AW, The Netherlands
e-mail: schomakercorrie@gmail.com

Jason D. Stanley, Florida Department of Agriculture and Consumer Services, Division of Plant Industry, The Doyle Conner Building, 1911 SW 34th St, Gainesville, FL 32608, USA
e-mail: jason.stanley@fdacs.gov

Sergei A. Subbotin, Plant Pest Diagnostics Center, California Department of Food and Agriculture, 3294 Meadowview Road, Sacramento, CA 95832-1448, USA
e-mail: sergei.subbotin@ucr.edu
ORCID: 0000-0001-6648-5889

Misghina G. Teklu, Wageningen University and Research Centre, Plant Research, PB 9101, 6700 HB Wageningen, The Netherlands
e-mail: misghina.goitomteklu@wur.nl
ORCID: 0000-0002-1686-1089

Tim C. Thoden, Corteva Agriscience, Riedenburger Str. 7, 81677 Munich, Germany
e-mail: tim.thoden@corteva.com
ORCID: 0000-0002-9167-7433

Peter E. Urwin, Centre for Plant Sciences, University of Leeds, Leeds LS2 9JT, UK
e-mail: P.E.Urwin@leeds.ac.uk
ORCID: 0000-0003-0398-2372

Nicole Viaene, Flanders Research Institute for Agriculture, Fisheries and Food (ILVO), Plant Sciences Unit, Crop Protection, Burg. Van Gansberghelaan 96, 9820 Merelbeke, Belgium
e-mail: nicole.viaene@ilvo.vlaanderen.be
ORCID: 0000-0002-7709-2578

Wim Wesemael, Flanders Research Institute for Agriculture, Fisheries and Food (ILVO), Plant Sciences Unit, Crop Protection, Burg. Van Gansberghelaan 96, 9820 Merelbeke, Belgium
e-mail: wim.wesemael@ilvo.vlaanderen.be
ORCID: 0000-0001-7960-797X

Denis J. Wright, Department of Life Sciences, Imperial College London, Silwood Park Campus, Ascot, Berkshire SL5 7PY, UK
e-mail: d.wright@imperial.ac.uk
ORCID: 0009-0007-9308-7342

Preface to the First Edition

Plant-parasitic nematodes are of considerable importance worldwide and their devastating effects on crops have major economic and social impacts. Control of these plant pests is imperative and with the banning or limitation of the use of many nematicides, alternative control strategies are required that, in turn, will have to be based on a sound knowledge of nematode taxonomy and biology. Such information is also a basic necessity for effective formulation and implementation of quarantine regulations. Molecular approaches to all aspects of nematology have already made a major contribution to taxonomy and to our understanding of host–parasite interactions, and will undoubtedly become increasingly important.

There have been several excellent specialized texts on plant-parasitic nematodes, aimed primarily at research scientists. However, there is no book on plant-parasitic nematodes aimed at a broader readership, especially one including students specializing in the subject at undergraduate and postgraduate levels. The driving force for this book was the need for a text to support the MSc course in nematology run by Ghent University, Belgium. The students on this course come from a wide spectrum of scientific backgrounds and from many different countries and, after obtaining their degree, will return to their own country to undertake various jobs, including advisory work, statutory and quarantine work, PhD degrees and teaching posts. Many of these students will return to countries where facilities for plant nematology work are basic. Thus, the book needed to provide a wide range of information on plant-parasitic nematodes and needed to be inclusive, appealing to workers from developing and developed countries. An excellent book, edited by John Southey and entitled *Plant Nematology*, provided this type of information but is now very dated and has long been out of print. We have used the general format of Southey's book as a template for the present volume. We hope that, as well as being informative, this book will stimulate interest in plant-parasitic nematodes.

Research on taxonomy, biology, plant–nematode interactions and control has generated an extensive volume of literature. In this book, we have deliberately limited the number of references, although key research papers have been included where these are of major significance. Important book chapters and reviews are cited so that a reader interested in a specific aspect can access these to obtain source references.

We are grateful to all the authors of the chapters for their time and effort in compiling their contributions. In addition, we wish to thank David Hunt (CABI, UK), John Jones (SCRI, UK) and Brian Kerry (Rothamsted Research, UK) for their advice and comments on various chapters, and Bram Moens (Wetteren, Belgium) for preparing some of the figures. This book is primarily for students and the impetus for it came from students; we would like to thank all those whose enthusiasm and interest in plant nematology made this book possible.

Roland Perry and Maurice Moens
May 2005

Preface to the Second Edition

We were delighted with the positive response to the first edition of *Plant Nematology*. Initially aimed at the MSc course in nematology at Ghent University, Belgium, the book has proved popular with students and staff worldwide. The first edition was written in 2005 and published in April the following year. Several chapters now need updating to meet the requirements of current nematology students.

In producing this second edition, we have taken the opportunity to revise all chapters, especially those, such as the molecular chapters, where a wealth of new information has accumulated over the intervening years since the first edition. We are grateful to the authors who have prepared the revised chapters; their time and effort are greatly appreciated. Also, we would like to acknowledge the constructive comments from users of the first edition; in particular, we thank Axel Elling (Washington State University, USA), Rick Davis (North Carolina State University, USA), David Hunt (CABI, UK), John Jones (The James Hutton Institute, UK), Charlie Opperman (North Carolina State University, USA) and Nicole Viaene (ILVO, Merelbeke, Belgium) for their important and useful comments.

The book is aimed at students, to introduce them to the delights and challenges of plant nematology, and the immense economic and social damage done by plant-parasitic nematodes. The need for young, enthusiastic nematologists to tackle the immense problems caused worldwide by plant-parasitic nematodes is paramount. We hope that the enthusiasm of the editors and chapter authors is contagious!

Roland Perry and Maurice Moens
December 2012

Preface to the Third Edition

The textbook *Plant Nematology* was initially conceived as a support volume for the MSc Nematology course at Ghent University, Belgium. It has evolved through two editions to be of use for students and researchers in many universities and institutes worldwide. It is now 10 years since the second edition was published and in many areas of plant nematology there has been considerable progress. The second edition needed updating. John Jones has joined as co-editor to bring his considerable expertise in molecular nematology. We have also taken the opportunity to include a completely new chapter on Ecology. Where possible, we have included new, young co-authors to the list of contributors to reflect the continuum of research and expertise in plant nematology. Many chapters have been expanded to reflect the vast new body of information that has accumulated since the last edition, particularly in relation to molecular aspects, and to management and control systems where there is a new assessment of the various options. The initial concept of *Plant Nematology* included limiting the number of references as much as feasible. However, in this edition this has changed somewhat so more details of references to recent work are included. CABI has also enabled many of the illustrations to be in colour where possible.

As with the previous two editions, this book would not have been possible without the time, dedication and efforts of the chapter authors. We are grateful for their contributions. We also appreciate the considerable assistance of the CABI staff for facilitating the publication of this edition. Their help is gratefully acknowledged.

The book is aimed primarily at students, to convey the spectrum of information about plant nematology and to introduce them to the delights and importance of working with plant-parasitic nematodes. We hope that in the context of the need for global food security and the need to limit the immense financial and social damage caused by plant-parasitic nematodes, this book will enthuse current and future nematologists.

Roland N. Perry, Maurice Moens and John T. Jones
October 2023

Online Supplementary Content

For both Chapter 11 *Plant Growth and Population Dynamics* and Chapter 12 *Distribution Patterns and Sampling*, supplementary files are available on the CABI website (https://www.cabidigitallibrary.org/doi/book/10.1079/9781800622456.0000).

For each example described in these chapters a zip file is available containing a number of files related to the example. Each zip file contains:

- An input file containing data – only when needed for the example.
- An 'R' file, containing the R script to run the described model and/or analysis.
- An .Rmd file, which is the mark down version of the R file.
- A pdf, which is 'knitted' from the .Rmd file. To knit, or convert, a mark down file into Word, PDF or HTML you need to download two additional libraries: library(knitr) and library(tinytex) to the R interface.

R is a freely available software package for statistical computing with graphics. R can be found on the website of *The R Project for Statistical Computing* https://www.r-project.org/. On the Intro page, under NEWS, links are available to download the latest version of R.

Several graphic user interfaces are available to enhance the usability of R, e.g. TinnR, Rstudio and many others. These are also freely available on the internet. They help the R user to get the most advantage of R by supplying user interfaces for scripting, data handling, wrangling and storage, and graphical techniques. We recommend https://posit.co/download/rstudio-desktop

This is all you need to run the .R script examples. The Rmd and pdf files are added for your convenience as they provide the correct result of the R script. There are four examples available for Chapter 11 and six for Chapter 12. Each Example zip is named as described in the text with the prefix of the Chapter it belongs to. An example:

'Chapter 11 Glasshouse.zip' contains

- DataGlasshouse.txt
- Glasshouse.R
- Glasshouse.Rmd
- Glasshouse.pdf

Questions about the R-scripts and their use can be addressed to the authors.

Why Rmd and pdf?

The PDF files are produced using R and Markdown (https://www.markdownguide.org/) an easy-to-use markup language that is used with plain text to add formatting elements (headings, text, bulleted lists, URLs) without the use of a formal text editor. Combined with R, markdown makes it possible to integrate formatted R code with output like graphs and tables into one document that is still responsive to R code changes (Rmd files). To create documents like Word, PDF or HTML for a web page, add the libraries knitr and tinytex to RStudio after which the Rmd is used to produce these documents.

Corrie Schomaker (schomakercorrie@gmail.com) and Thomas Been (thomas.been@wur.nl)

Part I Taxonomy, Systematics and Principal Genera

Knowledge of nematode morphology and life-cycle biology underpins all aspects of research, advisory work, implementation of quarantine legislation and selection of management strategies. Traditional, descriptive taxonomy is now routinely accompanied by molecular diagnostics, with the two approaches complementing one another. Molecular techniques may even supplant traditional identification methods. Sometimes this is because of a paucity of expert classical taxonomists, although where precision and rapid, reliable diagnostics are required, molecular techniques are replacing classical identification. This is particularly true for root-knot and cyst nematodes. Whilst the endoparasitic root-knot and cyst nematodes are the most devastating plant-parasitic groups worldwide, several species in the migratory endoparasitic and ectoparasitic groups of nematodes are also of considerable economic and social importance.

The chapters in Part I are intended to reflect all of these aspects by presenting the basic structure of nematodes and their functioning, followed by a chapter on molecular taxonomy, systematics and phylogeny. The subsequent four chapters focus on the major groups of plant-parasitic nematodes, presenting information on their morphology, taxonomy, basic biology and management.

Throughout this book, the systematic scheme follows the higher classification (i.e. family level and above) of De Ley and Blaxter (2002).

Structure and Classification*

WILFRIDA DECRAEMER[1]** AND DAVID J. HUNT[2]

[1]*Department of Biology, Ghent University, Belgium;* [2]*CABI Europe-UK, UK*

1.1. Introduction

Nematodes are pseudocoelomate, unsegmented worm-like animals, commonly described as filiform or thread-like, a characteristic reflected by the taxon name *nema* (Greek, *nema* = thread) and its nominative plural *nemata*. Zoologically speaking, the vernacular word 'nematode' is a corruption for the order name Nematoidea, one of the five historical orders of the class Helminthia, which embraced all thread-like forms or roundworms (gordians and nematodes). At present, nematodes are generally regarded as a separate

*A revision of Decraemer, W. and Hunt, D.J. (2013) Structure and classification. In: Perry, R.N. and Moens, M. (eds) *Plant Nematology*, 2nd edn. CAB International, Wallingford, UK.

**Corresponding author: wilfrida.decraemer@UGent.be

DOI: 10.1079/9781800622456.0001

phylum, the Nematoda or Nemata (De Ley and Blaxter, 2002). The systematic scheme presented is based on the higher classification proposed by De Ley and Blaxter (2002) and has been updated where appropriate to reflect new taxa proposals. However, recent molecular phylogenetic analyses recognize 12 clades within the Nematoda, with plant-parasitic taxa located in the basal clade I (Trichodoridae) and clade II (Longidoridae) and with the Tylenchomorpha in the more evolved clade 12 (Holterman *et al.*, 2006).

Nematodes are the most numerous Metazoa on earth (Ferris *et al.,* 2001). They are either free-living or parasites of plants and animals and although they occur in almost every habitat, they are essentially aquatic animals. Nematodes depend on moisture for their locomotion and active life and therefore soil moisture, relative humidity and other environmental factors directly affect nematode survival. However, many nematodes can survive in an anhydrobiotic state (see Chapter 7). Soil structure is influential as pore size affects the ease with which nematodes can move through the soil interstices (see Chapter 8). In general, sandy soils provide the best environment for terrestrial nematodes but saturated clay soils can be colonized successfully by certain specialized nematodes, including *Hirschmanniella* and some *Paralongidorus*. Soil pH may affect nematodes, but local variations in soil temperature are rarely a particularly important factor (see Chapter 8).

This review of the anatomy of plant-parasitic nematodes will also include mention of some free-living and animal-parasitic species for comparative purposes. Although currently only about 4000 species of plant-parasitic nematodes have been described (i.e. 15% of the total number of nematode species known), their impact on humans by inflicting heavy losses in agriculture is substantial. The maxim that 'where a plant is able to live, a nematode is able to attack it' is a good one. Nematodes are even able to attack the aerial parts of plants provided that the humidity is high enough to facilitate movement. Such conditions are provided in flooded rice fields where foliar species, such as *Aphelenchoides besseyi* and *Ditylenchus angustus*, can be devastating. Some *Bursaphelenchus* species, vectored by wood-boring insects, are found in the trunk of coconut palm or pines, where they locate the vector insects. Other nematodes, such as some *Hirschmanniella* and *Halenchus* spp., attack algae and can live in seawater. It has been estimated that a single acre of soil from arable land may contain as many as 3,000,000,000 nematodes and a single wheat gall formed by *Anguina tritici* typically contains approximately 11,000–18,000 nematodes, although as many as 90,000 have been recorded.

To constrain or even banish this limiting factor in agricultural production, it is vital to identify accurately the nematode pests and to understand their biology. At present, many nematode identifications still rely upon morphological characters, but an integrated approach is becoming increasingly common, including isozyme patterns and DNA sequences, and has become essential for many taxa, including root-knot nematodes (see Chapter 3) and cyst nematodes (see Chapter 4).

Despite their great diversity in lifestyle, nematodes display a relatively conserved body plan. The body consists of an external cylinder (the body wall) and an internal cylinder (the digestive system) separated by a pseudocoelomic cavity filled with fluid under pressure and containing a number of cells and other organs such as the reproductive system. About 99% of all known nematodes have a long, thin cylindrical body shape, which is round in cross-section and tapered towards both ends, although usually more so towards the posterior or tail end. The tail may be short or long and varies in form from broadly rounded to filiform. The tail may also differ in shape between the developmental stages or between sexes.

Nematodes usually crawl or swim with undulating movements operating in a dorso-ventral plane (see Chapter 8). Aberrant body shapes, for example a swollen female body, may indicate either a loss of locomotion, as in cyst nematodes, or be related to an atypical locomotory patterns, as associated with the epsilon-shaped body in epsilonematids. On a solid surface a nematode crawls on its lateral surface except, for example, in the free-living marine families Draconematidae and Epsilonematidae, which move on their ventral surface. Nematodes travel fastest in soil when pore space is about 0.3 times their body length. In plant-parasitic nematodes, all migratory ectoparasites of roots, including all Trichodoridae and Longidoridae and many Tylenchomorpha, are vermiform throughout their life cycle (Figs 1.1 and 1.2). Other, more highly evolved Tylenchomorpha have a sedentary endoparasitic lifestyle, one or more stages inciting specific feeding cells or feeding structures within the root tissue and becoming obese (Fig. 1.1). This type of life cycle is seen in the root-knot and cyst nematodes where the mature female becomes pyriform, globose or lemon-shaped in form.

The body shape, or habitus, assumed by nematodes on relaxation varies from straight through ventrally curved to spiral and this can be a useful diagnostic character, particularly under the stereomicroscope. Free-living and plant-parasitic nematodes are mostly

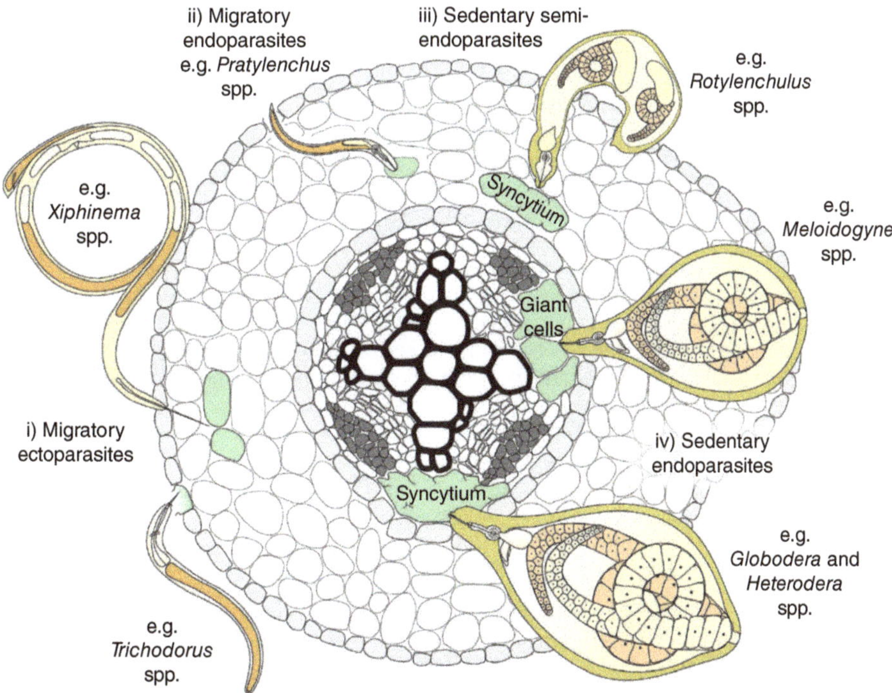

Fig. 1.1. Root in transverse section showing the diverse appearance and feeding modes of plant-parasitic nematodes: migratory ectoparasites (e.g. *Xiphinema*, *Trichodorus*), migratory endoparasites (e.g. *Pratylenchus*), sedentary semi-endoparasites (e.g. *Rotylenchulus*) and sedentary endoparasites (e.g. *Meloidogyne*, *Globodera*, *Heterodera*). Adapted from Eves-van den Akker *et al.* (2021) with permission from Elsevier.

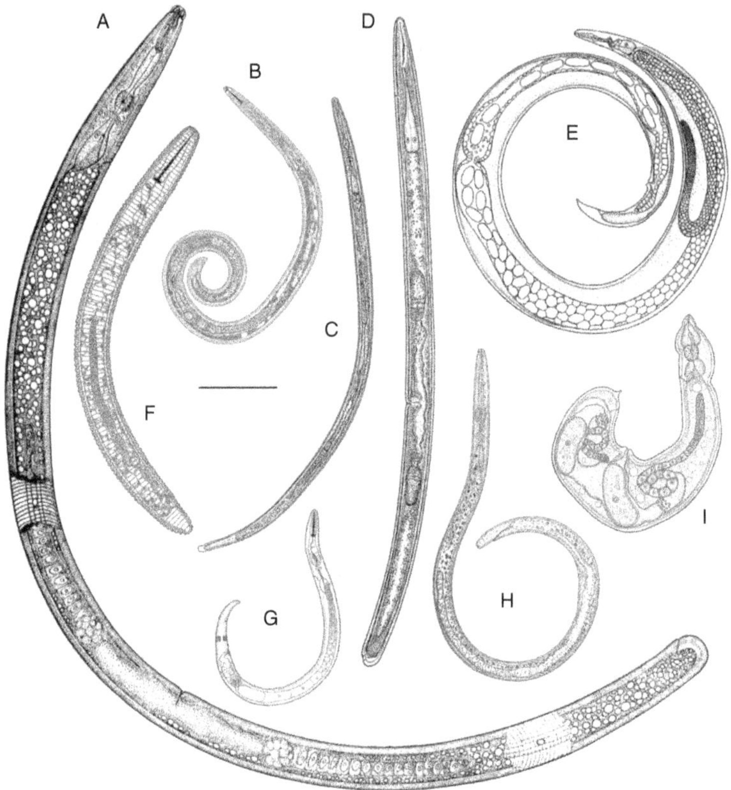

Fig. 1.2. Relative body size and range in form of mature females of some common plant-parasitic nematodes. A: *Hoplolaimus galeatus*. B: *Helicotylenchus dihystera*. C: *Tylenchorhynchus annulatus*. D: *Trichodorus primitivus*. E: *Anguina tritici* (note that, because of its length of 3–5 mm, this nematode is shown at a different scale to the others). F: *Criconemoides xenoplax*. G: *Paratylenchus bukowinensis*. H: *Aphelenchus avenae*. I: *Rotylenchulus reniformis*. Being typically rather long and often extremely slender, no members of the Longidoridae are figured due to problems of scale, but note that the smallest *Xiphinema* is about the same length as the *Hoplolaimus* and that the longest *Paralongidorus* is about six times as long. Figure digitally composed from drawings taken from *CIH Descriptions of Plant-parasitic Nematodes*. (Scale bar: A–D, F–I = 100 μm; E = 250 μm.)

less than 1 mm in length (Fig. 1.2), although some species, particularly in the Longidoridae, may greatly exceed this, *Paralongidorus epimikis*, for example, attaining a maximum length of over 12 mm. Animal-parasitic nematodes can be substantially longer and often achieve lengths of many centimetres, exceptionally even metres. Externally, the body shows little differentiation into sections apart from the tail region. The nematode body can be divided into a dorsal, a ventral and two lateral sectors. The secretory–excretory (S–E) pore, vulva and anus in the female or the cloacal opening in the male are all located ventrally (few exceptions) whereas the lateral regions may be identified by the apertures of the amphids (few exceptions), deirids and phasmids, when present. The mouth opening is usually located terminally at the anterior end. The body displays a bilateral symmetry although the anterior end also shows a radial symmetry. The body wall consists of the body cuticle, epidermis and somatic musculature.

1.2. General Morphology

The variation in size and body form of a selection of typical plant-parasitic nematodes is shown in Fig. 1.2, while the general morphology of a typical tylenchomorph nematode is shown in Fig. 1.3, major organ systems being depicted in relation to one another. Organ systems are usually tubular in form and are suspended within the pseudocoelomic cavity. The following sections deal with the various structures in more detail and also compare and contrast organ systems between tylenchomorph, longidorid and trichodorid nematodes, thereby facilitating diagnostics of the major groups of plant-parasitic nematodes.

1.2.1. Body cuticle

Most nematodes possess a cuticle, although some, such as *Fergusobia*, lack a cuticle in the adult insect-parasitic female. Cuticle structure may be extremely variable (Fig. 1.4; Box 1.1), not only between different taxa, but also intraspecifically between sexes and developmental stages or between different body regions of an individual (Decraemer *et al.*, 2003). The nematode cuticle varies from being simple and thin (Fig. 1.4H) to highly complex and multilayered (Fig. 1.4L). The body cuticle invaginates at the mouth opening, amphids, phasmids, S–E pore, vulva and anus or cloacal opening, forming the lining to the cheilostom, amphidial fovea (canal), terminal canal, part of the terminal duct of the S–E system, *pars distalis vaginae*, cloaca and rectum. As nematodes lack both a skeleton and a circular muscle system and possess a pressurized body cavity, the cuticle functions as an antagonistic system that prevents radial deformation of the body when the longitudinal muscles contract during undulatory locomotion. Initially, the cuticle plays a role in maintaining body shape after elongation of the embryo. The cuticle, together with the epidermis, also functions as a barrier to harmful elements in the environment and, being semipermeable, plays a role in secretion–excretion or in the uptake of various substances.

1.2.1.1. Outer cuticle structure and ornamentation

The body cuticle is often marked by transverse and/or longitudinal striae (Fig. 1.4). The transverse striae range from being very fine, superficial (i.e. restricted to the cortical zone) and close together, to deeper and wider apart, in which case they delimit the annuli. Annuli may either have a round or retrorse outline. Transverse striae are present in Longidoridae and Trichodoridae but are only visible by electron microscopy (Fig. 1.4A and L), being indistinct or difficult to observe by light microscopy. When both transverse and longitudinal striae are present all over the body, the cuticle has a tessellate or chequered appearance. Apart from striae, longitudinal elevations or ridges may also be present (with or without an internal support). Alae are thickened wing-like extensions of cuticle that are often found laterally or sublaterally on the body, but may also be localized in the caudal region of the male where they form the bursa or caudal alae and may play a role

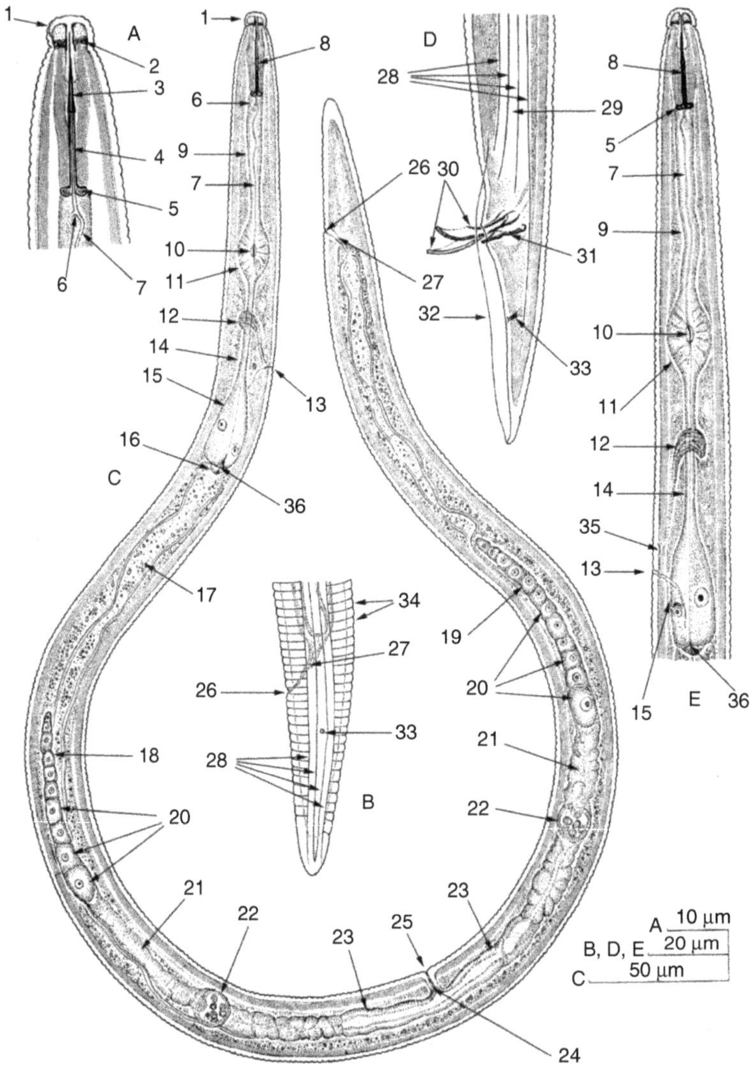

Fig. 1.3. Overview of the morphological structures in a female and male plant-parasitic nematode (*Tylenchorhynchus cylindricus*). A: Female anterior body region. B: Female posterior body region. C: Entire female. D: Male posterior region. E: Pharyngeal region. 1: Anterior body region. 2: Cephalic framework. 3: Stylet cone. 4: Stylet shaft. 5: Stylet knobs. 6: Outlet of dorsal pharyngeal gland. 7: Pharyngeal lumen. 8: Stylet protractor muscles. 9: Procorpus. 10: Valve. 11: Metacorpus. 12: Nerve ring. 13: Secretory–excretory pore. 14: Isthmus. 15: Pharyngeal bulb with gland nuclei: 16: Pharyngeal bulb (abutting intestine, not overlapping). 17: Intestine. 18: Ovary of anterior genital branch. 19: Ovary of posterior genital branch. 20: Developing oocytes in ovaries. 21: Oviduct. 22: Spermatheca with sperm inside. 23: Uterus. 24: Vagina. 25: Vulva. 26: Anus. 27: Rectum. 28: Four lines or incisures in lateral field. 29: Lateral field. 30: Spicules. 31: Gubernaculum. 32: Bursa or caudal alae. 33: Phasmid. 35: Hemizonid. 36: Cardia. Adapted from Siddiqi (1972a).

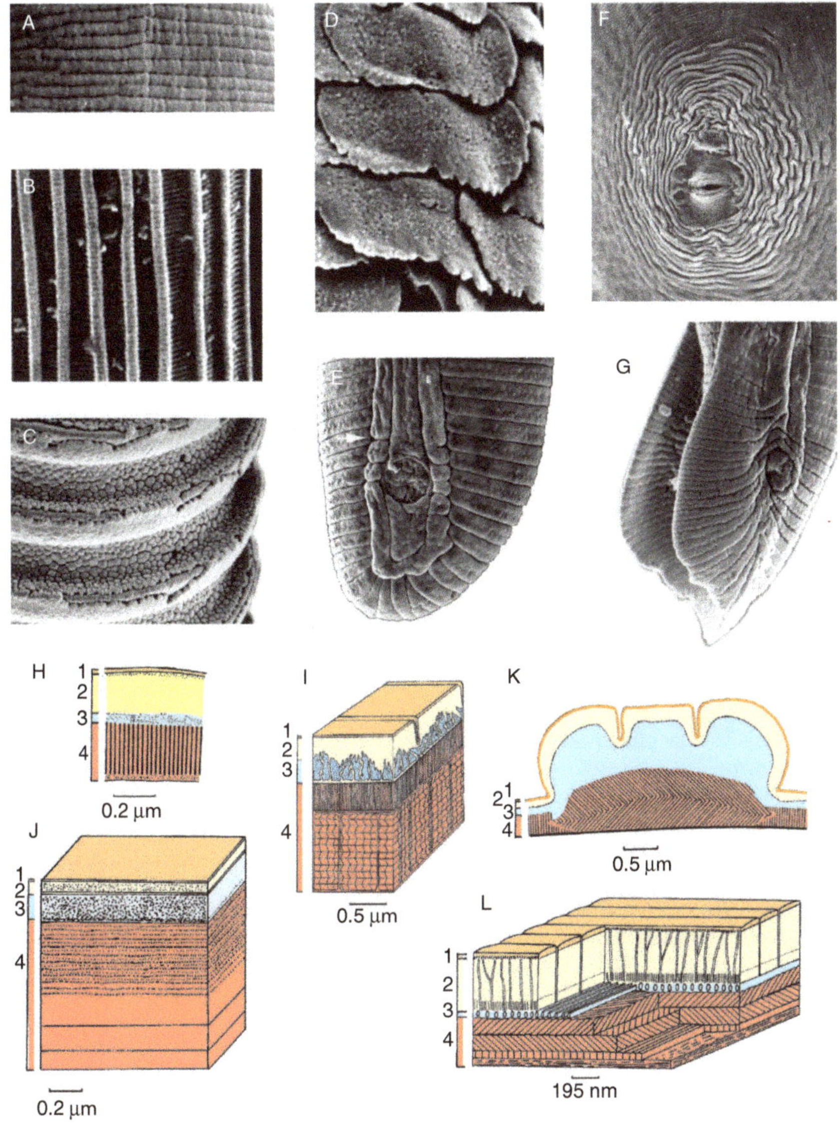

Fig. 1.4. Surface structure of the body cuticle. A: Transverse striae, scanning electron microscopy (Trichodoridae). B: Longitudinal ridges (Actinolaimidae, Dorylaimida) (Vinciguerra and Clausi, 2000. Courtesy Universidad de Jaén). C, D: *Criconema paradoxiger*. C: External cuticular layer in female. D: Scales in juvenile (Decraemer and Geraert, 1992. Courtesy Brill). E: Lateral field with longitudinal ridges and areolation (arrow) in *Scutellonema*. F: Perineal pattern in *Meloidogyne* (Siddiqi, 1986. Courtesy CABI). G: Caudal alae in *Scutellonema* male. H–L: Diverse ultrastructure of body cuticle of plant-parasitic nematodes. H: *Pratylenchus*. I: *Rotylenchus*. J: *Trichodorus*, adult (Trichodoridae). K: *Tylenchorhynchus*, lateral field. L: *Xiphidorus*, adult (Longidoridae). H, I, K: Tylenchomorpha, females. (Decraemer *et al.*, 2003. Courtesy Cambridge Philosophical Society (Biological Reviews)). See Box 1.1 for details of layers 1–4 in H–L.

Box 1.1. Ultrastructure of nematode body cuticle.

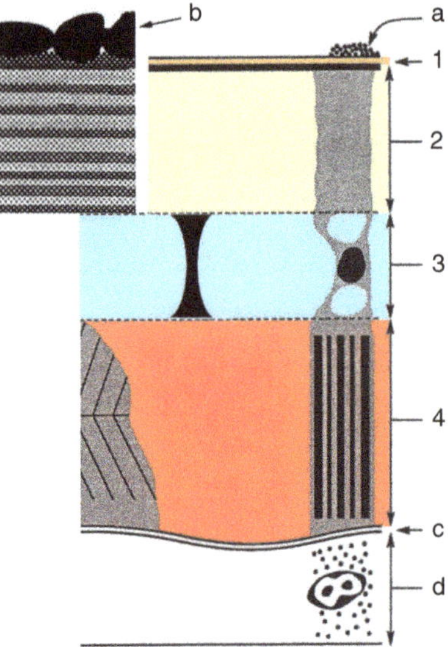

Scheme ultrastructure. 1: epicuticle, 2: cortical zone, 3: median zone, 4: basal zone, a: surface coat, b: extra- or non-cuticular material in criconematids, c: basal lamina, d: epidermis (based on Bird and Bird, 1991).

Epicuticle: Trilaminar outermost part of the cuticle, first layer to be laid down during moulting.

Cortical zone: Zone beneath the epicuticle, may be more or less uniform in structure, amorphous or with radial striae, or may show a subdivision into an outer amorphous part and an inner, radially striated layer, or be multilayered in criconematids. Cortical radial striae are unknown from plant parasites.

Median zone: Internal to the cortical zone, variable in structure: homogeneous or layered, with or without granules, globules, struts, striated material or fibres. The median zone may be absent.

Basal zone: Innermost zone of the cuticle, usually with the most complex structure of the three main zones; comprises outer sublayers of spiral fibres and inner layer with or without other fibres, laminae or radial striae.

Radial striae: Either cortical or basal in position, consist of longitudinal and transverse circumferential interwoven laminae which, at high magnification, appear as osmiophilic rods separated by electron-light material; the spacing and periodicity of these rods may vary between species, but in transverse sections of tylenchids it is about 19 nm.

Radial elements: Superficially resembling radial striae (in Longidoridae).

Basal radial striae: In basal cuticle zone, always interrupted at level of lateral chords and usually replaced by oblique fibre layers; mainly found in free-living terrestrial stages and most plant-parasitic Tylenchina.

Cortical radial striae: In cortical zone, not interrupted at level of lateral chords; characteristic of a free-living aquatic lifestyle.

Basal spiral fibre layers: Helically arranged fibre layers (angle 54°44′ or more), play a role in maintaining the internal turgor pressure.

Struts: Column-like supporting elements in the median zone (common in free-living aquatic nematodes and some animal parasites).

during copulation (Fig. 1.4G). Other cuticular outgrowths, such as transverse or longitudinal flaps, occur at the female genital opening where they may cover or guard the vulva. More elaborate cuticular ornamentation may also occur (spines, setae, papillae, tubercles, warts, bands, plates, rugae and pores). In plant-parasitic nematodes, cuticular ornamentations are important diagnostic features, especially in Criconematidae, and there may be an extra cuticular layer in *Criconema* (*Amphisbaenema*) and *Nothocriconema* (Decraemer *et al.*, 1996) (Box 1.1; Fig. 1.4C).

In many nematodes, the lateral body cuticle is modified to form the lateral fields (Fig. 1.4E and K). In Tylenchomorpha, the lateral fields are marked by longitudinal incisures and may be elevated above the body contour to form longitudinal ridges or bands. The widening of the body in the female and apparent elasticity of the body cuticle may be explained by unfolding of the ridges in the lateral field in the maturing female. These ridges may be intersected by the transverse striae, in which case the lateral field, which now has a block-like appearance, is described as being areolated (Fig. 1.4E). The number of longitudinal lines or incisures is of taxonomic importance but, as their number decreases towards the extremities, the number of lines should be counted in the mid-body region. It is important to differentiate between longitudinal ridges and lines or incisures, there being one more line or incisure than there are ridges; in older animals the lines gradually disappear and in some taxa (e.g. in *Malenchus*) additional fine ridges cannot be seen using light microscopy (LM). Sometimes, the lateral field shows anastomoses and occasionally lateral cuticle differentiation may be absent, as in obese females of *Heterodera* or all Longidoridae and Trichodoridae. Cuticular differentiations may also occur at or around the vulva and anus, as in the perineal patterns of mature females of root-knot nematodes (Fig. 1.4F).

1.2.1.2. Cuticle ultrastructure

The cuticle is secreted in layers and essentially consists of four parts: (i) a thin epicuticle at the external surface, which is provided with a surface coat of glycoproteins and other surface-associated proteins or, more rarely, with an additional sheath formed from cuticle or extra cuticular particles; (ii) a cortical zone; (iii) a median zone; and (iv) a basal zone. Certain zones may be absent. For example, in Tylenchomorpha, the body cuticle changes in structure at the base of the cephalic region, i.e. the median and radially striated basal zones disappear. The latter appears to continue as the electron-dense zone of the cephalic capsule. The trilaminar epicuticle acts as a hydrophobic barrier and is composed of non-collagenous proteins, cuticlins and lipids. In cyst nematodes, quinones and polyphenols in the epicuticle result, upon the action of phenoloxidase, in the tanning of the female cuticle to form a resistant cyst wall. The surface coat is a highly dynamic layer secreted by the epidermis and is part of the immune system (Davies *et al.*, 2008). In *Meloidogyne incognita*, *in vivo* root exudates triggered an increase in surface coat lipophilicity and allowed the root-knot nematodes to adapt to survive plant defence processes (Davies and Curtis, 2011). In some species (*Poikilolaimus*), the thickness of both an osmiophilic region of the median zone and total body cuticle thickness appear to be correlated with survival in the presence of a predator (Ichiishi *et al.*, 2022).

The most important structural elements of the cuticle morphology are the presence/absence of: (i) cortical radial striae; (ii) basal radial striae; (iii) spiral fibre layers in the basal zone; and (iv) supporting elements, e.g. struts, in the fluid matrix of the median zone (Box 1.1). All these features are thought to be responsible for the radial strength of the cuticle. At the level of the lateral chords, the cuticle may not only show an external differentiation in ornamentation or cuticular outgrowths, such as lateral alae, but also displays ultrastructural differences when compared to the dorsal and ventral regions of the body (Fig. 1.4K). Intracuticular canals have been observed in many species (e.g. Trichodoridae, Hoplolaimidae) and may be involved in transport of material from the epidermis to the other layers of the body cuticle. In Longidoridae, adults and juveniles have an identical cuticle structure composed of three main zones: (i) the cortical zone with radial filaments and radial elements at the inner base; (ii) the median zone with a layer of median thick longitudinal fibres; and (iii) the basal zone with two spiral fibre layers and either a layered or a homogenous inner region (Fig. 1.4L). Trichodoridae have a cortical zone without radial striae and a homogenous median zone, and a basal zone characterized by concentric layers, an apparent synapomorphy for *Trichodorus* and *Nanidorus* although lacking in several *Paratrichodorus* species (Karanastasi *et al.*, 2001) (Fig. 1.4J). The absence of radial striae in the cortical or basal zone, as well as the absence of spiral fibre layers, may be related to the low internal pressure in trichodorids (trichodorids do not burst open when punctured), as well as to their slow locomotion. In Tylenchomorpha, the cuticle structure is more diverse but cortical radial striae are always absent (Fig. 1.4H, I and K). The cortical zone is rarely subdivided but in females of the Criconematidae it is multilayered except in the cephalic region. The median zone in the majority of Tylenchomorpha is vacuolated, either with or without granular material or ovoid to globular structures, but may be absent resulting in the cortical zone abutting the radially striated basal zone (e.g. *Aphelenchus avenae*). In the majority of Tylenchomorpha, all developmental stages have the basal zone characterized by a radially striated layer; additional internal sublayers as part of the basal zone may be present in Hoplolaimidae and Heteroderidae females. In globose females of the Heteroderidae, the radially striated layer is discontinuous. Basal radial striae appear to induce some physical constraints, e.g. to growth, which may explain their absence under certain conditions or their breaking up into small patches in obese endoparasitic females of the Heteroderinae. In Tylenchomorpha, the cuticle at the level of the lateral field is differently structured compared to the rest of the body, resulting in replacement of the basal radial striae by fibre layers, an apparent functional requirement to accommodate small changes in body diameter. Basal radial striae also appear to be involved in locomotion because they disappear in second-stage juveniles (J2) of *Meloidogyne* shortly after the juvenile becomes a sedentary endoparasite.

Most juveniles of plant-parasitic Tylenchomorpha have a similar cuticle structure to the adults. Cuticular changes other than during the moulting process occur when changing lifestyle and in sedentary stages. Upon invasion of plant roots, the conspicuous radially striated basal zone of the cuticle of the pre-parasitic J2 of *Meloidogyne* is modified in the parasitic J2 into a thicker cuticle lacking basal radial striae (Jones *et al.*, 1993).

All stages of the plant-parasite *Hemicycliophora arenaria* possess an additional sheath covering the normal cuticle (Johnson *et al.*, 1970). This sheath is composed of a trilaminate outer layer and either four (female) or two (male) inner layers of cuticle.

Most nematodes moult four times (exceptionally three times, e.g. as in some Longidoridae) during their development. At each moult the cuticle is reconstructed, allowing adaptation to a possible changing environment, and to increase the nematode's length after the moult is completed and the old cuticle is shed. During moulting, the cuticle can either be shed completely (Trichodoridae) or partially resorbed (*Meloidogyne*), thereby recycling proteins in a confined space such as moulting within the egg. New cuticle formation is characterized by the occurrence of epidermal folds or plicae over which the new cuticle becomes highly convoluted (Bird and Bird, 1991). The epicuticle is the first layer to be laid down and is connected to the epidermis by hemidesmosomes.

In addition to the anterior cuticular sense organs, such as labial and cephalic sensilla and the amphids, there are also somatic sense organs that terminate in setae (not in plant-parasitic nematodes) or in pores, phasmids or deirids (see Section 1.2.5.3).

1.2.2. Epidermis

The epidermis secretes the cuticle and is responsible for the overall architecture, including elongation of the embryonic tadpole stage (Costa *et al.*, 1997). The epidermis is probably *the* limiting structure in homeostatic regulation. The epidermis consists of a thin layer and four main internal bulges that form the longitudinal chords, one dorsal, one ventral and two lateral, dividing the somatic muscles into four fields (Box 1.2). Anteriorly, it pervades the region of the cephalic framework and is responsible for its formation. The epidermis can be cellular, partly cellular or syncytial (Tylenchomorpha). The cellular condition is a primitive one occurring in free-living species and some parasitic species plus juveniles of parasites that possess a syncytial epidermis in adults. In some species there are no cell boundaries between the chords but cell walls exist within the chords, especially the lateral chords. The cell nuclei are usually located in the chords, although the dorsal chord only has nuclei in the pharyngeal region. The structure of the epidermis may show pronounced changes during development. For example, in the insect-parasitic stage of *Fergusobia* the cuticle and feeding apparatus are degenerate and the epidermis is convoluted into the numerous microvilli responsible for uptake of nutrients (Giblin-Davis *et al.*, 2001). The epidermis contains various specialized structures such as epidermal glands, caudal glands and ventral gland(s) of the S–E system. In some aquatic nematodes, such as *Halomonhystera disjuncta*, vacuoles in the epidermal chord may act as a compartmentalized hydrostatic skeleton (Van De Velde and Coomans, 1989).

1.2.3. Somatic musculature

Only a single layer of obliquely orientated and longitudinal aligned somatic muscle cells lies beneath the epidermis. The number of rows per quadrant between the chords varies from a few (up to five cells), known as the meromyarian condition, to the many rowed polymyarian condition. The general sinusoidal movement of nematodes is brought about by alternate contraction of the ventral and dorsal musculature, thereby giving rise to waves in a dorsoventral plane (see Chapter 8). In Criconematidae with strongly developed transverse cuticular annuli, contraction of the somatic muscles shortens, and relaxation extends the body, resulting in a creeping movement comparable to that of earthworms. A typical characteristic of a nematode muscle cell, a feature found in only a few other invertebrate taxa (e.g. some Gastrotricha), is that instead of the nerve process running towards the muscle, a process of the non-contractile portion of the muscle cell extends towards the dorsal or ventral nerve in the corresponding epidermal chord. The arrangement of

Box 1.2. Body wall and pseudocoel.

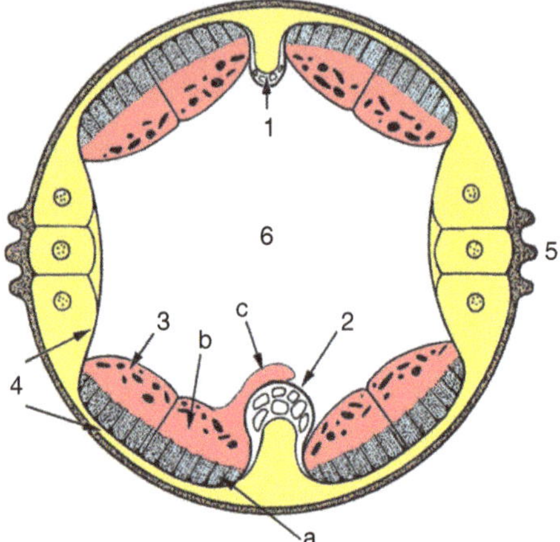

Diagram of transverse section depicting the epidermal chords, the somatic musculature with detail of muscle cell and pseudocoel. Internal organs have been omitted. 1: dorsal epidermal chord, 2: ventral chord with ventral nerve, 3: muscle cell, 4: basal lamina, 5: lateral alae, 6: pseudocoel, a: contractile part of muscle cell, b: non-contractile part, c: process of muscle cell (based on Bird and Bird, 1991).

Chord: The dorsal, the ventral and two lateral thickenings of the epidermis.

Somatic muscle cell: Mainly spindle-shaped, consists of (a) a contractile portion of the cell towards the epidermis, (b) a non-contractile portion towards the body cavity and (c) an arm or process that extends from the non-contractile portion of the cell toward the dorsal or the ventral nerve; muscle cells anterior to the nerve ring send processes directly into the nerve ring.

Platymyarian muscle cell: The whole contractile part of the muscle cell is flat and broad and borders the epidermis; common in small species.

Coelomyarian muscle cell: Spindle-shaped muscle cell; laterally flattened so that the contractile elements are arranged not only along the epidermis, but along the sides of the flattened spindle as well; coelomyarian muscle cells bulge into the pseudocoel; common in large species.

Circomyarian muscle cell: Muscle cell in which the sarcoplasm is completely surrounded by contractile elements.

Meromyarian musculature: Few (five or six) rows of muscle cells are present per quadrant.

Polymyarian musculature: More than six rows of muscle cells present per quadrant; spindle-shaped muscle cells laterally flattened.

the contractile portion (sarcomere) groups the muscle cells into three types: (i) platymyarian (flat contractile part bordering the epidermis); (ii) coelomyarian (muscle cell bulging into the pseudocoelom, contractile part not completely bordering the epidermis); and (iii) the circomyarian type (contractile elements surrounding the central sarcoplasm) (Box 1.2). The platy-meromyarian type is more common in small species such as plant-parasitic nematodes. The sarcomeres are separated from one another by dense bodies attached to the basal lamina at regular intervals, resulting in smooth bending of the body during locomotion. Specialized muscle cells are associated with the digestive system and the male and female reproductive systems.

1.2.4. Pseudocoelom

The pseudocoelom or body cavity is a secondary structure lacking mesentery and is lined by the somatic muscles and the basal lamina that covers the epidermal chords. Other authors consider the body cavity in nematodes as a primary body cavity instead of a pseudocoelom because it is not a persisting blastocoel. This fluid-filled cavity bathes the internal organs and contains some large amoeboid cells called pseudocoelomocytes. These vary in number, size and shape and their function includes osmoregulation, secretion and transport of material. The pseudocoelomic fluid acts as part of the turgor-pressure system, but also has some circulatory function. This high-pressure system also plays a role in feeding and egg-laying.

1.2.5. Cephalic region, sense organs and nervous system

1.2.5.1. Cephalic region and anterior sensilla

Nematodes lack a true head region although 'head' is commonly used. Other terms used for this region in the literature are labial and cephalic region. In this chapter the terminology used is as explained in Box 1.3. The basic pattern in nematodes is for there to be six lips around the mouth opening (two subdorsal, two subventral and two lateral) (Fig. 1.5A). The lips can be fused, for example two by two resulting in three lips, one dorsal and two

Box 1.3. Cephalic region terminology.

Cephalic region (CR) or head: The anteriormost body region; in several taxa this region is differentiated from the body by either a marked indentation and/or the possession of a thickened body cuticle that may form a capsule/helmet or by the possession of an inner cephalic framework (see Fig. 1.6B). The cephalic region is composed of a labial region and a postlabial region.

Labial region (LR): The anteriormost part of the cephalic region (CR), primitively, with six lips around the mouth opening; each lip with an inner (ILS) and an outer sensillum (OLS) on its radial axis; the two circlets of ILS and OLS are arranged in a hexaradial pattern (Fig. 1.5A). This labial pattern is also present when only three lips are present or when all lips are fused. However, the OLS may be described as anterior CS (Platt and Warwick, 1983) influenced by the ILS being small papilliform versus the OLS and CS setiform. LS do not possess the same neurotransmitters as somatic sensilla.

Postlabial cephalic region: In the primitive arrangement of anterior sensilla this region bears the third circlet of four 'cephalic' sensilla (CS), bilaterally arranged in a submedian position. Labial and cephalic sensilla have different origins (De Coninck, 1965). CS possess the same neurotransmitter as somatic sensilla do, whereas LS lack these substances (Voronov and Nezlin, 1994; Sawin *et al.*, 2000).

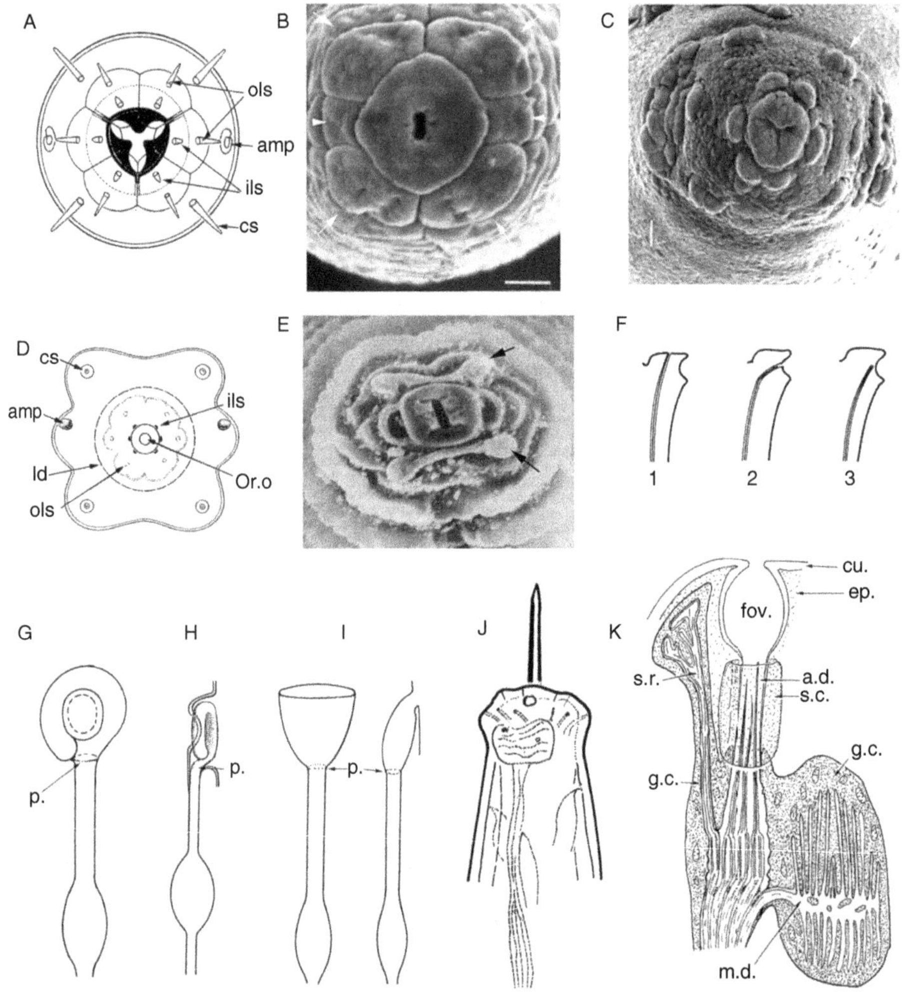

Fig. 1.5. Cephalic region and anterior sensilla. A: Basic scheme (de Coninck, 1965). B: *Scutellonema* with six lip areas (arrows); *en face* view. C: *Paratrichodorus* in *en face* view with outer labial and cephalic papillae in a single circlet (arrow). D: Arrangement in *Aphelenchoides*: amp: amphid, cs: cephalic sensilla, ils: inner labial sensilla, ld: labial disc, ols: outer labial sensilla, Or.o: oral opening. E: *Criconemoides* (= *Criconemella*), *en face* view showing pseudolips (arrows) (van den Berg and De Waele, 1989). F: Inner labial sensilla: 1: *Ditylenchus.* 2: *Merlinius.* 3: *Hemicycliophora* (after De Grisse, 1977). Amphid structure: G, H: Spiral amphid (not in plant-parasitic nematodes), lateral and ventral view; p. pore leading to sensillar canal. I: Pocket-like amphidial fovea in lateral (left) and ventral (right) view. J: *Xiphidorus*, pouch-like amphidial fovea with pore-like opening. K: Ultrastructure of amphid: a.d.: amphidial duct, cu: cuticle, ep: epidermis, fov: fovea, g.c.: gland or sheath cell, m.d.: multivillous dendrite, s.c.: socket cell (after Coomans, 1979).

ventrosublateral (*Ascaris*), or the lateral lips may be reduced or absent (*Pseudoacrobeles* (*Bonobus*) *pulcher*). The lips are either clearly separated or partially to completely fused (Longidoridae, Trichodoridae) (Fig. 1.5C). In Tylenchomorpha, the anterior end shows an amalgamated, usually hexagonal, lip region and so lip-like differentiations, when present, are better referred to as lip sectors or lip areas (Fig. 1.5B), e.g. there are six lip sectors in aphelenchs, but only four in *Belonolaimus* as the two lateral sectors have been reduced. Loof and De Grisse (1974) introduced the term 'pseudolips' for the six areas around the oral opening of some Criconematidae (Fig. 1.5E). The area around the oral opening may be differentiated into an oral and/or labial disc. In many groups of Tylenchomorpha there are two consecutive openings due to invagination of the cephalic cuticle, the outermost being the pre-stomatal opening and the innermost, the stomatal opening. The region between the openings, or anterior to the stoma opening when the prestoma opening is wide, is referred to as the prestoma.

In Tylenchomorpha, the cephalic region is internally supported by a variously developed cephalic cuticular framework (Fig. 1.3A) that may be well developed and heavily sclerotized. It is composed of six radial plates (one dorsal, one ventral, two sublateral on each side) running from stoma to the outside, the thickened posterior end forming the basal cephalic framework. The cephalic region can be continuous with the body contour, as in *Trophurus*, or more or less offset from the rest of the body, either by a depression or constriction (e.g. *Belonolaimus*, *Hoplolaimus*), or be broader than the adjoining body and therefore expanded (e.g. *Paralongidorus* spp. in the *Siddiqia* group; some *Xiphinema* spp.). In Tylenchomorpha, the true cephalic height is not always easy to establish as the cephalic region may be offset at a different level to the basal cephalic framework. The cephalic region may be smooth (e.g. *Trophurus*) or bear transverse striae (many genera), the annuli so formed sometimes being divided into blocks by longitudinal striae (e.g. *Hoplolaimus*, *Rotylenchus robustus*).

The cephalic region carries a concentration of anterior sensilla, each composed of a neuronal and non-neuronal section formed by two epidermal cells, the socket cell and the sheath cell (Coomans, 1979). In nematodes, there are primitively 12 labial sensilla and four cephalic sensilla arranged in three circlets to form six inner labial sensilla, six outer labial sensilla (both located on the lip region) and four cephalic sensilla posterior to the lip region (postlabial cephalic region). This is referred to as the 6 + 6 + 4 pattern. Two chemoreceptor sense organs, the amphids, are primitively located clearly posterior to the three circlets of anterior sensilla, but in more derived forms, such as in Tylenchomorpha, they have migrated forward onto the lip region. Each lip bears an inner and an outer sensillum on its radial axis, this hexaradiate pattern being maintained when lip number is secondarily reduced. The four cephalic sensilla are bilaterally arranged (two laterodorsal, two lateroventral) and represent the anteriormost somatic sensilla. In the plant-parasitic Longidoridae and Trichodoridae, the cephalic sensilla have migrated onto the lip region and are close to the outer labial sensilla, thereby forming a single circlet or 6 + (6 + 4) pattern. In Tylenchomorpha, the anterior sensilla are arranged in three circlets but, because of the small size of the cephalic region, the two posterior circlets are located close together.

In general, the six inner labial sensilla protrude from the surrounding cephalic cuticle via a terminal pore on top of a papilla. In a number of plant- and animal-parasitic nematodes, the inner labial sensilla either have pore-like openings around the oral opening or inside the pre-stoma (*Pratylenchus*) (Fig. 1.5F), or pores may be lacking entirely, the receptors ending blind in the cuticle of the oral disc (*Hemicycliophora*). Inner labial sensilla in open connection with the environment are chemoreceptive; those covered or embedded in the cuticle are mechanoreceptive. In most tylenchs, the inner labial sensilla possess two ciliary receptors and show a combined chemo- and mechanoreceptive function. In *Longidorus*, four such receptors can be found, whilst there are two or three in *Trichodorus*. Chemoreceptors enable the sensing of food, detection of volatile elements, avoidance of noxious conditions and navigation and mating.

The outer labial sensilla may protrude via papillae or setae, but in many plant- and animal-parasitic nematodes they end in simple pores or are embedded in the cephalic cuticle; the cuticle above each termination may show a slight depression. The lateral outer labial sensilla are often reduced, a reduction that may be related to the development of the amphids (*Meloidogyne*). In Tylenchomorpha, the outer labial sensilla have two receptors, a narrow and a large one (*Aphelenchoides*) or show different degrees of reduction to a complete lack of the fine receptor (Criconematoidea).

The cephalic sensilla are submedian in position and usually protrude from the surrounding cuticle as setae or papillae with a terminal pore. In many parasitic nematodes they are embedded in the cephalic cuticle.

The main constituent parts of an amphid are the aperture, the *fovea*, the *canalis* and the *fusus* or sensillar pouch (no spiral in plant-parasitic nematodes) (Fig. 1.5K). The distal part of the amphid, the *fovea*, is either an external excavation of the cephalic or body cuticle (as in many free-living Chromadorida species, Fig. 1.5G and H) or an invagination of the cuticle, thus forming a pocket connected with the exterior through an aperture (as in plant-parasitic nematodes, Fig. 1.5I and J). The amphidial aperture is typically located laterally, but may be shifted dorsad. The amphidial opening varies in form: in Trichodoridae (Fig. 1.5C) it is a post-labial transverse slit, whereas in Longidoridae it is of variable shape and size, being either a post-labial transverse slit or a pore (Fig. 1.5J). In Tylenchomorpha, the amphid openings have migrated onto the lip region and are usually close to the oral opening, their apertures being greatly reduced in size and slit-like to oval in form (Fig. 1.5B). The *fovea* is the most variable part of the amphid, varying in size and shape according to taxa and even within a taxon between sexes and may be completely or partially filled with a gelatinous substance (*corpus gelatum*) secreted by the amphidial gland and may protrude from the body. In nematodes with an internal fovea there is less variation in form. In Longidoridae, the *fovea* varies from stirrup- or goblet-shaped to a pouch (Fig. 1.5J), which may be bilobed. In tylenchs, there is no sharp demarcation between the often reduced *fovea* and *canalis* and the sensilla pouch may be located as far posterior as the stylet knobs and may be pouch-like as in *Malenchus*. The amphids are the largest of the chemoreceptors and possess a much greater number of receptors than other sensilla. In Tylenchomorpha, lip patterns and arrangement of the anterior sensilla and amphids are considered important diagnostic features in identification and for the analysis of relationships.

1.2.5.2. Somatic nervous system

The nerve ring usually encircles the isthmus of the pharynx, rarely the intestine near the pharyngo-intestinal junction (*Aphelenchoides*). The nerve ring is connected to several ganglia, longitudinal somatic nerves running anteriorly towards the anterior sensilla in the cephalic region while posteriorly four large nerves, the largest being the ventral nerve, run through each of the four epidermal chords, with four smaller nerves, two laterodorsal and two lateroventral, running adjacent. The cell bodies of the nerves of the anterior sensilla are located in six separate ganglia anterior to the nerve ring or in a single ganglion

(*Caenorhabditis elegans*). The amphidial nerves have an indirect connection with the nerve ring, their cell bodies being located in the paired lateral ganglia. Of the posterior longitudinal nerves, the lateral nerve contains a few ganglia, including the lumbar ganglia in the tail region. The ventral nerve has a chain of ganglia, the anteriormost being the retrovesicular ganglion whereas the dorsal and submedian nerves lack posterior ganglia; the dorsal rectal ganglion and the pre-anal ganglion are connected by commissures to the ventral nerve. The dorsal rectal ganglion locates a few unpaired ganglia such as the DVA interneuron in the tail, which informs the worm how its body bends as it moves (Robinson *et al.*, 2013). Posterior to the nerve ring, the longitudinal nerves are connected to each other by commissures. The presence and position of some of these commissures are of taxonomic importance. The most important commissure is the lateroventral commissure, which is also known as the hemizonid. This is visible as a refractive body near the S–E pore. At a short distance from the hemizonid lies the hemizonion. Caudalids in the tail region correspond to the paired anal–lumbar commissure, which links the pre-anal ganglion to the lumbar ganglion.

1.2.5.3. Peripheral and visceral nervous system

There may be numerous somatic sensilla in addition to the anterior sensilla in the cephalic region, arranged in dorsal, ventral and subventral longitudinal rows. In free-living marine nematodes, a peripheral lattice-like network of nerves connects somatic setae and papillae and coordinates impulses from these somatic sense organs, both with each other and with the somatic nervous system. In terrestrial nematodes, such as Longidoridae and Trichodoridae, these sensilla are mainly chemosensitive and open to the exterior as pores; they are often associated with glands (e.g. at lateral chord level in Longidoridae). In free-living and plant-parasitic Rhabditia, up to three pairs of laterally located deirids can be found in the pharyngeal region. The innervation of deirids (= cervical papillae common in Tylenchomorpha) is from the lateral nerves from the lateral ganglia. Other somatic sense organs common in Tylenchomorpha (although they may be lacking, as in criconematids) are the postdeirids and phasmids, which are usually situated in the tail region. Phasmids have the basic structure of a sensillum, the ciliary receptor being in open connection with the environment via a pore. The pore may be provided with a plug, which is probably secreted by the sheath cell, and which may be small, as in many Tylenchomorpha, or enlarged to a shield-shaped structure, the scutellum (e.g. in *Scutellonema*, Fig. 1.4E). In the posterior body region of males (see Section 1.2.8.2) specialized somatic sensilla or genital supplements may be present.

The pharynx possesses its own visceral nervous system with its own neurons, processes and receptors, but is also connected to the nerve ring. The detailed structure of this system is mainly known from work on *C. elegans*. Internal cephalic receptors have been detected in *Xiphinema*, e.g. in association with the amphidial sheath cell (Wright and Carter, 1980), although their function remains unclear. In *Longidorus* and *Xiphinema*, nerve endings have been observed in association with the cuticular lining of the pharynx at the level of the sinuses of the odontophore, the isthmus and the anterior bulb (Robertson, 1976, 1979). These may be mechanoreceptive, detecting the passage of food or regulating the flow of secretion. Photoreceptors or ocelli have not been observed in plant-parasitic nematodes.

1.2.6. Digestive system

The wide diversity of food sources and methods of ingestion is reflected in the diversity of the structure of the digestive system (Figs 1.6 and 1.7). In general, this system consists of three regions: the stomodeum, mesenteron and proctodeum. Only the mesenteron or mid-gut is of endodermal origin, the stomodeum being of mixed ecto-mesodermal origin and the proctodeum or rectum formed from the ectoderm. The stomodeum region is referred to as the pharyngeal, neck or cervical region.

1.2.6.1. Stomodeum

The stomodeum comprises the mouth opening, stoma *sensu lato* (composed of cheilostom and pharyngostom), pharynx and cardia. The oral aperture is mostly situated terminally. The cheilostom or lip cavity has a hexa- or triradial symmetry and is lined by body cuticle. The cuticle lining may be more or less sclerotized and is supported by simple rods (cheilorhabdia, a synapomorphy of the Chromadorida). Originally, the cheilostom was delimited by the lips but in derived conditions may extend further posteriorly, as in Longidoridae where it represents the stoma *sensu stricto*, extending from the oral opening to the guiding ring (Fig. 1.6A). In Tylenchomorpha, the cheilostom largely lines anteriorly the cavity of the cephalic framework, through which the stylet moves, and extends posteriad until the guide ring. The pharyngostom is triradially symmetrical, as is the pharynx, of which it is a specialized part. The pharyngostom is further subdivided (De Ley *et al.*, 1995) into a gymnostom (part lined by arcade cells) and in a stegostom (part surrounded by pharynx and differentiated into a pro-, meso-, meta- and telostegostom). Its structure reflects the method of feeding and food source. In plant-parasitic nematodes, some entomophilic and some predatory nematodes, the pharyngostom possesses a protrusible spear or stylet but in Trichodoridae, the spear is a protrusible dorsal tooth. The feeding apparatus in the majority of plant-parasitic nematodes is either a hollow stomatostylet (Tylenchomorpha) or odontostyle (Dorylaimida, Longidoridae). The stomatostyle consists of three parts: a conus with a ventral aperture, a shaft and a posterior region that may enlarge to form three basal knobs, these acting as attachments for the stylet protractor muscles. According to Baldwin *et al.* (2004) the stylet cone and shaft are formed by arcade syncytia and are homologous with the gymnostom while the stylet knobs are homologous with the prostegostom. Tylenchids have no stylet retractor muscles, the resulting tension in the alimentary tract causing retraction of the stylet when the protractor muscles are relaxed (Fig. 1.6B). The dorylaimid stylet is made up of two parts: an anterior odontostyle with a dorsal aperture anteriorly and with its posterior end furcate or simple, and a posterior extension or odontophore, a modification of the anterior pharyngeal region, which acts as a supporting structure for the odontostyle. The base of the odontophore may be enlarged to form three flanges for enhanced insertion of the stylet protractor muscles (Fig. 1.6A). Longidorids possess eight stylet protractor muscles (typical of dorylaims) but true stylet retractor muscles are poorly developed or absent. Pharyngeal retractor muscles at the level of the narrow part of the pharynx assist in retracting the odontostyle. Because of the extreme elongation of the cheilostom, *Xiphinema* species possess a special set of *dilatores buccae* muscles that counteract compression of the cheilostome wall on stylet protraction. In Longidoridae, the odontostyle is long and needle-like. When it is protruded, the anterior stomatal lining folds along the

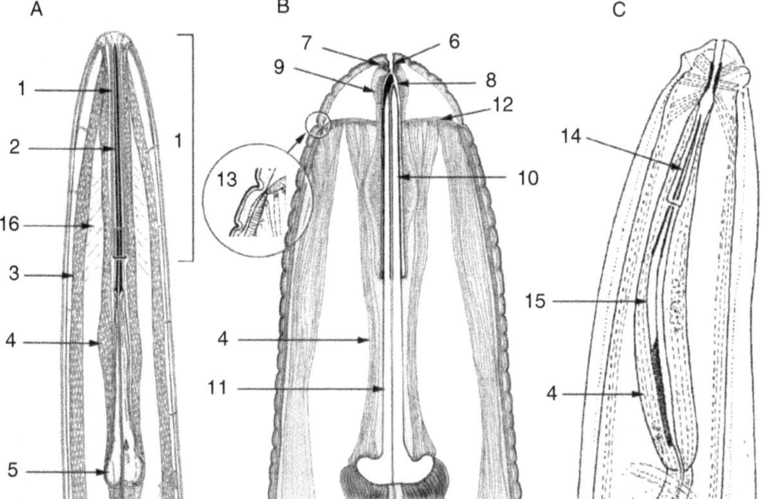

Fig. 1.6. Stoma region and types of feeding apparatus in plant-parasitic nematodes. A: Odontostyle and odontophore (Longidoridae). B: Stomatostylet with detail of body cuticle (inset) at base of cephalic framework (Tylenchomorpha). C: Onchiostyle (Trichodoridae). 1: Cheilostom. 2: Odontostyle. 3: Somatic muscles. 4: Stylet protractor muscles. 5: Odontophore with flanges. 6: Prestoma. 7: Thickening of cuticle around prestoma. 8: Stylet opening. 9: Stoma. 10: Stylet conus. 11: Stylet shaft and knobs. 12: Basal cephalic framework. 13: Body cuticle in detail, showing disappearance of median and striated basal zone in cephalic region. 14–15: Onchiostyle with onchium (14) and onchiophore (15). 16: *Dilatores buccae*. A: From Coomans (1985). B: Adapted from Endo (1980). C: From Maafi and Decraemer (2002).

odontostyle, thereby forming a guiding sheath, the inner cuticular lining being separated from the rest of the wall (except at the level of the guiding ring and at its posterior end) by a fluid-filled cavity that acts as a hydrostatic skeleton. In *Longidorus*, for example, the hydrostatic tissue forms four compensation sacs to regulate the pressure on stylet protraction. The odontostyle is secreted by a cell in the ventrosublateral sector of the pharynx, whereas the odontophore, together with the guiding sheath and the pharyngeal lining, is derived from pharyngeal tissue. In Longidoridae, as in all Dorylaimoidea, the replacement odontostyle in the first juvenile stage lies close to the functional one, i.e. its tip is within a sinus of the odontophore. During the first moult, the functional odontostyle is shed together with the body cuticle, cheilostome and inner lining of the anterior stomodaeum. In J2–J4 the replacement odontostyle upon formation moves posteriorly, turns around with the opening in ventral position and becomes enclosed in the wall of the pharyngeal isthmus until moulting begins, when the reserve odontostyle moves anteriorly, it returns to its normal position with the opening dorsal (Radivojević, 1998). The ventrally curved onchiostyle in Trichodoridae consists of an anterior stylet-like tooth or onchium with a solid tip and a posterior support, the onchiophore, which is formed by the thickened and sclerotized dorsal lumen wall of the pharyngostom (Fig. 1.6C). The protrusion of the onchiostyle occurs upon contraction of protractor muscles of the pharyngostom; there are no retractor muscles. In juveniles, the replacement onchium lies close behind the functional one and is obliquely inserted in the dorsal wall of the odontophore.

The structure of the pharynx itself is related to feeding mode. In Tylenchomorpha, the anterior part of the pharynx (or corpus) is subdivided into an anterior, muscular procorpus and a muscular, larger diameter and more robust metacorpus, which is located posteriorly. The corpus is followed by a non-muscular isthmus (may be very short or absent) and three pharyngeal glands arranged either in a terminal bulb (postcorpus) or in a lobe(s) overlapping the intestine predominantly dorsally, ventrally or laterally (Fig. 1.7C and D). The metacorpus shows a constant relative position in the pharynx, i.e. the distance posterior to the median bulb is more or less determined by the distance anterior to it. So when the pharyngeal glands become larger than the available space between nerve ring

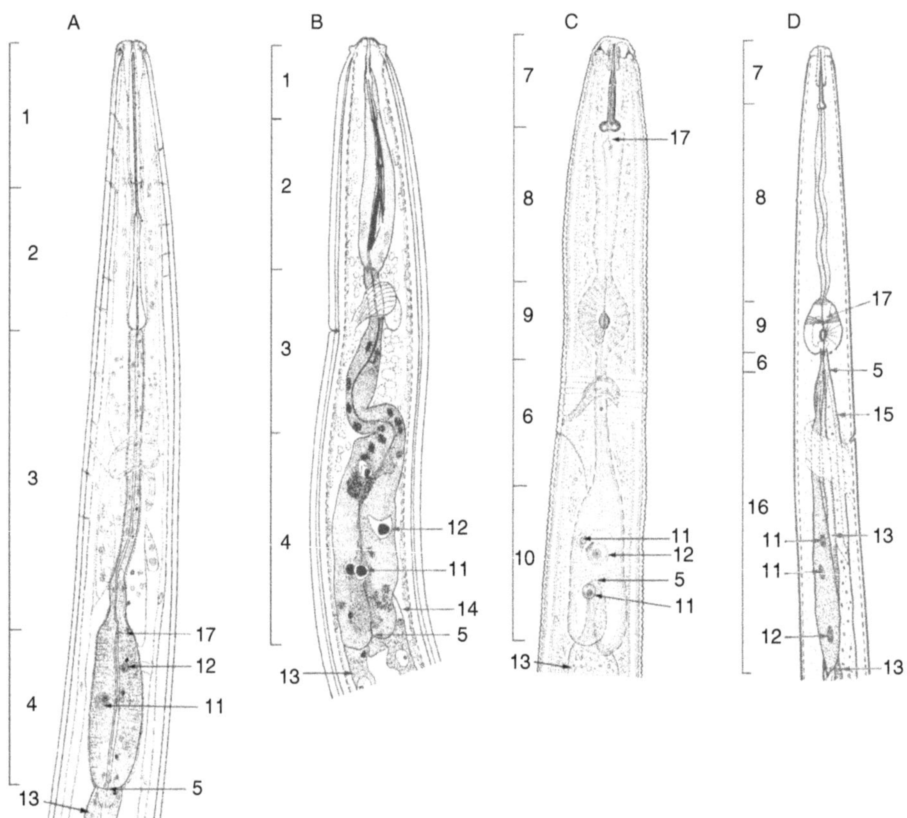

Fig. 1.7. Digestive system of plant-parasitic nematode taxa. A: *Paraxiphidorus* (Longidoridae). B: *Paratrichodorus* (Trichodoridae). C: *Pratylenchoides* (Pratylenchidae). D: *Aphelenchoides* (Aphelenchoididae). 1: Cheilostom. 2: Pharyngostom. 3: Narrow anterior region of pharynx. 4: Pharyngeal bulb. 5: Pharyngo-intestinal junction. 6: Isthmus. 7: Stomatostylet. 8: Procorpus. 9: Metacorpus. 10: Postcorpus. 11: Ventrosublateral pharyngeal gland nuclei. 12: Dorsal pharyngeal gland nucleus. 13: Intestine. 14: Intestine dorsally overlapping pharynx. 15: Pharyngo-intestinal junction valve cell. 16: Pharyngeal gland lobe. 17: Dorsal pharyngeal gland orifice. A: From Decraemer *et al.* (1998). B: From Decraemer and De Waele (1981). C: From Siddiqi (1986). D: From Shepherd *et al.* (1980).

and intestine, they overlap the intestine (Geraert, 2006). Some genera (e.g. *Ditylenchus*) can be characterized by the position of the median bulb. The arrangement of pharyngeal glands is of taxonomic and phylogenetic importance. In the Sphaerularioidea, a group of taxa combining fungus feeding and insect parasitism, the pharynx is simple and lacks differentiation into a procorpus. In addition, they have a low body pressure (e.g. *Hexatylus*), which is increased upon shortening of the body when penetrating the hypha; before ingestion, the body relaxes, reducing the body pressure and ingestion is passive. Passive feeding occurs also in primitive Tylenchina, such as *Ditylenchus*. Species consuming liquid food, such as plant-parasitic tylenchs, have a median bulb or metacorpus with a cuticularized triradiate valvate apparatus attached to well-developed musculature, thereby allowing a stronger pumping action. Within the Tylenchomorpha, the position of the outlet of the dorsal gland is of taxonomic importance and differentiates the Aphelenchoidea (where the outlet is in the metacorpus) from the other taxa of the infraorder (outlet in the procorpus and often close to the stylet base) (Fig. 1.7C and D). In Longidoridae, the pharynx is flask-shaped and posterior to the pharyngostome continues as a narrow flexible tube with a circular lumen followed by an offset, muscular and glandular bulb that is often cylindroid in form (Fig. 1.7A). The terminal bulb in Dorylaimida contains five pharyngeal glands, reduced to three in Longidoridae, and their respective orifi, the cuticular lining of the lumen being reinforced by six platelets. In Trichodoridae, the pharynx consists of a narrow isthmus that gradually expands into a largely glandular basal bulb with five gland nuclei and gland orifi (Fig. 1.7B). The pharyngo-intestinal junction may be directly abutting or characterized by various types of overlap (either a ventral overlap by the pharyngeal glands, or a dorsal overlap by the intestine over the pharynx, or both types of overlap together). In most virus vector trichodorid species, nearly the entire lining of the pharynx anterior to the outlets of the posterior ventrosublateral glands can act as a retention site for tobraviruses. In longidorid virus vectors, the virus particles adhere to the lumen wall of the odontostyle or inner surface of the guiding sheath (*Longidorus, Paralongidorus*) or lumen wall of the odontophore and pharynx (*Xiphinema*). The pharyngo-intestinal junction or cardia exists in many different types. In Tylenchomorpha, the cardia usually consists of two cells, which differ in size and position in relation to the anterior intestine, whereas in Longidoridae and Trichodoridae, the rounded cardia has a triradiate lumen and is small or only weakly developed (Fig. 1.7B and D).

1.2.6.2. Mesenteron

The intestine is entirely of endodermal origin. It is a simple, single layered tube, either cellular or syncytial (criconematids), with or without a clear lumen lined with microvilli. The function of the intestine is mainly absorption (its lumen being bordered with microvilli), storage and secretion of proteins and enzymes. In the evolutionary advanced root parasites, such as *Meloidogyne*, the intestine becomes a storage organ; the thin lumen wall lacks microvilli and loses connection with the anus. Anteriorly, the intestine may be differentiated into a ventricular part that may overlap the pharynx (Trichodoridae) or become modified to house symbiotic bacteria (entomopathogenic *Steinernema* species). Posteriorly, the intestine may be differentiated into a pre-rectum, which is separated from the intestine by a valve-like structure formed by columnar cells (Longidoridae).

1.2.6.3. *Proctodeum*

The rectum is of ectodermal origin and is lined with body cuticle. It is a very simple short tube and is apparently formed from only a few cells. The junction between the rectum and pre-rectum is guarded by an H-shaped sphincter muscle. Defecation is mediated by the H-shaped *dilator ani*. In criconematids, both rectum and anus are poorly developed and probably non-functional. In *Meloidogyne*, six large rectal glands produce the gelatinous matrix in which the eggs are deposited. In Mononchida and Dorylaimida, males possess rectal glands consisting of three to five pairs of cells located dorsolaterally or laterally from the posterior intestine (pre-rectum). Caudally or dorsocaudally from the spicules the ducts of these rectal glands form loops that run anteriorly between the spicules and open via the dorsal wall of the cloaca (Coomans and Loof, 1986).

1.2.7. Secretory–excretory system

The secretory–excretory (S–E) system is a system of variable complexity (Fig. 1.8). It is often called the 'excretory system' on morphological grounds, but physiologically the evidence supports more of a secretory and osmoregulatory function than an excretory one. The S–E system is either: (i) glandular, in general consisting of a renette cell or ventral

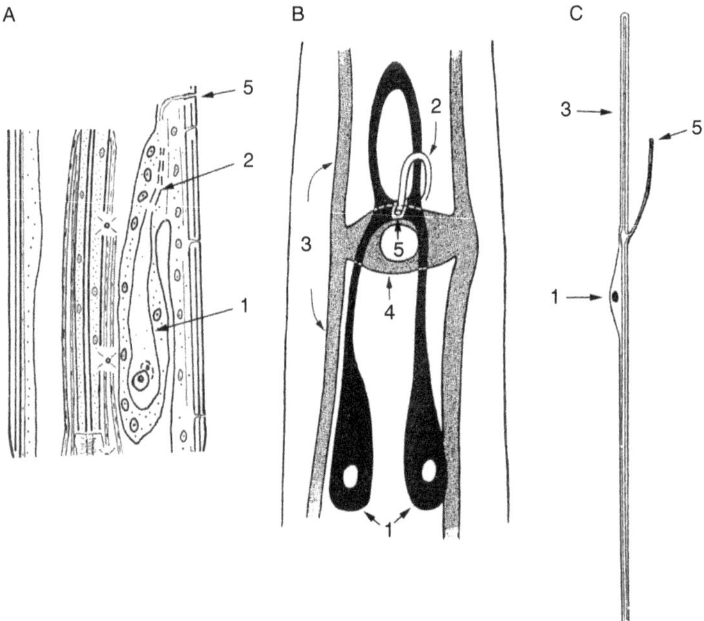

Fig. 1.8. Secretory–excretory (S–E) system. A: Glandular system in *Longidorus macrosoma* (Longidoridae). B: Complex tubular system in *Caenorhabditis elegans* (Rhabditoidea) ventral view. C: Simple tubular system in tylenchomorph nematodes. 1: S–E gland cell. 2: S–E duct. 3: Tubular cell. 4: Transverse duct. 5: S–E pore. A: From Aboul-Eid (1969). B: From Nelson *et al.* (1983) in Bird and Bird (1991). C: From De Coninck (1965).

gland connected to a ventral pore by a duct, the terminal part of which is lined by cuticle; or (ii) tubuloglandular, the most complex system consisting of an H-shaped cell with longitudinal canals running in the lateral chords between epidermis and basement membrane, and joined by a transverse duct connected to the S–E pore by a median duct, associated with a S–E sinus and an A-shaped binucleated gland cell (*C. elegans*, Fig. 1.8B). The canicular vesicles and lumen of the lateral canal(s) show a higher osmotic pressure than the body cavity collecting excess of water for excretion. In Tylenchomorpha, the S–E system is of the tubuloglandular type and is asymmetrical with a single renette cell situated laterally or lateroventrally, usually posterior to the pharynx (Fig. 1.8C). The S–E duct leads to the S–E pore, which is usually located posterior to the nerve ring, although exceptionally it may be close to the cephalic region or the vulval region (*Tylenchulus*). In *Tylenchulus*, the renette cell is enlarged and produces a gelatinous matrix into which eggs are deposited. In Trichodoridae, the S–E system is not developed and only a ventral pore and a very short duct are present. In Longidoridae it is not yet clear if a S–E system exists, a glandular structure with two cells being described only from *Longidorus macrosoma* (Fig. 1.8A). Other examples of secretions from the S–E system involve secretion of a protective surface coat and lubricating material facilitating movement, providing an immunosuppressive role in parasites and secretion of fluid at the start of the moulting process.

1.2.8. Type of reproduction and reproductive system

Most nematodes are dioecious (having the sexes separate) and gonochoristic (meaning they are either male or female) but bisexual individuals or protandrous hermaphrodites (e.g. *C. elegans*) exist; the latter usually have the appearance of females. Gonochoristic species reproduce by amphimixis or cross-fertilization. Uniparental reproduction or autotoky takes the form of either parthenogenesis, where development occurs through females producing female offspring (i.e. without fertilization), or automixis or self-fertilization in hermaphrodites (where male and female gametes are produced in the same individual). Parthenogenesis may be either meiotic (following meiosis the diploid chromosome number is restored by fusion with a polar body or by first doubling of the chromosomes), as in *Aphelenchus avenae*, *Pratylenchus scribneri*, *Xiphinema index* or *Meloidogyne hapla* race A, for example, or mitotic (i.e. without meiosis), examples being found in several species of *Meloidogyne* and *Pratylenchus*. In both types of autotoky, males may show up sporadically (androdioecious upon environmental stress) and amphimixis may then be possible in some species. In *Meloidogyne*, males arisen due to sex reversal rarely inseminate females and, even when they do, there is no fusion of male and female gamete (see Chapter 8). In *C. elegans*, transformed males showed a robust mating behaviour and 53% of them were fertile (Prahlad *et al.*, 2003). Some rhabditids (*Auanema*) are trioecious with male, female and hermaphrodites. Autotoky has arisen independently in several taxa of the phylum. Intersexes are found in some species and should not be confused with hermaphrodites as only one set of reproductive organs is functional, the other being vestigial. Pseudogamy, a way of reproduction intermediate between amphimixis and automixis/parthenogensis, where development of the egg is activated by a spermatozoon, which then plays no further role, is less common. Sex is mostly determined genotypically, mostly XX in female and hermaphrodites and XO or XX in male, but in some cases the genotype is changed under epigenetic influences.

The reproductive system is quite similar in both sexes and generally comprises one or two (rarely multiple) tubular genital branches (Figs 1.9 and 1.10). Apart from sexual characters, sexual dimorphism is not a common feature among nematodes, but when it occurs it is most evident among parasitic groups, e.g. a swollen saccate female contrasting with the vermiform male (*Verutus*, Tylenchomorpha).

1.2.8.1. Female reproductive system (Fig. 1.9)

The basic system is didelphic (i.e. composed of two uteri) and amphidelphic (referring to the uteri extending in opposite directions) and connected to a single vagina opening to the outside via a mid-ventral vulva (Fig. 1.9A). A derived system with a single uterus is called monodelphic. A monodelphic system with only the anterior branch developed is described as prodelphic; if it is only the posterior branch that is developed it is called opisthodelphic. The terms monogonic and digonic refer, respectively, to the presence of one or two ovaries. In didelphic systems, the vulva is located at about 50% of the entire body length from the anterior end, although it may be more anterior in monodelphic conditions or virtually subterminal in position as in some monodelphic species or certain obese females. Each genital branch consists of an ovary (= gonad) and a gonoduct. The gonoduct consists of an oviduct and uterus and may have one or two sphincters (valves) and a spermatheca. The spermatheca is either a specialized part of the oviduct or the uterus. Reduction of one of the genital branches is not uncommon, ranging from partial reduction in various degrees (the pseudomonodelphic condition, as in *Xiphinema*) or completely lost, apart from the possible retention of a small post-uterine sac (monodelphic system as in Criconematidae). In the didelphic condition, both branches may, in nematodes where the vulva is virtually subterminal (e.g. *Meloidogyne*, *Globodera*), be anteriorly outstretched, i.e. prodelphic.

The ovary is usually a tubular structure, either outstretched (Tylenchomorpha) or reflexed (Longidoridae, Trichodoridae), and consists of three main zones. The blind end usually functions as a germinal zone, the telogonic condition (although primary oocytes may be formed before the last moult), and is followed by the growth zone and ripening zone. In some taxa the oocytes are connected to a central protoplasmic core or rachis (e.g. *Anguina*, *C. elegans*). In species without a rachis, the oocytes are interconnected by protoplasmic bridges.

The oviduct may show great variation in structure and development between taxa, and is considered to be of fundamental importance in nematode systematics (Geraert, 1983, 2006). In Tylenchomorpha, the oviduct is generally formed by two rows of a few cells (e.g. three or four cells in Tylenchoidea, Criconematoidea and two cells in Aphelenchoididae), the number of cells corresponding to major taxa and being of diagnostic and phylogenetic importance. The tightly packed oviduct cells separate to form a tiny canal when a ripe oocyte is squeezed through. In Trichodoridae, the oviduct consists of two cells, while in Longidoridae the oviduct consists largely of flattened disc-like cells and has a collapsed lumen and a wider part that acts as a spermatheca. The presence or absence of a spermatheca(e) is of diagnostic importance (Trichodoridae). In Tylenchomorpha, the number of cells of the spermatheca, their shape and spatial arrangement is of taxonomic importance (Bert *et al.*, 2003).

The uterus may be a simple tube (Fig. 1.9A) but is usually more complex, being subdivided, for example, into a glandular, a muscular and a non-muscular portion (Fig. 1.9B).

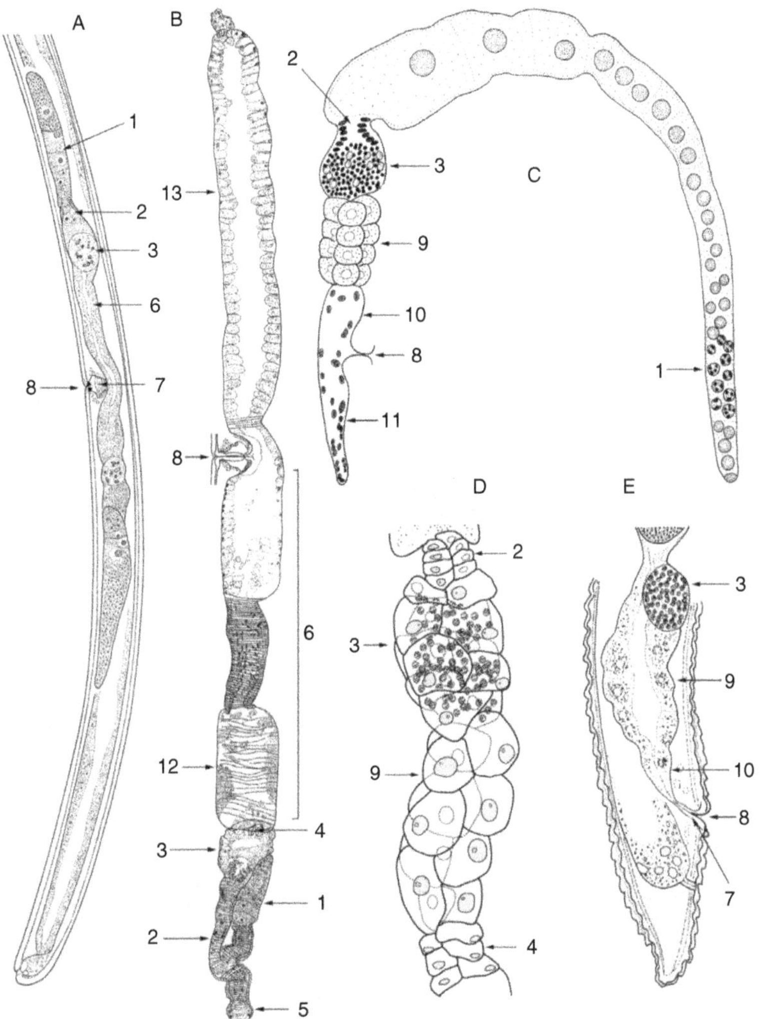

Fig. 1.9. Female reproductive system in plant-parasitic taxa. A: Didelphic-amphidelphic system (*Trichodorus*, Trichodoridae). B: Pseudomonodelphic system (*Xiphinema surinamense*, Longidoridae). C–E: Monodelphic system. C: Total female reproductive system of *Pratylenchus coffeae* (Pratylenchidae). D: Detail of oviduct–uterus region in *Rotylenchus goodeyi* (Tylenchoidea). E: Posterior body region (*Hemicriconemoides minor*, Criconematoidea). 1: Ovary. 2: Oviduct. 3: Spermatheca. 4: Sphincter/valve. 5: Ovarial sac. 6: Uterus. 7: Vagina. 8: Vulva. 9: Crustaformeria uterus. 10: Uterine sac. 11: Post-vulval uterine sac. 12: Uterus *pars dilata*. 13: Reduced anterior branch. A: From Decraemer (1991). B: From Decraemer *et al.* (1998). C: From Román and Hirschmann (1969). D: From Bert *et al.* (2003). E: From Decraemer and Geraert (1992).

In Tylenchomorpha, the uterus *sensu stricto* is restricted to the eggshell-producing region of the gonoduct, the crustaformeria, of which the number and arrangement of cells (tricolumella, tetracolumella, quadricolumella) are of taxonomic importance (Fig. 1.9C and D).

In Longidoridae the uterus may vary from very short and simple to very long and complex, with local uterine differentiations such as the Z-organ (*Xiphinema*). The Z-organ probably slows down the descent of the eggs towards the vulva, although its function is not entirely clear. Part of the uterus or uteri opposite the vagina may be differentiated into a muscular ovejector to assist with egg laying. The uterine structure is of taxonomic importance in Longidoridae. The uterus is connected to the vulva by the vagina, which may vary in size and shape (Trichodoridae, Longidoridae) and be of taxonomic importance. The vulva may also vary in shape from pore-like to slit-like (transverse or longitudinal) and may be occluded by a copulation plug (e.g. in *Trichodorus*). Various sets of muscles attach to the vagina, some serving to dilate the lumen while others constrict the vagina or suspend it in the body, thereby preventing it from prolapsing during egg laying. Egg laying is also mediated by the vulval dilators, which connect the vulva to the lateral body wall and is innervated by two specific motor neurons. Egg laying may be obstructed by absence of food or by aging and reduced plasticity of the body cuticle. Eggs may be laid singly or stuck together in masses in a gelatinous matrix secreted by the female. Such egg masses are associated with species where the females swell and become sedentary, although some obese genera retain all the eggs within the body, the female cuticle tanning on death to form a tough cyst. Egg sacs and cysts serve to protect the vulnerable eggs. Most nematode eggs are morphologically very similar and are cylindroid with rounded ends in shape with a transparent shell (except for some animal-parasitic nematodes), and are of similar dimension, irrespective of the size of the adult. Different-sized eggs may occur between free-living and parasitic forms of the same species. The eggshell consists of four main layers: an outer vitelline layer derived from the oolemma and the first layer formed after sperm penetration, middle chitinous and chondroitin layers, and an inner lipid layer (see Chapter 7, Fig. 7.3). The lipid layer is largely responsible for the impermeability of the eggshell, which is only permeable to chemicals before the lipid layer is formed during the passage of the egg down the uterus. The permeability of the lipid layer also alters before hatching (see Chapter 7).

1.2.8.2. Male reproductive system (Fig. 1.10)

Nematodes may have a single gonad or testis (monorchic) or two testes (diorchic). In Longidoridae, the male reproductive system comprises two testes, the posterior one being reflexed; the gonoduct consists of a single *vas deferens*, which may be differentiated into a strongly muscular *ductus ejaculatorius* (Enoplida) opening to the cloaca, a common cavity with the outlet of the digestive system. The posterior part of the testis (plural testes) and/or the anterior part of the *vas deferens* may form a *vesicula seminalis* or sperm storage zone. Ejaculatory glands may be associated with the *vas deferens*. Typically, four cells are present along each side of the intestine at pre-rectum level in dorylaims and the ducts merge posteriorly with the *vas deferens*. Such glands are absent in Trichodoridae. Trichodoridae males are monorchic, as are males of the Tylenchomorpha, except for some *Meloidogyne* males, which occasionally have two testes due to sex reversal during development. Sperm are continuously produced in most plant-parasitic taxa but in genera of the Criconematoidea and some Sphaerularioidea they are produced before the final moult. In such forms the testis appears to be degenerate, although well-developed sperm fill the gonoduct. Sperm size and shape may differ between taxa.

The copulatory apparatus generally consists of two equal, cuticularized, tubular structures called the spicules and a gubernaculum as a guiding structure. Spicules are rarely absent (e.g. the free-living rhabditid *Myolaimus*), may be dimorphic (*Pratylenchus penetrans*), partly fused (in rhabditids) or reduced to a single spicule (as in some mermithids) and are independently retractable. Spicules are formed by gradual invagination of the posterior wall of the spicular pouches, which originate from the spicular primordium, specialized cells of the dorsal wall of the cloaca. Spicules are sensory structures that serve

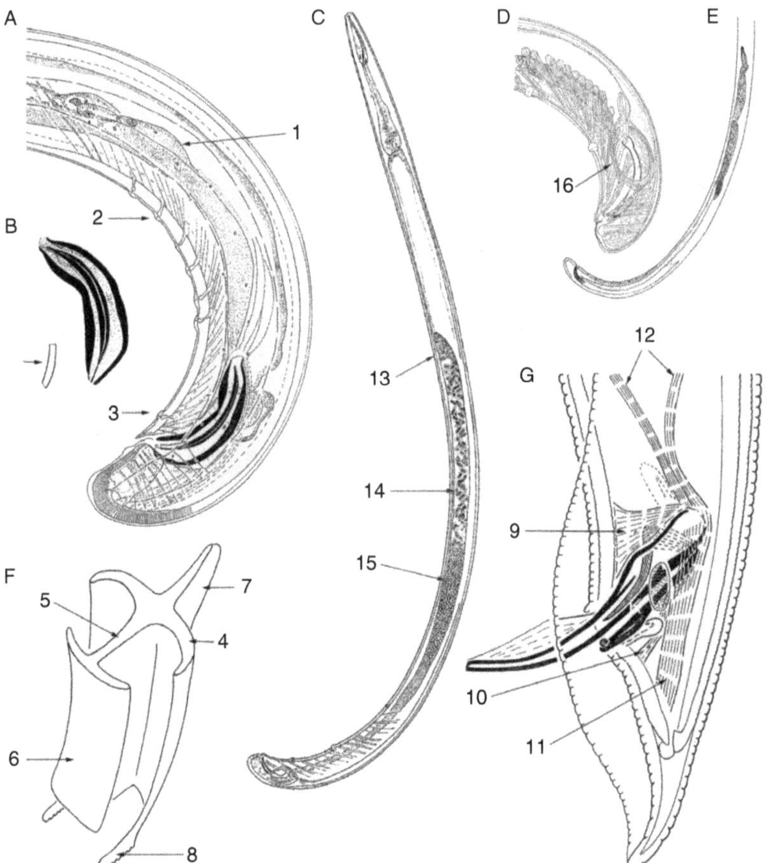

Fig. 1.10. Male reproductive system. A: Posterior body region (*Paraxiphidorus*). B: Spicule and guiding piece or crura (arrow). C: Monorchic system (*Trichodorus*). D: Copulatory apparatus with spicule retractor muscles and capsule of protractor muscles (arrow) in *Trichodorus similis*. E: Diorchic system (*Xiphinema*). F: Gubernaculum. G: Muscles and copulatory apparatus of *Hoplolaimus*. 1: Ejaculatory gland. 2: Mid-ventral supplement. 3: Adanal supplement. 4: Corpus. 5: Cuneus. 6: Capitulum. 7: Apophysis or apodeme. 8: Crura. 9: Anterior spicule protractor. 10: Gubernaculum protractor. 11: Posterior spicule protractor. 12: Spicule retractor muscles. 13: Germinal zone of testis. 14: *Vesicula seminalis*. 15: *Vas deferens*. A, B: From Decraemer *et al.* (1998). C: From Decraemer (1991). D: From Decraemer (1995). F: From Maggenti (1981). G: From Coomans (1962).

to detect the female vulva and help to channel sperm during copulation. Each spicule contains sensilla with one or two dendrites and dendritic process or receptor enclosed in a channel leading to a pore or pores at or near the spicule tip. In several taxa, the spicules are differentiated and show ornamentations such as striae, bristles, a ventral velum or subventral vela (flanges), the presence and form of which are of taxonomic importance. The gubernaculum is a cuticular thickening of the dorsal wall of the spicular pouch and acts as a guide during spicule protrusion. The gubernaculum can be very simple to complex; in Longidoridae only the crura (= lateral guiding pieces) remain (Fig. 1.10B). The copulatory apparatus functions by means of sets of protractor and retractor muscles. There are protractor and retractor muscles for each spicule as well as protractor and retractor muscles attached to the gubernaculum. In Triplonchida (Trichodoridae) the spicule protractor muscles form a capsule of suspensor muscles. Two sets of retractor muscles extend anteriorly and subdorsally from the tip of this capsule and one set runs from tip capsule to the spicule head. Contraction of the retractor muscles pulls the spicules inside the body, whilst contraction of the capsule of the protractor muscles protrudes the spicules. In the caudal region, accessory genital structures such as caudal alae, pre- and post-cloacal supplements, genital papillae or rays on the bursa, setae or suckers may be present. The arrangement of the pre- and post-cloacal supplements (see Longidoridae, Trichodoridae) or genital papillae (Rhabditomorpha) are of taxonomic importance. Paired genital papillae may also be present on the post-cloacal lip (= hypoptygma) as in Merliniinae (Tylenchomorpha). Trichodoridae possess in addition ventromedian cervical papillae (may be observed as pores) in the pharyngeal region. Nematode sperm cells are non-flagellate, non-ciliate and show amoeboid motility. They are diverse in size and shape and possess a major sperm protein associated with their unusual motility and correlated with a fibrillar skeleton in mature spermatozoa. Except for Enoplida, mature sperm lack a nuclear envelope. In Trichodoridae, sperm shape and size are used as diagnostic features.

1.3. Life Cycle Stages

Nematodes typically have an egg stage, four juvenile stages (J1–J4) and the adult male and female (Fig. 1.11). The egg is usually cylindroid with a chorion/eggshell of varying thickness (details of eggshell structure are given in Chapter 7, Section 7.6.5.1). Eggs may be deposited singly or in masses. In the latter case they may be held within a gelatinous matrix (root-knot nematodes) or protected within a tough cyst formed from the body of the dead female. Most nematodes moult four times before becoming adult, although there are nematodes, such as certain *Xiphinema* species, that have life cycles shortened to only three moults, probably due to epogenesis (maturation of gametes before completion of body growth), a phenomenon not uncommon in parthenogenetic organisms. In the Longidoridae and Trichodoridae, the juvenile that hatches from the egg is J1, whereas in Tylenchomorpha it is the J2, the first moult occurring within the egg. In some animal-parasitic nematodes, two moults occur within the egg.

Before reaching maturity, the juveniles usually resemble the adult female in morphology, differing in the absence of a mature reproductive system and in certain measurements and proportions. In some groups, one juvenile stage (usually the J3, but may be the J2 or the J4) is more resistant to environmental stress than the other stages (see Chapter 7). This juvenile stage is therefore specialized for dispersal and for surviving inhospitable conditions, and often represents the infective stage, usually the J3 in animal parasites, or the J2 in plant

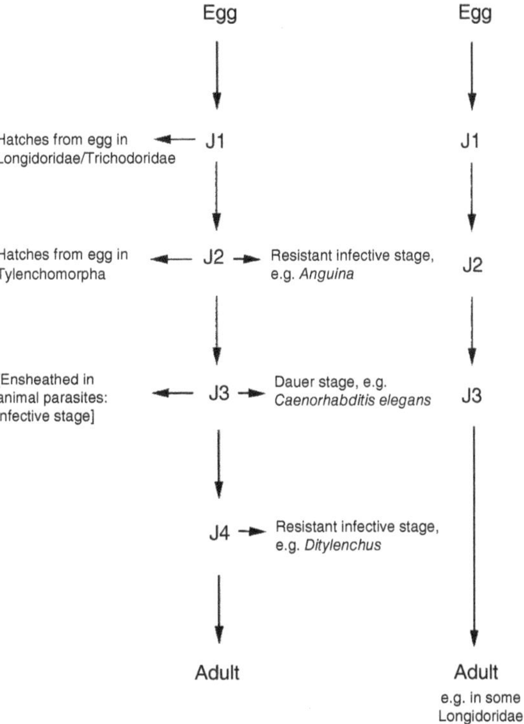

Fig. 1.11. Basic life cycles of plant-parasitic nematodes.

parasites, rarely the juvenile female (e.g. *Rotylenchulus*). This infective stage is non-feeding and may retain the cuticle from the previous stage as a protective sheath around its body. In free-living rhabditid nematodes, this modified stage is often called a dauer larva (from the German for endurance), dauer juvenile or dauer, and may represent an alternative pathway to the normal development process (see Chapter 7). Dauers possess a thicker cuticle, which is often hydrophobic, and are non-feeding, the oral aperture being closed, the non-functional pharynx and intestine reduced in development and the anterior sensilla modified. Many free-living species with dauers are dispersed by insects or other arthropods, the dauers either attaching themselves to the vector or congregating beneath the elytra or within the intersegmental folds. This is known as phoresy, dauers attached to the outside of the insect vectors being ectophoretic whilst those inside the vectors (for example in the Malpighian tubes, intestine, tracheal system or bursa copulatrix) are endophoretic. When conditions are suitable, the dauer stages can reactivate by absorbing water; the life cycle then continues. The term dauer has also been used for resistant stages of some plant-parasitic nematodes (e.g. J2 of anguinids) although here it does not seem to represent an alternative pathway to the formation of a normal J2 (see Chapter 7).

Nematodes are typically amphimictic and have separate males and females (see Chapter 7). Many species, however, lack males and reproduce either by parthenogenesis (the usual case) or, more rarely, by hermaphrodism. The generation time of nematodes can, depending on the species concerned, vary from a few days to a year or more. Females are usually oviparous, but in some groups, juveniles can hatch inside the body of the female (ovoviviparity), usually resulting in her death (this is known as endotokia matricida). Ovoviviparity can be induced in normally reproducing species by pollutants, e.g. sulphur dioxide (Walker and Tsui, 1968). In some specialized plant parasites, the female body swells and becomes greatly enlarged. The cuticle in some species thickens and tans/darkens on the death of the female to form a tough cyst that surrounds the retained eggs and protects them from deleterious factors, such as drought or extremes of temperature.

1.4. Feeding Groups

Nematodes display a wide range of feeding habits or trophisms. Some species of nematodes are microphagous/microbotrophic, feeding on small microorganisms, while others are saprophagous, feeding on dead and decaying organic matter. Many species of nematodes are phytophagous, obtaining nourishment directly from plants, whilst others are omnivorous or predatory. Parasitism of invertebrates or vertebrates is also common. There are three main types of plant parasitism: ectoparasitic, endoparasitic and semi-endoparasitic (Fig. 1.1).

1. **Ectoparasitic** – the nematode remains in the soil and does not enter the plant tissues. It feeds by using the stylet to puncture plant cells – the longer the stylet the deeper it can feed within the plant tissues. The majority of ectoparasitic species remain motile whereas others, e.g. *Cacopaurus*, become permanently attached to the root by the deeply embedded stylet.
2. **Semi-endoparasitic** – only the anterior part of the nematode penetrates the root, the posterior section remaining in the soil phase (e.g. *Helicotylenchus*).
3. **Endoparasitic** – in this type of parasitism the entire nematode penetrates the root tissue. Migratory endoparasites, such as *Pratylenchus* and *Radopholus*, retain their mobility and have no fixed feeding site within the plant tissue, whereas the more advanced sedentary endoparasites have a fixed feeding site and induce a sophisticated trophic system of nurse cells or syncytia (see Chapter 9). Establishment of a specialized feeding site enhances the flow of nutrients from the host, thereby allowing the females to become sedentary and obese in form and highly fecund. Sedentary endoparasites also have a migratory phase before the feeding site is established. In root-knot and cyst nematodes it is only the J2 and adult male that are migratory but in *Nacobbus*, for example, all juvenile stages, the male and the immature vermiform female are migratory, only the mature female being sedentary (see Box 3.1).

The above categories are not mutually exclusive as some genera may, depending on the host, be either semi-endoparasitic or migratory ecto-endoparasitic, e.g. *Hoplolaimus* (Tytgat *et al.*, 2000) or *Helicotylenchus*, whilst some sedentary parasites have only the anterior body embedded in the root (= sedentary semi-endoparasites), e.g. *Rotylenchulus, Tylenchulus.*

In ectoparasites and most migratory endoparasites, any vermiform stage may feed on, or penetrate, the root, but in those plant-parasitic nematodes where the female becomes obese and sedentary, the infective stage is usually the vermiform J2. This is true for *Heterodera*/*Globodera* and *Meloidogyne* species, for example, although in *Rotylenchulus* it is the immature vermiform female that is the infective stage, the non-feeding juveniles and males remaining in the soil. *Tylenchulus* has a similar life cycle, although here the female juveniles browse on epidermal cells, only the immature female penetrating deeper into the root cortex. In *Nacobbus*, adapted to a difficult and unpredictable environment, all vermiform stages, including the immature female, are infective and may enter and leave roots a number of times. In some Tylenchomorpha the males have a degenerate pharynx and do not feed. Such males are found in *Radopholus*, for example, and occur throughout the Criconematoidea. In *Paratylenchus*, it is the J4 female that lacks a stylet and hence cannot feed on plant cells.

1.5. Classification of Plant-parasitic Nematodes

Nematode classification is currently in a state of flux as molecular phylogenies become increasingly pertinent. For this reason, the higher classification is particularly fluid with proposals for Infraorders, etc., and the bringing together of groups that previously, under classical systematic systems, had been regarded as distantly related (De Ley and Blaxter, 2002). For plant-parasitic nematodes, the most recent classical schemes are those of Siddiqi (1986, 2000), Andrássy (2007) and Manzanilla-López and Hunt (2012) for the Tylenchina, except for the classification of *Meloidogyne* for which we follow the suggestion by Geraert (2013) to classify it as a separate family, Meloidogynidae, close to the Pratylenchidae (see Álvarez-Ortega *et al.*, 2019) and for the Criconematinae where we added *Mesocriconema* as a valid genus based on molecular analyses (Nguyen *et al.*, 2022), although *Mesocriconema* as well as *Criconema*, *Criconemoides*, *Lobocriconema*, *Ogma* and *Hemicriconemoides* are not monophyletic; Hunt (1993, 2008), Andrássy (2007, 2009), Kanzaki and Giblin-Davis (2012) and Manzanilla-López and Hunt (2012) for the Aphelenchoidea and/or Longidoridae, and Hunt (1993), Decraemer (1995), Andrássy (2009), Duarte *et al.* (2010) and Manzanilla-López and Hunt (2012) for the Trichodoridae.

In the following scheme, the higher systematic categories are as proposed by De Ley and Blaxter (2002), whilst at subfamily and generic level, a **simplification** of Siddiqi (2000) and Hunt (1993, 2008) is followed with some genera and many synonyms omitted. The most important plant-parasitic genera are indicated in **bold** type.

It is useful to note that the various systematic ranks have different suffixes attached to the stem:

Class	-ea
Subclass	-ia
Order	-ida
Infraorder	-omorpha
Suborder	-ina
Superfamily	-oidea
Family	-idae
Subfamily	-inae
Tribe	-ini

PHYLUM NEMATODA POTTS, 1932
CLASS CHROMADOREA INGLIS, 1983
Subclass Chromadoria Pearse, 1942
Order Rhabditida Chitwood, 1933
Suborder Tylenchina Thorne, 1949
Infraorder Tylenchomorpha De Ley & Blaxter, 2002

Superfamily Aphelenchoidea Fuchs, 1937[1]
Family Aphelenchidae Fuchs, 1937
Subfamily Aphelenchinae Fuchs, 1937
Aphelenchus Bastian, 1865
Subfamily Paraphelenchinae T. Goodey, 1951
Paraphelenchus Micoletzky, 1922
Family Aphelenchoididae Skarbilovich, 1947
Subfamily Aphelenchoidinae Skarbilovich, 1947
Aphelenchoides Fischer, 1894
Schistonchus Cobb, 1927
Subfamily Parasitaphelenchinae Rühm, 1956
Bursaphelenchus Fuchs, 1937
Superfamily Tylenchoidea Örley, 1880
Family Tylenchidae Örley, 1880
Subfamily Tylenchinae Örley, 1880
Aglenchus Andrássy, 1954
Tylenchus Bastian, 1865
Coslenchus Siddiqi, 1978
Filenchus Andrássy, 1954
Subfamily Boleodorinae Khan, 1964
Basiria Siddiqi, 1959
Boleodorus Thorne, 1941
Subfamily Thadinae Siddiqi, 1986
Thada Thorne, 1941
Subfamily Duosulciinae Siddiqi, 1979
Duosulcius Siddiqi, 1979
Malenchus Andrássy, 1968
Miculenchus Andrássy, 1959
Ottolenchus Husain & Khan, 1967
Subfamily Tanzaniinae Siddiqi, 2000
Tanzanius Siddiqi, 1991
Subfamily Ecphyadophorinae Skarbilovich, 1959
Ecphyadophora de Man, 1921
Subfamily Ecphyadophoroidinae Siddiqi, 1986
Ecphyadophoroides Corbett, 1964
Subfamily Atylenchinae Skarbilovich, 1959
Atylenchus Cobb, 1913
Subfamily Eutylenchinae Siddiqi, 1986
Eutylenchus Cobb, 1913
Subfamily Tylodorinae Paramonov, 1967
Tylodorus Meagher, 1964
Subfamily Pleurotylenchinae Andrássy, 1976
Cephalenchus Goodey, 1962
Pleurotylenchus Szczygiel, 1969

[1] This scheme only covers the plant-parasitic forms. For a complete overview of this group, see Chapter 5

Subfamily Epicharinematinae Maqbool & Shahina, 1985
Epicharinema Raski, Maggenti, Koshy & Sosamma, 1980
Gracilancea Siddiqi, 1976
Family Dolichodoridae Chitwood *in* Chitwood & Chitwood, 1950
Subfamily Dolichodorinae Chitwood *in* Chitwood & Chitwood, 1950
Dolichodorus Cobb, 1914
Neodolichodorus Andrássy, 1976
Subfamily Brachydorinae Siddiqi, 2000
Brachydorus de Guiran & Germani, 1968
Subfamily Telotylenchinae Siddiqi, 1960
Histotylenchus Siddiqi, 1971
Neodolichorhynchus Jairajpuri & Hunt, 1984
Paratrophurus Arias, 1970
Quinisulcius Siddiqi, 1971
Sauertylenchus Sher, 1974
Telotylenchoides Siddiqi, 1971
Telotylenchus Siddiqi, 1960
Trichotylenchus Whitehead, 1960
Trophurus Loof, 1956
Tylenchorhynchus Cobb, 1913
Uliginotylenchus Siddiqi, 1971
Subfamily Meiodorinae Siddiqi, 1976
Meiodorus Siddiqi, 1976
Subfamily Macrotrophurinae Fotedar & Handoo, 1978
Macrotrophurus Loof, 1958
Subfamily Merliniinae Siddiqi, 1971
Amplimerlinius Siddiqi, 1976
Geocenamus Thorne & Malek, 1968
Merlinius Siddiqi, 1970
Nagelus Thorne & Malek, 1968
Scutylenchus Jairajpuri, 1971
Subfamily Belonolaiminae Whitehead, 1960
Belonolaimus Steiner, 1949
Carphodorus Colbran, 1965
Ibipora Monteiro & Lordello, 1977
Morulaimus Sauer, 1966
Family Hoplolaimidae Filipjev, 1934
Subfamily Hoplolaiminae Filipjev, 1934
Aorolaimus Sher, 1963
Aphasmatylenchus Sher, 1965
Helicotylenchus Steiner, 1945
Hoplolaimus Daday, 1905
Scutellonema Andrássy, 1958
Rotylenchus Filipjev, 1936
Subfamily Rotylenchulinae Husain & Khan, 1967
Acontylus Meagher, 1968
Bilobodera Sharma & Siddiqi, 1992

Rotylenchulus Linford & Oliveira, 1940
Senegalonema Germani, Luc & Baldwin, 1984
Verutus Esser, 1981
Subfamily Heteroderinae[2] Filipjev & Schuurmans Stekhoven, 1941
Betulodera Sturhan, 2002
Cactodera Krall & Krall, 1978
Dolichodera Mulvey & Ebsary, 1980
Globodera Skarbilovich, 1959
Heterodera Schmidt, 1871
Paradolichodera Sturhan, Wouts & Subbotin, 2007
Punctodera Mulvey & Stone, 1976
Vittatidera Bernard, Handoo, Powers, Donald & Heinz, 2010
Subfamily Ataloderinae Wouts, 1973
Atalodera Wouts & Sher, 1971
Bellodera Wouts, 1985
Camelodera Krall, Shagalina & Ivanova, 1988
Cryphodera Colbran, 1966
Ekphymatodera Bernard & Mundo-Ocampo, 1989
Hylonema Luc, Taylor & Cadet, 1978
Rhizonema Cid del Prado, Lownsbery & Maggenti, 1983
Sarisodera Wouts & Sher, 1971
Subfamily Meloidoderinae Golden, 1971
Meloidodera Chitwood, Hannon & Esser, 1956
Subfamily Nacobboderinae Golden & Jensen, 1974
Bursadera Ivanova & Krall, 1985
Meloinema Choi & Geraert, 1974
Family Meloidogynidae Skarbilovich, 1959
Meloidogyne Göldi, 1887
Family Pratylenchidae Thorne, 1949
Subfamily Pratylenchinae Thorne, 1949
Apratylenchus Trinh, Waeyenberge, Nguyen, Baldwin, Karssen & Moens, 2009
Pratylenchus Filipjev, 1936
Zygotylenchus Siddiqi, 1963
Subfamily Hirschmanniellinae Fotedar & Handoo, 1978
Hirschmanniella Luc & Goodey, 1964
Subfamily Radopholinae Allen & Sher, 1967
Achlysiella Hunt, Bridge & Machon, 1989
Apratylenchoides Sher, 1973
Hoplotylus s'Jacob, 1960
Pratylenchoides Winslow, 1958

[2] The subfamily Heteroderinae has been raised to family level and the genus *Verutus* classified as member of the Verutinae Esser, 1981 within the family Heteroderidae by Subbotin *et al.* (2017); the subfamily Nacobboderinae based on *Meloinema* is considered by Subbotin and Kim (2021) to be a sister group of *Meloidogyne* and classified within the Meloidogynidae but so far no molecular data are available for *Bursadera*.

Radopholus Thorne, 1949
Zygradus Siddiqi, 1991
Subfamily Nacobbinae Chitwood *in* Chitwood & Chitwood, 1950
Nacobbus Thorne & Allen, 1944
Superfamily Criconematoidea Taylor, 1936
Family Criconematidae Taylor, 1936
Subfamily Criconematinae Taylor, 1936
Bakernema Wu, 1964
Criconema Hofmänner & Menzel, 1914
Lobocriconema De Grisse & Loof, 1965
Neolobocriconema Mehta & Raski, 1971
Ogma Southern, 1914
Subfamily Macroposthoniinae Skarbilovich, 1959
Criconemoides Taylor, 1936[3]
Discocriconemella De Grisse & Loof, 1965
Mesocriconema Andrássy, 1965
Xenocriconemella De Grisse & Loof, 1965
Subfamily Hemicriconemoidinae Andrássy, 1979
Hemicriconemoides Chitwood & Birchfield, 1957
Family Hemicycliophoridae Skarbilovich, 1959
Subfamily Hemicycliophorinae Skarbilovich, 1959
Colbranium Andrássy, 1979
Hemicycliophora de Man, 1921
Subfamily Caloosiinae Siddiqi, 1980
Caloosia Siddiqi & Goodey, 1964
Hemicaloosia Ray & Das, 1978
Family Tylenchulidae Skarbilovich, 1947
Subfamily Tylenchulinae Skarbilovich, 1947
Trophotylenchulus Raski, 1957
Tylenchulus Cobb, 1913
Subfamily Sphaeronematinae Raski & Sher, 1952
Sphaeronema Raski & Sher, 1952
Subfamily Meloidoderitinae Kirjanova & Poghossian, 1973
Meloidoderita Poghossian, 1966
Subfamily Paratylenchinae Thorne, 1949
Cacopaurus Thorne, 1943
Paratylenchus Micoletzky, 1922
Tylenchocriconema Raski & Siddiqui, 1975
Superfamily Sphaerularioidea Lubbock, 1861
Family Anguinidae Nicoll, 1935
Subfamily Anguininae Nicoll, 1935
Afrina Brzeski, 1981
Anguina Scopoli, 1777
Diptylenchus Khan, Chawla & Seshadri, 1969

[3] *Criconemella, Macroposthonia.*

Ditylenchus Filipjev, 1936
Heteroanguina Chizhov, 1980
Indoditylenchus Sinha, Choudhury & Baqri, 1985
Litylenchus Zhao, Davies, Alexander & Riley, 2011
Mesoanguina Chizhov & Subbotin, 1985
Nothanguina Whitehead, 1959
Nothotylenchus Thorne, 1941
Orrina Brzeski, 1981
Pseudohalenchus Tarjan, 1958
Pterotylenchus Siddiqi & Lenné, 1984
Safianema Siddiqi, 1980
Subanguina Paramonov, 1967

CLASS ENOPLEA INGLIS, 1983
Subclass Dorylaimia Inglis, 1983
Order Dorylaimida Pearse, 1942
Suborder Dorylaimina Pearse, 1942
Superfamily Dorylaimoidea Thorne, 1935[4]
Family Longidoridae Thorne, 1935
Subfamily Longidorinae[5]
Australodorus Coomans, Olmos, Casella & Chaves, 2004
Longidoroides Khan, Chawla & Saha, 1978
Longidorus Micoletzky, 1922
Paralongidorus Siddiqi, Hooper & Khan, 1963[5]
Paraxiphidorus Coomans & Chaves, 1995
Xiphidorus Monteiro, 1976
Subfamily Xiphinematinae
Xiphinema Cobb, 1913
Subclass Enoplia Pearse, 1942
Order Triplonchida Cobb, 1920
Suborder Diphtherophorina Coomans & Loof, 1970
Superfamily Diphtherophoroidea Micoletzky, 1922
Family Trichodoridae Thorne, 1935
Allotrichodorus Rodriguez-M., Sher & Siddiqi, 1978
Ecuadorus Siddiqi, 2002
Monotrichodorus Andrássy, 1976
Nanidorus Siddiqi, 1974
Paratrichodorus Siddiqi, 1974
Trichodorus Cobb, 1913

[4] There are plant-parasitic forms in other groups (e.g. *Californidorus*, *Longidorella*) although little is known about their importance and they are usually ignored.

[5] We do not include *Siddiqia* Khan, Chawla & Saha, 1978 as a valid genus (Peña-Santiago, 2021) because *Paralongidorus* is paraphyletic, with species arranged in two subclades, each of them clustering within *Longidorus* (Mwamula *et al.*, 2020).

1.6. Common Morphometric Abbreviations

Nematodes are characterized by a combination of measurements and ratios derived from the various body parts (Hooper, 1986; Ye and Hunt, 2021). Such morphometric characters are usually abbreviated, the most common being listed in Box 1.4. Measurements of, for example, the body, pharynx and tail are taken along the mid-line of the relevant structure (Fig. 1.12). Measurements of the spicule, a curved structure, are usually taken along the median line (a genuine indicator of actual length), although occasionally (and particularly in aquatic nematodes and in older descriptions) the chord, a straight line joining the two extremities of the spicule, is used. Body diameter should be measured perpendicular to the longitudinal body axis and care should also be taken that the nematode being quantified is not squashed (as a result, for example, of the coverslip not being properly supported with glass rods or beads) as this will produce a higher value than would otherwise be the case. Nematodes that have a long and often fragile tail that may be readily broken, are often measured from the cephalic region to the anal or cloacal aperture instead of to the end of the tail. By removing the error or variable element caused by a long and/or broken tail, ratios resulting from this modified measurement of body length, such as female vulval position, are more consistent and are therefore of greater utility in diagnostics.

Box 1.4. A list of the most commonly used morphometric abbreviations.

L = Total body length (anterior extremity to tail tip).
L′ = Body length from anterior end to anal or cloacal aperture (use when the tail is very long and/or frequently in a damaged state).
a = Total body length divided by maximum body diameter.
b = Total body length divided by pharyngeal length (the pharynx is measured from the anterior end to the pharyngo-intestinal junction, i.e. not to the posterior tip of the overlapping gland lobes).
b′ = Total body length divided by distance from anterior end of body to posterior end of pharyngeal glands.
c = Total body length divided by tail length.
c′ = Tail length divided by body diameter at the anal/cloacal aperture.
V = Position of vulva from anterior end expressed as percentage of body length. Superior figures to the left and right refer to the extent of anterior and/or posterior gonad or uterine sac, respectively, and are also expressed as a percentage of body length.
V′ = Position of vulva from anterior end expressed as percentage of distance from head to anal aperture.
T = Distance between cloacal aperture and anteriormost part of testis expressed as percentage of body length.
m = Length of conical part of tylenchid stylet as percentage of total stylet length.
o = Distance of dorsal pharyngeal gland opening posterior to stylet knobs expressed as a percentage of stylet length.
MB = Distance of median bulb from anterior end expressed as a percentage of total pharyngeal length.
Caudal ratio A = Length of hyaline tail divided by its proximal diameter.
Caudal ratio B = Length of hyaline tail divided by its diameter at a point 5 μm from its terminus.
μm = One-thousandth of a millimetre (micron).

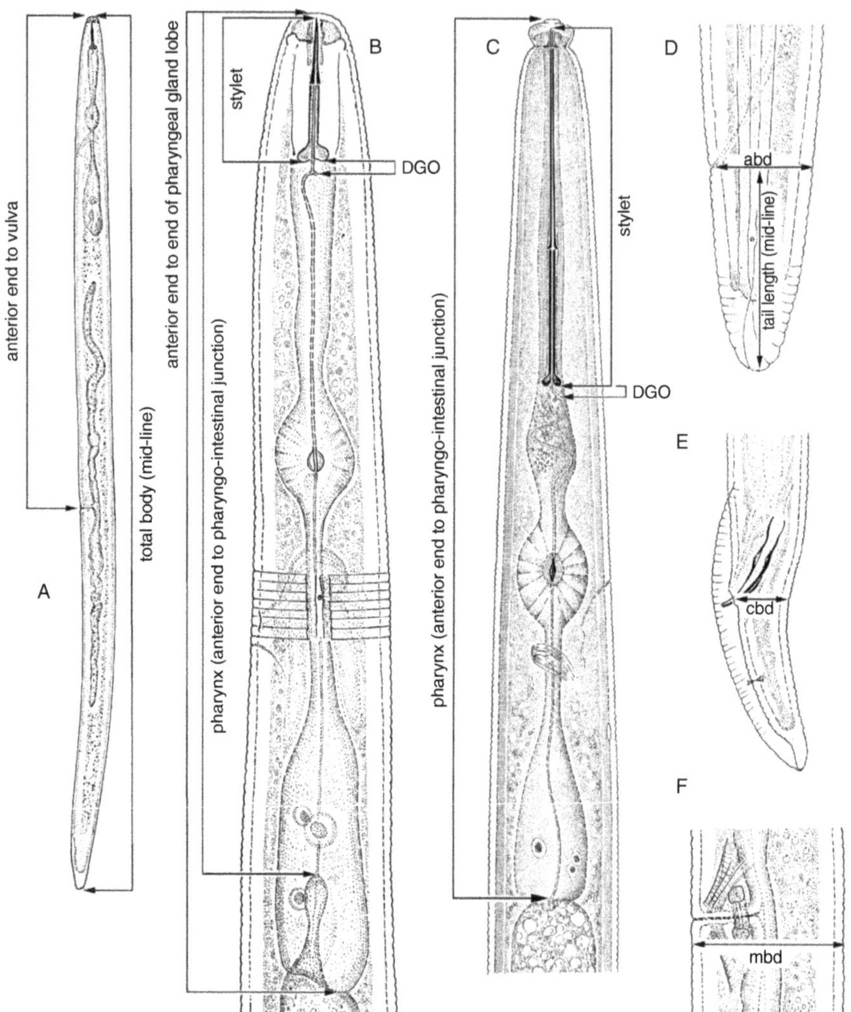

Fig. 1.12. How to take the basic measurements of a nematode. A: Entire nematode showing how to measure total body length along the mid-line from anterior end to tail tip, and distance of vulval aperture from the anterior end, also measured along the mid-line. B, C: Pharyngeal region showing how to measure stylet length, distance of dorsal gland orifice (DGO) from basal knobs, pharynx length (anterior end to pharyngo-intestinal junction) and anterior end to tip of pharyngeal gland lobes (this is not the same as the true pharynx length, but is usually taken in species where the glands overlap the intestine, the pharyngo-intestinal junction itself often being obscure). D: Female tail region showing how to measure tail length, and anal body diameter (abd) by extrapolating a line at 90° to the longitudinal axis. E: Male tail region showing corresponding measurement for body diameter at the cloacal aperture (cbd). Spicule length is measured from tip to tip along the curved median line. F: Mid-body region showing how to take the maximum body diameter (mbd), again at 90° to the longitudinal body axis. Figure digitally compiled from line drawings by Orton Williams (1974) and Siddiqi (1974a), *CIH Descriptions of Plant-parasitic Nematodes*, courtesy CABI.

Molecular Systematics*

Sergei A. Subbotin[1,2]** and Maurice Moens[3]

[1]California Department of Food and Agriculture, Sacramento, CA, USA; [2]Center of Parasitology of A.N. Severtsov Institute of Ecology and Evolution of the Russian Academy of Sciences, Moscow, Russia; [3]Flanders Research Institute for Agriculture, Fisheries and Food (ILVO), Merelbeke, Belgium and Laboratory for Agrozoology, Ghent University, Ghent, Belgium

*A revision of Subbotin, S.A., Waeyenberge, L. and Moens, M. (2013) Molecular systematics. In: Perry, R.N. and Moens, M. (eds) *Plant Nematology*, 2nd edn. CAB International, Wallingford, UK.
**Corresponding author

 Plant Nematology, 3rd edn (eds R.N. Perry, M. Moens and J.T. Jones)
DOI: 10.1079/9781800622456.0002

2.1. Phylogenetics and Phylogenomics

The tasks of systematics are: (i) to name, identify and catalogue organisms (**taxonomy**); (ii) to discover the ancestral relationships among organisms (**phylogenetics**); and (iii) to organize information about the diversity of organisms into a hierarchical system (**classification**). Molecular systematics is the application of knowledge of genome information, especially sequence and structure of DNA, RNA molecules and amino acid chains, for addressing questions regarding the phylogeny and taxonomy of organisms.

There are several reasons why molecular data are more suitable for phylogenetic studies than morphological ones. First, DNA and protein sequences are strictly heritable entities, whereas morphological characters can be influenced by various environmental factors. Second, the interpretation of molecular characters, such as the assessment of homologies, is generally easier than that of morphological characters. Third, molecular characteristics generally evolve much more regularly than morphological ones and, therefore, can provide a clearer picture of relationships. Fourth, molecular characters are much more abundant than morphological features, and many can be generated in a relatively short period of time. Various preserved, deformed and partly degraded materials can often be used for molecular studies. Using standard protocols and commercial kits, sequence information of certain genes or DNA fragments can be obtained from a single nematode or even only a part of one. Using specific primers, nematode DNA can be amplified from soil or plant extracts. Moreover, specialized methods enable the extraction of short DNA fragments from preserved, formalin-fixed and glycerin-embedded specimens. Recent achievements in molecular biology and the wide application of molecular techniques have revolutionized our knowledge in taxonomy and phylogeny of nematodes. The use of such techniques is becoming routine in nematology (Powers, 2004; Blok, 2005; Perry *et al.*, 2007; Subbotin *et al.*, 2010a,b; Subbotin, 2021a,b).

Molecular phylogenetics compares and analyses single or a few genes. However, systematics supported by DNA has entered a new era in which many thousands of nucleotides and whole genomes can be obtained inexpensively and in a relatively short period of time. The approach that involves genome data in evolutionary reconstruction is called **phylogenomics**. As a result, a major advance in the generation of genome data is happening, and studies on many more genomes are underway or planned, which will yield data that can be used to address critical questions in nematode phylogenetic systematics (Blaxter *et al.*, 2016).

2.2. Species Concepts and Delimiting Species in Nematology

There has been considerable debate concerning the definition of a species. Species were at first merely taxonomic units, i.e. the named categories to which Linnaeus and other taxonomists of the 18th century assigned organisms, largely on the basis of appearance. According to the **typological species concept**, the species is considered a community of specimens described by characteristic features of its type specimen. In the early 20th century, taxonomists had accumulated a great deal of evidence leading to the widely used modern concept of species. This species concept was based on two observations: (i) species are composed of populations; and (ii) populations are co-specific if they successfully interbreed with each other. This idea was articulated by Ernst Mayr (1942) in the **biological species concept**: "Species are groups of interbreeding natural populations that are reproductively isolated from other such groups". In the last 50 years several additional species concepts have been proposed. The most popular one in systematics is the so-called **phylogenetic species concept**. This concept does not emphasize the present properties of organisms or their hypothetical future, but rather points at their phylogenetic history. However, the applicability of this concept is debatable, for it proposes operational criteria of how to delimit species only as phylogenetic taxa, rather than describing the role that species play in the living world.

As with any concept, the biological species concept has its limitations. Application of the biological species concept is restricted to sexual, outcrossing populations over a short period of evolutionary time, and therefore excludes parthenogenetic organisms. Furthermore, in practice the diagnosis of biological species is seldom done by testing their propensity to interbreed and produce fertile offspring but is often made by examining the difference in morphology. This should not be a contradiction, because phenotypic characters often, although not always, serve as markers for reproductive isolation. Morphological distinctiveness is a good, but not infallible, criterion for separating species. Sibling species may remain undetected, even after careful morphological examination, unless allozymes or other genetic markers are studied. The genetic difference between related species appears to vary substantially but generally increases with the time elapsed since their reproductive isolation. The degree of the genetic distance among populations, estimated from allozyme frequency, nucleotide divergence, amplified fragment length polymorphism (AFLP), random amplified polymorphic DNA (RAPD), microsatellites or other markers, can be evidence to determine whether or not they should be assigned the status of species. These approaches to delimiting species are based on indirect inference of the presence or absence of gene flow.

Another approach for analysis of genetic data uses tree-based methods. In this approach species are delimited on the properties of a phylogenetic tree, which hypothesizes the relationships of groups bound by monophyly, or the shared presence of apomorphies. These methods can be used to detect asexual species. However, in practice, all of these methods can fail occasionally or be discordant with each other because in nature the process of speciation seems to create diffuse boundaries, or a hybrid zone, between diversifying species. As a result, groups of populations form, sometimes named as subspecies, that are not fully reproductively isolated from each other. Moreover, these methods have different sensitivities and can reflect different properties of speciation. The solution to the problem of delimiting species can be found in the concept of **polyphasic taxonomy and classification** or integrated approach in systematics.

2.3. Phylogenetics and Classification

Before the 1950s, taxonomists attempted to construct classifications on the dual criteria of common ancestry and their similarities. In the 1950s, the principles of numerical taxonomy were introduced basing classification not on a few important features but on multiple character data. Numerical methods of analysis were used to create diagrams of overall similarity among species. Such a diagram, called a phenogram, was intended to give an objective basis for classification. This approach was named as **phenetics**, giving rise to **phenetic classification**. This approach does not take into account the effects of parallel or convergent evolution in taxonomic interpretations. Another system was based on the argument that classification should rigorously reflect only phylogenetic relationships, not the degree of adaptive divergence or overall similarity. Classifications based on phylogenetic principles are named phylogenetic classifications; only shared, unique character states of similarity provide evidence for phylogenetic relationships. This approach to phylogenetic inference is known as **cladistics**. Branching diagrams constructed by cladistic methods are sometimes named as cladograms, and monophylic groups are called **clades**. A taxon should be a monophyletic group, originating from a single common ancestor, as opposed to a paraphyletic taxon, which includes only some of the descendants of a common ancestor, or a polyphyletic taxon, whose members share only a distant common ancestor and are usually circumscribed by other characteristics (i.e. **homoplasies**). Several terms are used to describe different character states for taxa under investigation: **plesiomorphy** (ancestral character state), **symplesiomorphy** (shared ancestral character state), **apomorphy** (derived character state), **synapomorphy** (shared derived character state) and **autapomorphy** (derived character state possessed by a single taxon). Within such a framework, the concept of **parsimony** is now widely applied to the reconstruction of phylogenetic relationships. It points out that among the various phylogenetic trees hypothesized for a group of taxa, the best one requires the fewest evolutionary changes, including the fewest homoplasies. Phylogenetic classification must always rely on an inferred phylogenetic tree, which is only an estimated part of a true history of the divergence of a species. In practice, creating a phylogenetic tree to resolve the phylogenetic relation between organisms is not a simple task.

The **polyphasic taxonomy**, or integrated approach, refers to classifications based on a consensus of all available data and information (phenotypic, genotypic and phylogenetic) used for delimiting taxa at all levels. Such analysis leads to a transition type of taxonomy in which a compromise can be formulated on the basis of results presently at hand.

2.4. Molecular Techniques

Almost all information from the genome and proteome at all levels, including the sequence of fragments of DNA, RNA or amino acids, structure of molecules, gene arrangement and presence versus absence of proteins or genes, can be applied to molecular systematics. Various biochemical and molecular techniques have been introduced to nematology for diagnostics, the estimation of genetic diversity of populations and the inference of phylogenetic relationships between taxa. The choice of the technique depends on the research question.

2.4.1. Protein-based techniques

These were the first of the molecular techniques to be applied in nematology. Soluble proteins extracted from nematodes are separated on polyacrylamide, starch, cellulose acetate or agarose gels under an electric field on the basis of their different molecular masses. Extracts from nematodes comprise thousands of different proteins but after total staining specific band patterns can be found for each sample. Differences in banding patterns between species or populations may be used as taxonomic markers. Isoelectric focusing (IEF), separating proteins on the basis of their charge in a pH gradient, enables more stable profiles to be achieved and resolves proteins into sharp bands. The application of enzyme-staining techniques for the characterization of a single protein or small subset of proteins on IEF gels provides another diagnostic method. Extensive characterization of isozymes has been carried out for *Globodera*, *Heterodera*, *Radopholus*, *Meloidogyne*, *Pratylenchus* and other nematode groups. For many groups these studies revealed a wide variation between populations of the same species; however, limited interspecific variation was detected for species of root-knot nematodes. IEF is used as a routine diagnostic technique for *Globodera pallida* and *G. rostochiensis* (Karssen *et al.*, 1995) as well as for the separation of other cyst nematode species (Fig. 2.1). Isozyme phenotypes of adult females, especially of esterase and malate dehydrogenase, are considered to be very useful as reliable markers for identification of root-knot nematodes (see Chapter 3). Because IEF differentiates root-knot nematode species only by specific isozyme patterns of young females, this technique can only be used to separate root-knot nematodes at this life stage.

Two-dimensional polyacrylamide gel electrophoresis (2D-PAGE) provides a better protein separation and fingerprint for any particular sample. In the first dimension, proteins are separated according to their charge; in the second dimension, they are separated on their mass. After staining, the position of individual proteins appears as spots of

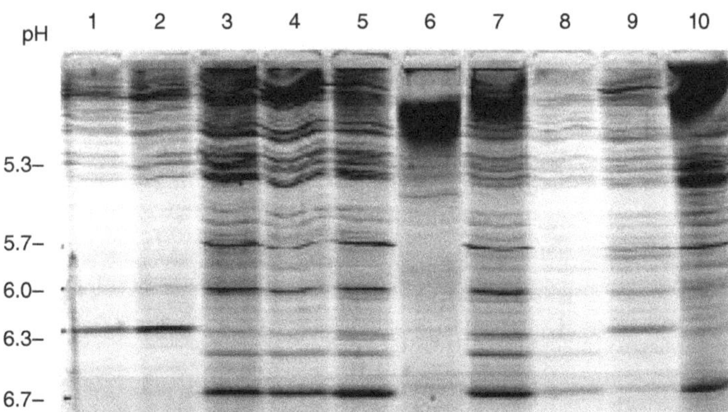

Fig. 2.1. Isoelectric focusing of proteins for species of the *Avenae* group. 1: *Heterodera avenae* (Rinkam, Germany). 2: *H. avenae* (Taaken, Germany). 3: *H. filipjevi* (Chabany, Ukraine). 4: *H. filipjevi* (Chernobyl, Ukraine). 5: *H. filipjevi* (Pushkin, Russia) 6: *H. pratensis* (Putilovo, Russia). 7, 8: *H. filipjevi* (Gorodets, Russia). 9: *H. filipjevi* (Vad, Russia). 10: H. *filipjevi* (Baimak, Russia). (From Subbotin *et al.*, 1996.)

various size, shape and intensity. This technique has been applied to separate species and populations of *Globodera* and *Meloidogyne*.

2.4.2. DNA-based techniques

Compared with the above approaches, analysis of DNA has several advantages. DNA profiles can be obtained rapidly from a few or even single nematodes and the clarity of the results enable species to be identified very easily without any effects of environmental or developmental variation.

2.4.2.1. DNA extraction

Extraction of DNA is the first step of molecular analysis. Using proteinase K is the most useful, cheap and rapid approach to extract DNA from nematodes (Subbotin, 2021a). It consists of two steps: (i) mechanical destruction of the nematode body and tissues in a tube using an ultrasonic homogenizer or other tools, or repeatedly freezing samples in liquid nitrogen; and (ii) chemical lysis with proteinase K in a buffer for 1 h or several hours with subsequent brief inactivation of this enzyme at high temperature. Various chemical treatments are subsequently applied to remove cell components and purify the DNA. Phenol or phenol/chloroform extractions are often employed to remove proteins, and ethanol is then used to precipitate and concentrate the DNA. Stanton *et al.* (1998) described an efficient method of DNA extraction from nematodes using chemical lysis in alkali solution without prior mechanical breaking of nematode bodies. Effective DNA extraction can also be achieved using commercial kits developed by various companies.

2.4.2.2. Polymerase chain reaction (PCR)

The polymerase chain reaction (PCR) technique has become one of the most widely used techniques for studying the genetic diversity of nematodes and their identification. PCR is a rapid, inexpensive and simple means of producing large numbers of copies of DNA molecules via an enzyme catalyst. Any DNA fragment can be amplified and detected by PCR. The PCR method requires a DNA template (starting material) containing the region to be amplified, two oligonucleotide primers flanking this target region, DNA polymerase and four deoxynucleotide triphosphates (dATP, dCTP, dGTP, dTTP) mixed in a buffer containing magnesium ions. A primer is a short oligonucleotide, containing about two dozen nucleotides, which is complementary to the 3′ end of each strand of the fragment that should be amplified. Primers anneal to the denatured DNA template and provide an initiation site for the elongation of the new DNA molecule. Universal primers are those complementary to a particular sequence of DNA present in a wide range of organisms; primers matching only to certain species are called species-specific primers. When sequences of the flanking regions of the amplified fragment are unknown, PCR with degenerate primers, containing a number of options at several positions in the sequence that allows annealing and amplification of a variety of related sequences, can be applied.

PCR is performed in a tube in a thermocycler with programmed heating and cooling. The procedure consists of a succession of three steps determined by temperature conditions: template denaturation (95°C for 3–4 min), primer annealing (55–60°C for 30 s to 2 min) and extension of the DNA chain (72°C for 30 s to 2 min). PCR is carried out for 30–40 cycles. As the result of PCR, a single target molecule of DNA is amplified into more than a billion copies. The resulting amplified products are electrophoretically separated according to their size on agarose or polyacrylamide gels and visualized using a fluorescent dye, which interacts with double-stranded DNA and causes it to fluoresce under UV radiation. Once identified, nematode target DNA generated by PCR amplification can further be characterized by various analyses including restriction fragment length polymorphism (RFLP), or sequencing. In some cases, the size of PCR amplicon may serve as a diagnostic marker for a nematode group or species. For example, it has been shown that primers amplifying the control; region of mitochondrial DNA (mtDNA) generate different amplicon sizes for different species of root-knot nematodes; primers amplifying nuclear ribosomal intergenic spacer generated species-specific size polymorphisms for *M. chitwoodi*, *M. hapla* and *M. fallax*.

2.4.2.3. PCR-restriction fragment length polymorphism (PCR-RFLP)

Variation in sequences in PCR products can be revealed by restriction endonuclease digestion. The PCR product obtained from different species or populations can be digested by a restriction enzyme, after which the resulting fragments are separated by electrophoresis. If differences in DNA sequence occur within restriction sites, the digestion of the PCR products will yield restriction fragment length polymorphism, i.e. different RFLP profiles. PCR-RFLP of the internal transcribed spacer (ITS) region of the ribosomal RNA (rRNA) gene is a very reliable method for identification of many plant-parasitic nematode groups including cyst (Fig. 2.2), root-knot, lesion and gall-forming nematodes as well as nematodes from the genera *Bursaphelenchus* and *Aphelenchoides*. Using six to nine restriction enzymes enables most of the economically important species of cyst nematodes to be distinguished from each other as well as from their sibling species. RFLP of the ITS rDNA obtained after restriction with several enzymes and their combination identifies important root-knot nematode species; however, it fails to separate species from the tropical group, including *M. javanica*, *M. incognita* and *M. arenaria*. PCR-RFLP of a mtDNA fragment between the cytochrome oxidase subunit II gene and large subunit (LSU) gene has been applied successfully for diagnostics of these nematodes (Powers and Harris, 1993).

2.4.2.4. Multiplex PCR

This type of PCR constitutes a major development in DNA diagnostics and enables the detection of one or several species in a nematode mixture by a single PCR test, decreasing diagnostic time and costs. In multiplex PCR, two or more unique targets of DNA sequences in the same sample are amplified by different primer pairs in the same amplification reaction. Multiplex PCR for detection of a single nematode species uses two sets of primers. One set is to amplify an internal control (e.g. universal primers for D2–D3 expansion regions of the 28S rRNA gene) confirming the presence of DNA in the sample and the success of PCR; the second set, including at least one species-specific primer, is targeted to nematode DNA sequences of interest (Fig. 2.3). Diagnostics using multiplex PCR with species-specific primers have been developed for a wide range of plant-parasitic nematodes.

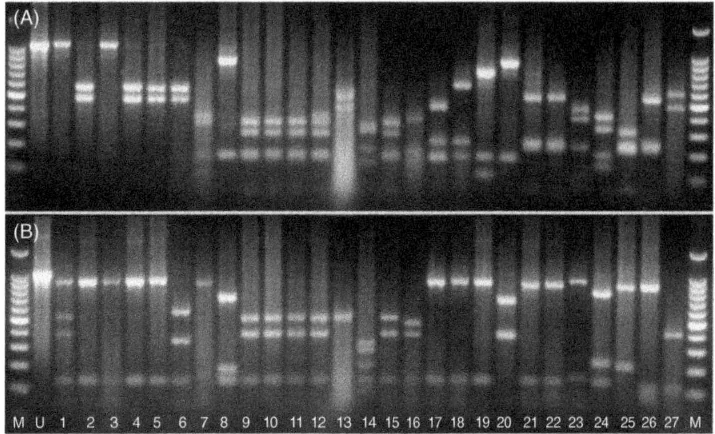

Fig. 2.2. RFLP profile of PCR-ITS-rRNA gene generated by *Alu*I (A) and *Bsh1236*I (B) for cyst-forming nematodes. M: l00 bp DNA ladder. U: unrestricted PCR product. 1 and 2: *Heterodera avenae*. 3: *H. arenaria*. 4: *H. filipjevi*. 5: *H. auklandica*. 6: *H. ustinovi*. 7: *H. latipons*. 8: *H. hordecalis*. 9: *H. schachtii*. 10: *H. trifolii*. 11: *H. medicaginis*. 12: *H. ciceri*. 13: *H. salixophila*. 14: *H. oryzicola*. 15: *H. glycines*. 16: *H. cajani*. 17: *H. humuli*. 18: *H. ripae*. 19: *H. fici*. 20: *H. litoralis*. 21: *H. carotae*. 22: *H. cruciferae*. 23: *H. cardiolata*. 24: *H. cyperi*. 25: *H. goettengiana*. 26: *H. urticae*. 27: *Meloidodera alni*. (From Subbotin *et al*., 2000.)

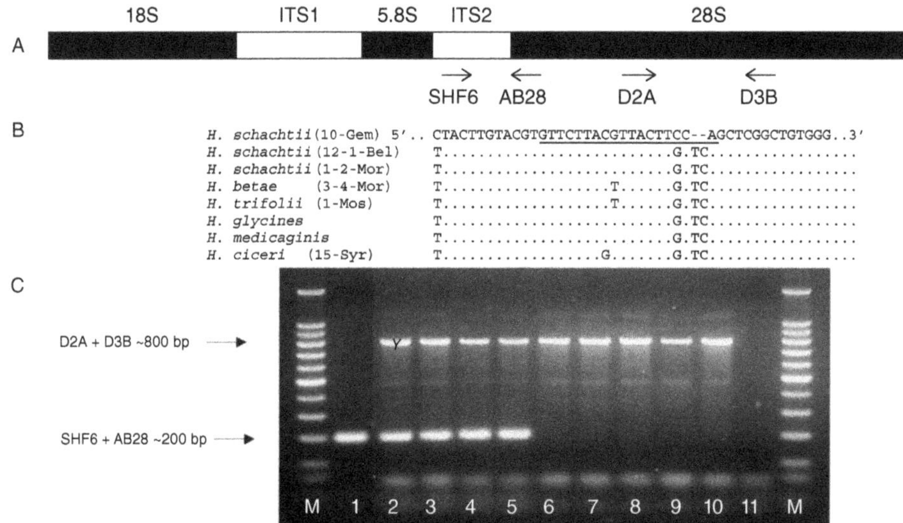

Fig. 2.3. PCR with a species-specific primer for the sugar beet cyst nematode *Heterodera schachtii*. A: Positions of specific (SHF6) and universal primers in the rRNA gene. B: Sequence alignment for *H. schachtii* and close related species with underlined sequence of specific primer. C: Agarose gel with PCR products generated with specific and universal primers (SHF6 + AB28) and universal (D2A + D3B) primers, or control band. 1–5: samples with *H. schachtii*. 6–10: nematode samples without *H. schachtii*. 11: a sample without nematode DNA. M: l00 bp DNA ladder. 1: resultant amplicon obtained with SHF6 + AB28 primer sets; 2-10: resultant amplicons obtained with SHF6 + AB28 and D2A + D3B primer sets. (Modified from Amiri *et al*., 2002.)

2.4.2.5. Random amplified polymorphic DNA (RAPD)

This method uses a single random primer of about ten nucleotides long for creating genomic fingerprints. This technique is often used for estimating genetic diversity between individuals, populations or closely related species. In this PCR-based approach, the short primer anneals to numerous similar sequences within the genome during the annealing step of the PCR cycle, which occurs at a lower temperature than does 'classical' PCR. If two complementary sequences are present on opposite strands of a genomic region in the correct orientation and within a close enough distance from each other, the DNA fragment between them can be amplified by PCR. Amplified DNA fragments obtained using different random primers from different samples are separated on gels and compared. RAPD polymorphisms result from the fact that if a primer-hybridization site in a genome differs by even a single nucleotide, the change can lead to elimination of a specific amplification product. The resulting individual bands are considered as equivalent independent characters (Fig. 2.4). The band polymorphisms can be binary scored and the data matrix is used for calculating the genetic distance between the samples under study and then presented as a dendrogram. Reproducibility of results is the most critical point for application of this technique.

The RAPD technique has been widely applied for separation of closely related species and studies of intraspecific variability of *G. pallida*, *H. glycines*, *Radopholus similis*, *D. dipsaci* and many other species (Powers, 2004; Blok, 2005). Specific sequences for certain species or races, called sequence characterized amplified regions (SCAR), can be derived from RAPD fragments. Specific pairs of SCAR primers have been designed for identification

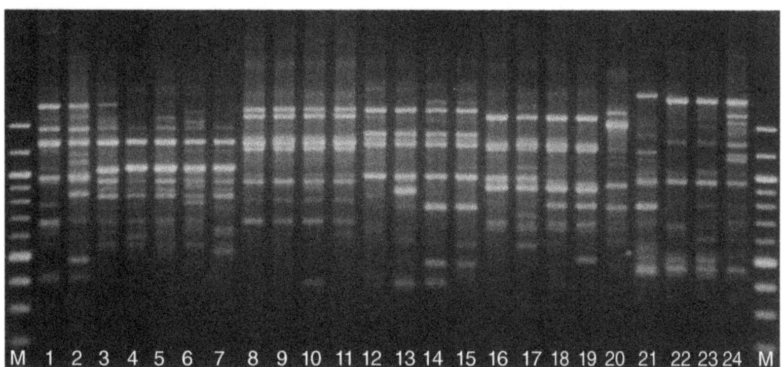

Fig. 2.4. Random amplified polymorphic DNA patterns for 26 populations of the *Heterodera avenae* complex. Primer G-10: 5′- AGGGCCGTCT-3′. 1: *H. avenae* (Taaken, Germany). 2: *H. avenae* (Santa Olalla, Spain). 3: *H. avenae* (Çukurova plain, Turkey). 4: *H. avenae* (Saudi Arabia). 5: *H. avenae* (Ha-hoola, Israel). 6: *H. avenae* (Israel). 7: *H. avenae* (near Delhi). 8: *H. australis* (South Australia, sample 3). 9: *H. australis* (Beulah, Australia). 10: *H. australis* (Victoria, Australia). 11: *H. australis* (Yorke Peninsula, Australia). 12: *H. mani* (Bavaria, Germany). 13: *H. mani* (Heinsberg, Germany). 14: *H. mani* (Andernach, Germany). 15: *H. mani* (Germany). 16: *H. pratensis* (Missunde, Germany). 17: *H. pratensis* (Östergaard, Germany). 18: *H. pratensis* (Lindhöft, Germany). 19: *H. pratensis* (Lenggries, Germany). 20: *H. aucklandica* (Auckland, New Zealand). 21: *H. filipjevi* (Saratov, Russia). 22: *H. filipjevi* (Akenham, UK). 23: *H. filipjevi* (Torralba de Calatrava, Spain). 24: *H. filipjevi* (Selçuklu, Turkey). M: 100 bp DNA ladder (Biolab). (After Subbotin *et al.*, 2003.)

of *M. chitwoodi*, *M. fallax*, *M. hapla* and other root-knot nematode species (Zijlstra *et al.*, 2000), as well as identification of stem nematodes *D. dipsaci* and *D. gigas*.

2.4.2.6. Amplified fragment length polymorphism (AFLP)

AFLP is also a random amplification technique, which does not require prior sequence information and it produces a higher number of bands than is obtained by RAPD. It is a much more reliable and robust technique, unaffected by small variations in amplification parameters; however, it is more expensive. The AFLP technique represents a conceptual and practical advance in DNA fingerprinting. It comprises the following steps: (i) restriction of the total DNA with two restriction enzymes; (ii) ligation of double-stranded adapters to the ends of the restriction fragments; (iii) amplification of a subset of the restriction fragments using two 17–21 nucleotide primers complementary to the adapter and one that is 1–3 nucleotides adjacent to the restriction sites; (iv) separation and visualization of the AFLP-PCR fragments with a variety of techniques, usually on denaturing polyacrylamide gels with further staining. A comparative study of *Globodera* species and populations using AFLP revealed greater inter- and intraspecific variability than obtained by RAPD, and enabled subspecies of *G. tabacum* to be distinguished. AFLP analysis also showed a clear distinction between species of the *D. dipsaci* complex (Fig. 2.5) (Esquibet *et al.*, 2003).

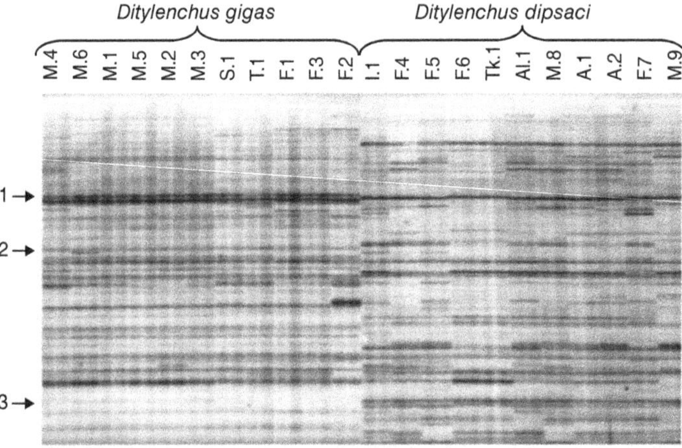

Fig. 2.5. Silver-stained 6% polyacrylamide gel showing AFLP amplification products generated using E-AA and M-CTG primers from 22 populations of *Ditylenchus dipsaci* and *D. gigas*. Two replicates were done for each population. Some polymorphic bands among races or populations are indicated by arrows: (1) giant-type-specific band; (2) population-specific band; (3) normal-type-specific band. (From Esquibet *et al.*, 2003.)

2.4.2.7. Real-time PCR

DNA technology also provides several methods for quantification of nematodes in samples. Real-time PCR requires an instrumentation platform that consists of a thermal cycler, optics for fluorescence excitation and emission collection, and computerized data acquisition and analysis software. The PCR quantification technique measures the number of nematodes indirectly by assuming that the number of target DNA copies in the sample is proportional to the number of targeted nematodes. Most of the difficulties with the PCR technique arise because only a very small number of the cycles (4–5 out of 40) contain useful information. The early cycles have an undetectable amount of DNA product; the final cycles, or the so-called plateau phase, are almost as uninformative. Quantitative information in a PCR comes from those few cycles where the amount of DNA grows logarithmically from just above background to the plateau (Fig. 2.6). The real-time technique allows continuous monitoring of the sample during PCR using hybridization probes (TaqMan, Molecular Beacons and Scorpions), allowing simultaneous quantification of several nematode species in one sample, or double-stranded dyes, such as SYBR Green, providing the simplest and most economical format for detection and quantification of PCR products in real-time reactions. Compared with traditional PCR methods, real-time PCR has advantages. It allows for faster, simultaneous detection and quantification of target DNA. The automated system overcomes the laborious process of estimating the quantity of the PCR product after gel electrophoresis. Real-time PCR has been used for detection and quantification of different species of *Heterodera*, *Globodera*, *Meloidogyne*, *D. dipsaci* and several other nematode species.

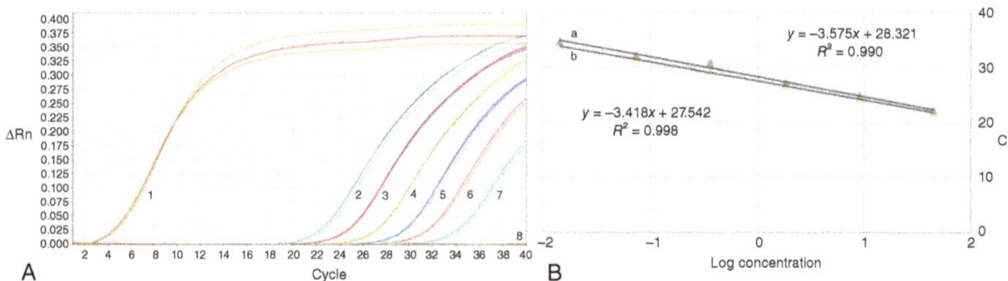

Fig. 2.6. Real-time detection of *Heterodera cajani*. Amplification of samples with serial dilution of *H. cajani* DNA and species-specific primer probe set targeting the *COI* gene. A: Amplification curve. 1: diluted COI PCR product of *H. cajani* (positive control). 2: *ca.* 8 eggs and J2. 3: *ca.* 1.6 eggs and J2. 4: *ca.* 0.32 eggs and J2. 5: *ca.* 0.06 eggs and J2. 6: *ca.* 0.01 eggs and J2. 7: *ca.* 0.003 eggs and J2 per tube. 8: negative control. B: Standard curves calculated with the log starting quantity and threshold cycle of the five-fold serially diluted DNA from *H. cajani* with (a) a species-specific primer-probe set targeting the *COI* gene and (b) universal nematode primer-probe set targeting the D3 of the 28S rRNA gene. (After Roubtsova and Subbotin, 2022.) © Brill, The Netherlands.

2.4.2.8. Loop-mediated isothermal amplification (LAMP)

LAMP is a novel approach to nucleic acid amplification. The LAMP reaction requires a DNA polymerase with strand displacement activity (*Bst* polymerase) and a set of 4–6 specially designed primers based on distinct regions of the target DNA. Due to the specific nature of the action of these primers, the amount of DNA produced in LAMP is considerably higher than PCR-based amplification. The reaction occurs under isothermal conditions (60–65°C) and yields large amounts of product in a short time (30–60 min). LAMP products can be visualized either by gel electrophoresis or the naked eye by adding DNA

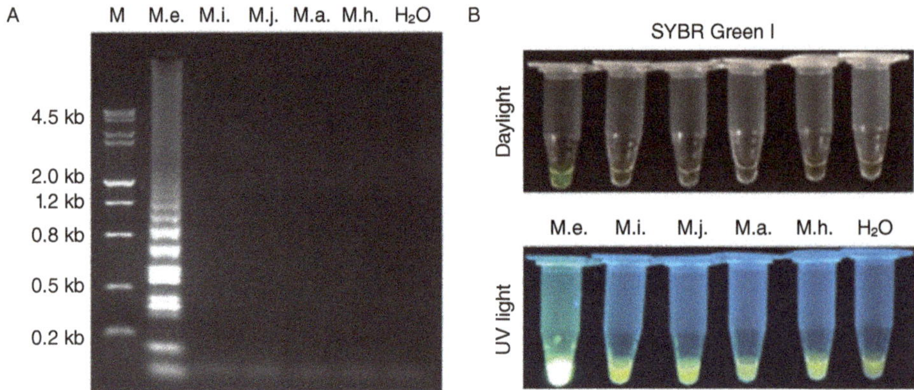

Fig. 2.7. LAMP detection of *Meloidogyne enterolobii*. A: LAMP assay products on agarose gel. B: LAMP assay products visualized by adding SYBR Green (top: direct visualization by the naked eye; bottom: observation under UV transillumination). Lane M: molecular marker. M.e, M.i, M.j, M.a and M.h represent *Meloidogyne enterolobii, M. incognita, M. javanica, M. arenaria* and *M. hapla*, respectively. (After Niu *et al.,* 2011.) Reproduced by permission of John Wiley & Sons, Inc.

intercalating dyes such as ethidium bromide or SYBR Green I in a tube. DNA concentration can also be detected by real-time detection methods. In addition, it has been improved by the use of a lateral flow dipstick (LFD) method to confirm visually the presence of amplicons. Because LAMP does not require an expensive thermal cycler and optical detection equipment and all LAMP steps are conducted within one reaction tube, this method clearly holds potential for testing in the field or in under-equipped laboratories. LAMP assays have been developed for detection of the pinewood nematode (Fig. 2.7), root-knot nematode species and other nematode species.

2.4.2.9. Recombinase polymerase amplification (RPA)

RPA is a relatively new isothermal methodology for amplifying DNA and represents a hugely versatile alternative to PCR. RPA uses a highly efficient displacement polymerase that amplifies a few copies of target nucleic acid in 20 min at a constant temperature (37–42°C). It does so by utilizing three core enzymes: recombinase, single-stranded binding protein (SSB) and strand-displacing polymerase. The recombinase enzyme forms a complex with a primer to facilitate binding to the targeted DNA template. Then, the SSB binds to the displaced strands of DNA and prevents the displacement of the recombinase–primer complex by branch migration. Thereafter, the strand-displacing polymerase recognizes the bound recombinase–primer complex and initiates DNA synthesis. Like PCR, RPA produces an amplicon constrained in size to the binding sites of the primers. The advantages of RPA include high efficiency and speed at a low constant operating temperature. RPA can be performed using fluorescent probes in real time or by detection after agarose gel electrophoresis or a lateral flow assay. Several tests using RPA have demonstrated high sensitivity and specificity for detecting small amounts of nematode DNA (Subbotin *et al.*, 2021a,b; Fig. 2.8).

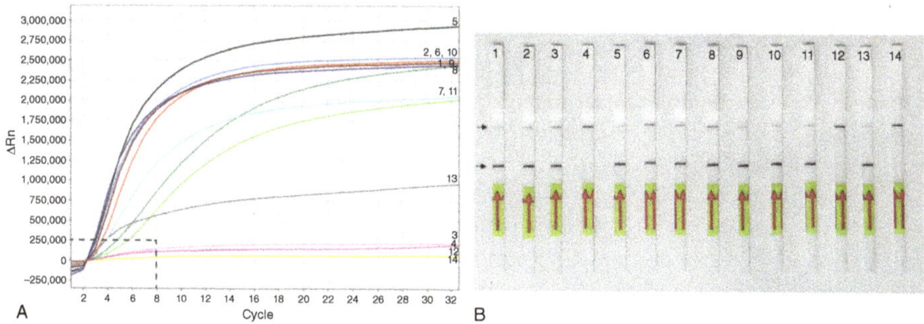

Fig. 2.8. RPA sensitivity assays of *Meloidogyne hapla* detection using A: real-time fluorescent detection and B: lateral flow strips. A dilution series of crude extracts of three young females (without egg masses) of *M. hapla*. Line (Strip): 1, 5, 9: 1/10 females per tube. 2, 6, 10: 1/100 females per tube. 3, 7, 11: 1/1000 females per tube. 4, 8, 12: 1/10,000 females per tube. 13: Positive control. 14: Negative control. Control (upper) and test (lower) lines are indicated by arrows. (After Subbotin and Burbridge, 2021.) CC BY 4.0.

2.4.2.10. DNA hybridization arrays

DNA hybridization arrays (also known as macro-arrays, micro-arrays and/or high-density oligonucleotide arrays) provide a powerful method for the next generation of diagnostics. The distinct advantage of this approach is that it combines DNA amplification with subsequent hybridization to oligonucleotide probes specific for multiple target sequences. DNA arrays can be used to detect simultaneously many nematode species based on differences in the rRNA gene. In general, arrays are described as macro-arrays or micro-arrays, the difference being the size and density of the sample spots, the substrate of hybridization and the type of production. Although the potential of DNA array methods for nematological diagnostics has been recognized, little progress had been made in their use, and only few research papers have been published on this technique (Blok, 2005). A reverse dot-blot assay has been developed for identification of several *Pratylenchus* species using oligonucleotides designed from the sequences of the ITS region of rRNA (Uehara *et al.,* 1999).

2.4.2.11. Sequencing of DNA

The process of determining the order of the nucleotide bases (adenine, guanine, cytosine and thymine) along a DNA strand is called sequencing. Several different methods have been developed for DNA sequencing.

2.4.2.11.1. Basic methods. In 1977, 24 years after the discovery of the structure of DNA, two methods for sequencing DNA were developed: the chemical degradation (Maxam–Gilbert) method and the chain-termination (Sanger dideoxy) method, the latter being more commonly used. The chain-termination method was developed by Frederick Sanger and coworkers (1977). The Sanger method was automated and used in the first generation of DNA sequencers. The chain-termination sequencing method is similar to PCR in that it involves the synthesis of new strands of DNA complementary to a single-stranded template. The sequencing reaction components are template DNA, DNA polymerase with reaction buffer, one primer and the mixture of all four deoxynucleotides (dNTP) and four dideoxynucleotides (ddNTP) labels, each with a different colour fluorescent dye. As all four deoxynucleotides are present, chain elongation proceeds until, by chance, DNA polymerase inserts a dideoxynucleotide. As the dideoxy sugar lacks a 3′-hydroxyl group, continued lengthening of the nucleotide chain cannot occur. Thus, the dideoxynucleotide acts analogously as a specific chain-terminator reagent. Therefore, the result is a set of new chains with different lengths. These fluorescently labelled fragments are separated by size, using capillary electrophoresis. As each label fragment migrates through the gel, a laser excites the fluorescent molecule, which sends out light of a distinct colour. The detection system records the chromatogram output on a computer. The computer software Chromas then presents the sequencing result as chromatogram sequence files (Box 2.1). The leader in automated Sanger sequencing is Applied Biosystems (AB) (now part of ThermoFisher). Currently commercialized AB sequencers all utilize fluorescent dyes and capillary electrophoresis. The machines vary in capacity, from 4 to 48–96 capillaries. All of these sequences can have an average accurate read of 600–1000 bp in length.

2.4.2.11.2. HIGH-THROUGHPUT METHODS. The high demand for low-cost sequencing has driven the development of several high-throughput sequencing (HTS) technologies that parallelize the sequencing process, producing thousands or millions of sequences at once. HTS includes next-generation short-read (Illumina, 454 pyrosequencing and others) and third-generation long-read (Single-Molecule Real-Time Sequencing, Nanopore Sequencing and others) sequencing methods. Illumina Sequencing or sequencing-by-synthesis technology is the most widely used next-generation sequencing technology in the world. Illumina sequencing is based on a technique known as 'bridge amplification' wherein DNA molecules with appropriate adapters ligated on each end are used as substrates for repeated amplification synthesis reactions on a solid support that contains oligonucleotide sequences complementary to a ligated adapter. The oligonucleotides on the slide are spaced such that the DNA, which is then subjected to repeated rounds of amplification, creates clonal clusters of oligonucleotide fragments. During the synthesis reactions, proprietary modified nucleotides, corresponding to each of the four bases, each with a different fluorescent label, are incorporated and then detected. The nucleotides also act as terminators of synthesis for each reaction, which are unblocked after detection for the next round of synthesis. The reactions are repeated for many rounds. The use of fluorescent detection increases the speed of detection due to direct imaging (Slatko *et al.*, 2018). In contrast to the short reads (150–300 bp) generated by most Illumina operations, long-read sequencing technologies are capable of reading longer lengths, between 5000 and 30,000 bp, or even more. Pacific Biosciences have developed a Single Molecule Real-Time sequencer that generates reads that can exceed 10,000 bp in less than 2 h. Oxford Nanopore Technologies' platform is capable of producing reads of up to 1 million bp. This method relies on changes in the ion flow as nucleotides pass through a nanopore. The goal of DNA sequencing technology development is to be faster and more accurate (lower error rates, minimal artifacts) and take lower amounts of input DNA at lower cost.

2.5. Genes used for Molecular Systematics

A gene is usually defined as a DNA segment that codes for a polypeptide or specifies a functional RNA molecule. Eukaryotic protein-coding genes consist of transcribed and

Box 2.1. Computer programs.

Packages for manipulation and align of sequences

Chromas (http://technelysium.com.au/wp/chromas/) is a program for displaying, editing and exporting chromatogram sequence files.
Clustal (http://www.clustal.org) is a package of multiple sequence alignment programs for DNA and proteins. It provides an integrated environment for performing multiple sequence and profile alignments and analysing the results.
BioEdit (https://bioedit.software.informer.com/) is a biological sequence alignment editor. An intuitive, multiple document interface with convenient features makes alignment and manipulation of sequences relatively easy. Several sequence manipulation and analysis options and links to external analysis programs facilitate a working environment that allows one to view and manipulate sequences with simple point-and-click operations.

General phylogenetic packages

PAUP* (http://paup.csit.fsu.edu) is the most sophisticated and user-friendly program for phylogenetic analysis, with many options. It includes parsimony, distance matrix and maximum likelihood methods.
PHYLIP (https://phylipweb.github.io/phylip/) includes programs to carry out parsimony, distance matrix methods and maximum likelihood, including bootstrapping and consensus trees. There are programs for data types including DNA and RNA, protein sequences, gene frequencies, restriction sites and restriction fragments, and discrete and continuous characters.
MrBayes (http://nbisweden.github.io/MrBayes) is a program for the Bayesian estimation of phylogeny. The program uses a Markov chain Monte Carlo (MCMC) technique to approximate the posterior probabilities of trees. One of the program features is the ability to analyse nucleotide, amino acid and morphological data under different models in a single analysis.
MacClade (http://macclade.org/) is a program for phylogenetic analysis with analytical strength in studies of character evolution. It also provides many tools for entering and editing data and many descriptive statistics as well as for producing tree diagrams and charts.

Package for selecting models of evolution

jModelTest (http://evomics.org/learning/phylogenetics/jmodeltest/) is a program for selecting the model of nucleotide substitution that best fits the data.

Packages for tree visualization

TreeView (https://treeview-x.en.softonic.com/) is a simple program for displaying and manipulating phylogenetic trees.
FigTree (http://tree.bio.ed.ac.uk/software/figtree/) is designed as a graphical viewer of phylogenetic trees and as a program for producing publication-ready figures.

untranscribed parts, called flanking regions. Flanking regions are necessary for controlling transcription and processing of pre-messenger RNA (pre-mRNA). A pre-mRNA consists of coding regions (exons), which encode amino acids, and non-coding regions containing information necessary for regulation of polypeptide production. Some segments of the non-coding regions (introns) are spliced out in the process of production of a mature mRNA.

The eukaryotic cell contains two different genomes: that of the nucleus and that of mitochondria. Molecular systematics use data from both genomes. Mitochondria are inherited maternally, whereas the nucleus is biparental

To determine true evolutionary relationships between organisms, it is essential that the correct gene fragment or molecule is chosen for sequence studies. This is important for several reasons: (i) the molecule should be universally distributed across the group chosen for study; (ii) it must be functionally homologous in each organism, i.e. the phylogenetic comparisons must start with molecules of identical function; and (iii) it is critical in sequence comparisons to be able to align the molecules properly to identify regions of sequence homology and sequence heterogeneity.

2.5.1. Nuclear ribosomal RNA genes

Historically, the only nuclear genes with a high enough copy number for easy study were ribosomal genes. These genes encode rRNAs, which are nearly two-thirds of the mass of the ribosome. The genes encoding rRNA are arranged in tandem, in several hundred copies, and are organized in a cluster that includes a small subunit (SSU or 18S) and a large subunit (LSU or 26 to 28S) gene, which are themselves separated by a small 5.8S gene. The whole set of genes is transcribed as a single unit. Another ribosomal gene, a ubiquitous component of large ribosomal subunits in eukaryotic cells, is 5S rRNA. The gene is found at different localizations in different organisms. The 5S rRNA gene linked to the intergenic spacer (IGS) regions of rRNA repeated units has been described for several root-knot nematode species.

In addition to these coding sequences, the rDNA array also contains spacer sequences that contain the signals needed to process the rRNA transcript: an external transcribed spacer (ETS) and two internal transcribed spacers, ITS1 and ITS2. A group of genes and spacer sequences together make up an rRNA transcript unit. These units are separated from each other by an IGS region, also known as a non-transcribed spacer (NTS).

rRNA (18S and 28S) genes evolve slowly and can be used to compare distant taxa that diverged a long time ago, whereas external and intergenic spacers have higher evolution rates and so have been used for reconstructing relatively recent evolutionary events and for the comparison of closely related species and subspecies. The IGS region contains many repeats and is more variable than the ITS region.

Although rRNA genes are present in many copies, their sequences are almost identical, because the highly repetitive sequences undergo homogenization processes known as concerted evolution. If a mutation occurs in one copy of a sequence, it is generally corrected to match the other copies, but sometimes the non-mutated copies are corrected to match the mutated one, so that nucleotide changes propagate throughout the arrays. However, this process may be disrupted, so that several different copies of this gene may be present in genome. The risk of incorporating ITS paralogues into phylogenetic studies should be considered with caution. Inspection of some basic features of the sequence, including the integrity of the conserved motifs and the thermodynamic stability of the secondary structures of the RNA transcripts, enables rRNA pseudogenes to be excluded from the dataset.

2.5.2. Nuclear protein-coding genes

The numbers of protein-coding genes predicted in plant-parasitic nematodes usually range from 14,000 to 19,000. Protein-coding genes have some advantages over rRNA genes and their spacers in that the alignment of sequences is less problematic. Protein sequences also lend themselves to different phylogenetic weighting of bases by codon position. The intron position patterns may also serve as decisive markers for phylogenetic analysis. Heat shock proteins, RNA polymerase II, actin, major sperm protein and other genes have been used for phylogenetic studies of cyst, root-lesion and other nematodes.

2.5.3. Mitochondrial DNA

mtDNA has been used to examine population structure and evolutionary relationships between different nematode groups. All nematode mtDNA is circular and double-stranded DNA. The mitochondrial genome of nematodes ranges in size from 12 to 25 kb and contains 36 (sometimes 37) genes: 12 (or 13) protein-coding genes (*cox1–cox3*, *cytb*, *nad1–nad6*, *nad4L*, *atp6* and rarely *atp8*), two rRNA genes (*lrRNA* and *srRNA*) and 22 tRNA genes (Fig. 2.9). In addition, there is usually a non-coding AT-rich region or a region with high levels of the nucleotides adenine and thymine in the mitochondrial genome containing an initiation site for replication and transcription. The remainder of the approximately 1000 mitochondrial proteins is nuclear-encoded and is imported in the organelle. The arrangement of genes in the mitochondrial genome is not consistent within Nematoda. Nematodes are characterized by a surprising variation in gene order. The Enoplia displays substantial gene rearrangement even among closely related species, whilst members of Chromadoria show far less rearrangement. A unique feature of the mitochondrial genome organization of nematodes is that some of them, e.g. *Globodera pallida* and *G. rostochiensis*, have at least five mini-circles, ranging in size from ~6 to 9 kb. *Globodera ellingtonae* comprises two large circles sharing a ~6.5 kb non-coding region.

Nematode mtDNA sequences accumulate substitution changes much more quickly than the ITS sequences and usually tend to show a strong nucleotide compositional bias toward A and T, which together account for between 63% and 85%. The T-content seems to be greater at the third codon position, compared with the first and second positions. Although the high rate of substitution makes mtDNA very useful for low-level phylogenetic applications, failure to correct for this severe substitution bias can potentially lead to phylogenetic error.

The relatively rapid rate of evolution and rearrangements that occur in mtDNA has limited the design of universal primers and, thus, mtDNA has not been as widely used as other markers for nematode phylogenetic or diagnostic purposes, except for root-knot nematodes. The region between the *coxI* and *lrRNA* gene containing an intergenic region with unique size and nucleotide polymorphism might be utilized for distinguishing different species and host races of *Meloidogyne* (Powers and Harris, 1993).

2.6. Microsatellites

Microsatellites or simple sequence repeats (SSRs) are short, 1–6 base nucleotide sequences (e.g. AAG) that are repeated many times in tandem (...AAGAAGAAG...) and are found

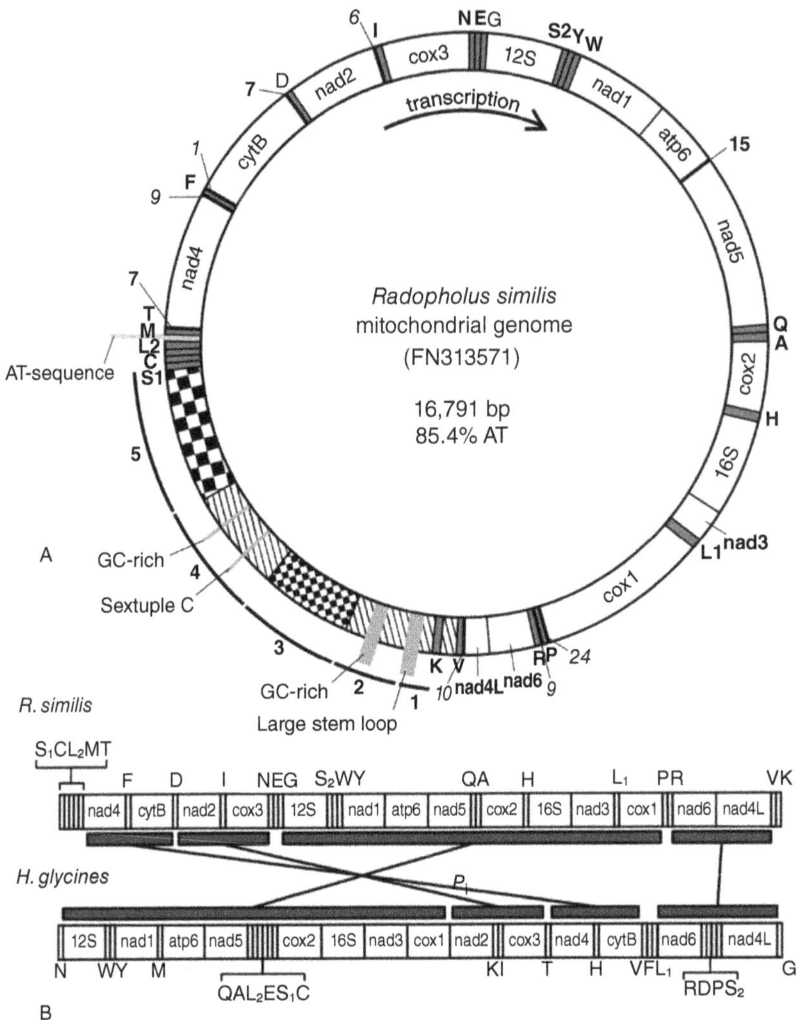

Fig. 2.9. A: Overview of the organization of the circular mitochondrial DNA of *Radopholus similis*. Genes and non-coding regions are indicated: in white, the protein-coding and rRNA genes, in grey, the tRNA genes called by their amino acid symbol (S_1: Ser-AGN, S_2: Ser-UCN, L_1: Leu-CUN, L_2: Leu-UUR). Bold and italic numbers indicate non-coding and overlapping nucleotides between neighboring genes, respectively. The pattern-filled part represents the large non-coding region. The repeat region of 302 bp is filled with large checkers and the 26 bp repeat region is filled with small checkers. (After Jacob *et al.*, 2009.) B: Partial mitochondrial genome organization of *R. similis* and *Heterodera glycines*. Bars joined by lines indicate regions of conserved genome organization. (After Gibson *et al.*, 2011.)

in all eukaryotic genomes. Overall, $(AT)_n$, $(AG)_n$, $(CT)_n$, $(AAT)_n$ and $(ATT)_n$ are the most frequent microsatellite motifs presently known in nematode genomes. They are present in both coding and non-coding regions, and found covering from 0.09 to 1.20% of the nematode genomes (Castagnone-Sereno *et al.*, 2010). A very high mutation rate, from 10^{-4} to 10^{-3} mutations per microsatellite and per generation, is usually associated with microsatellite loci, resulting in high heterozygosity and the presence of multiple alleles at a given locus. They are present in both coding and non-coding regions and are usually characterized by a high level of length polymorphism. Despite the fact that the mechanism of microsatellite evolution is still unclear, SSRs have been widely used as powerful markers for studies in population genetics. Microsatellites mutate over time, their alleles diverging in the number of sequence repeats. The flanking regions surrounding the microsatellites can be conserved across genera or even higher taxonomic levels and, therefore, are used as designs for PCR primers to amplify microsatellite loci. Based on analysis of the microsatellite variation in populations, inferences can be made about population structures and differences, genetic drift and the date of a last common ancestor (Jarne and Lagoda, 1996).

2.7. DNA Barcoding

DNA barcoding first came to the attention of the scientific community in 2003 when Paul Hebert's research group at the University of Guelph published a paper (Hebert *et al.*, 2003) entitled 'Biological identifications through DNA barcodes'. In the paper, they proposed a new system of species identification and discovery using a short section of DNA from a standardized region of the genome. DNA barcoding is a taxonomic method that uses a fragment containing the first half of the *COI* gene of mtDNA to identify it as belonging to a particular species. It is based on a relatively simple concept: most eukaryote cells contain mitochondria, and mtDNA has a relatively fast mutation rate and more differences than the ribosomal genes, which results in significant variance in mtDNA sequences between species and, in principle, a comparatively small variance within species. Molecular barcoding involves isolation of the nematodes (as individuals or in bulk), amplification of the target gene, cloning, sequencing and phylogenetic analysis, leading to the assessments of species content, abundance and diversity. However, DNA barcoding is only as good as the reference database: it cannot be used to identify species not already catalogued. Barcoding can potentially be used for identification of potential putative new species, but only for species groups whose genetic diversity has been well surveyed. DNA barcoding cannot replace the traditional methods of species description. Currently, there is insufficient information in databases for some nematode species identification based on the *COI* gene. However, the increasing deposition of DNA sequences in GenBank and BOLD databases will be beneficial for diagnostics. Participants in the DNA barcode initiative come in many configurations, including consortia, databases, networks, laboratories and projects that range in size from local to global. The largest consortium is International Barcode of Life (iBOL) (https://ibol.org/). The Barcode of Life Data System (BOLD) (https://www.boldsystems.org/) provides an integrated bioinformatics platform that supports all phases of the analytical pathway from specimen collection to tightly validated barcode library.

2.8. Phylogenetic Inference

Phylogenetic analysis is a complex field of study that embraces a variety of techniques that can be applied to a wide range of evolutionary questions. However, a complete understanding of all assumptions involved in analysis is essential for a correct interpretation of the results. A possible work flow of a molecular phylogenetic project could be presented as a flow diagram: (i) selection and sampling of a group of organisms;

(ii) choice of molecular markers; (iii) sequencing and assembling of sequence data; (iv) alignment, or establishment of homology between molecules; (v) construction of phylogenetic tree using distance or discrete methods and making an assessment of the reliability of its branches; and (vi) testing of different alternative hypotheses.

2.8.1. Alignment

The major step of any phylogenetic study is the construction of alignment or establishment of positional homology between nucleotides or amino acid bases that have descended from a common ancestral base. Errors incurred in this step can lead to an incorrect phylogeny. The best way to compare the homologous residues is to align sequences one on top of another in a visual display, so that, ideally, each homologous base from different sequences line up in the same column. Three types of aligned pairs are distinguished: (i) **matches** (same nucleotide appears for all sequences); (ii) **mismatches** (different nucleotides were found in same position); and (iii) **gaps** (no base in a particular position for at least for one of the sequences). A gap indicates that a deletion has occurred in one sequence or an insertion has occurred in another sequence. However, the alignment itself does not enable these mutational events to be distinguished. Therefore, insertions and deletions are sometimes collectively referred to as **indels**.

The optimal alignment is considered to be that in which the numbers of mismatches and gaps are minimized according to the desired criteria. The program Clustal (Box 2.1) is one of the most commonly used computerized alignment programs using a progressive alignment approach. Sequences are aligned in pairs to generate a distance matrix, which then is used for calculating a neighbour-joining guide tree. This tree gives the order in which progressive alignment should be carried out. Progressive alignment is a mathematical process that is completely independent of biological reality. The use of structural components of the given molecule can significantly improve estimations of homology, thus generating a better alignment.

2.8.2. Methods for inferring phylogenetic trees

The methods for constructing phylogenetic trees from molecular data can be categorized into two major groups: (i) distance methods; and (ii) discrete methods. In distance methods such as analysis by minimum evolution, sequences are converted into a distance matrix that represents an estimate of the evolutionary distances between sequences, from which a phylogenetic tree is constructed, by considering the relationships among these distance values, which are supposed to represent distances between taxa. In discrete methods, maximum parsimony, maximum likelihood, Bayesian inference methods map the history of characters onto a tree. Each method requires some assumptions about evolution.

2.8.2.1. Minimum evolution method

The minimum evolution method is very useful for analysing sequences. In this method, the sum of all branch lengths is computed for all plausible trees and the tree that has the smallest sum value is chosen as the best tree. The neighbour-joining (NJ) method applies the minimum evolution principle and estimates the tree based on data transformed into a pairwise distance matrix. This method does not examine all possible topologies but at each stage of taxon clustering a minimum evolution principle is used. The NJ algorithm is extremely popular because it is relatively fast and performs well when the divergence between sequences is low.

2.8.2.2. Maximum parsimony

Maximum parsimony is an important method of phylogenetic inference. The goal of parsimony is to find the tree with the minimum total tree length, or the minimum amount of evolutionary changes, i.e. the transformation of one character state to another. The better a tree fits the data, the fewer homoplasies will be required and the fewer number of character state changes will be required. Several different parsimony methods have been developed for treating datasets. The problems of finding optimal trees under the maximum parsimony criterion are twofold: (i) determining the tree length; and (ii) searching over all possible trees with the minimum length. When the number of taxa is small, it is possible to evaluate each of the possible trees, or to conduct an exhaustive search. An exhaustive search is carried out by finding each of the possible trees by a branch–additional algorithm. However, if the number of trees is large, the application of this approach is near impossible, and a heuristic strategy is used.

As with any method, maximum parsimony has its pitfalls. If some sequences have evolved much faster than others, homoplasies have probably occurred more often among the branches leading to these sequences than in others, so that parsimony tends to cluster these highly divergent branches together. This effect, called long-branch attraction, can be reduced by sampling additional taxa related to those terminating the long branches, so that the branches may be broken up into smaller ones.

2.8.2.3. Maximum likelihood

Maximum likelihood (ML) is the method that is generally considered to make the most efficient use of the data to provide the most accurate estimates of phylogeny. The likelihood is not the probability that the tree is the true tree; rather it is the probability that the tree has given rise to the data that were collected. The basic idea of the likelihood approach is to compute the probability of the observed data assuming it has evolved under a particular evolutionary tree and a given probabilistic model of substitution. The likelihood is often expressed as a natural logarithm and referred to as the log-likelihood. The tree with the highest likelihood is the best estimate of the true phylogeny. The main obstacle for the widespread use of ML methods is computing time, because algorithms that find the ML score must search through a multidimensional space of parameters to find a tree. Maximum likelihood requires three elements: a model of sequence evolution, a tree and the observed data.

2.8.2.4. Bayesian inference

Bayesian inference of phylogeny is based on a quantity called the posterior probability of a tree. The posterior probability of a tree can be interpreted as the probability that the tree is correct. The posterior probability involves a summation over all trees and, for each tree, integration over all possible combinations of branch length and substitution model parameter values. This method is almost impossible to complete by exhaustive analysis, so the Markov chain Monte Carlo (MCMC) is a search method used to approximate the posterior probabilities of trees. Maximum likelihood and Bayesian analysis are both based upon the likelihood function, although there are fundamental differences in how the two methods treat parameters.

2.8.2.5. *Evolutionary models*

In order to reconstruct an evolutionary tree, some assumptions about the evolutionary process for the studied molecules should be made. Evolutionary substitution models for DNA are implemented in a different way in distance, maximum likelihood and Bayesian analysis. The substitution model is a description of the way sequences evolved in time by nucleotide replacements. The nucleotide substitution process of DNA sequence can be described by a so-called homogeneous Markov process that uses the Q matrix, which specifies the relative rates of change of each nucleotide along the sequences. The Jukes–Cantor model (JC69) was one of the first proposed and is perhaps the simplest model of sequence evolution. It assumes that the four bases have equal frequencies, and that all substitutions are equally likely. The general time-reversible (GTR) model is the most general model where all eight free parameters of the reversible nucleotide rate Q matrix are specified. The best-fit model of evolution for the dataset can be selected through statistical testing. The fit to the data of different models can be compared through likelihood ratio tests (LRTs) or information criteria to select the best-fit model within a set of possible ones.

2.8.3. Phylogenetic tree and network

The result of a molecular phylogenetic analysis is expressed in a phylogenetic tree and a network. A tree consists of nodes, which are connected by branches. The branch length usually represents the evolutionary distance between two consecutive nodes. Terminal nodes (leaves) are often called **operational taxonomic units** (OTUs). Internal nodes represent hypothetical ancestors and may be called **hypothetical taxonomic units** (HTUs). The ancestor of all the taxa that comprise the tree is the root of the tree. An outgroup is a terminal taxon whose most recent common ancestor with any taxon within a given clade occurs at a node outside that clade. The OTUs within a given clade are called ingroup taxa. A group of taxa that belong to the same branch is called a cluster. Sister groups or sister taxa refer to two groups on a tree with the same immediate common ancestor, or are more closely related to each other than either is to any other taxon. The branching patterns, or the order and arrangement of nodes, are collectively called the topology of the tree. If three branches connect to an internal node, then the node represents a bifurcation, or dichotomy. If more than three branches connect to an internal node, then the node represents polytomy. A tree implicitly assumes that once two lineages appear they subsequently never interact with each other. However, in reality such interactions might have occurred, such as through hybridization (rare in animals) or reticulate evolution, and such relationships can be presented as a network. Compared to the phylogenetic tree approach, a phylogenetic network has many advantages. Labelled appropriately, the network can predict haplotypes, indicate most mutated sites and reveal sites where recombination and sequence errors are likely to have occurred. Since the network harbours all trees for the input data, it yields a more concise picture of relationships.

Parsimony analyses often arrive at multiple trees with the same length but with different branch order. Rather than choosing among these trees, systematists may simply want to determine what groups can be found in all the shortest trees. There are approaches to summarize information which are common to two or more trees in a single tree. The resulting tree is called a **consensus tree.** A **strict consensus tree** shows only those relationships that were hypothesized in all the equally parsimonious trees, whereas a **majority consensus rule tree** shows those relationships hypothesized in more than half the trees being considered.

2.8.4. Evaluation of the reliability of inferred trees

The estimation of phylogeny should be accompanied by an indication of its confidence limit. Phylogenetic trees should always be evaluated for reliability, which could be measured as the probability that the taxa of a given clade are always members of that clade. Bootstrap and jack-knife analyses are the techniques used most often for this purpose. Bootstrapping and jack-knifing are so-called re-sampling techniques, because they estimate the sampling distribution by repeatedly re-sampling data from the original dataset. These methods differ in their methods of re-sampling. Bootstrapping is the more commonly used approach for phylogenetic reconstruction. To estimate the confidence level by bootstrapping, or the bootstrap value of the clade, a series of pseudo samples or pseudo alignments are first generated by randomly re-sampling the sites in the original alignment with replacement. In such pseudo alignments some characters are not included at all, while others may be included twice or more. Secondly, for each pseudo alignment, a tree is constructed, and the proportion of each clade among all the bootstrap replicates is computed in a majority-rule consensus tree. If the value of support of the clade obtained as a result of these analyses is greater than 95%, the branch is considered to be statistically significant. Branch support of less than 70% should be treated with caution.

Confidence in maximum parsimonious trees can also be evaluated by calculating the decay index, or Bremer support, which expresses the number of extra steps required for each node not to appear in the tree, i.e. the length difference between the shortest trees including the group and the shortest threes that exclude this group. The higher the decay index the better the support for the group.

2.8.5. Testing of hypotheses

Once phylogenetic trees are obtained from a molecular dataset using different methods, they should be compared with each other or with trees generated from other, non-molecular datasets. There are several tests that allow the evaluation of alternative hypotheses and determine if one tree is statistically worse than another: Templeton test, Kishino–Hasegawa test, Shimodaira–Hasegawa test, approximately unbiased test and Bayes' factors.

2.9. Reconstruction of Historical Associations

Historical associations, when a lineage that tracks another lineage, can be divided into three basic categories: (i) genes and organisms; (ii) organisms and organisms; and (iii) organisms and areas (Page and Charleston, 1998). In each association, one entity tracks the other with a degree of fidelity that depends on the relative frequency of four events: co-divergence, duplication, horizontal transfer and a sorting event. The testing of hypotheses for co-divergence could be made using tree topology or data-based methods. Both methods are not without problems, and should rely on the estimation of inferred phylogenies or adequacy of models of sequence evolution.

Genes and organisms. Each gene has a phylogenetic history that is intimately connected with, but not necessarily identical to, the history of the organisms in which the gene resides. Processes such as gene duplication, horizontal gene transfer, gene loss and lineage sorting can produce complex gene trees that differ from organismal trees.

Organisms and organisms. Several associations between nematodes and other organisms are known, e.g. nematodes of the Heteroderidae or Anguinidae and their host plants; nematodes from the *Xiphinema americanum* group and symbiotic bacteria of the genus *Xiphinematobacter*; seed-gall-forming nematodes of the genus *Anguina* and host-plants and soil-inhabiting nematodes and parasitic bacteria of the genus *Pasteuria*, and plant-parasitic nematodes and bacterial intracellular endosymbiont *Candidatus* Cardinium. A comparison of the *Xiphinematobacter* 16S rRNA and *Xiphinema americanum COI* phylogenies revealed a high level of co-speciation events between host and symbiont (Fig. 2.10). Molecular data also suggest that anguinid groups are generally associated with host-plants from the same or related systematic groups. Although the strict co-speciation hypothesis for seed-gall nematodes and grasses was rejected, the analysis showed a high level of co-speciation events, which cannot be explained as a result of random establishments of host–parasitic association. Analysis of phylogenies of *Pasteuria* or *Candidatus* Cardinium, and their nematode hosts suggest that horizontal host-switching is the common event in this association.

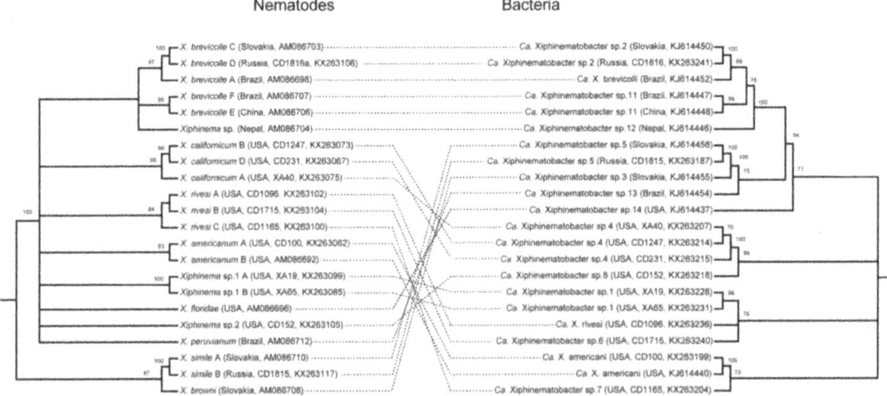

Fig. 2.10. Co-phylogenetic relationships between the endosymbionts *Candidatus* Xiphinematobacter (right, 16S rRNA tree) and their hosts *Xiphinema americanum*-group species (left, *COI* tree). Bayesian 50% majority consensus trees with posterior probability values more than 70% for appropriate clades. Dotted lines show the association between an *X. americanum*-group species and its symbiont (Orlando *et al.*, 2016). © Brill, The Netherlands.

Organisms and areas. Organisms can track geological history such that sequences of geological events are directly reflected in the phylogenies of those organisms. Phylogeography is the study of the historical processes (vicariance, sympatry, dispersal and extinction) that take place in this association (Page and Charleston, 1998). Studies of the geographical distributions of genealogical lineages within and between species of some genera give interesting views on the origin and dispersal of some nematode groups. Phylogeographic studies use analytical tools such as haplotype networks, mismatch analyses, genetic differentiation estimators, phylogenies, analyses of molecular variance (AMOVA), gene trees and coalescence analyses. As a result, phylogeography has been considered one of the most integrative disciplines in biology (Hickerson *et al.*, 2010). The number of phylogeographic studies on nematodes has grown in recent years owing to the increased accessibility of sequencing techniques and the emergence of genomic tools.

2.10. Databases

Phylogenetic analyses are often based on data accumulated by many investigators in different databases. All novel sequences have to be submitted in a public database. Databases are essential sources for modern bioinformatics, as they serve as information storage equipped with powerful query tools and a well-developed system of cross-references (Box 2.2).

Box 2.2. Databases.

Numerous genetic databases are spread out all over the world. The biggest public databases containing nucleotide sequence information are as follows: **GenBank** (National Center for Biotechnology Information, USA) (http://www.ncbi.nlm.nih.gov), **EMBL-ETI** (European Molecular Biology Laboratory's European Bioinformatics Institute) (https://www.ebi.ac.uk/) and **DDBJ** (DNA Data Bank of Japan) (http://www.ddbj.nig.ac.jp). Exchange of data between these international collaborating databases occurs on a daily basis.

There are several specialized nematode databases:

WormBase (http://www.wormbase.org) is the central data repository for information about the model organism *Caenorhabditis elegans* and related nematodes. As a model organism database, WormBase extends beyond the genomic sequence, integrating experimental results with an extensively annotated view of the genome. WormBase also provides a large array of research and analysis tools. WormBase is one of the organizations participating in the Generic Model Organism Database (GMOD) project.

Nematode.net (http://www.nematode.net) is a web-accessible resource for investigating gene sequences from nematode genomes. The database is an outgrowth of the parasitic nematode expressed sequence tag (EST) project. ESTs are usually shorter than the full-length mRNA from which they are derived and are prone to sequencing errors. The database provides NemaGene EST cluster consensus sequence, enhanced online BLAST search tools, functional classifications of cluster sequences and comprehensive information concerning the ongoing generation of nematode genome data.

WormBase ParaSite (https://parasite.wormbase.org) is an open access resource providing genome sequences, genome browsers, semi-automatic annotation and comparative genomics analysis for parasitic worms, aiming to serve parasitologists working on helminths. WormBase ParaSite leverages the infrastructure and expertise of the WormBase project.

2.11. Examples of Molecular Phylogenies

2.11.1. Position of Nematoda within metazoans

The relative position of nematodes in animal phylogeny remained uncertain for a long time. In the traditional, morphologically based view, bilateral organisms were subdivided according to their internal organization and emerged in a universal phylogenetic tree with following order: (i) the Acoelomata, lacking a body cavity (mainly the platyhelminths and nemertines); (ii) the Pseudocoelomata (nematodes and some other minor phyla), with an internal cavity outside the mesoderm; and (iii) the Coelomata, which have true coelomic cavities splitting the mesoderm. Another vision of the phylogenetic relationships between metazoan phyla was obtained after the analysis of molecular data. The concept is known as the Ecdysozoa hypothesis, grouping moulting organisms, including arthropods and nematodes, a superphylum Ecdysozoa of animals first defined through analyses of molecular markers (Aguinaldo *et al.*, 1997). The Ecdysozoa comprises two groups, the Panarthropoda (phyla Tardigrada, Onychophora and Arthropoda) and Cycloneuralia (Nematoda, Nematomorpha, Priapulida, Kinorhyncha and Loricifera). Nematoda are consistently placed as sisters to Nematomorpha in morphological and molecular analyses (Schmidt-Rhaesa, 1997; Dunn *et al.*, 2008; Blaxter and Koutsovoulos, 2015).

2.11.2. The phylum Nematoda

Since the first publications of general phylogenies of nematodes based on 18S rRNA in 1998, a massive nematode rRNA database has been gathered. The most up-to-date single gene analysis of phylogeny of the entire phylum consisted of more than 2700 sequences (Holterman *et al.*, 2019), The analysis of the phylum highlights a number of paraphyletic taxa and indicates new relationships between previously unconnected taxa. The Nematoda seems to have arisen from adenophorean ancestry: the classic split into Adenophorea and Secernentea is not supported. In the first phylum-wide molecular phylogeny of nematodes (Blaxter *et al.*, 1998), five major clades (clades I–V) were identified; however, Holterman *et al.* (2006) presented a subdivision of the phylum Nematoda into 12 clades.

Nematoda comprises the subclasses Enoplia, Dorylaimia and Chromadoria (De Ley and Blaxter, 2002, 2004). In SSU analyses the branching order of these three groups was unresolved, although there were hints that Enoplia may be the earliest-branching of the three (van Megen *et al.*, 2009; Ahmed and Holovachov, 2021). It was suggested that non-vertebrate parasitism arose independently at least ten times, five times for parasitism of vertebrate hosts across the three subclasses and at least three times for plant parasitism, i.e. Tylenchomorpha, Dorylaimida and Diphtherophorina (Fig. 2.11) (Blaxter *et al.*, 1998; De Ley and Blaxter, 2004; van Megen *et al.*, 2009; Blaxter and Koutsovoulos, 2015). However, the correctness of the phylogenetic reconstruction for nematodes using SSU might be influenced by two main factors: grouping of long branches occurring as a result of abnormally high evolution rate and a total deficit of informative characters. Because the SSU tree is reconstructed based on a single gene, effort is continuously made to sequence other genes. For the majority of nematode clades, mitogenome analyses have yielded results similar to nuclear SSU gene tree. In addition, mitochondrial data often clarified relationships where there was poor resolution by nuclear DNA sequence data

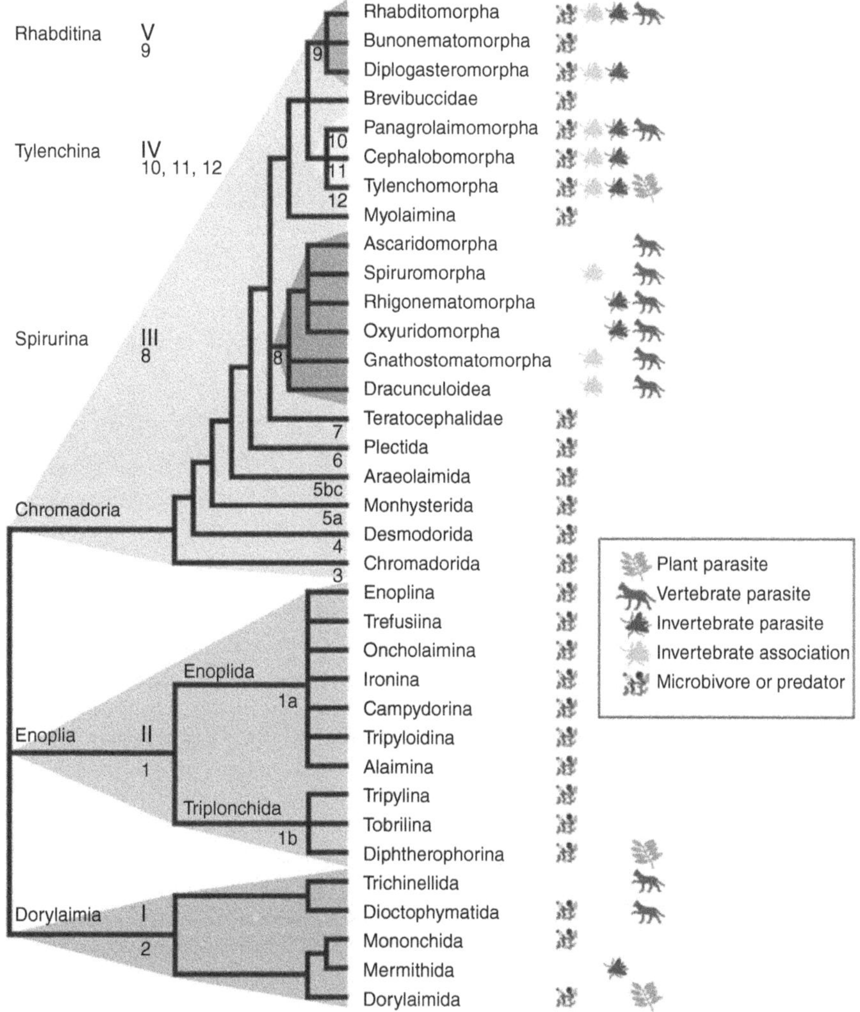

Fig. 2.11. An overview of phylogeny of the phylum Nematoda based on the small subunit ribosomal RNA gene. The systematic names given by De Ley and Blaxter (2004) are given, as is the 'clade' naming convention introduced by Blaxter *et al.* (1998). Helder and colleagues (Holterman *et al.,* 2006; van Megen *et al.*, 2009) introduced a numerical clade name scheme; this is given in numbers below the relevant branches. Feeding mode, and animal- and plant-parasitic and vector associations, are indicated by small icons (Blaxter, 2011).

(Kern *et al.*, 2020; Kim *et al.*, 2020). Recently, multigene phylogeny also became available when sequences of genes from nematode transcriptome and genome projects were obtained and this approch gave better supported relationships for some clades (Smythe *et al.*, 2019; Ahmed *et al.*, 2022).

2.11.3. The infraorder Tylenchomorpha

The evolutionary relationships of tylenchid and aphelenchid nematodes have been evaluated using sequence data of the 18S and the 28S rRNA genes. The order Tylenchida *sensu* Siddiqi containing plant-parasitic nematodes appears to be clearly monophyletic. The order Aphelenchida *sensu* Siddiqi comprising fungal-feeding Aphelenchidae and Aphelenchoididae is polyphyletic in all molecular analyses. Several studies have confirmed the sister relationship of tylenchids *sensu* Siddiqi (2000) with the bacteriovorous Cephalobidae (Blaxter *et al.*, 1998). The molecular datasets showed that the order Tylenchida *sensu* Siddiqi comprises lineages that largely correspond to two suborders, Hoplolaimina and Criconematina, and other taxonomic divisions by Siddiqi (2000). Molecular analysis supported the classical hypothesis of the gradual evolution of feeding types from simple forms of plant parasitism, such as root hair and epidermal feeding and ectoparasitism, towards more complex forms of endoparasitism. Sedentary endoparasitism has also evolved several times independently: (i) cyst and non-cyst nematodes of Heteroderidae probably evolved from migratory ectoparasitic nematodes; (ii) root-knot nematodes appear to be related to the false root-knot nematode Nacobbus and have evolved from migratory endoparasitic nematodes; and (iii) sedentary nematodes from Tylenchulidae and Sphaeronematidae (Criconematida).

2.11.4. Root-knot nematodes of the family Meloidogynidae

The genus *Meloidogyne* contains 98 valid species. The evolutionary relationships of root-knot nematodes have been inferred from several types of data: isozymes, DNA hybridization, DNA amplification fingerprinting, RAPD-PCR, sequencing of SSU rRNA, D2-D3 expansion segments of LSU rRNA, ITS rRNA and mtDNA. De Ley *et al.* (2002) were the first to use 18S rRNA gene sequences for a rigorous reconstruction of the *Meloidogyne* phylogeny. This analysis, which included 12 species of *Meloidogyne* and four outgroup taxa, revealed three clades (I, II and III) within the genus. Further analysis of five genes revealed that the root-knot nematode species studied are distributed among 11 highly or moderately supported clades, seven of which compose a 'superclade', containing 75% of the studied species. Clade I includes *Meloidogyne* species distributed in warmer climates and contains *Meloidogyne incognita*, *M. javanica*, *M. arenaria* and 17 other species that are commonly referred as the tropical root-knot nematode complex. Three species of this complex: *M. arenaria*, *M. incognita* and *M. javanica* belonging to *Incognita* group are globally distributed, polyphagous pests of many agricultural crops. *Meloidogyne nataliei* together with *M. indica* represents an early branching lineage of root-knot nematodes, which exhibit shared ancestral characteristics. The phylogenetic placement of the amphimictic *M. nataliei* with n=4 at a basal position to all other *Meloidogyne* species supports the hypothesis of a low chromosome number in ancestral species (Fig. 2.12) (Alvarez-Ortego *et al.*, 2019).

2.11.5. Cyst nematodes of the family Heteroderidae

Cyst nematodes are highly evolved sedentary plant parasites. Phylogenetic analysis of the ITS rRNA and D2–D3 expansion segment of 28S of rRNA gene sequences confirmed the monophyly of the subfamilies Punctoderinae and Heteroderinae with the genus

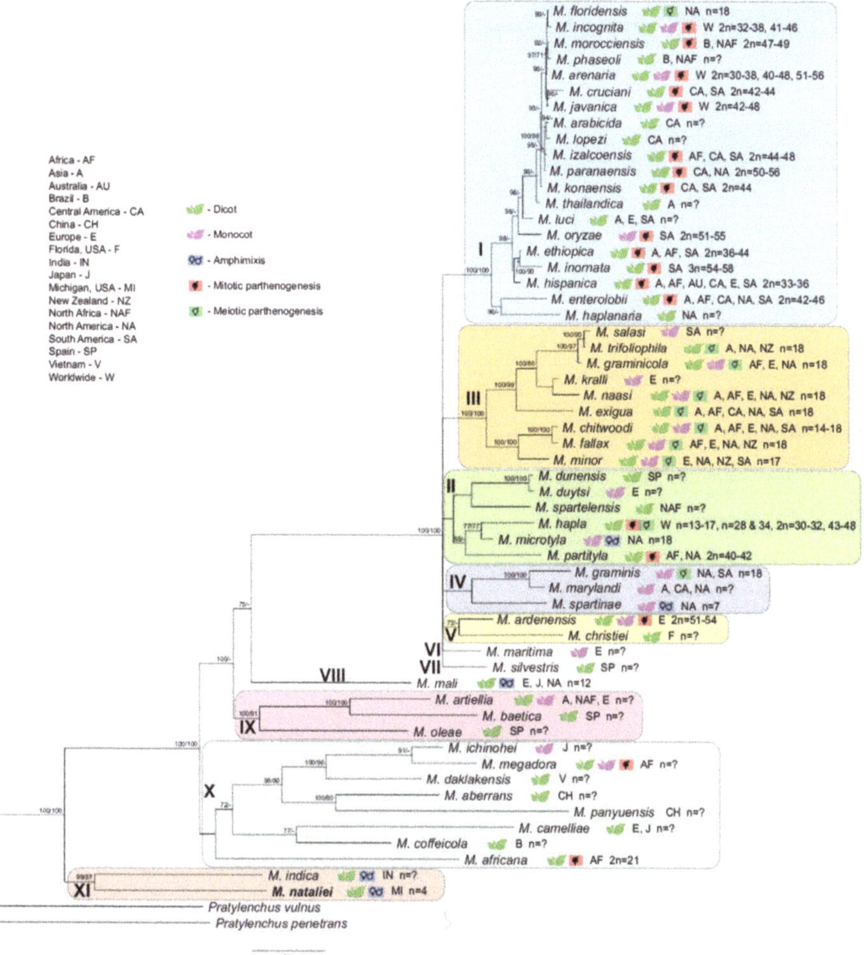

Fig. 2.12. Phylogenetic relationships within the genus *Meloidogyne*. Bayesian 50% majority rule consensus tree as inferred from 18S rRNA, ITS1 rRNA, D2–D3 expansion segments of 28S rRNA, *COI* gene and *COII*–16S rRNA sequence alignment under the GTR + I + G model. Branch support of over 70% is given for appropriate clades and it is indicated as: posterior probabilities value in Bayesian inference analysis/bootstrap value from maximum-likelihood analysis. Geographical distribution, plant host (monocots; dicots), reproduction mode and chromosome are given for each species. (After Álvarez-Ortega *et al.*, 2019.) CC BY 4.0.

Heterodera. The combination of molecular data with morphology of the vulval structure and the number of incisures in the lateral field of second-stage juveniles (J2) enabled nine main groups within *Heterodera*. Close relationships were revealed between the *Avenae* and *Sacchari* groups and between the *Humuli* group and the species *H. turcomanica* and *H. salixophila*. Some inconsistencies between molecular phylogeny and earlier proposed morphological groupings may be attributed to homoplastic evolution, e.g. a bifenestral vulval cone developed independently at least three times during the evolution of cyst nematodes. Likewise, the presence of three incisures in the lateral field of J2 seems to have arisen twice independently (Subbotin *et al.*, 2001). Molecular data suggested an early divergence of tropical and temperate heteroderid species and often revealed an association of nematodes with their host plants from related families (Fig. 2.13).

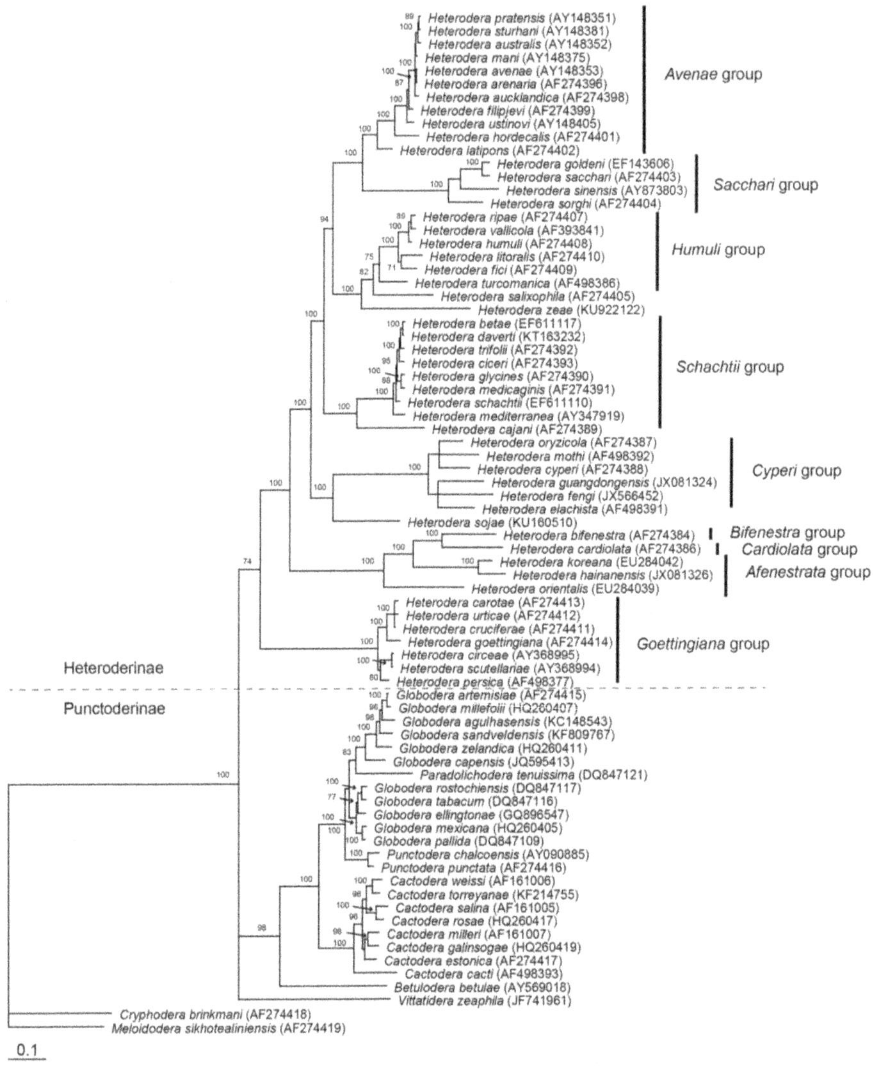

Fig. 2.13. Phylogenetic relationships within the genus *Heterodera*: Bayesian 50% majority rule consensus tree from two runs as inferred from ITS1–5.8S–ITS2 sequences of rRNA gene alignment. Posterior probabilities more than 70% are given for appropriate clades. (After Subbotin and Skantar, 2018.)

2.11.6. Stem and gall-forming nematodes of the family Anguinidae

The stem nematode, *D. dipsaci*, occurs as more than 20 biological races. Molecular approaches using RAPD-PCR, AFLP, PCR-RFLP and sequences of the ITS rRNA confirmed that *D. dipsaci* constitutes a complex sibling species. The phylogenetic analysis of the ITS sequences of plant-parasitic species of *Ditylenchus* revealed two main clades: (i) *D. dipsaci sensu stricto* with diploid chromosome numbers and comprising most isolates from agricultural, ornamental and several wild plants; and (ii) a complex of *Ditylenchus* species with polyploid chromosome numbers, including *Ditylenchus gigas* from *Vicia faba*, *D. weischeri* and several species parasitizing various Asteraceae and a species from *Plantago maritima*. Molecular methods failed to distinguish biological races within *D. dipsaci sensu stricto*.

Over 40 nominal species of gall-forming nematodes have been described. Testing of recognized anguinid classifications using the ITS sequences strongly supported monophyly of the genus *Anguina* and paraphyly of the genera *Mesoanguina, Heteroanguina sensu* Chizhov & Subbotin and *Subanguina sensu* Brzeski. Molecular data demonstrate that the main anguinid groups are generally associated with host plants belonging to the same or related systematic groups. The molecular analysis supports the concept of narrow host-plant specialization of seed-gall nematodes, and reveals several undescribed species infecting other species of grass (Fig. 2.14) (Subbotin *et al.*, 2004).

2.11.7. Pine wood nematode and other *Bursaphelenchus* species

A phylogeny of *Bursaphelenchus* species from Europe, North America, Central America and Asia representing much of the known biological diversity in this genus has been reconstructed using sequences of the 18S, 28S and ITS of rRNA and *COI* genes. Phylogenetic analyses using several methods of inference were congruent, with the greatest resolution obtained with combined datasets. Phylogenetic analysis revealed *B. abruptus* as the basal taxon among all investigated *Bursaphelenchus* species and a large number of significantly supported monophyletic groups that are largely consistent with morphological and life history variation in the genus (Fig. 2.15) (Ryss and Subbotin, 2017).

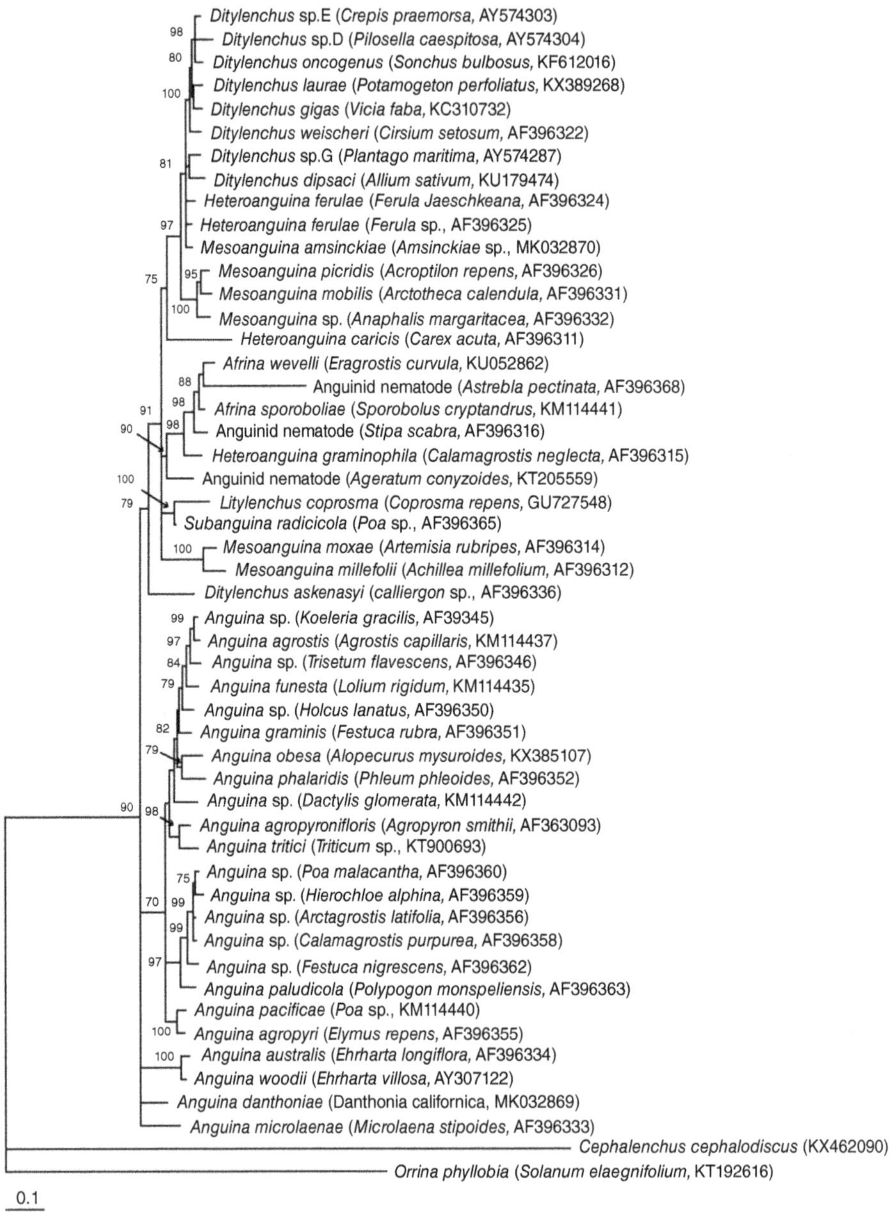

Fig. 2.14. Phylogenetic relationships within some plant-parasitic species from the family *Anguinidae* as inferred from the Bayesian analysis of the ITS rRNA gene sequences. Posterior probabilities are given on appropriate clades. (After Cid Del Prado Vera *et al.*, 2018, with modifications.)

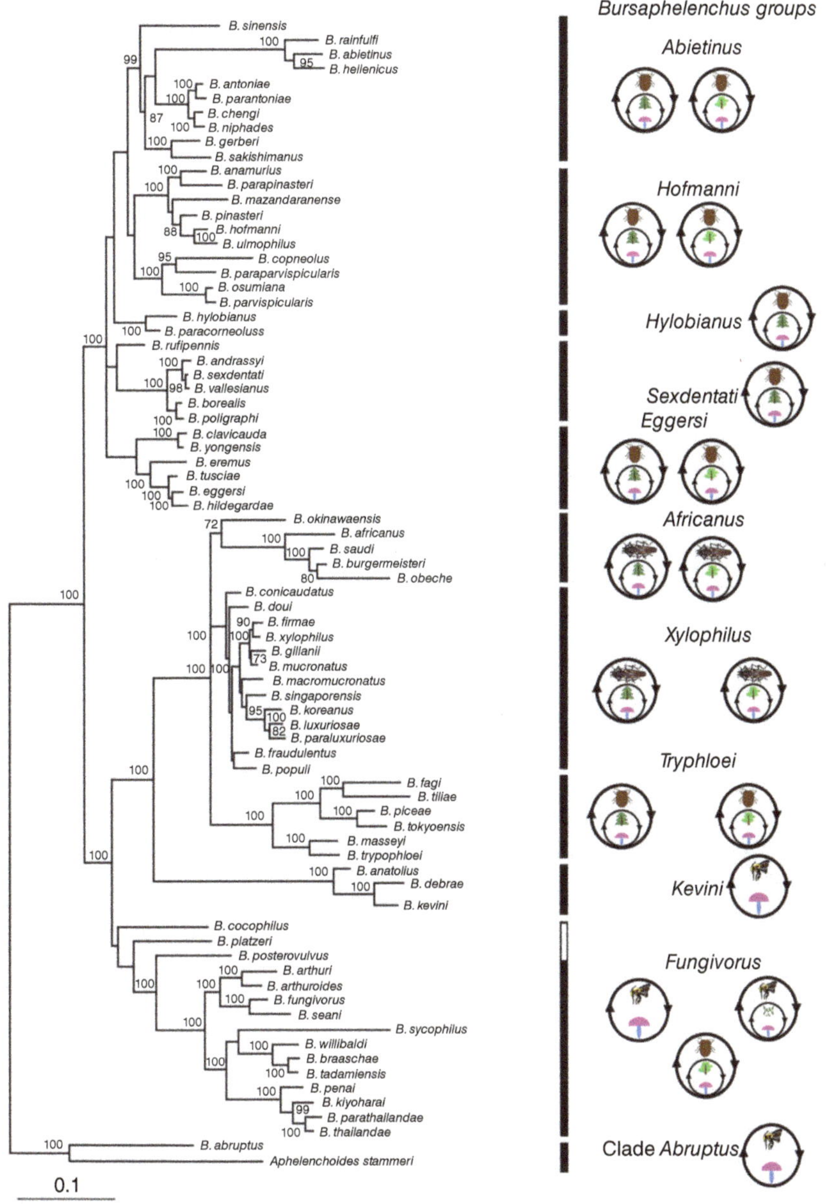

Fig. 2.15. Phylogenetic relationships within the genus *Bursaphelenchus*. Bayesian 50% majority rule consensus tree as inferred from 18S rRNA and D2–D3 expansion segments of 28S rRNA gene sequence alignment under the GTR + G model. Branch support of more than 70% is given for appropriate clades. Life cycles: dixenic (hosts: fungus and hymenoptera) and trixenic (hosts: fungus, plant – herbaceous, coniferous or deciduous – and insect – hymenoptera, bark beetle or longhorn beetle). (After Ryss and Subbotin, 2017.)

Root-knot Nematodes*

Wim Wesemael[1]**, Maurice Moens[1] and Stéphane Bellafiore[2]

[1]Flanders Research Institute for Agriculture, Fisheries and Food (ILVO), Plant Sciences Unit, Merelbeke, Belgium and Ghent University, Laboratory for Agrozoology, Department of Plants and Crops, Faculty of Bio-Science Engineering, Ghent, Belgium; [2]Plant Health Institute of Montpellier, IRD, CIRAD, University of Montpellier, l'Institut Agro, Montpellier, France and Institute of Technology of Cambodia (ITC), Phnom Penh, Cambodia

*A revision of Karssen, G., Wesemael, W. and Moens, M. (2013) Root-knot nematodes. In: Perry, R.N. and Moens, M. (eds) *Plant Nematology*, 2nd edn. CAB International, Wallingford, UK.
**Corresponding author: wim.wesemael@ilvo.vlaanderen.be

DOI: 10.1079/9781800622456.0003

3.1. Introduction to Root-knot Nematodes

Root-knot nematodes are included within the genus *Meloidogyne* Göldi, 1887 (*Meloidogyne* = apple-shaped female) and belong to a relatively small but important polyphagous group of highly adapted obligate plant pathogens. They are distributed worldwide and parasitize nearly every species of higher plant. Typically, they reproduce and feed within plant roots and induce small to large galls or root-knots (Fig. 3.1). Root-knot is one of

Fig. 3.1. Root-knot nematode galls of, respectively (starting top left, clockwise): *Meloidogyne incognita* on cucumber, *M. mali* on elm, *M. chitwoodi* on potato and *M. chitwoodi* on carrot.

the oldest known nematode diseases of plants. Other plant-parasitic nematodes, such as the false root-knot nematode *Nacobbus aberrans* (Thorne, 1935) Thorne & Allen, 1944 (Box 3.1) and *Subanguina radicicola* (Greef, 1872) Paramonov, 1967, also cause root-galls and can be confused with *Meloidogyne*.

Due to their endoparasitic way of living and feeding, root-knot nematodes disrupt the physiology of the plant and may reduce crop yield and product quality and, therefore, are of great economic importance. For reliable identification of root-knot nematodes, the best approach is to integrate morphological, isozyme and DNA data, together with information on mode of reproduction, chromosome number, host plants and distribution. Field samples often include mixtures of *Meloidogyne* species.

The systematic position of the root-knot nematodes at family level has been the subject of discussion for many years. Box 3.2 shows the classification of *Meloidogyne* after De Ley and Blaxter (2002). At genus level, root-knot nematodes were confused with cyst nematodes (*Heterodera* Schmidt, 1871) for a long period of time. Between 1884 and 1932 root-knot nematodes were generally named *Heterodera radicicola*, whilst between 1932 and 1949 the name *Heterodera marioni* was commonly used. Highlights of the taxonomical history of the genus are given in Box 3.3. A more comprehensive description of the taxonomical history of root-knot nematodes can be found in Karssen (2002) and aspects covered below are dealt with in greater detail in the book *Root-knot Nematodes* (Perry *et al.*, 2009).

Box 3.1. *Nacobbus* – the false root-knot nematode.

The species of the genus *Nacobbus* are sedentary root endoparasites. The parasitic behaviour of these species comprises similarities to both root lesion and root-knot nematodes. The migratory and vermiform juveniles and immature adults behave like lesion nematodes, causing cavities and lesions inside the root tissues, whereas the mature females are sedentary and obese and induce root galls as do true root-knot nematodes; hence their vernacular name 'false root-knot nematode'. The mature females induce specialized feeding sites similar to syncytia caused by cyst nematodes.

The genus *Nacobbus* has three species, *N. aberrans*, *N. dorsalis* and *N. bolivianus*. Of these, *N. aberrans* is the most widespread and has an extensive host range, among which are many cultivated plants (potato, tomato, sugar beet, carrot, lettuce, spinach, pepper and cucumber).

Nacobbus aberrans lays eggs in a gelatinous matrix secreted by the female. The first-stage juvenile develops inside the egg; the first moult occurs within the egg. After hatching, the second-stage juvenile (J2) penetrates a root, where it moves intracellularly causing necrotic lesions. The J2 moults either in the root or in the soil to the third-stage juvenile (J3). The moult to female or male fourth-stage juvenile (J4) occurs in the root or in the soil. Moulting of the J4 to the adult stage may take place inside or outside the root; immature females are vermiform. Mature females penetrate roots, become swollen and sedentary, and cause the formation of root galls and enlarged cells. Syncytia are associated only with adult females; hyperplasia, abnormal proliferation of lateral roots, and asymmetry of root structure are additional anatomical changes induced by adult females. Males leave the root swellings in search of the sedentary mature females in the galls. Obligatory amphimixis has been demonstrated in several South American populations. The egg sac is produced inside the gall and discharges to the root surface through a small channel, which is presumed to be formed during root penetration by the vermiform immature female.

Box 3.2. Classification of the genus *Meloidogyne* Göldi, 1887 within the phylum Nematoda, after De Ley and Blaxter (2002).

Phylum Nematoda Potts, 1932
Class Chromadorea Inglis, 1983
Order Rhabditida Chitwood, 1933
Suborder Tylenchina Thorne, 1949
Infraorder Tylenchomorpha De Ley & Blaxter, 2002
Superfamily Tylenchoidae Örley, 1880
Family Meloidogynidae Skarbilovich, 1959
  Genus *Meloidogyne* Göldi, 1887

Box 3.3. Highlights of the taxonomical history of the genus *Meloidogyne*.

1855 First illustrated report of root-knot disease by Berkeley (UK).
1879 First root-knot nematode description (*Anguillula marioni*) by Cornu (France).
1884 Müller (Germany) described and illustrated a perineal pattern for the first time.
1885 Treub (Indonesia) described *Heterodera javanica* (= *M. javanica*).
1889 Neal (USA) described *Anguillula arenaria* (= *M. arenaria*).
1887 Göldi (Brazil) described *Meloidogyne exigua*.
1919 Kofoid and White (USA) described *Oxyuris incognita* (= *M. incognita*).
1949 Revision of the genus (five species) and the name *Meloidogyne* revived by Chitwood (1949).
1968 *Meloidogyne* monograph (23 species) by Whitehead (1968).
1987 *Meloidogyne* monograph (51 species) by Jepson (1987).
2002 Revision of the European *Meloidogyne* species by Karssen (2002).
2021 *Meloidogyne* monograph (98 species) by Subbotin *et al.* (2021a).

3.2. Life Cycle and Behaviour

Root-knot nematode eggs are enclosed in gelatinous egg sacs that are usually deposited on the surface of galled roots (Figs 3.2 and 3.3). Sometimes they occur within the galls. Some species, such as *M. graminicola* and *M. spartinae*, deposit their eggs in the root cortex (Fig. 3.4). Following embryogenesis, the first moult occurs within the egg, giving rise to the second-stage juvenile (J2). Hatching of *Meloidogyne* eggs is temperature driven and occurs without requiring stimulus from plant roots; however, root diffusates sometimes stimulate hatching. Occasionally, the requirement for a hatching stimulus from the host plant is dependent on plant age. Hatching of J2 of *M. chitwoodi* produced on young plants does not require host root diffusate stimulus, whereas at the end of the plant growing season, egg masses contain a percentage of unhatched J2 that require host root exudate to cause hatch; the closely related species, *M. fallax*, does not show this difference (Wesemael *et al.*, 2006). The eggshell becomes flexible immediately prior to hatching and enzymes are thought to be involved in altering eggshell structure (see Chapter 7). When J2 leave the egg masses, they infect nearby galled roots or enter new roots. J2, i.e. the infective stage, and males are the stages of *Meloidogyne* that can be found freely in the soil. J2 can survive in the soil in a

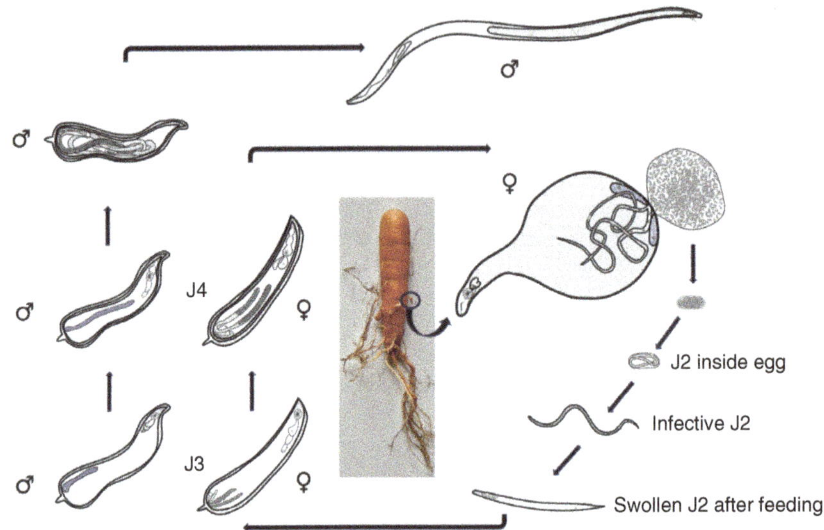

Fig. 3.2. Diagram of the life cycle of the root-knot nematode, *Meloidogyne*. J2: second-stage juvenile; J3: third-stage juvenile; J4: fourth-stage juvenile. (Courtesy of Wim Wesemael, partly created in BioRender.com.)

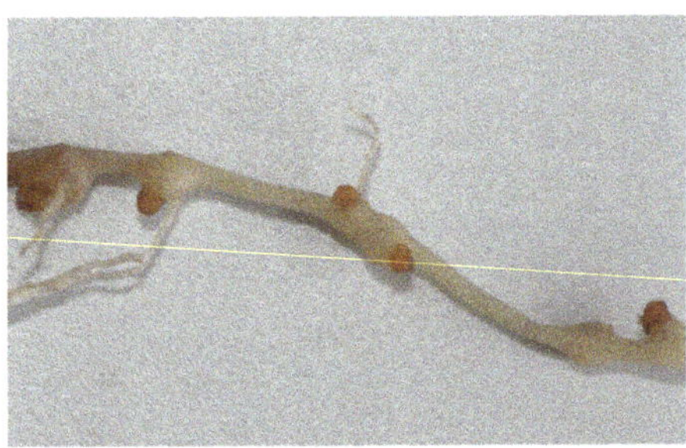

Fig. 3.3. Root piece of tomato with brownish egg masses of *Meloidogyne chitwoodi*.

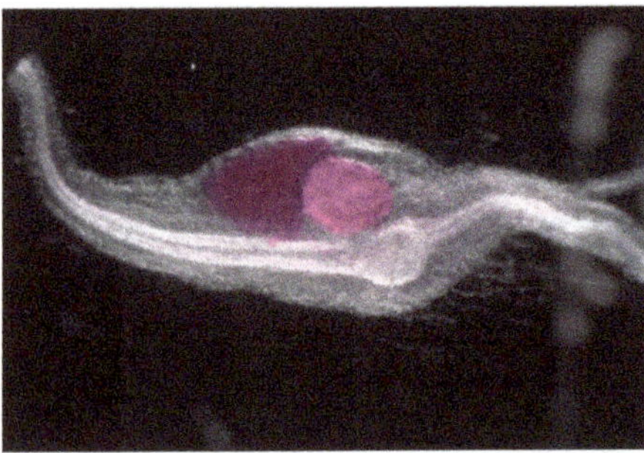

Fig. 3.4. *Meloidogyne graminicola* female and egg mass stained with acid fuchsin inside rice root. (Courtesy of Satish Chavan.)

quiescent state for an extended period. However, during that period they consume their food reserves stored in the intestine. As infectivity is related to food reserves (see Chapter 7), their infectivity will be reduced after long periods spent out of the roots. A comprehensive review on hatch and host location of *Meloidogyne* is given by Curtis *et al.* (2009).

J2 are attracted to plant roots and root location depends on perception of gradients of attractants emanating from the plant root. The infective J2 accumulate at the region of cell elongation just behind the root cap, even of plants resistant to root-knot nematodes. They are also attracted to apical meristems, points where lateral roots emerge, and penetration sites of other J2. There is little difference in the attractiveness of isolated susceptible and resistant plant roots; however, when both are present, the susceptible ones are preferred. The nature of the stimuli produced by the roots and perceived by J2 is not clear. Many organic and inorganic compounds excreted by roots form gradients from the root surface into the soil and may influence the nematodes. Carbon dioxide is frequently considered as being the most important factor for attracting plant-parasitic nematodes to the root area and specific attractants may be responsible for ensuring the J2 reaches the preferred invasion site (see Chapter 8). When root-knot nematodes come into contact with plant roots, they often penetrate immediately. Penetration usually occurs directly behind the root cap but it can occur at any site. J2 penetrate the rigid root cell walls by a combination of physical damage via thrusting of the stylet and breakdown by cellulyctic and pectolyctic enzymes secreted by the nematode (see Chapter 9). Following penetration, especially with multiple infections on the same root, the root tip may enlarge and root growth often stops for a short period. When egg masses are embedded in root tissue or freely deposited inside the root cortex, J2 can hatch and remain in the maternal gall or migrate intercellularly to new feeding sites within the same root, as seen for *M. graminicola*. This has also been reported for *Meloidogyne* spp. inside potato tubers (Teklu *et al.*, 2018).

After penetration into the root, J2 migrate intercellularly in the cortex to the region of cell differentiation. This migration causes cells to separate along the middle lamella. Cells along the path become distended but only rarely show signs of nematode feeding. Increases in levels of oxidoreductive enzyme activity have also been found; this is an indication of increased metabolic activity. To circumvent the barrier formed by the endodermis, the J2 migrate towards the root tip and turn around when they arrive in the apical meristimatic region. Subsequently, they move back up in the vascular cylinder towards the zone of differentiation. After migrating over a short distance, J2 become sessile in the cortical tissue in the zone of differentiation. The head of the J2 is embedded in the periphery of the vascular tissue where they feed on protoxylem and protophloem; the rest of the body is in the cortex parallel with the longitudinal axis of the root.

3.3. Host Response to Parasitism

Susceptible plants react to feeding by juveniles and undergo pronounced morphological and physiological changes. Giant cells (2 to 12, usually about 6), feeding sites for the root-knot nematode, are established in the phloem or adjacent parenchyma (Fig. 3.5). These cells are highly specialized cellular adaptations. The induction and maintenance of these giant cells has been the subject of several hypotheses. It is suggested that both induction and maintenance are controlled by stylet secretions originating from the subventral (mainly the early stages of giant cell formation, but not exclusively) and dorsal pharyngeal (mainly the later stages of giant cell formation and maintenance, but not exclusively) glands of the feeding juveniles (see Chapter 9). Removal of solutes from giant cells may be the stimulus necessary for maintaining the active metabolism of giant cells. If the host does not respond by the formation of giant cells, J2 fail to develop. The J2 will either starve or, if they have sufficient food reserves, may migrate out of the root and locate another root.

Giant cells induced by root-knot nematodes divide without forming new cell walls; they are most likely formed after repeated nuclear divisions without cytokinesis. Cell wall dissolution does not occur in the first 72 h of giant cell initiation. In normal divisions, cell plate vesicles coalesce during telophase to form a cell wall separating daughter nuclei. In giant cells around *Meloidogyne*, these vesicles are not consolidated; they disperse and are resorbed in the cytoplasm to yield binucleate cells. At later periods, cell wall stubs (plates, flanges) can be formed. The appearance of these flanges is different from that of wall fragments of syncytia in that they are not perforated (Fig. 3.6).

At the beginning of the giant cell formation the cells are occupied predominantly by the cell vacuole; the nuclei are located in the peripheral cytoplasm. Within a cell, mitosis of the different nuclei occurs synchronically (Fig. 3.7). In a more advanced giant cell,

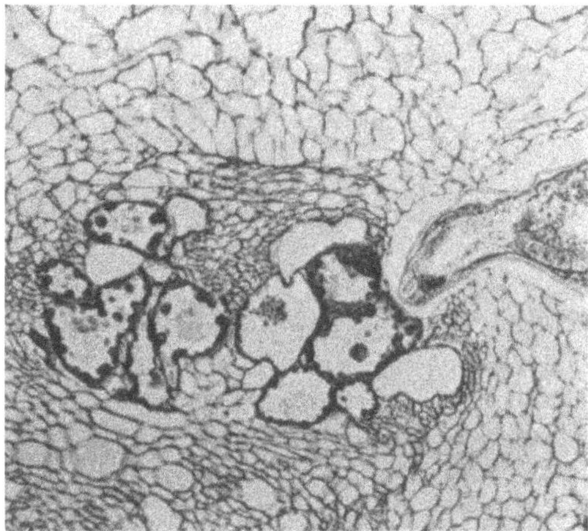

Fig. 3.5. Light micrograph of a cross-section of a root gall infected with a *Meloidogyne* female showing several giant cells. (After Eisenback and Hirschmann Triantaphyllou, 1991.)

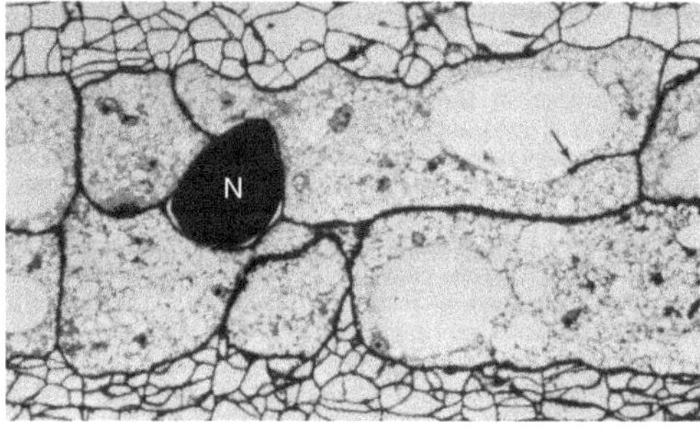

Fig. 3.6. Longitudinal section of giant cells within 6-day-old gall induced by *Meloidogyne incognita* in balsam root, and associated nematode (N). The cell vacuole is greatly reduced and the cells filled with cytoplasm and secondary vacuoles. A single cell wall stub (arrow) is present. (From Jones and Payne, 1978.)

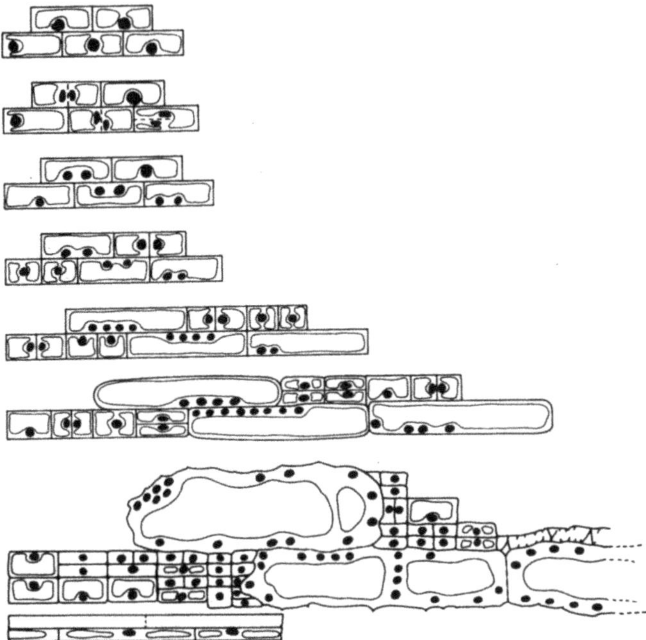

Fig. 3.7. Diagrammatic representation of the events which occur up to 72 h after the induction of three giant cells by *Meloidogyne* spp. Multinucleate cells develop after the failure of cytokinesis. Mitosis within one cell is synchronous, but not necessarily at the same rate as in other cells. Vascular elements start to differentiate outside the giant cells, and rapid division of neighbouring cells allows for their expansion. (From Jones and Payne, 1978.)

synchrony of the nuclear divisions may not be constant. As cytoplasmic contents (Golgi apparatus, endoplasmic reticulum, mitochondria, plastids and ribosomes) increase, the cells expand laterally. Walls that separate giant cells are apparently continuous without 'gaps' or 'perforations'. However, there are alternating thick and thin areas on the same wall. Plasmodesmata are found only in the thin portions of the wall. As the food demand of the nematode increases with increasing egg laying, the giant cell cytoplasm shows signs of intense metabolic activity. This is demonstrated by the presence of aneuploid nuclei with 14–16 times more DNA than nuclei of non-infected root tip cells. The increased metabolic activity mobilizes photosynthetic products from shoots to roots. Further details of giant cell induction and maintenance are given in Chapter 9.

At the same time as the establishment of giant cells, root tissues around the nematode undergo hyperplasia and hypertrophy, causing the characteristic root gall usually associated with *Meloidogyne* infections. Galls usually develop 1–2 days after J2 have penetrated the root. Their size is related to the host plant, the number of J2 and the nematode species. Gall formation is not essential for nematode development.

Plant growth regulators have been implicated in the development of giant cells and galls (see Chapter 7 for details). Auxins (promoters of cell growth) have been identified in higher concentrations in root-knot infected tissue than in non-galled tissue. Cytokinins (promoters of cell division) may also increase in *Meloidogyne*-infected plants. The application of these plant growth regulators to resistant plants reverses the resistance response and makes plants susceptible. These regulators have also been identified in different stages of *Meloidogyne*. Another plant growth regulator, ethylene, has been associated with gall formation. It may be involved in the hypertrophy of cortical parenchyma tissue during gall formation.

3.4. Post-infection Biology

Following penetration of host tissue and establishment of the feeding site, J2 stages undergo several morphological changes (Fig. 3.8). The nematode grows slightly in length. Simultaneously, enlargement of the pharyngeal glands and metacorpus occur. The cells of the genital primordial divide and six rectal glands, which secrete the ovisac, develop in the posterior end of the female juvenile stages. The genital primordium is V-shaped in the female, consisting of two limbs joined posteriorly to the vaginal rudiment that lies close to the posterior body wall. The male gonad develops as a single limb, the *vas deferens* differentiating at the hind end and eventually joining with the rectum. During their further development, the juveniles gradually assume a flask shape and undergo three moults. The last moult is a true metamorphosis for the male, which appears as a long filiform nematode folded inside the moulted cuticle of the fourth-stage juvenile, which is retained as a sheath. The adult female at first retains the same shape of the last juvenile stage but enlarges as it matures and eventually becomes pyriform. In the amphimictic species, the males mate after escaping from the last juvenile sheath. Females secrete a gelatinous matrix into which they extrude a large number of eggs.

The energy necessary for completion of the third and fourth moults must be obtained by the animal before the second moult because it is unable to feed on its host from the start of the second moult to the completion of the fourth moult. This happens because during this period the nematode has no stylet and remains within the sheath of the second cuticle. This cuticle retains a characteristic spike-like tail, although the enclosed juvenile is bluntly rounded posteriorly. By contrast, the developing juveniles of *Heterodera* and *Globodera* have no tail spike. After the fourth moult, the stylet is reformed and digestive and reproductive organs are developed. When the male is mature it uses its stylet to break through the cuticle and root tissue.

It is well known that the proportion of males in a population, at least in parthenogenetic species, varies according to the host plant and the environmental conditions. Food supply may be an important factor as males are more abundant under adverse conditions for development. It is known that under such conditions, female juveniles can become adult males. Sometimes true intersexes (males with vulva) appear.

The number of generations per year varies according to species and food availability. Usually there are many, but in some species there is only one (e.g. *M. naasi*). In a favourable host, several hundred eggs are produced by each female. Each female may lay 30–80 eggs per day, the number depending on the host plants and the environmental conditions. Eggs of *Meloidogyne* are mostly deposited in a gelatinous matrix, which is not retained in the body (as with most cyst-forming nematodes). If the female is deeply embedded in the plant tissue, the egg sac may also be embedded and the gelatinous matrix remains soft. If the egg sac is exposed on the root surface, the outer matrix layers may dry and become tough; sometimes they become orange–brown. In either case, the eggs are partially protected against drying and can survive in soil for a period that depends on temperature and moisture and on the nematode species.

Root-knot nematodes are known to have unbalanced sex ratios. Cross-fertilizing species (*M. carolinensis*, *M. spartinae*) usually have a 1:1 male-to-female ratio. Species

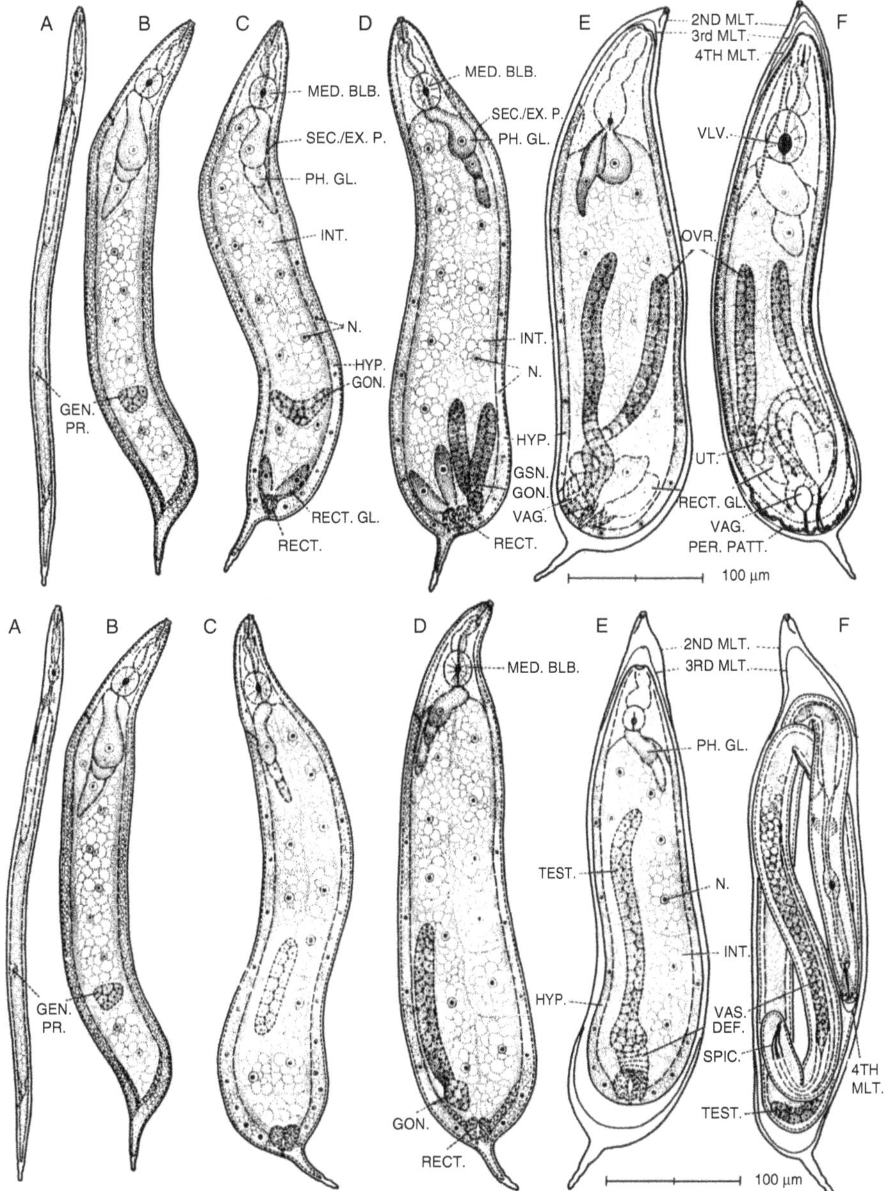

Fig. 3.8. Development of a female (top) and male (bottom) of *Meloidogyne incognita* from second-stage juvenile to adult. A: Second-stage infective juvenile. B: Swollen, sexually undifferentiated second-stage juvenile. C: Early second-stage juvenile differentiating into a female/male. D: Second-stage female/male juvenile shortly before second moult. E: Fourth-stage female/male juvenile. F: Adult female/male shortly after fourth moult. SEC./EX. P.: secretory–excretory pore; GEN. PR.: genital primordium; GON.: gonad; HYP.: hypodermis; INT.: intestine; MED. BLB.: median bulb; 2nd MLT.: second moult; 3rd MLT.: third moult; 4th MLT.: fourth moult; N.: nucleus; PH. GL.: pharyngeal glands; OVR.: ovary; PER. PATT.: perineal pattern; RECT. GL.: rectal glands; RECT.: rectum; SPIC.: spicule; TEST.: testis; UT.: uterus; VAG..: vagina; VAS. DEF.: *vas deferens*; VLV.: valve. (Modified from Eisenback and Hirschmann Triantaphyllou, 1991.)

that reproduce by facultative or obligatory parthenogenesis (*M. hapla*, *M. incognita*) have variable sex ratios. Depending on the environmental conditions, males may be absent or rare or abundant. Depending on the developmental stage at which sex reversal occurs, sex-reversed males may have one to two gonads of variable length.

3.5. Effect on Plant Growth and Yield

As described above, the response of roots of host plants to invasion by root-knot nematodes is very characteristic. The severity of the symptoms can be scaled with a root gall index (Greco and Di Vito, 2009). Crops from which tubers or tap roots are harvested can be worthless due to quality loss despite the normal total yield in tonnes (see Chapter 10). Above-ground symptoms observed on infected plants are similar to those produced on any plants having a damaged and malfunctioning root system. Symptoms include: (i) suppressed shoot growth and accompanying decreased shoot–root ratio; (ii) nutritional deficiencies showing in the foliage, particularly chlorosis; (iii) temporary wilting during periods of mild water stress or during the middle part of the day, even when adequate soil moisture is available; and (iv) suppressed plant yields. The importance of these symptoms is often related to the number of juveniles penetrating and becoming established within the root tissue of young plants. The common explanation for these above-ground symptoms is that *Meloidogyne* infection affects water and nutrient uptake and upward translocation by the root system.

Most of the studies focusing on the relationship between root-knot infection and plant nutrition demonstrated that *Meloidogyne* infection increased N, P and K concentrations in above-ground plant parts. However, this was not always the situation; sometimes nutrients appeared to accumulate in the roots. Supplying nutrients to infected plants often increases tolerance to *Meloidogyne*. The few studies on the water relationships of *Meloidogyne*-infected plants suggest that water consumption is not affected by nematode infection if soil moisture is not limiting, but under periodic stress, consumption is reduced. Plant development is suppressed by infections that inhibit root growth during moisture stress and prevent roots from extending into moist soil.

Meloidogyne infection of roots decreases the rate of photosynthesis in leaves. The nematodes interfere with the production of root-derived factors regulating photosynthesis. *Meloidogyne* functions as a metabolic sink in diseased plants. The increased metabolic activity of giant cells stimulates mobilization of photosynthetic products from shoots to roots and, in particular, to the giant cells where they are removed and utilized by the feeding nematode. Mobilization and accumulation of substances reaches a maximum when the adult females start egg laying and declines thereafter. Root branching and the degree of root extension are frequently affected by *Meloidogyne* infection. Abnormal root growth results in reduced root surface area and limits the capacity of a root system to explore the soil. As a consequence, the primary cause of poor nutrient uptake and suppressed growth of infected plants could be related to the reduced root system.

3.6. Survival

Meloidogyne eggs and juveniles do not survive as long as those of the closely related genera *Heterodera* and *Globodera*, where eggs in cysts can remain viable for several years. Eggs of *Meloidogyne* are deposited in a gelatinous matrix, which may be white or brown. Egg masses formed early in the host growing cycle contain eggs from which J2 hatch immediately, making several generations per growing cycle possible. Egg masses formed later or in adverse conditions by old or poorly nourished females are brown and contain eggs that are dormant and do not hatch immediately. This form of dormancy is termed diapause; more information on this topic is provided in Chapter 7. These latter egg masses ensure some carry over from one season to another.

Temperature is an important factor in several stages of the development of nematodes. As with all poikilothermic animals, temperature influences distribution, survival, growth and reproduction. Within the genus *Meloidogyne,* two distinct groups can be distinguished: thermophils and cryophils, which can be separated by their ability to survive lipid-phase transitions that occur at 10°C (see Chapter 7). *Meloidogyne hapla, M. chitwoodi, M. fallax* and probably *M. naasi* are cryophils and are able to survive soil temperatures below 10°C. *Meloidogyne chitwoodi* and *M. fallax* were found to stay infective for more than 300 days in soil at 5°C. *Meloidogyne incognita, M. javanica* and *M. exigua* are thermophils and do not have extended survival at temperatures below 10°C. Like survival, hatching is primarily controlled by temperature. Thermotypes exist within a species. Such populations can be distinguished from each other by small differences in the minimum temperatures needed for hatching, probably due to the adaptation of geographical populations to local temperature regimes.

Soil texture, moisture, aeration and osmotic potential are interacting factors and it is difficult to determine the effect of each one separately. Most *Meloidogyne* species are active in soils with moisture levels at 40–60% of field capacity. As the soils either dry or increase in moisture, nematode activity decreases. In drying environments, the gelatinous matrix of the egg sac appears to maintain a high moisture level and provide a barrier to water loss from eggs. Survival of eggs in egg masses at temperature extremes was found to be higher in dry soil compared with moist soil. Embryos and first-stage juveniles are more resistant to water loss than unhatched J2 because of changes in the egg membrane after the first moult. In drying soils, nematodes may be submitted to increased osmotic pressures, especially after fertilizer applications. Even small changes in osmotic potential may influence nematode behaviour. However, as discussed in Chapter 8, osmotic pressure within soil is unlikely to influence the water balance of plant-parasitic nematodes as the soil dries because most water will be extracted from nematodes by other forces before osmotic pressure is physiologically important. Increased crop damage is often associated with alkaline soils. This seems to be associated with stress on the host plant. As *Meloidogyne* prefer sandy soils, their agricultural importance is most frequently associated with sandy soils. *Meloidogyne artiellia* is an exception and causes severe damage in both sandy soils and in soils containing 30–35% clay.

3.7. Cytogenetics

The cytogenetic status of the genus *Meloidogyne* is complex but important in understanding the overall biology and evolution of these nematodes. The major root-knot nematodes have been well studied cytogenetically (Triantaphyllou, 1985). Within the genus *Meloidogyne* there are three modes of reproduction: amphimixis, automixis and apomixis (Table 3.1). Only a small number of species reproduce by **amphimixis,** i.e. with the obligatory fusion of a male and female gamete. The majority of species of *Meloidogyne* reproduce by parthenogenesis (automixis or apomixis). During maturation of the oocytes, the **apomictic** species undergo only a single mitotic division. Meiosis is completely absent in these

Table 3.1. Chromosome number and mode of reproduction of some root-knot nematodes.

	Chromosome number		
Meloidogyne species	*n*	*2n*	Mode of reproduction
M. carolinensis	18		
M. mali	12		
M. kikuyensis	7		
M. megatyla	18		
M. microtyla	18–19		Amphimixis
M. nataliei	4		
M. pini	18		
M. spartinae	7		
M. subartica	18		
M. chitwoodi	14–18		
M. exigua	18		
M. fallax	18		
M. floridensis	18		
M. graminicola	18		Facultative meiotic parthenogenesis (automixis)
M. graminis	18		
M. hapla (race A)	13–17		
M. minor	17		
M. naasi	18		
M. ottersoni	18		
M. trifoliophila	18		
M. africana		21	
M. ardenensis		51–54	
M. arenaria		30–38	
		40–48	
		51–56	
M. cruciani		42–44	
M. enterolobii		44–46	
M. ethiopica		36–44	
M. hapla (race B)		30–32	
		43–48	
M. hispanica		33–36	
M. incognita		32–38	
		41–46	
M. inornata		54–58	Obligatory mitotic parthenogenesis (apomixis)
M. izalcoensis		44–48	
M. javanica		42–48	
M. konaensis		44	
M. luci		42–46	
M. microcephala		36–40	
M. morocciensis		47–49	
M. oryzae		51–55	
M. paranaensis		50–52	
M. partityla		40–42	
M. petuniae		41	
M. platani		42–44	
M. querciana		30–32	
M. salasi		36	

After Triantaphyllou (1985) and updated by the authors.

obligatory mitotic parthenogens. In **automictic** species, oocytes undergo a meiotic division. When males are present, the sperm and egg nuclei fuse together. However, without the presence of a sperm nucleus, the nucleus of the second polar body fuses with the egg pronucleus and restores the diploid state (meiotic parthenogenesis). Therefore, automictic root-knot nematode species are facultatively parthenogenetic. Blanc-Mathieu *et al.* (2017) considered the success of asexually reproducing *Meloidogyne* species could be explained partly by their transposable element (TE)-rich composite genomes, facilitating plasticity and functional divergence between gene copies in the absence of sex and meiosis. *Meloidogyne* species with low chromosome count are ancestral traits, from which the mitotic parthenogenetic species have evolved (Triantaphyllou, 1985).

Most of the amphimictic and automictic species are diploid with a haploid chromosome number of 18. *Meloidogyne nataliei* is a diploid amphimictic species that has a haploid complement of only four chromosomes. The four chromosomes of *M. nataliei* are relatively large when compared to those of other *Meloidogyne* species. Moreover, *M. nataliei* reproduces exclusively by cross-fertilization. The majority of the apomictic species are polyploid or aneuploid and usually show a wide variation in chromosome number. *Meloidogyne africana* has the lowest number of chromosomes ($2n=21$) of root-knot nematodes known to reproduce by mitotic parthenogenesis. *Meloidogyne incognita* reproduces exclusively by apomixis. The majority of those populations of this species that have been studied have $2n=42$–44 chromosomes and are considered to be triploid or hypotriploid (i.e. derived from triploid forms through the loss of chromosomes); however, some populations have 32 or 38 chromosomes and represent a diploid form ($n=16$–18). *Meloidogyne hapla* is cytogenetically the most complex species of the genus. Most populations reproduce by automixis (termed race A) with a haploid chromosome number of 17; some populations with lower numbers (13–15) are known. Also, polyploid *M. hapla* females ($n=28$–34) have been observed within some diploid populations reproducing by automixis; even conversion from polyploid to diploid was noted. Other *M. hapla* populations reproduce by apomixis (race B), with usually 43–48 chromosomes and a few populations have 30–32 chromosomes. Direct polyploidization within the apomictic *M. microcephala* has also been recorded, i.e. through doubling of its somatic chromosome number from a diploid ($2n=37$) to a tetraploid ($2n=74$) state. Usually polyploidization mainly affects nematode morphometrics, resulting in an increased size of various structures, but rarely affects morphology.

3.8. General Morphology

Female. Sedentary adult root-knot nematode females are pearly white with a rounded to pear-shaped body and a protruding, sometimes bending, neck (Fig. 3.9). A cyst stage is not present. Females range in length from 350 µm to 3 mm and in maximum width from 300 to 700 µm. In full-grown females the cuticle annulation is only visible in the head region and the posterior part where a characteristic unique cuticular pattern or perineal pattern can be observed around the perineum (vulva–anus region). The shape of this pattern is often variable and influenced by several developmental factors. The perineum is terminal with the phasmids just above the anus. Usually the anus is covered with a small cuticular fold and the perineum can be slightly elevated. Some species have lateral lines or subcuticular punctuations above the perineum (Figs 3.10 and 3.11).

The head is usually not or only slightly set off with a distinct but delicate cephalic frame work present. The labial disc is not, to slightly, raised and fused with the medial

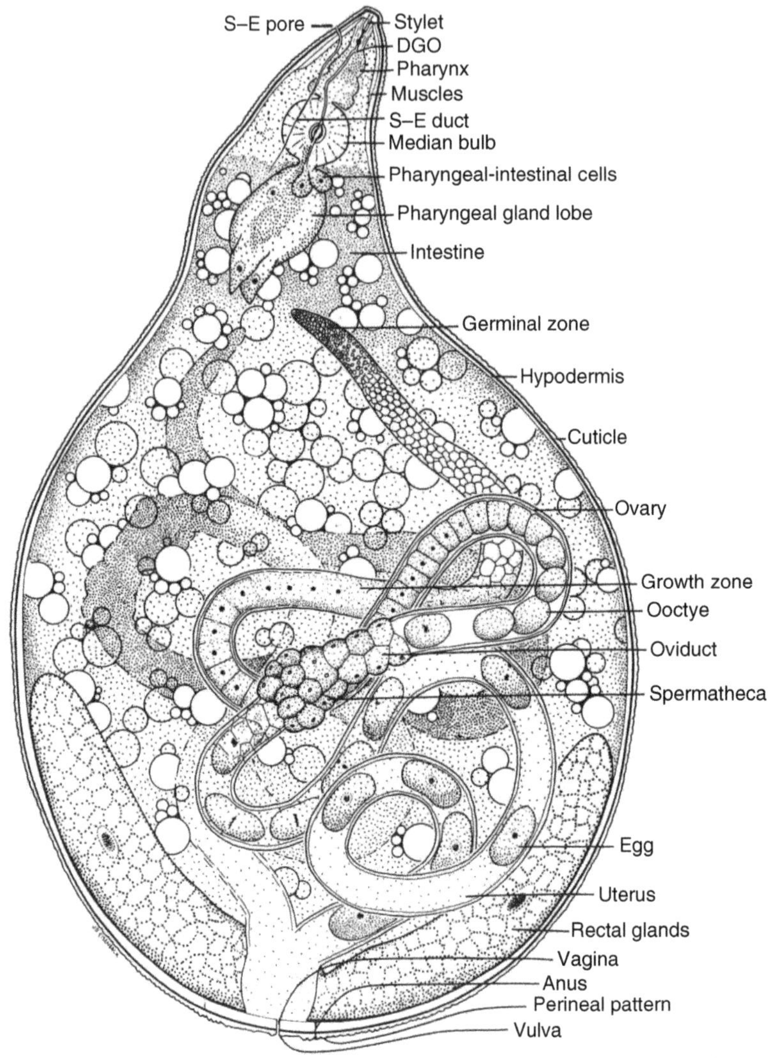

Fig. 3.9. Drawing of a female root-knot nematode. S–E: secretory–excretory; DGO: dorsal pharyngeal gland orifice. (From Eisenback and Hirschmann Triantaphyllou, 1991.)

and lateral lips. Two slit-like amphidial and ten small sensilla openings are present around the stylet stoma. The delicate stylet ranges in length from 10 to 25 μm; in most species the cone is slightly curved dorsally; the shaft is straight and includes three basal knobs. Stylet knob shape varies from rounded to transversely elongate and can be set off or is posteriorly sloping. The dorsal pharyngeal gland orifice (DGO) is located between 2.5 and 9.0 μm behind the stylet knobs. The secretory–excretory pore (S–E pore) is located usually between the stylet knobs and the metacorpial level. The metacorpus itself is relatively large and connected posteriorly with the pharyngeal glands. These glands are highly variable in size and shape and overlap the intestine ventrally. Two long didelphic, partly

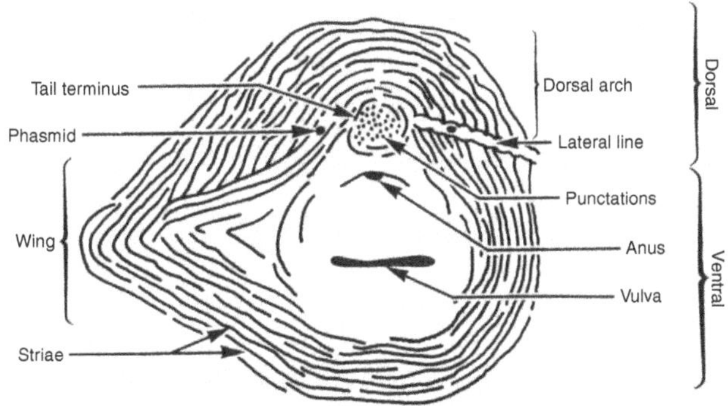

Fig. 3.10. Diagram of a root-knot nematode perineal pattern. (From Eisenback and Hirschmann Triantaphyllou, 1991.)

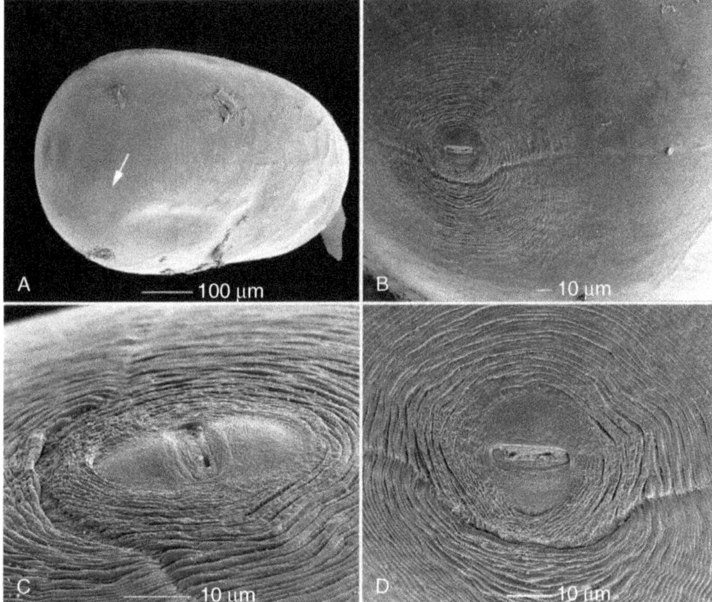

Fig. 3.11. Female root-knot nematode scanning electron microscope photographs. A: *Meloidogyne hapla* body (arrow = lateral line); B–D: *M. hapla* perineal patterns.

convoluted, gonads are present. Each gonad is composed of an ovary with germinal and growth zone, an oviduct, a large lobed spherical spermatheca and a very long uterus. Most of the unembryonated eggs are deposited in an egg sac, produced by six large unicellular rectal glands and secreted through the anus.

Male. The non-sedentary male is vermiform, clearly annulated and ranges in length from 600 to 2500 µm (Fig. 3.12). The head is composed of a head cap and head region (= post-labial annule). The head region can be set off and/or is partly subdivided by transverse incisures or annulations. The head cap has a relatively large rounded labial disc and is usually fused with four medial lips. Six inner labial sensilla are centred around the oral

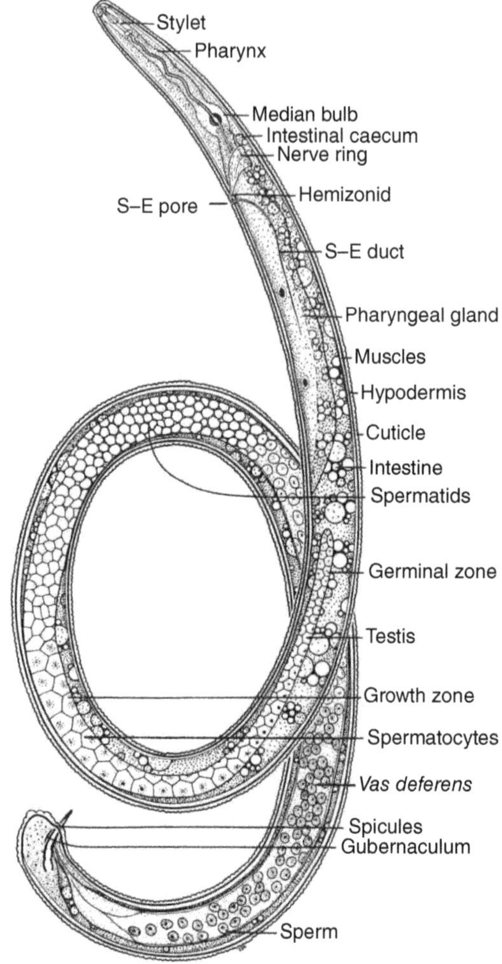

Fig. 3.12. Drawing of a male root-knot nematode. S–E: secretory–excretory. (After Eisenback and Hirschmann Triantaphyllou, 1991.)

opening and one cephalic sensillum is present on each medial lip. The two large slit-like amphidial openings are located between the labial disc and the lateral lip. In some species, these lateral lips are reduced or absent (Fig. 3.13). The cephalic framework and straight stylet are well developed; the latter ranges in length from 13 to 33 μm. The DGO is located 2–13 μm behind the three stylet knobs; stylet knob shape is as in females.

The metacorpus is much smaller compared to females. The S–E pore and the hemizonid are located between the metacorpial level and the ventral overlapping of the pharyngeal glands. The hemizonid is positioned anterior or sometimes posterior to the S–E pore. The pharyngeal gland nuclei are usually reduced to two. Usually one long testis is present; sometimes, however, two reduced ones can be observed in sex-reversed males. In most species, the lateral field has four incisures and the outer bands are often areolated. The tail is very short, bluntly rounded and without bursa. The small phasmids are positioned

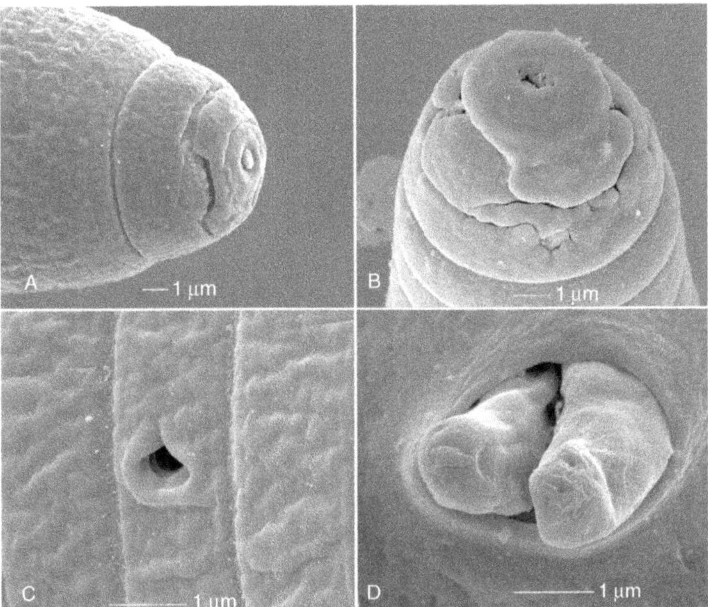

Fig. 3.13. Male root-knot nematode scanning electron microscope photographs. A: *Meloidogyne duytsi* head region (lateral view). B: *M. minor* anterior end. C: *M. minor* secretory–excretory pore. D: *M. duytsi* spicule.

near the cloaca. The spicules are slender and 20–40 µm long, while the crescentic gubernaculum is about 10 µm long.

Second-stage juvenile. The infective J2 is vermiform, annulated and ranges in length from 250 to 600 µm (Fig. 3.14). The head structure is as in males but much smaller and with weakly sclerotized cephalic framework. The delicate straight stylet is about 9–16 µm long, and the position of the DGO ranges from 2 to 12 µm behind the stylet knobs.

The metacorpus is relatively small with well developed valve plates. The hemizonid is usually positioned anterior or posterior to the S–E pore. Three pharyngeal glands are present and ventrally overlap the intestine. Usually the rectum is clearly inflated. The tail is 15–100 µm long and tapers towards the tail tip and ends into a hyaline tail part (tail terminus). Phasmids are very small and located about one third of the tail length below anus level. The lateral field has four incisures with the outer bands usually areolated.

The third- and fourth-stage juvenile stages are sedentary inside the root and swollen; they have no stylet and develop within the J2 cuticle. A detailed description of the ultrastructure of *Meloidogyne* is given by Bird and Bird (1991). Box 3.4 summarizes the criteria used for morphological identification of *Meloidogyne* species.

3.9. Principal Species

A list of nominal root-knot nematodes and synonyms is given in Table 3.2. At the end of 2023, nearly 100 nominal species of root-knot nematodes have been described. The number of descriptions of new species continues to increase but, unfortunately, they are not always

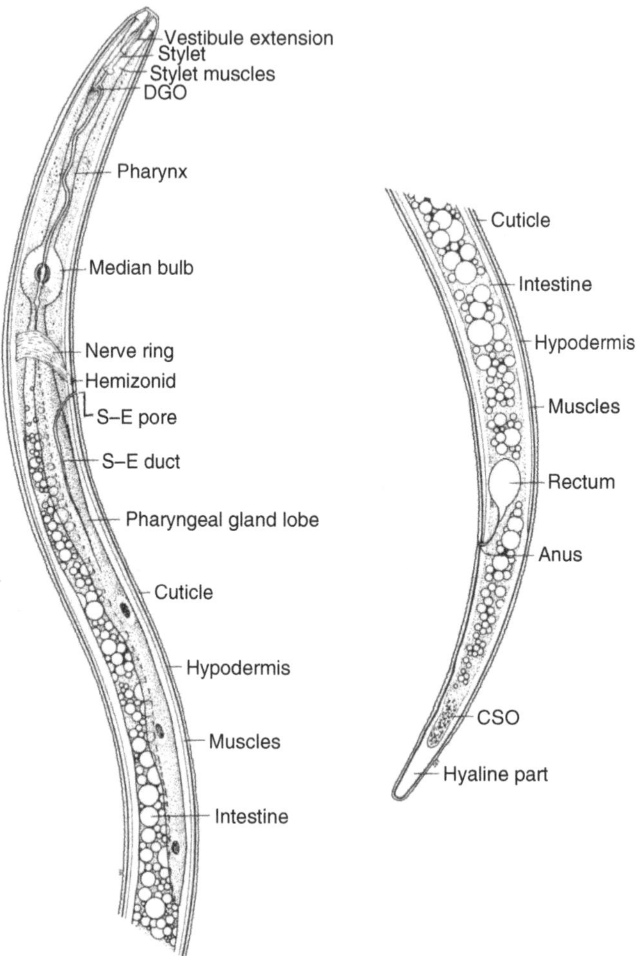

Fig. 3.14. Drawing of a second-stage juvenile root-knot nematode. DGO: dorsal pharyngeal gland orifice; S–E: secretory–excretory; CSO: caudal sense organ. (From Eisenback and Hirschmann Triantaphyllou, 1991.)

well described and compared with other species from the genus. It is not surprising that, at present, a satisfactory tool for the identification of all root-knot nematodes has not been developed. The following species of root-knot nematodes are economically important agricultural pests and described briefly (Figs 3.15 and 3.16); Jepson (1987), Eisenback and Triantaphyllou (1991) and Subbotin *et al.* (2021a) provide extensive information on these and other species. More information on isozyme characterization is given in Section 3.10.1.

3.9.1. *Meloidogyne arenaria*

Female. Body pear-shaped and without posterior protuberance. Stylet 13–17 (16) µm long, knobs rounded and backwardly sloping. Perineal pattern rounded with low dorsal arch and distinct striae, lateral lines sometimes weakly visible.

Box 3.4. Morphological identification.

Characters used for identification of root-knot nematodes using light microscopy are:

Female: Body shape, stylet length, knob shape and perineal pattern shape.
Male: Head shape, stylet length, knob shape and DGO–stylet knob length.
Second-stage juvenile: Body length, tail and hyaline tail length, DGO–stylet knob length, hemizonid position, tail and hyaline tail shape.

For a reliable morphological observation of *Meloidogyne* species, it is very important to use the correct methods for extraction, isolation, microscopical preparation and observation, as described in detail by Jepson (1987).

Table 3.2. Nominal root-knot nematode species (*Meloidogyne* Göldi, 1887).

Type species
M. exigua Göldi, 1887
  syn. *Heterodera exigua* (Göldi, 1887) Marcinowski, 1909

Species
M. aberrans Tao, Xu, Yuan, Wang, Lin, Zhou & Liao, 2017
M. acronea Coetzee, 1956
  syn. *Hypsoperine acronea* (Coetzee, 1956) Sledge & Golden, 1964
  Hypsoperine (*Hypsoperine*) *acronea* (Coetzee, 1956) Siddiqi, 1985
M. aegracyperi Eisenbeck, Holland, Schroeder, Thomas, Beacham, Hanson, Paes-Takahashi & Vieira, 2019
M. africana Whitehead, 1960
  syn. *M. decalineata* Whitehead, 1968
  M. oteifae Elmiligy, 1968
M. aquatilis Ebsary & Eveleigh, 1983
M. arabicida López & Salazar, 1989
M. ardenensis Santos, 1968
  syn. *M. deconincki* Elmiligy, 1968
  M. litoralis Elmiligy, 1968
M. arenaria (Neal, 1889) Chitwood, 1949
  syn. Anguillula arenaria Neal, 1889
  Heterodera arenaria (Neal, 1889) Marcinowski, 1909
  M. arenaria arenaria (Neal, 1889) Chitwood, 1949
  M. arenaria thamesi Chitwood, 1952
  M. thamesi (Chitwood, 1952) Goodey, 1963
  M. thamesi gyulai Amin, 1993
  M. gyulai Amin, 1993
M. artiellia Franklin, 1961
M. baetica Castillo, Volvlas, Subbotin & Troccoli, 2003
M. brevicauda Loos, 1953
M. californiensis Abdel-Rahman & Maggenti, 1987
  syn. *M. californiensis* Abdel-Rahman, 1981
M. camelliae Golden, 1979
M. caraganae Shagalina, Ivanova & Krall, 1985

Continued

Table 3.2. Continued.

M. carolinensis Eisenback, 1982
  M. carolinensis Fox, 1967
M. chitwoodi Golden, O'Bannon, Santo & Finley, 1980
M. chosenia Eroshenko & Lebedeva, 1992
M. christiei Golden & Kaplan, 1986
M. citri Zhang, Gao & Weng, 1990
M. coffeicola Lordello & Zamith, 1960
  syn. *Meloidodera coffeicola* (Lordello & Zamith, 1960) Kirjanova, 1963
M. cruciani Garcia-Martinez, Taylor & Smart, 1982
M. cynariensis Bihn, 1990
M. daklakensis Trinh, Le, Nguyen, Nguyen, Liebanas & Nguyen, 2018
M. donghaiensis Zheng, Lin & Zheng, 1990
M. dunensis Palomares Rius, Vovlas, Troccoli, Liébanas, Landa & Castillo, 2007
M. duytsi Karssen, van Aelst & van der Putten, 1998
M. enterolobii Yang & Eisenback, 1983
  syn. *M. mayaguensis* Rammah & Hirschmann, 1988
M. ethiopica Whitehead, 1968
  syn. *M. brasilensis* Charchar & Eisenback, 2002
M. fallax Karssen, 1996
M. fanzhiensis Chen, Peng & Zheng, 1990
M. floridensis Handoo, Nyczepir, Esmenjaud, van der Beek, Castagnone-Sereno, Carta, Skantar & Higgins, 2004
M. fujianensis Pan, 1985
M. graminicola Golden & Birchfield, 1965
  syn. *M. hainanensis* Liao & Feng, 1995
  M. lini Yang, Hu & Xu, 1988
M. graminis (Sledge & Golden, 1964) Whitehead, 1968
  syn. *Hypsoperine graminis* Sledge & Golden, 1964
  Hypsoperine (Hypsoperine) *graminis* (Sledge & Golden, 1964) Siddiqi, 1985
M. hapla Chitwood, 1949
M. haplanaria Eisenback, Bernard, Starr, Lee & Tomaszewski, 2003
M. hispanica Hirschmann, 1986
M. ichinohei Araki, 1992
M. incognita (Kofoid & White, 1919) Chitwood, 1949
  syn. Oxyuris incognita Kofoid & White, 1919
  Heterodera incognita (Kofoid & White, 1919) Sandground, 1923
  M. incognita incognita (Kofoid & White, 1919) Chitwood, 1949
  M. incognita acrita Chitwood, 1949
  M. acrita (Chitwood, 1949) Esser, Perry & Taylor, 1976
  M. kirjanovae Terenteva, 1965
  M. elegans da Ponte, 1977
  M. grahami Golden & Slana, 1978
  M. wartellei Golden & Birchfield, 1978
  M. polycephannulata Charchar, Eisenback, Vieira, Fonseca-Boiteux & Boiteux, 2009
M. indica Whitehead, 1968
M. inornata (Lordello, 1956) Carneiro, Mendes, Almeida, Santos, Gomes & Karssen, 2008
  syn. *M. incognita inornata* Lordello, 1956
M. izalcoensis Carneiro, Almeida, Gomes & Hernández, 2005

Continued

Table 3.2. Continued.

M. javanica (Treub, 1885) Chitwood, 1949
  syn. *Heterodera javanica* Treub, 1885
  Anguillula javanica (Treub, 1885) Lavergne, 1901
  M. dimocarpus Liu & Zhang, 2001
  M. javanica javanica (Treub, 1885) Chitwood, 1949
  M. javanica bauruensis Lordello, 1956
  M. bauruensis (Lordello, 1956) Esser, Perry & Taylor, 1976
  M. lucknowica Singh, 1969
  M. lordelloi da Ponte, 1969
M. jianyangensis Yang, Hu, Chen & Zhu, 1990
M. jinanensis Zhang & Su, 1986
M. karsseni Singh, Gitonga, Hajihassani, Verhage, van Aalst-Philipse, Couvreur & Bert, 2023
M. kikuyensis de Grisse, 1960
M. konaensis Eisenback, Bernard & Schmitt, 1994
M. kongi Yang, Wang & Feng, 1988
M. kralli Jepson, 1984
M. lopezi Humphreys Pereira, Flor es Chaves, Gomez, Salazar, Gomez Alpizar & Elling, 2014
M. luci Carneiro, Correa, Almeida, Gomes, Abbas, Castagnone Sereno & Karssen, 2014
M. lusitanica Abrantes & Santos, 1991
M. mali Itoh, Oshima & Ichinohe, 1969
  syn. *M. ulmi* Marinari-Palmisano & Ambrogioni, 2000
M. maritima (Jepson, 1987) Karssen, van Aelst & Cook, 1998
M. marylandi Jepson & Golden, 1987
M. megadora Whitehead, 1968
M. megatyla Baldwin & Sasser, 1979
M. mersa Siddiqi & Booth, 1992
M. microcephala Cliff & Hirschmann, 1984
M. microtyla Mulvey, Townshend & Potter, 1975
M. mingnanica Zhang, 1993
M. minor Karssen, Bolk, v. Aelst, v.d. Beld, Kox, Korthals, Molendijk, Zijlstra, v. Hoof & Cook, 2004
M. moensi Le, Nguyen, Nguyen, Liebanas, Nguyen & Trinh, 2019
M. morocciensis Rammah & Hirschmann, 1990
M. naasi Franklin, 1965
M. nataliei Golden, Rose & Bird, 1981
M. oleae Archidona Yuste, Cantalapiedra Navarrete, Liebanas, Rapoport, Casttillo & Palomares Rius, 2018
M. oryzae Maas, Sanders & Dede, 1978
M. ottersoni (Thorne, 1969) Franklin, 1971
  syn. *Hypsoperine ottersoni* Thorne, 1969
  Hypsoperine (*Hypsoperine*) *ottersoni* (Thorne, 1969) Siddiqi, 1985
M. ovalis Riffle, 1963
M. panyuensis Liao, Yang, Feng & Karssen, 2005
  syn. *M. panyuensis* Liao, 2001
M. paramali Gu, Fang, Ma, Shao & Zhuo, 2023
M. paranaensis Carneiro, Carneiro, Abrantes, Santos & Almeida, 1996
M. partityla Kleynhans, 1986
M. petuniae Charchar, Eisenback & Hirschmann, 1999
M. phaseoli Charchar, Eisenback, Charchar & Boiteau, 2008
M. pini Eisenback, Yang & Hartman, 1985

Continued

Table 3.2. Continued.

M. piperi Sahoo, Ganguly & Eapen, 2000
M. pisi Charchar, Eisenback, Charchar & Boiteau, 2008
M. platani Hirschmann, 1982
M. polycephannulata Charchar, Eisenback, Vieira, Fonseca-Boiteux & Boiteux, 2009
M. propora Spaull, 1977
  syn. *Hypsoperine* (*Hypsoperine*) *propora* (Spaull, 1977) Siddiqi, 1985
M. querciana Golden, 1979
M. salasi Lopez, 1984
M. sasseri Handoo, Huettel & Golden, 1993
M. sewelli Mulvey & Anderson, 1980
M. silvestris Castillo, Vovlas, Troccoli, Liébanas, Palomares Rius & Landa, 2009
M. sinensis Zhang, 1983
M. spartelensis Ali, Tavoillot, Mateille, Chapuis, Besnard, El Bakkali, Cantalapiedra Navarrete, Liebanas, Castillo & Palomares Rius, 2015
M. spartinae (Rau & Fassuliotis, 1965) Whitehead, 1968
  syn. *Hypsoperine spartinae* Rau & Fassuliotis, 1965
  Hypsoperine (*Hypsoperine*) *spartinae* (Rau & Fassuliotis, 1965) Siddiqi, 1985
M. subartica Bernard, 1981
M. suginamiensis Toida & Yaegashi, 1984
M. tadshikistanica Kirjanova & Ivanova, 1965
M. thailandica Handoo, Skantar, Carta & Erbe, 2005
M. trifoliophila Bernard & Eisenback, 1997
M. triticoryzae Gaur, Saha & Khan, 1993
M. turkestanica Shagalina, Ivanova & Krall, 1985
M. vandervegtei Kleynhans, 1988
M. vitis Yang, Hu, Liu, Chen, Peng, Wang & Zhang, 2021

Species *inquirendae*
M. actinidiae Li & Yu, 1991
M. cirricauda Zhang, 1991
M. marioni (Cornu, 1879) Chitwood & Oteifa, 1952
  syn. *Anguillula marioni* Cornu, 1879
  Heterodera marioni (Cornu, 1879) Marcinowski, 1909
M. megriensis (Poghossian, 1971) Esser, Perry & Taylor, 1976
  syn. *Hypsoperine megriensis* Poghossian, 1971
  Hypsoperine (*Hypsoperine*) *megriensis* (Poghossian, 1971) Siddiqi, 1985
M. pakistanica Shahina, Nasira, Salma, Mehreen & Bhatti, 2015
M. poghossianae Kirjanova, 1963
  syn. *M. acronea* apud Poghossian, 1961
M. vialae (Lavergne, 1901) Chitwood & Oteifa, 1952
  syn. *Anquillula vialae* Lavergne, 1901
  Heterodera vialae (Lavergne, 1901) Marcinowski, 1909

Nomina nuda
M. californiensis Abdel-Rahman, 1981
M. carolinensis Fox, 1967
M. goeldii Santos, 1997
M. panyuensis Liao, 2001
M. shunchangensis Wu, 2011
M. zhanjiangensis Liao, 200

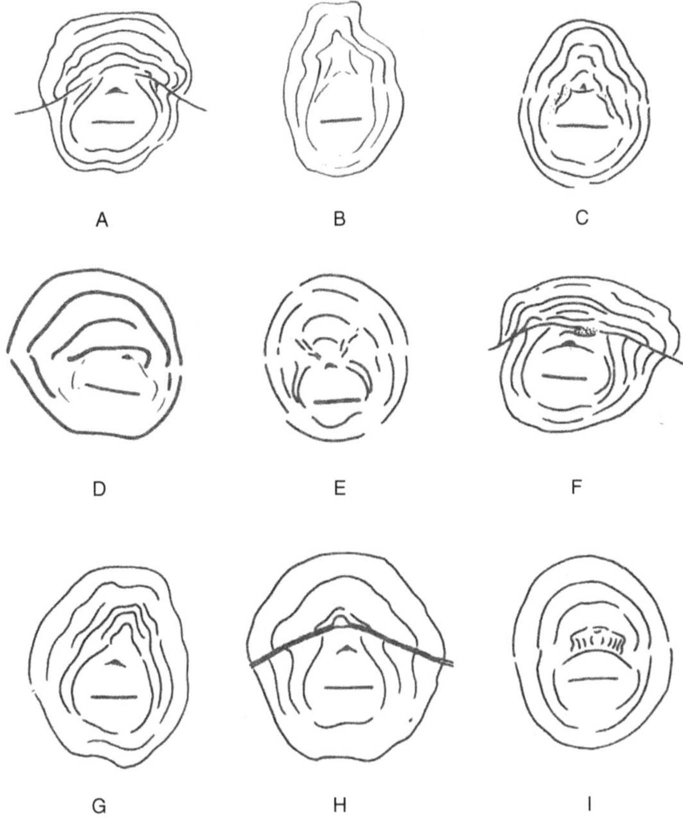

Fig. 3.15. Root-knot nematode perineal patterns. A: *Meloidogyne arenaria*. B: *M. artiellia*. C: *M. chitwoodi/M. fallax*. D: *M. exigua*. E: *M. graminicola*. F: *M. hapla*. G: *M. incognita*. H: *M. javanica*. I: *M. naasi*. (Redrawn after Jepson, 1987.)

Male. Head not set off from the body. Head cap rounded, labial disc not elevated, lateral lips absent. Stylet 20–27 (23) µm long, knobs rounded and backwardly sloping. DGO to stylet knobs: 4–8 (6) µm.

Second-stage juvenile. Body 400–600 µm long. Hemizonid anterior adjacent to the S–E pore. Tail long (45–70 µm) and slender, hyaline tail part 6–15 µm long, anterior region usually not clearly delimitated, tip finely rounded to pointed.

Isozymes. Esterase A1, A2 and A3 type and malate dehydrogenase N1 and N3 type (Fig. 3.17). A few atypical esterase phenotypes have been described.

Hosts. Able to reproduce on many monocotyledonous and dicotyledonous plants.

Distribution. *Meloidogyne arenaria, M. incognita* and *M. javanica* are the three most common species of root-knot nematodes, although *M. arenaria* is not as common as the other two species. Known worldwide; in temperate regions restricted to greenhouses.

Remarks. Induces large galls, which sometimes appear as a string of pearls along the root.

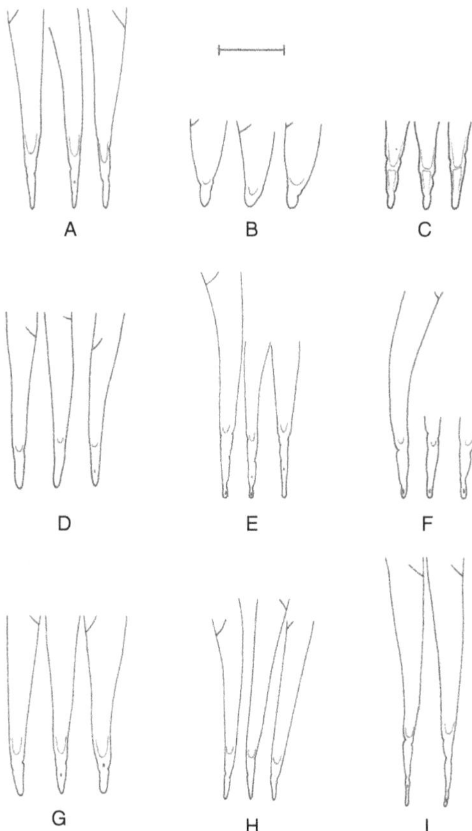

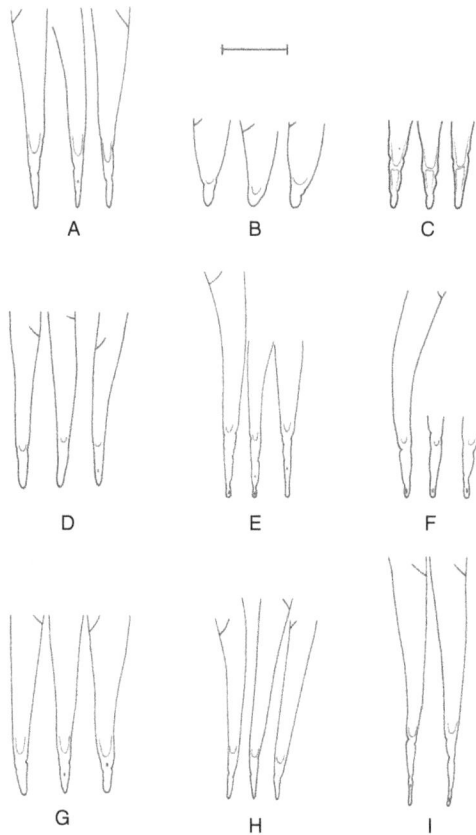

Fig. 3.16. Root-knot nematode second-stage juvenile tails. A: *Meloidogyne arenaria*. B: *M. artiellia*. C: *M. chitwoodi*/*M. fallax*. D: *M. exigua*. E: *M. graminicola*. F: *M. hapla*. G: *M. incognita*. H: *M. javanica*. I: *M. naasi*. Scale bar = 20 µm. (Redrawn after Jepson, 1987.)

3.9.2. *Meloidogyne artiellia*

Female. Body pear-shaped with short neck region and without posterior protuberance. Stylet 13–15 (14) µm long; knobs relatively small, ovoid and backwardly sloping. Perineal pattern striking, rounded with fine striae, lateral area with coarse ridges, dorsal arch angular and obscure lateral lines present.

Male. Head set off from the body. Head cap distinct, labial disc not elevated, lateral lips present. Stylet 15–19 (17) µm long, knobs backwardly sloping and ovoid shaped. DGO to stylet knobs: 4–6 µm.

Second-stage juvenile. Body 320–370 µm long. Hemizonid anterior adjacent to the S-E pore. Tail conical and short (19–25 µm), hyaline tail part 4–8 µm in length with a rounded tail tip.

Isozymes. Esterase M2-VF1 type and malate dehydrogenase N1b type.

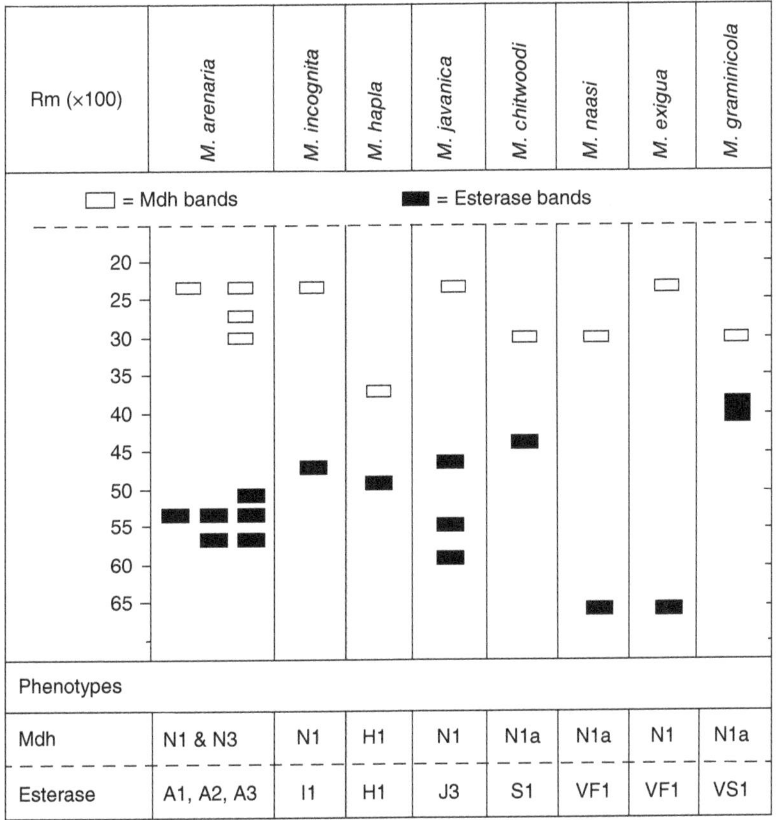

Fig. 3.17. Malate dehydrogenase (Mdh) and esterase isozyme phenotypes of several species of *Meloidogyne*. Rm: relative movement. (From Eisenback and Hirshmann Triantaphyllou (1991), plus information on codes for Mdh and esterase.)

Hosts. Detected on *Brassica napus*, *B. oleracea*, *Cicer arietinum*, *Vicia sativa*, *Avena sativa* and *Triticum vulgare*. Experimental studies indicate that nearly all members of the Brassicaceae, Fabaceae and Poaceae are good hosts for *M. artiellia*.

Distribution. In Europe reported from France, Greece, Italy, Spain and the UK. Outside Europe it has been found in Israel and Syria.

Remarks. The J2 are relatively small and can be overlooked easily in soil samples. Usually this species induces very small galls, sometimes with lateral root proliferation.

3.9.3. *Meloidogyne chitwoodi*

Female. Body rounded with short neck and slight posterior protuberance. Stylet 11–14 (13) µm long, knobs small, oval to irregularly shaped and backwardly sloping. S–E pore just below stylet knob level. Perineal pattern rounded to oval, striae relatively coarse and lateral lines weakly visible.

Male. Head not set off from the body. Head cap rounded, labial disc not elevated, lateral lips present. Stylet 17–19 (18) µm long, knobs small, oval to irregular shaped and backwardly sloping. DGO to stylet knobs: 2.5–4.0 µm.

Second-stage juvenile. Body 360–420 µm long. Hemizonid anterior and adjacent to the S–E pore. Tail conical and 40–45 µm long; hyaline tail part 9–12 µm long, anterior region clearly delimitated, tail tip bluntly rounded.

Isozymes. Esterase S1 type and malate dehydrogenase N1a type.

Hosts. Able to parasitize many mono- and dicotyledonous hosts, including economically important crops such as potato (tubers), wheat and maize. Usually it induces relatively small galls.

Distribution. A major pest in the Pacific Northwest region of the USA, also known from eight other states in the USA and Mexico. Also reported from Europe (Belgium, France, Germany, Lithuania, Portugal, Romania, Spain, Sweden, Switzerland, Türkiye and The Netherlands), Russia, Mozambique, South Africa and South America (Argentina and Chile).

Remarks. Listed in many countries as a quarantine organism.

3.9.4. *Meloidogyne enterolobii*

Female. Body pear-shaped without posterior protuberance. Stylet 13–18 (15) µm long, knobs oval, anteriorly often indented, slightly sloping backward to set off. Perineal pattern oval shaped, striae mostly fine, dorsal arch rounded to square, weak lateral line(s) sometimes present. General pattern shape as in *M. incognita*.

Male. Head slightly set off from the body. Head cap high and rounded, head region slightly set off, not annulated, lateral lips usually not present. Stylet 20–24 (22) µm long, knobs large, ovoid to rounded, slightly sloping backwards. DGO to stylet knobs: 3.5–6 µm.

Second-stage juvenile. Body 380–460 µm long. Hemizonid anterior and adjacent to the S–E pore. Tail slender and 45–60 µm long; anterior hyaline tail part not clearly deliminated, usually nearly straight and parallel, tapering to rounded tip.

Isozymes. Esterase VS1-S1 type and malate dehydrogenase N1a type.

Hosts. Able to parasitize many mono- and dicotyledonous hosts with potential to cause great economic damage.

Distribution. Recorded from Africa, Asia, North America and South America, and found in Europe (glasshouses).

Remarks. This species belongs to the so called *M. incognita* group, and is well known for its resistance-breaking behaviour. *Meloidogyne mayaguensis* is a junior synonym for *M. enterolobii*.

3.9.5. *Meloidogyne exigua* (type species)

Female. Body rounded, relatively small with short neck. Stylet 12–15 (14) µm long, knobs small, rounded and set off. Perineal pattern rounded to oval with coarse striae, lateral lines absent.

Male. Head not set off from the body. Head cap rounded, labial disk not elevated, lateral lips usually present. Stylet 15–19 (18) µm long, knobs small, rounded and set off. DGO to stylet knobs: 2–4 µm.

Second-stage juveniles. Body 330–360 µm long. Hemizonid anterior and adjacent to the S–E pore. Tail conical and 42–48 µm in length; hyaline tail part 12–14 µm long, narrow constricting annulations often present near anterior region, tail tip bluntly rounded.

Isozymes. Esterase VF1 type and malate dehydrogenase N1 type.

Hosts. The main host is coffee; a number of other dicotyledonous hosts are also known.

Distribution. Mainly known from South America, Central America and the Caribbean; incidentally reported from Europe and Asia.

Remarks. On coffee, this species usually induces elongated galls. Other root-knot nematode species described from coffee are: *M. africana*, *M. arabicida*, *M. coffeicola*, *M. decalineata*, *M. konaensis*, *M. megadora* and *M. paranaensis*. Also *M. arenaria*, *M. hapla*, *M. incognita*, *M. javanica*, *M. kikuyensis*, *M. enterolobii* and *M. oteifae* have been reported from coffee.

3.9.6. *Meloidogyne fallax*

Female. Body rounded with short neck and slight posterior protuberance. Stylet 13.5–15.5 (14.5) µm long; knobs large and rounded, slightly sloping posteriorly. S–E pore just below stylet knob level. Perineal pattern rounded to oval, striae relatively coarse and lateral lines weakly visible.

Male. Head slightly set off from the body. Head cap rounded, labial disc elevated, lateral lips present. Stylet 19–21 (20) µm long, knobs large, rounded and set off. DGO to stylet knobs: 3–6 µm.

Second-stage juvenile. Body 380–440 µm long. Hemizonid at S–E pore level. Tail conical and 40–45 µm in length; hyaline tail part 12–16 µm long, anterior region clearly delimitated, tail tip bluntly rounded.

Isozymes. Esterase '0' type (after prolonged staining a weak F3 type appears) and malate dehydrogenase N1b type.

Hosts. Like *M. chitwoodi* able to parasitize many mono- and dicotyledonous hosts, including economically important crops like potato (tubers) and wheat. *Meloidogyne fallax* and *M. chitwoodi* have similar hosts but species-specificity has also been found. Usually induces relatively small galls.

Distribution. Known from Europe (Belgium, France, Germany, Sweden, Switzerland, The Netherlands and the UK), South Africa, Chile, Indonesia, Australia and New Zealand (widely distributed).

Remarks. Closely related to *M. chitwoodi* and also listed in many countries as a quarantine organism.

3.9.7. *Meloidogyne graminicola*

Female. Body elongated with slight posterior protuberance. Stylet 8.1–15 (11.5) µm in length, knobs ovoid and set off. Perineal pattern rounded to slightly oval with smooth striae, without lateral lines.

Male. Head not set off from the body. Head cap rounded, labial disk not elevated, lateral lips usually present. Stylet 14.5–19.5 (16.7) µm long, knobs ovoid and set off. DGO to stylet knobs: 3–5 µm.

Second-stage juvenile. Body 380–510 µm long. Hemizonid anterior and adjacent to S–E pore. Tail slender and 60–90 µm long, small hyaline tail part 16–27 µm long. Tail tip finely rounded.

Isozymes. Esterase VS1 and malate dehydrogenase N1a type.

Hosts. Rice is the most important host, but many grasses and dicotyledonous weeds are known as good hosts.

Distribution. Common species found in all major rice growing areas except in Europe (only reported in Italy) and Africa (only reported in South Africa and Madagascar).

Remarks. Galls relatively large, often terminal and curved at root tips to form a hook. Other root-knot nematodes described from rice are: *M. hainanensis*, *M. lini*, *M. oryzae*, *M. salasi* and *M. triticoryzae*. Also *M. arenaria*, *M. incognita* and *M. javanica* have been reported from rice.

3.9.8. *Meloidogyne hapla*

Female. Body pear-shaped without posterior protuberance. Stylet ranging in length from 13–17 (15) µm, knobs relatively small, rounded and set off. Perineal pattern with fine striae, rounded with low dorsal arch, punctuations usually present above the anus, lateral lines present.

Male. Head clearly set off from the body. Head cap rounded, labial disc usually not elevated, lateral lips present. Stylet 19–22 (20.5) µm long; knobs relatively small, rounded and set off. DGO to stylet knobs: 4–5 µm long.

Second-stage juvenile. Body 360–500 µm long. Hemizonid anterior not adjacent to the S–E pore. Tail slender and 48–70 µm in length; hyaline tail part often irregular shaped, 12–19 µm long; anterior region not clearly delimitated, tail tip finely rounded.

Isozymes. Esterase H1 type and malate dehydrogenase H1 type.

Hosts. Many dicotyledonous hosts are known including important food crops and ornamentals. Records of monocotyledonous hosts are rare.

Distribution. Recorded worldwide from temperate regions; also found in tropical and subtropical regions at higher altitudes.

Remarks. Induces relatively small galls, often with secondary roots. Undoubtedly morphologically the most variable species of the genus.

3.9.9. *Meloidogyne incognita*

Female. Body pear-shaped without posterior protuberance. Stylet ranging in length from 15 to 16 µm, knobs rounded and set off. Perineal pattern usually with relatively high dorsal arch and without lateral lines.

Male. Head not set off from the body. Head cap with elevated labial disc, usually without lateral lips, head region often with incomplete head annulations. Stylet 23–26 (24.5) µm long, knobs rounded to oval shaped and set off. DGO to stylet knobs: 2–4 µm long.

Second-stage juvenile. Body 350–450 µm long. Hemizonid anterior or adjacent to the S–E pore. Tail slender and 43–65 µm in length; hyaline tail part 6–14 µm long, anterior region clearly delimitated, tail tip rounded.

Isozymes. Esterase I1 type and malate dehydrogenase N1 type.

Hosts. Able to reproduce on many monocotyledonous and dicotyledonous plants.

Distribution. *Meloidogyne incognita*, together with *M. arenaria* and *M. javanica*, are the most common root-knot nematodes. Known worldwide, but restricted in the temperate regions to greenhouses.

Remarks. Induces usually large galls. Some agricultural important populations, initially identified as *M. incognita*, have been described as new species, i.e. *M. brasilensis*, *M. enterolobii*, *M. hispanica*, *M. floridensis* and *M. microtyla*.

3.9.10. *Meloidogyne javanica*

Female. Body pear-shaped without posterior protuberance. Stylet ranging in length from 14 to 18 (16) µm, knobs ovoid and set off. Perineal pattern rounded with distinct lateral lines.

Male. Head not set off from the body. Head cap rounded and set off, usually labial disc not elevated and lateral lips not present. Stylet 19–23.5 (20.5) µm long, knobs well developed, ovoid and set off. DGO to stylet knobs: 3–5.5 µm.

Second-stage juvenile. Body 400–560 µm long. Hemizonid anterior and adjacent to the S–E pore. Tail slender and 47–60 µm in length; hyaline tail part 9–18 µm long, tail tip finely rounded.

Isozymes. Esterase J3 type and malate dehydrogenase N1 type.

Hosts. Able to reproduce on many monocotyledonous and dicotyledonous plants.

Distribution. Belongs together with *M. arenaria* and *M. incognita* to the most common root-knot nematodes. It is known worldwide, but restricted in the temperate regions to greenhouses.

Remarks. Usually induces relatively large galls.

3.9.11. *Meloidogyne naasi*

Female. Body rounded with slight posterior protuberance. Stylet ranging in length from 13 to 15 (14) µm; knobs ovoid and backwardly sloping. S–E pore near stylet knob level. Perineal pattern rounded to oval with prominent phasmids, relatively coarse striae, without lateral lines.

Male. Head set off from the body, labial disk not elevated and lateral lips not present. Stylet 16–18.5 (18) µm long, knob shape as in females. DGO to stylet knobs: 2.5–3.5 µm. Small vesicles are usually present in the anterior lumen lining of the metacorpus.

Second-stage juvenile. Ranging in length from 410 to 460 µm. Anterior metacorpial lumen lining with several (2–10) small vesicles. Hemizonid anterior adjacent to the S–E pore. Tail 60–75 µm in length and slender; hyaline tail part 16–25 µm in length with a fine rounded tail tip.

Isozymes. Esterase VF1 type and malate dehydrogenase N1a type.

Hosts. Reported from several grasses (including golf course and lawn grasses) and cereals. Rarely reported from dicotyledonous plants. Galls are relatively small, sometimes terminal, hooked or with lateral roots. Females usually completely embedded within the root-tissue.

Distribution. Widely distributed in Europe; also reported from North and South America and occasionally from Asia.

Remarks. The metacorpial vesicles of males and J2 can be observed easily in fresh material; in permanent slides they are hardly or not visible. These vesicles are also present in *M. sasseri* J2 but are smaller compared with *M. naasi*.

3.10. Biochemical and Molecular Identification

As with other genera, characterization of species of *Meloidogyne* includes the use of biochemical and molecular techniques. They allow a more rapid, reliable and cheaper identification than morphological approaches. In their monograph on 'Systematics of Root-knot Nematodes (Nematoda: Meloidogynidae)', Subbotin *et al.* (2021a) provide an elaborate overview of the methods developed for biochemical and molecular diagnostics.

3.10.1. Protein electrophoresis

Esbenshade and Triantaphyllou (1985) described a method to identify females of several *Meloidogyne* species with isozyme electrophoresis. In particular, esterase (EC 3.1.1.1) and malate dehydrogenase (EC 1.1.1.37) isozyme patterns discriminate most species clearly and, therefore, are used for routine identification of root-knot nematodes (Fig. 3.17). Esterase and malate dehydrogenase phenotypes are identified for 62 out of 100 *Meloidogyne* spp. Patterns of superoxide dismutase and glutamate-oxaloacetate are also available for some species. Additionally, 27 isozymes have been used for a robust study of the enzymatic relationship and evolution within the genus *Meloidogyne* (Esbenshade and Triantaphyllou, 1987).

Two-dimensional gel electrophoresis has also been used to study the total soluble protein patterns of root-knot nematodes. It was used with *M. hapla*, *M. chitwoodi* and *M. fallax* and confirmed these species as distinct biological groups. Antisera have also been used to discriminate *M. chitwoodi* and *M. fallax* from other root-knot nematodes, using a combination of two-dimensional gel electrophoresis, internal amino acid sequencing and serology.

3.10.2. DNA-based techniques

Another powerful method to identify *Meloidogyne* species is the use of nucleotide polymorphisms in DNA sequences among species. Initialy, genomic DNA was analysed using restriction enzymes to differentiate between species. However, this strategy is time consuming, needs a significant amount of DNA and lacks sensitivity. These issues were overcome with the development of the polymerase chain reaction (PCR). PCR-amplified DNA can be characterized by restriction fragment polymorphism (RFLP) or sequencing.

For the identification of *Meloidogyne* spp., different genes are targeted. Currently, diagnostics are based on the internal transcribed spacers (ITS1–5.8S–ITS2) of the D2–D3 expansion fragment of 28S ribosomal DNA (rDNA) gene, the intergenic spacer (IGS) ribosomal RNA (rRNA) and 18S rRNA (Subbotin *et al.*, 2021a). With respect to mitochondrial sequences (mitochondrial DNA (mtDNA)), the intergenic region between the *COII* and the large subunit (LSU) RNA gene containing an intergenic region have also been detected for distinguishing different species and host ranges of *Meloidogyne*. In addition, species-specific primers have been developed from fragments of mtDNA, rRNA gene or sequence characterized amplified regions (SCAR) of root-knot nematode species and are used as molecular markers for species identification. They detect a wide range of root-knot nematodes and can identify one or several species in a nematode mixture (see also Blok and Powers (2009) for a review of biochemical and molecular identification of *Meloidogyne* spp.). The first phylogenetic analysis, based on 18S rDNA sequences of 13 species, was published by De Ley *et al.* (2002). Since then, phylogenetic relationships within *Meloidogyne* clades have been refined using multiple DNA molecular markers and more species (Quist *et al.*, 2015; Álvarez-Ortega *et al.*, 2019). A schematic overview of the phylogeny of the Meloidogynidae derived from small subunit (SSU) rDNA sequence data of many more species is shown in Chapter 2. Real-time PCR, also called quantitative PCR (qPCR), provides a simple and elegant method for the quantification of nematodes in samples. qPCR is easy to perform and results are available after 45 min to 1.5 h (see Subbotin *et al.* (2021a) for a list of root-knot nematode species for which a procedure has been developed). In addition to the aforementioned techniques, two others were recently developed: loop-mediated isothermal amplification (LAMP) and recombinase polymerase amplification (RPA). LAMP holds potential for testing in the field (with no need for sophisticated equipment) and the sensitivity of some LAMP assays is 100 or more times higher that conventional PCR. Several tests using RPA have demonstrated high sensitivity and specificity; the method uses crude nematode extracts and results are available in 15–20 min.

3.11. Interactions with Other Plant Pathogens

In nature, plants are rarely exposed to the influence of a single pathogen. Roots usually host a great number of microorganisms, the actions of which are often combined to induce damage. *Meloidogyne* frequently plays a major role in disease interactions (Khan, 1993).

3.11.1. *Meloidogyne* as a wounding agent

Nematode parasitism of plants is associated with wounding their hosts. For a long time wounding was considered of paramount importance in facilitating the entry of fungal pathogens, with wounds acting as a passage or portal. However, the wound-facilitation concept in the fungus–nematode interaction, especially between *Fusarium* and *Meloidogyne*, was later disproved. Modification of host substrates was found to be primarily responsible for such interactions. There is evidence that predisposition to infection by *Fusarium* wilt fungi occurs through translocatable metabolites produced at the site of nematode infection.

Several studies indicated that root-knot nematodes cause wounds that facilitate entry of bacteria and that modify the host tissue to enrich the substrate nutritionally to the advantage of the bacteria. The role of *Meloidogyne* as a wounding agent has been observed for several combinations of bacterium and nematode; a well known example of this kind of interaction is the combination of *Ralstonia solanacearum* and *M. incognita* on tobacco. The interaction between these two pathogens makes the wilt symptoms more severe when nematodes and bacteria are inoculated together than when bacteria alone are inoculated on artificially wounded roots. *Meloidogyne*-modified tissue acts as a more favourable substrate for bacteria; the nematode causes physiological changes in the host substrate. On tomato cultivars susceptible or resistant to *Clavibacter michiganensis* subsp. *michiganensis*, the presence of *M. incognita* increased the bacterial canker only when the nematode was inoculated prior to the bacterium (de Moura *et al.*, 1975). Such an interaction did not occur when both pathogens were inoculated simultaneously or the bacterium was inoculated prior to the nematode.

3.11.2. Effects of *Meloidogyne* on host susceptibility

Root-knot nematode infection of *Fusarium*-infected plants increases the wilt symptom expression and death rate of the plants. Evidence indicates that the root diffusates from root-knot-infected plants stimulate the fungus in the rhizosphere. These diffusates also suppress actinomycetes, which are antagonists of *Fusarium*. It seems that the nature of interactions between root-knot nematodes and *Fusarium* wilt fungi is not physical, by the nematode puncturing the plant root, but is physiological.

The giant cells induced by *Meloidogyne* juveniles remain in a state of high metabolic activity through the continuous stimulation by the nematode. They are probably the major site of interaction between the nematode and *Fusarium*. The giant cells remain in a continuous juvenile state, which delays maturation and suberization of other vascular tissues, and thus *Fusarium* successfully penetrates and establishes in the xylem elements. Inhibition of tylose formation by root-knot nematodes has also been suggested as a possible mechanism for increased wilt severity (Webster, 1985; Seo and Kim, 2017).

Like wilt fungi, root-rot fungi are capable of causing disease on their own and have their own inherent mechanism of root penetration. The role of *Meloidogyne* in root-rot diseases in general is related to assisting the fungal pathogen in its pathogenesis and increasing host susceptibility. The invasion tracks formed by penetrating juveniles of root-knot nematodes provide a better substrate for the establishment of the fungal pathogens. Physiological alterations which ensure better nutrient availability for the penetrated fungal pathogen are the key factor of the synergistic damage caused to the host.

Fungi, which are non-pathogenic to a host plant, may become pathogenic in the presence of nematodes. Root-knot nematodes possess outstanding abilities to cause physiological changes in plants that can induce susceptibility in plants to attack by fungi present in the rhizosphere, whether pathogenic or non-pathogenic. Infection with *M. graminicola* rendered enhanced susceptibility of rice plants to rice blast (*Magnaporthe oryzae*) due to changes in above-ground plant immunity gene expression (Kyndt *et al.*, 2017).

Root-knot nematodes break the monogenic *Fusarium* wilt resistance in tomato cultivars. Histopathological changes in the host, caused by nematode infection, apparently are responsible for rendering the gene(s) for resistance ineffective and as a result the host is not able to express the resistant reaction. By contrast, there are reports indicating that *Fusarium* wilt resistance in cultivars of cabbage, peas and other crops, possessing a single gene for resistance against wilt fungi, is not altered by root-knot nematode infection. The multigenic resistance to *Fusarium* wilt in cotton is less effective when plants are infected by *Meloidogyne* but is not broken.

3.11.3. Interaction of *Meloidogyne* with other nematodes

Plant-parasitic nematodes frequently parasitize plants in mixed populations of two or more genera and species. Interaction may enhance or inhibit nematode reproduction, which is ultimately the driving force in interspecific competition. Competition is usually strongest between organisms that are most alike with respect to their physiological demands on the host (Eisenback, 1993).

Ectoparasites feeding on the preferred penetration sites of the infective J2 of *Meloidogyne* may be a limiting factor in the success of the latter. Ectoparasites may damage the root system and thus indirectly reduce the number of feeding sites available for *Meloidogyne*. By contrast, *Meloidogyne* can suppress ectoparasites even though they are separated by plant tissue, probably by physiological mechanisms. Sometimes *Meloidogyne* and ectoparasitic nematodes are mutually antagonistic. However, interactions between these two groups may be beneficial for one or both species.

Interactions between migratory endoparasitic nematodes and *Meloidogyne* species may be time dependent, i.e. inhibition of the development of one of the components of the system by the other may disappear after some time. Interaction between these two groups may also depend on the sequence of their infection. In experimental studies, the mutual effect of *Meloidogyne* and migratory endoparasitic nematodes depends on whether both groups are inoculated simultaneously or one after the other (Chitamber and Raski, 1984). In addition, the change in physiology of the host plants by *Meloidogyne* affects the suitability of the host for migratory endoparasites; root-knot nematodes produce a translocatable factor that inhibits the reproduction of *Pratylenchus* species (Estores and Chen, 1970). The effect of host suitability on nematode interactions is nematode density-dependent. Interspecific competition with *Radopholus similis* rendered frequency of populations and population levels of *M. incognita* very low in banana plantations (Moens *et al.*, 2006).

Two or more root-knot nematode species are commonly found in the same field, root system or gall. Factors other than competition for feeding sites may be important in the domination of a particular species. Temperature and other climatic factors may be important because certain species may be better adapted to cooler temperatures, whereas others are more common in warmer climates. Also, plant resistance to one of the *Meloidogyne* species may influence the interaction and even persistence in the soil.

3.12. Management and Control

Management systems are built around key pests (organisms that cause significant reduction in crop yield every year unless some pest control action is taken). *Meloidogyne* species frequently attain that status, even in the presence of other plant pathogens. The basic objective in any pest management is to increase both quantity and quality of crop yield. With respect to *Meloidogyne*, increasing yield quality is sometimes the most important objective. Successful management starts with a correct identification of the *Meloidogyne* problem.

3.12.1. Prevention

Meloidogyne species are not found in seeds but may be present in vegetative planting material such as corms, bulbs or roots. Obviously, planting materials are possible sources for passive dispersal. In plant parts, numbers of root-knot nematodes can be reduced by chemicals or hot water treatment; however, it is often better to discard infected material. It is recommended to use only certified nematode-free plants from reliable nurseries or seedlings produced in *Meloidogyne*-free seedbeds. Root-knot nematodes may also be disseminated via farm machinery or vehicles to which contaminated soil sticks. Cleaning machinery before and after use is recommended, but generally not practised. Waste soil at industrial processing companies is often returned to agricultural fields. To reduce risks via international trade, several root-knot nematode species have been listed as quarantine organisms (see Chapter 13).

3.12.2. Crop rotation

Meloidogyne species are obligate and specialized parasites and each has species and cultivars of plants where the susceptibility ranges between highly susceptible to immune. Populations of *Meloidogyne* in a field without host plants will become non-infective and die of starvation. In crop rotations for controlling *Meloidogyne* species, susceptible crops are rotated with immune or resistant crops. Susceptible hosts of the four most common species, *M. arenaria*, *M. hapla*, *M. incognita* and *M. javanica*, are numerous and belong to many plant families. However, rotations have been developed and several Gramineae are effective in reducing these *Meloidogyne* species (Nyczepir and Thomas, 2009). Other widely distributed species (e.g. *M. naasi* and *M. graminis*) have hosts that belong to only a few plant families, whilst other species (e.g. *M. partityla*, *M. kralli* and *M. ichinohei*) have a very restricted host range. With these species, the development of effective rotations for control is easier. The North Carolina Differential Host Test (Hartman and Sasser, 1985) not only separates the four most common species, it further distinguishes four races of *M. incognita* and two races of *M. arenaria*. Host assays separate the known two races of *M. chitwoodi* and their respective pathotypes (Brown *et al.*, 2009). Despite statistically significant morphological variation among adult females of *M. chitwoodi* from the USA, morphometrics were not able to distinguish between isolates representing these races and pathotypes, and molecular traits, determined by nuclear ribosomal genes, were stable across all isolates (Humphreys-Pereira and Elling, 2013). Crop rotations have been developed in relation to the presence or absence of these races. However, caution should be exercised in applying the results obtained in one location to another, as different populations of the same species of *Meloidogyne* may react differently on the same plant species (Moens *et al.*, 2009).

Trap crops allow the infective J2 to enter the roots but due to antagonistic plant responses *Meloidogyne* fail to complete their life cycle. The effect of *Tagetes* species on populations of *Meloidogyne* species is highly variable, depending on the combination of species and cultivar of *Tagetes* and the species and race of nematode. It appears that reduction of *M. incognita* by marigold (*Tagetes patula*) is due primarily to an antagonistic or trap crop effect. J2 enter roots but there is neither giant cell formation nor a hypersensitive reaction. *Eruca sativa* has shown potential as trap crop for *M. hapla*.

Fallow periods can be included in crop rotations and have been shown to be successful in vegetable production. However, lack of income and detrimental effects from soil erosion limit the use of fallowing. Finally, one should always be aware that *Meloidogyne* species can reproduce on many weeds, and the presence of weeds in the field can compromise the success of rotations.

3.12.3. Cultural methods and physical control

Root destruction and organic amendments can be practised to reduce the number and impact of root-knot nematodes (see Chapter 15). Removing infected roots after harvest reduces root-knot soil infection. Organic amendments, in addition to their positive influence on the soil physical structure and water-holding capacity, have the capability of reducing nematode population densities to varying degrees. Control of *Meloidogyne* is rarely the primary reason for their use (Nyczepir and Thomas, 2009). However, brassicaceous amendments with a high glucosinolate level can enhance nematicidal effects.

Where water is abundant and fields are level, it is sometimes possible to control *Meloidogyne* species by flooding land to a depth of 10 cm or more for several months. Flooding does not necessarily kill the eggs and juveniles of root-knot nematodes by drowning; it inhibits infection and reproduction on any plant that grows while the field is flooded.

In some climates, *Meloidogyne* populations can be reduced by ploughing at intervals of 2–4 weeks during the dry season. This exposes eggs and juveniles to desiccation and many in the upper layers of soil are killed.

Frost killing, steaming and solarization can be used to reduce root-knot nematode numbers. *Meloidogyne* infections of glasshouse soil can be lowered by exposing them to very low temperatures for a sufficient period of time. Steaming often fails to give satisfactory results for root-knot control, especially when survivors in the deeper layers of soil can build up infestations. Soil solarization (Gaur and Perry, 1991) is only adaptable to regions where sufficient solar energy is available for long periods of time.

3.12.4. Resistance

The identification in both woody and herbaceous plants of many varieties with natural resistance genes to *Meloidogyne* has allowed the control of these pests in many crops for decades. The first *Meloidogyne* major resistance gene to be cloned was *Mi-1.2*, which was originally discovered in *Solanum peruvianum* (Milligan *et al.*, 1998). This resistance (R)-gene has been used for more than 50 years in cultivated tomatoes to confer resistance to several species of root-knot nematodes. *Mi-1.2*-mediated resistance in tomato is due to a hypersensitive response (HR) that arises in root cells in contact with the anterior part of *Meloidogyne* where the feeding site is initiated, leading to the stoppage of nematode development in the early stages of infection (Williamson and Kumar, 2006). The *Ma* gene from plum (*Prunus cerasifera*) was the second major *Meloidogyne* resistance gene to be cloned (Claverie *et al.*, 2011). This *R*-gene confers a heat-stable, durable and broad resistance against several *Meloidogyne*. As with *Mi-1.2*, *Ma* induces the HR at the early stage of infection, shortly after J2 penetrate the root system. Other *R*-genes have been mapped and/or are commonly used in breeding programmes. For example, in bell pepper, the major resistance genes, *Me1*, *Me3* and *Me7*, have been shown to reduce parasite pressure from *M. incognita*, *M. javanica* and *M. arenaria* (Djian-Caporalino *et al.*, 2007) with *Me3*, just as for the *Mi-1.2* gene, inducing an early HR, whilst *Me1* allows the development of slightly swollen root tips with defective giant cells that subsequently collapse (visible symptoms of senescence with rupture of nuclear and cytoplasmic membranes in 7-day-old giant cells) (Bleve-Zacheo *et al.*, 1998). In 2021, the *MeR1* gene inducing resistance against *M. enterolobii* was patented (Frijters *et al.*, 2021).

The emergence of a virulent population (i.e. capable of bypassing resistant plant defence mechanisms) presents a challenge for control (Davies and Elling, 2015). Several virulent populations of *M. incognita*, *M. javanica* and *M. arenaria* that have overcome the resistance conferred by *Mi-1.2* in tomato have been sampled *in situ* in different geographical regions. Fortunately, this bypass does not always occur for all *R*-genes and, for example, the *Ma* and *Me1* genes appear to provide durable resistance.

In order to sustain crop resistance to pathogens by diversifying the combined putative resistance mechanisms, pyramiding of minor and major genes has been developed (Fuchs, 2017; Mundt, 2018). Pyramiding of major resistance genes is primarily used against pathogens with an asexual mode of reproduction. It has been suggested that pyramiding will be less effective for sexually reproducing pathogens, which would be able rapidly to combine virulence against multiple resistance genes (McDonald and Linde, 2002). However, *R*-gene pyramiding against the sexually reproducing potato cyst nematodes is successful. In the case of root-knot nematodes, only a few species are known to have a sexual mode of reproduction and those most damaging to agriculture reproduce essentially by parthenogenesis. Therefore, pyramiding should be a relevant and potentially sustainable approach to control root-knot nematodes. In order to accelerate the selection of durable resistant varieties containing multiple resistance genes, molecular marker-assisted selection of resistant varieties is essential. During the past few decades, sequencing methods have considerably reduced the cost of such analyses, speeding up the mapping of resistance genes and their introduction into elite varieties. Mutations at individual avirulence loci are assumed to be independent, and the probability of a pathogen overcoming the resistance gene pyramiding is theoretically very low. However, major *R*-genes do not confer absolute resistance and a small level of pathogen reproduction remains; therefore, it has the opportunity to accumulate virulence sequentially, rather than through simultaneous and independent mutations. *R*-genes recognize at first the corresponding avirulence products (*Avr*-genes) expressed by the pathogen and secreted into the plant upon infection, and then trigger the plant's specific defence reactions (Jones and Dangl, 2006). To acquire virulence, the parasite must be able to repress the expression of those *Avr*-genes or to modify their sequences, thus preventing recognition by the corresponding *R*-gene. Virulence can therefore arise from a simple mutation of the avirulence factor or from other phenomena such as the expression of suppressor genes that affect the avirulence loci, or the expression of new effectors that interfere with the plant's defence response. In any case, it must correlate with the appearance of molecular hallmarks in the genome and/or the epigenome. However, as none of the effectors recognized by the root-knot nematode genes has been identified to date, it is not possible to define exactly the mechanism responsible for this recognition evasion.

Genomic analyses of different populations of *M. incognita* have revealed low variability at the single nucleotide polymorphism (SNP) level among isolates collected worldwide, suggesting a slow rate of evolution or a recent expansion. Point mutations are not the only factors of genomic plasticity that may be involved in the adaptive evolution of *M. incognita*. Comparative genomics studies, in particular on virulent populations of root-knot nematodes, have revealed that the acquisition of virulence can take place in a few generations and that it is accompanied by modifications of the genome, in particular in the copy number variation (CNV).

Further studies on the incompatible interactions between plants and root-knot nematodes and, more specifically, on the identification of the *R*-genes and their pathogenic counterpart, the avirulent gene, should provide key information on how to manage the durability of resistance. While the identification of new resistance genes and their pyramiding must remain a major focus to mitigate parasitism, the adaptive capabilities of *Meloidogyne* and in particular their genomic plasticity support the need for a combination of control agents.

The use of resistant plants to combat *Meloidogyne* species and the associated molecular studies on resistance are detailed in Chapters 15 and 16.

3.12.5. Biological control using microorganisms

Live microbes (bacteria, fungi, etc.) and their gene products (mainly secondary metabolites), essential oils, plant extracts, individual and mixed acids such as organic and amino acids, and natural and bioactive substances have led to many studies in the integrated control of *Meloidogyne* (Forghani and Hajihassani, 2020). Bacteria and fungi remain the main antagonists for biocontrol of plant-feeding nematodes (see Chapter 15). However, their effect is not always consistent and depends on the environment where they are applied, e.g. *in vitro* versus *in vivo*, biotic and/or abiotic factors, soil structure and composition, and pH.

Strategies of using nematodes in the soil as prey to identify antagonistic microorganisms that bind to them or *a priori* approaches by bringing *Meloidogyne in vitro* into contact with microorganisms have led to the discovery of bacteria and fungi that can act as biocontrol agents and many can be found in suppressive soils. Until recently, the study of microorganisms present in suppressive soils was limited to 'cultivable organisms'. Thanks to the use of metabarcoding (*GyrA*, 16S and 18S rDNA) or metagenomics it is now possible to identify the community composition of suppressive soils (Topalović *et al.*, 2022). It is therefore potentially feasible to describe the full diversity of microorganisms present in suppressive soils and to detect new biological agents useful in the management of *Meloidogyne* (Masson *et al.*, 2022).

The discovery of bacterial biocontrol agents has made various products consisting of single species or mixtures of bacteria commercially available, including *Bacillus*, *Pseudomonas*, *Pasteuria*, *Arthrobacter*, *Serratia*, *Achromobacter*, *Rhizobium* and *Burkholderia*. These biopesticides have shown bionematicidal activity at different stages of pest development and have played an important role in the management of *Meloidogyne* (Stirling, 2014; Askary & Martinelli, 2015). Biocontrol of *Meloidogyne* can also be realized by fungal microorganisms. The most important are *Actylellina*, *Arthrobotrys*, *Aspergillus*, *Catenaria*, *Dactylellina*, *Hirsutella*, *Monacrosporium*, *Pochonia*, *Purpureocillium* and *Trichoderma*. *Arthrobotrys* and *Monacrosporium*, two hematophagous fungi that use sticky mycelia to capture nematodes, are particularly effective when they colonize the agrosystem. Arbuscular mycorrhizal fungi (AMF) are obligate root symbionts that can also help protect their host plants from *Meloidogyne* infection and reduce parasitism-related negative effects on plant development (Schouteden *et al.*, 2015). AMF compete with *Meloidogyne* for space and nutrition, induce systemic plant resistance, but also allow for more plant nutrient uptake. The ovicidal, nematicidal and nematistatic potential of crude extract and retained metabolites from liquid cultures of fungi has been evaluated against *M. javanica* and the results seem encouraging (Hahn *et al.*, 2019).

Natural and bioactive substances are also molecules that allow the biocontrol of *Meloidogyne*. Several parts of neem (*Azadirachta indica*) have been shown to be effective in nematode management. The effects of their products on nematode control occur after incorporation into the soil and during their decomposition, probably due to the release of nematotoxic compounds such as azadirachtins, phenols, fatty acids and tannins. Nematicidal activity has also been found in other plants such as *Camellia oleifera*, *Paenoia rockii* and *Vetiveria zizanioides*. Allicin (diallyl thiosulfinate), the key natural antimicrobial compound in garlic, has also showed notable effects against *Meloidogyne*. Saponin-rich extracts and plant biomasses of different *Medicago* spp. can be highly repressive against root-knot nematodes.

Various organic acids, including amino, acetic, butyric, formic and propionic acids, are known to have toxic effects on certain *Meloidogyne*. These are the result of either microbial decomposition of various compounds in the soil or metabolites produced by microorganisms (Oka, 2010). The ability of acids against nematodes is strongly influenced by soil conditions. Essential oils (EOs) acting as biopesticides for the control of *Meloidogyne* can be derived from different plants such as *Dysphania ambrosioides*, *Piptadenia viridiflora*, *Eucalyptus globulus* and *Pelargonium asperum*. The use of some of these EOs can significantly reduce nematode multiplication and/or root gall formation.

3.12.6. Chemical control

Nematicides (see Chapter 17) are effective against *Meloidogyne* and can give good economic returns on high-value crops; however, in low-yielding crops chemical control is uneconomical. Treatment may not be sufficient for a year and re-treatment usually will be required the following year if plants susceptible to root-knot nematodes are to be grown. Fumigants are commonly applied as pre-plant treatments to reduce nematode numbers. In addition to broad-spectrum fumigants, non-fumigant nematicides showed to be at least moderately effective for controlling root-knot nematodes. They are applied to the soil as granular or liquid formulations. As they are not toxic to plants, they are the only chemical options for established plants. Safer non-fumigant nematicides have been developed in recent years with high control effects on plant-parasitic nematodes. For root-knot nematodes, egg-laying, hatching, feeding and locomotion are targeted. Some compounds have effects on different developmental stages. Still under research but with promising results *in vitro* and in pot tests (i.e. to control *M. graminicola* on rice) are so-called priming agents. These compounds induce resistance, leading to an alert state in the plant protecting it from infection.

Cyst Nematodes*

Sergei A. Subbotin[1,2]** and Johannes Hallmann[3]

[1]*Plant Pest Diagnostics Center, California Department of Food and Agriculture, USA;* [2]*Center of Parasitology of A.N. Severtsov Institute of Ecology and Evolution of the Russian Academy of Sciences, Moscow, Russia;* [3]*Institute for Epidemiology and Pathogen Diagnostics, Julius Kühn-Institut, Braunschweig, Germany*

---

*A revision of Turner, S.J. and Subbotin S.A. (2013) Cyst nematodes. In: Perry, R.N. and Moens, M. (eds) *Plant Nematology*, 2nd edn. CAB International, Wallingford, UK.

**Corresponding author: sergei.subbotin@ucr.edu

DOI: 10.1079/9781800622456.0004

4.1. Introduction to Cyst Nematodes

The cyst nematodes are a major group of plant-parasitic nematodes and of great economic importance in many countries throughout the world. They cause considerable yield losses to many important crops, including cereals, rice, potatoes and soybean, with the most economically important species occurring within the genera *Heterodera* and *Globodera* (Table 4.1). *Heterodera* contains by far the largest number of species (Table 4.2), although several other cyst-forming species have been described within other genera (Table 4.3). Eight genera, *Heterodera* (87 species), *Globodera* (12 species), *Cactodera* (16 species), *Dolichodera* (one species), *Paradolichodera* (one species), *Betulodera* (one species), *Punctodera* (five species) and *Vittatidera* (one species), and a total of 124 valid species are presently recognized within this nematode group (Subbotin *et al.,* 2010a,b). The classification of cyst nematodes is given in Box 4.1. Cyst nematodes were originally considered to be largely a pest of temperate regions but many cyst nematodes are now known to be present in tropical and subtropical regions (Evans and Rowe, 1998).

It is impossible to evaluate the total economic losses caused by cyst nematodes throughout the world as many environmental, biological and cultural factors may confound these calculations, although some specific examples of calculated losses exist. Potato cyst nematodes (PCNs) have been well studied within Europe, and overall losses are estimated at about 9% of potato production; however, in other regions of the world, or when no control strategies are employed, total losses can occur. This range could equally be applied to all major crops that are hosts to cyst nematodes.

All cyst nematodes feed within the root system of their hosts and are characterized by the tanning and drying (cutinization) of the body wall of the sedentary adult female following fertilization and production of embryonated eggs (see Section 4.2). The resultant cyst, together with the eggshell and perivitelline fluid (see Chapter 7), allows the succeeding generation to survive for extended periods until a suitable host is growing in the near vicinity. It is this ability to persist for many years in the soil in the absence of a host that contributes to the economic importance of this group in agricultural situations.

Table 4.1. Cyst nematodes of major economic importance. (Adapted from Evans and Rowe, 1998 and Subbotin *et al.*, 2010a,b.)

Genus	Species	Main crops affected	Region
Globodera	*ellingtonae*	Potato	Temperate
	pallida	Potato, tomato, eggplant	Temperate
	rostochiensis	Potato, tomato, eggplant	Temperate
	tabacum	Tobacco, tomato	Temperate
Heterodera	*avenae*	Wheat, barley, oat, maize	Temperate
	filipjevi	Wheat, barley, oat, maize	Temperate
	cajani	Cowpea, pea, *Phaseolus* bean, pigeon pea, sesame, soybean, sweetcorn	Tropical
	cruciferae	Brussels sprout, broccoli, cabbage, cauliflower, radish, kohlrabi, pea, rape	Temperate
	glycines	Adzuki bean, broad bean, French bean, hyacinth bean, kidney bean, moth bean, mung bean, navy bean, rice bean, snap bean, soybean, blackgram, cowpea, sesame, white lupin, yellow lupin, tobacco	Temperate
	goettingiana	Broad bean, chickpea, lentil, pea, white lupin, yellow lupin, white clover	Temperate
	latipons	Barley, oat, rye	Temperate
	oryzicola	Rice, banana and plantain	Tropical
	sacchari	Rice, sugarcane	Tropical
	schachtii	Adzuki bean, beet, broccoli, Brussels sprout, cabbage, cauliflower, celery, chickpea, chicory, Chinese cabbage, cowpea, dill, kale, kohlrabi, lentil, pea, radish, rape, rhubarb, rutabaga, spinach, tomato, turnip, yellow lupin, tobacco	Temperate
	sorghi	Sorghum, maize, rice	Tropical
	trifolii	Carnation, chickpea, cucumber, gherkin, pea, pumpkin, red clover, rhubarb, spinach, squash, tomato, white clover, white lupin, zucchini	Temperate
	zeae	Maize, barley, rice, sorghum, wheat	Tropical

4.2. Life Cycle and Behaviour

The life cycle of cyst nematodes is shown in Fig. 4.1. After gastrulation, the embryo extends in length within the eggshell and movement begins, then folds develop in the embryo. After the first moult, the stylet forms at the anterior end of the second-stage juvenile (J2). This is the dormant stage of the life cycle and, depending on the species and environmental conditions, the J2 can remain unhatched within the egg and cyst for many years (see Chapter 7 for a discussion of dormancy as a survival strategy). The eggshell containing the J2 consists of three layers in cyst nematodes: the outer lipoprotein layer derived from the vitelline layers of the fertilized oocyte, the middle chitinous layer, which provides the eggshell with its structural strength, and the innermost lipid layer, which represents the main permeability barrier. The active part of the life cycle starts when the J2 hatches out of the egg, having used its stylet to cut a slit in the eggshell; the hatching process of cyst nematodes is discussed in detail in Chapter 7, Section 7.6.5.

Table 4.2. Cyst species (87) of the genus *Heterodera* Schmidt, 1871. (Adapted from Subbotin *et al.*, 2010b.)

Species	Main host plant family	Species	Main host plant family
africana	Poaceae	*leuceilyma*	Poaceae
agrostis	Poaceae	*litoralis*	Amaranthaceae
amygdali	Rosaceae	*longicolla*	Poaceae
arenaria	Poaceae	*mani*	Poaceae
aucklandica	Poaceae	*medicaginis*	Fabaceae
australis	Poaceae	*mediterranea*	Anarcardiaceae
avenae	Poaceae	*menthae*	Lamiaceae
axonopi	Poaceae	*microulae*	Boraginaceae
bamboosi	Poaceae	*mothi*	Cyperaceae
bergeniae	Saxifragaceae	*orientalis*	Poaceae
betae	Amaranthaceae	*oryzae*	Poaceae
bifenestra	Poaceae	*oryzicola*	Poaceae
cajani	Fabaceae	*pakistanensis*	Poaceae
canadensis	Cyperaceae	*persica*	Apiaceae
cardiolata	Poaceae	*phragmitidis*	Poaceae
carotae	Apiaceae	*plantaginis*	Plantaginaceae
ciceri	Fabaceae	*pratensis*	Poaceae
cireae	Onagraceae	*raskii*	Cyperaceae
cruciferae	Brassicaceae	*ripae*	Poaceae
cyperi	Cyperaceae	*riparia*	Urticacae
daverti	Fabaceae	*rosii*	Polygonaceae
delvii	Poaceae	*sacchari*	Poaceae
dunensis	Zygophyllaceae	*sacchariphila*	Poaceae
elachista	Poaceae	*salixophila*	Salicaceae
fengi	Poaceae	***schachtii****	Amaranthaceae
fici	Moraceae	*scutellariae*	Laimiaceae
filipjevi	Poaceae	*sinensis*	Poaceae
galeopsidis	Lamiaceae	*skohensis*	Poaceae
gambiensis	Poaceae	*sojae*	Fabaceae
glycines	Fabaceae	*sonchophila*	Asteraceae
glycyrrhizae	Fabaceae	*sorghi*	Poaceae
goettingiana	Fabaceae	*spinicauda*	Poaceae
goldeni	Poacea	*spiraeae*	Rosaceae
graminis	Poacea	*sturhani*	Poaceae
graminophila	Poacea	*swarupi*	Poaceae
guangdongensis	Poacea	*trifolii*	Fabaceae
hainanensis	Poacea	*turangae*	Salicaceae
hordecalis	Poacea	*turcomanica*	Amaranthaceae
humuli	Cannabaceae	*urtica*	Urticaceae
johanseni	Brassicaceae	*ustinovi*	Poaceae
kirjanovae	Betulaceae	*uzbekistanica*	Salicaceae
koreana	Poaceae	*vallicola*	Ulmaceae
latipons	Poaceae	*zeae*	Poaceae
lespedezae	Fabaceae		

*Type species.

Table 4.3. Cyst-forming species of genera other than *Heterodera*. (Adapted from Subbotin *et al.*, 2010a.)

Genus (number of species)	Species	Host plant family
Betulodera (1)	***betulae****	Betulaceae
Cactodera (16)	*acnidae*	Amaranthaceae
	amaranthi	Amaranthaceae
	cacti*	Cactaceae
	chenopodiae	Amaranthaceae
	eremica	Amaranthaceae
	estonica	Polygonaceae
	evansi	Caryophyllaceae
	galinsogae	Asteraceae
	milleri	Amaranthaceae
	radicale	unknown
	rosae	Poaceae
	salina	Amaranthaceae
	solani	Solanaceae
	thornei	Portulaceae
	tianzhuensis	Polygonaceae
	weissi	Polygonaceae
Dolichodera (1)	***fluvialis****	Poaceae
Globodera (12)	*agulhasensis*	Asteraceae
	artemisiae	Asteraceae
	capensis	Unknown
	ellingtonae	Solanaceae
	leptonepia	Solanaceae
	mali	Rosaceae
	mexicana	Solanaceae
	millefolii	Asteraceae
	pallida	Solanaceae
	rostochiensis*	Solanaceae
	tabacum	Solanaceae
	zelandica	Onagraceae
Paradolichodera (1)	***tenuissima****	Cyperaceae
Punctodera (5)	*achalensis*	Poaceae
	chalcoensis	Poaceae
	matadorensis	Poaceae
	punctata*	Poaceae
	stonei	Poaceae
Vittatidera (1)	***zeaphila****	Poaceae

*Type species.

Hatching represents the end of dormancy. Cyst nematodes exhibit diapause and quiescence, two types of dormancy (Jones *et al.*, 1998; Chapter 7). Diapause, a state of arrested development whereby hatching does not occur until specific requirements, including a time component, have been satisfied, enables the J2 to overcome environmental conditions that are unfavourable for hatch, such as extreme temperatures or drought. The extent of the diapause varies but several cyst nematodes, e.g. *Globodera rostochiensis* and *Heterodera avenae*, show obligate diapause during their first season of development. In *G. rostochiensis* and *G. pallida*, diapause is terminated in late spring, when the combination of rising soil temperature and adequate soil moisture is conducive for infection of the

Box 4.1. Classification of the cyst nematodes.

Phylum:	Nematoda Potts, 1932
Class:	Chromadorea Inglis, 1983
Subclass:	Chromadoria Pearse, 1942
Order:	Rhabditida Chitwood, 1933
Suborder:	Tylenchina Thorne, 1949
Infraorder:	Tylenchomorpha De Ley & Blaxter, 2002
Superfamily:	Tylenchoidae Örley, 1880
Family:	Hoplolaimidae Filipjev, 1934
Subfamily:	Heteroderinae Filipjev & Schuurmans Stekhoven, 1941
Genera:	*Heterodera* Schmidt, 1871 (Type Genus)
	Globodera Skarbilovich, 1959
	Punctodera Mulvey & Stone, 1976
	Cactodera Krall & Krall, 1978
	Dolichodera Mulvey & Ebsary, 1980
	Betulodera Sturhan, 2002
	Paradolichodera Sturhan, Wouts & Subbotin, 2007
	Vittatidera Bernard, Handoo, Powers, Donald & Heinz, 2010

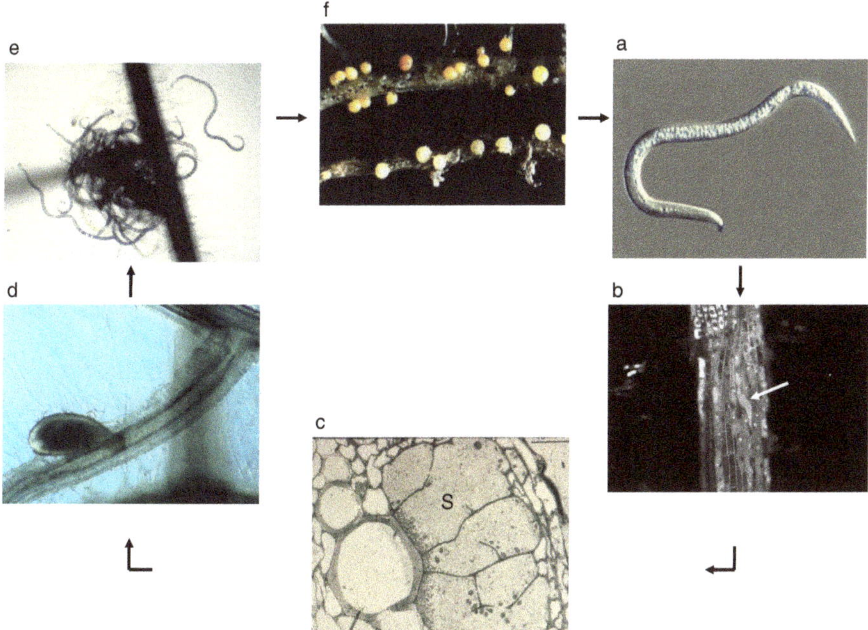

Fig. 4.1. Life cycle of a cyst nematode. Cysts contain 300–500 eggs on average but can reach up to 700 eggs under optimal conditions, each one containing a second-stage juvenile (J2). After hatch (a), the J2 moves through the soil, invades a host root (arrowed) (b) and moves through the root to establish a feeding site (syncytium (S)) (c) on which it feeds and develops. Juveniles develop either into females, which become saccate and rupture the root (d), or to vermiform males, which leave the root, locate the female and mate (e). The female then dies to form the cyst (f).

new potato crop. The duration of obligate diapause is affected by the photoperiod experienced by the infected plant, with unhatched J2 from plants grown under continuous light showing no obligate diapause. However, diapause is not always obligate, as shown for Kenyan populations of *G. rostochiensis* that completed up to three cycles in less than 1 year (Mwangi *et al.*, 2021). Facultative diapause is initiated by external factors, such as various environmental factors, from the second season onwards. Once diapause is completed, the J2 may enter into a quiescence state, which requires various environmental cues to effect further development of the life cycle. In temperate regions this usually occurs with an increase in soil temperature together with specific hatching stimuli produced by the host root system, termed root diffusate or root exudate. Whilst all species hatch in large numbers in response to appropriate host root diffusates, cyst nematodes can be classified into four broad categories based on their hatching responses: (i) low J2 water hatch, high J2 root diffusate hatch (*G. rostochiensis*, *G. pallida*, *H. cruciferae*, *H. carotae*, *H. goettingiana*, *H. humuli*); (ii) moderate J2 water hatch, high J2 root diffusate hatch (*H. trifolii*, *H. galeopsidis*, *H. glycines*); (iii) high J2 water hatch, high J2 root diffusate hatch of later generations (*H. schachtii*, *H. avenae*); and (iv) high J2 hatch in early generations, some dependence on root diffusate for J2 hatch in later generations (*H. cajani*, *H. sorghi*) (Moens *et al.*, 2018).

Cyst nematodes exhibit considerable variation in optimum temperature for hatch; for example, *G. pallida* is adapted to lower temperatures than *G. rostochiensis* (16°C and 20°C, respectively). Low optimum temperatures for hatching are characteristic of cyst nematodes that can invade during winter or early spring, such as *H. cruciferae*. As expected, nematodes adapted to warmer climates exhibit higher temperature optima, e.g. 30°C for *H. zeae*. Soil type can also affect rates of hatch. In general, coarse-textured soils favour hatching and subsequent invasion of root systems, providing suitable conditions for aeration and nematode migration. Maximum hatch usually occurs in soil at field capacity, whilst drought and waterlogging inhibit hatch.

Once hatched out of the eggshell, the J2 then leaves the cyst via either of the natural openings of the cyst, i.e. the fenestral region or the neck where the female's head has broken away. The J2 released into the soil will begin to search for a suitable host, relying primarily on gradients of chemicals released by the host's root system (see Chapter 8). As a survival strategy, not all juveniles hatch out at the same time. A proportion of J2 are retained either within the cyst body and/or in external egg masses. *Globodera* species do not produce egg sacs but occasionally exude a small droplet of moisture, whilst egg sac production varies between *Heterodera* species. While in *H. avenae* and *H. schachtii* all eggs are retained in the cyst, *H. glycines*, *H. carotae*, *H. cajani* and *H. cruciferae* lay a significant number of their eggs in an egg sac. The proportion of eggs laid in an egg sac can vary in individual species according to environmental conditions, e.g. *H. glycines* produces more eggs in egg masses under favourable conditions.

The J2 enters the root system of its host, usually directly behind the growing root tip, and then migrates to the pericycle and proceeds to select a suitable cell with which to form a feeding site (see Chapter 9). The hollow mouth stylet pierces a cell wall, being careful not to bridge the plasmalemma until a feeding tube is formed. Saliva from the pharyngeal glands is then injected and the cell contents are withdrawn into the nematodes by the action of the pharyngeal pump. The feeding tube acts as a particle filter to stop large molecules being ingested. This specific interaction induces enlargement of root cells and breakdown of their walls to form a large syncytial 'transfer cell' with dense, granular cytoplasm. The transfer cell develops cell-wall ingrowths adjacent to the conducting tissue, which greatly increase the internal surface area, facilitating the passage of nutrients into the syncytium. Provided that the J2 are able to stimulate the host plant to induce and maintain syncytia of sufficient size to receive all the nutrients they require, juveniles develop into both male and female adults. This stage of the life cycle will take approximately 7 days depending on the temperature, and the second moult to third-stage juvenile (J3) will then take place. The J3 has a well-developed genital primordial and rectum; the male has a single testis and the female has paired ovaries. The female at this point is about 0.4 mm long and its shape is becoming globular to facilitate the rapid growth of the developing ovaries. At the fourth moult, the female ruptures the root cortex and the formation of the vulva gives access to the reproductive system, which is being taken over by the formation of eggs.

Males develop at a similar rate in the same root as the females. They too emerge at the fourth moult but are still wrapped in the third-stage cuticle on emergence. Males are non-feeding, free-living and live for only a short time in the soil. The males are attracted to females, which exude sex pheromones (see Chapter 8) and may be the subject of multiple mating. After mating, the embryos develop within the egg as far as the formation of the J2 while still within the female's body. As indicated earlier, the female then dies and her cuticle tans to form a tough protective cyst containing several hundred embryonated eggs, the number depending on species and prevailing environmental conditions. In some of the *Heterodera* species an egg sac is exuded outside the cone region of the cyst. Eventually the cysts become detached from the roots as the plant dies and remain dormant in the soil until the next suitable host grows in the vicinity (Turner and Evans, 1998).

The time taken to complete the life cycle, from egg to egg, of a cyst nematode varies depending upon the co-evolution of the species with its host range and the environmental conditions. Typical life cycles are completed in about 30 days but this may be reduced in warmer climates; for example, *H. oryzicola* completes a life cycle in 23 days at 27°C, *H. glycines* 21 days at 25°C, whilst the temperate species *H. trifolii* requires 31 days at 20°C but 45 days at 15.5°C.

The number of generations per year varies between cyst nematode species It also varies with location of a given species; for example, *G. rostochiensis* has one completed generation per year in north-western Europe but Mwangi *et al.* (2021) considered that the Kenyan populations had the potential to complete up to three reproduction cycles in less than a year. Generally, as the soil temperature increases so does the number of generations, up to an upper threshold for each species. Under standard field conditions most temperate species of cyst nematodes will complete one or two generations, corresponding to the natural life cycle of its host combined with the length of the optimal temperature range. However, in tropical regions where favourable environmental conditions are more constant throughout the year, multiple generations are usual, with up to 11 generations being reported for *H. oryzicola*.

4.3. General Morphology of the Subfamily Heteroderinae

Nematode species of the subfamily Heteroderinae have similar gross morphology and are often distinguished from each other only by small details. The cyst and J2 are of the greatest importance in diagnosis as they are the stages most often found in soil extracts. Morphological identification using only the juveniles is not reliable and should be avoided, although they are the stage most likely to be obtained in soil extracts. If only juveniles are found in the soil it should be sampled again for cysts. Any live stages could be also used for molecular identification.

Mature females. The mature females are swollen into a spherical, sub-spherical or lemon shape to contain the developing ovaries (as well as the developing eggs) within the body cavity. Within some specimens, J2 can also be found. An egg sac may be extruded from the body but this depends on the species. In the mature female of the cyst nematode genera, annulations are restricted to the head region. The stylet and pharynx are strongly developed, with a prominent median bulb, and lie in the anterior part of the body, which forms a 'neck'. Posterior to the excretory pore, the swelling of the body is greatly developed and the excretory pore lies at the base of the neck. In most of the genera, the vulva is at the opposite pole of the body to the neck. The vulval slit runs transversely.

Cysts. Cysts are formed by the polyphenol oxidase tanning of the female cuticle, and they retain the female shape (Fig. 4.2A and B). The surface of the cyst is covered by a pattern of ridges derived from the pattern on the female cuticle. A thin-walled area surrounds the vulva and the cuticle can be lost, forming an opening, the fenestra (see Fig. 4.3 for the basic structures of terminal regions used for diagnosis). The fenestration (presence or absence; shape) is used in the diagnosis of genera (Fig. 4.4). Other measurements from the cysts are used in diagnosis of genera and species (Fig. 4.5).

Eggs. Eggs of most species fall within a similar size range and similar length width ratios, i.e. length (L) = 86–134 μm; width (W) = 36–52 μm; L/W ratio = 2.1–2.6. In general, eggs are usually unornamented and are not a reliable stage for diagnostic purposes. However, the eggs from species of *Cactodera* can be used for diagnostic purposes; in some species, such as *C. cacti* and

Fig. 4.2. Cyst and second-stage juvenile (J2) characteristics. A: Cysts of *Globodera*. B: Cysts of *Heterodera*. C: Anterior regions of J2 of *G. rostochiensis*, *H. schachtii*, *Punctodera punctata*. D: Posterior region of J2 of *G. rostochiensis*, *H. schachtii*, *P. punctata*.

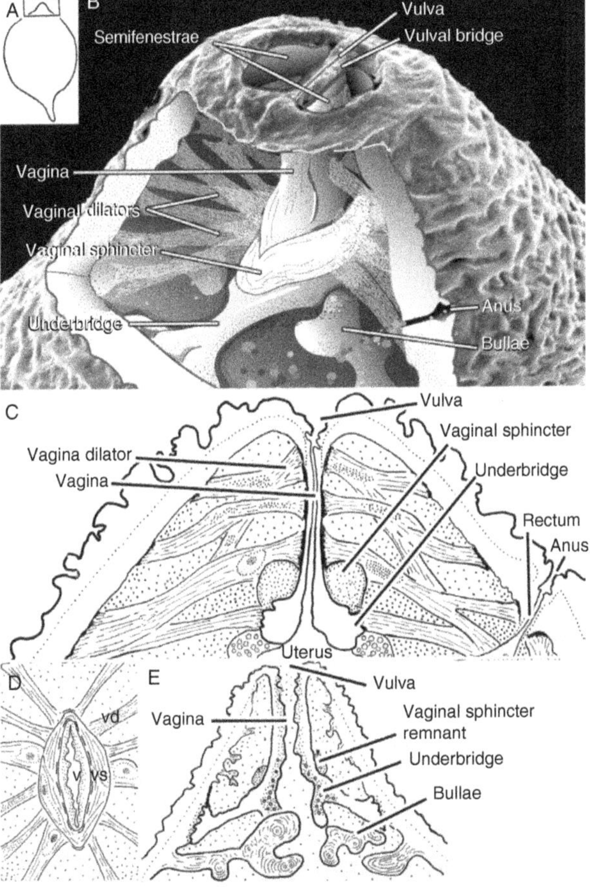

Fig. 4.3. Drawings depicting basic structures of terminal regions of cyst nematodes used for diagnosis. A: Overview of female. Box indicates terminal region as enlarged in B and C. B: Ventro-lateral 3D view with cutaway showing internal structures. C: Right lateral view of *Heterodera schachtii*. D: Transverse view through vulva (v) at level of vaginal sphincter (vs) and vaginal dilators (vd) of *H. schachtii*. E: Dorsal view of mature cyst. Main features for diagnosis are fenestrae type and measurements, underbridge and vulval features. (After Subbotin *et al*., 2010a.)

C. milleri, the eggshell is covered with small punctuations that resemble microvilli, whereas other members of the group, such as *C. weissi* and *C. amaranthi*, have smooth cuticles.

Second-stage juveniles. J2 are vermiform, with an offset, dome-shaped head and conical tail tapering to a point. On death, the body assumes a gentle curve with the ventral surface concave so that the nematode lies on its side. The cuticle is regularly annulated with the lateral field running from near the head to the tail; the number of incisures is three or four and they may be reduced anteriorly and posteriorly. The head skeleton, stylet and pharynx are well developed, the latter occupying approximately one-third of the body length (Fig. 4.2C). The median bulb is rounded in shape with a prominent valve. The pharyngeal glands overlap the intestine ventrally and subventrally; the single dorsal pharyngeal gland nucleus

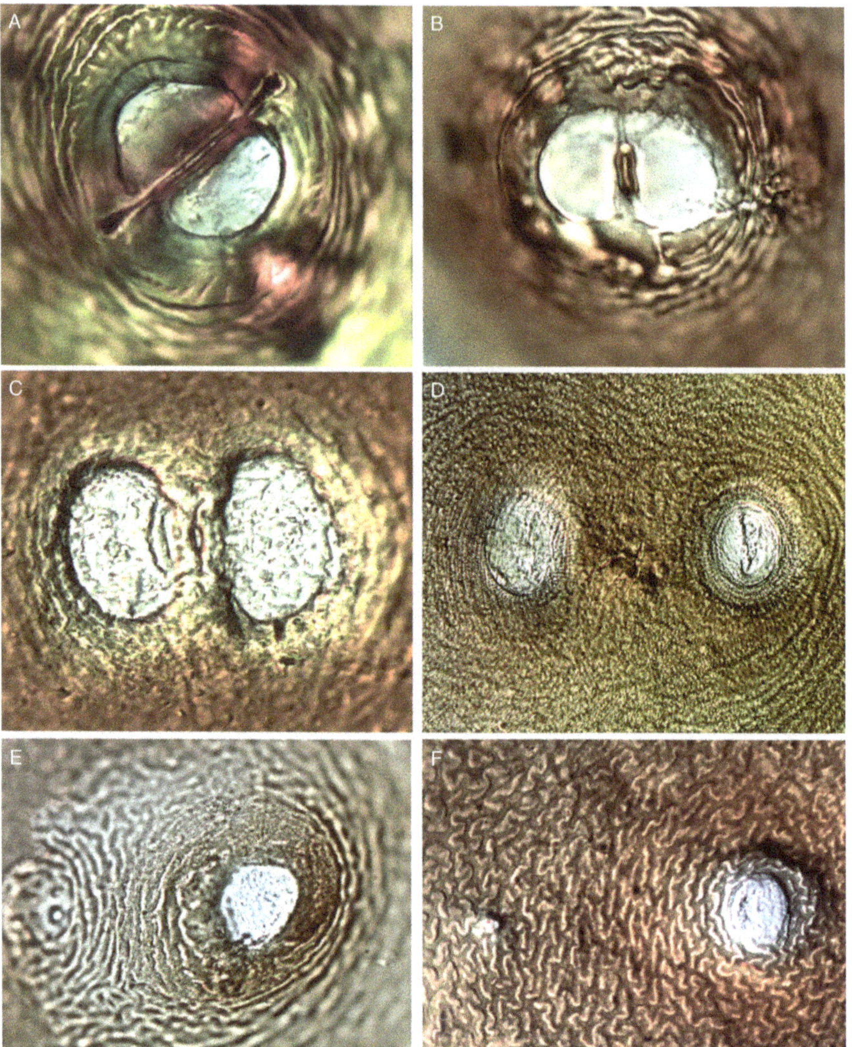

Fig. 4.4. Fenestration of cyst nematodes. The word fenestra (meaning window) refers to the thin-walled area on the vulval cone or perineal area of mature cysts. In young cysts the fenestral area is membranous but later decays, leaving a hole in the cyst wall. There are three main types of fenestration: circumfenestrate, bifenestrate and ambifenestrate. The fenestration shown by a particular cyst is an important feature in identification. A: *Heterodera trifolii* (ambifenestrate). B: *H. pratensis* (bifenestrate). C: *Heterodera* sp. (bifenestrate). D: *Punctodera punctata* (circumfenestrate). E: *Cactodera radicale* (circumfenestrate). F: *Globodera rostochiensis* (circumfenestrate). (Courtesy of V.N. Chizhov, Russia; after Subbotin *et al.*, 2010a.)

is more prominent than, and anterior to, the two subventral gland nuclei. The excretory pore is clearly visible on the ventral surface opposite the pharyngeal glands with hemizonid anterior to it. The anus may be marked by a small notch or step in the cuticle and the tail has a clear tip, the hyaline portion (Figs 4.2D and 4.6). Phasmids are visible in some species as

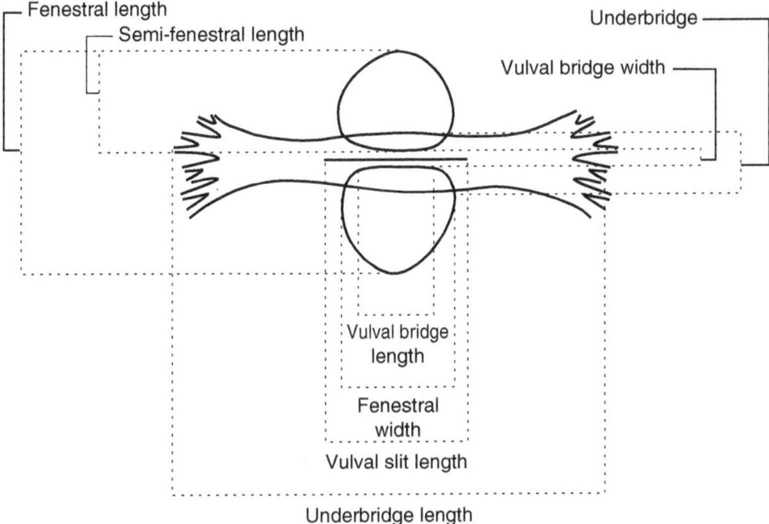

Fig. 4.5. Measurements of the fenestral area (fenestrae and underbridge) of a cyst, important for species and genus diagnostics.

small refractive points lying laterally on the tail surface, usually within the lateral field. Measurements from the J2 are used in diagnosis of genera and species (Fig. 4.6).

4.4. Genera and Principal Species

4.4.1. Genus *Heterodera* Schmidt, 1871

Females and cysts are usually lemon-shaped, the neck protruding from the anterior end and at the opposite pole the posterior usually ending in a cone (Figs 4.7 and 4.8). The cone carries the fenestra that internally is associated with the reproductive organs and, from the exterior, will provide diagnostic features used for identification. Some members of this genus, such as *H. schachtii*, have a very high conspicuous cone and others, such as *H. cruciferae*, have much lower cones. The mature female containing eggs dies, causing her cuticle to tan and dry out, thus protecting the eggs until the invasive J2 hatch. Eggs are retained in the body; in some cases (e.g. *H. cajani*) an egg sac is also present. The cyst can range in colour from light brown to dark brown or almost black. The cuticle surface displays folds and ridges often in specific patterns, e.g. zigzag or parallel, which are helpful in the diagnosis of species. Subcrystalline layer maybe present or absent. *Heterodera* cysts may or may not have an underbridge. Bullae may also be present and both features are very diagnostic of the groups contained within this genus. Molecular and morphological data support division of most *Heterodera* species into several groups: *Afenestrata*, *Avenae*, *Bifenestrata*, *Cardiolata*, *Cyperi*, *Goettingiana*, *Humuli*, *Sacchari* and *Schachtii*. Key features for study include the formation of the fenestra. These are classified as without fenestration (*Afenestrata* group), ambifenestrate (two openings divided by a narrow vulval bridge) or bifenestrate (two openings separated by a much wider vulval bridge). The

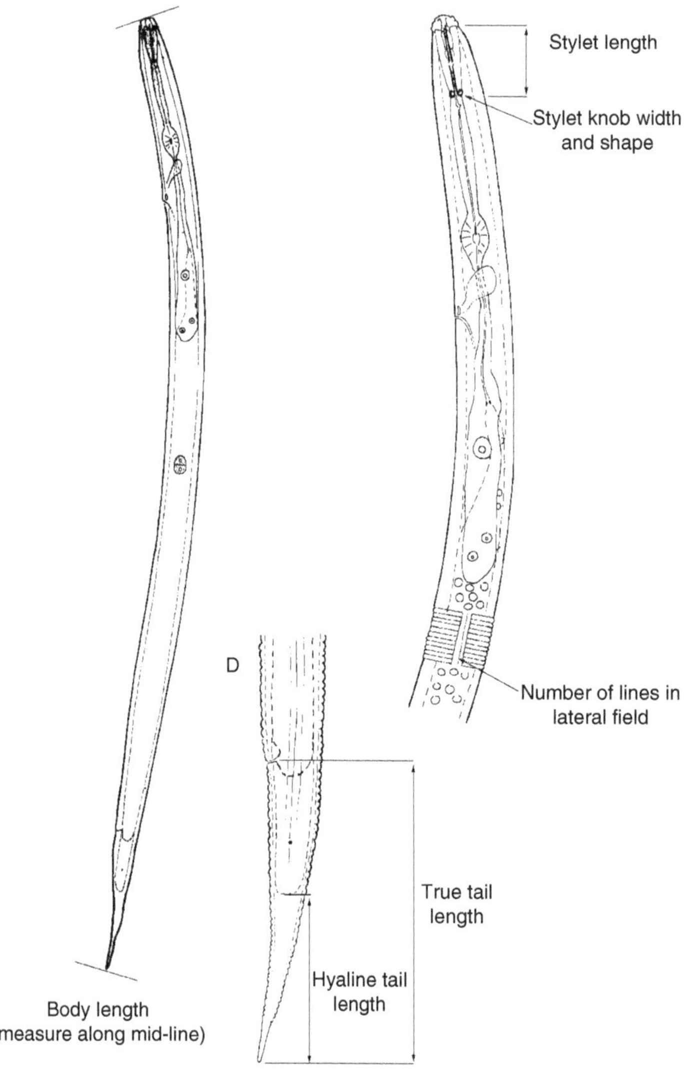

Fig. 4.6. Measurements of second-stage juvenile, important for species and genus diagnostics.

length of the vulval slit varies. In the *Avenae* group it is very short at 8–10 µm, whereas members of the *Schachtii* group have a much longer slit, averaging 65 µm in length.

The J2 are also used for diagnosis together with the cyst features. The stylet length and the position and shape of the basal knobs are important features. The number of lateral fields is usually three or four. The number of head annuli present, the width of the body at the excretory pore and the anus, and the length of both the true tail, i.e. from the anus to the tail tip, and the hyaline tail length are diagnostic (Fig. 4.6).

The type species of the genus is *H. schachtii*.

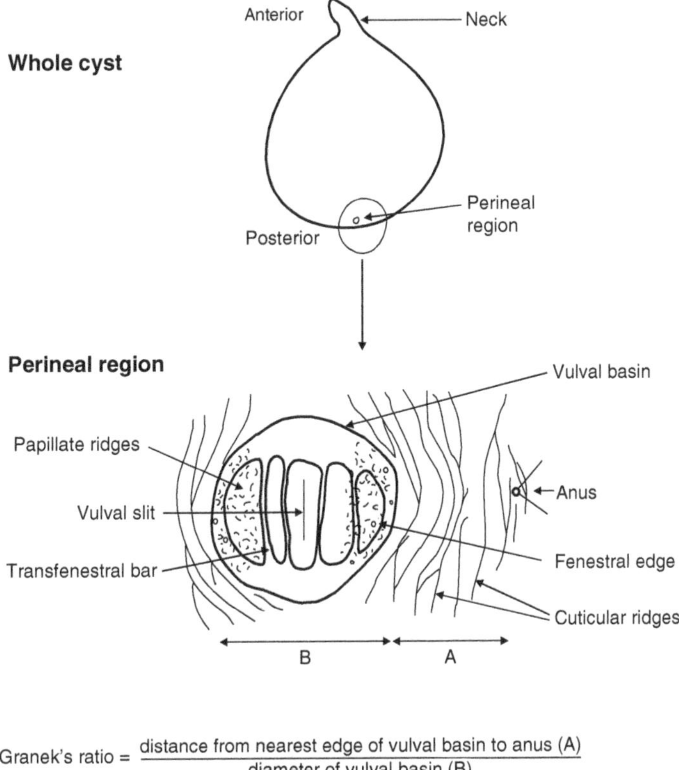

Fig. 4.7. The whole cyst shape and vulval area details for calculation of Granek's ratio.

4.4.1.1. Cereal cyst nematodes, Heterodera avenae *Wollenweber, 1924*

Presently, four species are collectively named as cereal cyst nematodes (CCNs) – the European CCN *H. avenae*, the Filipjev CCN *H. filipjevi*, the Sturhan CCN *H. sturhani* and the Australian CCN *H. australis* – and they are considered as major, economically important, nematode pests in cereal-growing areas. *Heterodera avenae* and *H. filipjevi* have been found in many countries of Europe, Asia and Northern America, where they occur often in mixed populations, whereas *H. australis* has a limited distribution and has only been reported to date in Australia, and *H. sturhani* has only been found in China (Smiley *et al.*, 2017).

In Europe, more than 50% of the fields in major cereal-growing areas are infected by the European CCN (Rivoal and Cook, 1993), with annual yield losses reaching £3 million (Nicol and Rivoal, 2008). In the USA, at least US$3.4 million is estimated to be lost annually in wheat production in the states of Idaho, Oregon and Washington because of CCN. The yield losses it causes on wheat are in the range 15–20% in Pakistan, and 40–92% on wheat and 17–77% on barley in Saudi Arabia. In China, yield losses of wheat crops induced by the CCN can reach 70%. Hosts of *H. avenae* include species of cereals and grasses from the following genera: *Agropyron*, *Agrostis*, *Alopecurus*, *Anisantha*, *Arrhenatherum*, *Avena*, *Brachypodium*, *Bromus*, *Dactylis*, *Echinochloa*, *Festuca*, *Hordeum*, *Koeleria*, *Lolium*, *Phalaris*, *Phleum*, *Poa*, *Polypogon*, *Secale*, *Setaria*, *Sorghum*, *Trisetum*, *Triticum*, *Vulpia*, *Zerna* and *Zea* (Williams and Siddiqi, 1972). CCN has only one generation per year, with J2 hatch from the eggs determined largely by temperature (Rivoal and Cook, 1993).

DESCRIPTION.

Cysts: L = 518–801 µm; W = 432–744 µm; L/W ratio = 0.8–1.8; fenestral length = 32–55 µm; vulval slit = 7–12 µm.

Male: L = 1020–1590 µm; stylet = 27–33 µm; spicules = 33–38 µm; gubernaculum = 10–13 µm.

J2: L = 505–598 µm; stylet = 24–27.5 µm; hyaline region = 34–50 µm; tail = 52–79 µm.

Cyst lemon-shaped, with prominent neck and vulval cone. Subcrystalline layer conspicuous, sloughing off with formation of dark brown cyst. Bifenestrate, bullae prominent, crowded beneath vulval cone. J2 vermiform, with a sharply pointed tail. Stylet well developed, with large, anteriorly flattened to concave basal knobs.

Species of CCN can be differentiated using morphometrical characters of J2 and cysts, isoelectric focusing (IEF) of proteins, polymerase chain reaction (PCR)-restriction fragment length polymorphism (RFLP) of internal transcribed spacer (ITS) ribosomal RNA (rRNA), sequences of the ITS rRNA and *COI* genes.

4.4.1.2. Sugar beet cyst nematode, Heterodera schachtii *Schmidt, 1871*

The sugar beet cyst nematode (Fig. 4.8) has been recognized as a plant pathogen since 1859 when it was associated with stunted and declining sugar beet in Germany. In the following years it became recognized as a pest of great importance in beet-growing areas of several European countries. *Heterodera schachtii* is found in all major sugar beet production areas of the world, favouring temperate regions but apparently tolerating a broad range of climates. It is widespread in Europe, the USA and Canada (Baldwin and Mundo-Ocampo, 1991). Annual yield loss in EU countries based upon world market sugar prices was estimated in 1999 at up to US$90 million (Müller, 1999). The optimum temperature for development is around 25°C. In some climates, three to five generations may complete development on sugar beet in one season (Franklin, 1972). *Heterodera schachtii* was described from the host *Beta vulgaris* and it parasitizes mainly plants of the families Amaranthaceae (Caryophyllales) (many species of *Beta* and *Chenopodium*) and Brassicaceae (Brassicales) (*Brassica oleracea, B. napus, B. rapa, Rhaphanus sativus* and many others including a diversity of common weeds) (Franklin, 1972). Some plants from Polygonaceae, Scrophulariaceae, Caryophyllaceae and Solanaceae are susceptible to nematode infection.

DESCRIPTION.

Cysts: L = 480–960 µm; W = 396–696 µm; L/W ratio = 0.9–2.0; fenestral length = 28–48 µm; vulval slit = 33–54 µm.

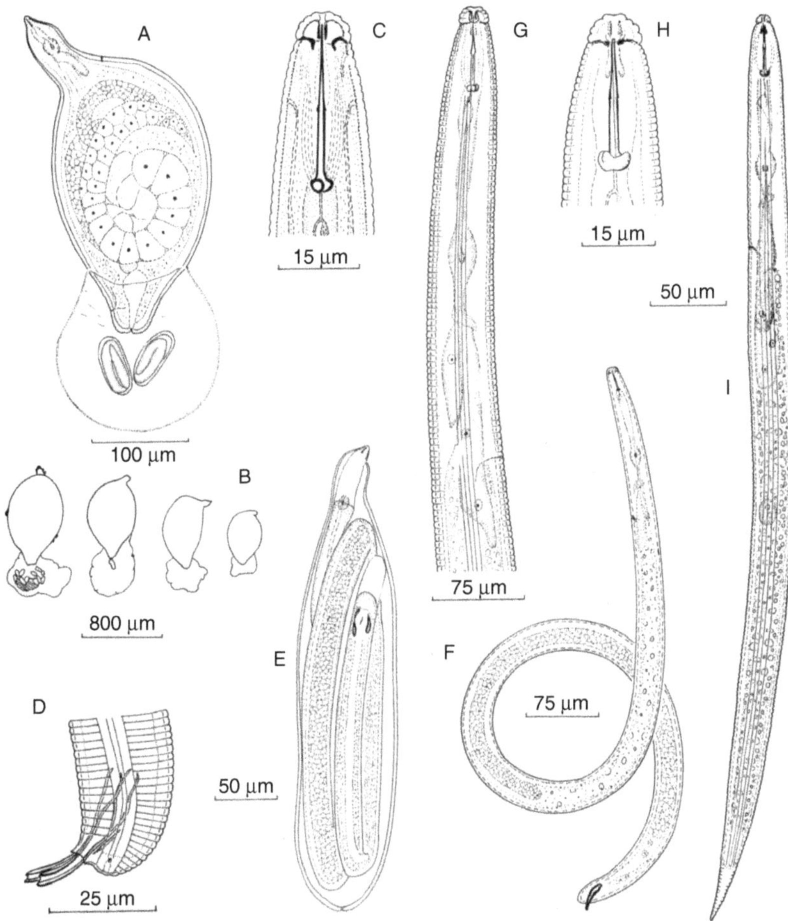

Fig. 4.8. *Heterodera schachtii*. A: Adult female with egg sac. B: Cysts and egg sacs. C: Anterior region of male. D: Male tail. E: Fourth-stage male moulting. F: Adult male. G: Male pharyngeal region. H: Anterior region of second-generation juvenile (J2). I: J2. (After Franklin, 1972.)

Male: L = 1038–1638 µm; stylet = 27–30 µm; spicules = 27–39 µm; gubernaculum = 10–11 µm.

J2: L = 400–512 µm; stylet = 23–28 µm; hyaline region = 17–33 µm; tail = 40–56 µm (Fig. 4.8).

Cyst colour light to dark brown. Ambifenestrate, within cone, remnants of vagina attached to side walls by underbridge and a number of irregularly arranged, dark brown molar-shaped bullae situated a short distance beneath the vulval bridge. J2 labial region offset, hemispherical, with four indistinct annuli. Stylet moderately heavy with prominent, forwardly directed knobs. Tail acutely conical with rounded tip, distinct hyaline terminal section 1–1.25 stylet lengths long.

Heterodera schachtii belongs to the *Schachtii* group and is distinguished from closely related species (*H. trifolii*, *H. glycines*, *H. betae* and others) by a combination of morphological and morphometric characteristics. The ITS rRNA and *COI* gene sequences clearly differentiate this species from all others. Diagnostics of this species using PCR with species-specific primers has also been developed.

4.4.1.3. Soybean cyst nematode, **Heterodera glycines** *Ichinohe, 1952*

A cyst nematode parasitizing soybean plants, *Glycine max*, and causing 'yellow dwarf' symptoms was recorded from Shirakawa, Fukushima Prefecture, Japan, in 1915. Ichinohe (1952) was the first to make careful morphological comparisons with other *Heterodera* species and to give a specific name and brief description of this nematode. In Japan, yield losses have been estimated at 10–70% (Ichinohe, 1988). Presently, *H. glycines* occurs in most countries of the world where soybean is produced. In a study of losses predicted in ten soybean-producing countries together accounting for 97% of the world crop, *H. glycines* appeared to be the most important constraint (losses of 8,969,400 t) on yield and estimated at US$1960 million (Wrather *et al.*, 2001). In these countries, total yield losses attributed to *H. glycines* were greater than those for any other pest of the crop. *Heterodera glycines* is widely distributed throughout the north-central USA where different maturity groups with the same source of resistance to *H. glycines* are grown. State surveys in the region report from 14% to 63% of fields are infested with *H. glycines*.

Heterodera glycines has a broad host range, especially Fabaceae, but also other families. More than 66 weed species of nine families are suitable hosts. Riggs (1992) provided a list of non-fabaceous hosts comprising 63 species in 50 genera from 22 families (e.g. Boraginaceae, Capparaceae, Caryophyllaceae, Chenopodiaceae, Brassicaceae, Lamiaceae, Fabaceae, Scrophulariaceae, Solanaceae). In field conditions, *H. glycines* was also found in several other plants, including henbit (*Lamium amplexicaule*), purple deadnettle (*Lamium purpureum*), mouse-ear chickweed (*Cerastium holosteoides*) and common chickweed (*Stellaria media*) (Riggs, 1992).

Three to five generations develop during the cropping season. Optimum temperature is 23–28°C; development stops below 14°C and above 34°C. In the absence of a host, J2 and eggs in cysts may remain viable in soil for 6–8 years.

Heterodera glycines disturbs root growth, interferes with nodulation and causes early yellowing of soybean plants. The above-ground symptoms of damage on individual plants and appearance of infested fields are usually not sufficiently specific to allow direct identification. Infected plants are predisposed to *Fusarium* wilt. Sudden death syndrome is a soilborne disease of soybean caused by the fungus *Fusarium solani* in association with *H. glycines*.

DESCRIPTION.

Cysts: L = 340–920 µm; W = 200–688 µm; L/W ratio = 1.0–2.4; fenestral length = 35–72 µm; vulval slit = 36–60 µm.

Male: L = 911–1400 µm; stylet = 24–27 µm; spicules = 28–45 µm; gubernaculum = 8–13 µm.

J2: L = 345–504 µm; stylet = 21–25 µm; hyaline region = 18–36 µm; tail = 35–59 µm.

Cyst mainly lemon-shaped, sometimes round with a protruding neck and cone. Ambifenestrate, bullae prominent, located at or anterior to underbridge, extending into vulval cone from interior of body wall cuticle. Shape varying from round to finger-like, round bullae differently sized, finger-like bullae of variable length and thickness. Underbridge well developed. J2 body vermiform with regularly annulated cuticle. Stylet robust with anteriorly protruding knobs. Tail tapering uniformly to a finely rounded terminus.

Heterodera glycines belongs to the *Schachtii* group and is distinguished from similar species (*H. medicaginis*, *H. schachtii*, *H. trifolii*, *H. daverti* and others) by a combination of morphological and morphometric characteristics. It differs from *H. schachtii* by the shape of the stylet knobs of J2 (slightly convex versus moderately or strongly concave), shorter average J2 stylet and longer average fenestral length. The ITS rRNA and *COI* gene sequences clearly differentiate this species from all others.

4.4.1.4. Pea cyst nematode, Heterodera goettingiana *Liebscher, 1892*

In 1890, G. Liebscher reported infection and yield loss of pea (*Pisum sativum*) and vetch (*Vicia sativa*) by nematodes identified as *H. schachtii* at fields of the Agricultural Institute at Göttingen, Germany. Two years later he described this species as *H. goettingiana*. Several researchers have reported diseases of peas, primarily in European countries, caused by this nematode; however, little was known about biology and pathogenicity of *H. goettingiana* until the mid-1900s (Franklin, 1951). Infected pea fields show sharply delineated patches with dwarfed, poorly branched and yellowing plants that die prematurely. Infected plants either fail to flower or flower too early. The root system is poorly developed. Development takes 3–15 weeks depending on soil temperature and moisture as well as host species. One or two generations occur during the growing season in the UK, and three generations may develop in southern Italy.

DESCRIPTION.

Cysts: L = 400–780 µm; W = 310–540 µm; L/W ratio = 1.3–2.2; fenestral length = 43–71 µm; vulval slit = 43–61 µm.

Male: L = 1270 µm; stylet = 27 µm; spicules = 27 µm; gubernaculum = 12 µm. (No ranges available.)

J2: L = 408–519 µm; stylet = 23–26 µm; hyaline region = 27–38 µm; tail = 54–74 µm.

Cyst lemon-shaped with light to dark brown cyst wall. Subcrystalline layer not visible. Vulval cone ambifenestrate. In some old cysts, vulval bridge ruptured, fenestrae joining to form a large oval fenestrum. Bullae absent, although bullae-like structures and vulval denticles present. Underbridge weak. J2 body vermiform, curved ventrally after fixation. Labial region hemispherical, with 2–5 annuli, slightly offset from body. Lateral field with four incisures, not areolated. Stylet knobs rounded, slightly projecting anteriorly. Tail tapering uniformly to a finely rounded terminus.

Heterodera goettingiana belongs to the *Goettingiana* group. It differs from several other representatives of the *Goettingiana* group (*H. cruciferae*, *H. carotae*, *H. circeae*, *H. scutellariae* and *H. persica*) by longer average J2 body, longer tail and longer hyaline region. The ITS rRNA and *COI* gene sequences clearly differentiate this species from all others.

4.4.1.5. Mediterranean cereal cyst nematode, Heterodera latipons *Franklin, 1969*

In the early 1960s, a cyst nematode similar to *H. avenae* was detected in Israel and Libya on the roots of stunted wheat and barley plants. It was morphologically studied and described by Franklin (1969) as a new species under the name *H. latipons*, based on characteristics of the Israel population. This nematode was later recorded in many countries, mainly from the Mediterranean and the Orient. *Heterodera latipons* often occurs in mixed populations with *H. avenae* in cereal cropping systems. Hosts include barley (*Hordeum vulgare*), oat (*Avena sativa*), rye (*Secale cereale*), *Phalaris minor*, *P. paradoxa* and *Elytrigia repens*. *Triticum durum* was considered to be a poor host of this nematode. Yield losses as high as 50% were reported on barley in Cyprus. In Syria, the nematode causes average yield losses of 20 and 30% in barley and durum wheat, respectively, and the nematode was more damaging under water stress conditions. Moreover, damage is more severe in fields infested concomitantly with *H. latipons* and the fungus *Bipolaris sorokiniana*, the causal agent of the common root rot and seedling blight of barley, i.e. the presence of the nematode increases the aggressiveness of the fungus (Scholz, 2001). In all areas studied, *H. latipons* completed only one life cycle during the growing season (Mor *et al.*, 1992).

DESCRIPTION.

Cysts: L = 300–700 µm; W = 320–560 µm; L/W ratio = 0.6–1.7; fenestral length = 52–76 µm; vulval slit = 6–11 µm.

Male: L = 960–1406 µm; stylet = 22–29 µm; spicules = 32–36 µm; gubernaculum = 8 µm.

J2: L = 401–559 µm; stylet = 22–25 µm; hyaline region = 20–36 µm; tail = 42–68 µm.

Cysts are dark to mid-brown covered with white subcrystalline layer. Bifenestrate, semifenestrae separated by a distance greater than fenestral width, vulval slit short. Strong underbridge with pronounced thickening in middle and with ends splayed. Bullae usually absent, sometimes present at underbridge level. J2 body slightly curved dorsoventrally when killed by heat. Stylet with well-developed, anteriorly concave knobs.

Heterodera latipons belongs to the *Avenae* group and morphologically closely resembles *H. hordecalis* and *H. turcomanica*. These nematodes share similar circular semifenestrae separated by a distance longer than the semifenestra diameter and a rather typical underbridge but with a pronounced enlargement underlying the vulval slit. The most important differentiating character between *H. latipons* and *H. hordecalis* is the vulval slit, which in *H. latipons* is much shorter. The ITS rRNA and *COI* gene sequences clearly differentiate this species from all others.

4.4.2. Genus *Globodera* Skarbilovich, 1959

Mature females and cysts are spheroid, lacking terminal cone. Vulval area is circumfenestrate. Vulva located in a cavity beneath outline of body, vulval slit <15 µm. There is no anal fenestra. Vaginal remnants, underbridge and bullae rarely present (Fig. 4.7). Cuticle with distinct D layer (Cordero and Baldwin, 1990). All eggs retained in body, egg sac absent. Colouration is helpful in making a diagnosis of species, especially separation of the PCN species *G. rostochiensis* and *G. pallida*. Egg surface is smooth. Male lateral field with four lines, spicules >30 µm, distally pointed. J2 has four incisures in lateral field. Tail conical, pointed, phasmids punctiform. *En face* pattern typically with six separate lips, sometimes with fusion of adjacent submedial lips. The type species of the genus is *G. rostochiensis*.

PCNs include three species, *G. rostochiensis, G. pallida* and *G. ellingtonae,* and the first two species have been reported from many countries and are considered to be one of the most economically important pests of potato. *Globodera rostochiensis, G. pallida* and *G. ellingtonae* are differently distributed in the Andes. Factors that may be responsible include day length, temperature, altitude, rainfall or the interaction of any of them with the host potato. Human activities over centuries may also have influenced distribution. In South America, these species are found mainly between 2000 and 4000 m above sea level, with the heaviest infestations between 2900 and 3800 m above sea level. These three species occupy different zones in the Andes. The demarcation line between *G. rostochiensis* and *G. pallida* is near 15.6°S. With few exceptions, populations north of this line are mainly *G. pallida*. Those from areas around Lake Titicaca and further south are predominantly *G. rostochiensis* with few *G. pallida* or mixtures of both species. *Globodera ellingtonae* was found in southern Bolivia, northern Argentina and Chile. *Globodera rostochiensis* and *G. pallida*, have been introduced in many parts of the world, particularly to Europe, and also to the USA, Canada, New Zealand and numerous other countries where potatoes are grown.

4.4.2.1. Golden potato cyst nematode, **Globodera rostochiensis** *(Wollenweber, 1923) Skarbilovich, 1959*

The golden cyst nematode associated with potato plants, *Solanum tuberosum*, from Rostock, Germany was first reported in 1881 and was considered to be *H. schachtii*, this being the only known species of cyst nematode at that time. During the early 1900s, the PCN became more widely known throughout Europe and was described in 1923. In temperate regions, *G. rostochiensis* usually completes only one generation, although a second generation may be initiated but not completed; J2 hatched from first-generation eggs, but they were unable to reach the adult stage. In subtropical regions two generations might occur. Development of one generation requires 6–10 weeks. The J2 can go into diapause and remain viable for many years, hatching continuing for 25 or more years. Heavily infected plants become yellow and stunted. Infected plants have reduced root systems, which are abnormally branched and brownish in colour. Symptoms in the field first appear in small patches. At low nematode densities tuber sizes are reduced, whereas at higher densities both number and size of tubers can be reduced. At 8 and 64 eggs g^{-1} of soil, yield losses of about 20% and 70%, respectively, can be expected. Seinhorst (1982) and Elston *et al.* (1991) proposed models that described the relationships of PCN population densities before planting with potato yield and post-harvest nematode populations (see Chapter 10). The damaging effect of PCN is not only determined by nematode density, but also by such factors as cultivar, crop husbandry and environmental conditions. PCNs are responsible for annual potato tuber losses of up to 9% in Europe. Information on economic importance in some South American countries is scarce or unavailable although yield losses in Bolivia and Peru have been estimated to be around US$13,000,000 and US$128,000,000, respectively (Franco and Gonzalez, 2010).

Hosts include potato, *S. tuberosum* (Solanaceae, Solanales), tomato, *S. lycopersicum* and eggplant, *S. melongena*. Other hosts include many *Solanum* spp., *Datura* spp., *Hyoscyamus niger*, *Nicotiana acuminata*, *Physalis* spp., *Physochlaina orientalis*, *Salpiglossis* spp., *Capsicum annuum* and *Saracha jaltomata*.

DESCRIPTION.

Cysts: L = 450–990 µm; W = 250–810 µm; L/W ratio = 0.9–1.8; fenestral diameter = 14–21 µm; number of ridges between anus and fenestra = 16–22; Granek's ratio = 2.3–7.0.

Male: L = 960–1406 µm; stylet = 22–29 µm; spicules = 32–36 µm; gubernaculum = 8 µm.

J2: L = 366–502 µm; stylet = 19–23 µm; dorsal gland orifice (DGO) = 2.4–6.7 µm; hyaline region = 18–30 µm; tail = 37–57 µm.

Female colour changing from white to yellow to light golden as female matures to cyst stage. Cyst brown, ovate to spherical in shape with protruding neck, circumfenestrate (Fig. 4.3F), abullate. Fenestra circular, anus conspicuous at apex of a V-shaped subsurface cuticular mark. J2 body tapering at both extremities but more at posterior end. Stylet well developed, with prominent rounded knobs as viewed laterally. Lateral fields with four lines extending for most of body length. Tail tapering to small, rounded terminus.

Globodera rostochiensis is morphologically similar to *G. pallida* and *G. tabacum*. It differs from *G. pallida* by yellow or gold versus cream-coloured maturing females, higher number of ridges between the vulva and anus, larger mean for Granek's ratio, stylet knob shape, shorter average stylet length and rounded versus more pointed J2 tail terminus (Box 4.2).

4.4.2.2. Pale potato cyst nematode, Globodera pallida *Stone, 1973*

The pale PCN, *G. pallida*, is considered to be a major pest of potato crops in cool temperate climates. It is reported from several counties in Europe, Asia, Africa and South America. In Central and North America, *G. pallida* has been reported in Panama, the USA and Canada, but in the last two countries *Globodera* species on potato have a rather restricted distribution with small infested areas, because of rigorous phytosanitary regulations and seed potato certification programmes, compared to the widespread infestations found in European countries. Recently, mitochondrial DNA (mtDNA) analysis has been used to study genetic relationships among Peruvian populations of *G. pallida*, thus identifying the origin of Western European populations of this species (Picard *et al.*, 2007; Plantard *et al.*, 2008; Subbotin *et al.,* 2020). Using the mtDNA gene, cytochrome *b* (*cytb*) sequences and microsatellite loci, Plantard *et al.* (2008)

Box 4.2. Morphometric comparison between *Globodera rostochiensis, G. pallida* and *G. ellingtonae* (measurements in mm; values for second-stage juveniles are ranges compiled from various sources).

Second-stage juvenile characters

	Body length	Stylet length	Stylet knob shape	Tail length	Hyaline region tail length
G. rostochiensis	366–502	19–23	Rounded	37–57	18–30
G. pallida	380–533	23–25	Robust; square hooked	40–57	20–31
G. ellingtonae	365–515	19–23	Rounded; flattened or forward projection	39–55	20–33

Cyst characters

	Distance from anus to vulval basin; mean (range)	Fenestral diameter	Number of ridges between anus and vulval basin	Granek's ratio
G. rostochiensis	66.0 (50.0–77.0)	Usually <19.0	16–22	2.3–7.0
G. pallida	45.0 (35.0–55.0)	Usually >17.0	7–17	1.2–3.6
G. ellingtonae	64.5 (50–85)	>17	10–18	1.7–3.0

showed that the *G. pallida* presently distributed in Europe derived from a single restricted area in the extreme south of Peru, located between the north shore of Lake Titicaca and Cusco. *Globodera pallida* develops one generation for a vegetation season. This species is adapted to cool temperatures and is able to hatch earlier in the year and develop at 2°C cooler than *G. rostochiensis* (Langeslag *et al.*, 1982). The symptoms of attack by *G. pallida* are similar to those for *G. rostochiensis* and the damage threshold is 1–2 eggs g^{-1} of soil. Hosts include potato (*S. tuberosum*), eggplant (*S. melongena*), tomato (*S. lycopersicum*), many other species of *Solanum*, and black henbane (*Hyoscyamus niger*).

DESCRIPTION.

Cysts: L = 420–748 µm; W = 400–685 µm; fenestral diameter = 17.5–25 µm; number of ridges between anus and fenestra = 7–17; Granek's ratio = 1.2–3.6.

Male: L = 1198 µm; stylet = 27 µm; spicules = 36 µm; gubernaculum = 11 µm.

J2: L = 380–533 µm; stylet = 22.5–25 µm; DGO = 2.7–5 µm; hyaline region = 20–31 µm; tail = 40–57 µm.

Female is white in colour, some populations passing, after 4–6 weeks, through a cream stage, turning glossy brown when dead. Cyst vulval region intact or fenestrated with single circumfenestrate (Fig. 4.3D) opening occupying all or part of vulval basin, abullate. J2 lateral field with four incisures but with three anteriorly and posteriorly, occasionally completely areolated. Stylet well developed, basal knobs with distinct anterior projection as viewed laterally. Tail tapering uniformly with a finely rounded point, hyaline region forming about half of tail region.

Globodera pallida is most closely related to *G. rostochiensis* and *G. tabacum*. It differs from *G. rostochiensis* by cream-coloured females versus yellow or gold, smaller number of ridges between the vulva and anus, smaller mean for Granek's ratio, stylet knob shape, longer stylet length, tail terminus and presence of refractive bodies on hyaline part of tail (usually 4–7 refractive bodies versus absence) in J2.

4.4.2.3. Tobacco cyst nematode, **Globodera tabacum** *(Lownsbery & Lownsbery, 1954) Skarbilovich, 1959*

Globodera tabacum is considered as a serious and important pest of shade and broadleaf tobacco. It is recorded from several countries in Europe, Asia, Africa, and South and North America. *Globodera tabacum* is a polytypic species containing the following subspecies: *G. tabacum tabacum* (Lownsbery & Lownsbery, 1954), *G. tabacum virginiae* (Miller & Gray, 1968) and *G. tabacum solanacearum* (Miller & Gray, 1972). All three subspecies develop on tobacco and horse nettle (*Solanum carolinense*), but otherwise differ in host preference. *Globodera tabacum* parasitizes *Nicotiana tabacum, S. carolinense*, tomato and other species of the genera *Nicotiana* and *Solanum*, as well as *Atropa belladona*, *Hyoscyamus niger*, *Nicandra physalodes* and *Capsicum annuum*. Two or more generations usually occur. Infected tobacco plants have small root systems and aboveground symptoms are similar to those associated with severe root-knot and lesion nematode infestations. Nematode infection is often associated with increased damage from bacterial wilt and black shank. Farmers in Virginia, USA, have recorded complete crop failures, but losses generally average 15%. A high density of nematode populations early in the growing season can reduce flue-cured tobacco yield by 25–50%, although tobacco may escape significant losses from moderate populations, especially under favourable growing conditions.

DESCRIPTION.

Cysts: L = 337–937 µm; W = 232–812 µm; L/W ratio = 0.9–1.5; fenestral diameter = 13–36 µm; number of ridges between anus and fenestra = 5–15; Granek's ratio = 1.4–4.2.

Male: L = 710–1450 µm; stylet = 24–29 µm; spicules = 26–35 µm; gubernaculum = 9–12 µm.

J2: L = 458–621 µm; stylet = 20–27 µm; DGO = 4.3–9 µm; hyaline region = 17–35 µm; tail = 34–64 µm.

Female body ovate to spherical with elongate neck, white, becoming yellow. Cyst light shiny brown, circumfenestrate, abullate. J2 with well-developed rounded basal knobs. Terminus of tail finely rounded.

Globodera tabacum differs from *G. rostochiensis* by J2 with longer mean values of body length, mean stylet and by cysts with smaller mean number of cuticular ridges. It differs from *G. mexicana* by J2 with longer mean body length and from *G. pallida* by cysts with a smaller mean number of cuticular ridges and by J2 with longer mean body length.

4.4.3. Genus *Punctodera* Mulvey & Stone, 1976

Mature females and cysts are spherical, pear-shaped or ovoid, with short projecting neck and heavy subcrystalline layer. Cuticle reticulate, subcuticle with punctations. D-layer present. Terminal region not cone-shaped; cyst light to dark brown. Vulval slit extremely short (<5 µm), anus at a short distance from vulval fenestra. Circumfenestrate, fenestra surrounding vulva 16–40 µm (approximately 20 µm in type species) in diameter, anus offset toward ventral margin of fenestra, an anal fenestra of similar shape and size to vulval fenestra present. Underbridge and perineal papilla-like tubercles absent. Bullae present or absent. Eggs retained in body, no egg sac. Males vermiform, less than 1.5 mm long. DGO 2.6–4.6 µm. Spicules 31–33 µm long, distally pointed. Tail less than 0.5 anal body diameter long, cloacal lips not forming a tube. J2 body length is 0.35–0.49 mm, stylet 24–26 µm and tail conical, 63–78 µm long, hyaline region in type species 38–41 µm long. Lateral field with four incisures. Phasmid openings punctiform, without a lens-like structure. Parasites of monocotyledonous plants. The type species of the genus is *Punctodera punctata*.

4.4.3.1. Grass cyst nematode, Punctodera punctata *(Thorne, 1928) Mulvey & Stone, 1976*

This species was described by Thorne based on specimens from heavily infected wheat roots from a field in the Humboldt area, Saskatchewan, Canada. Several further attempts to collect topotype specimens failed. Subsequently, *P. punctata* was also reported as a common species infecting grasses from Europe, the USA and Canada. However, all attempts to infect wheat or other cereals by these nematodes failed to give any positive results. Several authors suggested that *P. punctata* might represent a complex of several closely related species. Many grasses are good hosts of this nematode. Only a single generation occurs each year.

DESCRIPTION.

Female and cysts: L = 330–901 µm; W =170–720 µm; L/W ratio = 1.2–3.0; vulval fenestral diameter = 16–33 µm; anal fenestral diameter = 33 (25.2–42.0) µm.

Male: L = 910–1270 µm; stylet = 26–33 µm; spicules = 28–36 µm; gubernaculum = 8–10 µm.

J2: L = 520–643 µm; stylet = 23–32 µm; DGO = 3.5–6.5 µm; hyaline region = 38–64 µm; tail = 63–93 µm.

Females and cysts are ovoid, pear- or flask-shaped without vulval cone, white. Vulva slit approximately 4 µm long, bordered by thickened ridges, set in a subcircular translucent area of cuticle. Anal slit less than 4 µm long, positioned towards ventral side of a similar subcircular area. Newly formed cysts with conspicuous subcrystalline layer. J2 with well-developed projecting anteriorly basal knobs. Conspicuous hyaline region at least twice as long as stylet, distal third of tail tapering, ending in a rounded point.

Punctodera punctata differs from other *Punctodera* species by the pear-shaped cysts and the absence of bullae.

4.4.4. Genus *Cactodera* Krall & Krall, 1978

Mature females and cysts are lemon-shaped to spherical, with posterior protuberance. Vulva terminal, vulval slit <30 μm, fenestra circumfenestrate. Anus without fenestration. Bullae and underbridge absent, vulval denticles usually present. Cuticle with D-layer. The eggs are usually retained within the cyst body. The eggshell surface may be covered by tiny punctations. The surface structure of eggs is important diagnostically within *Cactodera* as some species have smooth surfaced eggs, e.g. *Cactodera weissi* and *C. amaranthi*, whilst others, such as *C. milleri*, *C. eremica* and *C. thornei*, are punctuated. J2 have a lateral field with four incisures, phasmid openings punctiform. The type species of the genus is *C. cacti*.

4.4.4.1. Cactus cyst nematode, Cactodera cacti *(Filipjev & Schuurmans Stekhoven, 1941) Krall & Krall, 1978*

A cyst nematode infecting cacti, *Discocactus akkermannii* and *Cereus speciosus*, both of which were expressing declining symptoms, was first recorded and described from Maartensdijk, near Utrecht, The Netherlands. The cactus cyst nematode is distributed worldwide, mainly on plants of the family Cactaceae grown in glasshouses as ornamentals. The dispersal of *C. cacti* from native regions in the Americas is beyond doubt associated with the international trade of infested ornamental cactus plants around the world. The cactus cyst nematode has been associated with or found to infect plants belonging to three families: Cactaceae (Caryophyllales): *Cereus*, *Cleistocactus*, *Coryphantha*, *Discocactus*, *Echinocactus*, *Echinopsis*, *Echinocereus*, *Epiphyllum*, *Gymnocalycium*, *Hatiora*, *Heliocereus*, *Hylocereus*, *Leuchtenbergia*, *Mammillaria*, *Melocactus*, *Notocactus*, *Nopalea*, *Notocactus*, *Opuntia*, *Oreocereus*, *Rebutia*, *Rhipsalis*, *Schlumbergera*, *Selenicereus*, *Thelocactus*; Umbelliferae (order Apiales): *Apium*; and Euphorbiaceae (order Malpighiales): *Euphorbia*. Infected plants may exhibit various symptoms including branched roots and increased numbers of rootlets. Plants become reddish-brown to yellow in colour, wilted and stunted, with reduced flower production and shortening of the flowering period. With high infection the host may die.

DESCRIPTION.

Cysts: L = 328–780 μm; W = 240–598 μm; L/W ratio = 1.1–2.0; fenestral diameter = 16–48 μm.

Male: L = 910–1113 μm; stylet = 22–29 μm; spicules = 30–37 μm; gubernaculum = 10–15 μm.

J2: L = 344–584 μm; stylet = 21–26 μm; hyaline region = 12–21 μm; tail = 34–60 μm.

Female body lemon-shaped to almost spherical, pearly white, yellow or golden, maturing to light brown. Cyst usually lemon-shaped, but may be rounded with protruding neck and vulva, light or medium brown, sometimes reddish-brown. Vulval denticles generally present, visible beneath fenestral surface. Cone tops abullate, circumfenestrate. J2 vermiform, body tapering anteriorly and posteriorly. Tail tapering, with hyaline region often shorter than stylet. Eggshells heavily punctuate as seen under optical microscope with oil immersion.

Cactodera cacti resembles *C. weissi*, *C. acnidae*, *C. milleri* and *C. galinsogae*. It differs from *C. weissi* and *C. acnidae* in having eggshells heavily punctate versus shells without visible markings, and J2 with larger stylet. It differs from J2 of *C. galinsogae* by a longer tail and from *C. milleri* by cysts with a larger fenestral diameter.

4.4.5. Genus *Dolichodera* Mulvey & Ebsary, 1980

Females and cysts: body elongate to oval, without terminal protuberance, white, swollen part 400–500 µm long, 140–270 µm wide, 2–2.8 times as long as wide, neck moderately long. Cuticle not annulated but with fine irregular striae. Vulval area terminal or just subterminal, circumfenestrate, fenestra approximately 20 µm in diameter, bullae present, perineal tubercles absent. Anus pore-like, lacking a fenestra, located 10–13 µm dorsal to vulval fenestral margin. Cyst with several large bullae. Perineal tubercles absent. Vulva circumfenestrate, underbridge absent. Anus lacking fenestra. Male not found. J2 with long tail (95–120 µm). Lateral field with three incisures, inner one faint. Labial region hemispherical, offset, with two annuli. Tail tip narrowly rounded. Phasmid openings lacking a lens-like ampulla, located about one anal body diameter posterior to anus. The type species of the genus is *Dolichodera fluvialis* Mulvey & Ebsary, 1980, parasitizing *Spartina pectinata*.

4.4.6. Genus *Betulodera* Sturhan, 2002

Cysts are lemon-shaped, pear-shaped or spheroid with insignificant, obtuse vulval cone. Cyst wall thick, with irregular network-like pattern, D-layer absent (no punctations in inner, deeper layers of cyst wall), subcrystalline layer heavily developed. Vulva terminal, surrounded by circumfenestration, vulval slit short (<10 µm), underbridge absent, denticles occasionally present, anus without fenestration. Male body twisted, no cloacal tube, spicules with bifid distal tips, phasmid openings punctiform. J2 has lateral field with three incisures, phasmid openings punctiform, without lens-like structure, labial region with three or four labial annuli and labial disc fused with submedial lips. The type and only species: *Betulodera betulae* (Hirschmann & Riggs, 1969) Sturhan, 2002.

4.4.7. Genus *Paradolichodera* Sturhan, Wouts & Subbotin, 2007

Mature female and cyst are elongate to ovoid, with rounded posterior end. Cuticle transparent, with faint transverse striations on anterior part of body and posteriorly mostly with faint irregular ridges superimposed on distinct punctations. Cuticle turning yellowish to light brown on death, covered by a subcrystalline-like film. Eggs retained in body, egg sac not observed. Labial disc squarish. Stylet well developed. Vulva terminal or subterminal, vulval slit short, circumfenestrate. Anus lacking fenestration. Male body not twisted, lateral field with four incisures. Cloacal tube present, spicules rounded at tip. Phasmids lacking. J2 long, extremely slender for family, lateral fields indistinct. Stylet short (<20 µm). DGO located more than half stylet length posterior to stylet base, pharyngeal glands long, filling body cavity. Tail long, slender, phasmid openings punctiform. Type and only species: *Paradolichodera tenuissima* Sturhan, Wouts & Subbotin, 2007.

4.4.8. Genus *Vittatidera* Bernard, Handoo, Powers, Donald & Heinz, 2010

Cysts are orange–brown to brown, lemon-shaped with short necks and vulval cone. Vulval aperture circular to rhomboid, circumfenestrate, with irregular denticle-like protuberances around the periphery of orifice. Bullae, vulval bridge and vulval underbridge absent. It resembles representatives of the genera *Cactodera* and *Betulodera* in having lemon-shaped cysts and circumfenestrate vulval area. Male variable length, stylet knobs rounded. J2 having conoid tail with narrowly rounded tip, phasmid apertures pore-like. Stylet length <18 µm. Stylet knobs rounded. Lateral field with four incisures. Eggshell smooth. The type and only species is *Vittatidera zeaphila* Bernard, Handoo, Powers, Donald & Heinz, 2010, parasitizing maize in north-western Tennessee, Indiana and Kentucky, USA. *Zea mays* and *Eleusine indica* are presently the known hosts for this species.

4.5. Pathotypes and Races

As resistant varieties were increasingly developed as a means of controlling cyst nematodes in several major crops (see Chapter 15), it became apparent that genetic variation existed within populations able to overcome such resistance (Cook and Rivoal, 1998). This led to the growing realization that within cyst nematode species that are morphologically identical, distinct virulent strains occur. Various pathotype schemes for the major cyst nematodes were proposed, with 'pathotype' being regarded as a group of individual nematodes with common gene(s) for (a)virulence and differing from gene or gene combinations found in other groups. Three cyst nematode groups have been extensively studied and pathotype schemes proposed, which are all based on the ability (or inability) of populations within each species to reproduce on a range of 'differential' host plants; these three groups are PCN (*G. rostochiensis* and *G. pallida*), cereal cyst nematode (*H. avenae*) and the soybean cyst nematode (*H. glycines*).

Two pathotype schemes for *G. rostochiensis* and *G. pallida* were proposed in 1977 that described the virulence of populations from Europe and South America (Table 4.4). In the pathotype/differential clone interactions, susceptible (+) indicated a multiplication rate (P_f/P_i) >1.0, and resistant (–) indicated a P_f/P_i <1.0, where P_i and P_f are the initial and final population sizes, respectively. This standardized disparate national schemes, especially those used within European countries, but it soon became clear that environmental

Table 4.4. Pathotype groups of potato cyst nematodes, *Globodera rostochiensis* and *G. pallida*. (Adapted from Cook and Noel, 2002.)

	Species	*G. rostochiensis*					*G. pallida*						
	Globodera species virulence group[a]	Ro1		Ro3		Ro5	Pa1			–	–	Pa2/3	
	European pathotypes[b]	Ro1	Ro4	Ro2	Ro3	Ro5	Pa1		–	–	Pa2	Pa3	–
Species and accession	South American pathotypes[c] Ploidy, resistance gene	R1A	R1B	R2A	R3A	–	P1A	P1B	P2A	P3A	P4A	P5A	P6A
Solanum tuberosum ssp. *tuberosum*	4×, (minor)	+/–	+	+	+	+	+	+	+	+	+	+	+
S. tuberosum ssp. *andigena* CPC 1673	4×, *H1* on chromosome 5	–	–	+	+	+	+	″	″	″	+	+	″
S. kurtzianum KTT 60.21.19	2×, *K1 K2* A and B	–	(+)	–	(+)	(+)	+	+	+	+	+	+	+
S. vernei GLKS 58.1642.4	2×, quantitative	–	+	–	–	+	+	+	+	–	+	+	+
S. vernei Vt 62.33.3	2×, quantitative		–	–	–	+	–	+	–	–	–	+	+
ex. *S. multidissectum* hybrid P55/7	2×, 1 + polygenes *H2*	+	+	+	+	+	–	–/+	–	+	+	+	+
S. tuberosum ssp. *andigena*	*H3* + polygenes	+	″	″	″	″	(–)	″	″	″	(–)	(–)	″
CIP 280090.10	Quantitative	″	″	″	″	″	″	″	″	″	–	–	+
S. vernei hybrid 69.1377/94	2×, polygenes	–	–	–	–	–	–	″	″	″	–	–	″
S. vernei hybrid 63.346/19	2×, polygenes	–	–	–	–	–	+	″	″	″	+	+	″
S. spegazzinii	2×, Fa = *H1*	–	–	+	+	+	″	″	″	″	″	″	″
S. spegazzinii	2×, Fb + 2 minor	+	–	+	–	–	″	″	″	″	″	″	″
	Glo1 on chromosome 7	(–)	+	+	+	+	″	″	″	″	″	″	″

[a]Trudgill (1985); [b]Kort *et al.* (1977); [c]Canto-Saenz and de Scurrah (1977).

Note: + = compatible interaction: nematode multiplication, potato susceptible; – = incompatible interaction: nematode no multiplication, potato resistant; () = partial or uncertain interaction; ″ = no information.

influences and the extensive heterogeneity of some populations, especially those of *G. pallida*, caused problems. Populations in the centres of origin of the two species in South America are more heterogeneous in virulence characteristics than those introduced and dispersed in the rest of the world. Some populations are relatively homozygous for virulence, e.g. Ro1 (R1A) and Pa1 (P1A). Others, such as most *G. pallida* populations, are heterogeneous and give varying results; thus, these populations cannot reliably be described as pathotypes and are increasingly referred to as virulence groupings within pathotypes. Such virulence is not a stable scenario but will change according to environmental conditions. For example, intensive cultivation of starch potatoes with resistance to pathotype Pa2/3 in Central Europe resulted in increased virulence that finally overcame the resistance (Mwangi *et al.*, 2019). Besides its virulence, this population also had a higher fitness on susceptible cultivars. The virulence remained unchanged even after several propagations on a susceptible cultivar, indicating that there was no trade-off following selection in the field.

The pathotype scheme for cereal cyst nematodes (*H. avenae*) is based on their multiplication on host differentials of barley, oats and wheat crops, with the major division into three pathotype groups based upon reactions of the barley cultivars with known resistance genes (*Rha1*, *Rha2*, *Rha3*). Each pathotype group is further subdivided by their reactions on other differentials. Resistance is defined as fewer than 5% new females on susceptible controls (Table 4.5). As with the PCN pathotype schemes, because the genetics of field populations are largely unknown and variability exists within them, the term 'virulence phenotype' has been proposed. Evidence suggests that the different species of cereal cyst nematodes (*H. avenae*, *H. filipjevi* and *H. australis*) have populations with different virulence phenotypes. There is limited evidence for loss of effectiveness of resistance genes used in widely grown cultivars and part of the reason for this may be that endemic biological control develops when cereals are intensively cultivated in moist temperate soils and, therefore, selection pressures for virulent strains are reduced.

Differences in virulence between soybean cyst nematode (*H. glycines*) populations are referred to as races rather than pathotypes. Such differences were recognized during breeding programmes in the USA for resistant soybean varieties. Using four soybean differentials ('Pickett', 'Peking', 'PI 88788' and 'PI 90763'), 16 races were characterized (Table 4.6). A resistant response (avirulence) is defined as a Female Index of <10% of that obtained on the susceptible cultivar 'Lee'. Although the soybean cyst nematode race scheme predicts whether a cultivar will control the nematode population within a field in a particular season, as with other schemes, it cannot predict the consequences of selection pressure with a heterogeneous population. According to recent studies, virulence is increasing, especially in dominating races (Lian *et al.*, 2022).

Despite the limitations of the various pathotype/race schemes for cyst nematodes, providing their limitations are recognized, they continue to give a useful indication of the virulence characteristics of particular nematode gene pools. As such, they can provide critical information necessary for effective management and the emergence of new virulent strains.

Variability in virulence up to pathotype formation is also known for other cyst nematodes, such as the beet cyst nematode *H. schachtii*. Control of this pest is mainly accomplished by growing resistant fodder radish or mustard cover crops, but also resistant sugar beet cultivars. While for cover crops the resistance seems to be very stable and no virulent populations have yet been reported, pathotype formation does occur in sugar beet. *Heterodera schachtii* populations that carry the virulence factor for $Hs1^{pro\text{-}1}$ are selected under

Table 4.5. Pathotype groups of cereal cyst nematodes, *Heterodera avenae, H. filipjevi* and *H. australis*. (Adapted from Cook and Rivoal, 1998.)

Species	*H. avenae*								*H. australis*	*H. filipjevi*	
Pathotype group	Ha1 group							Ha2 group	Ha3 group		
Pathotype	Ha11	Ha21	Ha31	Ha41	Ha51	Ha61	Ha71	Ha12	Ha13	Ha23	Ha33
Different species and cultivar											
Barley											
Varde	+	″	″	+	″	+	+	+	+	+	+
Emir	+	+	″	+	″	–	+	+	+	+	+
Ortolan/Drost	–	–	–	–	–	–	–	+	+	+	+
Morocco	–	–	–	–	–	–	–	–	–	–	–
Siri	–	–	–	+	+	+	–	–	+	+	+
KVL 191	–	–	–	″	+	+	+	–	″	″	″
Bajo Aragon	–	″	″	–	″	–	–	–	+	+	+
Herta	+	+	–	″	–	″	–	+	+	″	″
Martin 403	–	″	″	–	″	–	–	–	–	+	+
Dalmatische	(–)	″	″	+	″	–	(+)	+	+	(–)	+
La Estanzuela	″	″	″	″	″	″	+	″	″	(–)	″
Hartian 43	–	″	″	″	″	″	–	–	″	–	+
Oat											
Nidar	+	″	″	(+)	″	+	–	+	+	+	+
Sol II	+	–	–	–	–	+	–	+	+	+	+
Pura Hybrid BS1	–	–	″	–	–	–	–	–	+	–	+
Avena sterilis 1376	–	–	″	–	–	–	–	–	–	–	–
Silva	(–)	″	″	–	″	(–)	–	(–)	(–)	(–)	+
IGV.H. 72–646	–	″	″	–	″	–	–	–	+	+	+
Wheat											
Capa	+	+	″	+	″	+	+	+	+	+	+
AUS10894	–	″	″	–	″	–	+	–	(–)	+	+
Loros	–	–	″	–	″	(–)	–	–	(–)	+	+
Psathias	″	″	″	+	″	″	″	+	+	+	–
Iskamish K-2-light	+	″	″	–	″	(–)	″	+	+	+	+

Note: + = susceptible; – = resistant (<5% new females compared to numbers on susceptible control); () = intermediate; ″ = no information

Table 4.6. Races of soybean cyst nematode, *Heterodera glycines*. (Adapted from Cook and Rivoal, 1998.)

Differential cultivar	Race	3	6	13	9	1	5	11	2	8	10	12	14	7	15	1	4
	Virulence phenotype	0	1	2	3	4	5	6	7	8	9	10	11	12	13	14	15
'Pickett'		–	+	–	+	–	+	–	+	–	+	–	+	–	+	–	+
'Peking'		–	–	+	+	–	–	+	+	–	–	+	+	–	–	+	+
PI 88788		–	–	–	–	+	+	+	+	–	–	–	–	+	+	+	+
PI 90763		–	–	–	–	–	–	–	–	+	+	+	+	+	+	+	+

Note: – resistant (female index <10% cultivar 'Lee'); + susceptible (female index >10% that of susceptible control cultivar 'Lee').

repeated cultivation of resistant sugar beet cultivars and finally end up in a resistant-breaking pathotype. But also growing resistant cover crops will affect nematode virulence, as they usually allow for some nematode reproduction. Thus, the frequency of growing resistant cover crops will determine the selection pressure and shape the virulence of the nematode population.

4.6. Biochemical and Molecular Diagnosis

Biochemical approaches such as gel electrophoresis for separating protein and enzyme profiles have shown great potential for helping to identify cyst nematodes. Isoelectric focusing (IEF) is currently used in nematology laboratories for routine diagnostics of *G. pallida* and *G. rostochiensis*. By comparison with the biochemical approaches, analysis of DNA for diagnostics has several advantages. The main DNA regions targeted for diagnostics of cyst nematodes are the ITS1 and ITS2, which are situated between 18S and 5.8S and 5.8S and 28S rRNA genes, respectively. Genes of mtDNA, with their relatively higher rate of mutations relative to rRNA genes, have great potential for diagnostics of races and populations of cyst nematodes. PCR-RFLP and PCR with species-specific primer(s) are currently used for diagnostics of many cyst nematode species (Subbotin *et al.*, 2021; see Chapter 2).

4.7. Interactions with Other Plant Pathogens and Beneficial Microorganisms

Plant-parasitic nematodes interact with other soil organisms. This also applies for cyst nematodes. Those interactions can be additive, synergistic or antagonistic. Although the focus of this section is on the interaction of cyst nematodes with other plant pathogens, other interactions are also of relevance and therefore will be briefly mentioned. Plant mutualistic symbionts (e.g. rhizobia, arbuscular mycorrhiza fungi) compete with cyst nematodes for energy supply. In the case of rhizobia, infestation with cyst nematodes generally impairs nodule formation and nitrogen fixation (Lopez-Nicora and Niblack, 2018). In the case of arbuscula mycorrhiza fungi, the interaction with cyst nematodes is generally more variable depending on mycorrhizal strain, crop plant and growth conditions. Effects can reach from reduced mycorrhization rates in the presence of the nematode all the way to increasing plant tolerance towards nematode damage. Further highly relevant players in such types of interactions are other nematodes. Nematode communities usually contain many species that may potentially interact with each other. In most instances, where cyst nematodes are present, they are antagonistic to other plant-parasitic species. Examples of this include *G. tabacum* and *Pratylenchus penetrans* on tobacco, *H. avenae* and *P. neglectus* on wheat, *H. cajani* and *Helicotylenchus retusus* in pigeon pea, and *H. glycines* and *Meloidogyne incognita* on soybean. Very few examples of neutral or stimulatory responses between cyst and other nematode species have been documented. The potential interactions between cyst nematodes and bacteria, insects and other pests have received only limited study. One notable exception is that of *G. pallida* and the bacterium *Ralstonia solanacearum* on potato, in which the nematode enhances damage caused by the associated wilt.

The reality is, however, more complex than studying two players alone as all organisms within a certain environment interact with each other. Thus, future approaches need to focus more on the entire community of plant-associated organisms, including nematodes, bacteria, fungi and other soil organisms and their interactions (Helder and Heuer, 2021). Modern molecular tools are powerful enough and ready to explore large datasets on nematode interactions with other organisms within the soil microcosm.

As with all plant-parasitic nematodes, the mechanism of feeding on plant tissue results in wounding and provision of entry sites for other pathogens (Barker and McGawley, 1998). However, more specific associations, which can result in either synergistic or antagonistic responses by the host plant, demonstrate that more complex interactions have evolved. Most investigations of the interrelationships of cyst nematodes and other plant parasites have focused on those with fungi, especially those causing wilt and root rot (Table 4.7). A number of cyst nematode species interact with *Fusarium* wilt species, causing the wilt disease to be more severe than in the absence of the nematode, e.g. *G. tabacum* on tobacco, *H. cajani* on pigeon pea and *H. glycines* on soybean. Generally, these interactions involve synergism with regard to disease development but often result in restricted nematode reproduction because of the associated root damage. Only limited examples of interactions between cyst nematodes and root-rot fungi (*Rhizoctonia* spp.) have been documented, but those that have generally show an enhancement of the disease in the presence of the nematode, e.g. *G. rostochiensis* on potato, *H. avenae* on wheat, *H. glycines* on soybean and *H. schachtii* on sugar beet. Although such associations usually have an adverse effect on the host plant, the interaction of *H. cajani* with *Rhizoctonia bataticola* suppresses the associated damage caused by the fungus.

The economic effect of these interactions varies but their effect can be important with major high-value crops such as soybean. A major disease of soybean is 'sudden death syndrome' (SDS) caused by *Fusarium solani*, which is sometimes associated with the presence of *H. glycines*. Although experimental data indicate that the nematode is not necessary for the development of SDS, field observations have shown that soybean cultivars resistant to *H. glycines* were less affected by SDS than susceptible ones. The decreased *H. glycines* population levels correspond to the restriction in root and shoot growth attributable to the additive root damage by the fungus and nematode.

Thus, cyst nematodes show a wide range of important interrelationships with associated organisms in a wide range of habitats. As such, full evaluations of host–parasite relationships should be undertaken in the presence of other pathogens likely to be present in their natural habitat. Such evaluations are needed for effective management strategies.

Table 4.7. Summary of interrelationships of selected cyst nematodes of the genera *Globodera* and *Heterodera* and plant-pathogenic fungi. (Adapted from Barker and McGawley, 1998.)

Nematode	Associated fungi	Host	Comments
G. rostochiensis	*Rhizoctonia solani*	Potato	Yield loss, but only small interaction effect
	Pyrenochaeta lycopersici	Tomato	Fungus probably prevents syncytium formation by the nematode
G. rostochiensis and *G. pallida*	*Verticillium dahliae*	Potato	Results in 'early dying disease'
G. tabacum	*Fusarium oxysporum*	Tobacco	Wilt disease enhanced
	Fusarium oxysporum f. sp. *lycopersici*, *Verticillium albo-atrum*	Tomato	More *Verticillium* but less *Fusarium* wilt in the presence than absence of *G. tabacum*
H. avenae	*Rhizoctonia solani*	Wheat	In combination, greater reduction in tillering, height, weight, root number and length than with either alone
	Gaeumannomyces graminis	Wheat	Antagonism
H. cajani	*Fusarium udum*	Pigeon pea	Wilt enhanced
	Rhizoctonia bataticola	Cowpea	Damage suppressed
H. glycines	*Rhizoctonia solani*	Soybean	Limited nematode reproduction
	Calonectria crotalariae		Enhanced activity of both parasites
	Phytophthora megasperma var. *sojae*		Increased seedling disease (additive)
	Fusarium oxysporum		Increased wilt
	Fusarium solani		Variable; increased foliar symptoms of fungus; suppressed nematode reproduction
	Macrophomina phaseolina		Synergism
H. oryzicola	*Sclerotium rolfsii*	Rice	Variable, but usually synergistic
H. schachtii	*Fusarium oxysporum*	Sugar beet	Damage was less when fungus present; fungus inhibited nematode invasion
	Pythium ultimum		Synergistic or damping off when in combination
	Pythium aphanidermatum		Additive
	Pythium solani		Synergism
	Rhizoctonia solani		Synergism, especially at high inoculum levels

4.8. Management

Management of cyst nematodes presents a unique problem in the way that many or all of the eggs are produced inside the female body that, upon death, becomes a cyst with a hardened protective wall. This structure is resistant to invasion by potential parasites and protects the eggs inside from rapid desiccation, enhancing their ability to remain dormant for many years. In many cases (e.g. *Globodera* spp.) substantial hatch will only occur in the presence of a hatching factor produced by a potential host (see Section 4.3 and Chapter 7), so that any management strategy must be effective over a period of years or usable year after year. However, unlike root-knot nematodes (see Chapter 3), cyst nematodes have a relatively narrow host range, making appropriate crop rotation a viable option in most situations.

4.8.1. Prevention

Fundamental to the prevention of cyst nematodes spreading into non-infested regions is the use of certified planting material, and strict legislation for those commodities being traded both internationally and locally. This tactic has been the mainstay for controlling several major pests such as *G. rostochiensis, G. pallida* and *H. schachtii* (see Chapter 13). Efficient management and containment of an infestation may be compromised by the ease with which cysts can be dispersed by, for example, wind, in small aggregates of soil, on small roots attached to other parts of plants, by flood water run-off, or by adhering to machinery or animals passing through infested land. General hygiene practices should be adopted in higher risk situations when the pest is known to be present in the locality. Such measures would include cleaning machinery both before and after use, restricting movement of soil outside the field boundary and construction of natural wind breaks.

4.8.2. Crop rotation

In comparison to other plant-parasitic nematodes, cyst nematodes have a limited host range and, therefore, crop rotation is the economically and ecologically most important component for their management. Alternative non-host crops can safely be cultivated, during which time a combination of spontaneous hatch and natural mortality will reduce the field population to below threshold levels (see Chapter 12). However, weeds should be carefully controlled to avoid nematode survival on alternate hosts. Cyst nematodes that have only one to three cultivated host plants include *G. rostochiensis, G. pallida* (potato, eggplant and tomato), *H. avenae* (oat, barley and wheat), *H. zeae* (cultivated and wild maize) and *H. carotae* (cultivated and wild carrot). Even those cyst nematodes with broader host ranges, such as *H. schachtii* and *H. glycines*, have relatively few cultivated hosts, facilitating the potential for control by use of rotations. Further information on the host status and reproduction potential of a given nematode species can be retrieved from the Best4Soil support tool for nematodes (available at: https://www.best4soil.eu).

4.8.3. Resistance

Next to crop rotation, cultivar resistance remains the most economical practice for managing cyst nematodes (see Chapter 15), although these are not always available. Resistance of major crop hosts to *G. rostochiensis*, *G. pallida*, *G. tabacum tabacum*, *G. tabacum solanacearum*, *H. avenae*, *H. glycines*, *H. schachtii* and *H. cajani* have been found and attempts made to incorporate it into commercial cultivars. Only low level, or no, resistance is known in the major crop hosts of *H. cruciferae*, *H. oryzae*, *H. sacchari* and *H. oryzicola*. However, in many cases resistance is found only in wild species, with the accompanying inherent difficulties of transferring into commercial cultivars (Blok *et al.*, 2018). The inappropriate continuous planting of resistant cultivars is now known to increase selection pressure for virulent populations (e.g. potatoes and *G. rostochiensis* and *G. pallida*), limiting the durability of resistance in some cultivars, or resulting in the increase of other nematode problems (e.g. cereal cyst nematode with the associated build-up of *Pratylenchus neglectus*).

4.8.4. Tolerance

If resistant cultivars are not available, tolerant cultivars or crops might be an alternative. Tolerance (see Chapter 15) helps securing yield and thus generating income for the farmer. However, tolerance allows nematode reproduction and, as a result, the nematode population increases. The question then is what nematode densities are still acceptable before yield declines in the long term? Use of tolerant cultivars is a major issue for managing *H. schachtii* on sugar beets since resistant cultivars are not economic owing to their lower yield potential at low to medium nematode levels and susceptible cultivars fail at higher nematode densities. Currently available tolerant sugar beet cultivars show similar or even higher yield potential than susceptible cultivars even under non-infested conditions. For this reason, tolerant varieties are grown almost exclusively today in some regions.

4.8.5. Biological control

As early as 1877, Kühn indicated the enormous importance of natural antagonists occurring in the soil for the control of cyst nematodes. He observed numerous cysts whose eggs were parasitized by a fungus, which he described as *Tarychium auxiliarium*, today *Catenaria auxiliarum*. Cyst nematodes appear to be the perfect target for the use of biological agents in their management. Eggs of cyst nematodes are contained either inside the female's body/cyst or in a gelatinous sac attached to the cyst, so they are easily exposed to parasitism by fungi or bacteria. Most cyst nematodes have prolonged diapause stages making those eggs especially suitable to antagonism. Numerous studies using nematophagous fungi and bacteria against economically import cyst nematodes have been undertaken with varying degrees of success and failure (Stirling, 2014; see Chapter 14). This research has resulted in several biocontrol products that are now commercially available. Active compounds include, among others, the fungi *Purpureocillium lilacinum*, *Pochonia chlamydosporia* and *Myrothecium verrucaria* or the bacteria *Bacillus firmus* and *Pasteuria nishizawae* (Davies *et al.*, 2018). Although the effectiveness of such biological agents is generally lower than that of chemical nematicides, they are increasingly of interest, especially in sustainable or organic farming systems. Besides application of nematode antagonists, a different approach makes use of the naturally occurring antagonists in the soil to be managed in a manner to suppress the cyst nematodes. Examples of soil suppressiveness toward cyst nematodes are well documented for the control of *H. avenae*, *H. glycines* and *H. schachtii*. In case of *H. glycines*, suppressiveness was induced by continuous growing soybean for more than 15 years. However, suppressiveness failed to develop when soybean was rotated with wheat or maize (Wei *et al.*, 2015). Those suppressive soils were associated with higher levels of antagonistic fungi and bacteria such as *P. chlamydosporium*, *Nematophora gynophila*, *P. lilacinum*, *Trichoderma harzianum* or *Pasteuria penetrans* (see Chapter 14). Thus, different microbial antagonists operate in concert to control the cyst nematodes.

4.8.6. Chemical control

Chemical nematicides are an important tool in integrated nematode management (see Chapter 17) that can be combined with other tools such as crop rotation, resistant or tolerant cultivars or resistant cover crops (see Chapter 15 for details of resistance). Modes of action vary from acetylcholinesterase inhibitors such as carbamates and organophosphates, interfering with nematode nerve and muscles activity, to inhibitors of the mitochondrial complex II electron transport such as cyclobutrifluram or fluopyram, interfering with nematode respiration. In other cases, the mode of action is still unknown (e.g. fluensulfone, fluazaindolizine). Over the past decades, several chemical nematicides, especially of the carbamate and organophosphate group, have been phased out for human and environmental safety reasons. The new generation of chemical nematicides have a much better safety profile and required application rates are low, ranging from 400 g to 4 kg ha^{-1}. Furthermore, toxicity to non-target organisms is often negligible. However, if used multiple times on the same site, one should be aware that the effectiveness of nematicides might be reduced due to biological degradation by soil organisms. In general, species that produce several generations in a year, such as *H. glycines*, appear to be more difficult to control by chemical nematicides than cyst nematodes that produce only one or two generations a year, i.e. *G. rostochiensis* and *G. pallida* on potatoes, *H. goettingiana* on peas, and *H. avenae* on cereals.

4.8.7. Other methods

In general, any agricultural practice or amendment will affect cyst nematodes in one way or another, but only few are suitable for sustainable control of nematodes. For example, no- tillage tends to reduce damage by cyst nematodes, as shown for *H. avenae* (Roget *et al.*, 1996) on wheat and *H. elachista* on rice (Ito *et al.*, 2015), but effects are often inconsistent. Other methods promise much better control, such as anaerobic soil disinfection and inundation. Both methods rely on oxygen deficiency and toxic substances for killing the nematodes. In the case of anaerobic soil disinfestation, these conditions are achieved by incorporating fresh organic matter in moist soil followed by coverage with airtight plastic for several weeks, whilst for inundation the soil is flooded for a prolonged period of time (Runia *et al.*, 2014). Although both methods are commonly used in praxis, documented evidence on nematodes control is scarce. None the less, anaerobic soil disinfection as well as inundation were shown to reduce *Globodera* significantly, with the latter method more effective than the first one (Shrestha *et al.*, 2016). Weed management, biofumigation, solarization and organic amendments are further management tools that might support cyst nematode control. Their potential use and relevance have to be decided on a case-by-case scenario.

4.8.8. Integrated nematode management

The repeated use of a single control measure is likely to fail, sooner or later, from the selection of virulent biotypes, accelerated microbial degradation of nematicide or the selection of populations better able to overcome such management programmes; in general, selection of individuals unaffected by any control measure that may be applied will occur. The potential for managing cyst nematodes by combining two or more, most likely complementary, control strategies in an integrated programme has been widely demonstrated (Sikora *et al.*, 2021). Common control strategies might focus on nematode prevention (e.g. seed certification, quarantine regulation), reducing nematode populations (e.g. resistant cultivars, non-host plants, biological or chemical control), improving plant tolerance (e.g. tolerant cultivars, biostimulants) and/or using supportive tactics (e.g. nematode inventory, decision-support systems, remote sensing) (Desaeger *et al.*, 2021). The advantage of an integrated approach that includes partially effective strategies is seen in the better protection of highly effective ones from nematode adaptation or environmental risk. Examples include integrated control of *G. rostochiensis* and *G. pallida* in Europe, and *H. glycines* in the USA (Roberts, 1993).

5 Migratory Endoparasitic Nematodes*

ANTONIO ARCHIDONA-YUSTE, PABLO CASTILLO AND JUAN E. PALOMARES-RIUS**

Institute for Sustainable Agriculture, Department of Crop Protection, Córdoba, Spain

*A revision of Duncan, L.W. and Moens, M. (2013) Migratory endoparasitic nematodes. In: Perry, R.N. and Moens, M. (eds) *Plant Nematology*, 2nd edn. CAB International, Wallingford, UK.

**Corresponding author: palomaresje@ias.csic.es

DOI: 10.1079/9781800622456.0005

5.1. Introduction to Migratory Endoparasitic Nematodes

Migratory endoparasitic nematode species are among the top 10 economically significant plant-parasitic nematodes (Jones *et al.*, 2013). All life stages of migratory endoparasitic nematodes can be found within plant tissues. Unlike sedentary endoparasites, the nematodes in this group do not induce permanent feeding sites, but instead feed and reproduce while migrating between or through plant cells. The symptoms caused by migratory endoparasites vary. Some species are known for enzymatic degradation of host tissues, others for inducing hormonal imbalances that cause galling, swelling and other tissue distortions, and some cause visible lesions when phenolic compounds are released in and around wounded cells. Species in three nematode families are adapted to migratory endoparasitism. Those in the Pratylenchidae inhabit primarily cortical cells in roots and other below-ground plant parts. Species in the Anguinidae and Aphelenchoididae are unique among nematodes in parasitizing above-ground parts of plants, often exhibiting impressive survival and dispersal behaviours that permit them to live within stems, leaves, buds and seeds (see Chapter 7).

5.2. The Pratylenchids: Lesion, Burrowing and Rice Root Nematodes

Pratylenchus (root-lesion nematodes), *Radopholus* (burrowing nematodes) and *Hirschmanniella* (rice root nematodes) are among 11 genera in the family Pratylenchidae, and belong to the subfamilies Pratylenchinae, Radopholinae and Hirschmanniellinae, respectively (see Chapter 1; Box 5.1). All are obligate plant parasites, but economically important species of these subfamilies are known only in the three genera treated here. Quite a large number of morphological features differentiate these taxa but two broad groups can be distinguished. In *Pratylenchus, Apratylenchus*, *Hirschmanniella* and *Zygotylenchus* there is no pronounced sexual dimorphism and the pharyngeal glands overlap the intestine ventrally, whereas in *Radopholus* and the remaining genera (*Apratylenchoides*, *Achlysiella*, *Pratylenchoides*, *Hoplotylus*, *Zygradus*), male and female anterior morphology differs distinctly and the pharyngeal overlap is dorsal (Fig. 5.1). These morphological differences suggest that the family is polyphyletic. Molecular analyses have also shown a closer relationship between *Radopholus similis* and ectoparasitic and cyst-forming species in the Hoplolaimidae. Morphological similarities in the genera *Pratylenchus* and *Radopholus* probably resulted from convergence, due to their similar feeding behaviours.

Two other subfamilies occur in the Pratylenchidae, each with a single genus. Although the juvenile, male and vermiform female stages of *Nacobbus* (Nacobbinae) behave as migratory endoparasites, the group is conventionally treated among the sedentary plant-parasitic nematodes because the females eventually establish permanent feeding sites within induced root galls (Box 3.1) (Manzanilla-López *et al*., 2002). Similarly, the immature stages and males of a newly described genus, *Apratylenchus* (Apratylenchinae), resemble those of *Pratylenchus*; however, the mature female posterior assumes a distinctive club-shaped

Box 5.1. Classification of the subfamilies Pratylenchinae, Radopholinae, Hirschmanniellinae and Apratylenchinae.

Phylum Nematoda Potts, 1932
Class Chromadorea Inglis, 1983
Subclass Chromadoria Pearse, 1942
Order Rhabditida Chitwood, 1933
Suborder Tylenchina Thorne, 1949
Infraorder Tylenchomorpha De Ley & Blaxter, 2002
Superfamily Tylenchoidea Örley, 1880
Family Pratylenchidae Thorne, 1949
Subfamily Pratylenchinae Thorne, 1949
  Genus *Pratylenchus* Filipjev, 1936
Subfamily Radopholinae Allen & Sher, 1967
  Genus *Radopholus* Thorne, 1949
Subfamily Hirschmanniellinae Fotedar & Handoo, 1978
  Genus *Hirschmanniella* Luc & Goodey, 1964
Subfamily Apratylenchinae Trinh, Waeyenberge, Nguyen, Baldwin, Karssen & Moens, 2009
  Genus *Apratylenchus* Trinh, Waeyenberge, Nguyen, Baldwin, Karssen & Moens, 2009

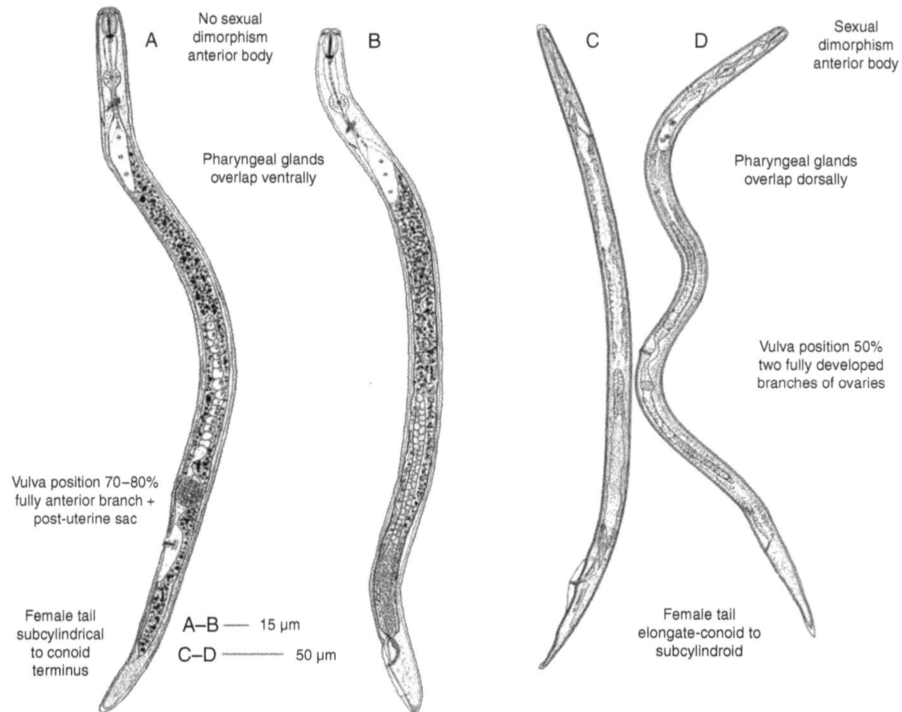

Fig. 5.1. Entire female and male body of *Pratylenchus dunensis* (A, B) and *Radopholus arabocoffeae* (C, D) showing differences in pharyngeal glands overlapping the intestine, in sclerotized lip region, in stylet morphology between male and female, in position of the vulva and in tail shape. (A, B after de la Peña *et al.*, 2006; C, D after Trinh *et al.*, 2004.)

form. Two species of *Apratylenchus* were described from roots of coffee in Vietnam but their economic significance is unknown.

5.2.1. Root-lesion nematodes, *Pratylenchus* spp.

Among the pratylenchids, *Pratylenchus* is easily the most cosmopolitan genus in terms of the variety of habitats occupied. The genus may also have the broadest host range among plant-parasitic nematodes. *Pratylenchus* is worldwide in distribution and while most species are of no or little economic importance, others are responsible for substantial yield losses in many agronomic and horticultural crops. Indeed, next to root-knot and cyst nematodes, *Pratylenchus* spp. cause the greatest crop damage by nematodes worldwide (Castillo and Vovlas, 2007).

5.2.1.1. Morphology and identification

Species of *Pratylenchus* (Luc, 1987) are stout, small to medium size nematodes with body length less than 0.9 mm and rarely strongly motile. The stylet and lips are heavily sclerotized. The lip region is usually not (or only slightly) offset from the body and is low and flattened anteriorly, with two, three or occasionally four labial annuli. The pharyngeal glands overlap the intestine ventrally for a medium distance and there is no sexual dimorphism in the anterior part of the body. The vulva is posterior and females are monoprodelphic with an anterior genital branch that may or may not contain a developed spermatheca and a posterior branch that is reduced to a post-vulval uterine sac. In species with males, the male tail is pointed with caudal alae extending to the tip. The gubernaculum is simple and does not protrude.

Of the more than 100 species in the genus, a large number are poorly described (Castillo and Vovlas, 2007; Geraert, 2013; Troccoli *et al.*, 2021). Some are known to be species complexes. *Pratylenchus* exhibits relatively little interspecific morphological variation when viewed with the light microscope. Some characters, such as tail shape, can be quite variable within a species. This makes it easy to recognize members of the genus but difficult to differentiate between species. The ranges of dimensions of most characters overlap between species, so it is often necessary to examine at least ten specimens to sample the range of morphometric and morphological variation in a population. The face morphology when viewed with scanning electron microscopy (SEM) is often useful to differentiate otherwise similar species of *Pratylenchus* (Fig. 5.2). Some characteristics for several of the more economically important species of *Pratylenchus* are given in Table 5.1.

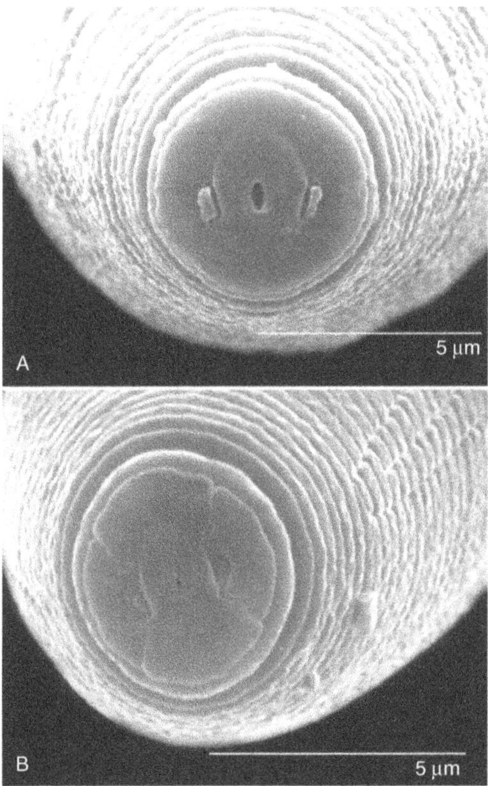

Fig. 5.2. Scanning electron micrographs of the first lip annulus face views showing a smooth pattern in *Pratylenchus coffeae* (A) isolated from citrus in Florida, and a divided pattern in *P. pseudocoffeae* (B) isolated from *Aster elliottii* in Florida.

Table 5.1. Some commonly encountered species of root-lesion nematodes, their geographic ranges, some of the crops they damage and morphological characters useful for their identification. (From Loof, 1991, and other sources.)

Species	Lip annuli	Males present	Face pattern	Additional characters	Geographic distribution and selected economically important hosts
Pratylenchus coffeae	2	Yes	Smooth	Tails of *P. coffeae* bluntly rounded, truncate or indented versus tail narrowly	Pantropic and subtropic. Banana, plantain, coffee, citrus, yam and many other crops.
P. loosi	2	Yes	Smooth	rounded for *P. loosi.*	South Asia and Japan. Tea, coffee, citrus.
P. brachyurus	2	No	Smooth	Very posterior vulva and tail with smooth rounded or truncate terminus.	Pantropic and subtropic. Peanut, potato, pineapple, peach, soybean, tobacco, coffee, rubber.
P. neglectus	2	No	Divided	*P. neglectus* vulva is usually more posterior (81–86%) than that of *P. scribneri* (72–80%).	Temperate and subtropical regions of most continents. Cereals, turf, crucifers, legumes, strawberry, tobacco, mint, maize.
P. scribneri	2	No	Divided		Widely distributed in subtropical and warmer temperate regions. Maize, tomato, sugar beet, onion, soybean, potato, bean, ornamentals.
P. penetrans	3	Yes	Divided	Smooth rounded tail terminus for *P. fallax*.	Cosmopolitan, chiefly temperate. Very wide host range. Fruit trees, conifers, ornamentals, potato, vegetables, maize, forage crops, sugar beet, fern, among many others.
P. fallax	3	Yes	Divided		Europe and Japan. Cereals, fruit trees, ornamentals, forage crops, strawberry.
P. thornei	3	No	Divided	Long vulva–anus distance and tail bluntly rounded or truncate.	Widespread in subtropical and warmer temperate regions. Cereals, legumes, fruit trees, woody and herbaceous ornamentals.
P. zeae	3	No	Smooth	Vulva 65–76% and tail tapered with sub-acute terminus in *P. zeae*	Pantropic and subtropic. Cereals, forage crops, turf, sugarcane, tobacco, peanut.
P. crenatus	3	No	Smooth	versus vulva 78–86% and crenate tail for *P. crenatus*.	Temperate, but occasionally subtropics and elevated tropics. Cereals, forage grasses, carrots.
P. vulnus	3–4	Yes	Divided	Tail is dorsally sinuate before terminus, vulva 73–76% in *P. goodeyi* versus narrowly rounded to sub-acute tail	Worldwide subtropical and Mediterranean climates. Mostly woody plants including rose, deciduous fruit and nut, citrus seedlings.
P. goodeyi	4	Yes	Divided	and vulva at 77–82% in *P. vulnus.*	East Africa, Canary Islands, Greece and Australia. Banana.

Note: Partial DNA sequences (D2–D3, ITS, *COI*, *hsp90*) for all of these species are available from GenBank (http://www.ncbi.nlm.nih.gov/gquery/gquery.fcgi).

5.2.1.2. Life cycle and behaviour

Parthenogenesis appears to be the mode of reproduction in about half of the known *Pratylenchus* species, based on absence of males and absence of sperm in the female spermatheca. Obligatory amphimixis (see Chapter 7) has been demonstrated in several of the remaining species. Eggs are oviposited at rates up to two per day, mainly in root tissue but also in soil along the root surface. Eggs are clustered in root tissue due to the gregarious behaviour of root-lesion nematodes. Eggs can also function in survival between hosts. *Pratylenchus penetrans* second-stage juveniles (J2) remain unhatched and quiescent, protected within eggs in the soil until stimulated to hatch by diffusates from roots, especially young host plant roots. This behaviour also reduces the amount of time that juveniles actively deplete energy reserves in search of food. All life stages of root-lesion nematode species can be isolated from both soil and roots, and the proportion of the population in the soil is distinctly seasonal in regions where temperature or moisture regulates root availability. As root-lesion nematodes readily migrate to fresh cortical tissue when resources become limiting, higher numbers of nematodes are often recovered from feeder roots that appear healthy than from diseased roots, suberized roots or soil.

Juvenile root-lesion nematodes sometimes feed ectoparasitically on plant root hairs, although cell death seldom occurs without repeated feeding. More commonly, all vermiform stages penetrate into and reside in the root cortex, but usually do not penetrate or feed on the endodermis or central stele. The nematodes are attracted to the zone of elongation or to root junctures. Root penetration attracts additional nematodes to the same entry site, resulting in clusters of nematodes within the root tissue. For amphimictic species this behaviour aids in locating a mate. Nematodes rupture cell walls by repeated stylet thrusting, after which they migrate through the cell or pause to feed. When feeding, the nematode secretes salivary enzymes for a short time before ingesting the cytoplasm. Feeding time may be brief during migration or may last for several hours. Cell death occurs when nematodes migrate through a cell and following a long feeding bout, but not following brief feeding (Fig. 5.3). Migration and long feeding bouts alternate with resting phases of several hours, during which the nematode coils within a cell. Between temperatures of 17°C and 30°C *P. penetrans* completed its life cycle in 548 degree days (base 5.1°C; 22–46 days) on clover roots. In general, tropical species at elevated temperatures complete a life cycle in 3–4 weeks, whereas temperate species at cooler temperatures require 5–7 weeks.

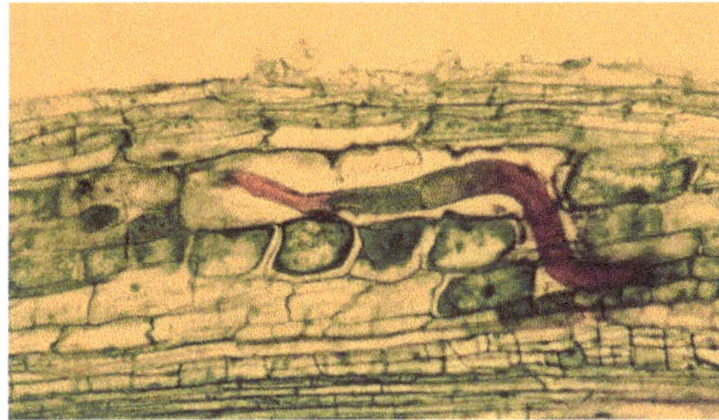
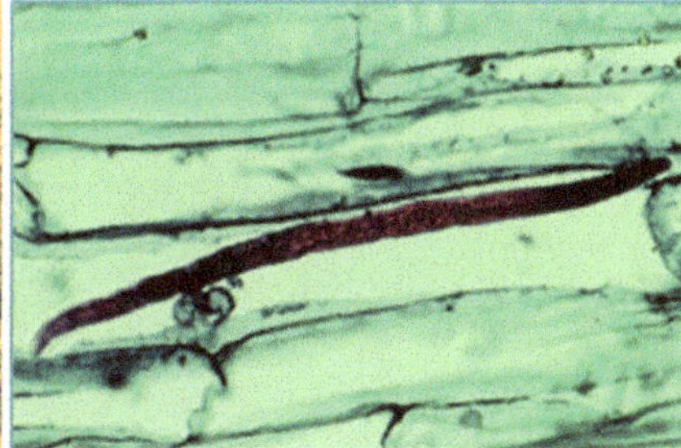

Fig. 5.3. Histopathological changes induced by the feeding and migration of the root-lesion nematode *Pratylenchus thornei* in chickpea showing the primary cellular reaction to nematode infection by enlargement of the nucleus and nucleolus, thickened cell walls, and granular cytoplasm of cells adjacent to the nematode pathway.

5.2.1.3. Host reaction

Brown to reddish necrotic lesions parallel to the root axis and eventual secondary decomposition characterize roots infected by root-lesion nematodes. The presence and extent of root lesions vary with the plant and nematode species, depending on the amount of polyphenol deposition caused by nematode damage. These symptoms differ from those caused by migratory stages of sedentary endoparasitic nematodes that travel between and through cortical cells without feeding until a permanent feeding site is established. By contrast, *Pratylenchus* species tend to migrate through the cortical cells, feeding frequently and causing extensive necrosis and visible lesions. Cells adjacent to those fed upon for long periods may also die. The propensity of migratory endoparasites to move continuously in and out of roots also increases entry points for secondary invaders. Heavily infected root systems are reduced in size and may exhibit additional symptoms such as 'witch's broom', irregular root swelling and stunted rootlets. Symptoms in tubers, such as potato, range from scabby or sunken lesions (caused by *P. scribneri*) to warty protuberances (*P. penetrans*). In stored yam tubers, *P. coffeae* causes dry rot in the surface tissues.

Above-ground symptoms result from root dysfunction – stunted, chlorotic plants appear in patches in the field. In woody perennials, these symptoms are associated with young trees when orchards are replanted on sites where the nematodes, along with other pathogens, have increased over many years. Symptoms that develop on older trees include canopy thinning, stem dieback and reduced fruit yield.

Despite the wide host range of *Pratylenchus*, host status varies greatly, often in proportion to nematode aggressiveness, and their reproductive fitness is generally related to damage. For high aggressive host–nematode combinations, such as *P. coffeae* on citrus, population densities can exceed 10,000 nematodes g^{-1} of roots. By contrast, *P. brachyurus* is widely associated with citrus, but population densities are less than 10% those of *P. coffeae* and the nematode causes little if any economic loss.

5.2.1.4. Dispersal

Endoparasitic nematodes are readily dispersed by movement of infected propagative plant material. Within fields, the nematodes move actively through soil when host roots are available and are passively moved with irrigation or precipitation run-off. Anhydrobiotic survival during dry periods permits the airborne dispersal of some *Pratylenchus* species during windstorms.

5.2.1.5. Ecology, host range and distribution

The geographic distribution of *Pratylenchus* species has been described as zonal (De Waele and Elsen, 2002). Species tend to occur worldwide in climates with suitable temperature ranges. Thus, *P. coffeae*, *P. loosi*, *P. goodeyi*, *P. brachyurus* and *P. zeae* are found throughout the tropics (Table 5.1). Most other economically important species occur in zones encompassed by temperate, subtropical or cooler (elevated) tropical regions. A few, such as *P. fallax* and *P. crenatus*, are prevalent only in temperate zones. The distribution and population development within a zone are affected by plant host suitability and physical factors such as soil texture and rainfall patterns.

Root-lesion nematode species characteristically reproduce on a wide range of hosts. *Pratylenchus penetrans* and *P. coffeae* have more than 350 and 130 known hosts, respectively. Most plants serve as hosts to some species in the genus, but the economic importance of root-lesion nematodes has been characterized for relatively few crops and regions.

Multiple species of root-lesion nematodes pathogenic to cereals are often present in fields and can reach damaging levels when cereals are grown intensively. In northern Europe, intensive maize cropping increases the likelihood of damage to winter wheat by *P. penetrans*, *P. crenatus* and *P. fallax*. Yield of winter wheat in the north-west USA is inversely related to cropping intensity and to numbers of *P. neglectus*, which may have a greater impact on yield than even soil moisture. Widespread damage to cereals by root-lesion nematodes in Australia has fostered variety screening and the discovery of wheat cultivars tolerant to *P. neglectus* and *P. thornei* that can increase grain yield by up to a third. Lesion nematodes are also the most important nematode pathogens of maize in the USA maize belt, where grain losses of more than 1 tonne ha^{-1} occur in heavily infested fields. Among many species that damage maize, *P. scribneri*, *P. penetrans* and *P. hexincisus* are important in temperate zones and *P. zeae* and *P. brachyurus* are most frequently encountered in warmer regions.

Pratylenchus affects harvestable below-ground plant parts directly and also by compromising the feeder root system. High numbers of *P. penetrans* can severely reduce potato yield; in addition, the marketable tubers are blemished by scabby or shallow sunken lesions. *Pratylenchus coffeae* is a pest of yam in the Americas and the Pacific Islands but, interestingly, not in Africa. Reduced vine growth and yield reduction are usually the result of planting infested seed tubers. However, soil populations are sufficient to reduce the quality of tubers due to dry rot that continues to develop during storage. Peanut in Australia, India and the southern regions of the USA and Africa are widely infected by *P. brachyurus*, which infects root, pegs and pods. In addition to reducing plant size, nematode infection of pegs reduces pod and seed development and pod infection results in necrotic lesions that can serve as entry points for secondary invaders that further reduce seed yield or quality.

Pratylenchus coffeae, *P. goodeyi* and *P. speijeri* are the primary root-lesion nematode pests of banana and plantain. The nematodes cause plant stunting and delayed maturity, but their major economic impact is a reduction of the plantation life and the ability to cause extensive toppling of the mature, fruit-bearing plants during storms. In eastern and central Africa, *P. goodeyi* is often absent at lower altitudes where *Radopholus similis* is the dominant nematode parasite of *Musa* spp. However, *P. goodeyi* predominates in the cooler, higher altitudes of Africa and the Canary Islands. *Pratylenchus coffeae* attacks *Musa* spp. throughout the tropics, causing greater losses in some regions than others. *Pratylenchus speijeri*, a species especially virulent to the staple food crop plantain in West Africa, was distinguished from *P. coffeae* based on molecular phylogenetic relationships.

Damage to roots of woody perennials by *Pratylenchus* spp. can reduce tree vigour and yield, but more importantly the nematodes contribute with other soilborne pathogens to orchard replant problems. *Pratylenchus penetrans* is a worldwide pathogen of pome and stone fruit and nut trees. The geographical range of *P. vulnus* is very similar but damage is restricted to warmer climates. *Pratylenchus coffeae* is the most common lesion nematode associated with coffee decline worldwide, but in West Africa, Brazil and Peru, *P. brachyurus* is widespread on coffee and in Central America coffee is attacked by a complex of described and undescribed species, all formerly considered to be *P. coffeae*. Similarly, citrus is a host and is severely damaged by *P. coffeae* in some, but not all, regions where both occur (Fig. 5.4).

Fig. 5.4. Citrus tree condition in a Florida orchard infested by *Pratylenchus coffeae* showing typical early decline symptoms of mature trees and the need for extensive replacement of non-profitable trees. Similar symptoms are caused by *Radopholus similis* in Florida citrus groves.

5.2.1.6. Molecular diagnosis

DNA sequences are often critical to species determination and identification (Fig. 5.5). Several molecular methods have been used for accurate diagnosis of *Pratylenchus*, including restriction fragment length polymorphism (RFLP) analysis of internal transcribed spacers (ITSs), species-specific primers, duplex and quantitative polymerase chain reaction (PCR) methods, and ribosomal (18S, ITS, D2–D3 of 28S ribosomal DNA (rDNA)) and *COI* mitochondrial DNA regions are being used extensively as molecular markers to identify to species level and to allow the detection of cryptic species (Palomares-Rius *et al.*, 2010). RFLPs of portions of the rDNA (ITS) discriminated multiple populations of 18 *Pratylenchus* species. Whereas populations of *P. goodeyi* and *P. vulnus* showed no intraspecific variation by this method, several putative *P. coffeae* populations were readily differentiated by combinations of restriction enzymes. Sequence analysis of a portion of the 28S rDNA gene revealed that half of 14 populations of *P. coffeae* from five continents are phylogenetic species, each possessing characteristics that are unique among the populations. Consequently, newly described species that differ from *P. coffeae* in Java (the species type locality) based variously on reproductive incompatibility, minor morphological differences and molecular phylogeny include *P. jaehni* from citrus in Brazil, *P. speijeri* from plantain in Ghana, and two species from native plants in Florida, USA. Presumably, additional species await description among the *P. coffeae* complex. Primers that selectively amplify rDNA of either *P. coffeae* or *P. loosi* were developed (Uehara *et al.*, 1998) and demonstrated the usefulness of molecular probes to discriminate between morphologically cryptic species in this genus. Species-specific, molecular probes are used in real-time PCR to quantify *P. neglectus* and *P. thornei* in soil samples, allowing grain growers in Australia to make management decisions before planting. A Moroccan

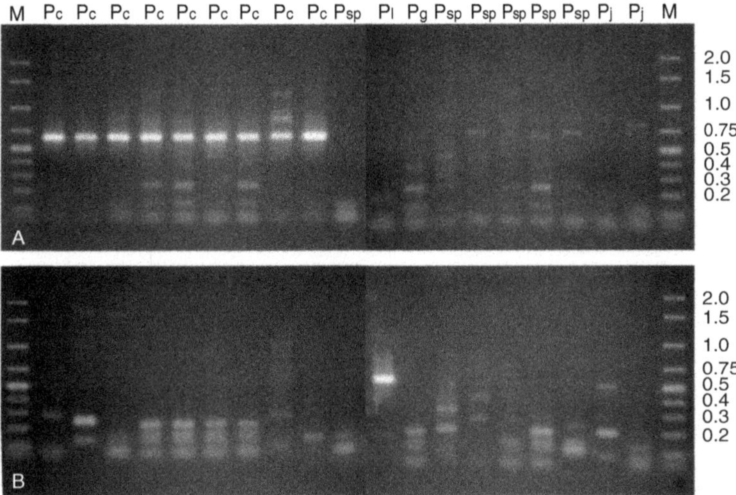

Fig. 5.5. Amplification of ribosomal DNA spacer regions of closely related *Pratylenchus* species from around the world. Populations of *P. coffeae* (Pc), *P. loosi* (Pl), *P. gutierrezi* (Pg), *P. jaehni* (Pj) and undescribed *Pratylenchus* species (Psp) were identified based on morphology and genetic sequences. DNA from these populations was then amplified using primers developed for populations of *P. coffeae* (A) and *P. loosi* (B) in Japan. Molecular weights (number of base pairs) are also indicated (M). (Adapted from Duncan *et al.*, 1999.)

study demonstrated that real-time PCR could be used to quantify *P. penetrans* and *P. thornei* in screenings for resistance of wheat lines. The systematic position of the genera within Pratylenchidae evaluated with maximum likelihood tests based on ribosomal RNA (rRNA) datasets showed that they do not conflict with the relationships that resulted from the present morphological classifications (Palomares-Rius *et al.*, 2010).

5.2.1.7. Interaction with other pathogens

Infection by root-lesion nematodes predisposes roots to numerous primary and secondary pathogens. Some *Pratylenchus* species appear to damage a host only in the presence of other disease-causing organisms, e.g. *P. brachyurus* and *Fusarium oxysporum* on cotton. Others, such as *P. penetrans* and *Rhizoctonia fragariae* on strawberry, do not interact but rather cause additive crop damage. The best-known interaction involves *Verticillium dahliae* in the disease syndrome called 'potato early dying'. Although the mechanism is unknown, the disease will occur at population densities below damaging levels for either pathogen individually (synergism). *Pratylenchus penetrans* and *P. thornei* can induce potato early dying, whereas *P. crenatus* and *P. neglectus* cannot. *Pratylenchus scribneri*, a species widely distributed throughout tropical and temperate regions, interacts with *V. dahliae* only at elevated temperatures. In wheat fields naturally infested by *P. thornei* and *F. pseudograminearum* (Fusarium crown rot) additive responses were detected (Smiley, 2021).

5.2.1.8. Management and control

Sanitation is important to reduce damage by species of *Pratylenchus*. Citrus nurseries in Florida, South Africa, Brazil and elsewhere have either mandatory or voluntary programmes to certify that young trees are free of nematode pathogens such as *P. coffeae* and *P. jaehni*. Pruning tissue with *Pratylenchus* lesions from banana and plantain corms followed by hot water treatment before planting greatly increases productivity, even in infested fields (Fig. 5.6).

Crop rotation is sometimes used to manage damage by root-lesion nematodes. In Australia, wheat that is susceptible to *P. thornei* can be rotated with several non-susceptible crops. However, the wide host ranges of most root-lesion nematodes can make successful crop rotation difficult. Numbers of *P. thornei* and *P. neglectus* were little affected by rotations of grains and commonly grown broadleaf crops in the north-western USA, whereas use of a summer fallow in the rotation decreased nematode numbers and increased grain yields. The use of sunn hemp (*Crotalaria* spp., notably *Crotalaria spectabilis*) as a leguminous rotation crop to increase soil fertility in sub-Saharan Africa is compromised by reports that it increases numbers of *P. zeae* to levels that can damage subsequent maize crops. Marigolds (*Tagetes patula*) produce a nematicidal chemical, thiophene α-terthienyl, and have been widely reported to provide effective control of *Pratylenchus* spp. when used as a cover crop (Pudasaini *et al.*, 2006a).

Germplasm resistant to root-lesion nematodes is less available than for sedentary endoparasites whose complex host–parasite relationships are more readily disrupted by the cell death response that characterizes the hypersensitive reaction typically induced by

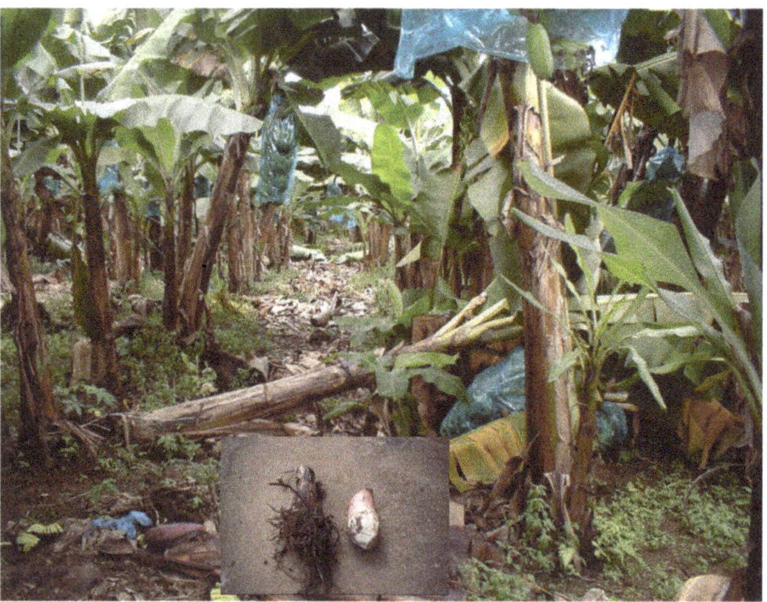

Fig. 5.6. Uprooted banana plants with reduced root systems caused by *Radopholus similis* (photograph courtesy of P. Quénéhervé) and (inset) banana corms before and after paring of the nematode-infected tissue (photograph courtesy of D. De Waele).

activation of resistance genes. Nevertheless, resistant germplasm has been identified in a number of important crops such as banana, potato and sweet potato, various cereals, legumes (chickpea) and forage crops, strawberry and some rootstocks of woody plants. Occasionally, varieties exhibit resistance to more than one nematode group. Several sweet potato cultivars resistant to *P. coffeae* are also resistant to the root-knot nematode *Meloidogyne incognita* and potato clones that are less susceptible to *P. penetrans* (quantitative resistance) are also resistant to some species and races of potato cyst nematode. Coffee varieties with resistance for *P. coffeae* and *Radopholus arabocoffeae* were detected in Vietnam (Trinh *et al.*, 2012). Conversely, resistance in a cultivar can vary among *Pratylenchus* species. Some wheat cultivars resistant to *P. thornei* in Australia are damaged by *P. neglectus*. Sources of resistance to *P. thornei* and *P. neglectus* in chickpea, wild *Cicer*, have been identified in breeding lines in India and Australia (Thompson *et al.*, 2011; Rostad *et al.*, 2022), and in wheat to *P. thornei* and *P. penetrans* (Linsell *et al.*, 2014; Mokrini *et al.*, 2018). New sources of resistance to *P. zeae* by using wild relatives of sugarcane were found in some introgression families (Foreman *et al.*, 2007), and in Asia and Brazil in rice genotypes (Biela *et al.*, 2015; Lahari *et al.*, 2020). In Ethiopia, new sources of resistance to *P. goodeyi* were found in the food crop enset (*Ensete ventricosum*) cultivars (Kidane *et al.*, 2021), and in California, USA, candidate *Juglans* genotypes have been identified as putatively resistant and/or tolerant to *P. vulnus* (Westphal *et al.*, 2021). Variation of cultivar resistance to different populations of the same species of *Pratylenchus* is also fairly common and further complicates the use of this tactic. Functional genomic approaches to developing root-lesion nematode resistant crops are promising. Analyses of the *P. coffeae*, *P. thornei* and *P. penetrans* transcriptomes have revealed potential parasitism genes that can be targeted, and reduced reproduction in *P. thornei* and *P. neglectus* was achieved using double-stranded RNA to silence genes (RNA interference (RNAi)) in these nematodes (Vieira *et al.*, 2019; Channale *et al.*, 2021).

Soil fumigation and use of nematicides can increase yield in many crops attacked by root-lesion nematodes but, due to the high cost and environmental concerns, chemical control is generally restricted to high-value crops such as coffee and banana. Where available, use of resistance is often as profitable as and more durable than chemical control. In addition, the interaction between arbuscular mycorrhizal fungi and *Pratylenchus* spp. showed variation in responses as a result of cultivar, crop species and mycorrhizal fungi species; the biocontrol effect of mycorrhizal fungi is likely to be a combination of increasing host tolerance, competition between organisms and systemic resistance (Gough *et al.*, 2020).

5.2.2. Burrowing nematodes, *Radopholus* spp.

Radopholus similis is the only one of more than 30 species in the genus recognized as a pathogen of widespread economic importance. *Radopholus* is probably indigenous to either the Indo-Malayan or Australasia regions, which are rich in burrowing nematode species and include the centres of origin of many of the important hosts of burrowing nematodes. Several new species of local importance have been described in the past decades. *Radopholus citri* and *R. bridgei* were found to be pathogens of citrus and turmeric, respectively, in Java. *Radopholus musicola* was described from northern Australia where it damages banana. *Radopholus duriophilus, R. arabocoffeae* and *R. daklakensis* were all discovered in Vietnam in association with declining durian (*Durio zibethinus*), Arabica coffee and Robusta coffee, respectively. By contrast, *R. similis* is widespread throughout the tropics where it is a serious limiting factor in the production of many crops. The pantropic distribution of the species probably resulted from the widespread movement of infected banana corms from South-east Asia (Marin *et al.*, 1998).

5.2.2.1. Morphology and identification

As described by Loof (1991) and Luc (1987), the genus is characterized by strong sexual dimorphism. For the female *R. similis*, the body is slender and ranges in length from 0.53 to 0.88 mm. The lip is strongly sclerotized internally, composed of three to four annuli and not offset. The short, stout stylet has well-developed knobs. The pharyngeal glands overlap the intestine dorsally. The vulva is post-equatorial, both branches of the ovaries are fully developed (some species in the genus have a post-uterine sac), and the round spermatheca contains rod-shaped sperm (identified as spermatids). The tail terminus is almost pointed and striated. The male lip is strongly offset, knob-shaped, unsclerotized, with four lobes and reduced lateral sectors. Male characters are useful to differentiate several *Radopholus* species from *R. similis*. For example, the main feature that distinguishes *R. similis* from *R. citri* is a thin stylet with reduced knobs in males of the former and a strongly developed stylet conus in males of the latter, while *R. musicola* differs from *R. similis* and *R. bridgei* for having a small stylet, but with well-developed knobs.

5.2.2.2. Life cycle and behaviour

The behaviour and life cycle of burrowing nematodes are, in the main, very like that of lesion nematodes. An interesting difference is that whilst burrowing nematodes normally reproduce sexually, *R. similis* females that do not mate after a period of time can reproduce as hermaphrodites. A further difference is that mature male burrowing nematodes do not feed. Males may comprise 0–50% of the population and differences in reproductive capacity between some populations is inversely related to the male:female sex ratio.

Burrowing nematodes infect at or near the root tip and reside almost exclusively in the root cortex, although they also damage the stele in banana. Migration and feeding behaviours are like that of *Pratylenchus*. As with most migratory endoparasites, the nematode remains within the root until forced by overcrowding and decay to migrate. Population development is host-dependent. Within roots, mature females begin to lay eggs at an average rate of nearly 2 day^{-1} on citrus and 5 day^{-1} on banana. At optimum temperature, juveniles may hatch in 2–3 days on some hosts or up to three times longer in others. The life cycle on citrus can be completed in 18–20 days under optimum conditions. Most populations of *R. similis* reproduce best at intermediate (25°C) or high (30°C), rather than lower (15–20°C) temperatures. Populations introduced into European parks and nurseries appear to have adapted to temperate conditions, reproducing at temperatures too low for most tropical populations. In Florida citrus groves, the optimum temperatures occur longest each year in the deeper soil horizons where root infection is greatest. Temperature extremes in the surface soils are nearer the limits for development of *R. similis*, which may explain low population development in surface roots. The nematode does not have a known resting stage, so recurring moisture deficits typical of surface soils may also inhibit development near the soil surface. The tendency of *R. similis* to inhabit deeper soil horizons in Florida affects its pest status and its management options (see Section 5.2.2.8).

Soil texture affects *R. similis* population growth and virulence. Population growth on banana is greater in coarse sand than in finer soils. The nematode is more pathogenic to citrus in sandy than loamy soils in controlled pot studies and in the field. Movement from tree to tree is also greatest in coarse textured soil.

Endosymbionts *Wolbachia*-like bacteria are widely associated with insects, filarial and plant-parasitic nematodes. A symbiosis has been detected between *Radopholus similis* and a *Wolbachia*-like bacterium. The bacterium appears to supply essential nutrients to filarial nematodes, but the nature of the symbiosis with *R. similis* is unknown (Haegeman *et al.*, 2009a). Further molecular and ultrastructural analyses confirmed the presence of other endosymbionts such as *Cardinium* in *Heterodera glycines* (Noel and Atibalentja, 2006), *Heterodera sturhani* (Yang *et al.*, 2017), *P. penetrans* (Wasala *et al.*, 2019), *Rotylenchus zhongshanensis* (Guo *et al.*, 2022) and *Bursaphelenchus mucronatus* (Yushin *et al.*, 2022).

5.2.2.3. Host reaction

Root damage by *Radopholus* is typical of that caused by *Pratylenchus* – reddish, brownish to black lesions caused by cell wall collapse as nematodes move inter- and intracellularly. The nematode usually penetrates the region of elongation but if terminals are penetrated, root tips can become swollen and stubby. Tissue rot occurs following secondary infections. Primary and secondary roots as well as corms can suffer severe damage in crops such as banana, whereas only non-lignified fibrous roots of woody plants are damaged. The quality of root crops such as turmeric is affected because rhizomes lose their bright yellow colour. Stunting, wilting and chlorosis characterize heavily infected plants. Tree damage includes smaller and sparser foliage and fruit. Branch ends become bare and eventually entire branches die. Disease progression is greatest during the dry season in affected tree crops.

5.2.2.4. Dispersal

The worldwide occurrence of burrowing nematodes is most likely due to the spread of contaminated plant material over many centuries. Use of infected planting material and movement of soil for construction purposes or on agricultural machinery routinely spreads the nematode within regions. The wide host range and virulent behaviour of the nematode results in efficient local dispersion at rates as high as 15 m year^{-1}. Reinfestation of disinfested banana fields in the French West Indies appears to be primarily the result of surface water run-off during heavy rainfall.

5.2.2.5. Ecology, host range and distribution

Radopholus similis is highly polyphagous, attacking more than 250 plants in 16 families. There are many reports of differential host preference among populations but two races of the nematode are commonly recognized. A citrus race exists in Florida that reproduces on most citrus species and varieties in addition to banana and the other known hosts of the genus. Populations that do not reproduce on citrus are collectively known as the banana race. The citrus race was given species status (*R. citrophilus*) in 1984, but subsequent research showed that *R. citrophilus* is a junior synonym of *R. similis,* based on karyotype identity, morphological and genetic identity, and reproductive compatibility (see Chapter 13).

The anthropic distribution of *R. similis* makes it one of the major species of economic importance. Following black Sigatoka and perhaps the banana weevil, burrowing nematodes are the most serious banana malady in many countries. The nematodes cause 'toppling disease', so named because fruit-laden, mature plants become top heavy when the root system is damaged beyond its ability to anchor the plant (Fig. 5.6). During the stormy rainy season losses can be very high. In addition, *R. similis* damage stunts plant growth, delays maturity, reduces bunch weight and shortens the productive life of the mother and daughter plants. Burrowing nematode populations vary in their capacity to damage banana. An inverse relationship between nematode population development and banana plant growth suggests that the degree of damage results from differences in nematode numbers rather than differing virulence (Fig. 5.7).

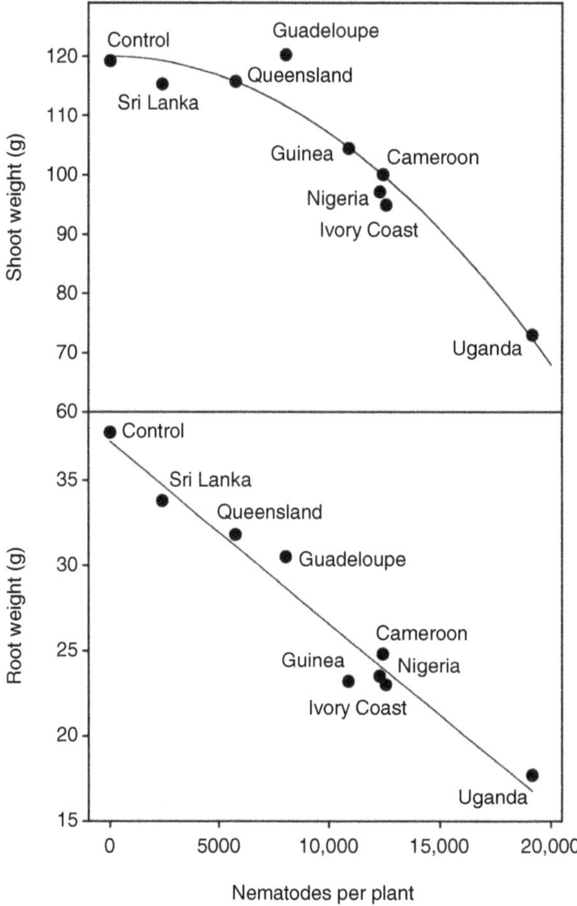

Fig. 5.7. Relationships between population growth of different populations of *Radopholus similis* and the growth of banana seedlings (*Musa* AAA 'Poyo'). These data support the idea that differences in virulence of burrowing nematode populations is caused by differences in the numbers of nematodes parasitizing the host. (Adapted from Fallas *et al.*, 1995.)

Populations with exceptionally high reproductive capacity may even reproduce well on some normally resistant cultivars. It has also become an important pathogen for ornamentals, such as *Anthurium*, *Calathea* and *Dracaena* (Uchida *et al.*, 2003).

During 20 years following the Second World War, 22 million black pepper vines in Indonesia were lost to 'yellows disease', eventually found to be the result of *R. similis*. The disease occurs throughout South-east Asia and India. Leaf yellowing is followed by drop and then branch dieback. Progression of decline during the dry season followed by recovery during the monsoons results in a gradual decline of the plants, called 'slow wilt' in India.

For several decades, 'spreading decline' was the most important disease of citrus growing on the deep sandy soil of central Florida. Eventually, the citrus race of *R. similis* was identified as the causal agent and aggressive management programmes were initiated. The race is known only in Florida and in addition to *P. coffeae* is the most virulent nematode pathogen of citrus. The name of the disease is descriptive of the very rapid spread of decline incidence from the location of the initial infestation. A prominent feature of the disease is that fibrous root density is relatively normal near the soil surface, but is severely reduced below depths of 25–50 cm. During the dry season, the trees lack the normally deep root system needed to avoid moisture stress. As with slow wilt of pepper, the tree condition alternates with the rainy and dry seasons, becoming progressively weaker over time. Above-ground symptoms are as those for yellows disease of black pepper. Without proper management, fruit yield can decline by more than half and orchards become economically non-viable.

Radopholus similis is increasingly a problem on tea in Sri Lanka where it tends to occur at lower elevations, whilst *P. loosi* is a problem at higher altitude. Other crops attacked by burrowing nematodes in their Indo-Malayan centre of origin and beyond include coconut palm, betel vine, ginger, turmeric, cardamom and nutmeg.

5.2.2.6. Molecular diagnosis

Populations of *R. similis* from many countries and host plants, as well as several of the recently described species in the genus, have been characterized by rDNA sequences (ITS and D2–D3 regions) and mitochondrial sequences (*COI* gene). PCR primers (ITS) specific to the genus and to just *R. similis* have been optimized for use in several assays to detect the nematode, and in some cases other pathogens, in soil and plant samples. Recently, new technologies, such as the loop-mediated isothermal amplification (LAMP) assay for a specific fragment of the rRNA gene (D2–D3 region), have been developed for nematode and plant samples (Peng *et al.*, 2012). Populations of *R. similis* appear to be remarkably homogeneous at these phylogenetically important gene regions, perhaps due to their relatively recent dissemination worldwide via major crops such as banana (Marin *et al.*, 1998). A 2.4 kb sequence tag site was used as a marker to demonstrate that hybrid progeny are produced when citrus and banana races of *R. similis* are mated. Significant variability of total DNA among *R. similis* populations (detected by random amplified polymorphic DNA (RAPD) analysis) has thus far failed to discriminate populations based on host range, pathogenicity or geographic origin.

5.2.2.7. Interaction with other pathogens

Burrowing nematodes can predispose plants to attack by other pathogens (e.g. black pepper to infection by *Fusarium solani*); however, unlike some lesion nematodes, *R. similis* is not known for its involvement in synergistic disease complexes. Nevertheless, secondary infections associated with burrowing nematodes greatly exacerbate damage to the root system. In addition to cortical rot, fungi may colonize the stele of plants such as banana, which is not affected directly by the nematode. Thus, the presence of the nematode exacerbates Panama wilt on banana caused by *F. oxysporum* f. sp. *cubense* or can lead to changes in the resistance of banana cultivars (de Jesus Rocha *et al.*, 2020).

5.2.2.8. Management and control

As with *Pratylenchus* spp., sanitation is one of the most important ways to manage burrowing nematode. Citrus nursery certification in Florida results in added crop value of more than US$17 million year per year by slowing the spread of *R. similis*. The use of tissue culture-derived banana plants ensures clean planting material and infected planting material such as banana corms can be treated profitably with hot water or by paring the lesions (Fig. 5.6). Ditching between banana fields is employed to prevent the dissemination of *R. similis* from infested to disinfested fields in surface run-off of rainwater.

Some cultural practices have proved useful in helping plants tolerate the nematode. *Radopholus similis* damage is restricted to the deeper citrus root system, so growers protect the shallow roots by mowing or using herbicides rather than cultivation for weed control. Use of frequent irrigation and fertilization maintains nutrients in the shallow horizon, allowing trees to grow reasonably well even during the dry season. Mulching of infested black pepper vines provides similar benefits during the dry season to help the vines tolerate nematode damage to the root system. Mulching also reduced soil temperatures, which was associated with lower population growth and damage to cooking bananas by *R. similis*. *Radopholus similis* does not survive for long periods in the absence of host roots, so crop rotation is an option. In Martinique, treatment of banana plants with herbicide to kill roots in soil before uprooting, in combination with non-host rotations and use of clean planting stock, allows banana to be grown without nematicides for several years and the programme has reduced the use of chemical nematicides by 60%. However, many of the crops commonly grown in heavily infested areas of South-east Asia are perennial (less amenable to rotation) and are hosts of the nematode.

Resistant germplasm is known in banana, citrus, coconut and betel nut, among other crops. Resistance-breaking biotypes of the citrus race of *R. similis* can develop on the commercially available resistant rootstocks. Two confirmed sources of resistance in banana are 'Pisang Jari Buaya' and 'Yangambi km5' and several potential sources have been reported, including lines resistant to nematodes in addition to *R. similis*. Breeding programmes to develop commercially acceptable cultivars are ongoing. These are complicated by the variable reactions among nematode populations. Banana and plantain have been transformed to express both a cystatin and a peptide that can provide a dual mode of action against *R. similis* by inhibiting digestion and chemoreception, respectively (see Chapter 16). No black pepper varieties that are resistant to *R. similis* have been found and favoured cultivars are sometimes grafted onto the rootstocks of *Piper colubrinum* (Brooks, 2008). Citrus and tea cultivars of high quality and yield are also grafted onto resistant rootstocks (Brooks, 2008).

Soil fumigants were formerly used heavily in citrus and banana orchards and resulted in serious environmental and health problems. Systemic nematicides have the potential to control profitably migratory endoparasitic nematodes, but their usage is becoming more restricted because of safety concerns and awareness of problems such as accelerated microbial degradation. Biological control has proved useful for partial control of the disease using mycorrhizal fungi (Shanthi and Rajendran, 2006; Koffi *et al*., 2013), parasitic fungi (Mendoza and Sikora, 2009) and bacteria (Shanthi and Rajendran, 2006; Aravind *et al*., 2010), and it could be enhanced via the combined application of antagonists with different modes of action that target different stages in the infection process (Mendoza and Sikora, 2009).

5.2.3. Rice root nematodes, *Hirschmanniella* spp.

This genus is aptly described by its common name because more than half of the 26 recognized species are parasites of rice and nearly 60% of the world's rice fields are infested by species of *Hirschmanniella*. These nematodes predominate in moist and even aquatic habitats on hosts such as blackberry, cabbage palm (*Sabal palmetto*), bulrush (*Scirpus robustus*), sagittaria (*Sagittaria subulata*) and hydrilla (*Hydrilla verticillata*). A small number of species occupy temperate zones but the majority occur in the tropics and subtropics. The female is long (1–4 mm) and slender. The lip region is not offset and may be hemispherical (e.g. *H. spinicaudata*) or low and flattened (e.g. *H. oryzae*). Other major diagnostic characters include stylet length, presence or absence of sperm in the spermatheca (and males in the population), and presence or absence of a mucro or a ventral, subterminal notch on the long-pointed tails.

In rice paddies, *Hirschmanniella* spp. are efficiently dispersed in water, so that almost all of the plants may be infected. Symptoms are non-specific and include reduced early growth rate, occasional chlorosis and decreased tillering. Infected roots are yellow initially, then they rot. The rot is caused in part by transcriptional reprogramming by the nematode that induces programmed cell death, oxidative stress and obstruction of normal root metabolic activity. By comparison, root-knot nematode infection of rice roots affects transcription in ways that suppress the local defence pathways while stimulating metabolic pathways and nutrient transport towards the induced root gall. Rice root nematode eggs are deposited within the roots and the life history is like that of other pratylenchids. Depending on the region and nematode species, from one to three generations per year have been reported. It is common to recover both *H. oryzae* and *H. spinicaudata* from a sample. These two species infect predominantly monocotyledons, but *H. oryzae* also infects some broadleaf plants such as cotton. The nematodes overwinter in roots of rice stubble or many species of native plants. In the absence of hosts, they persist in soil, and low soil moisture induces a quiescence that extends longevity for up to a year. Crop rotation is effective in reducing populations of rice root nematodes. The use of trap crops such as *Sesbania rostrata* and *Aeschynomene afraspera* that permit infection but not egression from the roots is very effective and they are good sources of nitrogen for the following crop, but may be uneconomical if used in place of a cash crop in the rotation.

5.3. Anguinids and the Stem and Bulb Nematode, *Ditylenchus dipsaci*

The family Anguinidae contains mycophagous nematodes and nematodes that attack plant aerial parts, bulbs and tubers (Boxes 5.2 and 5.3). Of several economically important genera, *Ditylenchus* spp. have the widest impact on agriculture. The genus has more

Box 5.2. First report of a plant-parasitic nematode.

The Reverend John Needham made the first known report of a plant-parasitic nematode in a 1743 communication to The Royal Society of London. When he added a drop of water to improve resolution of the microscopic bundles of dry, white fibres he discovered in 'small black grains of smutty wheat', he was greatly surprised to find that the fibres 'took life, moved irregularly, not with a progressive, but twisting motion; and continued to do so for the space of nine or ten hours, when I threw them away'. At the time, Needham's observation generated interest because of its bearing on the question of spontaneous generation. The nematode he observed, *Anguina tritici*, was described by Steinbuch in 1799.

Box 5.3. Classification of the subfamily Anguininae.

Phylum Nematoda Potts, 1932
Class Chromadorea Inglis, 1983
Subclass Chromadoria Pearse, 1942
Order Rhabditida Chitwood, 1933
Suborder Tylenchina Thorne, 1949
Infraorder Tylenchomorpha De Ley & Blaxter, 2002
Superfamily Sphaerularioidea Lubbock, 1861
Family Anguinidae Nicoll, 1935
  Subfamily Anguininae Nicoll, 1935 (1926)
    Genus *Anguina* Scopoli, 1777
      Ditylenchus Filipjev, 1936

than 80 described species but only a few parasites of higher plants, including *D. africanus, D. angustus, D. destructor, D. dipsaci* and *D. gigas*, while the majority are fungivorous nematodes. The numerous races of *D. dipsaci* probably represent a species complex, living mostly as endoparasites in aerial parts of plants (stems, leaves, flowers) but also attacking bulbs, tubers and rhizomes. From this complex some species have been described; for example, the 'giant race' multiplying on Fabaceae was singled out and described as *D. gigas* (Vovlas *et al.*, 2011). *Ditylenchus destructor* and *D. africanus* are important pests of potato tubers in Europe and North America and groundnuts in South Africa, respectively. These two species are also mycophagous, as is *D. myceliophagus*, which is a pest of the cultivated mushroom *Agaricus bisporus*. 'Ufra' is a serious disease of deep-water and lowland rice in India and South-east Asia, caused by *D. angustus*, which resides in water films, feeding ectoparasitically on the growing rice stems and leaves, or accumulates in the leaf sheaths and developing inflorescences. *Ditylenchus dipsaci*, known commonly as the 'stem' or 'stem and bulb' nematode, is equally as damaging as the former species, but is economically more important by virtue of its wide host range.

Seed gall nematodes, *Anguina* spp., inhabit the aerial parts of cereals and forage grasses. Thirteen species are recognized and it is thought that host specialization has acted as an isolating mechanism in the evolution of this genus. Restriction sites in the ITS1 region of rDNA have been used to discriminate eight anguinid species of regulatory importance. *Anguina tritici* invade ovules where they induce galls, mate, lay eggs and reside as J2. The J2 can remain as anhydrobiotes within dried seed galls for many years (see Chapter 7). When the infective J2 exit the moistened galls, they infect the stem growing point, which elongates carrying the nematodes upward with the developing ear. Up to 60% seed reduction can result from sowing infected seed, but modern seed cleaning methods have largely eliminated the problem, except in regions where farmers save the seeds they sow. *Anguina funesta* and *A. agrostis* infect seeds of different forage grasses. Both species can introduce the bacterium *Rathayibacter toxicus*, which can be fatal to grazing livestock. As with *A. tritici*, rotation and use of clean seed effectively control these pests.

5.3.1. Morphology and identification

The species of *Ditylenchus* can be categorized as members of the '*D. dipsaci* group' or the '*D. triformis* group'. Members of the former (including *D. angustus*) have sharply pointed tails and four incisures in the lateral field. They feed on plants almost to the exclusion of the more primitive trait of mycophagy (*D. dipsaci* is an obligate phytoparasite). The *D. triformis* group (including *D. destructor, D. myceliophagous* and *D. africanus*) have rounded tail tips, six lateral incisures and are mainly fungal feeders.

Female *D. dipsaci* are slender, less than 1.5 mm long and not curved when relaxed. Stylets are short and delicate, less than 15 µm long. The median bulb is muscular with distinct valve plates. A basal bulb may extend over the intestine. The ovary is outstretched with one or two rows of oocytes and post-uterine sac. The tail is elongate, conoid and acute. The male testes are outstretched and the bursa is adanal to subterminal, enveloping one-quarter to three-quarters of the tail. *Ditylenchus gigas* can be distinguished from all other *Ditylenchus* spp. by several morphological characteristics (body size, lateral field, stylet and post-vulval uterine sac). *Ditylenchus gigas* can be separated from *D. dipsaci* by its longer body length and longer vulva–anus distance.

5.3.2. Life cycle and behaviour

As a parasite of above-ground parts of plants, *D. dipsaci* is not buffered from changes in ambient weather conditions as are root parasites. Consequently, the nematode is highly resistant to desiccation and can survive in a state of anhydrobiosis for many years (see Chapter 7). Species adapted to more humid habitats such as *D. angustus* do not have this ability. Masses of anhydrobiotic *D. dipsaci* ('eelworm wool') frequently overwinter in dried plant debris in the field. The fourth-stage juvenile (J4) is the primary survival and infective stage. When conditions permit, the nematodes migrate to germinating host plants and invade hypocotyls or petioles, entering through stomata or penetrating the epidermis, where they moult to the adult stage. The nematode feeds on parenchymatous tissue where all life stages occur. Reproduction is by amphimixis and population growth can be very rapid; females lay 500 eggs from which the J2 hatch within 2 days and develop into females within 4–5 days. Females live more than 10 weeks and the life cycle requires 19–23 days at 15°C. The rapid population growth of this nematode can result in severe crop damage even when the initial population density is low.

5.3.3. Host reaction

Early infection of plants in the field causes high rates of seedling mortality. Surviving plants are stunted and deformed. The host response to *D. dipsaci* is due partly to enzymatic action that dissolves the middle lamellae and separates host cells. Plant hormone imbalance also occurs. Cell hypertrophy and formation of intercellular cavities causes local swelling. Affected leaves and petioles are often stunted and distorted and internodes are shortened. Stems of onion or garlic also become distorted with pimple-like spots called spikkles. Eventually stems become soft and puffy and often collapse. In plants grown from bulbs the nematodes migrate downward to infect the outer scales, eventually migrating throughout the bulb. Symptoms in infected bulbs progress downwards over time to reveal discoloured rings internally. Onion bulbs rot readily in storage whereas garlic bulbs tend to desiccate. Damage caused by *D. gigas* differs from that caused by *D. dipsaci* in a heavier distribution of symptoms through the main stem, leaves and pods, and in the greater percentage of infected seeds (Fig. 5.8).

Fig. 5.8. Symptomatology of damage caused by *Ditylenchus gigas* on broad bean. A: Apical stems and leaves necrotized and deformed by severe attacks. B: Symptomatic range of necrotic areas on stems. C: Longitudinal sections of healthy (first at left) and damaged (two at right) stems showing internal necrosis. D, E: Deformed and undersized pods. F: Deformed (middle and bottom) and uninfected (top) seeds. (Source: P. Castillo and N. Vovlas.)

5.3.4. Dispersal

Active dispersal of *D. dipsaci* is limited; however, its capacity for desiccation survival makes it readily disseminated (mostly as the resistant J4) as a contaminant of planting material (seeds and bulbs), plant debris or contaminated equipment. The nematode is seedborne in broad bean, lucerne, garlic, clover, for example, and is frequently transmitted in irrigation water, which can result in greater crop loss than when infestation is from a point source.

5.3.5. Ecology, host range and distribution

The host range of *D. dipsaci* includes about 500 plant species. More than 30 physiological races of the nematode are known, some being host-specific and others widely polyphagous. Races of the nematode are named after the crop from which they were identified or after a major host. *Ditylenchus dipsaci* is a serious pest of clover, pea, broad bean, celery, garlic, onion, potato, strawberry, oats, sugar beet and rye, and is the most important nematode pest of several crops, such as lucerne and the bulb ornamentals narcissus and tulip. In addition to broad bean, *D. gigas* parasitizes a number of *Lamium* spp., other dicotyledonous weeds and *Avena sterilis*.

Ditylenchus dipsaci is adapted to temperate conditions and occurs at higher elevation in some areas of the subtropics and tropics, providing humidity is adequate. The use of irrigation facilitates the nematode's survival in agriculture in hot, arid climates. The nematode's requirement for free moisture on plant surfaces makes it a seasonal pest in some regions. Winter crops in Italy are more susceptible to *D. dipsaci* damage due to the greater incidence of dew, fog and precipitation. Dense canopy formation in crops such as bean create humid conditions that permit heavy population development even during drier parts of the year. Fine-textured soils are most favourable for population growth and damage by *D. dipsaci*, perhaps due to their greater water-holding capacity.

Survival of *D. dipsaci* in soil is enhanced by low temperature and moisture deficit. The nematode is freeze-tolerant and will survive for many years in a frozen or desiccated state. In the absence of a host, the nematode survives for a relatively short time in warm moist soil.

5.3.6. Molecular diagnosis

The mainly morphologically indistinguishable races of *D. dipsaci* greatly complicate management based on crop rotation or use of resistance. Attempts to distinguish races of the nematode with esterase and catalase profiles and polyclonal antibodies were unsuccessful. Use of monoclonal antibodies proved to be specific to only one of several populations of the oat race. *Ditylenchus gigas* differs from related *Ditylenchus* spp. in the ITS1–5.8S–ITS2 region, the D2–D3 fragment of the 28S gene of rDNA, the small subunit 18S rDNA sequence, mt*COI* gene and *hsp90* gene sequences, real-time PCR, and chromosome numbers. Similarly, *Anguina* spp. can be also identified by these molecular markers (Vovlas *et al.*, 2011; Roubtsova *et al.*, 2020). LAMP-based diagnostic assays were also developed for *Anguina agrostis* using the ITS region (Yu *et al.*, 2020); another LAMP-based assay using the 28S rRNA gene allowed sensitive detection of the potato tuber nematode, *D. destructor*, from complex plant–nematode DNA mixtures within 50 min (Deng *et al.*, 2019).

5.3.7. Interaction with other pathogens

Although *D. dipsaci* has been reported to interact synergistically with other pathogens, such as *F. oxysporum* on lucerne and garlic, true disease interactions do not appear to be common (McDonald *et al.*, 2021). Nevertheless, secondary infections in many crops are an important cause of increased damage to above- and below-ground organs, particularly with ornamental and food bulb crops.

5.3.8. Management and control

Clean planting material is critical to avoid damage by stem and bulb nematodes. Some garlic and onion producers rely heavily on programmes to produce and distribute certified planting material. Fumigants and nematicides are uneconomical for management of *D. dipsaci* in most crops but can be used in nurseries to reduce the infection rate of planting material. Seed disinfestation by hot water or formaldehyde treatment is also practised. Disinfestation of equipment used between fields is important.

Short persistence of the nematode in warm moist soil makes it amenable to control by 2–3-year rotations and by soil solarization in suitable climates. The success of rotation depends on the host range of the race in question and availability of suitable non-host crops. Sources of resistance for many of the races are known and commercial cultivars are available in crops such as lucerne, clover, oat, garlic, strawberry and sweet potato. Use of resistance against this nematode can be highly profitable, more than doubling yields in many crops for which cultivars are available. Similarly, resistance to *D. angustus* has been found in different rice cultivars and wild rice species in Asia (Khanam *et al.*, 2018).

5.4. Plant-parasitic Aphelenchs

There are four trophic types within the Aphelenchoidea: (i) entomopathogenic; (ii) mycophagous; (iii) plant-parasitic; and (iv) predatory. Aphelenchs are predominantly free-living and mycophagous in habit. Only a small number of more than 400 species are known to damage plants. A few species of *Aphelenchoides* (Box 5.4) are plant parasites of economic

Box 5.4. Classification of the superfamily Aphelenchoidea.

Phylum Nematoda Potts, 1932
Class Chromadorea Inglis, 1983
Subclass Chromadoria Pearse, 1942
Order Rhabditida Chitwood, 1933
Suborder Tylenchina Thorne, 1949
Infraorder Tylenchomorpha De Ley & Blaxter, 2002
Superfamily Aphelenchoidea Fuchs, 1937
Family Aphelenchoididae Skarbilovich, 1947
  Subfamily Aphelenchoidinae Skarbilovich, 1947
    Genus *Aphelenchoides* Fisher, 1894
  Subfamily Parasitaphelenchinae Rühm, 1956
    Genus *Bursaphelenchus* Fuchs, 1973

importance: *A. fragariae* and *A. ritzemabosi* live in buds and leafs of higher plants, *A. besseyi* is an important parasite of rice worldwide and *A. composticola* can destroy mushroom beds. Among approximately 100 species within the genus *Bursaphelenchus* only two are considered plant parasitic; they are both vectored by insects. *Bursaphelenchus xylophilus*, vectored by *Monochamus* spp., is the cause of the pine wilt disease. *Bursaphelenchus cocophilus* is vectored by *Rhynchophorus palmarum* weevils and causes red ring disease in several palm species in the Caribbean and Latin America.

5.4.1. The bud and leaf nematodes, *Aphelenchoides fragariae* and *A. ritzemabosi*

These two species with similar life cycles attack hundreds of herbaceous and woody plant species. By feeding on leaf mesophyll, they cause typical leaf blotches and when feeding on buds, they may kill the growing point and prevent flowering.

5.4.1.1. Morphology

Aphelenchoides fragariae and *A. ritzemabosi* show the general characters of the genus *Aphelenchoides* (Hunt, 1993). They are usually between 0.4 and 1.2 mm long. Heat-relaxed females die straight to ventrally arcuate, whereas the males assume a 'walking stick' shape with the tail region curled ventrally. The stylet is often about 10–12 µm long; the ovoid or spherical median bulb with central valve plates is well developed. The vulva is post-median, usually between 60% and 75% of the body length, and the female genital tract is monoprodelphic. A post-uterine sac is usually present. The tail is conoid with a variable terminus. The male tail is strongly hooked ventrally to form the characteristic 'walking stick' form; spicules are thorn-shaped and a bursa is absent.

Aphelenchoides fragariae and *A. ritzemabosi* are separated from each other using morphological characteristics (Fig. 5.9) of both females and males (Table 5.2).

5.4.1.2. Life cycle and behaviour

Bud and leaf nematodes are facultative plant-parasitic nematodes and can readily reproduce on fungi. They are not true endoparasites, being essentially tissue-surface feeders. They are not attracted by host plants and their orientation is not affected by gravity or light but is positively influenced by CO_2. The nematodes enter leaves through stomata when the surface is covered with a thin film of water or by penetrating the epidermis of the under-surface. Within the leaves, the nematode destroys the spongy mesophyll cells and its movement seems to be delimited by leaf veins. On some plants, the nematode lives ectoparasitically within the folded crowns and runners (strawberry) or flower buds (violets, azalea).

Aphelenchoides fragariae and *A. ritzemabosi* are bisexual species for which amphimixis (n = 4) seems to be obligatory. Fertilized females of *A. ritzemabosi* go on reproducing for 6 months, without refertilization. It is suggested that caryogamy (hybridization) between both species is possible. Their life cycle is simple and very short (10–14 days at 18°C).

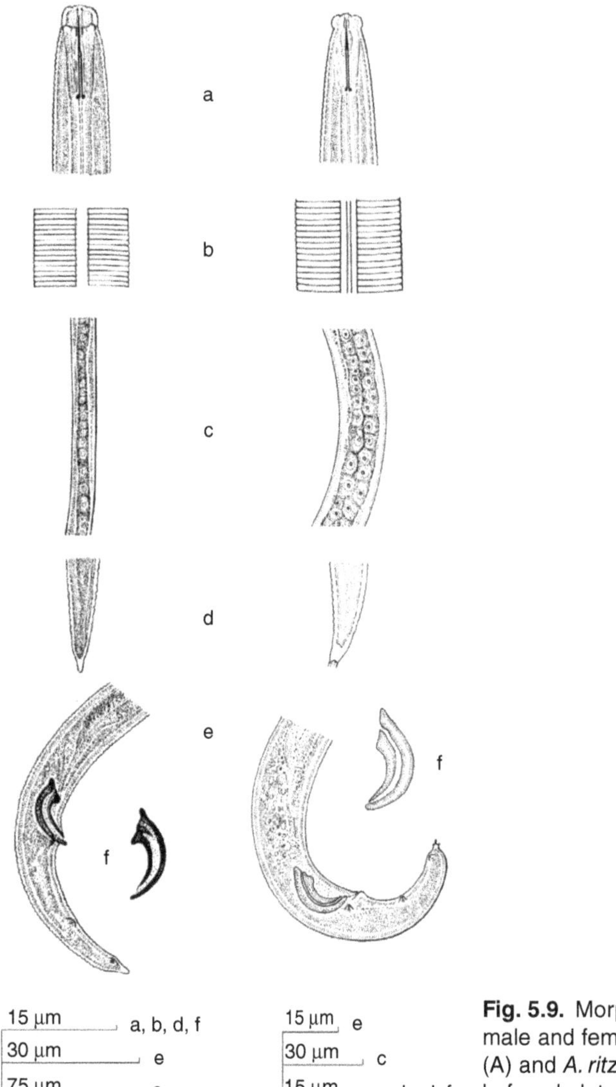

Fig. 5.9. Morphological comparison between male and female *Aphelenchoides fragariae* (A) and *A. ritzemabosi* (B); a: female lip end; b: female lateral field; c: oocytes; d: female tail tip; e: mail tail end; f: spicule. (From Siddiqi, 1974a, 1975.)

J2 hatch from eggs in 4 days and about 20–30 eggs are laid per female, the low fecundity being compensated by the short life cycle. It survives in soil for as little as 4 weeks.

5.4.1.3. Host reaction

By feeding on leaf mesophyll, bud and leaf nematodes cause typical leaf blotches initially limited by the veins (Fig. 5.10). Breakdown products of damaged tissues are nematotoxic and

Table 5.2. Differential diagnosis of *Aphelenchoides fragariae* and *A. ritzemabosi* based on female and male morphological characteristics. (From Siddiqi, 1974b, 1975.)

Characteristic	*A. fragariae*	*A. ritzemabosi*
Female		
Lateral field	Two incisures	Four incisures
Cephalic region	Smooth, elevated, anteriorly flattened with sides straight to rounded, almost continuous with body contour	Lip region spherical offset by a constriction, slightly wider with adjacent body
Excretory pore	With or close behind nerve ring	0.5–2 body widths posterior to nerve ring
Oocytes	In single file	In multiple rows
Tail end	Simple blunt spike devoid of any processes	Terminal peg with 2–4 minute processes pointing posteriorly, giving paintbrush-like appearance
Male		
Tail end	Arcuate through 45–90° not sharply curved like a hook	Usually curved through about 180° when relaxed
	Simple blunt terminal spine	2–4 processes, of variable shape
Spicules	Moderately developed apex and rostrum; dorsal limb 14–17 µm long	Lacking a dorsal or ventral process at proximal end; dorsal limb 20–22 µm long

the nematodes move progressively from the invasion site until barrier veins prevent access to fresh tissue. This discontinuous process produces discrete blotches with different stages of discoloration. Eventually, dead, shrivelled and hanging leaves extend from the tops of plants. Ferns infested with *A. fragariae* typically show narrow or linear patches perpendicular to the midrib of the frond. Sometimes blotches have an irregular shape and look water-soaked (*Begonia*). On strawberry, the nematode causes malformations such as twisting and puckering of leaves, discoloured areas with hard and rough surfaces, undersized leaves with crinkled edges, rending of petioles, short internodes of runners, reduced flower trusses with only one or two flowers and death of the crown bud. Malformations also occur when *A. ritzemabosi* infects lucerne seedlings. Infestations of *Lilium* spp. by *A. fragariae* cause dieback of the leaves from the base of the stem upwards. When young shoots are attacked, stunting and distortion result from the nematodes feeding externally around the growing point.

5.4.1.4. Dispersal

Bud and leaf nematodes can spread from plant to plant when tissues are touching and the humidity is high enough to create a water film needed by nematodes to move on the plant surface. Rain splash also helps local spread of the bud and leaf nematodes. Over longer distances, the nematodes can be spread by infected propagating material.

5.4.1.5. Ecology, host range and distribution

Bud and leaf nematodes are found in temperate and tropical regions of all continents and both species may occur sympatrically. The nematodes overwinter in dormant buds and

Fig. 5.10. Interveinal necrosis in a leaf of *Primula* sp. infested with *Aphelenchoides ritzemabosi*. (From Southey, 1993.)

growing points. In dry leaves, adults of *A. ritzemabosi* coil and can survive in a desiccated state for several months, mostly as late-stage juveniles and adults. They do not survive exposures to –20°C, but most populations survive from –2°C to 1°C.

The host range of both species is extremely large and composed of ferns and species of the Compositae, Liliaceae, Primulaceae and Ranunculaceae. *Aphelenchoides fragariae* is a major pest in strawberries and is found on more than 100 fern species. It occurs on common weeds in addition to cultivated plants.

5.4.1.6. Interaction with other pathogens

Aphelenchoides ritzemabosi was found in association with *Phytophthora cryptogea* on diseased gloxinia in Florida. In strawberry runners the 'cauliflower disease' was experimentally induced through co-inoculation of *A. ritzemabosi* and *Rhodococcus fascians*. In *Barleria cristata, A. fragariae* is associated with the bacterium *Pseudomonas chicorii*. When associated with *Xanthomonas begoniae*, the nematode causes leaf spot symptoms to appear earlier and to be more severe.

5.4.1.7. Management and control

Selecting nematode-free propagation material is of paramount importance to reduce the impact of bud and leaf nematodes; whenever possible certified stocks should be used. Regenerating plants from clean, dormant, excised axillary buds is effective for eliminating *A. fragariae* on strawberry. Planting stock can be disinfested by hot water treatment. Sometimes symptom-free stock may be infested and suspect material should be isolated from other plants to observe if symptoms develop.

Chemical control with insecticides (e.g. parathion) or non-fumigant nematicides used as foliar or soil applications are effective. Care should be taken when using these chemicals as phytotoxic thresholds are difficult to predict.

5.4.2. *Aphelenchoides besseyi*

Aphelenchoides besseyi is seedborne and causes the disease 'white tip' on rice. The species is widespread in rice-producing areas. The nematode lives mainly ectoparasitically on young tissue; at the end of the growing season it lives in a quiescent state under the hulls of the seed. This nematode was identified as the causal agent of 'black spot disease' on beans (*Phaseolus vulgaris*, Fabaceae) (Chaves *et al.*, 2013). *Aphelenchoides besseyi* also infects *Fragaria* species (strawberries), where it is the causal agent of 'summer dwarf' or 'crimp' disease.

5.4.2.1. Morphology

Within its genus, *A. besseyi* is characterized by lateral fields about one-quarter as wide as the body, with four incisures (Franklin and Siddiqi, 1992). The posterior part of the stylet has slight basal swellings 1.75 μm across. The secretory–excretory pore is usually near the anterior edge of the nerve ring. The spermathecae are elongate oval, usually packed with sperm. The ovary is relatively short and not extending to pharyngeal glands, with oocytes in 2–4 rows. The post-uterine sac is narrow, inconspicuous and does not contain sperm; the sac is 2.5–3.5 times the anal body width long but less than one-third distance from vulva to anus. The female tail terminus bears a mucro of diverse shape with 3–4 pointed processes; the male tail has terminal mucro with 2–4 pointed processes; spicules lack an apex at the proximal end.

5.4.2.2. Life cycle and behaviour

When infested grain is sown, the quiescent nematodes become active and move to the growing points of stems and leaves of the seedlings. There they feed ectoparasitically and lay eggs in the leaf axils and flower panicles. A rapid increase in nematode number takes place at late tillering and is associated with the reproductive phase of plant growth. *Aphelenchoides besseyi* is able to enter the spiklets before anthesis and feed ectoparasitically on the ovary, stamens, lodicules and embryo. However, *A. besseyi* is more abundant on the outer surface of the glumes and enter when these separate at anthesis. *Aphelenchoides besseyi* reproduces by amphimixis but can also reproduce parthenogenetically. There are several generations in a season.

As grain filling and ripening proceed, reproduction of the nematodes ceases, and the development of third-stage juveniles (J3) to adults continues until the hard dough stage. At this phase of grain development, the nematodes coil and aggregate in the glume axis. The nematodes (mostly adult females) are found beneath the hulls of rice grains. They may remain viable for 2–3 years on dry grain but die in 4 months on grain left in the field and do not survive in soil. Survival is enhanced by aggregation and a slow rate of drying (see Chapter 7) but the number and infectivity of nematodes is reduced as seed age increases.

5.4.2.3. Host reaction

The typical symptom is the emergence of the chlorotic tips of new leaves. Later, these tips become necrotic, while the rest of the leaf may appear normal. *Aphelenchoides besseyi* also causes crinkling and distortion of the flag leaf enclosing the panicle. In severe infections, the latter may be hindered from emerging. The size of the panicle and the number and size of grains is reduced. Viability of infected seed is lowered, germination is delayed and diseased plants have reduced vigour and height. On *Fragaria* species, symptoms include leaf crinkling and distortion, and dwarfing of the plant with an associated reduction in flowering.

5.4.2.4. Dispersal

Aphelenchoides besseyi is disseminated primarily by infected seed. On a local scale the nematode can be transmitted in flood water in lowland rice. High concentrations of seedlings in infested seedbeds also facilitate dispersal.

5.4.2.5. Ecology, host range and distribution

The optimum temperature for oviposition and hatch is 30°C; the optimum temperature for development is 21–25°C. At this temperature the life cycle is 10 ± 2 days. No development occurs below 13°C. Rice is the most important host of *A. besseyi*. The nematode is able to infect rice in most environments but infection and damage are generally greater in irrigated lowland and deep water than in upland rice. The nematode occurs in rice-growing regions of Africa, Asia, Caribbean, Europe and the USA. The nematode has hosts in more than 35 genera of monocotyledons and dicotyledons.

5.4.2.6. Interaction with other pathogens

In Bangladesh, *A. besseyi* is reported to occur with *D. angustus* and *Meloidogyne graminicola* but little is known of their association. The nematode appears to influence symptoms caused by some fungal pathogens of rice. It is reported that *A. besseyi* reduces the severity of *Sclerotium oryzae* symptoms, whereas the deterioration of *Pyricularia oryzae*-infected leaves is accelerated by the nematode reproducing in the blast lesions. In both cases, the concomitant infection of the fungus and *A. besseyi* reduced yield more than either organism separately.

5.4.2.7. Management and control

Prophylaxis is important for management of *A. besseyi*. Using nematode-free seed is of paramount importance; if unavailable, hot water seed treatment is probably the most cost-effective control measure.

Resistance to *A. besseyi* is reported from different countries and in the USA. *Aphelenchoides besseyi* is controlled principally through the use of resistant cultivars. Screening for resistance, based primarily on symptom expression, has commonly revealed symptomless but susceptible (tolerant) cultivars. Symptom expression in the field is particularly variable and variation occurs between plants of a cultivar. For this reason, efficient sampling in rice seed and good extraction procedures are important in order to detect false negatives. This genus of nematodes must be identified under an integrative approach as they are difficult to identify to species level. Several molecular tools, such as specific-PCR (Devran *et al.*, 2017) and real-time PCR (Rybarczyk-Mydłowska *et al.*, 2012; Devran *et al.*, 2017), have been developed and DNA sequences from conserved genomic or mitochondrial regions are available in GenBank. However, there are many sequences misidentified in GenBank and populations from rice and leguminous plants cluster in different clades from those of the strawberry populations in phylogenetic analysis, suggesting that they could be different taxa, as pointed by Oliveira *et al.* (2019).

Irrigating seedbeds or direct seeding into water reduces infection. Early planting in cooler conditions also reduces or eliminates *A. besseyi* infection. Various chemicals are used for seed treatments, soil applications or root dips; however, no economic assessment of the use of chemical control for *A. besseyi* has been made. Experiments studying the effect of *Pseudomonas fluorescens* on *A. besseyi* development and rice growth and yield demonstrated that biological control agents can be used to reduce the impact of the nematode.

5.4.3. The pinewood nematode, *Bursaphelenchus xylophilus*

Nearly all *Bursaphelenchus* species are associated with bark beetles. Many species have specialized J3, termed 'dauer larvae', which are usually ectophoretic and use the insect for transport. Several *Bursaphelenchus* species are associated with longhorn beetles. *Bursaphelenchus xylophilus*, the pinewood nematode (PWN), causes a serious disease of pines but also feeds on fungi (Mota and Vieira, 2008).

5.4.3.1. Morphology

Bursaphelenchus xylophilus shows the general characters of the genus *Bursaphelenchus* (Ryss *et al.*, 2005): the cephalic region relatively high and offset from the body by a constriction; a 10–20 µm long stylet with reduced basal swellings; the median bulb well developed, ovoid to elongate ovoid in shape. The dorsal pharyngeal gland opening is situated inside the median bulb; pharyngeal glands dorsally overlap the intestine. In the female, the post-uterine sac is long and the vulva is situated at 70–80% of body length. In the male, the tail is curved ventrally, conoid and has a pointed terminus. A small bursa-like cuticular structure is situated terminally; the spicules are well developed.

Within the genus *Bursaphelenchus* several groups of species are separated on the basis of their morphology (Braasch, 2004; Ryss *et al.*, 2005). The number of incisures in the lateral field, clearly visible by scanning electron microscope, is considered an important diagnostic feature, as are spicule shape, number and position of caudal papillae, presence and size of vulval flap and the shape of female tails. Within the *B. xylophilus* group, species can be distinguished by the shape of the female tail. *Bursaphelenchus xylophilus* is distinguished from other species in the genus (Fig. 5.11) by the simultaneous presence of: (i) in the male, spicules flattened into a disc-like structure (the cucullus) at their distal extremity; (ii) in the female, the anterior vulval lip is a distinct overlapping flap; and (iii) the posterior end of the body is rounded in nearly all individuals. This last character separates *B. xylophilus* from *B. mucronatus,* a non-pathogenic species in which the female has a mucronate posterior end. However, morphological differentiation between *B. mucronatus* and populations of *B. xylophilus* with mucronate tails in North America is difficult.

5.4.3.2. *Life cycle and behaviour*

Bursaphelenchus xylophilus has two different modes in its life cycle, a dispersal mode (primary transmission, phytophagous phase) and a propagative mode (secondary transmission,

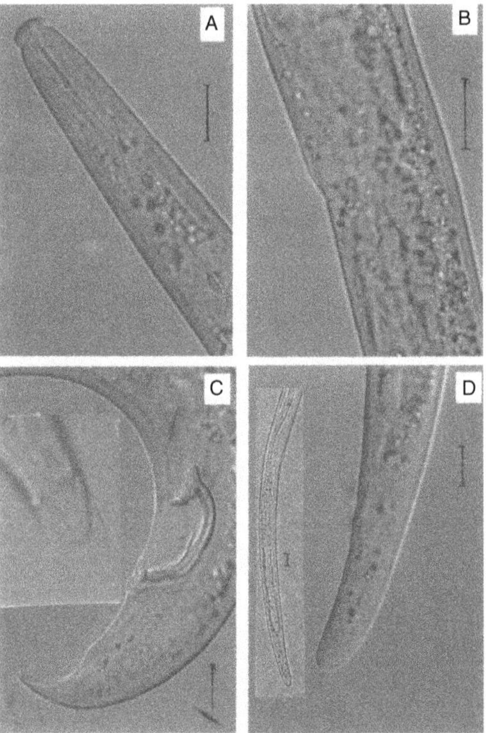

Fig. 5.11. Light microscope observations of *Bursaphelenchus xylophilus*. A: Female lip region with distinct labial region and stylet with reduced basal swellings. B: Female vulva and vulval flap. C: Male tail showing curved spicule with enlarged flattened cucullus (inset: ventral view showing bursa). D: Female round tail terminus (inset: female tail with long post-uterine sac). Scale bars = 10 μm. (From Mota *et al.*, 1999.)

mycophagus phase) (Fig. 5.12). During primary transmission, the PWN, phoretically associated in the tracheae and elytra of young adults of their longhorn beetle hosts (*Monochamus* spp.), are transmitted to young twigs of a susceptible conifer host (usually *Pinus* spp.). This happens when the insects, loaded with J4, feed on the bark of current or 1-year-old twigs. The initial feeding stage, lasting about 10 days to 3 weeks, is essential for the sexual maturation of the beetle. The nematodes enter the shoots through the feeding wounds.

In the young *Pinus* shoots, *B. xylophilus* spreads through the vascular system and resin canals, attacking their epithelial cells and living parenchyma. The nematode continues its life cycle, including four juvenile stages and an adult male and female stage. Under favourable conditions (20°C), the nematodes complete their life cycle in 6 days, each female laying between 80 and 150 eggs during a period of 28 days. As a result, nematodes block water transport in the xylem leading to plant death.

About 3 weeks later, the tree shows the first symptoms of 'drying out' or reduced oleoresin exudation. The nematodes now move freely throughout the dying tree. As a consequence of the reduction of its oleoresin defence mechanism, the tree becomes attractive to adult insects, which gather on the trunk to mate. At this stage, intensified wilting and yellowing of the needles is seen. In susceptible pines under the right conditions this can lead to classical pine wilt disease, resulting in the death of the tree within a year of infection (Giblin-Davis, 1993).

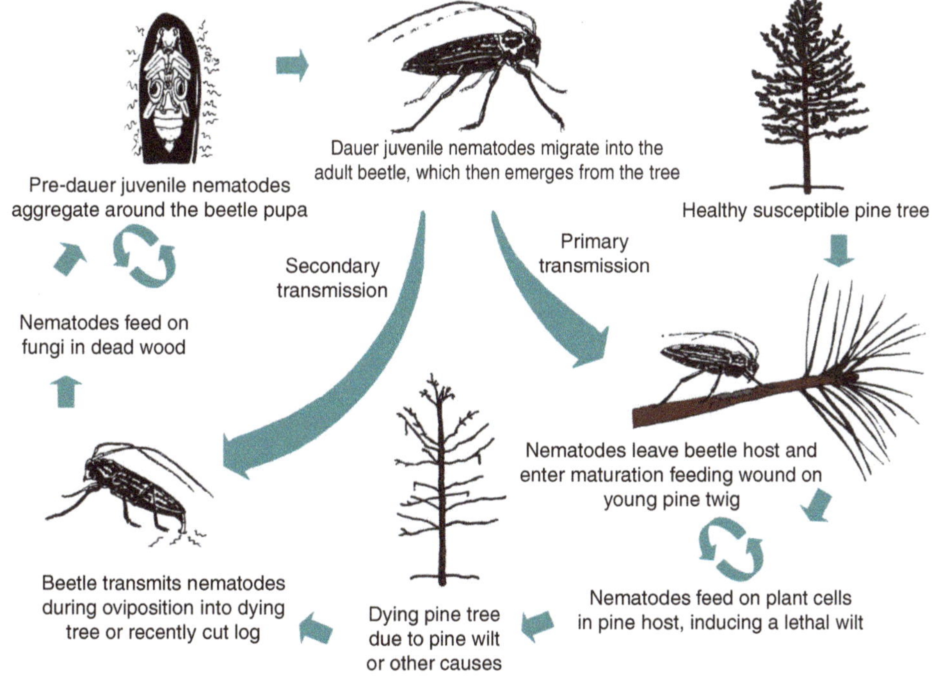

Fig. 5.12. Diagram representing the dispersal mode (primary transmission) and the propagative mode (secondary transmission) of the pinewood nematode, *Bursaphelenchus xylophilus*, by its *Monochamus* vector on a susceptible conifer host. (From Giblin-Davis *et al.*, 2003.)

The secondary transmission pathway occurs when conifers wilt, die and become suitable breeding hosts for *Monochamus*. During oviposition by the insect vector in the bark of dead or dying trees or recently cut logs, the nematodes migrate out of the beetle into oviposition slits. This means of transmission, however, is much less significant than the one by maturation feeding. In the tree the PWN feed, reproduce and greatly increase their population densities on fungi transmitted to the wood by ovipositing beetles (usually *Ceratocystis* spp.). The biology of the PWN is synchronized with the development of the longhorn beetle progeny. A different type of third-stage juvenile, DJ3 (pre-dauer juvenile), develops as conditions in the host conifer become less suitable. This is usually correlated with the development of late instar larvae or pupae of its *Monochamus* vector and results in large aggregations of DJ3 around the pupal chambers where they overwinter with the beetle. A large proportion of the pre-dauer juveniles of *B. xylophilus* moult to non-feeding dauer juveniles (fourth-stage, DJ4) and are attracted to CO_2 produced by newly enclosed adults. Nematodes also gather at the tips of perithecia formed by fungi growing on the sides of pupal chambers. When newly emerged beetles brush against the perithecial necks they acquire the nematodes, which settle below the elytra and in the tracheae. The immature adult beetle then flies from the wood carrying nematodes.

5.4.3.3. Host reaction

Two phases of symptom development occur after invasion of the wood by nematodes (Suzuki, 2002). In the early stage, cytological changes occur in the xylem parenchymatous cells, followed by cavitation and blockage within the tracheids. These internal symptoms are induced in both compatible and incompatible combinations of pine tree and PWN isolates. The first indication of the presence of nematodes in the tree is a reduction of oleoresin exudation rate, which marks the onset of advanced stage symptoms. Cambial death and cavitation in the outer xylem result in a water deficit that reduces transpiration and photosynthesis, causing the first obvious external symptoms, yellowing and wilting of the needles. The physiological water status of the tree and the nematode population density are both considered to be key factors in disease development. Experimental results suggest that pine seedlings do not wilt solely by virtue of the number of nematodes under favourable conditions such as a well-watered environment. Pine wilt disease seems to occur more frequently and to be more destructive in summers with little rainfall. The wilting may first appear on only one branch ('flag'), although the whole tree may later show symptoms. Infections with *B. xylophilus* eventually lead to death of the tree.

5.4.3.4. Movement and dispersal

The nematodes can move within the wood tissues and can leave one piece of wood to move into neighbouring pieces. However, without their vectors they are incapable of moving from one host tree to another. Adults of the vector beetles have a peak of flight activity about 5 days after emergence. Beetles are capable of flying up to 3.3 km, but, in most cases, dispersal is only for a few hundred metres. Infested wood is the most probable means of international transport of *B. xylophilus*, and the species has been intercepted on sawn wood, round wood and wood chips. The most serious pathway of introduction of *B. xylophilus* is when imported together with vector insects, which then carry the nematodes to coniferous trees.

Whether introduced with or without a vector insect, long-term establishment requires the nematode to find a means of coming into contact with a native vector, and this can probably be achieved only if the nematode first invades wood containing larvae or pupae of a potential vector.

5.4.3.5. Ecology, host range and distribution

Serious pine wilt disease is associated with higher temperature and occurs only where the mean summer temperatures exceed 20°C. In the laboratory, *B. xylophilus* can be maintained on fungal cultures. It reproduces in 12 days at 15°C, 6 days at 20°C and 3 days at 30°C. Egg-laying starts on the fourth day after hatching and the juveniles hatch in 26–32 h at 25°C. The temperature threshold for development is 9.5°C.

Bursaphelenchus xylophilus is found mainly on *Pinus* spp. Apparently, the dead wood of all species of *Pinus* can act as a substrate for *B. xylophilus* development. However, only a limited number of species of *Pinus* are susceptible to attack as living trees. The Far Eastern species *P. bungeana*, *P. densiflora*, *P. luchuensis*, *P. massoniana* and *P. thunbergii* in their native habitats), the European species *P. nigra*, *P. sylvestris* and *P. pinaster* are the only species known to be killed by pine wilt disease as mature trees in the field. Many other species have been found to be damaged or killed by the nematode only under experimental conditions (mainly as seedlings in glasshouses). Other conifers can also act as hosts (primarily *Larix*, *Cedrus*, *Tsuga*, *Pseudotsuga*, *Abies* and *Picea*) but reports of damage are rare. Isolated cases of death of *Picea* and *Pseudotsuga* due to this nematode have been reported in the USA. Some pine trees could act as asymptomatic carrier trees. Such trees survive 1 year or more after being infected with PWNs without displaying any visible symptoms, and then show wilting symptoms and release volatiles that attract *Monochamus* beetles. This could play an important role in the epidemic occurrence of pine wilt disease (Futai, 2003).

It is presumed that *B. xylophilus* originated in North America and was transported from there to Japan in infested timber at some time around the beginning of the 20th century. The fact that native American conifers are mostly resistant, while Japanese species are susceptible, supports this view. From Japan, *B. xylophilus* spread to other Asian countries including China, Korea Republic and Taiwan. In North America the nematode is present in Canada, Mexico and the USA, probably wherever *Pinus* occurs except Hawaii. In Europe, *B. xylophilus* was detected for the first time in Portugal on *P. pinaster*. The nematode was later reported several times from Spain, where eradication measures were implemented.

5.4.3.6. Molecular diagnosis

Bursaphelenchus xylophilus has a quarantine status in many countries (see Chapter 13) but is morphologically similar to *B. mucronatus*, so efficient diagnosis is critical. Both species (including the mucronate form of *B. xylophilus*) can be separated by sequencing or digesting fragments of DNA of several regions, such as the ITS and intergenic spacer (IGS) regions in the rDNA. Species-specific primers and probes have been designed for targeting ITS and IGS regions in rDNA, satellite DNA (satDNA) and real-time PCR technology based on a heat shock protein gene (*hsp70*). Satellite DNA technique can discriminate a single nematode spotted onto a filter. Because the *Msp*I satellite sequence is polymorphic within the species, this technique can also be used to characterize strains. Some of these identification tools could be adapted for wood and vector sampling after specific DNA extraction, or multiplexed to detect several species at the same time (Filipiak *et al.*, 2019). New technologies using isothermal amplification techniques have been included such as LAMP (Kikuchi *et al.*, 2009), denaturation bubble mediated strand exchange amplification (SEA) (Liu *et al.*, 2019) and recombinase polymerase amplification (RPA) (Cha *et al.*, 2020; Zhou *et al.*, 2022b). These techniques could be combined with the use of a lateral flow dipstick (LFD), and results could be directly interpreted by the naked eye in a few minutes.

5.4.3.7. Interaction with other pathogens

The interaction of *B. xylophilus* with *Monochamus* spp. is of paramount importance because the insect is the natural vector for the nematode. *Monochamus alternatus* is the major vector in Asian countries; in Portugal the major vector is *M. galloprovincialis*, but other *Monochamus* spp. are less effective in the transmission. *Monochamus* spp. are not only vectors but also pests of *Pinus* spp.

It is hypothesized that a third party, namely toxin-producing bacteria, is involved in pine wilt disease (Oku *et al.*, 1980). The disease is the result of an integrative double vector system, where the nematode transmitted the pathogenic bacteria within the tree, and the cerambycid beetle transmitted the nematodes from tree to tree. Later, evidence of these bacteria in plant tissues infected with PWN as well as in the body surface of the nematode was also reported (Vicente *et al.*, 2012). Studies have indicated that aseptic PWNs do not cause pine wilt disease in aseptic pine trees, while PWNs associated with bacteria cause wilting symptoms (Zhao *et al.*, 2003), while other authors found the contrary. Inoculation with bacteria alone did not lead to the development of disease symptoms (Zhao *et al.*, 2003). Some bacteria associated with *B. xylophilus* from different pine species might be helpful to adjust the parasitic adaptability and virulence of PWN (Tian *et al.*, 2015).

5.4.3.8. Management and control

Control measures against pine wilt disease aim at breaking the pine tree–nematode–insect disease triangle. Control must be concentrated on the activities that may represent risk of entry and dissemination. Wood trade between countries is nowadays highly monitored and all infested wood should be carefully treated before shipment or use in manufacturing.

In Japan, PWN control strategies have concentrated on a combination of removing dead or dying trees from the forest to prevent their use as a source of further infection, and control of vector beetles with insecticides. Research is ongoing to find alternative means of control, including natural and/or chemical nematicides/insecticides with low side effects, biological control of nematodes and vectors, use of insect attractants, and the more challenging and long-term approach, breeding programmes for resistance and tolerance of forest trees. In early-stage symptomatic trees or asymptomatic carrier trees, detection and removal are important for the successful and efficient management of the disease. Artificial intelligence techniques and unmanned aerial vehicle imagery have been used recently to detect these types of trees (Iordache *et al.*, 2020; Syifa *et al.*, 2020; Yu *et al.*, 2021a).

Ectoparasitic Nematodes*

Wim Bert**, Wilfrida Decraemer and Etienne Geraert

Department of Biology, Ghent University, Belgium

*A revision of Decraemer, W. and Geraert, E. (2013) Ectoparasitic nematodes. In: Perry, R.N. and Moens, M. (eds) *Plant Nematology*, 2nd edn. CAB International, Wallingford, UK.

**Corresponding author: Wim.Bert@UGent.be

DOI: 10.1079/9781800622456.0006

6.1. Introduction to Ectoparasitic Nematodes

Molecular research supports the hypothesis that plant parasitism in the phylum Nematoda evolved independently at least three times (see Fig. 2.9). Two of the three major plant-parasitic taxa, the infraorder Tylenchomorpha De Ley & Blaxter, 2002 and the Longidoridae, show a similar adaptation to plant parasitism by the possession of a hollow protrusible stylet as a tool for penetrating the plant cell wall, whilst in the Trichodoridae, the stylet is a modified stylettiform mural dorsal tooth.

The Tylenchomorpha (belonging to the suborder Tylenchina) are the largest group of plant-parasitic nematodes and include the tylenchs or tylenchids and the aphelenchs (superfamily Aphelenchoidea Fuchs, 1937). This group comprises the economically important endoparasitic root-knot nematodes (Chapter 3), cyst nematodes (Chapter 4), migratory endoparasites (Chapter 5) and ectoparasites (this chapter) (Fig. 1.1, Chapter 1).

Ectoparasitic tylenchs *sensu stricto* can be grouped according to their parasitic strategies: (i) the migratory ectoparasites, which stay vermiform throughout their life cycle and feed for short periods along the root system (e.g. *Trichodorus*, *Belonolaimus*, *Dolichodorus*); and (ii) the sedentary ectoparasites, which are those species that may feed for several days on the same cell, either a cortical or an epidermal cell (Tytgat *et al.*, 2000). Other tylench taxa are able to also act as semi- or facultative ecto-endoparasites (*Hoplolaimus*, *Helicotylenchus*, *Scutellonema*) but are still considered as ectoparasites.

The Longidoridae and Trichodoridae are migratory root ectoparasites. They are responsible for substantial direct damage to a wide variety of plants, but their major pest status is as virus vectors, despite the rather low number of vector species in both families. Virus–vector association in Nematoda is only known from these two families; both families acquired the association with the non-related viruses independently, with Nepoviruses being associated with members of the family Longidoridae and Tobraviruses with Trichodoridae.

6.2. Definition of Ectoparasites/Ectoparasitism

The term 'ectoparasite' comes from the Greek: *ecto* (= outer), *para* (= with, at), *siteo* (= feeding). An ectoparasite is a parasite that lives on the outer surface of a host; nematode ectoparasites do not enter the plant tissues with their body but use their stylet to puncture plant cells and feed upon the cytoplasm. The longer the stylet, the deeper the nematode ectoparasite can feed. In general, short stylet-bearing forms, e.g. trichodorids and tylench taxa, such as *Tylenchorhynchus* and *Helicotylenchus*, usually feed on root hairs and/or epidermal cells; ectoparasites with long stylets, such as longidorids and the tylench genera *Belonolaimus* and *Dolichodorus*, exploit deeper tissues. All stages of ectoparasites generally feed on the root. As ectoparasites *sensu stricto* do not enter the plant, the damage they cause is usually limited to necrosis of those cells penetrated by the stylet, and this may cause galling. However, cells are not always killed; in trichodorids, for instance, cells can be punctured without being consumed and cytoplasmic streaming may be restored once the nematode moves away. It is worth noting that, according to Yeates (1971), the various migratory ecto- and endoparasites are not considered true parasites but rather a highly adapted group of plant browsers.

6.3. Classification

The classification (Box 6.1) follows the scheme proposed by De Ley and Blaxter (2002) as used throughout this book. The classification is visually represented by schematic drawings that display the most economically important ectoparasitic families (Fig. 6.1).

Box 6.1. Classification of major ectoparasitic taxa.

Phylum Nematoda Potts, 1932
Class Chromadorea Inglis, 1983
Subclass Chromadoria Pearse, 1942
Order Rhabditida Chitwood, 1933
Suborder Tylenchina Thorne, 1949
Infraorder Tylenchomorpha De Ley & Blaxter, 2002
Superfamily Tylenchoidea Örley, 1880
Family Dolichodoridae Chitwood *in* Chitwood & Chitwood, 1950
  Genera *Dolichodorus* Cobb, 1914
    Belonolaimus Steiner, 1949
    Tylenchorhynchus Cobb, 1913
    Amplimerlinius Siddiqi, 1976
Family Hoplolaimidae Filipjev, 1934
  Genera *Hoplolaimus* Daday, 1905
    Scutellonema Andrássy, 1958
    Helicotylenchus Steiner, 1945
    Rotylenchus Filipjev, 1936
Superfamily Criconematoidea Taylor, 1936 (1914)
Family Criconematidae Taylor, 1936
  Genus *Criconemoides* Taylor, 1936
Family Tylenchulidae Skarbilovich, 1959
  Genus *Paratylenchus* Micoletzky, 1922
Family Hemicycliophoridae Skarbilovich, 1959
  Genus *Hemicycliophora* de Man, 1921

Class Enoplea Inglis, 1983
Subclass Dorylaimia Inglis, 1983
Order Dorylaimida Pearse, 1942
Suborder Dorylaimina Pearse, 1942
Superfamily Dorylaimoidea Thorne, 1935
Family Longidoridae Thorne, 1935
 Subfamily Longidorinae
  Genera *Longidorus* Micoletzky, 1922
    Longidoroides Khan, Chawla & Saha, 1978
    Paralongidorus Siddiqi, Hooper & Khan, 1963
 Subfamily Xiphinematinae
  Genus *Xiphinema* Cobb, 1913
Subclass Enoplia Pearse, 1942
Order Triplonchida Cobb, 1920
Suborder Diphtherophorina Coomans & Loof, 1970

Continued

Box 6.1. Continued.

Superfamily Diphtherophoroidea Micoletzky, 1922
Family Trichodoridae Thorne, 1935
Genera *Nanidorus* Siddiqi, 1974
Paratrichodorus Siddiqi, 1974
Trichodorus Cobb, 1913

6.4. Tylenchina (Chromadorea, Chromadoria): selected genera

Morphological descriptions are based on females, unless otherwise stated.

6.4.1. *Dolichodorus* Cobb, 1914 (Fig. 6.2A–H)

Dolichodoridae. Body long (1–3.5 mm) and slender. Lateral field with three incisures, areolated. Head offset, rounded, striated, roughly quadrangular to prominently four lobed in *en face* view; labial disc more often prominent; subdorsal and subventral lip sectors distinct; lateral lip sectors reduced or absent. Sclerotization of basal plate prominent. Stylet very long (50–160 μm), strong. Median and basal bulbs well-developed, joined by a short, slender isthmus. Vagina variously sclerotized; no epiptygmata. Tail hemispherical-spiked, rarely conoid. Males with large, trilobed bursa; spicules most generally with prominent flanges; gubernaculum also robust, protrusible (Luc and Fortuner, 1987).

The genus *Dolichodorus* contains 17 species, known as 'awl' nematodes (Geraert, 2011). The pathogenicity of *D. heterocephalus* Cobb, 1914 and the symptoms expressed by its host plants (see below) are well known. Those of some other species of *Dolichodorus* are less well known but from the information available appear to be very similar to those of *D. heterocephalus*. The genus has been recorded principally from widely separated states in the USA and from widely separated countries all over the world, predominantly in subtropical and tropical areas. The host range of *D. heterocephalus* is extensive and includes agronomic crops, vegetable crops (Hallmann and Meressa, 2018), fruit trees, ornamental trees, shrubs, grasses and weeds. All stages of the nematode feed from the surface of roots by injecting its long stylet into cells mostly at root tips. If feeding occurs in the piliferous area for 1–4 h, the root becomes constricted in that area. Normal growth of the root tip is inhibited in the vicinity of a feeding site and the elongation of cells of the non-parasitized areas causes the root to curve. Root tips may appear enlarged because the tissues mature right to the tip, and occasionally galls appear on root tips. When an infected plant produces new root initials, the nematode feeds on them, stopping their growth as well. The result is a skeleton root system with almost no feeder roots and few, if any, secondary roots. This is often referred to as stubby root or coarse root or both. Root tissue surrounding the feeding site is destroyed and discoloured. Above ground, the plants are stunted and chlorotic due to the lack of an adequate root system. Control of these nematodes has been achieved with nematicides, with resistant cover crops, by rotating susceptible crops with resistant varieties and by using organic soil amendments (Smart and Nguyen, 1991).

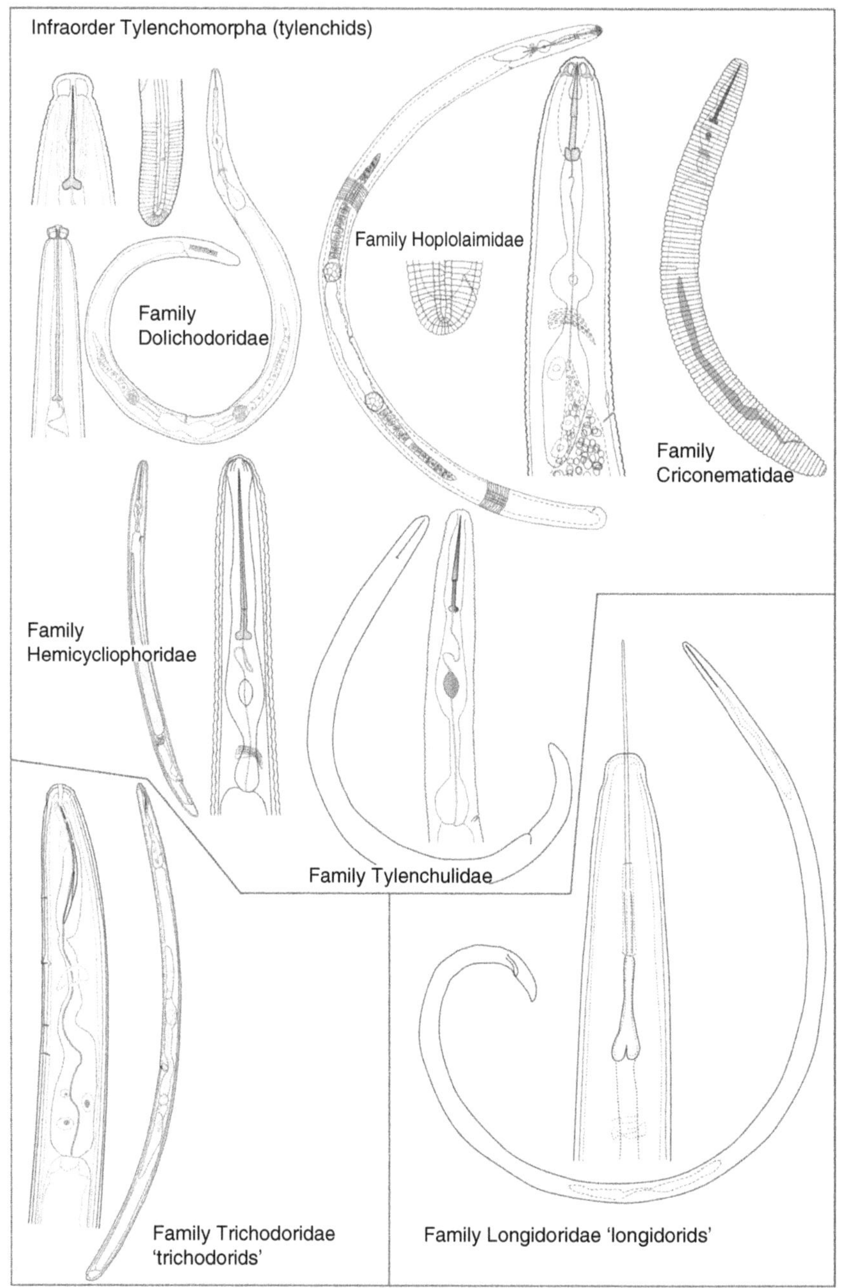

Fig. 6.1. Schematic representation of the major ectoparasitic families, belonging to the Order Rhabditida, Infraorder Tylenchomorpha ('tylenchids'); Order Dorylaimida, family Longidoridae ('longidorids'); and Order Triplonchida, family Trichodoridae ('trichodorids'). Not to scale.

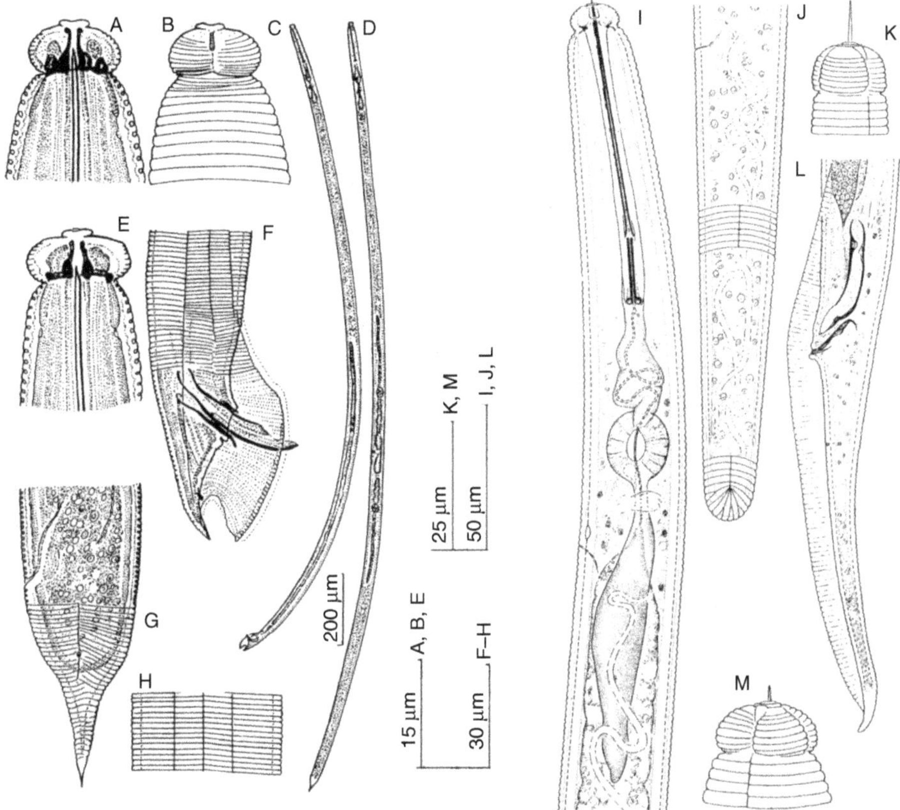

Fig. 6.2. A–H: *Dolichodorus heterocephalus* Cobb. I–M: *Belonolaimus longicaudatus* Rau. A, B and M, head ends of females; C, male; D, female; E and K, head ends of males; F and L, tail ends of males; G and J, tail ends of females; H, lateral field; I, pharyngeal region of female. (A–H after Siddiqi, 1976, courtesy *Nematologica*; I–M after Orton Williams, 1974, courtesy CABI.)

6.4.2. *Belonolaimus* Steiner, 1949 (Fig. 6.2I–M)

Belonolaimidae: Belonolaiminae. Large (2–3 mm) and slender (a = 50–80) tylenchs. Lateral field with a single, two or four incisures. Head large, rounded, mostly offset, divided by longitudinal grooves into four well-separated sectors; framework with basal plate moderately and arches lightly sclerotized. Stylet very long (90–160 µm), with conus more than twice the shaft length. Median bulb round, isthmus short; pharyngo-intestinal junction close to median bulb, anterior to main body of glands; pharyngeal glands overlapping intestine lateroventrally. Vulva equatorial, with epiptygmata; spermatheca axial, slightly offset; ovaries paired. Tail elongate, in female cylindroid, three to six anal body widths long; with post-anal extension of intestine and annulated terminus. Spicules robust, slightly arcuate, flanged. Gubernaculum large, not protrusible. Male tail elongate–conoid, enveloped by a low bursa. (Fortuner and Luc, 1987).

The genus *Belonolaimus* contains about ten species, of which *B. longicaudatus* Rau, 1958 is well known as an ectoparasite (or 'sting' nematode). Sting nematodes are predominantly found in sandy soils in the USA, but they have also been reported in Middle and South America and Australia. Although ectoparasitic, these nematodes feature a very long stylet that allows them to feed deep within plant root tissue, penetrating the cortex to reach the vascular cylinder. A large number of plants, including cereals, potato, fruit trees, cotton, peanut, corn, various fruits, vegetables, forage grasses and especially the citrus family (Shokoohi and Duncan, 2018) as well as turf grasses have been shown to be suitable hosts (Crow *et al.*, 2001; Hunt *et al.*, 2012). Whilst the effects of sting nematode parasitism vary somewhat depending on such factors as inoculum level, host plant variety and plant age, in general the affected plants display a greatly reduced root system and exhibit a combination of stubby and coarse roots, with dark lesions apparent along the length of the root as well as at the tips. Sometimes roots are girdled completely. Studies suggest that the salivary secretions of the feeding nematodes affect plant cells beyond those that were actually fed on (Willett *et al.*, 2020). Shoot symptoms consist of stunted growth, premature wilting and leaf chlorosis (Smart and Nguyen, 1991). Attacks involving *Belonolaimus* also increase the susceptibility of plants to *Fusarium*. Seasonal vertical migration allows the nematodes to escape from nematicides, although owing to the nematode's lack of a long-term survival stage, fallow periods or crop rotation with a non-hosting variety has been found to be effective in eradicating *Belonolaimus longicaudatus* (Crow, 2015). The management of *B. longicaudatus* in Florida strawberry has been extensively covered by Noling (2021).

6.4.3. *Tylenchorhynchus* Cobb, 1913 (Fig. 6.3)

Belonolaimidae: Telotylenchinae. Body about 1 mm or less. Lateral field with two, three, four or five lines, sometimes areolated. Longitudinal ridges sometimes present on body. Head offset from body or continuous, annulated or rarely smooth; head framework light to moderately sclerotized. Scanning electron microscopy (SEM) shows various end-on patterns, the most typical with labial disc fused with first lip annulus, and with lateral sectors regressed; the remaining sub-median sectors give a distinctive square appearance to the face view; papillae often present on sub-median sectors. Stylet 15–30 µm long, thin to slender, with cone about as long as shaft, sometimes needle-like; knobs prominent. Median bulb round to oval, with distinct refractive thickenings; basal bulb usually offset from intestine, sometimes dorsally overlapping. Vulva near middle of body, generally at 50–54%, lips not modified; spermatheca round, axial; ovaries paired, outstretched. Tail conoid to sub-cylindroid (exceptionally clavate), about three times as long as wide ($c' = 2–4$), sometimes with thicker cuticle in the distal end. Male with caudal alae well developed, rarely lobed. Spicules flanged, gubernaculum about half as long as spicule (Fortuner and Luc, 1987; Geraert, 2006).

The genus *Tylenchorhynchus* contains more than 130 species. Several characters, such as the number of lines in the lateral field, presence of longitudinal ridges and thickening of tail cuticle, can be used to separate groups of species and they are useful for identification. Nomenclatorial status for such groups by naming them as genera is often proposed (Siddiqi, 2000).

Tylenchorhynchus together with representatives of several related genera constitute a group of more than 250 species, generally known as 'stunt nematodes'. Few have been proven as pathogens in the strict sense but about 8% are known ectoparasites. *Tylenchorhynchus claytoni* Steiner, 1937 is present in the north temperate zone, where various plants are attacked, in particular azaleas, strawberry, holly and conifers, causing

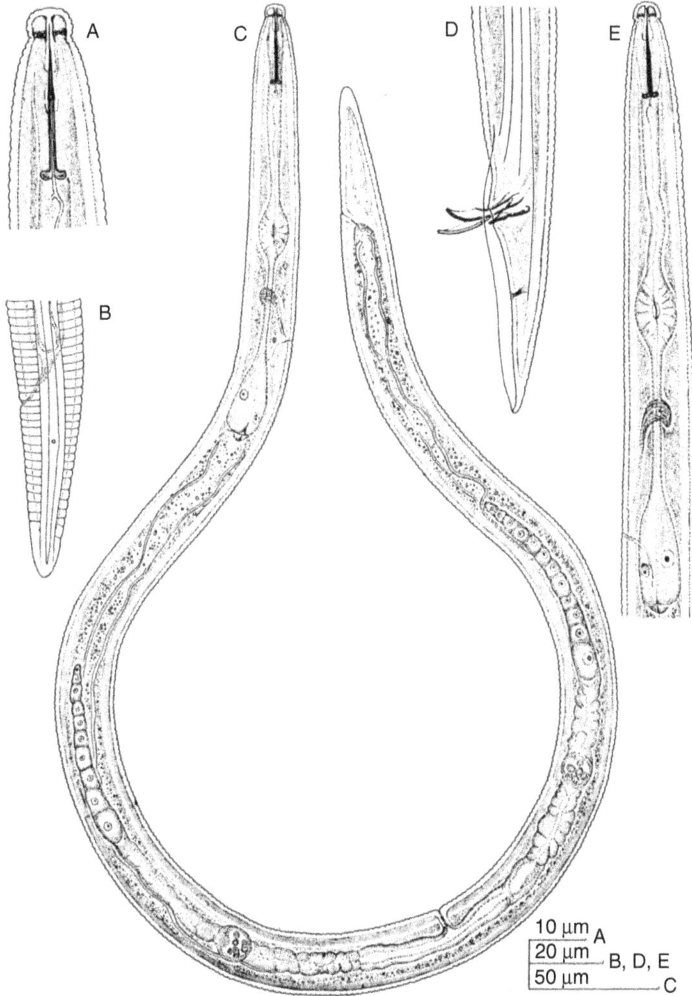

Fig. 6.3. *Tylenchorhynchus cylindricus* Cobb. A–C: neotype female. D and E: male. A, head end; B, tail end of female; C, entire female; D, tail end of male; E, pharyngeal region of female. (After Siddiqi, 1972c, courtesy CABI.)

reduced growth. *Tylenchorhynchus annulatus* (Cassidy, 1930) Golden, 1971 is found in the subtropical and tropical regions of all continents, except Europe. It is a pathogenic ectoparasite of sugar cane, Bermuda grass, sorghum and especially rice. *Tylenchorhynchus nudus* Allen, 1955 has been reported mainly from the Midwest USA, from the Orient and from India, and caused growth reduction in Kentucky bluegrass, spring wheat and sorghum, whilst sugar cane yield increased by 10–20% when nematodes were controlled by fumigation. *Tylenchorhynchus cylindricus* Cobb, 1913 (Fig. 6.2) is mainly found in North America where it causes moderate stunting in cotton and bean crops. *Tylenchorhynchus capitatus* Allen, 1955 is widespread across all continents and has a broad host range including hollyhock, apple, plum, strawberry, citrus, potato, wheat, tobacco and maize. *Tylenchorhynchus dubius* (Bütschli, 1873) Filipjev, 1936 is present throughout the northern hemisphere and is especially common in Europe. Many of the reports of *T. dubius* association with plants have indicated stunted growth and weight reduction as their main detrimental effects (Anderson and Potter, 1991). The presence of *Tylenchorhynchus brassicae* Siddiqi, 1961 has been shown to have economic significance for vegetable harvests, although studies indicate that certain cauliflower cultivars appear to be resistant to such attacks (Hallmann and Meressa, 2018).

6.4.4. *Amplimerlinius* Siddiqi, 1976 (Fig. 6.4)

Belonolaimidae: Telotylenchinae. Body medium to large (1–2 mm). Lateral field with six lines over most of the body, generally with areolations. Head continuous with body contour; labial framework robust. SEM face view broadly rounded, and laterally elongated, with labial disc partially or completely fused with first lip annulus, lip annulus sectors also partly or completely fused together. Stylet robust, 19–47 µm long; knobs large. Median bulb well developed, at or behind middle of pharynx; dorsal pharyngeal gland sometimes overlapping the beginning of the intestine for a short distance. Vulva with epiptygmata. Female tail cylindrical, occasionally sub-clavate, with hemispherical, annulated terminus; cuticle thickened at tip. Spicules robust, blunt and notched at tip; gubernaculum trough-shaped in lateral view (Fortuner and Luc, 1987).

The genus *Amplimerlinius* contains about 22 species (Geraert, 2011); two species are known to be parasitic: *Amplimerlinius icarus* (Wallace & Greet, 1964) Siddiqi, 1976 and *A. macrurus* (Goodey, 1932) Siddiqi, 1976. Both species are morphologically very similar; they differ mainly in measurements, *A. icarus* being much larger. Both species are known mainly from Europe. They are ectoparasites and semi-endoparasites, feeding on oat and ryegrass roots, with the anterior third of the body embedded. They feed as fairly sedentary ectoparasites but also penetrate roots and feed on internal tissues. However, they cause little observable damage, except for darkening of the cells surrounding the embedded head; root growth appeared normal (Anderson and Potter, 1991).

6.4.5. *Helicotylenchus* Steiner, 1945 (Fig. 6.5)

Hoplolaimidae: Hoplolaiminae. Small to medium-sized (0.4–1.2 mm), spirally coiled or rarely arcuate. Lateral field with four lines. Head continuous to slightly offset, rounded or anteriorly flattened, generally annulated but almost never longitudinally striated; anterior lip annulus generally not divided into sectors. Cephalic framework well developed. Stylet robust, about three to four times maximum width of cephalic region. Orifice of dorsal pharyngeal gland 6–16 µm from stylet base. Median bulb rounded. Pharyngeal glands overlapping intestine on all sides with the position of the pharyngeal lumen situated between the dorsal gland and one of the ventrosublateral glands, longest overlap being lateroventral. Two genital branches, the posterior one sometimes degenerated or reduced to a post-vulval uterine sac. Epiptygmata present but folded inwards, into the vagina. Tail 1–2.5 anal body diameters long, typically more curved dorsally, with or without a terminal ventral process, sometimes rounded. Males sometimes with slight secondary sexual dimorphism

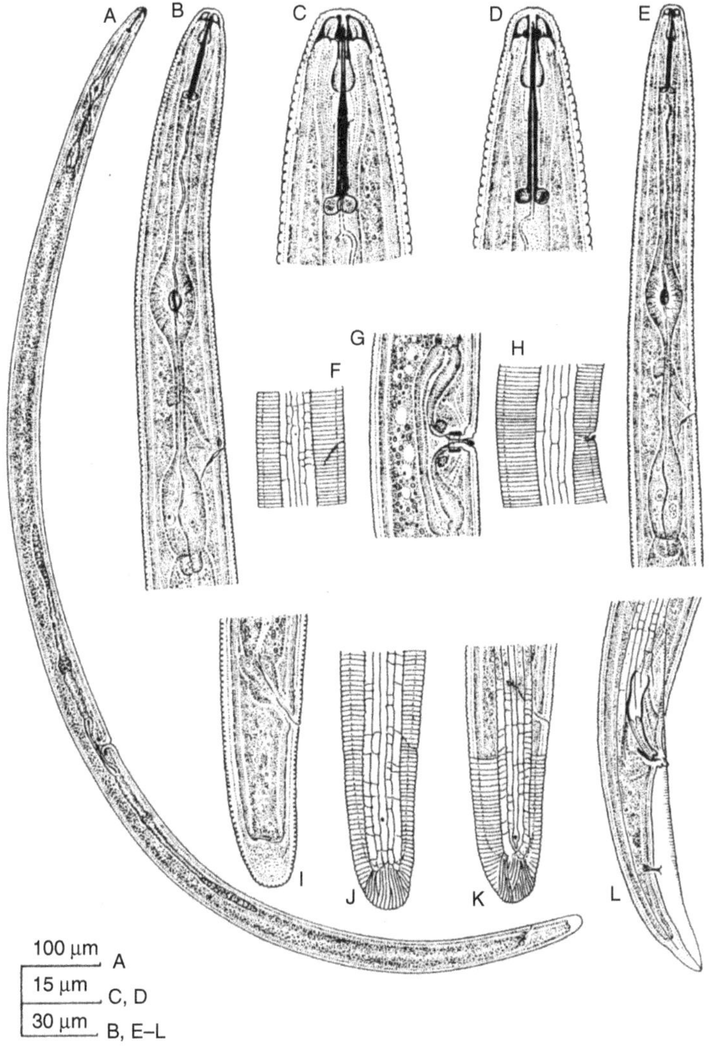

Fig. 6.4. *Amplimerlinius amplus*. A, female; B, pharyngeal region of male; C and D, head end of female and male, respectively; E, anterior region; F and H, lateral fields; G, vulval region; I–K, tail ends of females; L, tail end of male. (After Siddiqi, 1976, courtesy *Nematologica*.)

in smaller anterior end. Caudal alae enveloping tail end. Gubernaculum trough- or rod-shaped, fixed. Male tail short, conical (Fortuner, 1987).

The genus *Helicotylenchus* is one of the largest genera in the tylenchs, at present thought to contain more than 160 species. *Helicotylenchus dihystera* (Cobb, 1893) Sher, 1961 is a cosmopolitan species with an extremely broad host range, present as an ectoparasite or semi-endoparasite on the roots of many plants. The nematodes are found either partially or fully embedded in the root, where they feed on a single cell over several days. Such feeding results in cortical root lesions of the plants. *Helicotylenchus multicinctus*

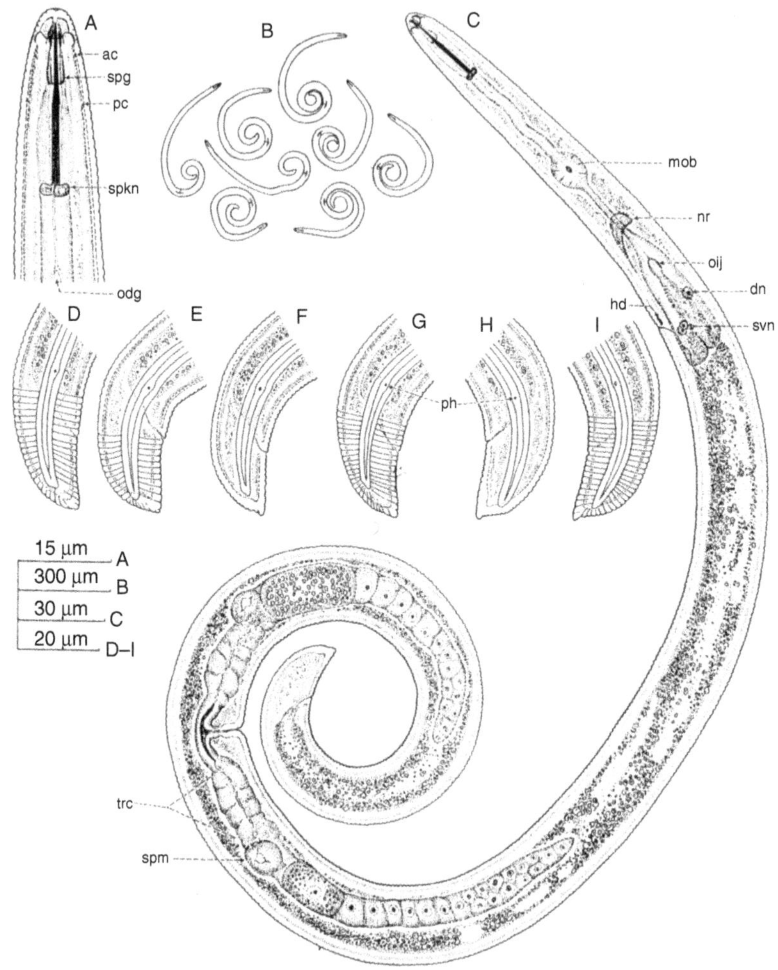

Fig. 6.5. *Helicotylenchus dihystera*. A–I: topotype females. A, head end; B and C, females; D–I, tail ends; ac, anterior cephalid; dn, dorsal pharyngeal gland nucleus; hd, hemizonid; mob, median bulb; nr, nerve ring; odg, orifice of dorsal gland; oij, pharyngo-intestinal junction; pc, posterior cephalid; ph, phasmid; spg, stylet guide; spkn, stylet knobs; spm, spermatheca; svn, subventral gland nuclei; trc, tricolumella (After Siddiqi, 1972a, courtesy CABI.)

(Cobb, 1893) Golden, 1956 is a known pest found in virtually all banana-growing areas, and is recognized as a key nematode causing reduced *Musa* production, especially that of plantain in West Africa (Coyne, 2009; Sikora *et al.*, 2018). It is an endoparasite in the cortex of the roots where it feeds and produces small superficial lesions. However, other *Helicotylenchus* species are also known to cause damage to banana plantations (Sikora *et al.*, 2018) as well as to other plants; for example, *Helicotylenchus brevis* (Whitehead, 1958) Fortuner, 1984 has been found on both cultivated plants (banana, mango) and uncultivated bush, fern and bulbous plants in Southern Africa (Fortuner, 1991). *Helicotylenchus dihystera* (Cobb, 1893) Sher, 1961 is known to be pathogenic to several crops, such as peanut, millet and tea (Baujard and Martiny, 1995; Gnanapragasam and Mohotti, 2018). Remarkably, it has been reported that the density of the more harmful root-knot nematodes inversely correlates with that of *Helicotylenchus*, leading to the suggestion that if the nematode population density can be skewed in favour of that of *Helicotylenchus*, nematode damage can therefore be reduced (personal communication, M. Daneel).

6.4.6. *Rotylenchus* Filipjev, 1936 (Fig. 6.6)

Hoplolaimidae: Hoplolaiminae. Usually rather large (1–2 mm) ectoparasites, body spiral to C-shaped. Lateral field with four lines, with or without scattered transverse striae. Head offset or continuous with body contours, anteriorly rounded or flattened, generally annulated, with or without longitudinal striae on basal lip annulus. Labial framework well sclerotized, stylet and stylet knobs well developed; stylet knobs with rounded to indented anterior surface. Opening of dorsal pharyngeal gland often close to stylet base (6 µm) but sometimes more posteriorly (up to 16 µm). Median bulb well developed; pharyngeal glands overlapping intestine dorsally and laterally; dorsal gland more developed than subventral glands. Two genital branches outstretched, equally developed; posterior branch rarely degenerated. One or two epiptygmata present. Phasmids pore-like, small, near anus level.

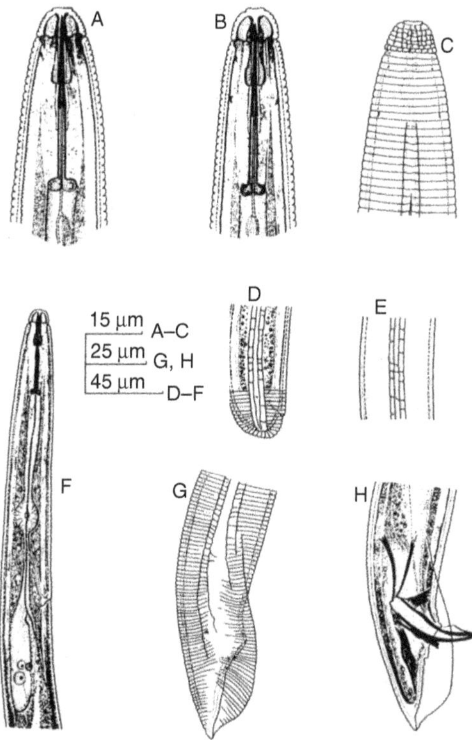

Fig. 6.6. *Rotylenchus robustus* (de Man). A and C, female head ends; B, male head end; D, female tail end; E, female lateral field at mid-body; F, female pharyngeal region; G and H, male tail ends. (After Siddiqi, 1972b, courtesy CABI.)

Tail short, hemispherical, rarely with small ventral projection. Males with caudal alae enveloping tail, not lobed. Spicules robust, flanged. Gubernaculum titillate (Fortuner, 1987).

Rotylenchus species have a worldwide distribution occurring on every continent, including Antarctica and more than 100 species have been recognized. *Rotylenchus* species are root ectoparasites to semi-endoparasites on a wide variety of economically important plants. They are able to penetrate deep into root tissues to feed on cortical parenchyma, resulting in destruction of epidermal and cortical cells (Castillo and Vovlas, 2005; Cantalapiedra-Navarrete *et al.*, 2012; Singh *et al.*, 2022). The most damaging *Rotylenchus* species include *Rotylenchus buxophilus* Golden, 1956, *Rotylenchus pumilus* (Perry in Perry, Darling & Thorne, 1959) Sher, 1961, *Rotylenchus robustus* (de Man, 1876) Filipjev, 1936 and *Rotylenchus uniformis* (Thorne, 1949) Loof & Oostenbrink, 1958, which can cause, for example, severe necrotic lesions in the root cortex of boxwood bushes; severe destruction in the root system of carrots; yield reduction in pea; stunting, yellowing and yield loss in lettuce crops; growth reduction in marguerite daisies; reducing root development in juniper; and severe parasitizing of ornamental Liliaceae plants (references in Castillo and Vovlas, 2005 and Singh *et al.*, 2022). *Rotylenchus laurentinus* Scognamiglio & Talamé, 1973 was found on carrots in Italy feeding as a semi-endoparasite and causing lesions and cavities in epidermal and cortical tissues. Other *Rotylenchus* species have been recorded in association with sugar cane in several countries (Fortuner, 1991), and cause reduction in root and top weights of *Pinus* seedlings in the savannah area of Nigeria and cause damage to fruit trees in China (Castillo and Vovlas, 2005).

6.4.7. *Hoplolaimus* Von Daday, 1905 (Fig. 6.7)

Hoplolaimidae: Hoplolaiminae. Body straight, large (1–2 mm). Lateral field with four lines or less, generally areolated at level of phasmids and anteriorly, sometimes with striae irregularly scattered over entire field, not areolated. Lip region offset from body, wide, anteriorly flattened, with clearly marked annuli and with longitudinal striae. Labial framework and stylet massive; stylet knobs anchor- or tulip-shaped. Dorsal pharyngeal gland opening 3–10 µm from stylet base. Pharyngeal glands overlap intestine dorsally and laterally; sometimes gland nuclei duplicated to a total of six nuclei. Two genital branches outstretched, equally developed. Tail short, rounded. Phasmids enlarged to scutella erratically situated on body, anterior to anus level and sometimes anterior to vulva level, not opposite each other. Male with caudal alae enveloping tail, regular. Secondary sexual dimorphism visible in labial region; pharyngeal structures smaller in males. Spicules massive with distal flanges; gubernaculum large, protrusible, titillate (Fortuner, 1987).

The genus *Hoplolaimus* contains more than 30 species, of which several are known endoparasites. *Hoplolaimus galeatus* (Cobb, 1913) Thorne, 1935 is widely distributed in the USA and has been reported from Canada, Central and South America, and India. It has a large variety of hosts, in particular cotton, trees (pine, oak, sycamore, etc.), maize, turf grasses and other graminaceous plants. It lives as an endoparasite on cotton, causing considerable damage to cortex and vascular tissue. Several other *Hoplolaimus* species are also parasites of cotton (Davis *et al.*, 2018). On pine, most of the cortex of infected roots is destroyed. In sycamore it causes extensive root necrosis but it is unable to penetrate completely within the roots and its body partly protrudes out of the root. *Hoplolaimus pararobustus* (Schuurmans-Stekhoven & Teunissen, 1938) Coomans, 1963 is found in Africa, mostly within the roots of banana (Sikora *et al.*, 2018), but it can also parasitize

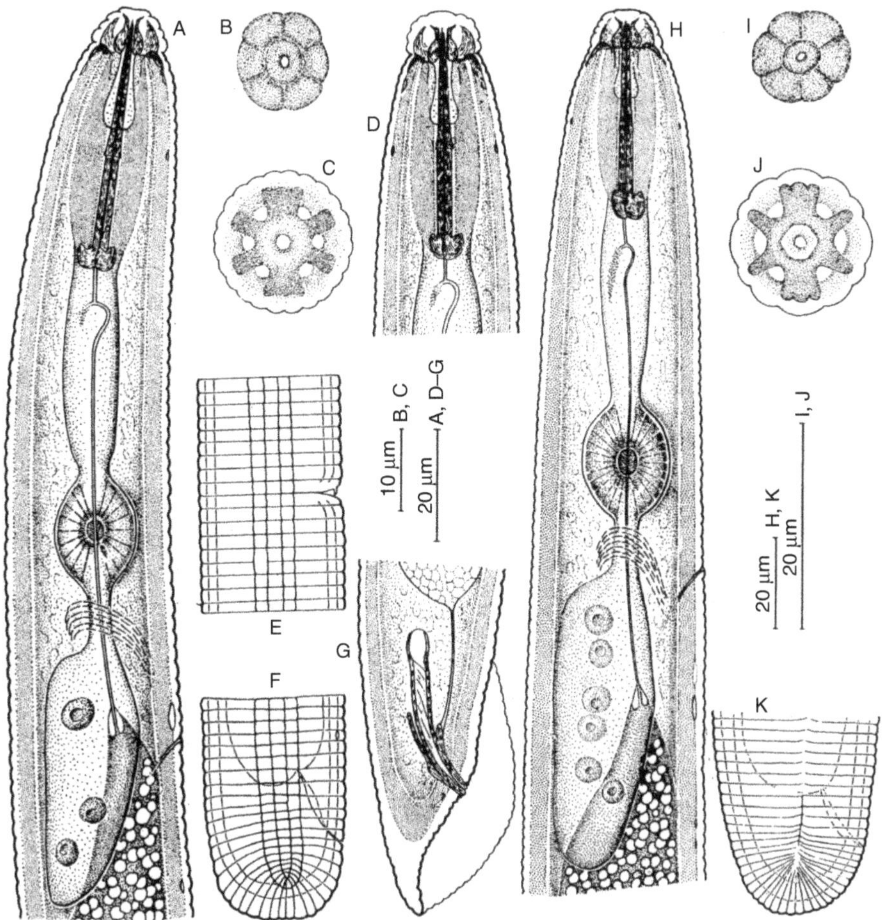

Fig. 6.7. A–G: *Hoplolaimus tylenchiformis* Daday. H–K: *H. seinhorsti* Luc. A and H, pharyngeal regions of females; B and I, end-on views of females; C and J, cross-sections through basal annulus of head of females; D and G, head and tail end of male, respectively; E, surface view at vulval region; F and K, tail ends of female showing lateral field. (After Sher, 1963, courtesy *Nematologica.*)

other plants, including coffee, tea, sugar cane, palm trees, various tropical fruit trees, rice and yam. It has been described from grass in India and from various plants in Pakistan. On banana, it feeds mostly endoparasitically but it has occasionally been observed only partially embedded within the roots. On coffee, it feeds semi-endoparasitically. Cortex penetration results in cavities and ruptured cells, and numerous irregular brown necrotic lesions develop on the roots of infested coffee plants. *Hoplolaimus columbus* Sher, 1963 is an important parasite of soybean and cotton in the USA (Mueller, 2021). It acts as an endoparasite of soybean roots, penetrating the endodermis, pericycle and phloem, whereas on cotton, it functions as a semi-endoparasite where it penetrates the cortex but not the endodermis (Fortuner, 1991). Unfortunately, a wide range of crops such as maize, grain, sorghum, millet, wheat, and many grasses and forage legumes also prove to be excellent hosts for *H. columbus* (Mueller, 2021), making carefully planned crop rotations and efficient weed control important factors for the successful control of this species. A comprehensive overview of the symptoms of *H. columbus*, its interaction with other pathogens and possible integrated nematode management strategies is provided by Mueller (2021). Further reported important species include *Hoplolaimus indicus* Sher, 1963 and *H. clarrissimus*, which are known parasites of rice crops (Peng *et al.*, 2018).

6.4.8. *Scutellonema* Andrassy, 1958 (Fig. 6.8)

Hoplolaimidae: Hoplolaiminae. Small to large nematodes, habitus spiral, C-shaped or almost straight. Lateral field with four lines, areolated anteriorly, usually areolated at level of scutella and anteriorly, and sometimes throughout length. Labial region set off or continues, rounded, rarely conical, and usually annulated with basal annulus often divided

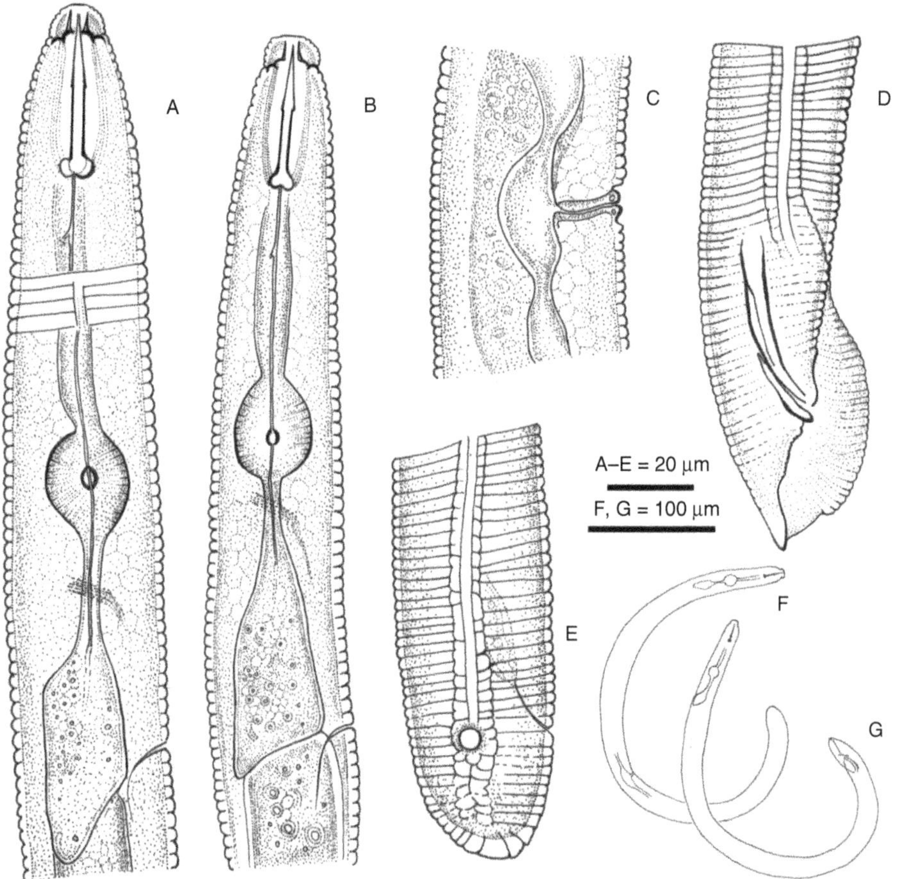

Fig. 6.8. *Scutellonema* sp. A, pharyngeal region of female; B, pharyngeal region of male; C, vulva region; D, male tail region; E, female tail region showing scutellum (two scutella at the same level, the other not visible); F, G, total body of female and male, respectively (X. Qing, unpublished).

into blocks by longitudinal lines. Labial framework sometimes strongly developed but mostly average sized. Stylet knobs rounded or indented anteriorly. Pharyngeal gland with greatest overlap dorsally and dorsolaterally. Two genital branches outstretched. Epiptygma present, sometimes well protruding but usually folded into the vagina or covering the vulval opening. Tail short, rounded or dorsally convex conoid, rarely conical or with ventral projection. Phasmids greatly enlarged to form scutella, opposite each other, pre- or postanal. Male similar to female with spicules and gubernaculum distinct. Bursa large and enveloping the tail tip (after van den Berg and Quénéhervé, 2012).

The genus *Scutellonema* contains some 50 species of which the majority are primarily ectoparasites of roots, notwithstanding certain species that are known to be semi-endoparasitic or endoparasitistic of roots and/or tubers. A key to the species is provided by Kolombia *et al.* (2017). This genus is most prevalent in Africa and most species are polyphagous, although *Scutellonema bradys* or 'the yam nematode' is a highly damaging migratory endoparasite of yam in Africa, Central and South America, India and the southern USA. Feeding by the nematode results in necrotic lesions beneath the yam's outer skin, and in combination with the colonization of deeper tissue by fungi can lead to the disease 'dry rot of yam', characterized by a persistent decline of tuber quality and even the complete destruction of tubers during storage. *Scutellonema brachyurus* has a worldwide distribution, is especially prevalent in Africa, and is found associated with a large variety of crops and wild plants. They are primarily ectoparasites of roots, but instances of semi-endoparasitism and endoparasitism have also been reported. It feeds, for example, both ecto- and endoparasitically on African violets and semi-endoparasitically on the roots of sycamore, with only the anterior end of the body being embedded in the cortex (van den Berg and Quénéhervé, 2012). Interestingly, it has also been reported that a high density of *S. brachyurus* can lead to stimulation rather than reduction in root growth of longleaf pine. *Scutellonema brachyurus* is the most common species of the genus in South Africa, recorded as parasitizing virtually every crop plant, fruit tree, and almost all other ornamental and natural vegetation. Very frequently found on tobacco in South Africa and Zimbabwe where it causes root lesions and reduction of top weight and height, this species is also the cause of poor sugar cane growth in Madagascar (van den Berg and Quénéhervé, 2012). The general approach for *S. bradys* management involves: (i) interrupting the nematode spread with seed tubers; and (ii) breaking the nematode life cycle in soil (Claudius-Cole, 2021). Management options for *S. bradys* therefore include the use of nematode-free planting material, hot water treatment of yam tubers, treatment of soil by chemical, biological and cultural means, and treatment of tubers after harvest (Coyne and Affokpon, 2018; Claudius-Cole, 2021). Thus far, cultivars showing resistance to *S. bradys* infection have remained elusive (Claudius-Cole, 2021).

6.4.9. *Criconemoides* Taylor, 1936 (Fig. 6.9)

Criconematidae: Macroposthoniinae. Sexual dimorphism. Females with a body of variable length (0.2–1 mm). Annuli large, distinct; mostly retrorse; 36–219 annuli; posterior edge smooth to finely crenate. Submedian lobes generally well developed, but may be poorly developed or even absent in some species; separated or connected in different ways; first annulus may be reduced or even divided into plates. Vulva closed to open; anterior lip may be ornamented. Spear strong, stylet knobs mostly anchor-shaped; stylet rarely thin and flexible or exceptionally short with rounded basal knobs. Male with head

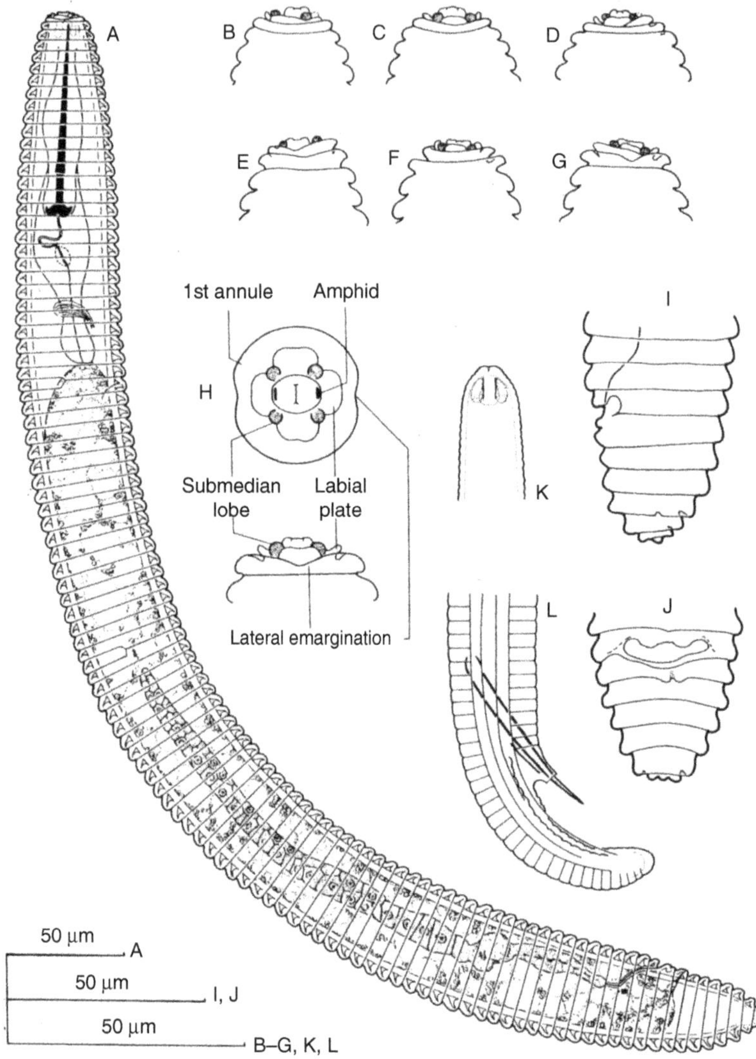

Fig. 6.9. *Criconemoides xenoplax*. A, entire body; B–H, head ends of females; I and J, female; K and L, male, head and tail, respectively. (After Orton Williams, 1972, courtesy CABI.)

end rounded to conoid; generally four lateral lines (rarely three or two); caudal alae distinct, exceptionally absent (Raski and Luc, 1987).

The genus *Criconemoides* (including *Mesocriconema*; some authors regard *Mesocriconema* as a valid genus based on characters of the labial region and whether the vagina is open or closed, see Geraert, 2010) contains more than 140 species. Several characters, such as head structure, vulva structure and structure of the lateral field in males, separate groups of species and they are useful for identification. Nomenclatorial status for such groups by naming them as genera is often proposed (Geraert, 2010). However, phylogenetic analyses revealed that almost all criconematid genera, including *Criconema*,

Ogma, *Crossonema*, *Discocriconema*, *Hemicriconemoides*, *Criconemoides*, *Mesocriconema* and *Lobocriconema*, are not monophyletic groups (Nguyen *et al.*, 2022). Four *Criconemoides* species have been reported as important ectoparasites: *Criconemoides onoense* Luc, 1959, *C. ornatus* Raski, 1958, *C. xenoplax* Raski, 1962 and *C. curvatus* Raski, 1952. *Criconemoides onoense* causes root decay on rice in tropical regions; the secondary roots become stunted with lesions appearing near the club-shaped root tips. *Criconemoides ornatus* affects groundnut and other fruit and nut crops in subtropical and tropical regions worldwide, characterized by the death of lateral root primordia and extensive lesions and pits on roots. *Criconemoides xenoplax* is injurious to almond, apple, apricot, carnation, cherry, grapefruit, guava, lettuce, peach, plum, wine grapes and walnut plants, producing interveinal chlorosis, leaf drop and also resulting in stunted plants. It is typically less prevalent in medium-textured soils but can cause plant damage in highly porous clay loam soils (McKenry, 2021). Damage caused by ring nematodes to stone fruits is twofold: first due to direct damage to the roots (especially provoking a lack of feeder roots), and second, but more importantly, owing to damage caused by interactions with other pathogens. In the USA, *C. xenoplax* association with *Pseudomonas syringae* in peach can lead to the disease complex called 'Peach Tree Short Life' or 'Bacterial Canker Complex' (Cid Del Prado Vera and Talavera, 2012; McKenry, 2021). Integrated management of *C. xenoplax* in wine grape production and *Prunus* spp. has been discussed by McKenry (2021) and Zasada and Forge (2021), respectively.

6.4.10. *Hemicycliophora* de Man, 1921 (Fig. 6.10A–C)

Hemicycliophoridae, Hemicycliophorinae. Extra cuticular sheath always present, loosely or closely fitting, never membranous. Female labial region two to three labial annuli, conoid or rounded, nearly always continuous with body. Body usually recessed immediately posterior to vulva. Tail elongate-conoid, filiform or cylindroid, rarely hemispherical. Vulva lips rounded, pointed or elongated. Male with labial region marked by discontinuity in body annulation, usually offset and labial framework in lateral view appearing as 'spectacle mark'. Spicules arcuate, semicircular or hook-shaped. Cloacal lips elongated to form a penial tube.

Hemicycliophora spp., also known as 'sheath nematodes', are migratory ectoparasites of agricultural crops, fruit and nut trees, and ornamental plants. To date, 134 species have been described (Nguyen *et al.*, 2021), albeit with only a limited number of species having been reported as causing damage to agricultural crops. These species include *H. arenaria* Raski, 1958, *H. conida* Thorne, 1955, *H. parvana*, *H. poranga*, *H. similis* Thorne, 1955, *H. typica* de Man, 1921 and *H. ripa* van den Berg, 1981. *Hemicycliophora* spp. can be temporarily sedentary during their feeding process and can induce the formation of root galls, causing stunted growth and the thickening of root termini (Chitambar and Subbotin, 2014). *Hemicycliophora arenaria* is a notorious citrus parasite (Shokoohi and Duncan, 2018) that also parasitizes pumpkin, tomato and other cultivated and non-cultivated plants. In infested field plots, tomato and squash yields have been reported as 10–20% less than in their non-infested counterparts. *Hemicycliophora parvana* has been reported as pathogenic to celery (Cid Del Prado Vera and Talavera, 2012). Comprehensive information of sheath nematodes can be found in the invaluable book by Chitambar and Subbotin (2014), which includes comprehensive

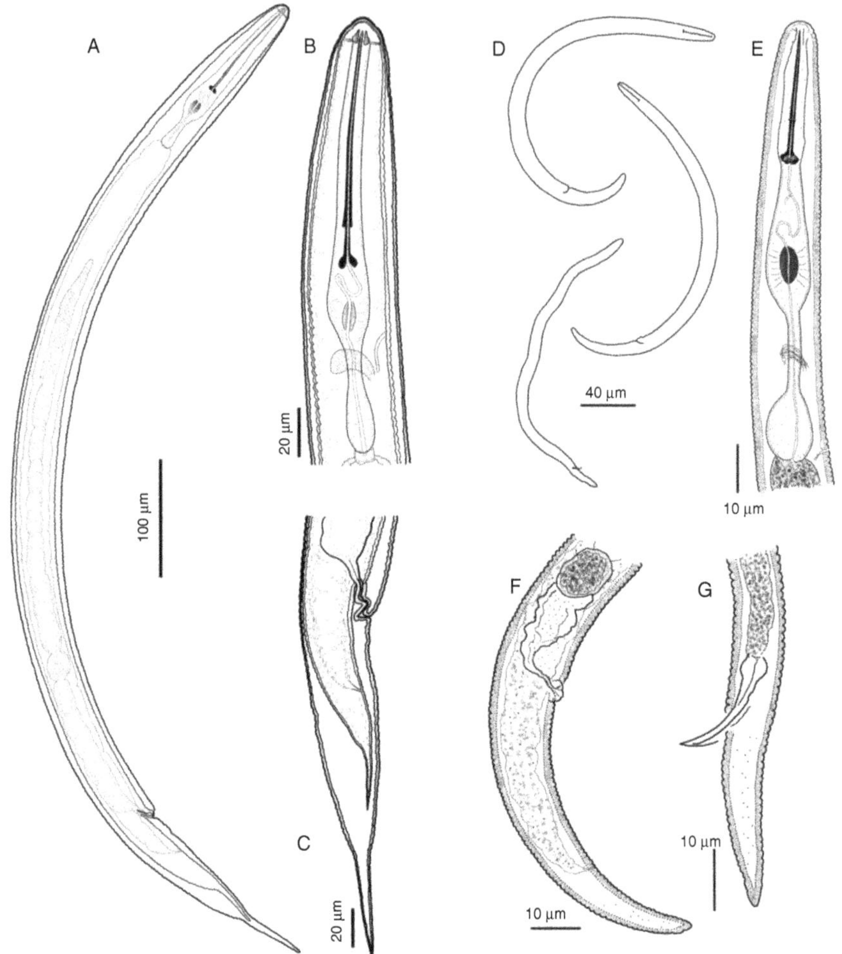

Fig. 6.10. A–C: *Hemicycliophora cardamomi.* A, entire body; B, pharyngeal region; C, vulva and tail region. D–G: *Paratylenchus enigmaticus.* D, entire bodies of females and male; E, pharyngeal region; F, vulva and tail region; G, male tail region. *H. cardamomi* modified from Nguyen *et al.* (2021) and *P. enigmaticus* modified from Maosa *et al.* (2024).

descriptions of 132 *Hemicycliophor*a species combined with both a dichotomous and a tabular key. Nguyen *et al.* (2021) have also provided an online key.

Hemicycliophora spp. tend to be polyphagous, making management by crop rotation difficult, although the application of organic matter to sandy soils has been known to inhibit population growth. Fumigation of the soil before planting can also be effective in reducing the number of *Hemicycliophora. Hemicycliophora arenaria* can be eradicated from root systems with warm water immersion and several citrus rootstocks are resistant to the nematode (Cid Del Prado Vera and Talavera, 2012). The host specificity, pathogenicity, symptomatology and ecology of this species (and other *Hemicycliophora* spp.) are provided by Chitambar and Subbotin (2014).

6.4.11. *Paratylenchus* Micoletzky, 1922 (Fig. 6.10D–G)

Tylenchulidae: Paratylenchinae. Sexual dimorphism, small size (<0.7 mm). Females vermiform to obese; C, J or 6 shapes when heat relaxed; two to four lateral lines; cuticle with or without ornamentations. Cephalic regions of rounded to conoid, truncate or trapezoid; protruding or non-protruding submedian lobes. Stylet length ranging between 10 and 120 μm. Well-developed valves of median bulb, slender isthmus and rounded to pyriform end bulb. Secretory–excretory (S–E) pore often at the level between median bulb and end bulb. Spermathecae with or without sperm cells; commonly swollen prevulval region; vulva with or without lateral flaps; presence or absence of a short post-vulval uterine sac. Tail ranging from conoid to hemispherical with variable tail termini. The diagnostic traits of juveniles and males are less frequently used for identification, except for the presence of a stylet and the length of the spicules of males.

Several related genera, such as *Gracilacus* Raski, 1962, *Paratylenchoides* Raski, 1973, *Gracilpaurus* Ganguly & Khan, 1990 and *Cacopaurus* Thorne, 1943, are considered junior synonyms of *Paratylenchus* (Singh *et al.*, 2021, 2022). The genus *Paratylenchus* contains more than 120 species, commonly known as pin nematodes. *Paratylenchus* species identification is notoriously difficult because of overlapping morphometrics, high levels of intra-specific variability, the presence of cryptic species and the frequent occurrence of more than one pin species in the same soil sample. Some species have a long stylet (>40 μm) that may become swollen by feeding from deeper layers in the root cortex when acting as sedentary ectoparasites (Clavero-Camacho *et al.*, 2021). However, the majority of pin nematode species feed as migratory ectoparasites on epidermal cells and root hairs. They are obligate ectoparasites of a wide variety of plants, including herbs, shrubs and trees, with a higher abundance in the rhizosphere of trees and perennials. They are distributed worldwide and cause various symptoms in their host plants. Clavero-Camacho *et al.* (2021) list 10 species with known pathogenicity on different crops, although more species are likely to be revealed as being pathogenic as studies continue. *Paratylenchus bukowinensis* is known to cause severe damage to vegetable host plants such as celery, carrot and cabbage. Hallmann and Molendijk (2021) state that this species is one of the most underestimated plant-parasitic nematodes in practical agriculture and they discuss its importance, biology, symptomatology and integrated management.

6.5. Enoplea

6.5.1. The family Longidoridae

Dorylaimida. Dorylaimoidea. Adults long to very long, slender, usually ventrally arcuate to C-shaped or spiral upon relaxation by gentle heat, rarely straight; body length (L) ranging between 1.5 and 13 mm. Tail usually short, hemispheroid or conoid, with or without peg, but may be elongate conoid to filiform. Body cuticle smooth by light microscopy, finely transversely striated by SEM; body pores present along the lateral, dorsal and ventral sides with the dorsal pores restricted to the anterior end, the lateral pores arranged in one (neck region) to two or three longitudinal rows (posteriorly) and ventral pores arranged along the body, but may be absent in vulva region or completely as in *Xiphidorus*. Lip region more or less rounded, continuous with body contour or offset to a variable extent (depression or constriction), with or without expansion of the lip region. Six lips largely or completely amalgamated. Anterior sensilla arranged in two circlets respectively of six inner labial sensilla and six outer labial sensilla + four cephalic sensilla; all papilliform or as small pores surrounded or not by a cuticular rim. Amphidial fovea, post-labial, with variable shape and size ranging from wide stirrup-shaped, funnel-shaped, goblet-like to pouch-like; amphidial opening either a pore or ranging from a minute to a large transverse slit. Cheilostome with posterior end marked by the guide ring and varying in length. Stylet greatly elongated, 95–350 µm long, consisting of a 50–220 µm long anterior odontostyle (forked or not) and a posterior supporting structure or odontophore, provided or not with flanges. Pharynx dorylaimoid, with an anterior slender flexible tube and a flask-shaped posterior bulb with enforced lumen wall and three pharyngeal glands, one dorsal and two ventrosublateral glands. Intestine with pre-rectum. Female reproductive system typically didelphic–amphidelphic with vulva at mid-body but monodelphic and pseudomonodelphic conditions also occur with shift of vulva; uterus varying in length and in structure, from simple to complex, with or without various inclusions such as a Z-differentiation, granules, spines, crystalloids. Male reproductive system diorchic, posterior testis reflexed. Spicules relatively large, dorylaimoid in shape; gubernaculum restricted to two lateral guiding pieces or crura. Pre-cloacal supplements consisting of an adanal pair and a medioventral series of 1–20 papillae, rarely staggered or arranged in a double row. Type genus: *Longidorus* Micoletzky, 1922 (Filipjev, 1934).

The family Longidoridae consists of two subfamilies: Longidorinae Thorne, 1935 and Xiphinematinae Dalmasso, 1969.

6.5.1.1. Longidorinae Thorne, 1935

Longidoridae. Guide ring single; dorsal pharyngeal gland nucleus (DN) smaller than ventrosublateral gland nuclei (VSN) and located at some distance posterior to its orifice. The subfamily Longidorinae contains two tribes: the Longidorini and the Xiphidorini. The latter tribe is not further discussed herein as it consists of genera that are largely restricted to Central and South America, which do not currently appear to be of major economic importance and from which viral transmission has not been detected. For more details, see Decraemer and Chaves (2012).

6.5.1.1.1. *LONGIDORUS* MICOLETZKY, 1922 (FIG. 6.11). Longidorini. Lip region rounded or more or less flattened, continuous or marked by depression or constriction. Odontostyle non-forked; base odontophore without flanges. Cheilostome relatively long; guide ring single, anteriorly situated (at most at 40% of odontostyle length from anterior end); compensation sacs present. Amphidial fovea a pouch, its aperture a pore. Male with an adanal pair of pre-cloacal supplements (some species have two or three adanal pairs) plus a ventromedian series of up to 20 pre-cloacal supplements without hiatus between the adanal pair and the series; in some species ventromedian row forms a double staggered row. Female reproductive system didelphic–amphidelphic, dorylaimoid; uterus without sclerotizations. Tail short, dorsally convex-conical with a finely or broadly rounded terminus. Mostly four juvenile stages present, but some species with three. The latter are mainly restricted to Asia and occur less frequently in Africa and Europe. Type species: *Longidorus elongatus* (de Man, 1876) Thorne & Swanger, 1936.

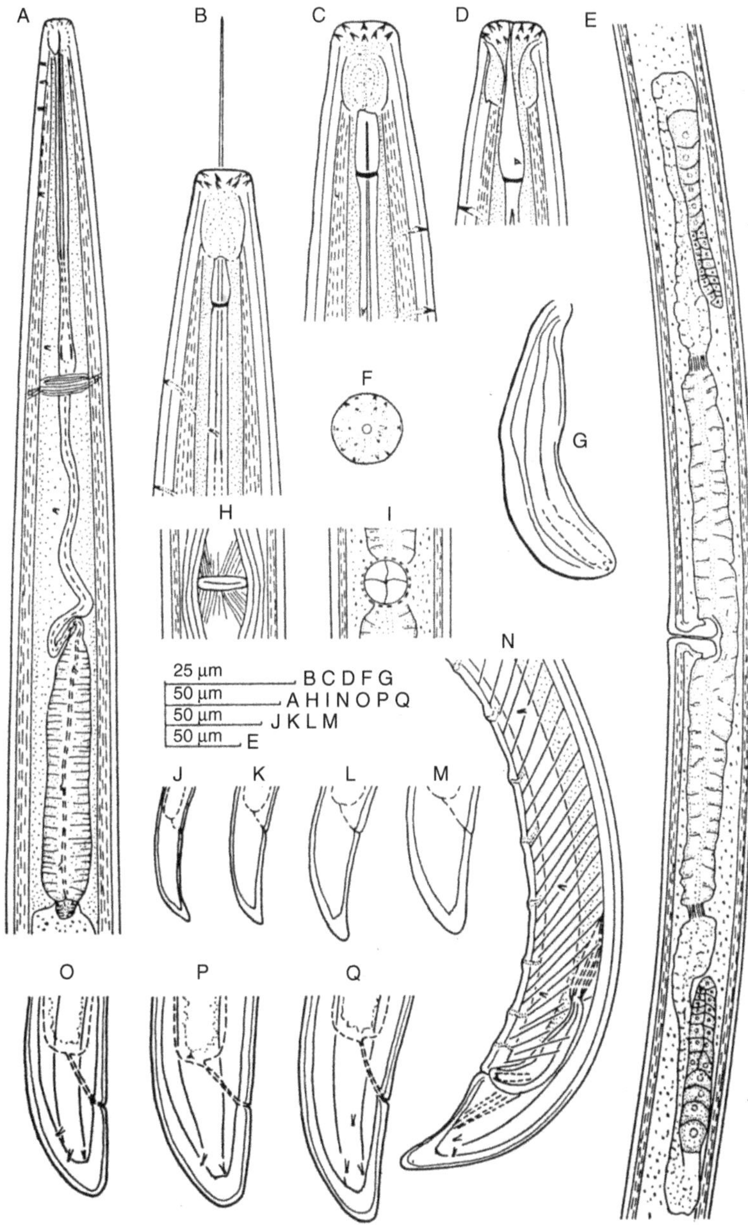

Fig. 6.11. *Longidorus elongatus*. A, pharyngeal region, dorsolateral; B–D, head region in surface view: lateral view of neotype (B), Scottish female, in lateral (C) and ventral view (D), respectively; E, female reproductive system; F, lip region, *en face*; G, spicule and lateral guiding piece; H and I, ventral view of vulva and vagina, respectively; J–M, juvenile tails from first-stage juvenile (J) to fourth-stage juvenile (M); N, posterior body region male; O–Q, female tails, respectively of neotype, topotype and other specimen. (After Hooper, 1961, courtesy *Nematologica*.)

Approximately 180 *Longidorus* species are known. Species discrimination in *Longidorus* is difficult because of the large number of species and because of the conserved morphology and overlapping morphometric characters. Furthermore, the use of molecular markers in species identification of *Longidorus* has indicated that some species actually comprise multiple genetically divergent and morphologically similar cryptic species (Liébanas *et al.*, 2022). Species identification is based upon the following features: (i) shape of pouch-like amphidial fovea; (ii) shape of lip region (rounded, flattened; offset by constriction, depression or continuous) and width of lip region; (iii) tail shape; (iv) length of odontostyle; (v) body length; (vi) tail length; (vii) ratios a and c′; (viii) distance of guide ring from anterior end; (ix) length of spicule; (x) number of ventromedian pre-cloacal supplements; and (xi) males known or unknown (Chen *et al.*, 1997).

Longidorus species are of economic importance owing to their ability to cause serious damage to a wide range of crops. Such damage is caused by these species directly feeding on root cells and, in addition, eight *Longidorus* species are also known vectors of Nepoviruses. Parasitism by *Longidorus* spp. has a detrimental effect on root growth by inducing hypertrophied highly metabolically active uninucleate cells, followed by hyperplasia with synchronized cell division (Palomares-Ruis *et al.*, 2017b). Virus transmissions show a marked specificity between plant viruses and their specific *Longidorus* vector species, except for *L. elongatus*, which not only transmits tomato black ring virus (TomBRV), but also raspberry ringspot virus (RRSV). Such considerations highlight the need for accurate identification of *Longidorus* species (Archidona-Yuste *et al.*, 2019). *Longidorus elongatus* is an important plant parasite and widely distributed throughout temperate regions worldwide. Generally limited to one generation per year, on good hosts such as strawberry, the life cycle may be half as long or less. Males are usually rare and reproduction is apparently parthenogenetic, although bisexual reproduction is possible in populations where males are common. *Longidorus elongatus* is widespread in temperate regions worldwide and has been found in various types of soil but mainly has a preference for coarse, preferably stable, well-drained soils. It has a wide host range, among them many crops and weeds (De Waele and Coomans, 1991). As a root ectoparasite, *L. elongatus* aggregates around growing roots and feeds just behind the root tips, causing a characteristic swelling or galling and a general stunting of the root system. Severe attacks on sugar beet results in a fang-like, distorted root system. Studies show that *L. elongatus* can survive more than 2 years in moist soil sealed in a polythene bag and in the absence of host plant material. *Longidorus*, *Paralongidorus* and *Xiphinema* have been shown to cause severe damage to vegetables, especially in sandy soils, and are likely often to be overlooked wherever root-knot nematodes predominate (Hallmann and Meressa, 2018). Although the nematode does not retain viral particles for more than 2 months, during which time they are lost from the inner surface of the guiding sheath during the moulting process, reinfection of the nematode is possible from many weed hosts that carry the virus (Taylor and Brown, 1997).

6.5.1.1.2. *PARALONGIDORUS* SIDDIQI, HOOPER & KHAN, 1963 (FIG. 6.12).

Syn. *Paralongidorus* (*Siddiqia*) Khan, Chawla & Saha, 1978, *Siddiqia* Khan, Chawla & Saha, 1978 and *Inagreius* Khan, 1982.

Longidorini. Lip region continuous with body contour or expanded and offset; in some species with a second depression behind the groove. Odontostyle non-forked; base of odontophore slightly swollen or non-swollen but without flanges. Cheilostome relatively

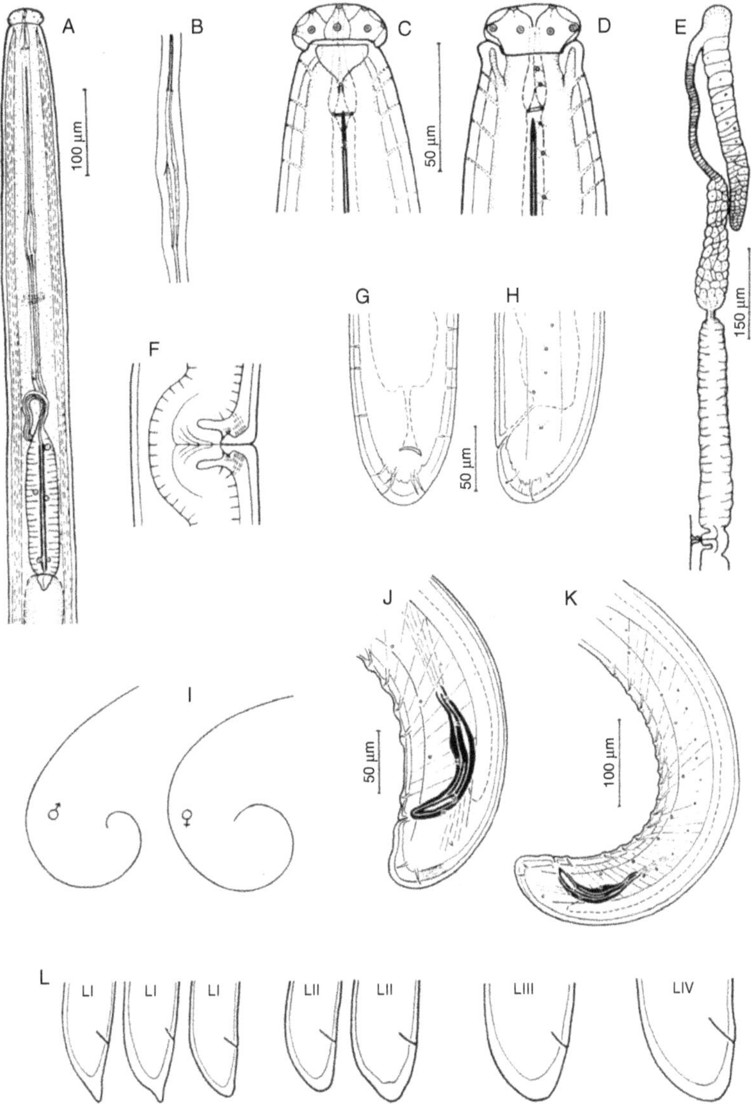

Fig. 6.12. *Paralongidorus maximus*. A, pharyngeal region; B, odontophore; C and D, head region in lateral view and dorsoventral view, respectively; E, anterior genital branch of female reproductive system; F, vagina region; G and H, female tail ventral view and view of left side, respectively; I, habitus of male and female upon thermal death; J and K, posterior body region of male; L, tails in lateral view of the four juvenile stages. (After Sturhan, 1963.)

long; guide ring single, anteriorly situated (at most at 60% of odontostyle length from anterior end); compensation sacs present. Amphidial fovea variable: elongate funnel-shaped or stirrup-shaped, its aperture usually a large transverse slit. DN most generally some distance from its orifice; VSN most generally more developed than dorsal one. Male with an adanal pair of pre-cloacal supplements plus a ventromedian series of supplements without hiatus between the adanal pair and the series. Female reproductive system didelphic–amphidelphic; uterus without sclerotized structures. Tail short, rounded; may be conoid or hemispheroid. Four juvenile stages with appearance largely similar to females except for tail. First-stage juveniles with rather long conoid tail, becoming relatively shorter (higher c-value) in later developmental stages but not in absolute length. In one species, *P. duncani* Siddiqi, Baujard & Mounport (1993) only three juvenile stages recognizable by relative body length and length of functional and replacement odontostyles.

Type species: Paralongidorus sali *Siddiqi, Hooper & Khan, 1963*

The genus currently contains about 90 species (15 *Longidoroides* species included). *Paralongidorus* can be distinguished from *Longidorus* mainly by the slit-like amphidial aperture and the stirrup-shaped fovea. Differentiation of *Paralongidorus* species is based on the following features: (i) shape of amphidial fovea; (ii) width of amphidial aperture; (iii) shape and width of lip region; (iv) tail shape in female; (v) length of odontostyle; (vi) body length; (vii) tail length; (viii) ratios a and c′; (ix) distance of guide ring from anterior end; (x) length of spicule; (xi) number of ventromedian pre-cloacal supplements; and (xii) males known or unknown (polytomous identification key of Escuer and Arias, 1997). *Paralongidorus* species parasitize a wide range of agronomic crops, ornamentals, turfgrass, herbaceous plants and forest trees. They are of global interest because they cause direct damage to the roots of the host plant that is attributable to their ectoparasitic feeding. Several *Paralongidorus* species are important parasites of rice (Peng *et al.*, 2018). *Paralongidorus maximus* is able to transmit damaging nepoviruses (Jones *et al.*, 1994). A known vector of RRSV, *Paralongidorus maximus* is considered as a major pest in Europe and other parts of the world, and is included on the A2 list of pest recommended regulation of the European and Mediterranean Plant Protection Organization. It can be locally widespread (Germany) where it is often associated with vineyards or found in association with pine and river banks. This species was also recorded from South Africa in association with grapevine (Swart *et al.*, 1996).

Longidoroides *Khan, Chawla & Saha, 1978*

Longidoroides Khan, Chawla & Saha, 1978, is characterized by: (i) the lip region usually continuous with the body, exceptionally expanded; (ii) a pouch-like usually bilobed amphidial fovea (the main difference from *Paralongidorus sensu stricto*); and (iii) its aperture a medium-sized to very small (1 µm) transverse slit. *Longidoroides* was considered a junior synonym of *Paralongidorus* by Siddiqi *et al.* (1993) but was accepted as valid by Hunt (1993), Coomans (1996), Decraemer and Coomans (2007) and the present authors until additional morphological and molecular data become available. Fifteen species have been described, but none of them is known to transmit a nepovirus.

6.5.1.2. *Xiphinematinae Dalmasso, 1969*

This subfamily has only one genus, *Xiphinema*, which is the oldest and most diversified genus of the family Longidoridae (Coomans, 1996; Coomans *et al.*, 2001). Dagger nematodes of the genus *Xiphinema* comprise plant-parasitic species that damage a wide range of both wild and cultivated plants either directly by feeding on root cells or indirectly by transmission of plant viruses. Within the genus, some 300 species are recognized as valid. They are divided into two species groups: the *Xiphinema americanum* group, which comprises a complex containing roughly 61 species, and the non-*americanum* group. Both species groups include vectors of several important plant viruses, in particular nepoviruses, which cause significant damage to a wide range of crops: seven such virual vectors belong to the *X. americanum* group (*X. americanum* Cobb, 1913, *X. bricolense*, *X. californicum*, *X. rivesi*, *X. intermedium*, *X. inaequale* and *X. tarjanense*) and three vector species to the *Xiphinema* non-*americanum* species group (*X. diversicaudatum*, *X. index* and *X. italiae*).

6.5.1.2.1. *XIPHINEMA AMERICANUM* GROUP (FIG. 6.13). The type species of the genus has long been considered to have a worldwide distribution. However, it is now regarded as a species complex of 61 putative species (Jeger *et al.*, 2018), of which *X. americanum sensu stricto*

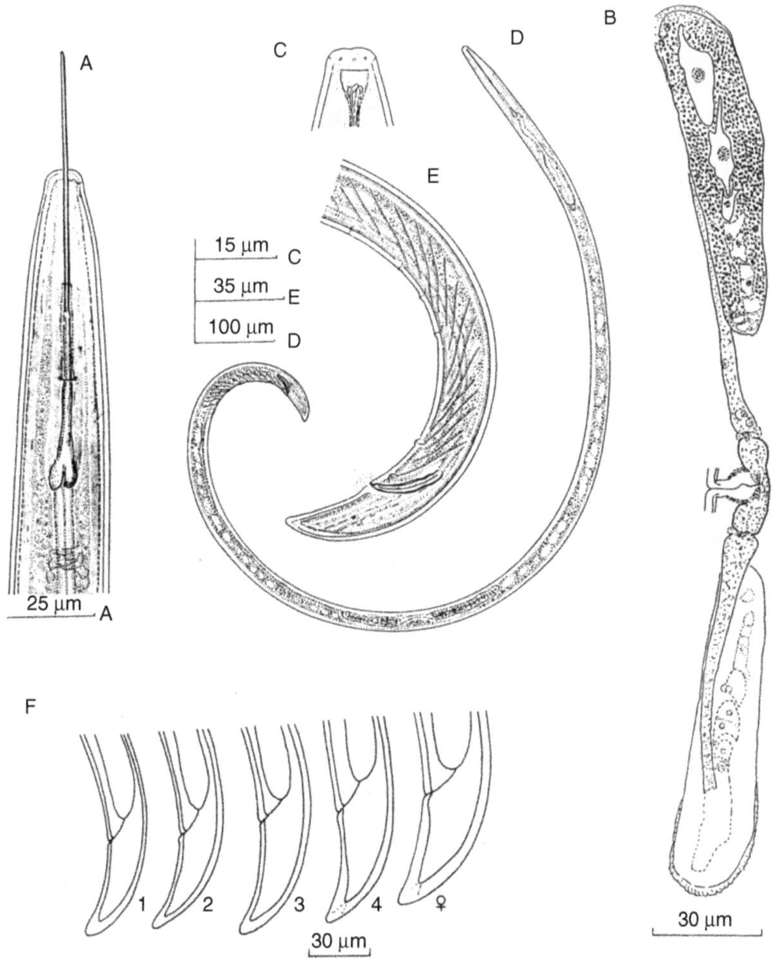

Fig. 6.13. *Xiphinema americanum*. Female: A, pharyngeal region; B, reproductive system. Male: C, head in surface view; D, total view; E, posterior body region. F, tail ends of the four juvenile stages (1–4) and female. (A, C–E, after Siddiqi, 1973; B, after Coomans and Claeys, 2001; F, after Coomans *et al.*, 2001.)

is widespread in eastern USA, and also occurs in Arkansas, California and South Africa. Identification within the *X. americanum* group is of particular importance for phytosanitary regulation (see Chapter 13), as these nematodes are vectors of economically important plant viruses that cause substantial damage to a wide range of crops.

The species of *Xiphinema americanum sensu lato* are characterized by: (i) adults with a more or less open C to spiral upon fixation or heat killing; (ii) body length small (<3 mm); (iii) lip region rounded, continuous with the body and marked by a depression, minor expansion or a constriction; (iv) odontostyle rarely longer than 150 µm; (v) female reproductive system didelphic, amphidelphic with rather short non-differentiated uteri, weakly developed sphincter muscles and oocytes usually with endosymbiotic bacteria, vulva between 42 and 65% from anterior end; (vi) tail short conoid with a more or less acute terminus, sometimes subdigitate; and (vii) male rare, female devoid of sperm. Four juvenile stages but species with three stages occur. The latter are widely distributed in the Americas but occur rarely in Africa, Europe and Asia (Lamberti *et al.*, 2000). Lamberti and Ciancio (1993) distinguished five species subgroups based on a hierarchical cluster analysis of morphometrics, among them a *Xiphinema pachtaicum* group; the latter also included *X. pachydermum*. There is no general agreement on the definition of the *Xiphinema americanum* species group morphologically as well as molecularly, especially with respect to the position of *X. pachydermum* and related species. The latter differ from the typical *X. americanum* group in females which possess ovaries without associated symbiotic bacteria (except in *X. mesostilum*), a well-developed sphincter muscle versus weakly developed in the *X. americanum* species group, clearly longer uteri and males common (Luc *et al.*, 1998; Coomans *et al.*, 2001). Comprehensive molecular phylogenetic analyses showed two main clades separating most of the species of *X. americanum*-subgroup *sensu stricto* from the *X. pachtaicum* subgroup (Archidona-Yuste *et al.*, 2016).

In the polytomous identification key of the *X. americanum* species group (Lamberti *et al.*, 2004), the following characters are used: (i) odontostyle length; (ii) V%; (iii) value of ratio c′ and c; (iv) body length; (v) tail length; (vi) shape of tail terminus; and (vii) lip region continuous or offset from the body. Two additional characters could be added: (i) lip region marked by a depression; and (ii) the length of the genital branch/uterus in the female. Attention should be paid to the different use of characters and coding in the different polytomous identification keys (see also Lamberti *et al.*, 2000). The identification of species within the *X. americanum* group is problematic as a result of general similarity of the morphology of the putative species. It has also been suggested that the genera *Longidorus* and *Xiphinema* have a greater tendency than other plant-parasitic genera for cryptic speciation (Cai *et al.*, 2020).

The *X. americanum* group includes the following vector species for four economically important nepoviruses and naturally occurring in the USA: cherry rasp leaf virus (CRLV), peach rosette mosaic virus (PRMV), tobacco ringspot virus (TRSV) and tomato ringspot virus (ToRSV): *X. americanum sensu sricto* (CRLV, TRSV, ToRSV), *X. americanum sensu lato* (CRLV, PRMV, TRSV, ToRSV), *X. bricolense* (ToRSV), *X. californicum* (CRLV, TRSV, ToRSV), *X. inaequale* (ToRSV), *X. rivesi* (CRLV, PRMV, TRSV, ToRSV), *X. intermedium* (TRSV, ToRSV) and *X. tarjanense* (TRSV, ToRSV) (Jeger *et al.*, 2018). Although only these seven species are known to transmit plant viruses, the entire *Xiphinema americanum sensu lato* was formerly considered as an A2 quarantine species in Europe because it is difficult to separate the species with certainty based on morphology. Nowadays, *X. americanum sensu stricto*, *X. bricolense* and *X. californicum* are listed as A1 and *X. rivesi* is listed as A2 quarantine organisms by the European and Mediterranean Plant Protection Organisation (http://www.eppo.int/).

Xiphinema americanum has been recorded worldwide, albeit often as *X. americanum sensu lato*, meaning that these identifications should be reconfirmed. *Xiphinema americanum sensu stricto* is widespread in North America, particularly in the eastern region, and has also been recorded from South Africa. The species has a wide host range including woody and herbaceous plants, cereals, cherry, prune, peach, citrus, grapevine, strawberry, maize, tomato, tobacco, soybean and many others. Marigold was found to support a 75-fold increase in population of *X. americanum* in 5 years, whilst millet, hairy indigo and cotton resulted in a less than 10-fold increase. By contrast, Sudan grass, crotalaria and beggar weed prevented multiplication (Brodie *et al.*, 1970). Possible integrated management programmes for *X. americanum sensu lato* in wine grape production have been put forward by Zasada and Forge (2021) and Halbrendt (2021).

6.5.1.2.2. *XIPHINEMA* NON-*AMERICANUM* GROUP (FIG. 6.14). Xiphinematinae. Body length 1.2–7.5 mm, body straight to spiral. Lip region varying from well offset and knob-like to continuous with body contour, high to low. Amphids with stirrup- or funnel-shaped, rarely goblet-shaped, fovea with wide (40–90% of central body width) slit-like aperture, rarely narrower. Base odontostyle forked; base odontophore with sclerotized flanges. Cheilostome long, ending around posterior third of odontostyle; guide ring always double due to folding of guiding sheath anteriorly beyond guide ring. *Dilatores buccae* or cheilostome retractor

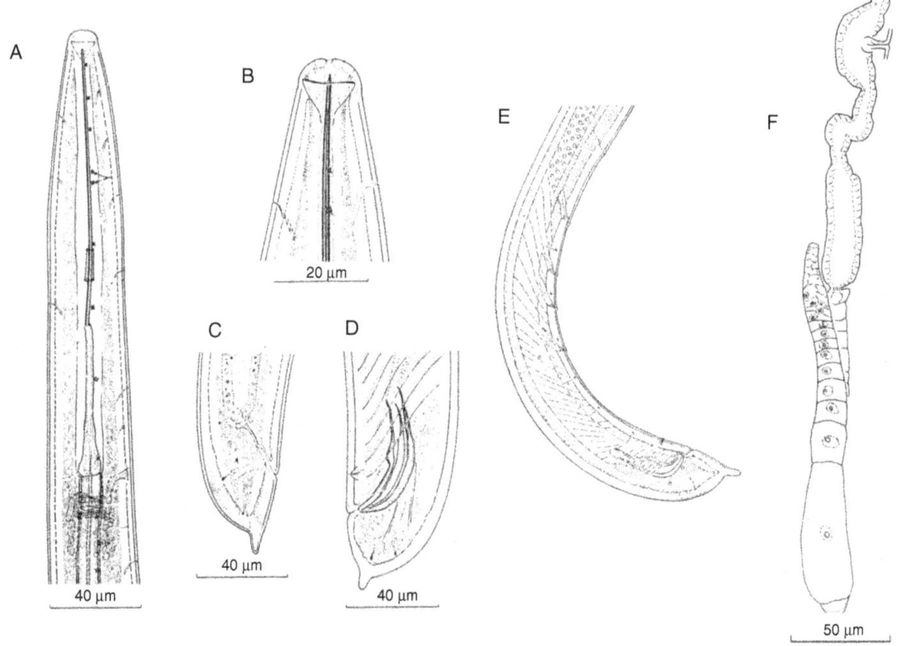

Fig. 6.14. *Xiphinema index.* A, pharyngeal region; B, head region, surface view; C, female tail; D, male tail region and copulatory apparatus; E, male posterior body region; F, female reproductive system, posterior branch. (From Siddiqi, 1974c.)

muscles connecting cheilostome to dorsal and ventral body wall. True stylet retractor muscles degenerated, pharyngeal retractor muscles composed of three bands extending from near base of odontophore and splitting into four muscles attaching to body wall opposite anterior part of pharyngeal bulb. DN large, located at same level as its dorsal orifice, more developed than the two nuclei of VSN. Pre-rectum well developed. Female reproductive system varying from typical didelphic–amphidelphic to various transitions (pseudomonodelphic) to complete mono-opisthodelphic; uterus varying from short and undifferentiated to long and differentiated into several parts and in many species provided with sclerotized structures (crystalloids, spines, Z- or pseudo Z-differentiation). Males diorchic, spicules dorylaimoid and gubernaculum reduced to sclerotized lateral guiding pieces or crura. Male genital pre-cloacal supplements consisting of an adanal pair followed by a hiatus and a ventromedian series varying from 1 to 11 (usually 2–5), mostly situated outside the range of retracted spicules, rarely absent. Tail shape varying from elongate filiform to short bluntly rounded; tails of both sexes similar or showing sexual dimorphism in some species. In about 40% of the species, males are abundant and reproduction is amphimictic. In 25% of *Xiphinema* species males are rare and in 35% of the species males are unknown; in both cases reproduction probably occurs through parthenogenesis. Four juvenile stages but species with three stages occur.

Species in the non-*americanum* group are subdivided into eight species groups based upon the following characters: (i) absence of anterior genital branch; (ii) absence of anterior ovary; (iii) anterior genital branch present but reduced; (iv) two equally developed genital branches with presence of a Z-organ; (v) as in (iv) but with a pseudo Z-organ with or without uterine spines; (vi) similar as in (iv) but without Z-organ or other uterine sclerotized structures; (vii) as in (iv) but uterus not differentiated and tail elongate to conical; and (viii) as in (vii) but tail short, rounded. Additional to these characters, species identification is also based on: (i) ratios c′ and c; (ii) the presence or absence of a terminal canal in the cuticle at the tail tip; (iii) V%; (iv) body length; (v) length of complete stylet; (vi) outline of lip region; (vii) habitus; (viii) presence of males; and (ix) tail shape in first- and fourth-stage juveniles (polytomous key by Loof and Luc, 1990; Coomans *et al.*, 2001; Peraza-Padilla *et al.*, 2018).

The non-*americanum* species group includes several economically important species, in particular the three virus vector species *X. diversicaudatum*, *X. index* and *X. italiae*. *Xiphinema index* has a high economic impact in vineyards in many parts of the world including Africa, Australia, Europe, South America and the USA. During its feeding on the plant, *X. index* is known to transmit grapevine fanleaf virus (GFLV) to grapevines. GFLV is considered to be the most severe viral grapevine disease and is responsible for fanleaf degeneration, which occurs in temperate regions of vine cultivation (Nguyen *et al.*, 2019). *Xiphinema index* has an eastern origin and an east-to-west nematode dissemination appears to match the spread of the domesticated grapevine throughout Antiquity, presumably being mainly dispersed by the Greeks and the Romans (Nguyen *et al.*, 2019). It is especially widespread in Mediterranean environments and temperate climates where grapevines are prevalent. It has a limited host range, and the domesticated grapevine is by far its major host. Its presence has not been reported in native forests or climax vegetation except for in the Middle East, where it has been reported in the Iranian natural woodland of the forests lying along the Caspian Sea where wild grapevines may be common (Nguyen *et al.*, 2019). It occurs in a wide range of soil types (including heavy soils), but has a preference for light- to medium-textured soils. Apart from grapevine, the species also attacks various woody plants including fig, mulberry, *Prunus* spp. and *Pyrus* spp. *Xiphinema index* and several other *Xiphinema* species are also pathogenic to citrus (Shokoohi and Duncan,

2018). As a result of direct feeding, *X. index* causes necrosis, terminal galling and lack of lateral root development. Males are extremely rare and reproduction is by meiotic parthenogenesis. The life cycle of *X. index* is variable and has been recorded as being as short as 22–27 days at 24°C in California and up to 7–9 months at 20–23°C in Israel.

6.5.2. The family Trichodoridae (Fig. 6.15)

Triplonchida. Diphtherophoroidea. Body straight cigar-shaped or J-shaped. Lip region rounded, continuous with body; lips amalgamated. Anterior sensilla papilliform and arranged in two circlets, an anterior circlet of six inner labial papillae and an outer circlet

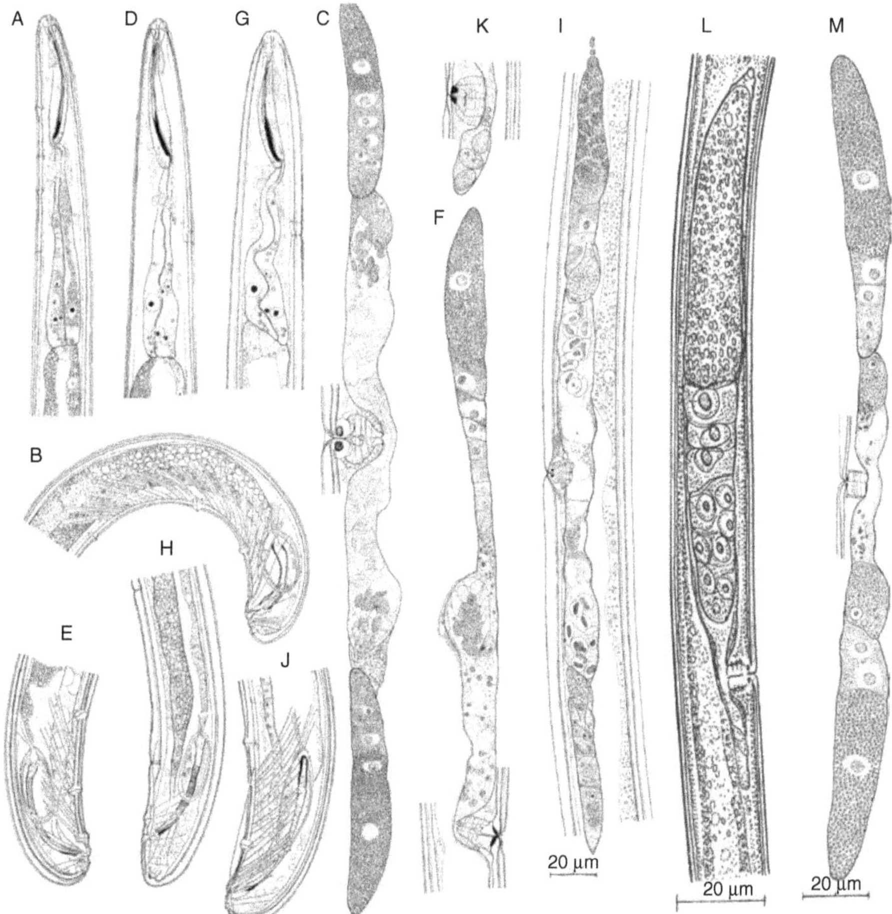

Fig. 6.15. The genera of the Trichodoridae. Anterior body region in male: A, *Trichodorus velatus*; D, *Monotrichodorus m. vangundyi*; G, *Paratrichodorus weischeri.* Male posterior body region: B, *T. velatus*; E, *M. m. vangundyi*; H, *P. pachydermus*; J, *Allotrichodorus campanulatus*. Female reproductive system: C, *T. taylori*; F, *M. m. vangundyi*; K, *A. campanulatus*, vulva region; I, *P. weischeri*; L, *Ecuadorus mexicanus*; M, *Nanidorus minor*. (A–K, after Decraemer, 1991, courtesy of Marcel Dekker Inc.; L, after Siddiqi, 2002; M, after Decraemer, 1995.)

of ten papillae, the two subventral and two subdorsal outer labial papillae and four cephalic papillae form a kind of protruding double papillae. Amphidial fovea post-labial, cup-shaped and aperture a medium-sized transverse slit. Onchiostyle relatively long, ventrally curved stylettiform dorsal tooth; guide ring simple, at level of posterior end of onchiostyle. Pharynx with swollen posterior glandular bulb; five gland nuclei present: one posterior ventrosublateral pair, one smaller anterior ventrosublateral pair and a single large dorsal nucleus. Intestine without pre-rectum, rectum almost parallel to longitudinal body axis; anus subterminal in female. Male with a single outstretched anterior testis, sperm cells variable in shape and size; 1–4 pre-cloacal supplements, one pair (rarely two) of post-cloacal papillae; weakly developed caudal alae present or absent; spicules straight to ventrally curved, variable in shape and ornamentation. Spicule protractor muscles transformed in a capsule of suspensor muscles surrounding the spicular pouch. Female reproductive system didelphic–amphidelphic or monodelphic–prodelphic, ovary (ovaries) reflexed; oviduct consisting of two finely granular cells; spermatheca(e) present or absent. Tail short, maximum length one anal body width in male. Apparently four juvenile stages but first-stage juveniles have not been recovered from field samples and are known only from laboratory cultures. Type genus: *Trichodorus* Cobb, 1913.

The Trichodoridae, or stubby root nematodes, are polyphagous root ectoparasites distributed worldwide. Plants that are known to favour the multiplication of at least some of the most common trichodorid species include potato, sugar beet, onion, carrot, maize, beans, cabbage, wheat, barley, rye and ryegrasses, strawberries, a range of tree and fruit tree species and many wild plants that occur as arable weeds (Brinkman and Hallmann, 2021). Trichodoridae can cause direct damage to a wide range of crops and natural vegetation and have gained some notoriety since some species are vectors of three viruses belonging to the tobravirus group – namely tobacco rattle virus (TRV), pea early browning virus (PEBV) and pepper ring spot virus (PRV) tobraviruses (see Section 6.7). The Trichodoridae comprise some 123 species and six genera: *Trichodorus* Cobb, 1913 (type genus), *Allotrichodorus* Rodriguez-M, Sher & Siddiqi, 1978, *Ecuadorus* Siddiqi, 2002, *Monotrichodorus* Andrassy, 1976, *Nanidorus* Siddiqi, 1974 and *Paratrichodorus* Siddiqi, 1974 (Fig. 6.14). The status of *Atlantadorus,* introduced at subgenus level within *Paratrichodorus* and subsequently raised to genus level by Siddiqi in 1974 and 1980, respectively, is still under discussion (Decraemer, 1995; Duarte *et al.*, 2010).

The three largest genera, *Trichodorus* (74 species), *Paratrichodorus* (29 species) and *Nanidorus* (seven species), are didelphic and occur worldwide. The three remaining genera have fewer species, *Monotrichodorus* (five species), *Allotrichodorus* (six species) and *Ecuadorus* (two species). These genera are monodelphic–prodelphic. They have only been recorded from Central America and the northern part of South America.

The main diagnostic characters for genus identification are: (i) female reproductive system (didelphic or monodelphic); (ii) length of vagina and size of vaginal sclerotized pieces; (iii) presence of advulvar lateral body pores; (iv) presence or absence of caudal alae in male; (v) degree of development of copulatory muscles and related habitus and capsule of spicule suspensor muscles; and (vi) type of sperm.

6.5.2.1. **Trichodorus** *Cobb, 1913 (Fig. 6.16)*

Trichodoridae. Cuticle usually not or slightly swollen after fixation, heat killing or fixatives. Pharynx usually with offset bulb, more rarely with short ventral overlap of pharyngeal

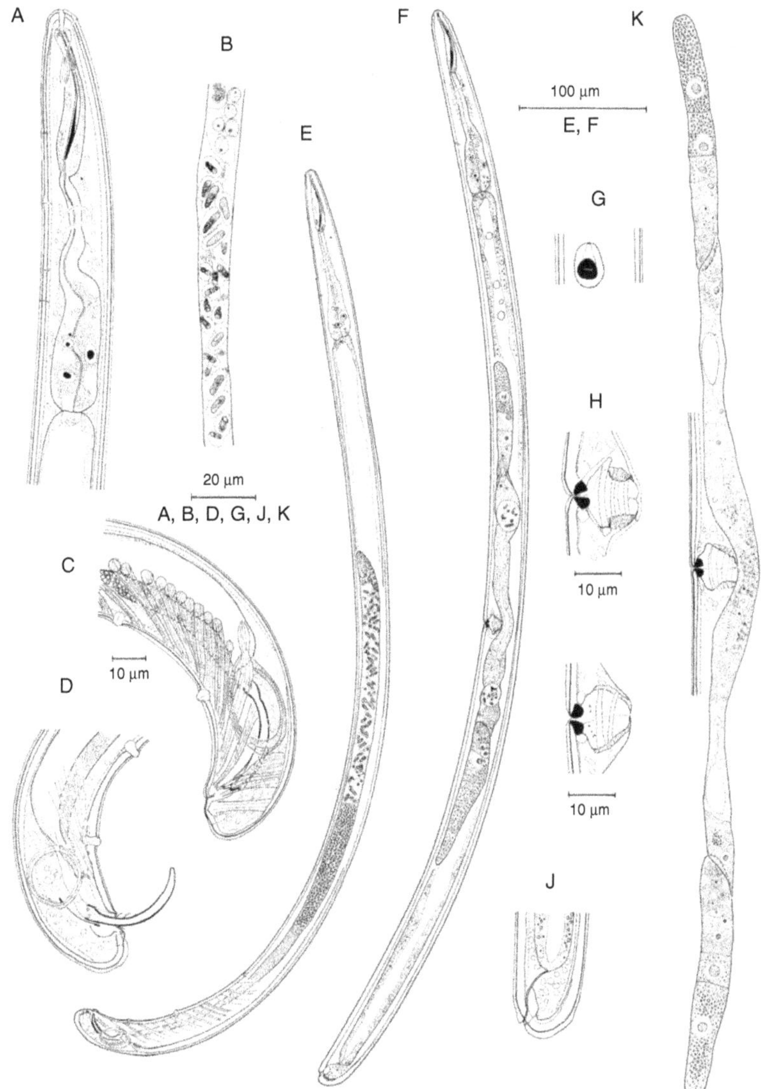

Fig. 6.16. *Trichodorus similis*. Male: A, pharyngeal region; B, sperm in testis; C and D, posterior body region; E, total view. Female: F, total view; G, ventral view; H and I, vaginal region; J, tail region; K, reproductive system. (After Decraemer, 1991, courtesy Marcel Dekker Inc.)

glands or dorsal intestinal overlap. S–E pore usually at level of narrow part of pharynx or anterior part of pharyngeal bulb. Female reproductive system didelphic–amphidelphic; spermathecae present, rarely absent; vagina well developed, usually about half the corresponding body width long, variable in shape; vaginal constrictor muscle well developed; vaginal sclerotized pieces in lateral view conspicuous, variable in shape and size. Vulva a pore, a transverse slit, or rarely, a longitudinal slit. One to four lateral body pores present on each side, rarely absent; one post-advulvar pair (i.e. within one body width posterior to the vulva) present, rarely absent. Male common. One to three ventromedian cervical papillae present in neck region; rarely four or none. Lateral cervical pores usually present, rarely absent; usually one pair near onchiostyle base or just behind nerve ring. Sperm cells large, with large sausage-shaped or rounded nucleus. Spicules ventrally arcuate, rarely straight, usually with ornamentation (bristles, transverse striae or velum) or smooth. Capsule of spicule suspensor muscles well developed. Oblique copulatory muscles well developed, extending far anterior to spicular region, causing upon contraction the typical J-shaped ventral curvature of posterior end in male. Caudal alae absent, exceptionally present (*T. cylindricus* Hooper, 1962 and *T. paracedarus* Xu & Decraemer, 1995). Three pre-cloacal ventromedian supplements, rarely two, four or five supplements, more or less evenly spaced, usually one within range of retracted spicules. Tail with one pair of post-cloacal subventral papillae and one pair of caudal pores. Type species: *T. obtusus* Cobb, 1913. Syn. *T. proximus* Allen, 1957.

The main morphological features for species identification are, in the female, the shape and size of the vaginal sclerotized pieces and, in the male, the spicule shape; other characters are body length, onchiostyle and spicule length, number and position of pre-cloacal supplements and ventromedian cervical papillae in the male. However, identification to species level in Trichodoridae is hampered by the largely conserved morphology, overlapping morphometrics, the difficulty in properly fixing samples, and co-occurrence of more than one trichodorid species, often in low numbers, in the same soil sample (Decraemer *et al.*, 2019).

Only four of the more than 70 *Trichodorus* species naturally transmit TRV and/or PEBV (see Section 6.7). *Trichodorus similis* is an important plant pathogen that occurs in temperate and cooler regions of Europe (west, east and north); few records are known from the USA. The species is most widespread in sandy or sandy loam soils, with highest densities at between 20 and 30 cm depth. Reproduction is amphimictic and under optimum conditions the life cycle is completed in 45 days. *Trichodorus similis* causes direct damage to the root system of carrot, potato, onion, sugar beet, tobacco and trees such as *Prunus* (plums, cherry), *Juglans* (walnut) and *Picea* (spruce). It also acts as a vector of TRV to potatoes and *Gladiolus*.

6.5.2.2. Paratrichodorus *Siddiqi, 1974 (Fig. 6.17)*

Syn. *Atlantadorus* Siddiqi, 1974.

Trichodoridae. Cuticle usually clearly swollen when fixed or heat-killed. Ventral pharyngeal and/or dorsal intestinal overlap usually present, bulb rarely offset. Female reproductive system didelphic–amphidelphic; spermathecae present, or absent with sperm throughout uterus. Lateral body pores present in about 50% of the species but rarely located advulvar; body pores exceptionally ventral or subventral in position. Vagina short (about one-third of corresponding body diameter), constrictor muscles inconspicuous, vaginal sclerotization small to inconspicuous in lateral view. Vulva a pore or a longitudinal slit. Male rare or unknown. Ventromedian cervical papillae in male usually one papilla near the S–E pore, exceptionally with two. Lateral cervical pores absent or, if present, usually one pair near onchiostyle base or S–E pore. Male tail region straight; caudal alae present, obscure to distinct. Sperm cells variable, large with large sausage-shaped nucleus, medium-sized with rounded nucleus, small with small oval to rounded

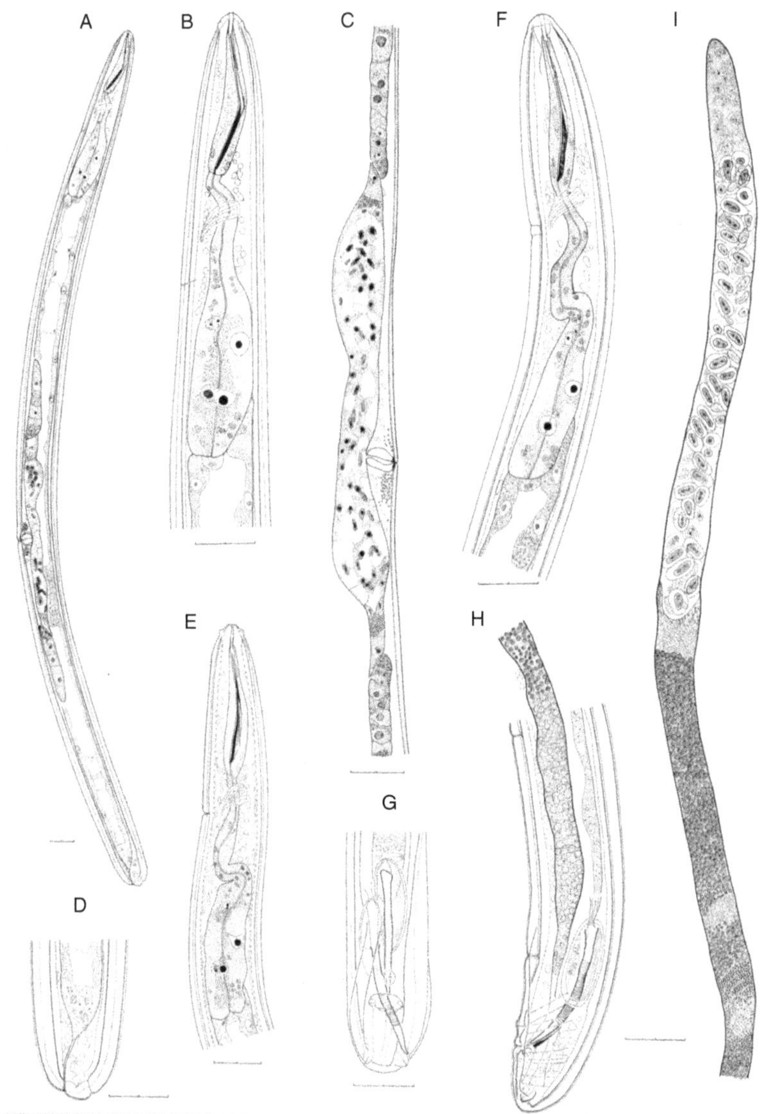

Fig. 6.17. *Paratrichodorus pachydermus*. Female: A, total view; B, anterior body region; C, reproductive system; D, tail. Fourth-stage juvenile: E, anterior body region. Male: F, anterior body region; G, ventral view of copulatory apparatus and tail; H and I, reproductive system and tail. Scale bars = 20 μm. (After Decraemer, 1991, courtesy Marcel Dekker Inc.)

nucleus or thread-like (degenerated) with nucleus obscure, sperm exceptionally long, fusiform with elongated nucleus. Oblique copulatory muscles poorly developed and restricted to spicule region; capsule of spicule suspensor muscles inconspicuous. Spicules usually straight and blade about equally wide and often finely striated; no bristles or velum. Two or three ventromedian pre-cloacal supplements, exceptionally four; usually two supplements within spicule region. Usually one pair, rarely two, subventral post-cloacal papillae. A pair of caudal pores present. Type species: *Paratrichodorus tunisiensis* (Siddiqi, 1963) Siddiqi, 1974. Species identification is based mainly upon the same characters as for *Trichodorus*. Additional features are the presence or absence of a spermatheca, the size and shape of sperm cells and nuclei and to a lesser extent the type of pharyngo-intestinal junction.

Within the genus *Paratrichodorus*, nine of the 29 known species are vectors of tobraviruses (see Section 6.7), and among them, *P. pachydermus* acts as a vector of both TRV and PEBV. Reproduction is amphimictic. *Paratrichodorus pachydermus* is polyphagous, and has not only been associated with a wide range of crops, trees and flowers, but is also known to withstand both drought and waterlogged conditions.

6.5.2.3. **Nanidorus** *Siddiqi, 1974 (Fig. 6.18)*

Based upon recent additional molecular support, *Nanidorus* is now generally accepted as a valid genus (Duarte *et al.*, 2010). Although molecularly more close to *Trichodorus,* morphologically it is largely similar to *Paratrichodorus* but can be differentiated from the latter in the female by: (i) the vulva with more posterior position (60–64% versus <60% in *Paratrichodorus*); (ii) the transverse slit-like vulva opening versus a pore or longitudinal slit; and (iii) non-functional striated sperm, rarely small rounded, spread along the uteri instead of more diverse in shape and size and either in a spermatheca or spread along the uteri in *Paratrichodorus* and in male by the single pre-cloacal supplement versus 2–4 in *Paratrichodorus*. Males are rare in *Nanidorus* (only known in three out of seven species), lack ventromedian cervical papillae and caudal pores are absent in male and female versus present in *Paratrichodorus*.

The genus *Nanidorus* includes one of the most important trichodorid plant pathogens, *N. minor*, which has been recorded mainly from tropical and subtropical regions worldwide, causing not only substantial damage by direct feeding on the root system but also as a vector for TRV in the USA and PRV in South America; it has occasionally been found in southern Europe. *Nanidorus minor* is common in sandy and sandy loam soils in warm climates. It has been associated with over 100 plant species including economically important crops such as citrus, sugar cane, walnut and a wide range of vegetables. On tomato, its life cycle is complete in 21–22 days at 22°C and in 16–17 days at 30°C.

6.6. Molecular Diagnostics

Various biochemical and molecular techniques are known for diagnostics of nematodes, including estimation of genetic diversity and inference of phylogeny (see Chapter 2). However, it is clear that sequence-based identification, or DNA barcoding, has become the most popular technique, and its use is growing rapidly, not only for the characterization of ectoparasites, but also for nematodes in general. However, it must be remembered that results of DNA barcoding are only as good as the reference database, i.e. useful molecular barcodes are only obtainable based on thorough 'traditional' taxonomy. Unfortunately, taxonomic expertise is waning, resulting in numerous sequences being based on incorrect identifications in GenBank. Logically therefore, forging a strong and lasting link between morphology and DNA sequences, including the use of voucher species and topotype specimens, is of crucial importance in order to prevent, or at least minimize, sequence-based misidentifications (Janssen *et al.*, 2017a; Palomares-Rius *et al.*, 2017a).

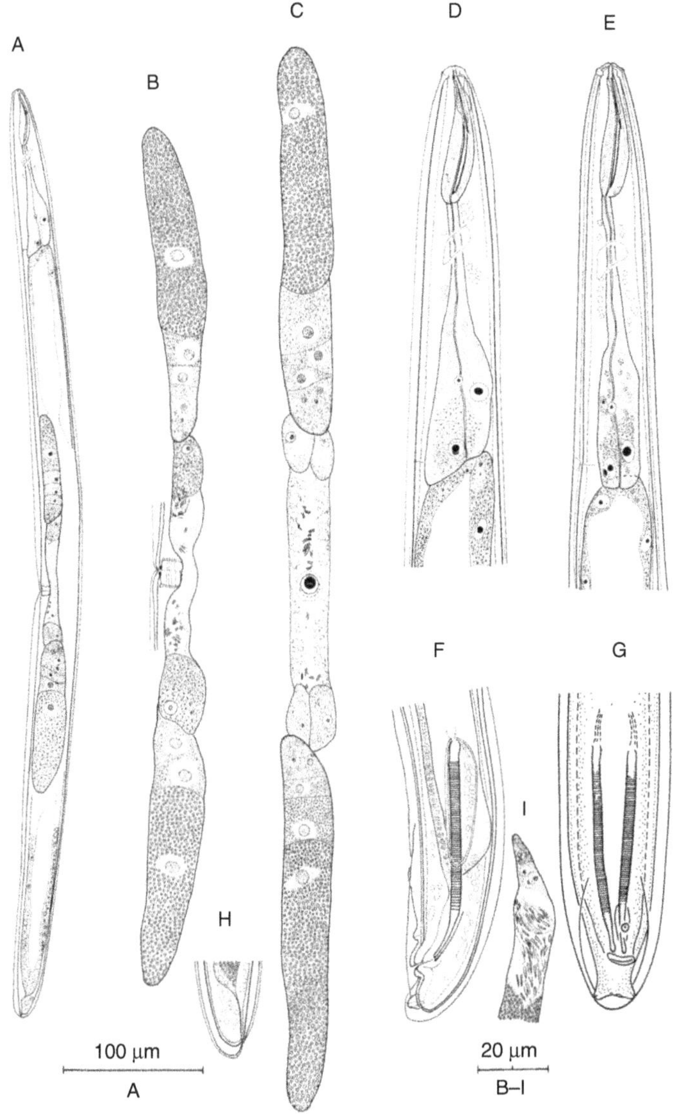

Fig. 6.18. *Nanidorus minor.* Female: A, total view (paratype *P. christiei*); B and C, reproductive system in lateral view and ventral view, respectively; D and E, anterior body region; F and G, male tail region and copulatory apparatus in lateral view and ventral view, respectively; H, female tail region; I, testis and *vesicular seminalis*. (After Decraemer, 1991, courtesy Marcel Dekker Inc.)

In general, nuclear D2–D3 expansion region of 28S ribosomal RNA (rRNA) and the mitochondrial *COI* gene sequences are most used and useful to characterize plant-parasitic nematodes, and especially ectoparasites. A case study in *Paratylenchus* (Singh *et al.*, 2020) indicated that internal transcribed spacer (ITS) and 18S rRNA genes provide, respectively, a likely overestimation and underestimation of the number of putative species based on

DNA-based species delimitation interferences, while *COI* and D2–D3 of 28S rRNA sequence information showed a better agreement with traditional methods. The D2–D3 domains of 28S genes display significant interspecific sequence differences that make them well-suited for use in nematode diagnostics, with the result that this is by far the most commonly used DNA barcode in modern plant-parasitic diagnostics. Conversely, use of the *COI* gene is advantageous over ribosomal genes as it shows not only a high interspecific variability within short lengths, but also an excellent barcoding gap, and furthermore provides an improved resolution in phylogenetic relationships among many closely related species. Although it is the most well-known barcode for most animals, GenBank and Barcode of Life Data Systems (BOLD) have only relatively recently been populated with *COI* data of plant-parasitic nematodes, and while *COI* is certainly promising as a barcode for plant-parasitic nematodes, it has also some shortcomings, e.g. mitochondrial heteroplasmy (i.e. presence of more than one type of mitochondrial genome) as was observed in *Rotylenchus goodeyi* (Singh *et al.*, 2022).

The ITS1 and ITS2 of rRNA genes usually contain more mutations compared to the former two subunits of the rRNA gene and therefore have proved useful in efficiently separating closely related species in many plant-parasitic groups. However, ITS DNA barcodes are also associated with serious shortcomings for some species of plant-parasitic nematodes because of large insertions and deletions, and can suffer from substitution saturation at a relatively low taxonomic level. Such mutations can result in extremely high variations both at inter- and intra-specific levels and thus obscure the boundaries between them, thereby creating confusion in ITS-based identification (Janssen *et al.*, 2017a). The 18S gene evolves relatively slowly and provide a good resolution above species level, and has therefore proved very useful in phylogenetic studies, although the low resolution of 18S at species level has been demonstrated in various plant-parasite groups. Nevertheless, it is still important to generate this gene sequence alongside other barcodes because of its widespread use as a barcode in traditional nematological research. For tylenchid ectoparasites, ITS, 18S RNA and especially the D2–D3 regions of 28S rRNA and mitochondrial *COI* sequences represent the preferred molecular barcodes for their diagnosis. This approach holds true for the majority of plant-parasitic nematodes, with the exception of tropical root-knot nematodes, for which more quickly evolving mitochondrial coding genes such as NAD5 are needed (Janssen *et al.*, 2016).

For the Longidoridae, sequences of nuclear rRNA genes, particularly D2–D3 and ITS1, have proven to be a powerful tool for providing accurate species identification and in assessing phylogenetic patterns necessary for molecular characterization and reconstruction of phylogenetic relationships (Gutiérrez-Gutiérrez *et al.*, 2013). A broad study of the variability of molecular markers used for phylogenetic relationships and the identification of Longidoridae was conducted by Palomares-Rius *et al.* (2017a), and in this study, despite some difficulties with obtaining an unequivocal alignment, the ITS1 region showed great promise for barcoding and species identification because of the clear molecular variability among species. Although the use of *COI* and D2 and D3 expansion segments of the 28S rRNA gene did not clarify the phylogeny at the genus level, they were found to be useful for species identification, with the exception of certain species in the *X. americanum* group. Although (partial) 18S rRNA sequences are known to have species resolution constrains for plant-parasitic species, the partial 18S rRNA gene showed potential for differentiating between species in Longidoridae. Palomares-Rius *et al.* (2017a) also stressed the need for more than one molecular marker to permit the unequivocal identification of Longidoridae unless integrative taxonomical approaches are also employed.

For the Trichodoridae, sequence information is available for the didelphic genera but molecular data are virtually non-existent for the monodelphic–prodelphic genera (*Monotrichodorus*, *Allotrichodorus* and *Ecuadorus*), except for *Monotrichodorus* (*M. vangundyi* and two unknown species). Available data for species of the didelphic genera are represented by D2–D3 of 28S rRNA sequences (some 30 *Trichodorus* species) and to a lesser extent by 18 28S rRNA (approximately 18 *Trichodorus* species), with very few ITS rDNA sequences currently available. Remarkably, *COI* sequence representation of trichodorids remains almost non-existent, despite the fact that Decraemer *et al.* (2019) have demonstrated the usefulness of *COI* for delimiting species in the *P. hispanus* group. An overview of other molecular methods to identify, differentiate and quantify Trichodoridae has been provided by Huang *et al.* (2019) and includes polymerase chain reaction followed by a restriction fragment length polymorphism analysis (PCR-RFLP), duplex PCR with species-specific primers, and quantitative real-time PCR. This paper also presents the development of a one-step multiplex PCR assay for rapid detection of the four major trichodorid nematodes in the USA (*Paratrichodorus allius, P. minor, P. porosus* and *Trichodorus obtusus*).

In conclusion, currently the most useful techniques available for the characterization and diagnosis of the majority of ectoparasitic nematodes and plant-parasitic nematodes in general appear to stem from the use of both D2–D3 of 28S and *COI* gene sequences, originating from both nuclear and mitochondrial genomes. However, while such molecular techniques represent informative and powerful investigational tools, their effectiveness can be severely hampered due to many factors, including inappropriate barcoding gene selection and data interpretation, difficulties encountered in linking cryptic species to informative barcodes, as well as the known flaws and limitations of existing sequence databases. Furthermore, the compilation and interpretation of molecular data involve a significant amount of manual intervention, and sole reliance on DNA-similarity searches such as BLAST can give misleading results, as there is no certain intra- or interspecific threshold available. To help mitigate such complications, Qing *et al.* (2019) have presented a reference database and molecular identification pipeline for plant-parasitic nematodes.

In the future, the increased affordability of next-generation sequencing is expected to lead to diagnostic techniques that will be progressively based on genomic data rather than individual sequences, in conjunction with metabarcoding-based identification rather than simple Sanger sequencing. Nevertheless, it is clear that the role of nematode taxonomists will remain crucial in assuring the vital link between classical systematics and multigene approaches, which will allow databanks to be usefully populated with professionally audited and validated data.

6.7. Ectoparasitic Nematodes as Vectors of Plant Viruses

Plant viruses represent the most important pathogens associated with ectoparasitic nematodes, and this nematode-based viral transmission process involves different phases, such as ingestion, acquisition, adsorption, retention, release, transfer and establishment. For a virus to be transmitted by a nematode, virus particles must be acquired by the nematode and then retained at a specific place in its food channel, with the transmission of the virus then occurring through the detachment of the retained virus particles from the retention sites, followed by the direct injection of those viruses into the root cells of the plant during nematode feeding. Longidoridae and Trichodoridae are currently the only families of plant-parasitic nematodes known to transmit plant viruses, and the relationship between virus and nematode is specific;

only certain species of nematodes transmit certain viruses. Plant-parasitic nematodes transmit two groups of viruses: nepoviruses and tobraviruses. As an example of the symptomatology of tobraviruses, the TRV disease symptoms in crops range from yellowing and necrosis of leaves to colour break in flowers and necrotic arcs in the tuber flesh of potatoes. The best-known example of a symptom associated with nepoviruses is the 'hen-and-chick' effect on grape clusters resulting from an infection of the ToRSV vectored via *Xiphinema* spp. These are grape clusters that retain a few normal-sized berries, although the majority are small and tasteless (Halbrendt, 2021). Of the 41 known nepoviruses, 13 species of nepoviruses are transmitted by 11 species of *Xiphinema*, 10 species by 11 species of *Longidorus* (Singh *et al*., 2020) and a single species of nepovirus is transmitted by one *Paralongidorus* species (Gutiérrez-Gutiérrez *et al*., 2017). All three tobraviruses (PEBV, PRV and TRV) are transmitted by trichodorid nematodes; within the Trichodoridae, four out of 74 *Trichodorus* species (5%), nine out of 29 *Paratrichodorus* species (32%) and two out of seven *Nanidorus* species (29%) are vectors of tobraviruses. Compared to the rest of the world, European species demonstrate greater percentages of virus vector versus non-vector species.

Nepoviruses and tobraviruses both consist of bipartite positive-sense single-stranded RNA (ssRNA) genomes, the larger being referred to as RNA1 and the smaller RNA2. Nepoviruses are isometric in particle makeup, while the particles of the tobraviruses have straight tubular particles occurring in two sizes. In nepoviruses, the large RNA1 molecule carries the genetic determinants for host-range, seed transmissibility and some types of symptom expression, whilst the small RNA2 contains the coat protein (involved in serological specificity), nematode transmissibility and some symptom expression (Taylor and Brown, 1997). In tobraviruses, every molecule contains genetic determinants for symptom expression and the smaller RNA2 contains determinants for serological specificity (coat protein) and vector transmissibility. Nepoviruses and tobraviruses transmitted by nematodes are primarily pathogens of wild plants, and most are usually seed transmitted. They have developed specific relationships with their vector species and have become associated with cultivated plants, although native wild plants infected with nematode-transmitted viruses typically do remain asymptomatic. It is possible that co-evolution of some of some of these viruses with native vegetation has resulted in plants that are tolerant of infection and thereby serve as a reservoir of the virus in nature (for related information concerning ToRSV, see Halbrendt, 2021).

The marked specificity between virus and vector is most well-known in Europe where a particular serological virus strain is transmitted by a single nematode species. For example, *X. diversicaudatum* only transmits arabis mosaic virus (ArMV), and the PRV serotype of TRV from Britain, Sweden and The Netherlands is transmitted only by *P. pachydermus* (Brown *et al*., 2004). The acquisition and transmission of viruses is linked to their methods of feeding. In *Longidorus*, virus particles are adsorbed in a single layer onto the inner surface of the odontostyle, and in *L. elongatus* virus particles of RRSV may also be located between the odontostyle and the guiding sheath (Taylor and Brown, 1997). In *Xiphinema* vector species, virus particles are associated with the cuticular lining of the odontophore and the pharynx. In trichodorid vectors, the tobravirus particles are associated with the lining of the complete pharyngeal region but are not attached to the onchiostyle. However, Karanastasi and Brown (2004) observed that the sites of virus particle retention differed between the individual tobravirus–trichodorid combinations. In addition, virus particles of several strains of tobraviruses were acquired and retained by trichodorids that are not natural vectors of the particular virus. In these instances, the particles were retained within the pharyngeal tract at locations from where, when released, they cannot be transferred anteriorly along the pharyngeal tract into plant cells but can move only posteriorly into the nematode intestine. It should be noted that, owing to their ability to induce post-transcriptional gene silencing (known as virus-induced gene silencing or VIGS), tobraviruses are employed as reverse-genetics tools for the functional characterization of plant genes. (Dubreuil *et al.*, 2009).

Although the relationship between virus and nematode is specific, virus vector nematodes are polyphagous in nature and the viruses may have a wide host range, making management of virus vectors challenging.

6.7.1. Management of virus vector nematodes

In addition to being vectors for tobraviruses and nepoviruses, many species of the families Longidoridae and Trichodoridae cause direct damage to a wide variety of plants when population densities are high (Fig. 6.19), but some species, such as *N. minor*, are considered more economically important than others possessing some of the widest host ranges. The amount of damage to the root system varies with the crop attacked and also appears greatly influenced by season, soil type and fertility. For example, 'docking disorder' of sugar beet in the UK resulting from feeding of *P. anemones* on seedlings was more severe after a prolonged period of rain in spring. In practical terms, management of these ectoparasites depends on suppression of their population densities to an acceptable level, as determined by the costs of control measures in relation to the financial return of the resultant increased crop yield. Although most of these nematode species are polyphagous, not all of their host plant species promote nematode reproduction to the same extent, meaning that selective crop rotation can still be effective as a measure to control their populations.

Fig. 6.19. Direct damage caused by Trichodoridae. A, stubby root symptoms in chicory caused by *Paratrichodorus teres*; B, galling on wheat caused by *Paratrichodorus* spp.; C, enlarged multinucleate cells in root gall caused by *Xiphinema index* on fig roots. (A, courtesy Plant Protection Service, The Netherlands; B and C, courtesy CABI.)

Whilst the agricultural importance of trichodorid and longidorid nematodes can result from their direct damage to host plants, the major pest status attributed to Longidoridae and Trichodoridae is due mainly to the ability of some species to transmit plant viruses, in which case a great amount of indirect damage can occur even at low nematode densities. However, in most situations, nepovirus and tobravirus infections in the field are apparent only in crop plants where symptoms are severe and may induce plant death.

As it is very difficult to eradicate trichodorids and longidorids and their associated viruses once they are present in a field, it is of vital importance that measures are taken to prevent them from establishing a presence in the first place. Such prevention includes the use of certified planting material, ensuring that machinery is cleaned adequately before moving it from one field to the other, and employing effective weed management, as many arable weed species are often a source of trichodorids and associated viruses. Complementary measures may include techniques such as tissue culture micropropagation that can result in the production of virus-free plants. Regular soil sampling and assessment of the plant-parasitic nematodes that are present are also essential to create a sound crop rotation (Brinkman and Hallmann, 2021), although, in many cases, crop rotation options are limited since most polyphagous virus nematode vectors and their associated viruses have a wide host range. As supplementary measures, the use of fumigants as well as non-fumigant nematicides (see Chapter 17), can help to a certain extent to prevent or, in the case of perennial plants, to delay, viral infection of newly planted crops. Knowledge of previous crops that have been harvested in a given field is occasionally of great importance, as some crops such as seed potatoes are more sensitive to viral infection than others. Soil solarization using clear plastic mulch or a photoselective (black) polyethylene film has been proven to reduce populations of *N. minor* on tomato as effectively as fumigation in warm climates. The presence of fresh organic matter also seems to hamper the transmission of TRV, potentially as a result of decreased nematode activity (Brinkman and Hallmann, 2021). To solve the complexity of plant virus infestation by nematodes in the long term, at present the most effective solution would appear to involve the incorporation of either virus tolerance or its complete resistance into the crop through breeding programmes or the development of transgenic plants. Such breeding programmes have already resulted in hybrid grapes that have proven to be resistant to ToRSV infection (Halbrendt, 2021). In contrast to the increasing reliance on plant resistance to viruses, knowledge of an equivalent resistance to the vector nematode remains very limited. Taking the example of longidorids, vineyard rootstocks have been reported to possess resistance to *X. index* but have not currently been assessed regarding their resistance to *X. americanum* (Zasada and Forge, 2021). Furthermore, transgenic resistance is primarily restricted to viruses and has not been applied to the nematode vectors. Despite notable exceptions, such as the research by Danchin *et al.* (2017) revealing the independent acquisition of parasitism genes through horizontal gene transfer in *X. index* and *L. elongatus*, the relationship between ectoparasitic nematodes and their host plants remains an area with much yet to be explored. Unlike the wealth of information available on sedentary endoparasitic nematodes, the mode of interaction between ectoparasitic nematodes and their host plants remains largely to be investigated in order to understand and evaluate possible ways that they might acquire resistance to the nematode vector.

Part II Nematode Biology and Plant–Nematode Interactions

Information on the functional biology and ecology of nematodes is central both to fundamental science and to practical aspects, such as targeting various management strategies effectively. This knowledge is also vital for identification and development of novel control targets. This is an exciting period for studies on nematode biology and host–parasite interactions because the availability of genome and transcriptome sequences, expressed sequence tag libraries and tools for functional analysis, such as RNAi, provides immense potential for functional genomic studies. The first two chapters in this section focus on nematode biology and lead into the third chapter, which reflects the advances in molecular information by focusing on molecular aspects of plant–nematode interactions. The fourth chapter deals with ecology of plant-parasitic nematodes, and includes information on the effects of climate change.

7 Reproduction, Physiology and Biochemistry*

Roland N. Perry[1**], Denis J. Wright[2], David J. Chitwood[3] and James A. Price[4]

[1]*School of Life and Medical Sciences, University of Hertfordshire, UK and Biology Department, Ghent University, Belgium;* [2]*Department of Life Sciences, Imperial College London, UK;* [3]*Formerly Mycology and Nematology Genetic Diversity and Biology Laboratory, USDA, ARS, USA;* [4]*CMS, The James Hutton Institute, UK*

*A revision of Perry, R.N., Wright, D.J. and Chitwood, D.J. (2013) Reproduction, physiology and biochemistry. In: Perry, R.N. and Moens, M. (eds) *Plant Nematology*, 2nd edn. CAB International, Wallingford, UK.

**Corresponding author: r.perry2@herts.ac.uk

DOI: 10.1079/9781800622456.0007

7.1. Reproduction and Development

Nematodes exhibit a variety of sexual and asexual reproductive methods and associated behaviours. Where there are two separate sexes, sexual reproduction, or **amphimixis**, can occur where male gametes (spermatocytes) and female gametes (oocytes) fuse to form the zygote. Reproduction can occur asexually, termed parthenogenesis or **amixis**, where males are not involved. A further type of reproductive method is **hermaphroditism**, where both egg and sperm are produced by the same individual. As will be evident in the following sections, there are several types of these basic reproductive forms and some genera can exhibit more than one type of reproduction (Table 7.1). Most information is available for *Meloidogyne* spp. (reviewed by Chitwood and Perry, 2009) and details are summarized in Chapter 3, Table 3.1.

7.1.1. Amphimixis

Female nematodes are the homogametic sex (genetically XX) and males are heterogametic (sometimes XO, usually XY). This is found in *Anguina tritici* and *Ditylenchus dipsaci*, for example, and the sex ratio is determined genetically. In genera such as *Globodera*, *Heterodera* and *Meloidogyne*, sex chromosomes are absent and the sex ratio may be influenced by environmental factors, such as food availability and temperature.

As with many animal groups, development of gametes includes meiotic division that results in the diploid ($2n$) complement of chromosomes being reduced by half (n). Fusion (fertilization) of the haploid gametes restores the diploid complement and ensures that the offspring receive genetic material from both parents. The following outline of oogenesis is based on observations of several species of nematodes, summarized by Triantaphyllou (1985). When the gonad is mature, oogonia are produced, which increase in size as they progress down the growth zone of the ovary. The young oocytes enter the spermatheca where development ceases unless they are entered by sperm. After copulation, the sperm are stored in the spermatheca and occasionally the uterine sac, if present. When young oocytes pass through the spermatheca they are individually fertilized with a sperm, thus

Table 7.1. Reproduction method and chromosome numbers of some plant-parasitic nematodes. The haploid chromosome number is given first followed by the diploid number in brackets. (Modified from Evans, 1998.) For details relating to *Meloidogyne* spp., see Chapter 3, Table 3.1.

Nematode group	Chromosome number Haploid (diploid)	Type of reproduction
Class Enoplea		
Order Dorylaimida		
Longidorus	7	Parthenogenesis
Xiphinema pachtaicum (*X. americanum* group)	5	Meiotic parthenogenesis
X. index	10 (20)	Meiotic parthenogenesis
Class Chromadorea		
Order Rhabditida		
Family Anguinidae		
Anguina tritici	19 (38)	Amphimixis
Ditylenchus dipsaci	12 (24)	Amphimixis
D. dipsaci	(54)	Amphimixis
Family Dolichodoridae		
Tylenchorhynchus claytoni	8	Amphimixis
Family Pratylenchidae		
Nacobbus aberrans	5–8	Amphimixis
Pratylenchus brachyurus	(30–32)	Mitotic parthenogenesis
P. coffeae	7	Amphimixis
P. neglectus	(20)	Mitotic parthenogenesis
P. penetrans	5 (10)	Amphimixis
	6	Amphimixis
P. scribneri	6 (12)	Meiotic parthenogenesis
	(25–26)	Mitotic parthenogenesis
P. vulnus	6	
P. zeae	(26)	Parthenogenesis
	(21–26)	Parthenogenesis
Family Hoplolaimidae		
Helicotylenchus dihystera	6–8 (30, 34, 38)	
H. erythrinae	5 (10)	Amphimixis
Hoplolaimus galateus	10, 11 (20)	Amphimixis
Rotylenchus buxophilus	8	Parthenogenesis?
Rotylenchulus reniformis	9 (18)	Amphimixis
Globodera mexicana	9	Amphimixis
G. pallida	9 (18)	Amphimixis
G. rostochiensis	9–11 (18–24)	Amphimixis
G. solanacearum	9	Amphimixis
G. tabacum	9	Amphimixis
G. virginiae	9	Amphimixis
Heterodera avenae	9 (and rarely, 19)	Amphimixis
H. betulae	12, 13 (24)	Meiotic parthenogenesis
H. carotae	9	Amphimixis
H. cruciferae	9, 10	Amphimixis
H. galeopsidis	(32)	Mitotic parthenogenesis

Continued

Table 7.1. Continued.

Nematode group	Chromosome number Haploid (diploid)	Type of reproduction
H. glycines	9 (18)	Amphimixis
H. goettingiana	9	Amphimixis
H. lespedezae	(27) triploid	Mitotic parthenogenesis
H. oryzae	9 (27) triploid	Mitotic parthenogenesis
H. sacchari	(27) triploid	Mitotic parthenogenesis
H. schachtii	9 (18)	Amphimixis
	9 (19)	Amphimixis
	9	Amphimixis
H. trifolii	(24–27) triploid	Mitotic parthenogenesis
	(26–34)	Mitotic parthenogenesis
Cactodera weissi	9	Amphimixis
Family Criconematidae		
Hemicriconemoides spp.	5	Amphimixis
Family Tylenchulidae		
Tylenchulus semipenetrans	5	Amphimixis, facultative meiotic parthenogenesis
Cacopaurus pestis	5, 6 (10, 11, 12)	Amphimixis
Family Aphelenchidae		
Aphelenchus avenae	8	Meiotic parthenogenesis, and amphimictic pseudogamy
	9	Meiotic parthenogenesis
Family Aphelenchoididae		
Aphelenchoides composticola	3	Amphimixis

stimulating further development. Soon after penetrating the oocyte, the sperm becomes temporarily inactive near the oocyte membrane it penetrated. Two meiotic divisions occur in the nucleus of the activated oocyte and the eggshell begins to form. One of the diploid nuclei from the first division is reduced to a polar body and is extruded from the oocyte. The second division of the remaining oocyte nucleus results in a haploid pronucleus and a second polar body, which is also extruded. At about the time the egg is expelled from the uterus, the sperm pronucleus, also haploid, is formed and the gamete fusion occurs to produce the diploid zygote that commences development.

In many genera, including *Heterodera*, *Pratylenchus* and *Radopholus*, the female can be fertilized by several males, thus enhancing the genetic diversity of the offspring. In a few genera (e.g. *Trichodorus*), copulatory plugs deposited by a male nematode within the female vulva may interfere with subsequent mating (Decraemer, 2012). Amphimixis is often a secondary option in species where parthenogenesis is predominant.

7.1.2. Parthenogenesis

In species of nematodes that reproduce asexually, males are absent or occur only rarely. Among nematodes, parthenogenesis has been found to be particularly common in plant-parasitic forms. There are two main types of parthenogenesis, meiotic parthenogenesis (or **automixis**), which involves a reduction division, and mitotic parthenogenesis (or **apomixis**), where there is no reduction division.

7.1.2.1. Meiotic parthenogenesis

In this type of reproduction, there is a first meiotic division in the oocytes, although there are variations between species. This meiotic division allows the possibility of some genetic reorganization. *Meloidogyne hapla* race A exhibits facultative meiotic parthenogenesis, oogenesis and spermatogenesis proceeding as in amphimictic species to yield one haploid nucleus and two polar bodies per oocyte. If sperm are present, amphimixis occurs. However, if the egg is not fertilized by sperm, the diploid state is achieved by the egg pronucleus recombining with the second (haploid) polar body. In monosexual populations of *Aphelenchus avenae*, females produce only female progeny and reproduction is by obligate meiotic parthenogenesis. Meiosis produces only one polar body, with the egg nucleus having the $2n$ chromosome number that then develops into the zygote. Other species reproducing by meiotic parthenogenesis include *Xiphinema index* and *Pratylenchus scribneri* (Table 7.1).

7.1.2.2. Mitotic parthenogenesis

This is the most common method of asexual reproduction (Table 7.1). The only division that occurs is mitosis and the oocytes retain the diploid chromosome number. Mitotic parthenogenesis is always obligate. This prevents any genetic reorganization and, except for mutations, the ova produced should be clones. Frequently, mitotic parthenogenesis is associated with polyploidy (Table 7.1), which may increase the likelihood of mutation. There is no direct evidence for this, although variability in pathogenicity exists and can be selected for (Blok *et al.*, 1997). Mitotic parthenogenesis occurs in several species of *Meloidogyne* and *Pratylenchus*; in *Meloidogyne* it has evolved at least four times (Janssen *et al.*, 2017a). Polyploidization may facilitate divergence among gene copies in mitotically parthenogenetic *Meloidogyne* species (Blanc-Mathieu *et al.*, 2017).

7.1.3. Hermaphroditism

In hermaphrodites, both egg and sperm are produced in the same individual. Usually the sperm is produced first and is stored in the spermatheca, then the gonad produces oocytes, which are fertilized by the sperm until the sperm supply is exhausted. Hermaphroditism is a common method of reproduction amongst rhabditids and other free-living nematodes but is relatively rare in plant-parasitic nematodes, being found in some members of the Criconematoidea, species of *Paratrichodorus*, and one population of *M. hapla* (Triantaphyllou, 1993). Pseudogamy is a type of hermaphroditism where sperm penetration activates oocyte development but the nuclei do not fuse. The sperm nucleus degenerates and the diploid state of the oocyte is restored after the first meiotic division. However, to our knowledge this form of reproduction has not been reported in plant-parasitic nematodes.

7.1.4. Development

Development includes embryogenesis and growth, about which there is a very large amount of information for *Caenorhabditis elegans* (see Priess and Seydoux eds, http://www.wormbook.org) but little for plant-parasitic nematodes. With the availability of genome and transcriptome sequences (e.g. Da Rocha *et al.*, 2021; Siddique *et al.*, 2022) for a wide range of plant-parasitic nematodes, as well as tools for functional analysis such as RNA interference (RNAi) (reviewed by Rosso *et al.*, 2009), there is now potential for functional genomic studies on development of plant-parasitic nematodes. Embryogenesis has been described for many species of plant-parasitic nematodes (Hope, 2002) but most information derives from observations of fixed material. Most of the cell division occurs in the first half of embryogenesis. In the second half, the embryo elongates markedly, starts to move and synthesizes the cuticle. The number of cells in most nematodes is presumed to be similar to that in *C. elegans* (959 somatic cells) despite differences in size between species and juvenile and, especially, adult stages.

Juveniles hatch from eggs that are laid by the adult female. Each egg is ovoid-shaped and contains a single juvenile. Despite the vast difference in size between adults of various species of nematodes, the majority of eggs are of similar size (50–100 μm long and 20–50 μm wide) and morphology. In most plant-parasitic nematodes the juvenile moults within the egg and the resulting second-stage juvenile (J2) hatches. The egg and the hatching responses are part of the survival attributes of many species of nematodes, and hatching is discussed in Section 7.6.5.

7.1.4.1. Moulting

The primary signal for moulting in nematodes is unlikely to be an increase in body size as growth takes place between moults. The cuticle is shed and replaced four times during the life cycle, although there is evidence that some species of *Longidorus* and *Xiphinema* have only three juvenile stages.

There are differences between species in the moulting pattern (Bird and Bird, 1991; Lee, 2002) but, in general, the process involves three phases: (i) the separation of the old cuticle from the epidermis (apolysis); (ii) the formation of a new cuticle from the epidermis; and (iii) the shedding of the old cuticle (ecdysis). Wright and Perry (1991) studied the moulting process in *Aphelenchoides hamatus* from fourth-stage juvenile (J4) to adult. The process took 12–13 h to complete and showed features typical of most moulting patterns. Initially, *A. hamatus* did not move (a period termed lethargus) while the cuticle separated from the epidermis, which thickened and showed evidence of increased metabolic activity as the new cuticle was formed. There was also a loss of the knobs and shaft of the stylet. The water content of the nematode decreased, causing a reduction in body volume that enabled the adult to retract within and away from the old cuticle as the new stylet and head skeleton were formed. The entire body of the adult then moved within the old cuticle. The pharynx became active as the adult took up water and expanded to burst the old cuticle. It is possible that pharyngeal secretions and associated enzymic activity might aid in degrading the old cuticle before ecdysis. In some endoparasitic species, such as *Meloidogyne javanica* and *Heterodera glycines*, cuticles are partially or completely reabsorbed. During adverse conditions, the moulted cuticles of *Rotylenchulus reniformis* are retained as sheaths to aid survival (see Section 7.6.3).

7.2. Musculature and Neurobiology

Information on the musculature and nervous system of plant-parasitic nematodes is confined almost completely to morphological details (see Chapter 1). The information here on physiology and biochemistry derives mainly from work on animal-parasitic and free-living nematodes, especially *C. elegans*, the assumption being that the basic features are likely to be the same for plant-parasitic forms.

7.2.1. Musculature

There are two types of muscle: **somatic** and **specialized**. Somatic, or body wall, muscles are obliquely striated (multiple sarcomere). The somatic muscle cell comprises three parts: the spindle that contains the contractile elements (called sarcomeres), a non-contractile muscle cell body that projects into the pseudocoelom and the arm, which extends from the cell body to the longitudinal nerve chord or to the nerve ring. In many species of nematodes muscle cells have more than one arm. Before each arm reaches the nerve chord, it divides into several 'fingers', which subdivide into multiple fine processes that receive synaptic input from excitatory and inhibitory motor neurons. The fingers are interconnected by gap junctions, probably facilitating electronic spread of impulses between neighbouring muscle cells. Gap junctions occur between muscle cells and between neurons but are rare or absent between muscles and neurons. Specialized muscles are non-striated (single sarcomere) muscles and they include pharyngeal muscle cells, intestinal muscles and others associated with defecation, mating, fertilization and egg laying.

Each sarcomere consists of thick filaments flanked and interdigitating with thin filaments; the striated appearance of somatic muscle results from the alternation of regions containing thick and thin filaments. At the centre of each thick filament, a specialized region crosslinks the thick filaments to maintain their alignment. Thin filaments are anchored at one end to dense bodies. Thick filaments contain myosin and paramyosin, and thin filaments contain actins, tropomyosins and troponins. Actin forms the core component of thin filaments and binds and activates myosin. The process of muscle contraction depends on ATP and requires direct interaction of myosin head and actin. Cross-bridges, originating on the thick filaments, attach and detach cyclically to specific sites on the thin filaments, causing the filaments to be pulled past each other, resulting in shortening of sarcomeres and, thus, contraction of muscles.

Somatic muscles are arranged in longitudinal rows in a single layer divided by the epidermal chords into four nearly symmetrical quadrants; circular somatic muscles are absent in nematodes. Individual motor neurons of the ventral chord innervate either dorsal or ventral muscles, thus restricting body flexures to the dorsoventral plane only. Alternate contraction of the dorsal or ventral longitudinal somatic musculature results in sinusoidal locomotion. The more complex innervation of the anterior end permits the lateral and dorsoventral movements of the head that are components of oriented and unoriented behaviour (see Chapter 8).

7.2.2. Nervous system

The organization of the nervous system appears to be essentially conservative. The neurons have a simple, relatively unbranched morphology and a single gap junction is sufficient for functional coupling between neurons. The majority of nerve processes run longitudinally, as ventral and dorsal nerve chords, or circumferentially, as commissures. The nervous system in *C. elegans* comprises two essentially independent parts, the pharyngeal system and the much larger extrapharyngeal (somatic) system, which are connected via a bilateral pair of gap junctions between a pair of pharyngeal interneurons and a pair of ring/pharynx interneurons. Most neuronal cell bodies of the extrapharyngeal nervous system occur around the pharynx, along the ventral mid-line and in the tail.

The circumpharyngeal nerve ring is the main mass of the central nervous system. Neuronal processes from the anterior sensilla (sense organs) run posteriorly as six cephalic nerve bundles, four of which have their cell bodies just anterior to the nerve ring in a region termed the 'anterior ganglion'. The other cephalic nerve bundles, which contain processes from the lateral pair of amphid sensilla, run past the nerve ring to cell bodies in two lateral ganglia. Processes from the sensilla eventually synapse at the nerve ring. Processes from cells in the tail ganglia connect with the posterior (caudal) sensilla. Details of sensilla structure and function are given in Chapter 8.

The majority of the nerve processes leaving the nerve ring form the ventral nerve chord. The interneuron processes in this chord synapse either with the pre-anal ganglion or with excitatory motor neurons in the nerve chord. The motor neurons are arranged in a linear sequence and some have processes that run under the epidermis as commissures to the dorsal mid-line where they pass anteriorly or posteriorly forming the dorsal nerve chord. The latter is simpler in structure than the ventral nerve chord with few interneurons and no cell bodies.

7.2.3. Neurotransmission

Neurotransmission in nematodes has been reviewed by Wright and Perry (1998) and Holden-Dye and Walker (2011); for specific reviews on *C. elegans* see Jorgensen (2023). Classical synaptic transmission involves the arrival of an action potential at a pre-synaptic nerve ending, which leads to the opening of voltage-gated Ca^{2+} ion channels near the synapse and the influx of Ca^{2+} ions into the nerve cell. This results in the secretion, by exocytosis from the pre-synaptic nerve ending, of neurotransmitter molecules, which diffuse across the synaptic cleft and bind reversibly to specific receptor proteins on the post-synaptic membrane of a nerve or muscle cell. This causes a conformational change in the receptor proteins that are linked directly (ionotropic receptors) or via secondary messenger systems (metabotropic receptors) to ion channels. The release of the neurotransmitter and the threshold response of the receptors can be regulated by other chemicals ('neuromodulators') released in the area of the synapse. Whether the response is excitatory or inhibitory depends on the type of receptor and, thus, which ion channel is activated. Thus, the same neurotransmitter can be excitatory and inhibitory.

Classical transmitters include acetylcholine, probably the primary excitatory transmitter, several amino acids and various biogenic amines. Characterization of different molecular forms of acetylcholine has been conducted with several species of nematodes. There is extensive information on the amino acid transmitter gamma-amino butyric acid (GABA) in *C. elegans*, including evidence for a novel excitatory GABA receptor as well a classic inhibitory GABA receptor (Schuske *et al.*, 2004). GABA has also been reported in J2 of *Globodera rostochiensis* and *Meloidogyne incognita* (Stewart *et al.*, 1994), and the expression of a putative GABA synthesis gene as well as the number of GABA immunoreactive neurons is lower in immobile developmental stages than in J2 (Han *et al.*, 2018). Other putative neurotransmitters in nematodes include the biogenic amines dopamine and 5-hydroxytryptamine, and various neuropeptides (see Section 7.3.1). Dopamine has been detected in a wide variety of nematodes, including *M. incognita* (Stewart *et al.*, 2001), and genes involved in the biosynthesis or utilization of dopamine and six other neurotransmitters occur in *Globodera pallida* (Cotton *et al.*, 2014). An essential feature of all neurotransmitter systems is a mechanism for the rapid removal of neurotransmitter from the synaptic cleft. With acetylcholine, the mechanism is enzymatic and inhibition of acetylcholinesterase is the target site for the control of plant-parasitic nematodes by organophosphate and carbamate nematicides (see Chapter 17).

7.3. Biosynthesis

This subject was comprehensively reviewed for plant-parasitic and free-living nematodes by Chitwood (1998), who pointed out that much more is known about the chemical composition of nematodes than about their biosynthetic pathways. Increased knowledge of the latter may lead to the design of novel control strategies for plant-parasitic species by exploiting targets unique to nematodes. Again, functional genomics and metabolomics (e.g. Assefa *et al.*, 2021) can be expected to provide substantial information in the future. Moreover, as our knowledge of the extent of microbial endosymbionts increases, their role in biosynthesizing compounds critically important to their hosts must be considered (Brown, 2018). A somewhat more in-depth presentation of cyst nematode lipid, carbohydrate and protein biochemistry was provided by Chitwood and Masler (2018).

7.3.1. Amino acids and related compounds

Work on *Caenorhabditis* spp. has shown a dietary requirement for arginine as well as nine typical mammalian essential (i.e. dietary) amino acids, histidine, isoleucine, leucine, lysine, methionine, phenylalanine, threonine, tryptophan and valine, although the requirement for methionine can be satisfied by a known precursor, homocysteine. Studies on various species of plant-parasitic nematodes (*Meloidogyne* spp., *Aphelenchoides* spp.) showed that they could all synthesize a number of typical mammalian non-essential amino acids. In addition, several amino acids that are essential in mammals could be synthesized, including tryptophan (in *Meloidogyne* spp.) and lysine, phenylalanine, leucine/isoleucine, threonine and valine (in *Aphelenchoides* spp.), although not necessarily in sufficient quantities to support reproduction.

Many putative neurotransmitters in nematodes, including biogenic amines and GABA, are related to amino acids and evidence for the pathways required for their synthesis has been obtained for *C. elegans* and to a more limited extent in some other nematodes (Wright and Perry, 1998; Perry and Maule, 2004). There has been little work on plant-parasitic nematodes, other than the localization of GABA and several biogenic amines within the nervous system (see Section 7.2.3). Various small polypeptides are also putative neurotransmitters or neuromodulators in nematodes. Typically, these are members of the FMRFamide-related peptide (FaRP) or FMRFamide-like (FLP) family, and they have been localized in the nervous system of several species, including *H. glycines*, *G. pallida* and *M. incognita* (Johnston *et al.*, 2010a). Insect neuropeptides are derived from large parent polypeptides. Many genes encoding FLPs and some other neuropeptides have been isolated from *C. elegans* and various plant-parasitic nematode genera, including *Bursaphelenchus*, *Globodera*, *Heterodera*, *Meloidogyne* and *Pratylenchus*.

The plant auxin indoleacetic acid (IAA), a tryptophan metabolite, and several related compounds, have been identified in species of *Meloidogyne* and *Ditylenchus* but there is no direct evidence of auxin biosynthesis by nematodes (Chitwood, 1998). Although most animals, including *C. elegans*, possess a nutritional requirement for B vitamins, *H. glycines* appears to have acquired the enzymatic machinery for biosynthesizing the amino acid derivatives biotin, thiamine and pantothenate via horizontal gene transfer (Craig *et al.*, 2009). Genomic studies indicate that endosymbionts within *Xiphinema* species may provide host nematodes with components essential for the biosynthesis of thiamine, other vitamins and amino acids (Myers *et al.*, 2021).

7.3.2. Nucleic acids and proteins

Like most organisms, nematodes are capable of synthesizing the purine and pyrimidine bases of nucleotides, although studies on the synthesis of nucleoside monophosphates are lacking. Work on *C. elegans* has provided evidence for the synthesis of cyclic nucleotides from their corresponding nucleoside triphosphates. The two enzymes at the start of the pentose pathway, glucose 6-phosphate dehydrogenase and 6-phosphogluconate dehydrogenase, have been demonstrated in several free-living and plant-parasitic nematodes (Barrett and Wright, 1998). However, during active nucleic acid synthesis in mammals the pentose pathway ends at the next step with the formation of D-ribose-5-phosphate, which is used for nucleotide synthesis. So the presence of the starting enzymes does not necessarily mean there is a complete cycle (see Section 7.3.4.).

Studies on nucleic acid polymerases in nematodes have focused on *C. elegans* and a large number of transcription factors have been identified in this species (Reinke *et al.*, 2013). Translational mechanisms involved in protein synthesis in *C. elegans* have been reviewed by Rhoads *et al.* (2006).

7.3.3. Lipids

The total lipid content of the free-living and plant-parasitic nematodes examined ranges from 11% to 67% of dry weight and in general they have a greater lipid content than animal-parasitic nematodes, which rely more on glycogen energy reserves and, at least partially, anaerobic metabolism (Barrett and Wright, 1998). Neutral lipids are the primary lipid energy reserves and are predominantly triacylglycerols (containing three fatty acid molecules). These usually make up >70% of total lipid (up to 94%). Polar phospholipids generally comprise 6–16% of total lipid. Glycolipids have been reported to comprise 2.5% of total lipid in *M. javanica* females. Sterols occur in much smaller quantities, comprising 0.01–0.06% of dry weight in plant-parasitic species (Chitwood, 1998).

The fatty acid composition of nematodes is distinctive at the genus level; some qualitative differences have also been detected at the species level (Sekora *et al.*, 2010). Nematodes contain a much greater variety of fatty acids than their hosts. In general, the majority of fatty acids in most plant-parasitic nematodes contain 18 or 20 carbon atoms (see Box 7.1 for fatty acid nomenclature), and are mostly unsaturated (75–92%), although

Box 7.1. Fatty acid nomenclature.

Fatty acid nomenclature uses a numeric system. The general type of acid is designated as C_x with *x* as a number denoting the number of carbon atoms; for example, C_{18} (unsaturated acids) and C_{20} (polyunsaturated acids). They are further characterized as, for example, 18:1 or 20:4. The number after the colon represents the number of double bonds. When characterized with the double bond, the numbers are usually not written as subscripts and the C is omitted. In addition, the specific acid can be identified by including the location of the double bond, which is represented by, for example, *n*-3 (this is usually put in brackets). The number is the number of carbon atoms from the methyl end of the fatty acid. Thus 18:1 (*n*-7) is vaccenic acid and 18:1 (*n*-9) is oleic acid.

this can be influenced by the culture media and temperature, and even by the age of nematode juveniles (Lu *et al.*, 2022). Between 19 and 24 C_{12}–C_{20} fatty acids have been identified in *P. penetrans*, *Aphelenchoides ritzemabosi*, *Tylenchorhynchus claytoni*, *D. dipsaci*, *D. triformis*, *M. arenaria* and *M. incognita*; *D. triformis* also contained a 22:1 fatty acid. The major fatty acid group in each species was 18:1 and in species of *Ditylenchus* and *Meloidogyne* its major component was vaccenic acid, C18:1 (*n*-7). In *G. rostochiensis*, *G. pallida* and *G. tabacum solanacearum* the major fatty acid groups were 20:4 > 20:1 > 18:1. The differences in the fatty acids of host and nematodes indicate that *G. rostochiensis* is capable of fatty acid chain elongation and/or desaturation, an observation reflected by the discovery of genes involved in both processes in *G. pallida* (Cotton *et al.*, 2014). In mycophagous *Aphelenchoides* spp., carboxyl-directed desaturation is indicated as the major pathway for the synthesis of long-chain polyunsaturated fatty acids (Ruess *et al.*, 2002). Some polyunsaturated fatty acids (20:4, 20:5) are known to be phytoalexin elicitors in potato and their presence in *G. rostochiensis* has been linked to pathogenesis. In insects, the insecticide–nematicide spirotetramat inhibits acetyl-CoA carboxylase; it appears to have the same mode of action in *H. schachtii* (Gutbrod *et al.*, 2020).

Phospholipids are the major components of cellular membranes and are typically derived from a molecule of glycerol, two fatty acids, phosphoric acid and another alcohol such as choline, ethanolamine or inositol. Another phospholipid, sphingomyelin, contains a long-chain nitrogenous base (sphingosine), esterified to a phosphorylcholine molecule and amide-linked to a fatty acid. Ethanolamine and choline phosphoglycerides have been identified as the major phospholipids in *M. javanica*, *A. tritici* and *G. rostochiensis*. As in most organisms, phospholipids regulate membrane fluidity in nematodes. Electron spin resonance (ESR) spectroscopy of spin-labelled nematode polar lipid extracts reconstituted in water yielded plots that were linear between 0 and 30°C for cold-tolerant species (*M. hapla* and *A. tritici*) but had a sharp change in slope for *M. javanica* at 10°C, suggesting a physiologically important lateral phase separation or other localized event in the latter, more tropical, species. Phospholipids are also likely to be involved in signal transduction cascades in nematodes, including protein kinase C-mediated cascades. Incubation of *G. rostochiensis* eggs with tritiated inositol resulted in the formation of various inositol phosphates, which may reflect the involvement of inositol phosphoglyceride-mediated signal transduction in nematode hatching.

Glycolipids vary in structure from simple diacylglycerols attached to one sugar residue to glycosphingolipids, which consist of long chains of sugar residues attached to a lipophilic sphingosine base connected to a fatty acid moiety. Glycolipids often anchor proteins in cell membranes and also have roles in facilitating recognition events, and as precursors for lipoidal secondary messengers involved in signal transduction. The surface coat of nematodes contains glycolipids and can change in response to plant signals (Davies and Curtis, 2011); it is also implicated in the attachment of endospores of *Pasteuria penetrans* (see Chapter 14). Large amounts of a cerebroside, a simple glycosphingolipid, are synthesized *de novo* in *C. elegans*, and similar glycosphingolipids have been identified in *M. incognita*. The sphingoid bases of these glycolipids have predominantly branched structures, unlike those of other organisms examined; the fatty acids are hydroxylated and much longer (20–26 carbon atoms) than in typical nematode glycerolipids. The dauer stage of *C. elegans* contains unique glycolipids comprising two fatty acids esterified to a trehalose molecule (Penkov *et al.*, 2010).

Sterols (tetracyclic polyisoprenoids) are biosynthesized in plants and higher animals via the hydroxymethylglutaryl Coenzyme A pathway. Nematodes possess a wide variety of sterols but cannot biosynthesize them *de novo*, unlike plants and higher animals. Sterols are one of the few essential nutritional dependencies that plant-parasitic nematodes have upon their host plants, a dependency that could possibly be exploited for their control. Comparison of the sterol composition of plant-parasitic nematodes with their plant hosts has indicated that two major metabolic transformations can occur in these species: dealkylation of the C-24 side chain and the saturation of double bonds in the four-membered ring system to yield stanols (Chitwood, 1998). Sterols generally have two major functions. They act as modulators of cell membrane fluidity and as precursors of biologically active molecules, such as steroid hormones. In plant-parasitic nematodes, low membrane phase transition temperatures due to the predominance of unsaturated phospholipid fatty acids, plus the relatively small quantities of sterols present, suggest that the overall effects of sterols such as cholesterol on membrane fluidity are likely to be small, although effects upon specific domains cannot be ruled out (Chitwood, 1998). Studies on *C. elegans* suggest that nematode steroid hormones may be quite different from the ecdysteroid hormones of insects, but information is lacking for plant-parasitic species. In *C. elegans*, entry into the dauer stage is regulated by 3-keto bile acid-like steroids termed dafachronic acids (Gerisch *et al.*, 2007). Orthologues of the *daf-9* gene encoding the enzyme catalysing dafachronic acid synthesis in *C. elegans* occur in *M. hapla* (Gilabert *et al.*, 2016) and *Bursaphelenchus xylophilus* (Wang *et al.*, 2013). The biochemical precursors to dafachronic acid in *C. elegans* are regulated by a sterol 4-methyl transferase unique to nematodes (Hannich *et al.*, 2009) and responsible for the surprising occurrence of 4-methylsterols in many nematodes (Chitwood, 1998).

7.3.4. Carbohydrates and related compounds

As in most other organisms, glycogen is the major storage carbohydrate in nematodes (3–20% of dry weight). Radiotracer studies have demonstrated its biosynthesis in at least two free-living species (Barrett and Wright, 1998; Chitwood, 1998). Specific sugars and related compounds, including the disaccharide trehalose, ribitol, inositol and glycerol, are synthesized by some nematodes as they undergo the process of anhydrobiosis (see Section 7.6.2.2). This synthesis is typically at the expense of glycogen or lipid depletion. The pathway for trehalose synthesis in nematodes involves, as in other organisms, the transfer of glucose by trehalose 6-phosphate synthase from uridine diphosphate glucose to glucose

6-phosphate to form trehalose 6-phosphate and the conversion of the latter to trehalose by trehalose 6-phosphate phosphatase, the former process being the rate-limiting step (Behm, 1997; Chitwood, 1998). Genes encoding the synthase occur in several plant-parasitic nematode species. Hatching in cyst nematodes is associated with the release of trehalose, which is found in high concentrations in the perivitelline fluid (see Section 7.6.5).

An apparently unique feature of nematodes among the metazoa is the presence of the glyoxylate cycle, which synthesizes succinate from two acetyl-CoA molecules (see Section 7.4.4). This pathway is used by many plants and bacteria to generate carbohydrates from acetate, and enables lipids to be converted into carbohydrates via β-oxidation of fatty acids.

Chitin, a polymer of *N*-acetylglucosamine, is a major component of nematode eggshells (Bird and Bird, 1991) and in *C. elegans* it is involved in the prevention of fertilization by multiple sperm (Johnston *et al.*, 2010b). Evidence for chitin synthases has been reported in several nematode species, including *M. artiellia* (Veronico *et al.*, 2001). The *M. incognita* genome contains genes encoding three different enzymes involved in chitin biosynthesis (trehalase, glucose 6-phosphate isomerase and chitin synthase); RNAi-based targeting of these genes via transgenic tobacco reduced egg mass production (Mani *et al.*, 2020).

Ascarosides are nematode-specific glycosides obtaining their name from the unique ascarylose sugar, 3,6-dideoxy-L-mannose. Although first identified in eggshells of *Ascaris* spp., ascarosides have become central to understanding dauer formation and pheromone signalling in *C. elegans* (Butcher *et al.*, 2007; Ludewig and Schroeder, 2013). There are relatively few publications directly identifying the presence of ascarosides in species of plant-parasitic nematodes. However, ascarosides have been identified in exo-metabolome samples from root-knot species *M. incognita*, *M. javanica*, *M. hapla*, cyst nematode *H. glycines*, pinewood nematode *B. xylophilus* and lesion species *Pratylenchus brachyurus* (Manosalva *et al.*, 2015; Zhao *et al.*, 2020). Of the ascarosides identified from these species, ascaroside #18 had implications in host expression of pattern-triggered immunity (PTI) genes as it is detected as a pathogen-associated molecular pattern. The specificity of ascarosides to the phylum Nematoda raises the proposition for distinct nematode-associated molecular patterns (NAMPs) (Manosalva *et al.*, 2015).

7.3.5. Inferring biochemical features from genomics

Since the previous edition of this book (2013), the availability of genomic data for plant-parasitic nematodes has grown exponentially. Consequently, the opportunity to identify and infer biochemical pathways and features no longer relies solely on wet-laboratory research. Once genomes have been produced, transcriptomic data can be used in addition to metabolism pathway information (e.g. KEGG) to identify biosynthesis possibilities for various molecules, allowing comparisons to be drawn between species. Examples of how this information can be used to identify different aspects of the parasite–host relationship have been outlined below and were interpreted from the online resource 'NemaPath' through Nematode.net. It is important to highlight that metabolic pathway predictions will only be as strong as the genomic data supporting them.

Globodera pallida transcribes the required proteins to convert threonine and pyruvate into isoleucine and valine. Strong orthologues for three of the required enzymes in this pathway are missing in *M. incognita* (threonine dehydratase, ketol acid reductoisomerase and branched-chain amino acid aminotransferase). Similarly, *M. incognita* is missing L-amino acid oxidase suggesting that both phenylalanine and tyrosine cannot be cycled into phenylpyruvate and 4-hydroxyphenylpyruvate, respectively, to be resynthesized into functional amino acids when necessary. This suggests that *M. incognita* must acquire these four amino acids from its host plant and could partially explain the 149% increase in tyrosine production by roots infected with the nematode as seen by Hanounik and Osborne (1975).

Lipids and fatty acids present in various nematode species differ at the genus level. Earlier in this chapter, differences between fatty acid groups for *Globodera* spp. and *Meloidogyne* spp. were highlighted. The differences in common fatty acid chain lengths between these genera indicates that *Globodera* spp. is capable of chain elongation and/or desaturation. However, a more complete picture can be obtained by combining this information with biosynthesis pathways inferred from genomic data. For example, *G. pallida* has orthologues for fatty acid synthase (FAS) and enoyl-acyl carrier protein reductases I, II and III (FabI, FabK and FabL), which catalyse the final step in the fatty acid elongation cycle, completing various fatty acid synthesis pathways. *Meloidogyne incognita* has fewer fatty acid synthase orthologues and only one reductase (FabL), suggesting a severe reduction in fatty acid biosynthesis and chain elongation. The ectoparasitic *X. index* has no strong matches for any enoyl-acyl carrier protein reductases and so it would be reasonable to predict that common fatty acid groups have shorter chains and that any longer chain fatty acids are almost exclusively host derived.

7.4. Respiration and Intermediary Metabolism

There have been relatively few studies on the respiratory physiology of nematodes since the mid-1980s and the review by Wright (1998). Studies on intermediary metabolism show that free-living and plant-parasitic nematodes all appear to catabolize energy reserves by glycolysis/β-oxidation and the tricarboxylic acid (TCA) cycle. Most biochemical studies on nematodes have been on animal-parasitic species and *C. elegans*. The relatively limited number of studies on plant-parasitic nematodes were reviewed by Barrett and Wright (1998). Absolute proof for the presence of a particular metabolic pathway is generally lacking in nematodes, although it is likely that nematodes possess the same pathways as those found in other metazoans. One unusual feature is the presence of the glyoxylate cycle in many nematode species.

7.4.1. Respiratory physiology

The cylindrical body plan of nematodes, together with their lack of a circulatory system, places severe restrictions on their body radius for a given level of aerobic metabolism (Atkinson, 1980). However, almost all plant-parasitic nematodes are sufficiently small for diffusion to supply sufficient oxygen for full aerobic respiration at partial pressures of oxygen >15 mmHg. There are some habitats where microaerobic (<15 mmHg) conditions can occur. For example, waterlogged soils, or organic soils with high levels of microbial activity, can have sufficiently low levels of oxygen to affect nematode activity. The roots of plants in lowoxygen environments (e.g. mangroves, paddy and deep-water rice) may also have low oxygen tensions.

Metabolically, nematodes are typical poikilotherms and show a marked change in their oxygen consumption with changes in ambient temperatures. Other factors that can influence their rate of oxygen consumption include ageing, osmotic and ionic regulation and locomotory activity, although the energy cost of movement appears low compared with metabolic maintenance and reproduction.

7.4.2. Amino acid catabolism and nitrogenous waste products

There is some evidence that starved nematodes, including J2 of *M. javanica* and *Tylenchulus semipenetrans*, can use endogenous amino acids as an energy source (Barrett and Wright, 1998). However, nitrogenous waste products in animals, usually in the form of ammonia, must either be excreted directly or converted to less toxic metabolites, such as urea. Nematodes, like most aquatic organisms, are 'ammonotelic', continually excreting ammonia (approximately 40–90% of non-protein nitrogen eliminated) into the environment as it is produced, thus avoiding the toxic storage problem. Urea is excreted by some nematodes (up to around 20% non-protein nitrogen) but not by others, although whether urea is produced via the ornithine–arginine (urea) cycle in nematodes is more problematic (Wright, 1998, 2004). The excretion of waste nitrogenous products into the soil environment may be a significant contribution to nitrogen availability to plants (Ferris *et al.*, 1998).

In amino acid catabolism, the amino group is removed first, followed by catabolism of the carbon skeleton. The two main routes for the removal of the amino group are transamination and oxidative deamination. There is also non-oxidative deamination of specific amino acids. All such enzyme systems have been identified in the free-living nematode *Panagrellus redivivus* and, in general, the ability of this species to transaminate and deaminate amino acids is similar to other organisms. Following transamination or deamination, the carbon skeletons of amino acids are converted to glycolytic or TCA cycle intermediates and eventually to carbon dioxide.

7.4.3. Lipid catabolism

Neutral lipids are the major energy reserves in plant-parasitic nematodes and the lipids usually contain a high percentage of unsaturated fatty acids (see Section 7.3.3). The aerobic utilization of lipid stores has been demonstrated in a wide variety of nematode species and where investigated it is the triacylglycerol fraction that is catabolized. In some species of plant-parasitic nematodes, e.g. *G. rostochiensis*, a decline in infectivity has been correlated with a decline in lipid reserves. Organophosphate and carbamate nematicides cause reversible inhibition of nematode movement, and this reduces neutral lipid consumption in J2 of *G. rostochiensis*, enabling them to remain infective on recovery.

Under aerobic conditions, fatty acids are broken down by β-oxidation to give acetyl-CoA, NADH and reduced flavoprotein. The presence of β-oxidation enzymes has been reported in a number of plant-parasitic nematodes and, in several species, radiotracer studies have demonstrated a functional pathway. The nematicide fluensulfone inhibits the consumption of lipid reserves by *G. pallida* J2 (Kearn *et al.*, 2017) and also inhibits the expression of every gene involved in the fatty acid β-oxidation pathway in *M. incognita* (Wram *et al.*, 2022).

7.4.4. Carbohydrate catabolism

The main carbohydrate reserve in nematodes is glycogen (3–20% dry weight); significant amounts of trehalose (0.1–5%) have also been reported in some species but whether it has a role in energy metabolism as well as in anhydrobiosis has yet to be established. Glycogen utilization under aerobic and anaerobic conditions has been demonstrated in free-living and plant-parasitic nematodes as well as in animal-parasitic species (Barrett and Wright, 1998). In most nematodes, glycogen content is greatest in the non-contractile regions of the muscles, epidermis, intestinal cells and the epithelial cells of the reproductive system. Glycolytic enzymes have been demonstrated in nematodes together with a range of glycolytic intermediates. The regulatory properties of glycolytic enzymes in plant-parasitic nematodes have not been investigated. TCA cycle enzymes have also been demonstrated in a wide range of nematode species and radiolabel studies are consistent with a functional TCA cycle.

The glyoxylate cycle enables organisms to use acetyl-CoA from the β-oxidation of fatty acids for gluconeogenesis (see Section 7.3.4). This cycle has been found in bacteria, algae, fungi, plants, protozoans and nematodes. It consists of two enzymes, isocitrate lyase and malate synthase, which effectively 'short circuit' the TCA cycle by bypassing the two decarboxylation steps (isocitrate dehydrogenase and 2-oxoglutarate dehydrogenase) and catalyse the net conversion of isocitrate and acetyl-CoA to succinate and malate. Both glyoxylate cycle enzymes have been demonstrated in a range of plant-parasitic nematodes, including J2 of *M. incognita*; additional evidence for a functional glyoxylate cycle in nematodes has come from radiotracer studies and molecular genetic studies on *C. elegans* (Barrett and Wright, 1998; McCarter *et al.*, 2003; Jeong *et al.*, 2009). RNAi-based gene silencing utilizing parts of the *M. incognita* isocitrate lyase gene and transgenic tobacco plants decreased egg production by 77% (Lourenço-Tessutti *et al.*, 2015).

Plant-parasitic nematodes produce ethanol as a major end product under anaerobic conditions. Other compounds that may be produced and excreted under anaerobic conditions include acetate, lactate, succinate and glycerol. In organisms that are periodically exposed to anaerobic conditions, resynthesis of carbohydrate from anaerobic end products is important in substrate conservation. For example, the presence of an active glyoxylate cycle would allow the resynthesis of carbohydrate from ethanol.

7.4.5. Cytochrome chains

NADH and reduced flavoprotein are oxidized by mitochondrial cytochrome chains, resulting in the reduction of oxygen to water and the formation of ATP. Cytochromes $a+a_3$ and c have been identified in *D. triformis* and $a+a_3$, b and c in *A. avenae* and there is some evidence for a non-mammalian type of terminal oxidase in *M. javanica*. Oxidation of iron (Fe^{2+}) centres in the cytochrome chain of nematodes has been suggested as one mode of action of halogenated aliphatic hydrocarbon fumigant nematicides (Wright, 1981). The cytochrome *b* and *c* oxidase 1 (*cox 1*) genes have been used in molecular taxonomy for many plant-parasitic nematode genera (e.g. Plantard *et al.*, 2008; Kumari *et al.*, 2010; see Chapter 2).

7.5. Osmotic and Ionic Regulation and Excretion

Nematodes develop and reproduce in an environment where water is freely available and like all aquatic organisms must deal with the opposing problems of osmosis and diffusion.

Osmotic and ionic regulation in nematodes and techniques to study these aspects have been reviewed by Wharton and Perry (2011). In soil water, the free-living, infective stages of endoparasitic nematodes and all stages of ectoparasitic species are normally in a very dilute osmotic (hypo-osmotic) and ionic environment compared with their body fluids, although the osmotic potential of soil water may become hyper-osmotic under some conditions.

Some nematodes, notably euryhaline estuarine species, are regularly exposed to fluctuating osmotic conditions and in some cases may survive as osmoconformers. Species that live in hypo- or hyper-osmotic conditions for any appreciable period must regulate their water content and volume to retain normal locomotory activity (Wright, 1998, 2004). Whilst this would not apply to sedentary stages of endoparasitic species, such as root-knot and cyst nematodes, the maintenance of an osmotic and ionic balance might be expected to be important in other aspects of nematode biology (see Section 7.5.1).

7.5.1. Physiological ecology

Little is known about water and ion balance in plant-parasitic nematodes compared with animal-parasitic species or *C. elegans*, and the information available relates almost entirely to free-living soil stages. The soil-living infective juveniles of plant-parasitic nematodes (e.g. J2 of *H. schachtii* and *G. rostochiensis*) tend to be very good hyper-osmotic regulators but are less good hypo-osmotic regulators. For such stages, the ability to maintain locomotory activity under a range of soil water solute concentrations is critical because almost all species actively infect their host. Marked increases in the solute concentration can occur in the upper layers of soils, where agricultural amendments and changes in the net evaporative rate have the greatest impact, and where irrigation can lead to 'saline' soils (Wright, 2004). However, the relative importance of the solute concentration compared with the soil matric potential on the water balance and movement of nematodes is uncertain. The matric potential of soil water can be tenfold greater than the solute component, increasing at a much greater rate than osmotic pressure as soils dry (Burr and Robinson, 2004), and is the more important factor for plant–water relations and as a predictor of nematode movement in soils (Robinson, 2004).

There have been no physiological studies on osmotic and ionic regulation by plant-parasitic stages of nematodes, although they would be expected to tolerate marked fluctuations in water potentials within plants, particularly at times of drought or nematode-induced stress. In the case of endoparasites, such as species of *Meloidogyne, Heterodera* and *Globodera*, osmotic and ionic balance could be important for sedentary juvenile and adult female stages, particularly for feeding and reproduction.

J2 of *M. javanica* migrate away from areas of high soil salinity. *Meloidogyne incognita* and *R. reniformis* respond to the constitutive cations and anions of salts, respectively; the former species is strongly repelled by ammonium salts, whilst the latter species is repelled by chloride salts (Le Saux and Quénéhervé, 2002). How such differences in chemotactic responses relate to the behaviour and distribution of nematodes in soils of varying ionic content remains to be investigated. The amphids and phasmids of nematodes, which have chemosensory functions, are linked to osmotic avoidance behaviour in *C. elegans* and may play a similar role in other species.

7.5.2. Mechanisms

Two complementary osmoregulatory mechanisms are found in animals: **isosmotic intracellular regulation**, where the osmolarity is adjusted to conform with the extracellular osmotic pressure, and **anisosmotic extracellular regulation**, where the extracellular fluid is maintained hypo- or hyper-osmotic to the external environment. The latter mechanism, which may involve a non-ionic component, has been demonstrated in several animal-parasitic nematode species (Wright, 2004). The pseudocoelom is the principal extracellular fluid compartment in nematodes, and in actively moving nematodes sinusoidal waves of contraction and accompanying internal pressure changes may result in some mixing of pseudocoelomic fluid (Wright, 2004).

The body wall, including various epidermal glands, intestine and excretory–secretory system have been suggested as sites of urine production in nematodes capable of volume regulation in hypo-osmotic environments, but there is little direct evidence. Passive factors associated with osmotic and ionic regulation in hypo-osmotic environments are thought to involve a relatively impermeable nematode cuticle or underlying epidermal membrane (Wright, 1998). In *X. index*, the continuity of outer epidermal membrane invaginations with the endoplasmic reticulum suggests a transport function. Most nematodes retain a functioning gut, and the regular removal of material by defecation in actively feeding nematodes suggests the intestine has an important role in fluid excretion. The structure of the excretory–secretory system has been studied in detail in *C. elegans*, where it consists of three unicellular tubes forming the excretory canal, duct and pore, together with a secretory gland, and two associated neurons (Sundaram and Buechner, 2016). In *C. elegans*, aquaporin water channel gene (*aqp*) homologues are expressed in the apical cell membrane of the intestine and the excretory cell. Hypo-osmotic preconditioning of wild-type (N2) nematodes has been shown to involve downregulation of *aqp-4*, suggesting the adaptive response to hypo-osmotic environments involves an active reduction in membrane water permeability (LaMacchia and Roth, 2015).

The ionic composition of the pseudocoelomic fluid in several animal-parasitic species suggests that ionic regulation must occur. However, there is only limited, largely indirect, physiological and biochemical evidence for the ion channels and pumps that would be required to maintain electrochemical gradients across nematode epidermal and intestinal cells (Thompson and Geary, 2002). There is evidence for K–Cl co-transporter proteins in *M. incognita*, which could be involved in the ionic and osmotic regulation (Neveu *et al.*, 2002). In the euryhaline marine species, *Litoditis marina*, transcriptomic and proteomic studies have associated various genes families, including osmolyte biosynthetic, vacuolar-type H^+-transporter ATPase and potassium channel genes, with short-term low and high salinity stress responses, and long-term acclimation processes (Xie and Zhang, 2022).

7.5.3. Nitrogen excretion

In nematodes, the importance of the body wall, intestine and excretory–secretory system in the excretion of waste nitrogen (primarily as ammonia; see Section 7.4.2) will vary between species and between stages of a particular species, and is likely to correspond closely with the relative importance of these structures for the removal of water (Wright, 1998, 2004). All plant-parasitic nematodes have a small body radius (<20 μm) and large surface area to volume ratio, with most cells adjacent to the external environment or the alimentary canal. This suggests that the body wall (cuticle/underlying hypodermis) and, where functional, the intestine are major routes for the removal of freely diffusible nitrogenous waste products, such as ammonia and urea. Physiological and molecular studies with *C. elegans* suggest that ammonia is at least partially excreted through the hypodermis; with functional expression analysis showing the ammonia transport capabilities of the Rhesus protein CeRhr-1 (Adlimoghaddam *et al.*, 2015). The significance of urea excretion in nematodes is uncertain (Section 7.4.2) but at least one nematode aquaporin homologue can transport urea as well as water (Thompson and Geary, 2002).

7.6. Survival Strategies

7.6.1. Terminology

To survive unfavourable conditions, some nematodes are able to suspend development and survive in a dormant state until favourable conditions return. Dormancy is usually subdivided into **quiescence** and **diapause**. Quiescence is a spontaneous reversible response to unpredictable unfavourable environmental conditions and release from quiescence occurs when favourable conditions return. Quiescence can be facultative or obligate. Obligate quiescence occurs when the environmental cue affects a specific receptive stage of the nematode life cycle. For example, quiescence of unhatched juveniles is a common survival strategy among soil nematodes, including species of cyst and root-knot nematodes, *R. reniformis* and *P. penetrans*. By contrast, facultative quiescence is not stage-specific. Adverse environmental conditions and the types of quiescence they induce include cooling (cryobiosis), high temperatures (thermobiosis), lack of oxygen (anoxybiosis), osmotic stress (osmobiosis) and dehydration, or desiccation (anhydrobiosis).

Cryptobiosis is a further term that has been used in connection with a quiescent state. Cryptobiosis is defined as a state where no metabolism can be detected, whereas dormancy involves lowered metabolism. In practice, it is frequently difficult to separate states on the basis of metabolic activity. The cause of the arrest in development is a more relevant criterion for separating states, and on this basis cryptobiosis can be viewed as the same kind of phenomenon as quiescence. Evans (1987) further distinguished between dormancy affecting ontogenetic development and that affecting somatic development.

In contrast to quiescence, diapause is a state of arrested development whereby development does not continue until specific requirements have been satisfied, even if favourable conditions return. It is either programmed into the life cycle (obligatory diapause) or is triggered by environmental stimuli (facultative diapause), such as day length.

The term 'dauer' was once rarely used in relation to plant-parasitic nematodes but with the availability of numerous genomes, its use has increased greatly (Perry, 2011; Vlaar *et al.*, 2021). 'Dauer' describes an alternative developmental stage for surviving unfavourable conditions for long periods and is a central component of the survival strategy of many free-living and entomopathogenic nematodes.

7.6.2. Quiescence, particularly anhydrobiosis

Separating different types of quiescence is somewhat artificial as many of the environmental stresses involve removal or immobilization of water. For example, desiccation concentrates body solutes and increases internal osmotic stress, exposure to hyper-osmotic conditions causes partial dehydration of a nematode, and freezing may involve dehydration through sublimation of water from the solid phase.

The ability of some species of plant-parasitic nematodes to withstand dehydration for periods considerably in excess of the duration of the normal life cycle is a feature often associated with a dispersal phase, such as the cysts of *Globodera* and *Heterodera* species. Most research on the physiological and biochemical aspects of dormancy has focused on anhydrobiosis (Perry, 1999). Adaptations associated with desiccation survival serve primarily to reduce the rate of drying, either to prolong the time taken for the nematode's water content to reach lethal low levels or, in true anhydrobiotes, to enable the structural and biochemical changes required for long-term survival to take place. Nematode anhydrobiotes can be grouped into those that depend on environmental factors to control water loss and those that have intrinsic abilities to control water loss; Perry and Moens (2011) termed these two groups **external dehydration strategists** and **innate dehydration strategists**, respectively. Both groups require controlled drying in order to survive, the first group to prolong the time to lethal low water content and the second group to enable biochemical changes to take place to facilitate long-term survival. Control of the rate of drying is the first phase; successful entry into long-term anhydrobiosis by the innate dehydration strategists depends on subsequent biochemical and molecular adaptations.

7.6.2.1. Behavioural and morphological attributes that enhance anhydrobiotic survival

The gelatinous matrix of species of *Meloidogyne*, which consists of an irregular meshwork of glycoprotein material, shrinks and hardens when dried, thus exerting mechanical pressure on the eggs to inhibit hatching during drought conditions and ensuring that the infective J2 are retained within the protection of the eggs and matrix. When exposed to desiccation, the permeability characteristics of the surface layers of the cyst wall and of the eggshell of *G. rostochiensis* change. The resulting control of water loss is a major factor in the survival of this species, as the hatched J2 is susceptible to environmental extremes. This susceptibility is offset by a sophisticated host–parasite interaction whereby the J2 does not hatch unless stimulated by host root diffusates (see Section 7.6.3). The ability of cyst nematodes to survive severe desiccation varies considerably between species and long-term anhydrobiosis seems to be associated primarily with those species, such as *G. rostochiensis*, that have a very restricted host range (see Chapter 4). In addition to its role in desiccation survival of unhatched J2, the eggshell enables J2 of *G. rostochiensis* to supercool in the presence of ice as a freeze-avoidance strategy for cryobiotic survival (Wharton *et al.*, 1993).

Rotylenchulus reniformis retains its moulted cuticles to protect the infective stages. J2 hatch in the soil and the subsequent moults are completed without feeding, resulting in a decrease in body volume. The young adults are enclosed in the three cuticular sheaths from the previous stages and remain inactive in dry soil over the summer months until favourable moist conditions return, allowing movement and exsheathment. The sheaths aid survival in the dry soil by slowing the rate of drying of the enclosed adult. The reduced rate of water loss assisted individuals to survive only for periods over which water loss was controlled; control of water loss merely prolongs the time taken for the nematode's water content to reach lethal low levels. Thus, control of the rate of drying does not, of itself, guarantee long-term survival but with protection of the sheaths plus the soil environment, where even if the soil water potential falls below −1.0 MPa the relative humidity in soil pores is still above 99%, survival in the dry season between crops is enhanced. Knowledge of the survival biology of *R. reniformis* formed the basis of a control strategy using alternating wetting and drying of soil; wetting the soil activates the adults and causes exsheathment, and drying the soil is lethal to the exsheathed adults.

Coiling and clumping (or aggregation) are two behavioural responses by some species of nematodes to removal of water (see Chapter 8). Nematodes will also coil in response to osmotic stress and increase in temperature. Coiling and clumping reduce the surface area exposed to drying conditions. The galls induced by *Anguina amsinckiae* contain hundreds of desiccated adults and juveniles of all stages, many of which are coiled. However, not all nematodes need to coil to survive drying. The galls induced by *A. tritici* contain tightly packed aggregates of J2 only, each of which remains uncoiled when dry. The classic image of anhydrobiotic nematodes is clumps of coiled, desiccated J4 of *D. dipsaci*, termed 'eelworm wool', associated with infected bulbs or inside bean pods. In infected narcissus bulbs, set out to dry at the end of the growing season, development is arrested at the J4 stage and hundreds emerge from the basal part of the bulb and aggregate before drying. The death of the peripheral J4 apparently provides a protective coat that, in a manner similar to the cyst wall and eggshell, aids survival of the enclosed nematodes by slowing their rate of drying. The cuticle of J4 is also able to resist water loss; the cuticular permeability barrier is heat labile and is destroyed by brief extraction with diethyl ether, indicating that an outer lipid layer, possibly the epicuticle, is involved.

A slow rate of water loss appears to allow orderly packing and stabilization of structures to maintain functional integrity during desiccation. This aspect has been examined in detail in J4 of *D. dipsaci*, where an initial rapid loss of water is followed by a period of very slow water loss before the third phase of rapid water loss to leave individuals with no detectable water content (Fig. 7.1). The first two phases are separated by a permeability slump during which the permeability of the cuticle, and hence the subsequent rate of water loss, is reduced. During the first phase, there is a rapid shrinkage of the cuticle, the lateral hypodermal chords and the muscle cells, followed by a slower rate of shrinkage during the second phase. The contractile region of the muscle cells appears to resist shrinkage until desiccation becomes severe during the third phase (Fig. 7.1). The mitochondria swell and then shrink during desiccation, which may indicate disruption of the permeability of the mitochondrial membrane. A decrease in thickness of the hyaline layer, caused by shrinkage of its constituent muscle cells and epidermis, results in a decrease in diameter that is of a much greater magnitude than the accompanying change in length. By contrast, the reduction in the rate of water loss of *Rotylenchus robustus* is achieved by controlled contraction of cuticular annuli resulting in decreased length, but not diameter, of the nematode. Desiccation of J4 of *D. dipsaci* did not result in any appreciable denaturation of metabolic enzymes and intestinal cells changed little, possibly because the large lipid droplets they contain resist shrinkage and may prevent structural damage. Lipid reserves are maintained at high levels in many nematodes and provide a food source for the nematode after termination of dormancy and before they are able to feed on a host.

The dynamics of rehydration are also important for survival. With *A. avenae*, successful revival depends on slow rehydration in saturated atmospheres (approximately 100% relative humidity). By contrast, J4 of *D. dipsaci* rehydrate very rapidly on direct transfer to water, although there is a delay, or 'lag phase', of several hours before the onset of locomotory activity (Fig. 7.2). The water permeability characteristics of the cuticles of *D. dipsaci* and

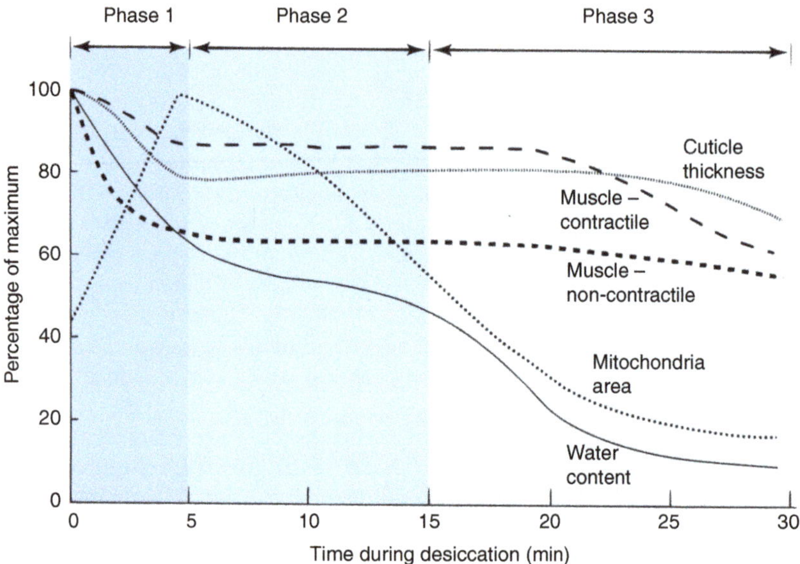

Fig. 7.1. Changes accompanying desiccation of J4 of *Ditylenchus dipsaci* following placement of hydrated individuals in 50% relative humidity at time zero. The three phases reflect differences in the rate of water loss. (Redrawn from Perry, 1999.)

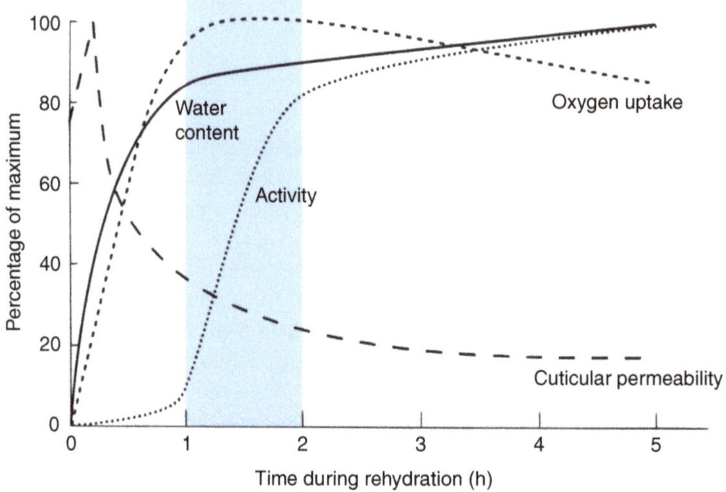

Fig. 7.2. Changes accompanying rehydration of J4 of *Ditylenchus dipsaci* following placement of desiccated individuals in water at time zero. Activity is defined as the percentage of J4s showing movement. The time difference between water uptake and activity is the 'lag phase' (highlighted). (Redrawn from Barrett, 1991.)

A. agrostis are restored during the lag phase and this process can be prevented by inhibitors that block enzyme activity and post-transcriptional protein synthesis, indicating an active repair mechanism. Leakage of inorganic ions during rehydration of *A. tritici* ceases during the lag phase, indicating the repair of damaged membranes or the restoration of the permeability barrier.

Morphological changes occur gradually throughout the lag phase. Muscle cells of J4 of *D. dipsaci* increase in thickness, small lipid droplets coalesce within the intestine to form large droplets and there is a decrease in body length of *D. dipsaci*, which may indicate a contraction of the muscle cells as they recover. Metabolism of J4 of *D. dipsaci* begins immediately after hydration. The metabolite profiles recover quickly, with noticeable changes after 10 min rehydration and completion by 1 h. However, the ATP content does not recover as rapidly as those of the other metabolites; after 10 min there is little change and even after 1 h it is still low. The slow trehalose depletion (up to 48 h to return to pre-desiccation levels) may be associated with the slow recovery of ATP levels. Mitochondria swell during rehydration before adopting a normal morphology; immediately after hydration, the mitochondria are essentially uncoupled and there is no oxidative phosphorylation.

7.6.2.2. Biochemical adaptations of anhydrobiosis

Only a few species are able to survive beyond the period during which water loss is controlled. With these innate dehydration strategists, additional biochemical adaptations, reviewed by Barrett (1991, 2011), are required for long-term survival. At water contents below about 20%, there is no free water in the cells. This 20%, usually referred to as 'bound water', is involved in the structural integrity of macromolecules and macromolecular structures, such as membranes. The water content of desiccated, anhydrobiotic nematodes is estimated to be about 1–5%, so it is probable that the bound water has been lost, although there is no experimental evidence that nematodes can survive the complete loss of structural water. Some molecules might replace bound water and preserve structural integrity but, in nematodes, there is only limited information on biochemical mechanisms of desiccation survival. The accumulation of the disaccharide trehalose, the only naturally occurring non-reducing disaccharide of glucose, during water loss of anhydrobiotic nematodes has been reported frequently. In *A. avenae*, glycogen and lipid reserves are converted to trehalose and glycerol, respectively. J4 of *D. dipsaci* and J2 of *A. tritici* also sequester trehalose, but not at the expense of lipid reserves; in these stages, other carbohydrates, such as myo-inositol and ribitol, may be involved. There has been much speculation on the role of trehalose in desiccation protection. It may replace bound water by attaching to polar side groups on proteins and phospholipids, thus maintaining the balance between hydrophilic and hydrophobic forces acting on the molecules and preventing their collapse. Preventing crosslinkage of molecules and fusion of membranes as bulk water is removed also preserves membrane stability. Trehalose may prevent protein denaturation. Glucose reacts with the amino acid side chains of proteins to form brown pigments called melanoidins. By contrast, trehalose does not react with proteins in this way and also appears to suppress this adverse reaction of other sugars with proteins. Trehalose can act as a free radical scavenging agent to reduce random chemical damage.

Synthesizing trehalose during dehydration may indicate preliminary preparation for a period in the dry state, but it does not necessarily mean that survival during subsequent severe desiccation is assured. It appears that, following trehalose synthesis, other, at present unknown, adaptations are required at the cellular and subcellular levels for nematode survival, and rate of drying still has to be controlled.

7.6.3. Diapause

Diapause has been documented for cyst and root-knot nematodes as a strategy to overcome cyclic long-term conditions, such as seasonal conditions and/or the absence of the host, that are not conducive to hatch and infection (see Chapters 3 and 4). The incidence of diapause varies greatly between species and between populations of the same species. For example, among species of *Meloidogyne*, the percentage of unhatched J2 that enter diapause varies from less than 10% for the predominantly tropical *M. arenaria* to 94% for *M. naasi*, which has a temperate distribution. Molecular studies have shown differences in gene expression changes between quiescence and diapause in unhatched J2 of *G. rostochiensis* after hydration or exposure to root diffusate (Palomares-Rius *et al.*, 2016).

Obligate diapause is initiated by endogenous factors and can be relieved by the J2 receiving exogenous stimuli for a required period of time. Nematodes can undergo obligate diapause only once in their life; in *G. rostochiensis* it is initiated by signals, such as photoperiod, via the host plant to the female nematode and thence to the developing juvenile. Temperature is the most important environmental cue for the termination of obligate diapause, with a fixed period of exposure to low temperatures relieving the arrested development.

Facultative diapause is initiated by exogenous, rather than endogenous, stimuli and terminated by endogenous factors after a critical period of time. This type of diapause is illustrated by the predictable periods (between autumn and spring from their second season onwards) of non-responsiveness to root diffusates of cyst nematode J2. Initiation of facultative diapause in *G. rostochiensis* is affected by both day length and low temperatures, whilst a range of environmental stresses on the female nematode initiates diapause in *H. avenae* and some *Meloidogyne* spp.

The different types of dormancy provide nematodes with a range of strategies with which to synchronize hatching with unpredictable and seasonal environmental changes. For example, the newly formed unhatched J2 of *G. rostochiensis* immediately enter obligate diapause, which is broken by the chilling stimulus of autumn and winter. In early spring, the unhatched J2 then enter obligate quiescence, which will be terminated by increasing soil temperature and potato root diffusate (PRD) to stimulate hatching of the J2. If no host is present, quiescence continues followed by facultative diapause as the unstimulated eggs enter their second winter. This combination of diapause and quiescence enables *G. rostochiensis* to persist in the soil for more than 20 years in the absence of the host plant.

7.6.4. Dauers

Dauers are a morphologically, behaviourally and physiologically adapted alternative juvenile stage for surviving unfavourable conditions (Grant and Viney, 2011). The term dauer has been defined differently by various groups of nematologists studying groups of nematodes with different life habits. Vlaar *et al.* (2021) provided a hybrid definition of dauer as a state in which the juvenile is: (i) non-feeding; (ii) essentially non-ageing; and (iii) the species has a conserved dafachronic acid biosynthetic pathway that produces (a chemical analogue of) dafachronic acid. The dauer stage of *C. elegans* is enclosed by a dauer-specific cuticle and exhibits several characteristic resistance adaptations, including reduced metabolism, elevated levels of several heat shock proteins, an enhanced resistance to desiccation and a several-fold increase in trehalose levels (Erkut *et al.*, 2011).

In *C. elegans*, the transforming growth factor (TGF)-β, insulin/insulin-like growth factor (IGF)-1 pathways and the mostly nematode-specific *gpcr* genes all have implication in dauer formation due to their interconnectivity with the dafachronic acid signalling pathway. The concentration of dafachronic acid is 30 times higher following dauer evasion (Li *et al.*, 2013). This corresponds with dafachronic acid binding the nuclear receptor DAF-12 (DAuer Formation), preventing dauer entry and promoting reproductive growth (Wang *et al.*, 2015). With the evolution of parasitism, it is possible that dauer formation has been lost. For example, species in clade 2 do not have an orthologue for the essential *daf-12*. Quiescent stages in this clade share the dauer phenotype but are probably the result of convergent evolution rather than dauer. The decision of *C. elegans* to enter the dauer stage is influenced by pheromonal signals involving ascarosides, esters of short-chain fatty acids to the dideoxyhexose ascarylose; over 300 ascarosides have been discovered thus far (Machado and von Reuss, 2022; see also Section 7.3.4). Pheromonal and other behavioural responses in *C. elegans* are modulated by ascarosides (Butcher, 2017). Ascarosides #1, #2, #3 (previously named daumones) and #5 are interpreted by G-protein-coupled receptors (GPCRs). However, due to ascarosides being highly species-specific, GPCRs are not all conserved and cannot be used as markers for dauer formation and/or ascaroside presence. Mixtures of ascarosides are produced by a wide range of nematode species, including *P. penetrans* (Choe *et al.*, 2012), and the discovery of pheromonal and other signalling roles for these molecules in plant-parasitic nematodes and in nematode–plant interactions is underway (Shivakumara *et al.*, 2019; Manohar *et al.*, 2020).

The evidence for dauers in plant-parasitic nematodes is fragmentary. Kikuchi *et al.* (2007) identified 31 homologues of 18 *C. elegans* genes, including nine homologues for *daf* (dauer formation) genes. *Meloidogyne hapla* has 14 orthologues of *C. elegans daf* genes (Abad *et al.*, 2008; Abad and Opperman, 2009) but it does not carry the *daf-28* orthologue, which is key in the signal transduction pathway. Abad and Opperman (2009) concluded that basic development mechanisms are conserved, although signalling is not. A similar conclusion was reached by Zhang *et al.* (2022), who found that the types of differentially expressed genes accompanying the moult from *B, xylophilus* dauers to adults (including four sterol hydroxylases) indicated that the signal transduction pathway was also quite different in this species. Comparison of expression profiles of dauer genes in *C. elegans* and in survival and other stages of parasitic nematodes (Elling *et al.*, 2007; Vlaar *et al.*, 2021) reveals marked differences and similarities in expression patterns between *C. elegans* and other nematodes, and there is insufficient information to be able to link individual *daf* genes to specific survival traits. Under adverse conditions, *D. dipsaci* development stops at the J4 stage and, compared with J4 in a population feeding and developing under ideal conditions, the survival form of J4 has more lipid reserves and shows a propensity to aggregate (Perry and Moens, 2011), properties that reflect the dauer state. Dauers of *B. xylophilus* survive on beetles of the genus *Monochamus*, using them for transport to susceptible hosts. In species such as *Radopholus similis* it is possible that dauer signalling has a wider function in dormancy as upregulation of several genes in the insulin and dafachronic acid pathways can be seen (Huang *et al.*, 2019).

The majority of evidence for dauer pathways in plant-parasitic nematodes has come from bioinformatic studies (Vlaar, 2021). While studies of this type allow a broad look at the presence of key genes, incomplete read mapping and analysis of differentially expressed genes from fragmented genomes weakens this tool. Orthologues of the crucial *daf-12* have been identified in most investigated clade 12 nematodes except for *G. pallida*, *M. hapla* and *M. enterolobii*. In the case of plant-parasitic nematodes that share almost identical life cycles, such as *G. pallida* and *G. rostochiensis*, it is difficult to believe that one may have the ability to form dauers whilst the other does not.

7.6.5. Eggshell structure and hatching

7.6.5.1. Eggshell structure

Eggshells primarily function as a safehouse for the developing embryo, retaining all contents for development whilst providing a physical barrier against external contamination. Almost all molecular research on nematode eggshell has used *C. elegans* as a model and is reviewed by Stein and Golden (2018). Due to the similarity in eggshell structures across the Nematoda it would be reasonable if much of the basic biological and structural components were shared between this model and plant-parasitic nematodes. A generalization of eggshell structure is given in Fig. 7.3.

Nematode eggshell formation typically begins as the oocyte passes through the oviduct producing the outer most layer, the vitelline membrane, formed from the oocytes' oolemma. The vitelline membrane controls sperm binding, preventing polyspermy upon exposure to sperm in the spermatheca (Johnston *et al.*, 2010b). As the egg passes through the spermatheca, towards the uterus, chitin synthesis and deposition begin. Although permeable, the chitin layer provides the eggshell with rigidity and structural tensile strength. To date, a chitinous layer has been found in all nematode eggshells. The thickness and relative percentage of chitin present in the eggshell varies between different species; for example, *G. rostochiensis* eggshells consist of 9% chitin (Clarke *et al.*, 1967), whereas *M. javanica* eggshells have a larger chitin content at 30% (Bird and McClure, 1975). Across several examples of plant-parasitic nematodes (*R. reniformis*, *T. semipenetrans*, *P. minyus* and *M. javanica*), width of the chitin layer varies between 50 and 400 nm.

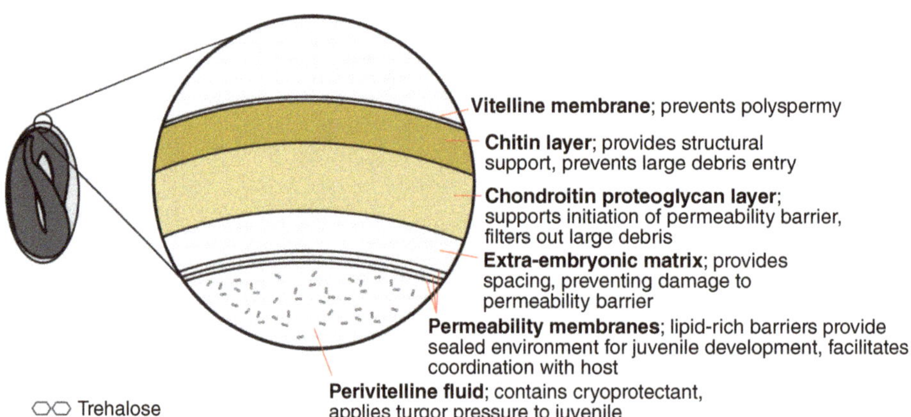

Fig. 7.3. Generalized plant-parasitic nematode eggshell schematic. Layers of the eggshell are not to scale and thickness will vary between different nematode species; however, each layer shown in the schematic is predicted to be present across all nematode eggshells. (Courtesy James Price.)

Within the last 10 years there have been large advances in understanding the next layer of the nematode eggshell. Initial reports separately in *Ascaris* (Foor, 1967) and *Heterodera* spp. (Perry and Trett, 1986) suggested that there was a lipid-rich layer beneath the chitin layer of the eggshell based on lipid composition and transmission electron microscopy. In *C. elegans* Olsen *et al.* (2012) showed that this layer is instead formed from chondroitin proteoglycans and is deposited during cortical granule exocytosis in anaphase I. More recently Price *et al.* (2023) identified chondroitin proteoglycans (CPG) in mass spectrometry data from *G. rostochiensis* eggshell protein extracts, supporting the presence of this layer being common in both free-living and plant-parasitic nematode eggshells. The CPG layer supports initiation of the permeability barrier through the hierarchical formation of the eggshell and potentially also reduces larger granules causing physical damage to the permeability barrier after entering through the porous chitin layer.

Beneath the CPG layer is a fluid-filled space separating the structurally rigid eggshell from the permeability barrier. This area has previously been called the perivitelline or perimembrane space. However, to bring plant-parasitic and free-living nematology together, the new naming convention of extra-embryonic matrix should be used (Johnston and Dennis, 2012). Without this matrix, the embryo's plasma membrane and the outer eggshell layers can interact, leading to cytokinesis failures.

Internal to the extra-embryonic matrix is the permeability membrane, which physically and chemically separates the developing embryo from the porous outer eggshell layers and the wider environment. This lipid-rich layer is produced during meiosis II. Until this point there is little reason that free-living and plant-parasitic nematode eggshells should differ as from a biological and evolutionary viewpoint these layers are functionally required for protection of the developing embryo and hierarchical formation of the eggshell. However, the need for coordination with a host seen in many parasitic nematode species consequently implies the need for a more controlled and durable permeability barrier to increase resistance against external stresses and, for cyst nematodes, permitting longevity in diapause before hatching. Multilaminar (predominantly tetra- and penta-laminae) permeability barriers can be seen in *H. schachtii* and *H. glycines* (Perry and Trett, 1986).

On a molecular level, all nematode eggshells share a large amount of homogeneity. As previously mentioned, all nematode eggs have chitinous and lipid components. Identification of eggshell lipids in plant-parasitic nematodes highlights similarities to animal-parasitic nematode eggshells. In *Meloidogyne* and *Globodera* spp., eggshells have an abundance of neutral lipid (55–90% of total lipids). Of these neutral lipids, triacylglycerols are most predominant. Fatty acids have mostly longer chains with low degrees of unsaturation, which aids function when forming strong membranes. Ascarosides were first identified in eggshells of *Ascaris* spp. The long, saturated, hydrocarbon chains help *Ascaris* regulate permeability against a harsh host environment. Therefore, it would be reasonable for ascarosides similarly to be found in eggshells of other parasitic nematode species that are also subject to strong abiotic stresses. However, to date, ascarosides have not been found in any other nematode eggshell. Using our current eggshell knowledge, it is possible that ascaroside layer described in the ultrastructure of *A. lumbricoides* eggshell may have been the CPG layer (Lýsek *et al.*, 1985). To date only one parasitic nematode eggshell protein has been identified (Price *et al.*, 2023). This calcium-dependent lipid-binding annexin contains a unique peptide insertion at one of the calcium-binding sites. This motif is only found in *Globodera* spp. and potentially enables their specific hatching behaviours. Further molecular analysis of eggshells will help gain a greater understanding of hatching mechanisms and function, and consequently could highlight control strategies for many plant-parasitic species.

7.6.5.2. Hatching

Once hatched, the J2 is susceptible to abiotic pressures and must find a host to commence feeding before lipid reserves are depleted. In some species environmental susceptibility is offset by a sophisticated host–parasite interaction whereby the J2 does not hatch unless stimulated by host root diffusates, synchronizing hatching with the availability of nearby host plants and reflecting classical host–parasite co-evolution. Hatching in this way is typically common in cyst nematodes (Box 7.2), although the reliance on root diffusates can vary between species. A few non-cyst species (e.g. *M. hapla*, *M. ottersoni* and *R. reniformis*) also hatch in response to host root diffusates. Species such as *G. rostochiensis* and *G. pallida* that have a very restricted host range are almost completely dependent on host diffusates for hatch. By contrast, species such as *H. schachtii* hatch well in water and its survival is correlated with a very wide host range (some 218 plant species, including many weeds). *Heterodera avenae* also has a large hatch in water but a relatively narrow host range; however, the hosts are very common. Most information on hatching mechanisms derives from research on *G. rostochiensis* and *H. glycines* (Jones *et al.*, 1998; Perry, 2002).

Host root diffusates can be separated into subclasses, hatching factors (HFs), hatching inhibitors (HIs) and hatching stimulants (HSs). HIs are produced earlier than HFs, inhibiting hatching while the roots of new plants become established, ensuring that the host plant and food source will be able to sustain the infection. HIs vary between host and nematode. Lipid peroxides such as L-ascorbyl 2,6-dipalmitate can inhibit *M. incognita* suppressing 86% of hatching at 2 mmol l^{-1} (Yang *et al.*, 2016), whereas thiazoles such as fluensulfone will inhibit *G. pallida* hatching (Feist *et al.*, 2020). By contrast, HFs are active chemicals that encourage hatching. Production of HFs is restricted to relatively short periods of plant growth. *Heterodera goettingiana* hatches only in diffusates from 4- and 6-week-old pea plants while *H. carotae* hatching is largely restricted to diffusates of 5- and 7-week-old carrots. Juveniles of *G. rostochiensis* hatch in response to a rise in

Box 7.2. Grouping of some species of cyst nematodes into four broad categories, based on their hatching response to host root diffusates. (From Perry, 2002.)

Group 1	Very large numbers of juveniles hatching in response to host root diffusates; few hatching in water.	e.g. *Globodera rostochiensis*, *G. pallida*, *G. ellingtonae*, *Heterodera cruciferae*, *H. carotae*, *H. goettingiana*, *H. humuli*
Group 2	Very large numbers of juveniles hatching in response to host root diffusates; moderate hatch in water.	e.g. *H. trifolii*, *H. galeopsidis*, *H. glycines*
Group 3	Very large numbers of juveniles hatching in response to host root diffusates; large hatch in water.	e.g. *H. schachtii*, *H. avenae*
Group 4	Hatching of juveniles induced by diffusates only in later generations produced during the host growing season; very large hatch in water for all generations.	e.g. *H. cajani*, *H. sorghi*

the HF:HI ratio (Byrne *et al.*, 1998). HSs synergize the effect of HFs to increase the number and/or rate of hatching but are inactive by themselves.

HFs have very high specific activities, stimulating hatch of *H. glycines* and *G. rostochiensis* at concentrations as low as 10^{-10} g ml^{-1}. The first HF purified was the terpenoid, glycinoeclepin A, which induced hatch of *H. glycines*; subsequently two more HFs, glycinoeclepins B and C, were purified. Potato steroid glycoalkaloids α-solanine and α-chaconine initiate the hatch of *G. rostochiensis* and *G. pallida* (Jones *et al.*, 1998). Hatching of *G. rostochiensis* varies between root diffusates from different potato varieties. Faster hatching in response to glycoalkaloids may relate to the adaptation of *G. rostochiensis* to the bitter (i.e. high glycoalkaloid) potatoes of the Andes. Solanoeclepin A is a hatching stimulant for potato cyst nematodes and the total chemical synthesis of solanoeclepin A has been reported (Tanino *et al.*, 2011). The related triterpenoid HF solanoeclepin B is less active than solanoeclepin A but is metabolized by soil organisms to solanoecepin A (Shimizu *et al.*, 2023). While total chemical synthesis for solanoeclepin A and glycinoeclepin A have been achieved, their use for nematode control through initiation of suicide hatch remain prohibitively expensive.

Although Gautier *et al.* (2020) considered that microbial HFs have no role in cyst nematode hatching, there is increasing evidence for a role for microbial HFs. A comprehensive, species-specific review of HFs and more than 200 other compounds that stimulate hatching of plant-parasitic nematodes has been provided by Čepulytė and Būda (2022).

Hatching can be affected by other biotic factors, such as microbial HFs. However, abiotic factors can contribute to hatching as HSs would. For example, hatching of the hop cyst nematode, *H. humuli*, increases as temperature increases from 15°C to 20°C. Similarly, at favourable temperatures and soil moisture *H. schachtii* is able to achieve up to 60% hatch in the complete absence of root diffusates (Ngala *et al.*, 2021).

There are considerable variations between species of nematodes in the sequence of events, the overlap of individual responses and their significance during the hatching process. Essentially, the hatching process can be divided into three phases: changes in the eggshell, activation of the juvenile, and hatching from the egg (eclosion). In many species, such as *Meloidogyne* spp., activation of the juvenile precedes, and may even cause, changes in eggshell structure; in others, such as *G. rostochiensis*, alteration of eggshell permeability characteristics appears a necessary prerequisite for metabolic, and consequent locomotory, changes in the juvenile. Exposure of cyst nematodes to host root diffusates induces a cascade of interrelated events leading to eclosion. Most research has centred on cyst and root-knot nematodes, with the hatching mechanism of *G. rostochiensis* being best understood. The hatching processes for *Globodera* spp. and *Meloidogyne* spp. are outlined in Box 7.3.

Inside the egg, the J2 of *Globodera* and *Heterodera* spp. is surrounded by perivitelline fluid, which contains trehalose at a concentration of 0.34 M in *G. rostochiensis* (Clarke *et al.*, 1978) and 0.5 M in *H. goettingiana* (Perry *et al.*, 1980), for example. The osmotic pressure generated by the trehalose reduces the water content of the unhatched J2 and this partial dehydration inhibits J2 movement because the turgor pressure is insufficient to antagonize the longitudinal musculature.

To 'activate' the unhatched J2, the pressure needs to be removed. In *G. rostochiensis* and some other species, this is achieved by the first event in the hatching sequence, a change in permeability of the inner lipoprotein membranes of the eggshell. This is caused by HF binding or displacing internal Ca^{2+} (Box 7.4). In both *G. rostochiensis* and *G. pallida*, a 5-min exposure to PRD is sufficient to stimulate hatch, suggesting the involvement

Box 7.3. Flow diagram showing events in the hatching process of second-stage juveniles (J2) of *Globodera* spp. and *Meloidogyne* spp. The main components are in chronological order but there will be overlap in the sequence of some events. Before emergence from cysts, J2 may become quiescent leading to subsequent 'spontaneous' hatch.

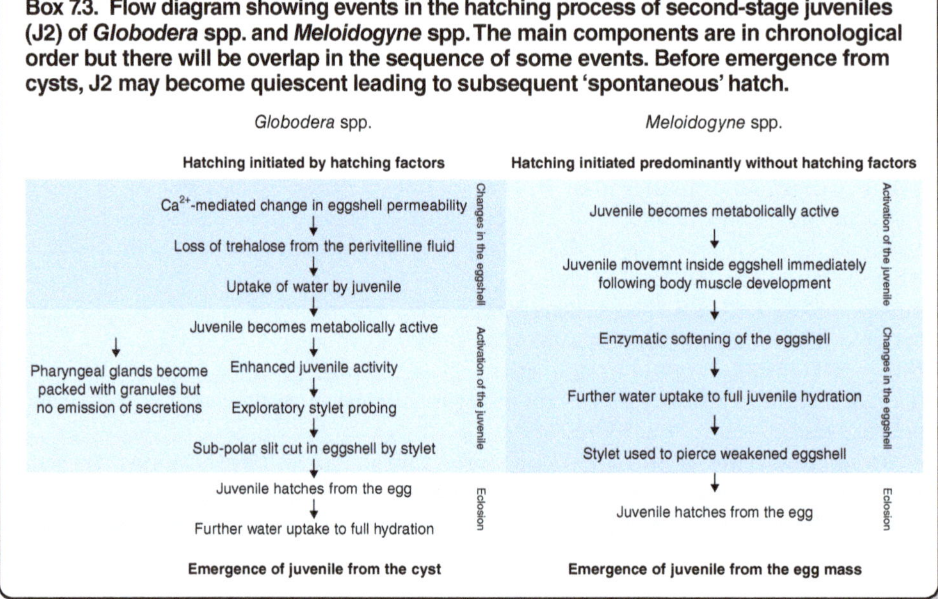

Box 7.4. Three classes of Ca^{2+}-binding site have been distinguished in the *G. rostochiensis* eggshell (Clarke and Perry, 1985):

(i) Sites in the outer layers that bind Ca^{2+} tightly; these sites are not involved in the hatching process.
(ii) Sites in the lipoprotein layer, from which Ca^{2+} can be removed by hatching factors (HFs); these are proposed to be the sites associated with HF-stimulated hatch.
(iii) Sites on the lipoprotein that bind additional Ca^{2+} ions in the presence of HF.

Atkinson and Taylor (1983) reported a sialoglycoprotein with high Ca^{2+} affinity that was involved in the hatching process, which may correspond to Ca^{2+}-binding site class (ii).

of a receptor–ligand interaction between the HF and the eggshell lipoprotein membrane. The change in eggshell permeability results in trehalose leaving the egg, permitting an influx of water and subsequent rehydration of the juvenile to a water content commensurate with movement. The presence of trehalose in eggs containing unhatched, viable J2 has been used as the basis for a viability assay for *G. rostochiensis* and *G. pallida* (van den Elsen *et al.*, 2012; Ebrahimi *et al.*, 2015).

The involvement of enzymes in eggshell permeability change has been postulated in several species. A Zn^{2+}-dependent enzyme mediates hatching of *H. glycines* and leucine aminopeptidase activity was found in the egg supernatant, although root diffusate does not increase its activity. Apparent softening of the eggshell before eclosion occurs in *X. diversicaudatum*, *A. avenae* and *M. incognita*, and in *M. incognita* lipase activity has been correlated with hatch. Transcriptomic investigation identified upregulation of the chitinase gene (*cht-2*) following exposure to host root diffusates in *Globodera* spp. (Duceppe *et al.*, 2017). However, the potato cyst nematode eggshell remains rigid during the hatching process suggesting no enzymatic activity on the eggshell, therefore the role of this chitinase remains unclear. Devine *et al.* (1996) demonstrated that the potato steroidal glycoalkaloids, α-solanine and α-chaconine, induce hatch of *G. rostochiensis*; glycoalkaloids are known to destabilize lipid membranes, during which leakage of trehalose is possible.

Several other events accompany rehydration. Within 24 h of root diffusate exposure, unhatched J2 of *G. rostochiensis* begin O_2 consumption and utilization of lipid reserves commences, the adenylate energy charge falls while the content of cAMP (a possible secondary messenger in receptor–ligand interactions) rises. These effects on the rehydrated J2 are due in part to removal of osmotic pressure and hydration and in part to direct stimulation of the J2 by root diffusate and mark the recommencement of metabolic activity. Changes in gene expression appear to occur during or immediately after the hatching process.

Vigorous movement of the J2 does not begin until at least 3 days after initial exposure to root diffusate, when the J2 starts local exploration of the inner surface of the egg, using its lips and stylet, and then begins thrusting movements with the stylet. This causes a regular pattern of perforations in the sub-polar region of the eggshell, which the J2 extends to a slit through which it emerges. Such use of the stylet to produce holes in the eggshell relies on the eggshell remaining rigid. J2 of *D. dipsaci* use a similar approach, except that the stylet thrusts are more random and the J2 uses its head to force open the slit in the eggshell. In *P. penetrans* and *H. avenae*, a single stylet thrust penetrates the eggshell and the head extends this into a tear. J2 of *Heterodea iri* (now *H. ustinovi*) use the tail tip to make the initial penetration.

The integration between host and nematode to ensure survival and invasion has progressed furthest in cyst nematodes and the sophisticated hatching mechanism is one aspect of this integration. Hatching patterns of cyst nematodes with multiple generations during the host-growing season are further survival adaptations. In several species of tropical cyst nematodes, with many generations during a crop growing season, there are changes in the hatching biology (Perry and Gaur, 1996). For example, under favourable conditions, *H. glycines* produces most eggs in egg sacs and J2 in these eggs hatch readily in water to invade the host and result in a rapid population build-up. Under less favourable conditions, there is a shift towards production of encysted eggs, which have a greater requirement for hatch stimulation by host root diffusates. In *H. cajani*, this shift occurs at the end of the host-growing season and is accompanied by increased energy reserves in the unhatched J2 to further enhance survival in the absence of a host.

8 Behaviour and Sensory Perception*

Roland N. Perry[1]** and Lindy Holden-Dye[2]

[1]*School of Life and Medical Sciences, University of Hertfordshire, UK and Biology Department, Ghent University, Belgium;* [2]*School of Biological Sciences, University of Southampton, UK*

*A revision of Perry, R.N. and Curtis, R.H.C. (2013) Behaviour and sensory perception. In: Perry, R.N. and Moens, M. (eds) *Plant Nematology*, 2nd edn. CAB International, Wallingford, UK.

**Corresponding author: r.perry2@herts.ac.uk

  Plant Nematology, 3rd edn (eds R.N. Perry, M. Moens and J.T. Jones)
DOI: 10.1079/9781800622456.0008

8.1. Overview of Perception and Behaviour

In order to enact behaviours that are commensurate with their survival, all animals must be capable of detecting changes in their external environment, monitoring their own internal status, and optimally interpreting this in the context of past events. These vital behaviours are achieved through neural circuits, including sensory receptors that respond to a diverse range of chemical, mechanical, light and thermal stimuli. The neural circuits have the potential to integrate multiple sensory inputs and encompass motor circuits that drive the appropriate behavioural response. Nematodes are no exception to this. This is best exemplified from studies on the free-living nematode *Caenorhabditis elegans* for which the connectome (wiring diagram) of the 302 neurones in the adult hermaphrodite has been described (White *et al.*, 1986). In this seminal paper, which had the running title 'The Mind of the Worm' the authors assigned a code to identify each neurone, e.g. VA1 is one of 12 motoneurones in the ventral nerve cord. Subsequently, the 387 neurones in the male and the circuitry of juvenile (= larval) stages have all been delineated (Cook *et al.*, 2019; Witvliet *et al.*, 2021). All of this information is available from an extensive database of *C. elegans* anatomy: https://www.wormatlas.org. Despite this anatomical simplicity, *C. elegans* can enact relatively complex behaviours (de Bono and Villu Maricq, 2005) that include aversive learning (Zhang *et al.*, 2005), behavioural preference (Barrios, 2014) and social behaviour (de Bono and Bargmann, 1998; Scott *et al.*, 2017).

To what extent these observations in the free-living bacteriovore *C. elegans* can provide broader insight into perception and behaviour of the Nematoda, where species have very varied lifestyle and habitats, is open to question. Whilst much less is known about the neural basis of behaviour in parasitic nematodes, there is sufficient evidence from comparative neuroanatomy to argue that these general principles may often extend to parasitic species that have similarly organized nervous systems. However, where there has been a detailed comparative analysis of *C. elegans* with another divergent species, *Pristionchus pacificus*, the gross anatomical similarities in terms of neuronal morphology belie discrete differences in sensory processing (Hong *et al.*, 2019). Given *C. elegans* is a bacterivore whilst *P. pacificus* is a predatory nematode found in association with insects this is not surprising. Unfortunately, such analyses on similarly divergent species is limited and for the time-being the extraordinary detail on the neural substrates of behaviour for *C. elegans* (de Bono and Villu Maricq, 2005; Taylor *et al.*, 2021) provides a useful framework on which to gain insight for other species in the phylum but with cautionary notes (Bumbarger *et al.*, 2009). This is especially pertinent as the genomes of many nematode species are sequenced and annotated, providing a great opportunity for mining for *C. elegans* orthologues as a gateway to understanding perception and behaviour more broadly.

In this chapter the discrete behavioural responses of different life stages of the nematode to specific stimuli are discussed. The movements of the nematode encompass those that result in: (i) travel or rotation of the nematode through space (locomotion); (ii) events of physiological or developmental importance (hatching, feeding, copulation, egg laying, defecation, root penetration); (iii) postures or integrated movements of groups of nematodes (coiling, clumping, swarming and nictating) that enhance survival and phoresis; and (iv) quiescence.

8.1.1. Fundamental concepts and terms in nematode behaviour

The terminology used here to define responses of organisms to environmental cues, i.e. context-dependent behaviour, is that of Burr (1984) as modified by Dusenbery (1992) (Box 8.1). The most important concepts are **migration, taxis** and **kinesis**. All are exhibited by nematodes. **Migration** is movement of an individual or population in a direction oriented with respect to a stimulus field and can be accomplished by taxis or kinesis. **Taxis** results from orienting the body to the stimulus direction, whereas **kinesis** results from undirected responses to changes in stimulus intensity, e.g. a change in forward speed (orthokinesis) or in random turn frequency (klinokinesis). Orientation can result from comparing stimulus intensity sequentially at two or more points in time (klinotaxis) or simultaneously at two points in space (tropotaxis) or many points in space (teleotaxis). Orientation can be positive, negative or transverse. When applying these concepts to nematodes, one should remember the dorsal–ventral orientation of body waves as well as the bilateral placement of amphids and the hexaradiate placement of papillae at the anterior end (see Section 8.2).

8.2. Sensory Transduction

Nematodes have sense organs that can transduce signals from their environment including chemical, mechanical, thermal and light stimuli and from their internal environment relating, for example, to gas tension and nutritional status. They also respond to noxious stimuli with escape responses, e.g. quinine avoidance (Hilliard *et al.*, 2004). In order to perform these behaviours, nematodes need to assimilate information from their external and internal environment via sense organs or sensilla. The sensors for external stimuli are exteroceptors and for internal stimuli are interoceptors.

There are two basic types of exteroceptor, internal and cuticular sensilla. The fundamental structure of a sensillum (see Fig. 1.5K) comprises three basic cell types: a glandular

Box 8.1. Behaviour terminology.

Kinesis. Behaviour comprised of undirected responses that are dependent on the intensity or temporal change in intensity of a stimulus. Undirected responses are unrelated to the orientation of the stimulus or stimulus field.
Orthokinesis. A kinesis in which translational motion (i.e. speed) is affected.
Klinokinesis. A kinesis in which rotational motion (i.e. direction) is affected.
Taxis. Migration oriented with respect to the stimulus direction or gradient that is established and maintained by directed turns. Directed turns are those that are biased in some way with respect to the orientation of the stimulus field.
Klinotaxis. A taxis which results from directed responses to sequential samples of stimulus intensity or direction.
Tropotaxis. A taxis which results from directed responses to two (or in three dimensions, three) simultaneous samples of stimulus spatial distribution or directional distribution.
Teleotaxis. A taxis due to directed response to information gathered by a raster of many receptors. Directed turning occurs until the fixation area of the raster is exposed to the directional stimulus. An extreme form of tropotaxis. (Added by Dusenbery, 1992.)

sheath cell, which is deeply folded with a very large surface area, a supporting socket cell, surrounding the duct that encloses the distal regions of the receptors, and a number of bipolar neurons, or dendritic processes, which are bathed in secretions. The main concentration of cuticular sensilla is on the head of the nematode (Fig. 8.1), consisting of a hexaradiate pattern comprising six lips containing 12 labial sensilla, four cephalic sensilla and two amphids. These are arranged in three circles consisting of an outer circle of four cephalic sensilla, a middle circle of six outer labial sensilla and an inner circle of six inner labial sensilla (see Chapter 1).

The function of interoception is performed by individual neurones. These interoceptors detect internal sensory cues. For example, they may detect body posture, i.e. proprioception, gas tension or nutritional status.

8.2.1. Chemosensilla

Chemosensilla encompass those responding to olfactory, airborne and gustatory, soluble, chemical cues. Nematodes detect and respond to a vast array of chemical cues including gases, inorganic and organic ions, and a variety of the constituents of root exudates as well as volatiles released from roots. Chemicals that cause interactions between organisms are called **semiochemicals**, which include **allelochemicals**, mediating interspecific responses such as nematodes responding to diffusate from host roots, and **pheromones**, mediating intraspecific responses such as male nematodes responding to sex pheromones from female nematodes. Semiochemicals may have volatile and non-volatile components. Sensilla that open to the environment have long been assumed to have a function in detecting semiochemicals.

The amphids are considered to be the primary chemosensilla but evidence from work on the free-living nematode *C. elegans* shows that they also have thermosensory (Mori

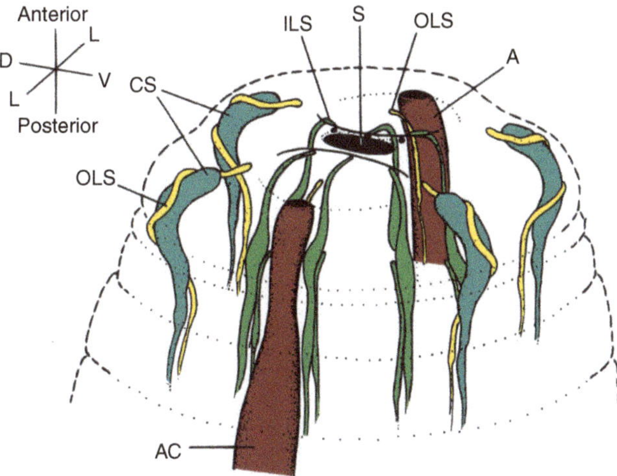

Fig. 8.1. Diagrammatic reconstruction summarizing the form and arrangement of the anterior sensilla of *Pratylenchus* spp. A, amphid; AC, amphidial canal; CS, cephalic sensillum; D, dorsal; ILS, inner labial sensillum; OLS, outer labial sensillum; S, stoma; V, ventral. (From Perry and Aumann, 1998.)

and Ohshima, 1995) and mechanosensory roles (Kaplan and Horvitz, 1993). Amphids are paired organs, positioned laterally with cephalic or cervical external openings, and contain several sensory receptors (the number varying among taxa) (Ashton *et al.*, 1999). The socket cell surrounds the amphidial duct; often the socket cell secretes copious material into the duct. The ciliary region of the receptors is enclosed by the sheath cell and within the sheath cell, and near the base of the duct, gap junctions occur between adjacent receptors. Typically, the amphid sheath cell is deeply folded or otherwise modified, resulting in a very large surface area. In *Meloidogyne* males and to a lesser extent in second-stage juveniles (J2), for example, the sheath cell has many extracellular, fluid-filled caverns continuous with a larger pouch surrounding the receptors (Baldwin and Hirschmann, 1973).

Phasmids are specialized pairs of sensory organs situated in the posterior lateral field (see Fig. 1.3D). Typically, the structure is that of a simple chemoreceptor, including one or two neuron receptors, each with a typical ciliary region surrounded by a sheath cell; the receptors terminate in a duct that opens to the exterior. The duct is surrounded by one or two secretory socket cells. Although there is much information on the ultrastructure of the phasmids of many species of nematodes, there is less information on their function. In *C. elegans* they are implicated in chemical avoidance (Hilliard *et al.*, 2002) and in mate-searching behaviour in males (Barrios *et al.*, 2008).

The socket and the sheath cells produce secretions, which may be specialized for parasitism. However, the secretions in the amphidial duct may derive only from the socket cell, where there is continuity between secretory vesicles and the secretions in the duct, rather than from the sheath cell, where there seems to be no direct contact with the amphidial duct as the receptor cavity of the sheath cell appears closed (Baldwin and Perry, 2004). The secretions may protect the dendritic ending of the nerve cells from desiccation or microbial attack, or may maintain electrical contact between the tips and the bases of the dendritic processes. Secretions from the posterior supplementary sensilla are likely to aid adhesion in copulation. The compounds present in the secretions are undoubtedly important in chemoreception and include *N*-acetylgalactosamine and fucose in J2 of *Meloidogyne incognita* and O-glycans in J2 of *Heterodera schachtii*, with mannose or glucose, *N*-acetylglucosamine and galactose and/or *N*-acetylgalactosamine forming the oligosaccharide chains. There is evidence that the composition of amphidial secretions differs between species. A glycoprotein associated with amphidial secretions of J2 of six species of *Meloidogyne* was not present in representatives from eight other genera that included *Globodera* and *Heterodera*, indicating a more specialized function for this protein in *Meloidogyne*. Secretions also may change at different stages of the life cycle of an individual species, perhaps reflecting different functional requirements. Amphid secretions may also contain 'effector' proteins that mediate interactions with the host (Semblat *et al.*, 2001; Eves-van den Akker *et al.*, 2014a).

The genetic basis of chemosensation has been extensively explored in *C. elegans* resolving a large panel of genes that are expressed in chemosensilla and play a role (Dusenbery *et al.*, 1975; Bargmann, 2006). This has provided a platform to investigate whether there is a conserved role in other species through tissue specific location of gene expression to chemosensory amphids or phasmids and the impact of RNA interference (RNAi) knockdown on responses to chemical stimuli. For example, in *M. incognita* four functional orthologues of *C. elegans* chemosensory genes have been identified and found to be expressed in the amphids and phasmids (Shivakumara *et al.*, 2019).

8.2.2. Mechanosensilla

The mechanosensilla detect gentle and harsh touch. Touch receptors, running along the body length (Jones, 2002), are pre-synaptic to many other neurones, including motor neurones, and detect gentle mechanical stimuli. A model for the functioning of these receptors (Driscoll and Kaplan, 1997) in *C. elegans* proposes that the cuticle is linked, via proteins, to a movement-sensitive ion channel. Microtubules within the receptor cell may also be linked, via other proteins, to ion channels. In *C. elegans* these sensory neurones express a characteristic portfolio of genes encoding mechanosensitive ion channels (Huang and Chalfie, 1994). The mechanosensilla neurones terminate beneath the cuticle surface and the ducts do not open to the exterior, although they may contain secretions that might aid in mechanical transmission. Many mechanosensilla are expressed at the cuticle surface as raised areas, or bumps, termed papillae. In addition, they may be directly exposed to the pseudocoelomic fluid. In general, the inner, outer and cephalic papillae (Fig. 8.1) are the main concentration of anterior mechanoreceptors, although in Tylenchomorpha the inner labial papillae are chemoreceptors.

Mechanosensilla are also involved in responses to external stimuli, such as tactile responses during copulation mediated by papillae in the tail region, and internal processes, such as egg laying and pharyngeal pumping. The involvement of mechanosensory perception (as well as chemoreception) in hatching and selection by endoparasitic nematodes of root penetration sites has been inferred from observations on pressing and rubbing of the anterior end accompanied by stylet probing.

8.2.3. Interoceptors

Interoceptors are neurones that express chemical or mechanosenory receptors and are key for controlling body posture, locomotion, energy homeostasis and the internal milieu.

Mechanosensory perception is required for coordinated locomotion. In *C. elegans* it has been demonstrated that a large number of neurones are proprioceptive, i.e. have the ability to detect body posture and position (Krieg *et al.*, 2022). Given the observation that many species of nematode adopt their posture and pattern of movement according to the viscosity of the environment it is likely this plays out more broadly in the phylum (see Section 8.2.4). Distinct processing for proprioception has been defined in *C. elegans* with neurones DVA and PVD assigned key roles (Krieg *et al.*, 2022).

Interoceptors also regulate energy homeostasis, e.g. by regulating feeding and by detecting nutritional status. An interoceptor expressed in *C. elegans* pharyngeal muscle cells has been attributed with the role of detecting the ingestion of food and regulating food intake through the expression of the ion channel PIEZO-1 (Millet *et al.*, 2022). A link between nutritional status and feeding rate has also been identified in which interneurones sensitive to the metabolite kynurenic acid mediate a post-starvation increase in feeding rate (Lemieux *et al.*, 2015). These observations open the way to the investigation of similar sensory regulation of behaviours in other species of nematodes by highlighting genes encoding proteins with a key role. A combination of bioinformatics to identify orthologues combined with studies to localize gene expression patterns has the potential to address questions relating to the molecular and cellular basis of context-dependent behaviours in other species of interest. However, it is prudent to be cautious when inferring functional equivalence if functional experiments are not possible.

Oxygen and CO_2 are long recognized as key regulators of nematode behaviour (see Section 8.5). In this regard it is of note that evidence from *C. elegans* indicates neurones that are specialized to detect oxygen tension, e.g. AQR, PQR and URX (Busch *et al.*, 2012). The neurone URX is also implicated in sculpting the behavioural response to CO_2 (Carrillo *et al.*, 2013). The behaviour of many nematode species is influenced by gas tension gradients and a parsimonious suggestion would be that there is a similar neural basis for interoception in other nematodes.

8.2.4. Other sensory receptors

One feature of nematode sensilla is that many of these sensory neurones are polymodal. There are some excellent examples of this in *C. elegans*. For example, PVD is a distinctively multi-dendritic neurone, sometime described as 'menorah-like', that responds to noxious harsh touch and cold temperature (Chatzigeorgiou *et al.*, 2010). Similarly, the amphidial neurone AFD has a unique morphology and is sensitive to temperature and CO_2 (Mori and Ohshima, 1995; Bretscher *et al.*, 2011). To what extent these neuronal morphologies and functions are conserved in the Nematoda is not known but these observations provide an informative starting point for investigation.

Nematodes can exhibit extremely sensitive thermotaxis (see Section 8.4.2). It is not known which sensilla are thermoreceptors in plant-parasitic nematodes, although *C. elegans* amphids can detect thermal cues. Photoreception in certain marine and insect-parasitic nematodes is mediated by photoreceptive sensilla termed ocelli, which are associated with a pair of red/brownish pigmented spots on either side of the pharynx, or lie within the central pharyngeal region and are shaded by a hollow cylinder of haemoglobin pigment that provides directional resolution. Further insight into the molecular basis of light detection has been provided by studying light avoidance in *C. elegans*. Ultraviolet light increases the locomotor rate of *C. elegans* by stimulating a gustatory-like receptor LITE-1 to facilitate an escape response (Edwards *et al.*, 2008). Gustatory-like receptors also mediate light-dependent inhibition of *C. elegans* feeding through the generation of hydrogen peroxide (Bhatla and Horvitz, 2015). Taken together, these sensors enable the nematode to evade a potentially toxic environment.

It has also been suggested that nematodes may be sensitive to sound despite the lack of a conventional ear-like organ for its detection (Iliff *et al.*, 2021). A key observation in this study was that airborne sound vibrated the nematode cuticle and the authors suggest this may be an additional mechanism required for nematode escape responses.

8.2.5. The sensory nervous system

The organization of the nematode nervous system (see Chapter 1) is essentially conservative, and information from the three best-studied species, *C. elegans*, *Ascaris suum* and *Strongyloides stercoralisis*, is likely to predict neuronal organization in other species (Fine *et al.*, 1998; Martin *et al.*, 2002; Baldwin and Perry, 2004). Counterparts of some of the well-characterized *C. elegans* neurones have been identified in other species of nematode and for the limited number of reports for which this detail is provided there would seem to be conservation of neuronal types, neurotransmitter phenotype and location of the neuronal cell body, e.g. for *Pratylenchus penetrans* (Han *et al.*, 2017). However, despite conservation of neuronal types in the anterior nervous system of *C. elegans* and *Pristionchus pacificus*, there are differences in the neural connectivity that suggest differences in the processing of sensory information (Bumbarger *et al.*, 2012; Hong *et al.*, 2019).

The main mass of the central nervous system is the circumpharyngeal nerve ring, and the majority of nerve processes run longitudinally, as ventral and dorsal nerve cords, or circumferentially, as commissures. With a couple of notable exceptions, the neurones have a simple, relatively unbranched morphology. Neuronal signalling uses small-molecule neurotransmitters, neuropeptides and gap junctions (see Chapter 7).

Much of our understanding of neurotransmission comes from work on either animal-parasitic nematodes or from *C. elegans* (Perry and Aumann, 1998; Perry and Maule, 2004). The initial pre-interactive events include passage of signals from the external environment through the sensilla secretions to the membrane-bound receptors of the nerve cells. On stimulation of a nerve ending, a neurotransmitter is released that triggers a downstream signalling event/cascade that results in a behavioural change. Further details of this process are given in Chapter 7.

Stimulation of a particular chemosensory neurone of *C. elegans* results in a distinct behavioural response (attraction or avoidance) demonstrating that the neurones are 'hard wired' and it is the neurone that determines the response, not the nature of the receptor molecule being stimulated (Bargmann and Horvitz, 1991). The chemoreceptors are G-protein-coupled receptors and consist of a large family of approximately 1000 proteins (Bargmann, 2006). Some receptors are expressed in a single sensory neurone, whilst others are present in up to five sensory neurones.

8.3. Locomotory Behaviours

Nematodes can express a range of locomotory behaviours that are either probabilistic or directed in response to a given stimulus. These behaviours include forward movement, backward movement, reversals, bending and turning, coiling and nictating. They can also cease movement, a state termed quiescence and purported to be analogous to sleep. The precise sequence of distinct locomotor patterns of behaviour can also exhibit temporal organization. The term 'dwelling' has been used to describe *C. elegans* that are on a bacterial lawn and actively feeding and it is manifest as a high frequency of switching between forward and backward movement. When *C. elegans* are moved to a food-free environment, they will first adopt a 'local area search' involving frequent reversals and turns. In a very stereotyped manner, after 15 min this behaviour switches to long periods of forward movement thus extending the food search into the wider arena (Hills *et al.*, 2004).

8.3.1. The importance of the hydrostatic skeleton

Microscopic nematodes are not thought to be strong enough to dislodge most soil particles, thus restricting movement to channels within the soil. However, their bodies are clearly stiff and can transmit sufficient axial force to burrow through pluronic gel and dilute (0.5–1.5%) agar. This stiffness derives from a high internal pressure, or turgor, which provides a hydrostatic skeleton. High internal pressure appears to be built up and maintained by pumping action and valves of the metacorpus. Crossed fibres within the basal zone or layer of the cuticle offer a mechanical explanation for the observation that upon osmotically induced swelling, increases in circumference were restricted and most volume increase resulted primarily from increased length (Alexander, 2002). Anatomical evidence indicates additional factors contribute to body stiffness (Burr and Robinson, 2004). The sinusoidal waveform is produced by alternate contraction of the dorsal and ventral longitudinal muscle trunks of the body wall antagonized by the hydrostatic skeleton. Other antagonistic factors include stiffness of the muscle fibres themselves, elastic restoring force of the cuticle and compartmentalization of turgor, especially between the so-called muscle bellies and the gut. The relative proportions of cuticle thickness, muscle belly diameter and body diameter vary more than tenfold across nematodes. Thus, it is reasonable to assume that the relative importance of these factors also varies.

Undulatory locomotion in nematodes presents six requirements related to neuromuscular control (Box 8.2). Although most is known about creation and propagation of the dorsal–ventral wave, understanding how all requirements might be met requires familiarity with nematode sensory structures (see Section 8.1) and elementary knowledge of nematode neuromuscular anatomy and function, much of which derives from work on *C. elegans* and the animal-parasitic nematode *A. suum* (Wright and Perry, 1998). Muscles consist of bundles of parallel elongated cells called muscle fibres. Each fibre in turn contains, in addition to essential cell organelles and a highly specialized endoplasmic reticulum, a stack of contractile units (sarcomeres) positioned end to end along the fibre's length and separated by so-called z-lines, which serve as connective tissue within and between contractile units. Each sarcomere contains parallel and overlapping molecules of the linear contractile proteins myosin (also called thick filaments) and actin (called thin filaments) within each sarcomere so that each thick filament is positioned centrally and thin filaments are attached on one end to a z-line with the other end projecting into and partly overlapping the thick filaments. During muscle contraction, membrane depolarization events create a strong molecular affinity between thick and thin filaments, causing them to slide across each other, maximizing the overlap and shortening the sarcomere. Taken together, the alignment of z-lines and overlapping regions of thick and thin filaments gives rise to the striated appearance; somatic muscles in nematodes are obliquely striated.

With only few exceptions, nematodes move by dorsoventral undulations of the body that are propagated backwards from the anterior end. As a result, on flat surfaces they tend to follow sinusoidal paths, as do snakes, but while lying on their sides rather than their ventral surface. *In vitro*, vermiform stages of plant-parasitic nematodes in water and

Box 8.2. Requirements of neuromuscular control for nematode undulatory locomotion (Burr and Robinson, 2004).

1. Creation of a reciprocating pattern of tension and relaxation on dorsal and ventral sides of the body.
2. Propagation of this pattern along the body posteriorly for forward locomotion and anteriorly for backward locomotion.
3. Control of the switching from forward to backward locomotion based on sensory input.
4. A rhythmic pattern generator to generate the waves cyclically.
5. Regulation of wave frequency, amplitude, wavelength and rate of propagation according to environmental requirements.
6. Utilization of sensory feedback to adjust waveform according to the location of objects along the body.

on agar typically exhibit rhythmic generated body waves that cease or reverse only when an obstacle or some other discontinuity in an otherwise homogeneous substrate is encountered. The typical response to an obstacle is to stop, withdraw one or two waves, probe in one or more directions with the anterior end, and then resume rhythmic forward movement in a new direction. In nature, plant-parasitic nematodes are usually on or within highly heterogeneous substrates, such as soil, plant tissue or foliar surfaces, where rhythmic propulsion is likely to be interrupted frequently.

The undulatory waveform varies with environment and species. Substrate resistance reduces wave amplitude, wavelength and speed of wave propagation, and in water where resistance is minimal some species propagate waves ten or more times as fast as others. However, the rate of forward progression, ignoring substrate-dependent slippage, is strictly determined by the speed of wave propagation in almost all nematode species utilizing undulatory propulsion (Fig. 8.2). Moreover, in at least some species, observations show that the waveform can be differentially modulated along the length of the body in response to external forces or stimuli, thus allowing efficient passage over irregular surfaces; this may also occur in nematodes within soil, plant tissues or other habitats where they are rarely or never observed.

Somatic muscles are oriented parallel to the body axis as two dorsal and two ventral muscle trunks. They are unusual in that each sarcomere is connected perpendicularly to the site of force application, the cuticle, by z-line functional equivalents consisting of a geometric arrangement of dense bodies and basal lamina that project into the fibre. The basal lamina, epidermis and cuticle are thought to comprise three tightly bound layers. Thus, the nematode body wall, consisting of somatic muscle, epidermis and cuticle, functions as a single differentially contractile organ (Burr and Robinson, 2004). As there are no circumferential muscles in nematodes, local constrictions in volume do not occur. Cuticular annuli allow for compression during bending of the body, but the magnitude of the longitudinal restoring force within the cuticle has not been measured, nor has its longitudinal elastic

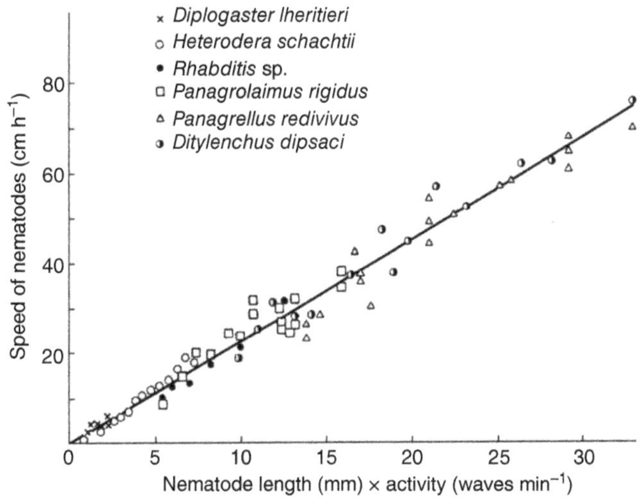

Fig. 8.2. Speed of propulsion by morphologically and ecologically diverse nematodes on a moist glass surface. (From Wallace, 1958.)

modulus during stretching in most species. However, it is clear that the nematode cuticle has little circumferential elasticity and longitudinal elasticity appears limited to a relatively narrow increase in length.

8.3.2. The motor nervous system

Sinusoidal waveforms are enabled by separate innervation of dorsal and ventral muscle trunks by their respective nerve cords along most of the body length. Innervation is achieved via somatic muscle arms that extend to and synapse only with their respective nerve cords (dorsal or ventral). Thus, contraction along most of the body is possible only in the dorsal–ventral plane and there is evidence that commissures between the dorsal and ventral nerve cords provide reciprocally inhibitory innervation. However, in the anterior end, additional innervation permits complex behaviours involving localized movement of the anterior in other planes, as observed during feeding, hatching, mating and tissue penetration. The most comprehensive information on sensory neurons and the circuitry by which forward and backward movements are triggered and effected derives from work on *C. elegans*, where numerous neurons and synapses have been carefully mapped and studied (see Section 8.6).

8.3.3. Transfer of forces and propulsion

The primary variables influencing nematode propulsion are body waveform geometry and forces arising from the nematode and its surroundings (Box 8.3).

Each sinusoidal body wave can be considered a locomotory unit (Fig. 8.3) exhibiting reciprocal inhibition of opposing dorsal and ventral sections of the longitudinal body

Box 8.3. Variables influencing nematode undulatory propulsion.

(A) Geometric variables

1. Body wavelength.
2. Body wave amplitude.
3. Wavelength-to-amplitude ratio.
4. Body diameter-to-wavelength ratio.

(B) Forces

1. Net contractile force of body wall and speed of contraction (determines power).
2. Axially transmitted resistance (determines ratio of power to speed and energy lost to heat of friction):
- (a) from friction along body:
  - (i) substrate resistance;
  - (ii) cuticular resistance; includes changes in coefficient of resistance of cuticle when compressed and expanded;
- (b) from force against head penetrating substrate.

3. Substrate resistance locally perpendicular to body axis (determines slippage and energy lost to work done to surroundings).

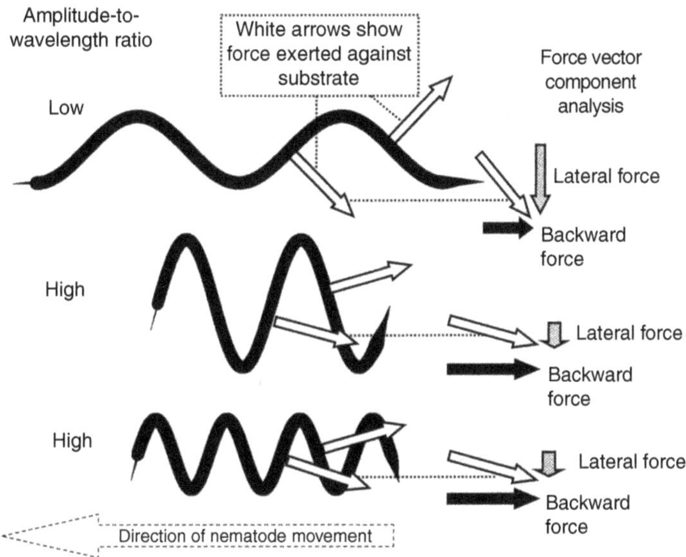

Fig. 8.3. Effect of amplitude-to-wavelength ratio on backward force exerted by nematodes when moving forward.

muscles (Alexander, 2002; Burr and Robinson, 2004). Comparisons of cuticular folds show that dorsally convex regions of curvature along the body axis are dorsally extended and ventrally contracted, and conversely for ventrally convex regions. During forward movement, when waves are propagated backwards, the relatively straight regions of longitudinally oriented body trunk muscles just behind those that are maximally contracted ventrally are undergoing ventral contraction, while those immediately ahead are ventrally distending. Thus, although the precise way that wave propagation is controlled neurologically is uncertain, reciprocal anterior–posterior as well as dorsal–ventral motor inhibition is indicated, with different inhibitory relationships for forward and backward movement, and candidate neural processes have been identified.

Vector analysis can be applied to understand essential relationships between waveform, external resistance and internal contractions (Fig. 8.3). To move, a nematode must overcome surface tension and friction by pushing its angled body against the substrate laterally (more precisely dorsally and ventrally) relative to the sinusoidal body axis but obliquely backwards to either side of the direction of travel of the nematode's centre of mass. When a vector representing the obliquely backwards force applied perpendicular to the body axis is resolved into components perpendicular and parallel to the direction of travel, the relative magnitude of the backward component useful to overcome resistance clearly varies directly with the amplitude-to-wavelength ratio. Thus, it is not surprising that nematodes characteristically increase the amplitude-to-wavelength ratio upon encountering increased substrate resistance. The angle of leverage against obstacles or tracks in the substrate is thereby increased. However, the speed of propulsion is typically decreased so that overcoming greater substrate resistance with perhaps no additional strength (strictly, power) is achieved at the cost of speed, much like shifting a bicycle into a lower gear. Conversely, most nematodes that normally move on solid or semi-solid substrates exhibit body waves with greatly exaggerated wavelength and speed when suspended in water, which offers relatively little resistance to movement.

Microscopic nematodes are at the limit of size of organism at which the viscosity of water surrounding the body becomes a major force and they differ greatly in ability to negotiate free water and water-saturated porous substrates, as discussed in Section 8.3. The primary factors affecting efficiency of swimming by any animal are size, shape, speed, inertia and viscosity. The dimensionless index known as the Reynolds number predicts the relative importance of inertia and viscosity for a given shape, size and speed; for small nematodes in general the value has been estimated as 0.01. When the Reynolds number is this small, animals swimming in water are so strongly influenced by viscosity that they must use non-reciprocal motions, such as undulatory propulsion, or they will continually pull themselves back and forth with no net change in position (Dusenbery, 1996). When moving while held by surface tension within a film of water on a solid substrate, viscosity contributes to substrate resistance and purchase. Additional factors potentially influencing nematode propulsion but little investigated include the body diameter-to-wavelength ratio, wrinkling and smoothing of the outer layer of the cuticle during undulations and differences in general cuticular topology among species. The latter will be discussed briefly in relation to peristaltic movement in *Criconemoides* spp. (see Section 8.5).

8.4. Movement in Soil

8.4.1. Physiological factors influencing activity

Probably the most important physiological factors influencing the rate of nematode movement are temperature, oxygen availability, toxins, water and ionic status, energy reserves and the nematode's state of activation as influenced by developmental triggers. As poikilotherms, nematodes tend to increase activity with increased temperature up to some limit; however, specific thermal optima for activity differ greatly among species and appear ecologically tuned (Fig. 8.4). In a study comparing nematodes from a common geographic region (southern Texas, USA), infective juveniles of two root parasites (*Rotylenchulus reniformis* and *Tylenchulus semipenetrans*) were found to have thermal optima for motility 10°C higher than those of two foliar parasites (*Ditylenchus dipsaci* and *D. phyllobius*), consistent with the difference between the warm median soil temperature in the summer, when most root growth occurs, and the need for foliar parasites to be maximally active during the cool rainy period of the summer, when the foliar moisture films required for infection of foliar buds are present. Ecological adaptation in speed of movement is afforded by the low energy cost of locomotion in microscopic nematodes, due to their small size (Alexander, 2002), and further enhanced in the infective stages of plant-parasitic species through the space-efficient storage of food reserves as lipid rather than glycogen (Barrett and Wright, 1998; Barrett, 2011; see Chapter 7).

In vitro, solutions with high osmotic pressure rapidly desiccate most nematodes. However, as will be discussed, except in highly saline soils, osmotic pressure within soil and plant tissue is unlikely to influence the water balance of plant-parasitic nematodes as the soil dries because most water will be extracted from nematodes by other forces before osmotic pressure is physiologically important. Moreover, the surface tension of soil water

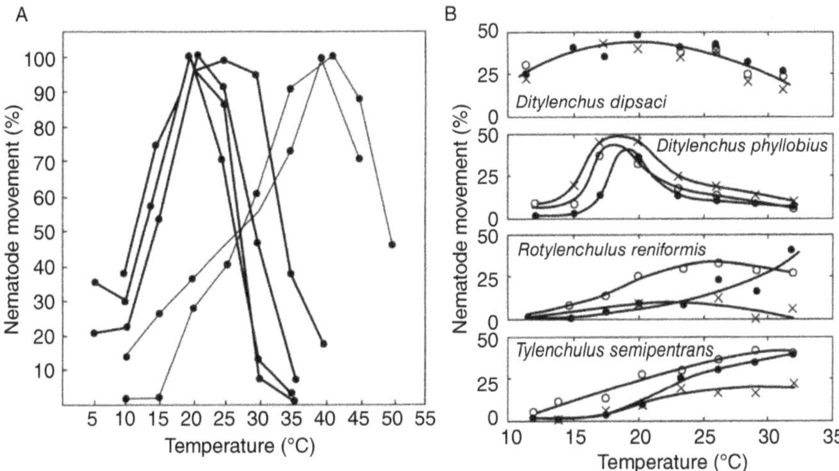

Fig. 8.4. Effects of temperature on nematode activity. A: Motility of skin-penetrating stages of parasites of warm-blooded hosts (thin lines) contrasted with plant parasites (bold lines). (From Croll, 1975.) B: Random dispersal on agar by foliar and stem parasites (*Ditylenchus dipsaci* and *D. phyllobius*) contrasted with root parasites (*Rotylenchulus reniformis* and *Tylenchulus semipenetrans*) from the same region of Texas. (From Robinson, 1989.)

will bind nematodes tightly to the soil before physiologically critical levels of osmotic pressure in the soil develop. Limited data available for effects of low oxygen tension on plant-parasitic nematodes indicate that activity of most species appears unaffected until the ambient partial pressure (normally 20%) drops below 1–2%. Thus, over most of the range of oxygen tension that plant-parasitic nematodes are likely to encounter within plant tissue and agricultural soil, activity is unlikely to be suppressed or stimulated. However, symptoms of CO_2 intoxication in plant-parasitic nematodes have been reported at concentrations as low as 5%, which is well within the range that can be expected in waterlogged soils with a high biological oxygen demand. Although many saprophytic nematodes appear able to withstand strongly hypoxic, reducing conditions, plant-parasitic nematodes as a general rule are quickly intoxicated and rendered immobile by oxygen deprivation or high ambient concentrations of CO_2 and nitrogenous waste products.

8.4.2. Physical factors influencing the occurrence and rate of movement

Substrates through which nematodes move may be porous or non-porous and porosity is scale-dependent. For example, freshly prepared agar is non-porous from a nematode's perspective but porous at the molecular level. Soil is particulate and is effectively solid for a nematode if pores are too small or too dry to permit nematode movement. Pores may be partly or completely filled with water. Yeates (2004) noted that nematodes can be considered **interstitial** or **pellicole**, based on the water content of the substrates they typically occupy. Interstitial nematodes occupy interstices completely filled with water or other fluid (marine sediments and many animal tissues), whilst pellicole nematodes occupy pores or channels filled only partly with water or other fluid (moist soil, spongy foliar tissue, leaf galls). Substrates with partially filled interconnected pores lined with water films allow rapid diffusion of gases and volatiles over relatively long distances through the substrate and expose nematodes to powerful surface tensions within water films.

Any substrate on which nematodes move can be considered an interface, which on either side may be solid, semi-solid, porous, liquid or gas. As movement along an interface constrains motion to one or two dimensions, the rate of random dispersal along an interface should be faster than random dispersal in three dimensions. When moving on an interface, such as the root epidermis or the surface of stems and leaves, topology is very important. Generally, a moisture film must also be present, and nematodes can be drawn into depressions such as embryonic leaf folds where moisture films recede as the surface dries. Obviously, some regions constrain motion largely but not entirely to one or two dimensions. For example, a moist stem surface covered with trichomes (leaf hairs) through which nematodes thread themselves is a two-dimensional interface but might more accurately be considered a narrowly bounded porous substrate where space for movement is less than 1 mm radially, and 10 mm circumferentially and 100 mm axially. Thus, the expected magnitude of random dispersal in those directions is greatly influenced by topology.

All soil classification systems recognize that soil particles fall into three classes: (i) those that are colloidal (stay suspended by Brownian movement) and have electrically charged surfaces largely due to laminated aluminium silicates (clay); (ii) those that are relatively large and primarily uncharged silicon dioxide particles (sand); and (iii) those that are between these two in size and charge characteristics. The quantity of clay and silt in most agricultural soils predicts that if small particles did not bind together only soils composed largely of medium to coarse sand would offer sufficiently large pores between particles to permit passage by most kinds of soil-inhabiting nematodes. However, this is clearly not the case for reasons that we shall examine next.

The energy required by plants to extract water from soil depends on the Gibb's free energy of soil water, which is a cumulative measure of the osmotic pressure of soil water, the pull of gravity and the powerful attraction of water to charged soil particle surfaces, known by plant physiologists as matric potential. For plants, the soil water matric potential is the primary factor affecting plant-water status. In agricultural soils it typically ranges from 0 to negative 15×10^5 Pa, the so-called permanent wilting point for plants. When soil is irrigated, a wetting front moves down through the soil profile saturating the soil interstices. In time, gravity partially drains the soil pores and air channels form, connecting the soil with the atmosphere. At this point, the soil is said to be at field capacity water content, which usually occurs at a matric potential near 5×10^3 Pa.

Classical experiments by Wallace in the 1950s and 1960s (summarized by Robinson, 2004) demonstrated, firstly, that it was not the soil texture that determined the suitability of soil for nematode movement but rather the soil crumb size, i.e. the porosity of the soil from a nematode's perspective (Fig. 8.5). He found, secondly, that the total amount of water per se in soil was not critical; what really mattered was the matric potential and the optimum matric potential was that observed at field capacity (Fig. 8.5). This is very important because finely textured soils can retain several times as much water as sandy soils at the same matric potential. Thus, the truly useful predictor of the suitability of the soil-moisture level for nematode movement was not the water content but the energy required to extract water, which is directly related to the surface tension and thickness of water films within the soil. Wallace also found that nematode movement within soil was

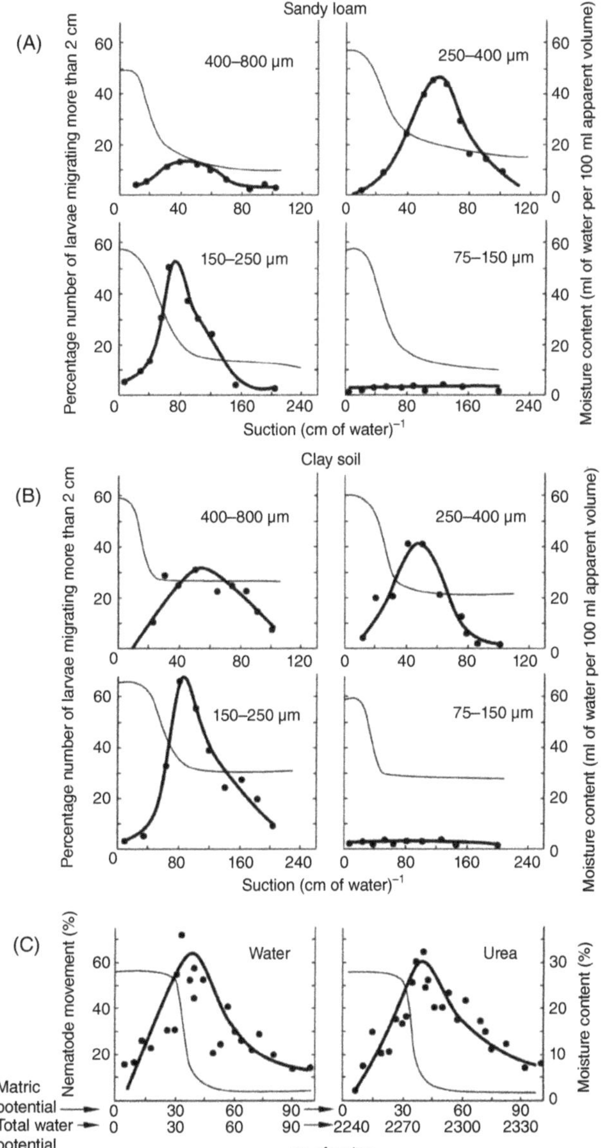

Fig. 8.5. Factors affecting nematode movement through soil. A, B: Soil texture is only indirectly important when compared with soil structure and matric potential. Similar effects in texturally contrasting sandy and clay soils are observed when movement by *Globodera rostochiensis* is compared in relation to soil structure (crumb diameter in μm) and water potential (cm of water). C: Matric potential is far more important than osmotic pressure over ranges that occur in natural soils. The marked influence of soil matric potential on movement by *Ditylenchus dipsaci* in sand wetted with water (left) is unaltered when 0.1 M urea is substituted for water to elevate osmotic pressure independently of soil matric potential (the energy required to extract water mechanically). The moisture characteristic of sand is shown as a thin line. (A, B from Wallace, 1968; C from Blake, 1961.)

restricted to a narrow range of matric potentials compared with the range over which plants can extract water and survive, and it was already well established that, except in highly saline soils, increases in osmotic pressure in soil water as a result of evaporation is physiologically insignificant to plants compared with the soil matric potential. Subsequently, other work demonstrated that osmotic pressure was of no importance to nematode movement in natural soils (Fig. 8.5).

Electrical charge separation on soil particles, particularly in the clay fraction, is of great significance to agricultural soil fertility and nematodes. Surface charges bind ionic species in the soil, providing a reservoir of plant nutrients and facilitating the establishment and collapse of gradients of ionic nematode attractants and repellents because specific ions are removed and added to different regions of the soil by plant roots, biodegradation, leaching, irrigation and fertilizer applications. Many nematodes are attracted and repelled by various salts, and attraction of root-knot nematodes to roots can be enhanced by the addition of small amounts of clay to pure sand. Surface charges on soil particles also favour plants and nematodes by binding soil particles together to create a structure consisting of larger diameter aggregates than would exist otherwise, thus increasing aeration, decreasing resistance to water percolation, and providing larger pores for nematodes to move through.

The pH of soil is typically buffered by ammonium, carbonates, sulphates and phosphates often bound to or comprising soil particles. Their concentrations are changed by plant roots and the soil microflora and vary markedly with depth and distance from roots. Thus, pH and any substance whose concentration is pH-dependent (such as CO_2, see Section 8.5.2) are likely to establish concentration gradients in the soil, providing information regarding depth, vertical orientation and the location of roots.

8.5. Movement in Response to Stimuli

In Section 8.4, we discussed physiological and physical factors that influence the activity of plant-parasitic nematodes in the soil. In this section, the response of plant-parasitic nematodes to various stimuli that they are exposed to in the soil is examined (Box 8.4). While in the soil, plant-parasitic nematodes are not feeding and therefore are dependent on their food reserves and need to locate a host rapidly in order to feed and develop. For example, under optimal conditions for movement, J2 of *Globodera rostochiensis* have an infective life of only 6–11 days after hatching. Around actively growing roots there exist several gradients of volatile and non-volatile compounds, including amino acids, ions, pH, temperature and CO_2. It is evident that nematodes orientate towards the roots using at least some of these gradients (Fig. 8.6) and this enhances the chances of host location and reduces the time without food. There is also increasing evidence for a role for semiochemicals in regulating their behaviour. This topic has seen a wealth of additional knowledge since the discovery of ascarosides (Butcher *et al.*, 2007; Srinivasan *et al.*, 2008). These are an extensive family of small molecules assembled in a modular fashion from the sugar ascarylose, fatty-acid-like side chains, amino acid derivatives and other primary metabolites. For more detail on ascarosides, see Chapter 7.

Evaluating the reality of the attractiveness or otherwise of an individual compound is difficult. Much of the available information is based on *in vitro* behavioural bioassays,

Box 8.4. Prefixes denoting stimuli investigated for effects on nematode behaviour.

Chemo-	Chemical	Photo-	Light
Galvano-	Electric field	Rheo-	Fluid flow
Geo-	Gravity	Thermo-	Temperature
Magneto-	Magnetic field	Thigmo-	Touch

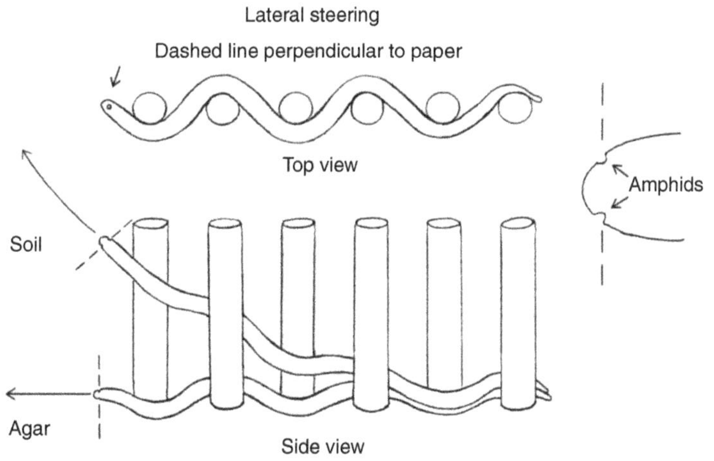

Fig. 8.6. Lateral movement of the anterior end of a nematode within a three-dimensional matrix, in theory, can permit orientation to gradients perceived by comparing sensory input of the two amphids, whereas amphids are necessarily positioned at the same concentration of chemicals in gradients on the surface of agar, because nematodes on the surface of agar lie on their sides.

which bear little if any resemblance to the situation in the soil; care must therefore be exercised in extrapolating from such assays to the field situation (Spence *et al.*, 2009). However, some generalizations can be made and certain gradients are strongly implicated in orientating nematodes to the roots. Perry (2005) separated gradients into three types: (i) **long-distance attractants** that enable nematodes to move to the root area; (ii) **short-distance attractants** that enable the nematode to orientate to individual roots; and (iii) **local attractants** that are used by endoparasitic nematodes to locate the preferred invasion site. In the following sections, some of the major stimuli in soil will be discussed and related to these three types.

8.5.1. Orientation to temperature

It is well established that plant-parasitic nematodes migrate in response to temperature but it is unclear why, although some interesting theories have been proposed. For example, metabolic heat from roots could be a local attraction for nematodes (Perry, 2005).

In addition, temperature might be the most consistent cue to differentiate up from down within soil where there is no light and gravitational effects are minuscule compared with surface tension.

Migration to a preferred temperature (temperature preferendum) has been well characterized in *C. elegans* (Hedgecock and Russell, 1975) and a similar phenomenon has been described for at least nine species of nematodes in which they migrate when placed on a temperature gradient plate *in vitro*. In most cases where tested, the preferendum is shifted partly or completely in the direction of a new adaptation temperature within several hours. One of these nematodes is the J2 of *M. incognita*. The threshold ambient temperature change eliciting a detectable change in the rate of movement of these nematodes is less than 0.001°C (Robinson, 2004), which is near the theoretical limit of temperature sensation in animals. In migration experiments, gradients sufficient to achieve a maximal response by *M. incognita* and other species, including *G. rostochiensis*, *D. dipsaci*, *T. semipenetrans* and *D. phyllobius*, have been found to lie between 0.01 and 0.1°C cm^{-1}.

Whilst the capability to detect thermal gradients for parasitic nematodes that seek mammalian hosts is intuitive (Bryant *et al.*, 2022) the survival value of this for plant-parasitic nematodes is less clear. Can they assist them to find roots? El-Sherif and Mai (1969) found two species in Petri dishes to be attracted to heat released by germinating alfalfa seedlings, demonstrating that this was possible. However, in soil, root metabolism is a small heat source compared with solar radiation, which in many latitudes can elevate the soil surface temperature by 20°C or more in several hours. Diurnal surface heating and cooling also sends a heat wave down through the soil every day (Fig. 8.7). The wave characteristically starts with maximum amplitude at the surface during the afternoon, dampening as it moves downwards 2–3 cm h^{-1}. Robinson (1994) measured magnitudes of vertical gradients in a cotton field in Texas during the afternoon on a typical sunny day and found them to be 0.3 and 0.15°C cm^{-1} at depths of 10 and 25 cm, respectively. Thus, behaviourally effective gradients extended deep into the soil. In addition, hourly collection of temperature data at 2.5 cm increments down to 60 cm across most months of the year clearly showed gradient inversions and other perturbations indicative of rainy periods, cold fronts and other weather patterns throughout most of the root growth zone (Fig. 8.7).

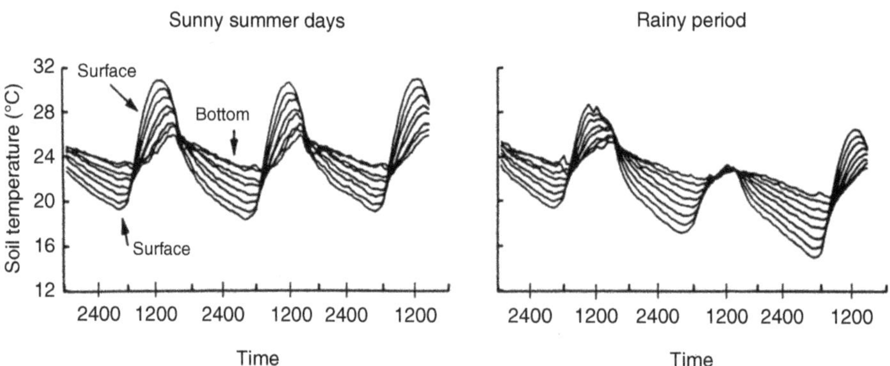

Fig. 8.7. Diurnal heat wave propagation down through soil in a cotton field during late summer in eastern Texas. Temperatures were measured with thermocouples planted 5–25 cm deep at 2.5 cm intervals.

As soil temperatures change constantly and gradients invert daily while nematodes are constantly adapting and migrating in response to gradients, it is exceedingly complex to predict the net result. Robinson (1994) subjected two root parasites, *M. incognita* and *R. reniformis*, and one foliar parasite, *D. phyllobius*, in the laboratory to gradient fluctuations in soil that precisely mimicked those measured previously in cotton fields. Nematodes in the laboratory were equilibrated for 24 h under fluctuating temperatures identical to those at the same depth in soil where they were released; at intervals following release their distributions were determined. The two root parasites consistently moved in opposite directions; moreover, movement of *R. reniformis* was down and *M. incognita* was up, consistent with their known vertical distributions in cotton fields. The foliar parasite appeared to move toward cool regions regardless, consistent with previous observations on agar, and consistent with its need to be attracted to the soil surface during rainy periods to ascend cool, moist foliage.

8.5.2. Response to carbon dioxide

CO_2 strongly attracts a number of species *in vitro* and may be the most common and potent nematode attractant (Perry, 2005). It is released abundantly by living and decaying plant and animal tissues, providing an obvious cue to the possible presence of food. Plant-parasitic nematodes are confronted with the problem of distinguishing roots from decaying material from which they would be unable to obtain food. It is possible that other substances modulate nematode attraction to CO_2. For example, *M. incognita* is repelled by ammonia and several nitrogenous salts released by decaying material. It has also been proposed that CO_2, like temperature, may serve as a collimating stimulus, providing a vertical directional reference to soil organisms.

The attraction of plant-parasitic nematodes to CO_2 *in vitro* was first observed in the 1960s. Not everyone who looked for the response observed it and failures to detect it in some cases may have resulted from CO_2 toxicity because pure CO_2 at high flow rates was used. Additionally, some species are repelled by CO_2. For example, fourth-stage juveniles (J4) of *Bursaphelenchus xylophilus* avoid CO_2 and this is thought to be an important cue for its transition from its beetle vector into the pine tree (Wu *et al.*, 2019). Higher concentrations of CO_2 exceeding 5% suppress motility. Robinson (1995) used infusion pumps to deliver small amounts of CO_2 over extended periods of time through needles positioned within tubes of moist sand. The optimal release rate for attracting *M. incognita* and *R. reniformis* to a point source was 15 µl min^{-1}, a miniscule rate indeed. Intervals of 40 and 29 h were needed to attract most nematodes within 52 mm of the source and the total amount of gas that was needed to be released over this very long period of time (about 10 cm^3) was calculated to be equivalent to that released by a germinating sunflower seed.

Pline and Dusenbery (1987) elegantly showed that the threshold at which *M. incognita* responds to CO_2 varies with ambient concentration; that is, nematodes become more sensitive as the ambient concentration drops and, thus, can detect about the same relative change at any ambient concentration. This effect is very important ecologically because it means that nematodes can detect gradients at far greater distances from the source than would be possible with a fixed concentration differential threshold. The threshold for *M. incognita* corresponded to a relative change of about 3% cm^{-1}. It is very important to understand that this does not mean 3% CO_2 in air but rather 3% of the concentration of CO_2 in air, whatever that concentration might be, but typically less than 1%. Results from other studies are consistent with Pline and Dusenbery's conclusions. The minimum effective gradient measured in the first study on *D. dipsaci* on agar was approximately 0.1% cm^{-1}. *Meloidogyne incognita* and *R. reniformis* were strongly attracted to CO_2 in sand, migrating up a 0.2% cm^{-1} gradient at a mean ambient CO_2 concentration of 1.2%, which was a 16% cm^{-1} relative change or 1% per nematode body length. Plant-parasitic nematodes attracted to CO_2 include *Aphelenchoides fragariae*, *D. dipsaci*, *H. schachtii*, *M. hapla*, *M. incognita*, *M. javanica*, *P. penetrans*, *P. neglectus* and *R. reniformis*.

However, interpreting the significance of attraction to CO_2 released by roots is confounded by effects of CO_2 on gradients of redox potential, pH, carbonic acid, bicarbonate and carbonate in the soil. CO_2 attracts nematodes to roots and either dissolved CO_2 or carbonic acid is the attractive species. Wang *et al.* (2009a) considered that the observed attraction of root-knot nematodes to CO_2 may be due to acidification of solutions by dissolved CO_2 rather than to CO_2 itself.

The neural basis of the nematode response to CO_2 has been carefully defined in *C. elegans*. However, it is clear that whilst this is probably a ubiquitous environmental cue for regulation of nematode behaviour, the precise response, attraction or repulsion, is species, and even life stage, dependent (Banerjee and Hallem, 2020).

8.5.3. Response to plant attractants

Long-distance attractants, such as CO_2, may enable the nematode to move to the root area but other gradients are important to ensure the nematodes locate individual roots. These short-distance attractants have received most attention and there is evidence that, in some instances, the attractiveness of a host to the pest species is correlated with its efficiency as a host.

Research on *G. rostochiensis* has demonstrated that diffusates from the roots of the host plant, potato, increase the activity of the infective J2 and also attract them to the roots. This potato root diffusate (PRD) is required to stimulate hatching of the majority of J2 of the potato cyst nematodes *G. rostochiensis* and *G. pallida* (see Chapter 7), but work by Devine and Jones (2002) has shown that the chemicals in PRD responsible for hatching differ from those responsible for attracting the J2 to the root. Electrophysiological analysis of sensory responses (Perry *et al.*, 2004) demonstrated that spike activity of J2 of *G. rostochiensis* increased on exposure to PRD but not to root diffusate from the non-host sugar beet, thus indicating that responses to diffusates may be host-specific. Two *Solanum* species, *Solanum tuberosum* and *S. sisymbriifolium*, respectively a susceptible and a trap crop for *G. rostochiensis* and *G. pallida*, exude different hatching factors but send common signals to attract the J2 towards the root area (Sasaki-Crawley *et al.*, 2010).

Research evaluating the attraction of *M. graminicola* and *M. incognita* for cereals or dicotyledonous roots supports the role of root diffusates in regulating host specificity. This research showed that these species take the most direct route to their preferred host but often take the longest route towards the poor host (Dutta *et al.*, 2011). When in the rhizosphere, plant-parasitic nematodes have to navigate their way to roots amongst a complex of repellent and attractant chemicals and gradients of these chemicals might influence the interaction between *Meloidogyne* spp. and their hosts, and Dutta *et al.* (2011) suggested that chemical differences in complexes of 'short-distance attractants' seem to define host preference for these root-knot nematodes. Attraction of *M. hapla* to roots of *Arabidopsis* is altered in plants with mutations affecting ethylene signalling, indicating that the root diffusates to which the nematode responds are modulated by ethylene signalling (Williamson and Čepulytė, 2017).

The role of amphids in chemoreception has been demonstrated in experiments where exposure of J2 of *M. javanica* and *G. rostochiensis* to antibodies to amphidial secretions blocked the response to host root allelochemicals. However, responses were not permanently blocked as, after a period of between 0.5 and 1.5 h, turnover of sensilla secretions presumably 'unblocked' the amphids (Perry and Maule, 2004; Curtis, 2008).

The orientation of J2 of endoparasitic nematodes to the preferred invasion site, the root tip, is well established but the active factors in root diffusates that constitute these 'local attractants' are unknown. It is possible that specific allelochemicals are responsible but it is also possible that the nematodes orient to an electrical potential gradient at the elongation zone of the root tip. For example, low pH amongst other local attractants might contribute to gather root-knot nematodes at the zone of elongation as *M. hapla* is attracted to, and aggregates between, pH 4.5 and 5.4, which corresponds to the acidic pH gradient encountered in the zone of elongation (Wang *et al.*, 2009a). Although this pH range was attractive to all tested root-knot nematode isolates and species, the level of aggregation depended on the species/strain assessed.

Root diffusates change in response to plant growing conditions. For example, the form of nitrogen supplied to rice plants directly affected the movement and invasion of *M. graminicola*. Rice plants grown in hydroponics containing ammonium nitrate or ammonium chloride repelled J2 and invasion was reduced, whereas plants grown in calcium nitrate attracted J2, resulting in increased invasion (Patil *et al.*, 2013a,b).

The relative importance of electrical and chemical attractants for location of roots and, especially, root tips is unknown. Local attractants from root cap exudates from peas increased motility and attracted J2 of *M. incognita*. However, when in contact with root border cells and mucilage, J2 of *M. incognita* entered a reversible state of immobility. These nematodes are able to recover fully and can find and penetrate root tips within 24 h after recovery from immobility (Curtis *et al.*, 2011). This ability of the root cap to deliver products that temporarily immobilize nematodes may have an important role in protecting the root tip from infection by acting as a natural trap for pathogenic organisms.

Once nematodes are near or at the root surface, local root signals induce the nematode to adopt an exploratory behaviour with rhythmic stylet movements, aggregation, increase of motility, production of pharyngeal secretions and changes to their surface cuticle (Curtis, 2008). Before root invasion, root-knot nematodes are able to perceive signals in root diffusates by changing behaviour and gene expression (Teillet *et al.*, 2013). These events are likely to be important for preparing nematodes for root invasion and possibly allowing their survival inside the host tissue (Curtis *et al.*, 2009, 2011). *Caenorhabditis elegans* also detects environmental signals present in root diffusates that lead to a change in the surface cuticle. Detection of plant signals by chemosensory organs of plant-parasitic nematodes might be important to protect the nematode surface from biological or chemical attack in the rhizosphere (Davies and Curtis, 2011).

Understanding the complexity of the molecular signal exchange and response during attraction and invasion, as well as the nature of root diffusates affecting nematode behaviour, is important. This knowledge can reveal targets for chemical or genetic intervention to control plant-parasitic nematodes that can be used to disorientate nematodes and disrupt host recognition. Once the nematode migratory phase in the rhizosphere is prolonged with a concomitant increased depletion of food reserves, there will be a reduction in their infectivity and an increase in their exposure to natural enemies. As well as identifying and purifying attractants from root diffusates, chemicals from other sources may be important. For example, Tsai *et al.* (2021) successfully purified and identified an attractant for *M. incognita* from flaxseed mucilage. The attractant was identified as rhamnogalacturonan-I (RG-I), a flaxseed cell wall component, and the linkages between rhamnose and L-galactose were shown to be essential for the attraction. Oota *et al.* (2020) screened a chemical library of synthetic compounds and determined that *M. incognita* is likely to be attracted by polyamines to locate the appropriate host plants. Naturally occurring polyamines may form the basis to protect crops from *Meloidogyne* infection.

8.5.4. Response to small molecule signals

Of the wide variety of small molecules known to influence behaviour of invertebrates, ascarosides and sex pheromones have been studied most extensively. Studies on ascarosides have focused on *C. elegans* demonstrating their control of dauer entry and exit, and sex-specific and social behaviours (Ludewig and Schroeder, 2013). Information on plant-parasitic nematodes is limited (see Chapter 7). Most studies on sex pheromones of plant-parasitic nematodes have been on cyst nematodes and have established that sex pheromones have volatile and non-volatile components.

Electrophysiological analysis of the responses of males of the potato cyst nematodes to non-volatile components of female sex pheromones showed that *G. rostochiensis* males exhibited specific mate recognition, whilst *G. pallida* males responded to the sex pheromones from females of both species. However, the homospecific response of *G. pallida* was much greater and may be dominant in the soil environment. The sex pheromone of *G. rostochiensis* is composed of several polar, weakly basic compounds. Chemical characterization of these compounds, and the two polar compounds in the sex pheromone of *H. schachtii*, is required. One male attractant has been identified for a plant-parasitic nematode: vanillic acid for *H. glycines* (Jaffe *et al.*, 1989).

Wang *et al.* (2009a) found that nematodes aggregated when suspended in pluronic gel without roots; when a coverslip was placed on the gel it served as a focus for the aggregation, suggesting that lower oxygen or a volatile attractant is involved in this aggregation behaviour. However, the possibility of a nematode pheromone being involved in this aggregation phenomenon should not be excluded; aggregation pheromones have been reported previously for some animal-parasitic nematodes and *C. elegans*.

8.6. Feeding and Movement within Plant Tissue

Plant-parasitic nematodes have adapted in diverse ways to feed on almost every part of the plant. When the life cycle of a given nematode species is compared on various plant hosts, a strongly characteristic pattern is typically observed in its migration route and feeding behaviour irrespective of the host (Wyss, 2002). Evidence accumulated across numerous taxa in all major taxonomic groups has shown that it is the specific combination of behavioural, anatomical, developmental, biochemical and molecular features that makes the particular feeding pattern and pathology induced by each plant-parasitic species unique. Anatomical variables influencing migration and feeding behaviour include the dimensions of the body, stylet strength, lumen and length, and pharyngeal glands. Developmental variables include the migratory potential of the adult female (vermiform and motile as in *Pratylenchus*, vermiform but sedentary as in *Criconemoides*, or saccate and sedentary as in *Meloidogyne*), the type of reproduction (amphimictic versus parthenogenetic because the enclosure of females within galls may prevent copulation), the number of eggs produced, and the occurrence of metamorphosis during male development. Biochemical and molecular variables include timing and capacity of dorsal and subventral pharyngeal glands to produce enzymes involved in plant cell wall degradation, cytoplasm pre-digestion or feeding tube formation, and elicitors triggering development of nurse cells and syncytia.

The behavioural repertoire of plant-parasitic nematodes moving through and feeding on plant tissue is restricted to a small number of simple activities that are exhibited to varying degrees by all species. However, the manner in which they are timed and coordinated with other events results in behaviours and effects on the plant that are highly consistent within nematode taxa across many plant hosts, yet sufficiently variable across nematode groups to enable utilization of flowers, foliar galls, stems, root hairs, root cortex and various cell layers within the root vascular system as food sources. Simple and consistent patterns of behaviour within a species or taxonomic group in response to an environmental trigger have been termed releaser responses of fixed action patterns (Sukhdeo *et al.*, 2002).

A number of discrete behavioural patterns are exhibited by plant-parasitic nematodes while moving towards, entering and moving through plant tissues (Box 8.5). The sequence and duration of these activities tends to be highly consistent within species; however, across species activities may be elaborated, sequences may be completely different or cyclic, and the duration of each activity may range from a few seconds or minutes to several days. Plant-parasitic nematodes can be placed in at least ten categories based on their migration and feeding patterns within plant tissue and usually the groups are taxonomically consistent. Since stylets evolved independently several times in plant-parasitic

Box 8.5. Behavioural patterns exhibited by plant-parasitic nematodes while moving toward and through plant tissues.

1. Undulatory locomotion through the soil, either toward specific regions of roots or to stem bases and then onto stem surfaces.
2. Lip rubbing of host cell surfaces, often with gentle stylet probing.
3. Orientation of the anterior body perpendicular to the host epidermal wall.
4. Rapid forceful thrusting jabs of the stylet until the cell wall is penetrated, often along a line of holes that merge to form a slit (this behaviour occurs during hatching, epidermis penetration and penetration of cells during intracellular migrations through the root cortex).
5. Migration within the host, either intracellularly or intercellularly, depending on the nematode species; some species move more or less randomly in the cortex and others follow a specific migration route through certain tissues to come to rest at a specific cell layer or region of tissue.
6. Periods of total inactivity.
7. Periods of inactivity with partial protrusion of the stylet and release of pharyngeal gland secretions.
8. Basal or metacorpal bulb pumping and ingestion of cellular constituents.
9. Withdrawal from the feeding site.

nematodes, it is prudent to examine the behaviour of different groups separately. Therefore, first we will examine the trichodorid and then longidorid nematodes. After that, we will compare cyst and root-knot nematodes, followed by discussions of other sedentary root parasites, foliar parasites and insect-vectored plant parasites.

8.6.1. Trichodoridae

The trichodorid nematodes are moderate-sized (<1.5 mm long), strictly epidermal feeders (Fig. 8.8, 1A) that can consume several hundred cells in a few days (Hunt, 1993; Wyss, 2002). The trichodorids exemplify the consistency of nematode feeding patterns within taxa. The trichodorid stylet is a ventrally curved, grooved tooth, termed the onchium, which uniquely permits them to ingest entire cell organelles. They feed on cells quickly (<4 min cell^{-1}), always exhibiting characteristically continuous stylet thrusting while feeding. Moreover, for every cell fed upon, five distinct feeding stages are seen (Box 8.6).

8.6.2. Longidoridae

In contrast to trichodorids, the longidorids are large (up to 12 mm long) with long slender stylets, including species of *Xiphinema* and *Longidorus*; only a few of either group have

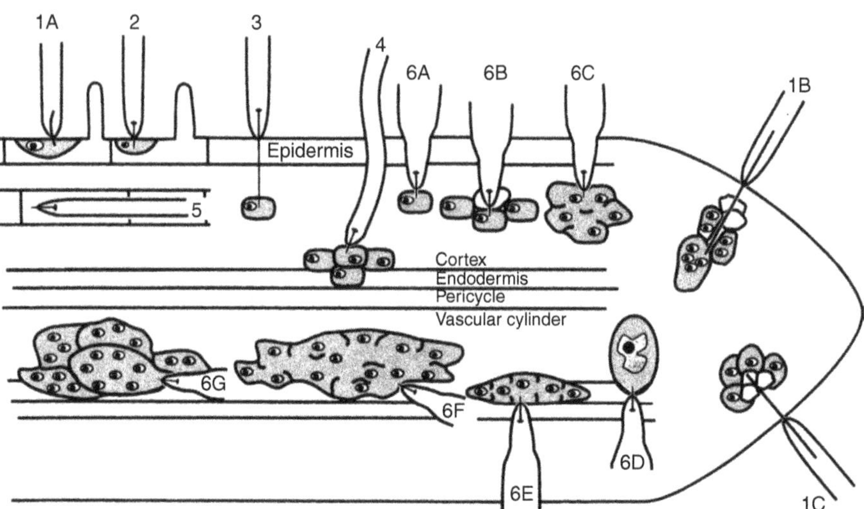

Fig. 8.8. Feeding sites of selected root parasitic nematodes (from Wyss, 1997). (Note: 1A, 1B and 1C are all dorylaimid migratory ectoparasites and all others are tylench nematodes.) 1A: *Trichodorus* spp. 1B: *Xiphinema index*. 1C: *Longidorus elongatus*. 2: *Tylenchorhynchus dubius* (migratory ectoparasite). 3: *Criconemoides xenoplax* (sedentary ectoparasite). 4: *Helicotylenchus* spp. (migratory ecto-endoparasite. 5: *Pratylenchus* spp. (migratory endoparasite). 6A: *Trophotylenchulus obscurus*. 6B: *Tylenchulus semipenetrans*. 6C: *Verutus volvingentis*. 6D: *Cryphodera utahensis*. 6E: *Rotylenchulus reniformis*. 6F: *Heterodera* spp. 6G: *Meloidogyne* spp. (6A–6G are all sedentary endoparasites).

Box 8.6. Feeding stages of trichodorids (Wyss, 2002).

1. Cell wall exploration with lip rubbings.
2. Wall perforation consisting of 1 min of forceful rapid stylet thrusting concurrent with contraction of radial muscles of the basal bulb.
3. Salivation or release of pharyngeal gland secretions through the stoma while thrusting continues and cytoplasm 'liquefies' (the longest phase).
4. Food ingestion (20% of total feeding behaviour).
5. Withdrawal from feeding site.

Box 8.7. Feeding stages of longidorids (Wyss, 2002).

1. Exploration via lip rubbings.
2. Insertion of stylet 2–3 cells deep via quick repeated jabs.
3. Progressive insertion of stylet into deeper cells, withdrawing food from each for several minutes, followed by withdrawal of stylet and reinsertion at a new point, with the same process then repeated.
4. Long quiescent period with no bulb pumping invariably followed by a prolonged period of food removal lasting up to 30 min.
5. Continued feeding leading to binucleate cells surrounding necrotic spaces that transform into multinucleate feeding cells (only in susceptible plants).

been studied extensively. Several distinct feeding patterns have been noted in longidorids (Fig. 8.8, 1B and 1C) but they are generally similar in many ways and all are markedly different from that in trichodorids. Several *Xiphinema* spp. feed on differentiated vascular tissue but those studied the most feed on root tips where they cause galls that eventually attract other nematodes. Root tip feeding has been carefully studied for *X. index* and *X. diversicaudatum* on *Vitis* spp. (grape) and *Ficus carica* (fig), where five feeding stages are evident (Box 8.7).

In *Longidorus* species, which only feed on root tips, *Longidorus caespiticola*, *L. elongatus* and *L. leptocephalus* have been studied *in vitro* on perennial ryegrass (*Lolium perenne*). Five feeding phases are noted: (i) as in *X. index*; (ii) as in *X. index* except rapid thrusting continues until the stylet is fully protracted; (iii) always a period of inactivity up to 1 h when pharyngeal secretions are thought to be released; (iv) continuous food ingestion for several hours (i.e. longer than in *X. index* and up to 40 cell volumes may be removed); and (v) continued feeding leading to galling and cell wall dissolution; however, nuclear divisions without cytokinesis as observed in *X. index* are generally not characteristic.

8.6.3. General aspects of tylench tissue migrations and feeding behaviour

Tylench nematodes can be divided into those that feed on roots and those that feed on above-ground plant parts. All nematodes in both groups are vermiform as juveniles and those feeding above ground generally are vermiform throughout life. Root parasites may be vermiform or saccate as adults (Fig. 8.8). Gravid females as a general rule are sluggish and, during the physiological events of moulting, preceding ecdysis (the physical shedding of the cuticular remnant), all nematodes exhibit quiescent behaviour termed moulting lethargus. Aside from these periods, most plant-parasitic nematodes that are vermiform as adults are motile throughout the life cycle, with the arguable exception of Paratylenchinae and Criconematinae (Siddiqi, 2000; Wyss, 2002).

8.6.4. Tylench root parasites with vermiform adults

The browsing ectoparasite *Tylenchorhynchus dubius* feeds on root hairs and epidermals cells (Fig. 8.8, 2). Feeding lasts 10–12 min cell^{-1} and is marked by four phases: (i) lip rubbings with gentle stylet probings; (ii) vigorous stylet thrusts; (iii) salivation; and (iv) ingestion (85% of total, or about 9 min). Thus, the period of ingestion for *T. dubius* lasts about ten times as long as in *Trichodorus* spp., perhaps because the former feeds through a narrow stylet lumen, whereas the latter feeds via the buccal cavity. The ecto-endoparasites (e.g. *Helicotylenchus* spp., *Hoplolaimus* spp. and *Rotylenchus* spp.) move partly or, in some cases, entirely into the cortex (Fig. 8.8), feeding on individual cells for long periods. By comparison, migratory endoparasites (e.g. *Pratylenchus* spp., *Hirschmanniella* spp. and *Radopholus* spp.) feed on root hairs when in the J2 and third-stage juvenile (J3) stages and later stages continuously migrate intracellularly in the cortex as they develop (Fig. 8.8), most often parallel to the root axis, leaving extensive tracks of destroyed cells and necrotic tissue. Cortical migrations are periodically interrupted by quiescent periods.

8.6.5. Tylench root parasites with saccate adults

By far the most commonly encountered plant-parasitic nematodes with sedentary, saccate adults are the root-knot nematodes (see Chapter 3) and the cyst nematodes (see Chapter 4). Our knowledge regarding feeding behaviour of root-knot nematodes comes mainly from studies on *M. incognita* and of cyst nematodes from studies on *G. rostochiensis*, *H. schachtii* and *H. glycines*; thus, generalizations should be made with caution.

In both cyst and root-knot nematodes the J2 is the motile, infective stage, which after invading the plant moves to the vascular region where it elicits highly specialized plant cell responses resulting in the development of feeding sites comprising enlarged hyperactive cells upon which the nematode permanently feeds (Fig. 8.8). Behaviourally, the two groups differ in the migration root taken by J2 through root tissue, the type of feeding site induced and, most importantly perhaps, in the structure and function of the stylet and pharyngeal glands utilized to enable root migrations and feeding cell establishment (see Chapter 9). The ability to induce feeding sites has evolved independently in cyst and root-knot nematodes. Thus, the behaviours associated with feeding site induction will also have evolved independently, presumably as a modification of the ancestral migratory endoparasitic behaviour patterns.

Cyst nematode J2 have a relatively robust stylet that allows them to enter the root, mainly in the zone of elongation, via repeated forceful jabs of highly coordinated stylet thrusts that produce a line of merging holes to form a slit, thereby cutting through the tough epidermal wall and subsequently other cell walls that they encounter as they move intracellularly to the vascular tissue. Here they initiate feeding on cells that, in *Arabidopsis thaliana*, have been confirmed consistently to be cambial or procambial. Root invasion behaviour for the cyst nematode *G. pallida* has been studied using nematodes restrained in a microfluidic device. Stylet thrusting is robustly activated by the neurotransmitter serotonin (Hu *et al.*, 2014) and inhibited by a root secondary metabolite reserpine (Crisford *et al.*, 2020).

Root-knot nematode J2 have a more delicate stylet and enter the root closer to the meristematic region. Root penetration involves continuous head rubbings and stylet movement, including occasional stylet protrusions followed by metacorpal bulb pumpings of a few seconds, indicating cell wall degradation enzymes are probably involved. The epidermal cells eventually soften and lyse. Once inside, the J2 initially migrate acropetally and intercellularly (in contrast to cyst nematodes, which migrate intracellularly) through the relatively soft tissues of the cortical parenchyma until reaching the apical meristematic region, where they turn around and enter the newly formed vascular cylinder, moving basipetally several body lengths before coming to rest and initiate feeding. There is biochemical and molecular evidence for cell wall degrading enzymes in the subventral gland secretions of both groups (see Chapter 9).

The destructive behaviour pattern of the cyst nematode J2 during migration changes into a subtle exploratory pattern upon reaching the vascular cylinder. A single cell, termed the initial syncytial cell (ISC) is then perforated by careful stylet thrusts, after which the stylet tip remains protruded a few micrometres deep for 6–8 h (the so-called preparation period); no obvious changes in the ISC are observed nor are there any metacorpal bulb pumpings. However, in addition to the gradual increase in the dorsal and decrease in the subventral glands, there are a few sudden shrinkages of the body, when the J2 defecates. Subsequently, the ISC enlarges and feeding begins, which occurs in repeated cycles, each consisting of three distinct phases: (i) nutrients are withdrawn by continuous rapid pumping of the metacorpal bulb with no forward flow of pharyngeal gland secretory granules; (ii) stylet retraction and reinsertion; and (iii) continuous forward movement of secretory granules.

When a J2 of root-knot nematodes reaches the cambial region where the eventual multinucleate nurse cells (called giant cells) form, forward motion stops but the head continues to move in all directions, exhibiting the same stylet tip protrusions and metacorpal bulb pumping as during migration; however, these periods increase until it is apparent the J2 are extracting food from the enlarging cells surrounding the anterior end.

Additional tylench nematodes with saccate adults are found in the families Hoplolaimidae and Tylenchulidae. Knowledge regarding tissue migrations in these nematodes comes primarily from histopathological observations rather than direct observations of the infection process *in vitro*, and much more information is available for some than others due to their greater economic importance. In species of *Rotylenchulus* and *Tylenchulus* the adult vermiform female is the infective stage and penetrates the root cortex of a largely differentiated zone, eliciting formation of a feeding cell and (or) syncytium without migrating longitudinally through the root. Thus, the initial and final orientation of the nematode is perpendicular to the root axis. The male of *R. reniformis* has a weak stylet and does not feed. However, the vermiform female of *R. reniformis*, which has a robust stylet similar to that of cyst nematode J2, penetrates the root cortex intracellularly, perpendicular to the root axis in a largely differentiated zone, coming to rest and feed, usually on an endodermal cell, less commonly on a pericycle or cortical cell on which it feeds permanently. A syncytium interconnecting the feeding cell with a curved sheet of hyperactive pericycle cells extending part way around the vascular cylinder through partial dissolution of common radial walls is characteristic in many plant species. In *T. semipenetrans*, the vermiform female also is infective and infection is similar to that in

Rotylenchulus, but feeding almost invariably involves cortical cells. *Nacobbus aberrans* (see Chapter 3) differs in that the J2 and J3 feed as migratory endoparasites, much like *Pratylenchus* spp. The adult females, however, are saccate as in *Rotylenchulus* and *Tylenchulus* and establish a syncytium where they feed as sedentary parasites in the cortex.

8.6.6. Above-ground parasites of plants

Nematodes parasitizing aerial plant parts occur in the infraorder Tylenchomorpha, and include species in the Sphaerularioidea and Aphelenchoidea. Within Sphaerularioidea, aerial parasites are common among the 150 species in the family Anguinidae, and particularly among the 100 or so species of *Anguina*, *Subanguina* and *Ditylenchus*. Within Aphelenchoidea, aerial parasites of plants include at least three species of *Aphelenchoides*, namely *Aphelenchoides besseyi* (white tip of rice nematode), *A. fragariae* (strawberry crimp nematode) and *A. ritzemabosi* (chrysanthemum nematode).

Aphelenchoidea also includes at least two insect-vectored wood inhabitants of major economic importance that have life cycles completely different from other aerial plant parasites in Tylenchomorpha. Both feed facultatively on fungi in the plant host and elicit the production by the tree of tyloses and autotoxins that contribute to tree death. These nematodes are *Bursaphelenchus cocophilus* (coconut red ring disease nematode) and *B. xylophilus* (pine wilt nematode).

Bursaphelenchus cocophilus is vectored endophoretically by the palm weevil *Rhynchophorus palmarum* (see Chapter 5). The J3 of *B. cocophilus* enter weevil grubs burrowing through palm tissue through the spiracles and mouth and migrate to the haemocoel, where they remain without developing until the insect undergoes metamorphosis and emerges as an adult female weevil. The J3 then concentrate in the ovipositor and when the weevil moves to another tree to oviposit they are transferred with the eggs. The nematodes migrate intercellularly through all the parenchymatous tissue of the palm, but in the trunk are largely restricted to the 2–4 cm wide 'red ring', which is about 3–5 cm beneath the trunk surface (Hunt, 1993).

Bursaphelenchus xylophilus is vectored ectophoretically primarily on two longhorn beetles, *Monochamus alternatus* in Japan and *M. carolinensis* in North America, although many other related beetles can carry the nematode. The J4 are carried from tree to tree under the elytra and in the tracheae. In susceptible tree species, juveniles enter feeding wounds made by the insect and gain access to the resin canals of the shoots where they feed on epithelial cells. Feeding inside the resin canals causes a pronounced decrease in oleoresin production allowing the nematodes to migrate throughout the dying tree and reproduce. Dying trees in turn attract insect vectors. Experiments *in vitro* showed the J4 to be attracted to CO_2 from the vector and to terpenes predominant in the vector's diet, especially β-myrcene. *Bursaphelenchus xylophilus* can nictate (standing vertically out from the substrate) and jump, behaviours generally not found among plant-parasitic nematodes. With some entomopathogenic nematodes, nictation often is stimulated by vibration, CO_2 or other cues from the insect.

The infective stages of plant-parasitic nematodes that attack aerial plant parts move faster than infective stages of root parasites. Body undulations in J2 of *D. phyllobius*, for example, exceed 1 wave s^{-1} and direct comparisons showed *D. phyllobius* and *D. dipsaci* J4 to disperse on agar 6–20 times as fast as infective stages of *T. semipenetrans* and

R. reniformis (Robinson, 2004). Aphelenchoidids in general are also considered highly active nematodes, and *A. besseyi* along with *D. angustus* are considered likely to be the best swimmers among plant-parasitic nematodes (Hunt, 1993). Such fast movement undoubtedly increases the chances that infection sites on aerial parts can be reached before moisture films required for migration dry.

Aphelenchoides besseyi, *A. fragariae* and *A. ritzemabosi* are usually described to feed both ecto- and endoparasitically on buds, stems, leaves and inflorescences. There are at least four reports of direct observations of entry and exit of *A. fragariae* and *A. ritzemabosi* through leaf stomata, and CO_2 has been demonstrated to lure *A. fragariae* through stomata-size pores in plastic film. However, it is clear that many nematodes commonly penetrate soft regions of the foliar epidermis directly, especially in the tender folds between embryonic leaves and floral structures within buds. For example, in a comparison of *D. dipsaci* and *A. fragariae* on alfalfa seedlings in pots, *D. dipsaci* penetrated through the epidermis at the base of cotyledons forming cavities in cortical parenchyma within 12 h of inoculation, whereas *A. fragariae* fed primarily ectoparasitically. *Ditylenchus phyllobius* similarly penetrates the epidermis of *Solanum elaeagnifolium* directly, but only in the leaf folds in buds above ground or on shoots produced by rhizomes as they grow up through the soil. This nematode induces massively galled leaves weighing up to 10 g, but these develop from leaves infected when at the embryonic stage and infection appears never to occur after leaves have expanded. Direct observations of nematode feeding within foliar galls are lacking but a number of histological studies have shown induction of putative nurse cells and associated hyperplastic tissue similar to that observed in root parasites and it can be surmised that the types of feeding behaviours directly observed for root parasites *in vitro* are probably shared by these foliar feeding species.

8.7. Other Types of Movement and Behaviour

Adverse environmental conditions engender behavioural responses of nematodes, frequently associated with survival and entry to a dormant state (see Chapter 7). For example, coiling and clumping by some species of nematodes, such as *D. dipsaci*, occur in response to gradual removal of water. Coiling and clumping both reduce the surface area of the nematode that is exposed to drying conditions and reduces the rate of water loss, a slow rate of dehydration being a critical factor for survival. In *Anguina amsinckiae* and *A. tritici*, aggregations occur within galls. Within the galls induced by *A. amsinckiae* are hundreds of desiccated adults and juveniles of all stages, many of which are coiled. By contrast, the galls induced by *A. tritici* contain tightly packed aggregates of J2 only, each of which remains uncoiled when dry. Mass movement or swarming found, for example, in *Aphelenchus avenae* is probably a behavioural response to lack of food or toxic products from decaying hosts.

Several hundred species within the Criconematidae have backwardly angled cuticular outgrowths, such as scales or spines, or retrorse annules in juvenile stages, which probably provide purchase for a peculiar form of travel. The criconematids are noted for moving forwards via sluggish, anteriorly propagated peristaltic contractions, in contrast to the backwardly propagated undulations in other nematodes (Siddiqi, 2000). This behaviour has received little study relative to its probable ecological importance (Burr and Robinson, 2004).

9 Molecular Aspects of Plant–Nematode Interactions*

GODELIEVE GHEYSEN[1]** AND JOHN T. JONES[2]

[1]*Department of Biotechnology, Ghent University, Belgium;* [2]*Cell and Molecular Sciences Department, James Hutton Institute, UK and School of Biology, University of St Andrews, UK*

9.1. Nematode Parasitism of Plants

The relationships between plant-parasitic nematodes and their hosts are diverse. Plant-parasitic nematodes may be sedentary or browsing and can be ectoparasites or endoparasites. For some nematodes, including most migratory ectoparasites, plants simply provide an ephemeral food source and the interaction between the nematode and the plant is limited.

*A revision of Gheysen, G. and Jones, J.T. (2013) Molecular aspects of plant–nematode interactions. In: Perry, R.N. and Moens, M. (eds) *Plant Nematology*, 2nd edn. CAB International, Wallingford, UK.
**Corresponding author: godelieve.gheysen@UGent.be

  Plant Nematology, 3rd edn (eds R.N. Perry, M. Moens and J.T. Jones)
DOI: 10.1079/9781800622456.0009

For other nematodes the interactions are far more complex and long-lasting. The most economically damaging plant-parasitic nematodes are biotrophic and induce changes in the roots of their hosts to form feeding sites that serve to supply the nematode with a rich and continuous food source.

Nematodes are well equipped for parasitism of plants. Plant parasitism has arisen independently at least three times in the Nematoda (see Chapter 2, Fig. 2.11) but despite the fact that plant parasites are not always directly related in phylogenetic terms, some structural features are always present. All plant-parasitic nematodes have a strong, needle-like structure that is used to pierce plant cells (Fig. 9.1). This structure (stylet, onchiostyle or odontostylet in the various taxonomic groups) is used to break or disrupt host tissues, to introduce nematode secretions into plant tissues (see below) and to remove plant cell contents during feeding. Stylets vary in shape and size according to the feeding strategy of the nematode; for example, nematodes such as *Trichodorus* that feed on epidermal cells have short onchiostyles, whereas those such as *Xiphinema* or *Longidorus* have considerably longer odontostylets (Fig. 9.1) and can feed on cells deeper within the plant. In many plant-parasitic nematodes the secretions of the pharyngeal gland cells play important roles in the host–parasite interaction. The active components of these secretions (which are mostly – but not always – proteins) that manipulate the host to the benefit of the parasite are referred to as effectors. Plant-parasitic nematodes have two sets of pharyngeal gland cells, dorsal and subventral. The precise number and arrangement of each cell type varies. In Tylenchoidea (including the major sedentary endoparasites) there are two subventral gland cells and one dorsal gland cell (Fig. 9.2). The subventral gland cells are large and full of secretory granules in the invasive second-stage juvenile (J2) after hatching and during the earliest stages of parasitism. These cells become smaller in the third- and fourth-stage juveniles and adult females. By contrast, the dorsal gland cell increases in size throughout the life cycle. These structural studies are mirrored by studies using antisera that recognize the various gland cells and by genomic and transcriptomic studies. Antibodies that bind to the subventral gland cells give a strong signal in J2 and little or no signal in adult females. By contrast, antibodies that recognize the dorsal gland cell give a weak signal in J2 but a much stronger signal in adult females. Analysis of expression patterns of potato cyst nematode genes encoding effectors show that these fall into two broad categories – one cluster of genes that includes known subventral gland cell expressed genes is expressed mainly in the J2 while another cluster that includes known dorsal gland cell genes is expressed at later stages. These observations show that the gland

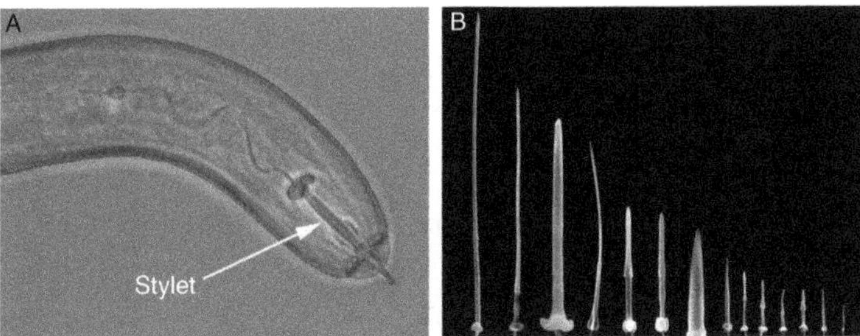

Fig. 9.1. Stylets of plant-parasitic nematodes. A: Stylet at the anterior tip of a second-stage juvenile of the beet cyst nematode *Heterodera schachtii*. B: Scanning electron micrograph of stylets dissected from various plant-parasitic nematodes.

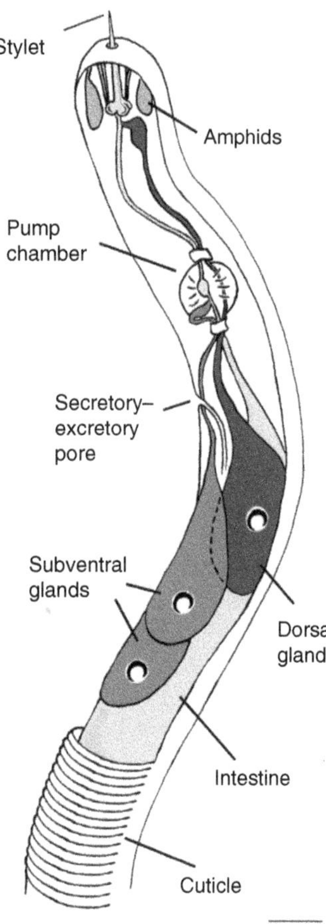

Fig. 9.2. Sketch of the anterior end of a tylench plant-parasitic nematode showing the relative positions of various nematode structures including the sub-ventral and dorsal pharyngeal gland cells.

cell products are developmentally regulated and suggest that the products of the subventral gland cells are important in the early stages of parasitism, during invasion and perhaps during feeding site induction, while the products of the dorsal gland cell play a role later in the parasitic process, probably in feeding site development or maintenance.

Nematode feeding sites (NFS) are incredibly diverse and can be induced in a variety of root tissues (see Fig. 1.1) but have some conserved features (Wyss, 2002). All have structural characteristics of metabolically active tissues including cytoplasm highly enriched in subcellular organelles. Many show signs of DNA replication and may have enlarged or multiple nuclei. The simplest feeding sites are composed of single modified cells. *Trophotylenchulus* feeds from a single cell that appears metabolically active and contains one enlarged nucleus. Other nematodes, such as *Tylenchulus*, feed on clusters of such cells. *Cryphodera* feeds from a single greatly enlarged cell that contains one enlarged nucleus. More complex feeding sites fall into two categories: syncytia and giant cells. **Syncytia** are large multinucleate cells formed by breakdown of plant cell walls and fusion of adjacent protoplasts. The most complex syncytia are induced by cyst nematodes (*Heterodera* and *Globodera*). **Giant cells** of root-knot nematodes are formed through repeated rounds of nuclear division and cell growth in the absence of cytokinesis.

The nematodes that induce these complex feeding sites cause severe damage to plants and are the most economically important of the plant-parasitic nematodes. These nematodes will form the basis for most of the rest of this chapter.

9.2. Invasion and Migration

Endoparasitic nematodes need to invade host tissues and migrate to where they feed. Plants have evolved complex defences to protect themselves against attack by nematodes and other pathogens (see Sections 9.3 and 9.8). One of the major physical barriers encountered by any invading pathogen is the plant cell wall. Each plant cell is surrounded by a cell wall and the middle lamella forms the interface between adjacent cell walls (Fig. 9.3). The plant cell wall is a complex but highly organized composite of polysaccharides and proteins. Cellulose, formed by β1–4-linked chains of glucan sugars, is the most abundant polymer in the plant cell wall. Cellulose exists in the cell wall in the form of microfibrils that are formed from dozens of individual polysaccharide chains, hydrogen bonded to one another along their length. These microfibrils are crosslinked to one another by hydrogen-bonded crosslinking glycans, the most important of which are the xyloglucans and glucuronoarabinoxylans. This network of cellulose and crosslinking glycans is embedded within a matrix formed from pectins. Pectins serve a variety of other roles in the plant cell wall including control of cell wall porosity and cell–cell adhesion at the interface with the middle lamella and are formed from highly branched chains of polysaccharides rich in D-galacturonic acid. Although the carbohydrates described above form the major part of the plant cell wall, two major classes of structural protein are also present. Hydroxyproline-rich glycoproteins such as extensin and glycine-rich proteins may form networks in the plant cell wall. The interrelationships of the cell wall components are summarized in

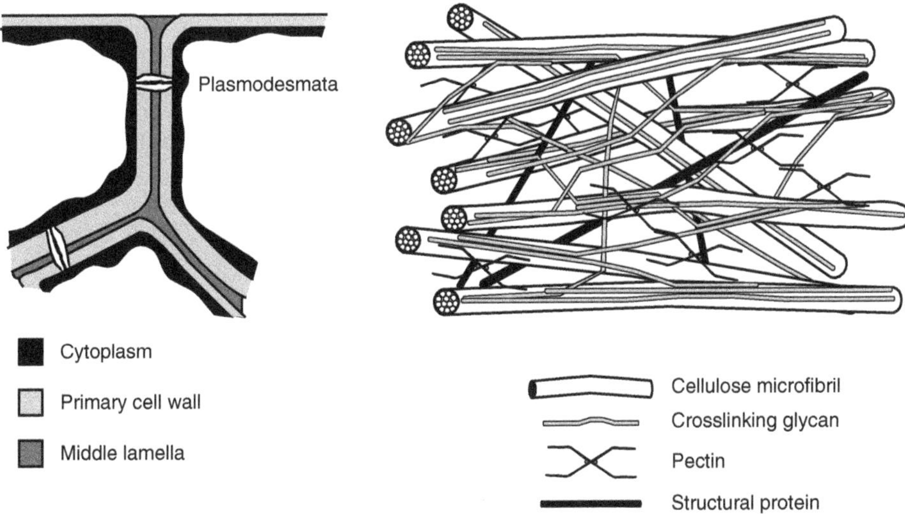

Fig. 9.3. Schematic diagrams of the plant cell wall (left) and of the molecules that make up the primary cell wall (right).

Fig. 9.3. The plant cell wall presents a formidable barrier and any invasive pathogen needs to have mechanisms that allow it to overcome this obstacle.

Plant-parasitic nematodes, particularly the endoparasites, migrate for considerable distances within the plant before settling to feed. Intriguingly, there are differences in the way in which various nematodes migrate within the plant. Cyst nematodes enter the root, mainly at the zone of elongation, and migrate destructively through cortical cells until they reach a position near the differentiating vascular cylinder. Root-knot nematodes enter the root and migrate intercellularly, first heading towards the root tip until they reach the root apex. They then turn around and migrate (again intercellularly) back up the root until they reach a site near the vascular cylinder suitable for feeding site induction. It is possible that this behaviour of migrating between plant cells, which prevents extensive damage to the host during migration, is partly responsible for the huge host range of root-knot nematodes, which encompasses many thousands of flowering plant species. Both the cyst and root-knot nematodes use the stylet during invasion and migration as a cutting tool. This physical disruption of plant tissues is thought to be more important for cyst nematodes (which are equipped with a strong stylet) than root-knot nematodes.

Detailed examination of the behaviour patterns of nematodes migrating through plant roots led to the suggestion some time ago that cell wall degrading enzymes might be used by plant-parasitic nematodes during migration. For example, root-knot nematodes within roots show stylet movements and protrusions that are followed by pumping of the metacorpal bulb. This prediction was proved correct when the first nematode effector was identified as a cellulase (see Box 9.1). Cellulases break the β1–4 links within the cellulose polysaccharide chain and a family of these proteins, which has probably arisen by gene duplication, is now known to be present within plant-parasitic nematodes. The genes that encode these proteins are expressed specifically in the subventral gland cells and the cellulase proteins are secreted into the plant from the stylet during migration. Subsequent studies have shown that this is not the only plant cell wall degrading enzyme produced by plant-parasitic nematodes. Pectate lyases, xylanases, polygalacturonases, arabinases and arabinogalactan galactosidases have been reported from various phylogenetic groups. It has also been shown that plant-parasitic nematodes secrete expansins during migration. Expansins are thought to disrupt the non-covalent interactions (hydrogen bonds) between cellulose chains and between cellulose and crosslinking glycans. The action of expansins greatly enhances the accessibility of the other polysaccharide chains to the activity of the cellulases and pectate lyases. Analysis of the genome of the root-knot nematode, *Meloidogyne incognita*, shows that six different gene families representing a total of up to 60 proteins are present. A similar range of cell wall degrading enzymes are also encoded in cyst nematode genomes. Plant-parasitic nematodes therefore secrete a cocktail of enzymes that degrade the plant cell wall and allow migration to where the feeding site is induced (Kikuchi *et al.*, 2017).

The presence of these cell wall degrading enzymes in nematodes is remarkable for several reasons. Plant cell wall degrading enzymes had previously only been shown to be produced by plants themselves, and by pathogenic bacteria and fungi, with no reports of their production by animals. Almost all other animals use endosymbiotic bacteria for the production of the enzymes and, before 1998, when the first nematode cellulases were described, it was believed that no animals contained endogenous genes of this type. The nematode genes, when compared to other sequences in databases, are most similar to cell wall degrading enzymes from bacteria. However, several convincing lines of evidence

Box 9.1. How do we identify nematode effector genes?

Plant-parasitic nematodes are extremely awkward experimental animals. They are inconveniently small, difficult to produce in large numbers and many of the most interesting parasitic stages live deep within plant roots. However, nematode genes and proteins have been identified using several different approaches.

Perhaps the most direct approach has been to collect nematode secretions and analyse the proteins present. This approach requires many millions of nematodes and was facilitated by the discovery that certain chemicals and plant root diffusates stimulate release of secreted materials. Once secreted proteins have been collected, they can be analysed directly using two-dimensional gel electrophoresis. In this technique, proteins are first separated on an immobilized pH gradient on the basis of their isoelectric point. The separated proteins are then run on a large gel, which separates the proteins on the basis of their size. The result is a large rectangular gel in which individual proteins appear as spots. Spots can be excised from the gel and short sequence tags can be obtained using mass spectrometry. This allows protein identification. Secreted proteins from *M. incognita* have been collected and chromatographically separated before being directly analysed by mass spectrometry (Bellafiore *et al.*, 2008). As with all protein-based approaches, this method requires extensive genome information to give the best chance of identifying the genes encoding the proteins found on the gels.

Antibodies raised against secreted proteins can also be used to identify nematode parasitism genes, either by screening of a complementary DNA (cDNA) library or by purifying the protein from nematode homogenate. This approach was used for identification of nematode cellulase encoding genes.

The most common method for identifying nematode effectors is now through direct analysis of genome or transcriptome sequences. If an annotated dataset is available, candidate effectors can be identified using a combination of bioinformatic and lab-based approaches. We know that effectors will be secreted proteins that can be identified by the presence of a predicted signal peptide at the N-terminus and their upregulation in life stages associated with parasitism. Candidates can then be studied in the laboratory using *in situ* hybridization to confirm expression in the gland cells.

A more direct approach is to isolate RNA from excised nematode gland cells and to sequence this directly. This is technically challenging but will provide a far greater enrichment of effector sequences than when analysing a whole transcriptome or genome.

Functional analysis of nematode genes, including effectors, has been greatly facilitated by the development of RNA interference (RNAi). Exposure of an organism to double-stranded RNA (dsRNA) generated from a gene of interest causes rapid degradation of the endogenous messenger RNA (mRNA) from which the dsRNA was made. This causes a drop in the level of the corresponding protein. The effects of the removal of the protein on the normal biology of the nematode can then be observed, allowing the function of the original gene product to be investigated.

Most effector proteins are likely to interact with host proteins when secreted into plants. These host target proteins can be identified using yeast two-hybrid screening of plant cDNA libraries. The role of the identified host proteins can subsequently be investigated in the plant. For more information see Eves-van den Akker *et al.* (2021).

show that these are endogenous nematode genes. The genes are expressed solely in the subventral gland cells, which do not contain symbiotic bacteria. Analysis of genomic sequences of the genes shows that they contain introns, which are not present in bacterial DNA. In addition, these introns have sequence features found in other nematode introns and are spliced in the way that is expected for nematode introns. Since these genes are not present in other (non-plant-parasitic) nematodes or almost all other invertebrate animals, the most logical explanation for their presence in plant-parasitic nematodes is that they have been acquired by horizontal gene transfer from bacteria.

The acquisition of genes encoding plant cell wall degrading enzymes by an ancestor of the plant-parasitic nematode species present today was a key event in the evolution of plant parasitism. Many of the Tylenchoidea plant parasites investigated to date have these genes present, including migratory endoparasites such as *Pratylenchus* and *Radopholus*. At a broader level, horizontal gene transfer may have driven the evolution of plant parasitism by nematodes on more than one occasion. Another plant-parasitic nematode, *Bursaphelenchus xylophilus*, a member of the Aphelenchoidea rather than the Tylenchoidea, has been shown to contain cellulase and pectate lyase (Kikuchi *et al.*, 2011). However, in this nematode the cellulase appears to have been acquired from fungi. In addition, a cellulase from an entirely unrelated gene family has been identified in *Xiphinema index* (Danchin *et al.*, 2017). These studies support the concept of multiple independent horizontal gene transfer events from a range of sources.

9.3. Defence Responses of the Plant

Plants recognize and react to the presence of pathogens by activating various defence responses. Defence responses include the production of toxic oxygen radicals and systemic signalling compounds as well the activation of defence genes that lead to the production of structural barriers or other toxins designed to harm intruding pathogens (see below). The plant defence (immune) system is thought to consist of two layers of responses and is summarized in both functional and evolutionary terms by the 'zigzag' model (Jones and Dangl, 2006; Fig. 9.4). Plants detect highly conserved pathogen molecules (pathogen-associated molecular patterns (PAMPs)) through cell surface pattern recognition receptors (PRRs) and trigger pattern-triggered immunity (PTI), also referred to as basal defences. PAMPs are highly conserved, specific to the pathogen and are essential for pathogen survival. PTI is relatively durable as the pathogens are not readily able to evolve changes to PAMPs. All successful biotrophic pathogens need to be able to suppress PTI in order to infect plants. This is achieved using effectors that target the signalling pathways invoked during PTI. However, plants have subsequently evolved a further layer of defences known as effector-triggered immunity (ETI). ETI is mediated by the products of resistance genes and these recognize the effectors or, more frequently, the changes to host metabolism induced by the effectors. ETI is mediated by a strong localized cell death response (the hypersensitive response), which targets the tissues on which the pathogen is trying to feed. Further details of resistance genes and the effectors that they recognize are provided in Section 9.8.

Changes in gene expression correlated with defence responses have been studied in several plant–nematode interactions. Just 12 h after inoculation of tomato roots with root-knot nematodes, general plant defence genes are upregulated (Williamson and Hussey, 1996). Most of these genes are induced in both the compatible and the incompatible interaction, albeit with differences in levels and timing. Genes encoding direct defence proteins that are activated include peroxidase, chitinase, lipoxygenase, extensin and proteinase inhibitors. Genes that encode enzymes in pathways that result in synthesis of other

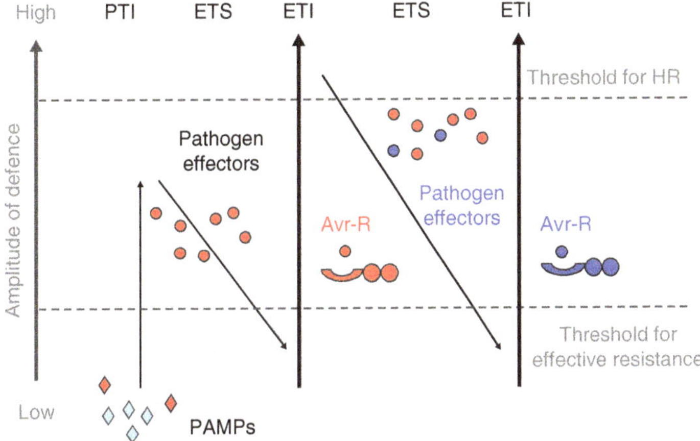

Fig. 9.4. The zigzag model (Jones and Dangl, 2006). Conserved pathogen molecules (pathogen-associated molecular patterns (PAMPs)) are detected by host cell surface pattern recognition receptors (PRRs), which activate pattern-triggered immunity (PTI). Adapted pathogens deliver effector proteins to suppress these defences, thus inducing effector-triggered susceptibility (ETS). To counter pathogen-induced PTI suppression, plants possess a second layer of immune receptors (resistance genes - *R*) to detect the presence of effectors, leading to effector-triggered immunity (ETI). Effectors that are recognised by *R*-genes are termed avirulence (*Avr*) genes. An evolutionary battle then follows with pathogens evolving new effectors that evade detection and plants evolving new resistance genes. HR, hypersensitive response.

defence compounds are also activated during plant defence responses. For example, genes encoding enzymes that lead to synthesis of phytoalexins (such as glyceollin in soybean), or deposition of physical barriers such as callose and lignin, are induced in the early phases of the nematode infection process.

The application of high-throughput molecular analyses (micro-arrays, RNA sequencing) to plant–nematode interactions has enabled a more comprehensive overview of the plant response to nematode invasion, indicating that the molecular defence response to nematode infection is very similar to the response to many other pathogens. These responses include activation of genes encoding pathogenesis-related (PR) proteins and defence signalling compounds, as well as general stress response genes and genes associated with the HR. However, specific plant defence pathways may be locally suppressed in the feeding sites, e.g. genes for the biosynthesis of the defence hormones salicylic acid (SA) and jasmonic acid (JA) are downregulated in giant cells induced by root-knot nematodes in rice (Ji *et al.*, 2013).

In addition to these local responses at the infection site, defence responses are often systemically induced in remote plant tissues after pathogen attack. As a result of this systemically acquired resistance (SAR), the plant is better prepared for attack by nematodes on other roots. Nevertheless, in some cases plant defence in systemic tissues is suppressed, possibly causing the plant to be more susceptible to leaf pathogens after nematode attack. During *H. schachtii* infection of *Arabidopsis*, the expression of *PR-1, PR-2* and *PR-5*, considered to be markers for SA-dependent SAR, was induced in roots and leaves. The expression of *PR-3* and *PR-4*, which are used as markers for JA-dependent SAR, was not altered in roots, but *PR-3* was induced, whereas *PR-4* was downregulated in the leaves of *H. schachtii*-infected plants (Hamamouch *et al.*, 2011). However, all tested *PR* genes (*PR-1* to *PR-5*) were downregulated in the leaves of *M. incognita*-infected plants, suggesting the suppression of both SA- and JA-dependent SAR. Similarly, many defence-related genes were suppressed in shoots of root-knot nematode infected rice at 3 days after infection, but not in shoots of rice infected by migratory nematodes (Kyndt *et al.*, 2012). These results suggest that root-knot nematodes are not only specialized in avoiding plant defence responses (see above) but actively suppress the host defence.

9.4. Suppression of Host Defences and Protection from Host Responses

Plant-parasitic nematodes, like other plant pathogens, have evolved a series of physical and biochemical adaptations that help them to either avoid eliciting a host response, suppress host defences or reduce the toxic effects of any plant defence response.

9.4.1. Suppression of host defences

All the evidence obtained to date suggests that the principles of the zigzag model (Section 9.3) apply to the interactions between biotrophic nematodes and their hosts. The first PAMP from plant-parasitic nematodes has been identified as an ascaroside signalling molecule, ascr18. Ascr18 induces defence responses in the host indicative of activation of PTI including activation of kinase signalling pathways. As seen for other PAMPs, exposure of plants to ascr18 increases resistance to diverse pathogens suggesting activation of general defence responses. There is also some evidence that nematodes produce other PAMPs – a heat-sensitive, proteinaceous molecule that is secreted from nematodes has been detected that activates PTI, although the nature of this 'Nemawater' remains unknown. The first PRR that detects a nematode PAMP, a leucine-rich receptor-like kinase called NILR1, has also been identified (Mendy *et al.*, 2017). NILR1 is essential for PTI responses mounted against nematodes and, like many other PRRs, requires a co-receptor (BAK1) for its function.

There are numerous examples of plant nematode effectors that suppress PTI or ETI. The root-knot nematode, *M. incognita*, secretes a calreticulin into the apoplast of host plants that suppresses PTI, possibly by preventing calcium influx from here into host cells (Jaouannet *et al.*, 2013). A modified ubiquitin-like effector protein from *G. rostochiensis*, which carries a 12 amino acid extension at the C-terminus, suppresses PTI induced by a bacterial PAMP, although no information on the mechanism by which this is achieved is available (Chen *et al.*, 2013). Similarly, plant nematode proteins have been identified that suppress ETI. Most notably, several members of a large family of proteins, the SPRYSECs, from the potato cyst nematodes *G. rostochiensis* and *G. pallida* suppress ETI (Postma *et al.*, 2012; Mei *et al.*, 2015).

Venom allergen-like proteins (VAPs) are a family of effector proteins that are required for the onset of parasitism. GrVAP1 from *G. rostochiensis* suppresses plant immune responses by targeting the extracellular protease RCR3. This protease is a target for many other plant pathogens including *Phytophthora infestans* and *Cladosporium fulvum* and is guarded by the R-protein Cf-2 (Lozano-Torres *et al.*, 2014). This demonstrates that diverse pathogens frequently target the same host defence signalling nodes and that plants seek to detect pathogens by monitoring these nodes. The 10A06 protein from *H. schachtii* suppresses host defences by disturbing the SA pathway. Constitutive expression of 10A06 in *Arabidopsis thaliana* enhances susceptibility to the nematode but also to unrelated pathogens such as the bacterium *Pseudomonas syringae* and cucumber mosaic virus, indicating that this effector targets a general component of the plant immune system. This effector protein stimulates polyamine biosynthesis through its interaction with spermidine synthase 2 causing upregulation of antioxidant genes and the disruption of SA-mediated defence signalling (Hewezi *et al.*, 2010). An effector from *M. javanica* that is similar to transthyretins interacts with a ferredoxin:thioredoxinreductase catalytic subunit. This is a key component of the host antioxidant system and the interaction of the effector with this protein reduces the host capacity to produce reactive oxygen species and thus reduces host defences (Lin *et al.*, 2016). Plant-parasitic nematodes thus suppress host defences using a wide range of different effector proteins targeting many different parts of the host defence signalling pathway.

9.4.2. Protection from host defence responses

The proteins that are present within the secretions of plant-parasitic nematodes (reviewed in Haegeman *et al.*, 2012) include proteins designed to protect the nematode from host defence responses. For example, plant-parasitic nematodes produce glutathione-S-transferase (GST) and one GST of *M. incognita* is expressed in the pharyngeal gland cells from where it is presumably secreted into the host. Knockout of the transcript encoding this gene led to a decrease in parasitism, suggesting it plays an important role in the nematode life cycle. In many animal-parasitic organisms, including nematodes, GSTs detoxify endogenous and xenobiotic compounds using a range of biochemical mechanisms. It is possible that GSTs may have a similar role in the interactions between nematodes and plants, detoxifying some of the wide range of secondary metabolites that the plant uses to deter invaders.

One of the major responses of plants to nematode infection is production of damaging free radicals, including hydrogen peroxide. In order to overcome this threat, plant-parasitic nematodes secrete enzymes that can break down this hydrogen peroxide in their surface coat. A peroxiredoxin (thioredoxin peroxidase) that specifically metabolizes hydrogen peroxide has been shown to be present in the surface coat of *G. rostochiensis*. In addition a secreted glutathione peroxidase is expressed in the hypodermis of this nematode and may also be secreted to the parasite surface. This protein does not metabolize hydrogen peroxide but has a preference for larger, lipid hydroperoxides. These compounds may be produced by the plant as direct anti-pathogen factors and can form part of plant defence signalling pathways. *Globodera rostochiensis* also secretes other proteins that may inhibit defence signalling pathways. Linoleic and linolenic acids are metabolized by lipoxygenase in the first steps of the JA signalling pathway. Activation of this pathway leads to changes in gene expression associated with systemic defence responses. The surface coat of *G. rostochiensis* contains a secreted lipid binding protein (GpFAR-1) that can bind linoleic and linolenic acids and it has been shown that the binding activity of GpFAR-1 can inhibit lipoxygenase-mediated peroxidation of these compounds. Plant-parasitic nematodes, therefore, secrete proteins that inhibit defence signalling pathways as well as proteins that can metabolize the toxins produced by plants as part of defence responses.

The presence of peroxiredoxin, glutathione peroxidase, GST and secreted lipid-binding proteins within the secretions of plant-parasitic nematodes shows that there are parallels in the ways in which plant and animal parasites protect themselves from their hosts. Whilst defence mechanisms of plants and animals are different, some common components are present. Both plants and animals respond to an invader by producing free radicals and both have signalling pathways that include lipid molecules. Like plant parasites, some animal-parasitic nematodes produce peroxiredoxins, glutathione peroxidases, GST and lipid-binding proteins in their secreted proteins. These substances are not present in secretions of non-parasitic nematodes such as *Caenorhabditis elegans*. Their presence in secretions of a range of parasites is therefore a remarkable example of convergent evolution, where the site of expression of similar standard nematode housekeeping proteins is changed in order to provide protection from similar host defences.

9.5. Molecular and Cellular Aspects of the Development of Nematode Feeding Cells

9.5.1. Cellular changes

Nematodes that feed for a prolonged period from the same cell induce cytological modifications that increase the metabolic and transport capacities of the cell (see Fig. 1.1). The precise mechanism underlying feeding cell formation is unknown but there is growing evidence that effectors have a pivotal role in this process (Mejias *et al.*, 2019). When a cyst nematode selects a competent root cell, this cell responds by gradually widening some of the plasmodesmata that link it to neighbouring cells. The protoplasts of the initial syncytial cell and those of its neighbours fuse through these developing wall openings. At later stages, cell wall openings are formed *de novo* and the syncytium continues to expand by incorporating hundreds of adjacent cells (Grundler *et al.*, 1998). The first sign of giant cell induction by a root-knot nematode is the formation of several binucleate cells (Jones and Payne, 1978; see Chapter 3, Fig. 3.7). Additional nuclear divisions (mitoses) uncoupled from cell division result in several large multinucleate cells being formed. At the site of cell plate formation in developing giant cells, vesicles accumulate but the formation of the cell plate is aborted, preventing cell division (Jones and Payne, 1978). These acytokinetic mitoses result in polyploid cells with up to 100 nuclei. In addition, cells that surround the giant cell divide and swell to form the typical gall or root knot.

Despite their different ontogeny, many of the physiological functions (such as nutrient supply to the nematodes) and underlying cellular features of these two types of feeding cells are similar. Both have multiple enlarged nuclei, small vacuoles and show proliferation of smooth endoplasmic reticulum, ribosomes, mitochondria and plastids. In both giant cells and syncytia the high metabolic activity of the feeding cells is reflected in the upregulation of mitochondrial and ribosomal genes, genes encoding components of the proteasome pathway (for protein turnover) and genes involved in carbohydrate metabolism (Box 9.2; Gheysen and Mitchum, 2009).

Box 9.2. Identification of plant genes that are responsive to nematode infection.

Cellular and biochemical changes in plants are caused by changes in a variety of proteins including structural proteins such as cytoskeleton compounds or enzymatic proteins such as kinases involved in cell cycle regulation. Proteins are made by genes through the production of an intermediate messenger, the mRNA, which is transcribed from the gene after activation of the promoter. A study of the molecular response of the plant can therefore be performed at different levels: analysis of promoter activity, mRNA differences or protein patterns. Indeed, the study of the plant response to nematode infection has been done at these different levels and using a variety of methods (Cabrera *et al.*, 2016). Examples include studies:

1. At the promoter level: a reporter gene was used to analyse activity in nematode feeding sites (NFS) of known or unknown promoters.
2. At the mRNA level: mRNA has been extracted from infected roots or from NFS and comparisons were made with mRNA from non-infected roots. These methods (see also Box 9.1) comprise analysis of cDNA libraries, quantitative reverse-transcriptase polymerase chain reaction (PCR), micro-arrays and RNA sequencing.
3. At the protein level using two-dimensional gel electrophoresis (see Box 9.1) and antibody labelling of plant tissue sections.

Data collection and interpretation on plant gene expression after nematode infection has profited from the use of the model plant *Arabidopsis thaliana* (Gheysen and Fenoll, 2011). A model organism is one studied by many scientists because it has features that make it easy to study and it is similar enough to other organisms for extrapolation of conclusions. *Arabidopsis thaliana* is a small weed of the cabbage family (Brassicaceae) and it has some key characteristics that make it suitable as a model organism for plant studies:

- Small genome size, 120 million base pairs, it was the first plant genome to be completely sequenced. For comparison, e.g. soybean has 1100 million, maize 2500 million and wheat 17,000 million base pairs.
- Small size, many plants can be grown in a growth chamber.
- Short life cycle, from seed to next generation seed in 6–8 weeks.
- Production of mutants is easy by chemical, radiation or insertion mutagenesis.
- Transformation is easy and rapid, e.g. the floral dip method does not require tissue culture facilities or sophisticated equipment.
- The thin transparent roots facilitate following the nematode infection process.

Since the first use of *A. thaliana* as a model species for plant nematode research, huge quantities of molecular data on the plant response have been obtained from this model species. Nevertheless, it is still necessary to analyse crop species to verify important conclusions, and sometimes different plants can yield very different results. Luckily, with the advancement in sequencing technologies, crop species such as rice, soybean, tomato, potato and wheat can now also be studied (Cabrera *et al.*, 2016).

Some nice examples of the advantages of working on a model species include the analysis of cell cycle or hormone mutants (see Section 9.5.6).

Although many different methods can be used to study plant gene expression (see Box 9.2), comprehensive analyses have only become feasible with the use of micro-arrays containing thousands of genes. Next-generation sequencing has opened up even more possibilities to analyse in detail differences in gene expression patterns between control and infected plant tissues or between resistant and susceptible plant responses. With sequencing methods becoming less expensive, there has been a huge expansion in the data available for crop plants, even those with huge genomes that were previously viewed as intractable. For example, a transcriptome analysis of wheat infected with the cereal cyst nematode *Heterodera avenae* was recently reported (Qiao *et al.*, 2019).

Other omics techniques to analyse the plant response to nematode infection have been less frequently used, but have also yielded new insights. For example, metabolomic analysis has shown an accumulation of sucrose, specific amino acids and phosphorylated metabolites in syncytia. For a historical review on the use of omics in plant–nematode interactions, see Cabrera *et al.* (2016).

9.5.2. Cell wall and membrane modifications

Nematode-induced feeding sites develop through cell expansion, which requires cell wall loosening and remodelling. This is reflected in the upregulation but also sometimes suppression (reviewed in Gheysen and Mitchum, 2009) of endogenous plant genes encoding cell wall modifying proteins including expansins, β1,4-endoglucanases, polygalacturonase and pectin acetylesterase. It is evident that feeding cell formation is a complex process that requires the coordinated regulation of several different types of plant cell wall modifying proteins, as well as the differential expression of individual gene family members. In syncytia, cell wall modifying proteins are also involved in the cell wall breakdown that occurs between adjacent cells, allowing the cytoplasm to move freely throughout the syncytium. Nematodes appear to enhance this process by secreting proteins that activate cell wall modifying proteins from the plant. A cellulose binding protein (CBP) from *H. schachtii* was shown to interact with a pectin methylesterase from *A. thaliana*, thereby activating and potentially targeting this enzyme to aid cyst nematode parasitism (Hewezi *et al.*, 2008).

In both syncytia and giant cells, the cell wall adjoining the xylem increases its thickness by forming finger-like wall invaginations lined with plasma membrane (Jones and Northcote, 1972). A similar process occurs in plant transfer cells and it is likely to facilitate water transport from the xylem into the feeding cell. It has been estimated that each developing juvenile of *H. schachtii* withdraws an amount of solute equivalent to four times the total syncytial volume from the syncytium every 24 h. Genes encoding water channel proteins are upregulated in giant cells but similar genes do not appear to be upregulated in syncytia. However, a syncytium is one huge cell that may consist of up to 250 fused root cells, whereas giant cells are interconnected separate cells. Therefore, giant cells may require water channels to redistribute the water rapidly, even though they are interconnected with each other through plasmodesmata.

While water is taken from the xylem, nutrients are retrieved from the phloem (reviewed in Grundler and Hofmann, 2011). However, it has been shown that syncytia are initially symplastically isolated from surrounding tissues, including the phloem. Assimilates must therefore be imported into the syncytium via the apoplast. The sucrose carrier AtSUC2 is responsible for the active import of sucrose into sink tissues and is highly expressed in syncytia. Therefore, it may have a role in forming or maintaining the metabolic sink activity in syncytia. The activity of sugar or other transporters is dependent on the energy generated by H^+-ATPases, which have also been shown to be upregulated in NFS. At later stages, functional plasmodesmata are being established and at 15 days after infection all syncytia are symplastically connected to the phloem. The supply of sucrose from phloem to giant cells induced by root-knot nematodes is mediated by sucrose transporters and/or plasmodesmata, their relative importance apparently being different depending on the specific plant host, the age of the plant and the infection stage.

9.5.3. The cytoskeleton

Sedentary endoparasitic nematodes induce long-term rearrangements of the cytoskeleton during the infection process. Actin genes are highly expressed in NFS, as would be expected in large expanding cells that need extensive internal transport (de Almeida Engler *et al.*, 2004). Tubulin genes are slightly upregulated in syncytia and highly upregulated in giant cells, probably reflecting the need for a mitotic cytoskeleton for the rapidly dividing nuclei in the latter. Analysis of the cytoskeleton by immunolocalization of actin and tubulin proteins and by the use of GFP fusions revealed that the cytoskeleton was strongly disrupted in syncytia (de Almeida Engler *et al.*, 2004). In giant cells, although disturbed compared to normal root cells, actin and microtubular fibres are visible. A functional mitotic apparatus, consisting of microtubules forming spindles (for separating the chromosomes during nuclear division) and phragmoplasts (which initiate the cell plate), is also present in developing giant cells (de Almeida Engler *et al.*, 2004). As nematode feeding involves the retrieval of large volumes of cytoplasm, a degree of cytoskeleton fragmentation may facilitate uptake during nematode feeding. It has been shown that nematodes can actively disturb the polymerisation of actin. A nematode effector from *M. incognita* was identified as a profilin protein. This effector disrupts actin polymerisation in an *in vitro* assay and when expressed in *A. thaliana* protoplasts, it shrinks the filamentous actin network (Leelarasamee *et al.*, 2018).

9.5.4. Nuclear changes and cell cycle activation

Several cell cycle genes are among those that are induced in the early stages of NFS formation (reviewed in Kyndt *et al.*, 2013). Both types of feeding cells undergo multiple rounds of shortened cell cycles leading to genome amplification but do so in different ways (Fig. 9.5). Thus, giant cells go through repeated (acytokinetic) mitoses, whereas nuclear divisions have never been observed inside syncytia, which repeatedly go through S (synthesis) phase (repeated DNA synthesis or endoreduplication), apparently shunting mitosis. The absence of clear differences in expression of cell cycle-related mRNAs between giant cells and syncytia, in spite of the differences in ontogeny of these structures, can be explained in several ways. First, both types of feeding cells need cell cycle activation and endoreduplication, although the latter process is essential for syncytium initiation, while nuclei in giant cells appear to go first through repeated (acytokinetic) mitosis before entering endoreduplication. It is possible that the genes involved specifically in acytokinetic mitosis have not yet been identified, and that endoreduplication in development of syncytia versus giant cells is controlled by subtle differences in expression time or levels. It is also possible that regulation occurs primarily at the post-transcriptional level, by regulation of protein turnover or activity.

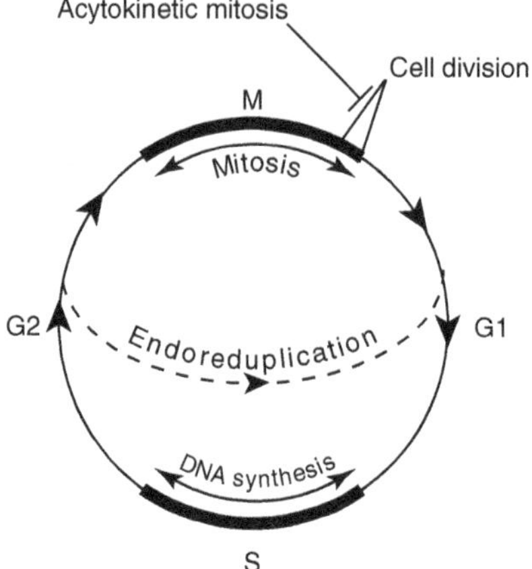

Fig. 9.5. The cell cycle. The standard eukaryotic cell cycle consists of four successive stages. The nuclear DNA is replicated during S (synthesis) phase and the nucleus divides in the M (mitosis) phase. The interval between the completion of mitosis and the beginning of DNA synthesis is called the G1 phase (G = gap) and the interval between the end of DNA synthesis and the beginning of mitosis is the G2 phase. At the end of the mitotic phase, the cell divides by a process called cytokinesis. Cyst nematodes activate the cell cycle in the feeding cells they induce, but no mitosis occurs. Instead the cells repeatedly go through the S phase (endoreduplication), probably also through (part of) G1 and G2 phases but bypass mitosis. Root-knot nematodes induce repeated nuclear division but no cell division; this is called acytokinetic mitosis.

9.5.5. Epigenetic control of feeding site development

Evidence has accumulated in the past decade of the importance of epigenetic processes in plant-nematode interactions. Epigenetic control of gene expression includes DNA methylation, chromatin modifications and the actions of small non-coding RNAs (comprehensively reviewed by Hewezi, 2020). An example of a microRNA that is a key regulator of syncytium development is miR396, first discovered in *A. thaliana*, but recently confirmed in soybean. The miR396 gene is strongly downregulated during initiation of a syncytium and upregulated at the transition to the syncytium maintenance phase. MicroRNAs are 21 nucleotides long and regulate gene expression by degrading their complementary target mRNAs. Two of the targets of miR396 are the growth-regulating transcription factors GRF1 and GRF3. Interference with expression of GRF1/3 or their modulating miR396 caused abnormal syncytium and nematode development. How the nematode influences miR396 is not known, but nematode effectors affecting other epigenetic processes have been discovered (see Section 9.6.1).

9.5.6. Functional analysis

Cataloguing plant genes that are up- or downregulated upon nematode infection is only the first step in gaining insight into NFS development. Many genes may be activated because they happen to have the appropriate regulatory signals that are recognized in developing NFS and not because they play an essential role. Furthermore, genes that are not differentially expressed can equally be important in NFS development. Recent additions to such lists of genes have been made by the identification of host proteins interacting with nematode effectors (see also Section 9.6). To understand the molecular basis of NFS ontogeny, functional analysis of the plant genes expressed in NFS is crucial. However, the consequence of knocking out a gene (or overexpressing it) on NFS formation has been investigated in only a small percentage of potentially involved plant genes (reviewed in Gheysen and Mitchum, 2009 and Mejias *et al.*, 2019). Some particular examples are highlighted below.

A transcription factor that is highly expressed in both early giant cells and syncytia is WRKY23. This gene is induced by auxin and is important for primary root development and nematode infection (Grunewald *et al.*, 2008, 2012). Knocking down the expression of WRKY23 resulted in lower infection of the cyst nematode *H. schachtii* (Grunewald *et al.*, 2008). Mutants in cell wall modifying proteins such as the *Arabidopsis Cel2* mutant resulted in a reduced number of developing *H. schachtii* females (Wieczorek *et al.*, 2008).

Knockdown or overexpression of several cell cycle genes have proven their importance in NFS development. A notable example is the *RHL1* gene encoding a DNA topoisomerase VI subunit. This topoisomerase is crucial in entangling DNA copies caused by endoreduplication, a process that is important in giant cells and syncytium development. In the *rhl1* mutant, root-knot nematodes could only form very small giant cells and cyst nematodes were unable to trigger syncytium formation (de Almeida Engler *et al.*, 2012).

Another method for functional analysis of the cell biology of the NFS is the use of chemical inhibitors that interfere with a specific process. Blocking the cell cycle with specific inhibitors, for example, has been shown to lead to an arrest in NFS development.

For more details on the involvement of hormones in NFS development, see below and Box 9.3.

Several types of plant hormones have been studied since the middle of the last century, with cytokinins, auxins and ethylene among the best characterized. These hormones (see Box 9.3) have also long been proposed to be relevant in establishment of NFS.

Box 9.3. Plant hormones play an important role in feeding site development.

Plant hormones are biochemicals that act specifically on plant cells and tissues to regulate growth and development at low concentrations. Plant hormones can be classified into auxins, cytokinins, salicylic acid (SA), jasmonates (JA), gibberellins, ethylene (ET), abscisic acid (ABA), polyamines, brassinosteroids, strigolactones, peptides and oligosaccharins, each of these regulating a variety of processes.

The classical defence hormones are JA and SA and also ET. Genes involved with biosynthesis and signalling of SA and JA are mainly downregulated in NFS, enabling these feeding cells to develop and sustain nematode feeding.

Evidence has accumulated for the involvement of auxins, cytokinins – and also in some cases ET or ABA – in supporting a successful nematode infection.

Auxins influence cell elongation, cell division and differentiation, and apical dominance. They are mainly produced in the plant apex and are transported to the root tip. This polar auxin transport is important for growth and development, for example in lateral root formation. Cytokinins promote cell differentiation and division, often acting in association with auxin. Their synthesis occurs in various tissues and they are inactivated by oxidation. Ethylene is involved in fruit ripening (a process that includes cell wall degradation), senescence and abscission, and its biosynthesis is stimulated by auxin.

Despite their strong record of scientific analysis and recent molecular evidence for their role in NFS formation, it is still very difficult to propose a unifying model, because of pleiotropic effects and interactions of these hormones and contradictory results about their hypothetical roles in nematode infection. For a comprehensive overview of this topic, see Gheysen and Mitchum (2019).

For the involvement of auxin in syncytium formation, however, a clear picture has emerged. The auxin-responsive promoter *DR5* is rapidly and transiently activated during NFS initiation by *Meloidogyne* and *Heterodera* (Karczmarek *et al.*, 2004). Disturbance of auxin gradients with a polar auxin transport inhibitor results in abnormal feeding sites, and strong auxin-insensitive mutants barely support cyst nematode reproduction (Goverse *et al.*, 2000). The importance of polar auxin transport and of the PIN efflux transport proteins in syncytium formation have been elucidated (Grunewald *et al.*, 2009). PIN1 is necessary for transporting auxin from the plant shoot to the infection site, and its expression is downregulated in the initial syncytial cell, preventing removal of auxin from the early syncytium. This, together with the local upregulation/activation (see also Section 9.6.2) of auxin import proteins AUX1 and LAX3 results in accumulation of auxin in the early syncytium (Lee *et al.*, 2011). Radial expansion of the syncytium is facilitated by the movement of PIN3 proteins to the lateral sites of the syncytium, from where they export auxin to the neighbouring cells preparing them for incorporation in the syncytium (Grunewald *et al.*, 2009).

Similarly, an important role for cytokinins in syncytium development has been described. Mutants in cytokinin biosynthesis or signalling are less susceptible to cyst nematode infection. Furthermore, a typical cytokinin biosynthesis gene (isopentyltransferase) has been discovered in the genome of *H. schachtii* and silencing this gene impedes nematode infection (Siddique *et al.*, 2015).

The role of ethylene might be different for giant cell versus syncytium development. Ethylene-overproducing *A. thaliana* mutants attract more cyst nematode juveniles and result in larger syncytia and females. The enhanced syncytial cell wall breakdown in these mutants indicates that ethylene-induced cell wall degradation is involved in syncytium development, pointing to similarities between cell wall alterations in syncytia and during fruit ripening. Ethylene can also induce endoreduplication. By contrast, ethylene biosynthesis inhibition and impaired ethylene signalling result in a higher susceptibility of rice toward root-knot nematodes (Nahar *et al.*, 2011). The fact that ethylene was not detected in galls at 1–2 days post-infection (Glazer *et al.*, 1983), in combination with the transcriptional patterns observed (Barcala *et al.*, 2010), argue against a crucial role for ethylene-activated pathways during early giant cell differentiation.

9.6. Nematode Signals for Feeding Site Induction and Other Processes

Understanding how nematodes induce the changes in plant root cells that lead to feeding site formation has long been one of the holy grails of plant nematology. In recent years the number of nematode molecules identified that may play a role in host–parasite interactions has increased dramatically (Ali *et al.*, 2017), mainly due to the development of new technologies for genome sequencing and effector gene identification (Box 9.1) and molecular methods for characterization (Eves-van den Akker *et al.*, 2021). However, many of the effector sequences identified are pioneers – proteins that have no matches to other sequences in the databases and that often contain no domains that provide an indication as to their function. Determining the function of these proteins is often extremely challenging.

9.6.1. Effectors manipulating gene expression in nematode feeding sites

Nematode feeding sites are fundamentally different from normal root cells in their morphology and physiology, this being reflected in differential gene expression. Remarkably, effectors have been identified that have a direct effect on plant gene expression (Eves-van den Akker *et al.*, 2021).

The effector GLAND4 of *H. schachtii* was shown to bind to specific DNA sequences and regulate expression of the nearby lipid transfer protein (LTP) genes involved in plant defence. Expression of GLAND4 in *A. thaliana* resulted in downregulation of these LTP-genes and LTP overexpression decreased infection by *H. schachtii.*

Both the root-knot nematode effector MiEFF18 and the cyst nematode effector 30D08 alter mRNA splicing in NFS by interaction with a plant splicing protein. Plants expressing those effectors resulted in altered mRNA patterns of genes implicated in hormone pathways, cell cycle and development.

The *H. schachtii* effector 32E03 induces chromatin alterations by interaction with a histone deacetylase in *A. thaliana*. This results in elevated levels of ribosomal RNA (rRNA) and higher development of female nematodes. More rRNA could contribute to the high metabolism in syncytia.

9.6.2. Effectors promoting elevated auxin in nematode feeding sites

The characterization of the Hs19C07 effector protein secreted by *H. schachtii* indicates that the nematode can modify the cellular partitioning of auxin in part by targeting one of the auxin importers of the AUX1/LAX family in *A. thaliana* (Lee *et al.*, 2011). Hs19C07 has been shown to interact with the LAX3 auxin transporter and can stimulate its activity. This work is important as it represented the first direct link between a nematode effector and the plant signalling pathways that control auxin in roots. This work also uncovered parallels between the mechanisms underlying syncytium formation and those that control formation and emergence of lateral roots. Auxin signalling is also hijacked by nematode effectors. The cyst nematode effector 10A07 binds to the transcription factor IAA16, and as a consequence, several auxin response factors are upregulated in *A. thaliana*. In addition, an effector from *H. schachtii* increases auxin levels and nematode susceptibility when expressed in *A. thaliana* (Habash *et al.*, 2017). The effector contains a tyrosinase functional domain, but the exact mechanism of how this effector affects hormone homeostasis is unknown.

9.6.3. Chorismate mutase

Cyst nematodes and root-knot nematodes both secrete chorismate mutases. The proteins are produced in one or both of the pharyngeal gland cell types and, like cell wall degrading enzymes, are not usually present in animals and are thought to have been acquired by horizontal gene transfer from bacteria. Chorismate mutase converts chorismate to prephenate. Compounds derived from chorismate include auxin, an important plant hormone that has been implicated in early feeding site development (see Section 9.5 and Box 9.3), whilst a variety of secondary metabolites are derived from prephenate. Expression of the nematode chorismate mutase in soybean hairy roots gives rise to a remarkable phenotype. Normal vascular tissue development is suppressed and lateral roots do not appear to develop normally. Both these phenotypes are indicative of an auxin defect, and exogenous application of auxins reverses these observed phenotypes. It has been proposed that nematode chorismate mutases may deplete levels of auxins in the early stages of feeding site development by removal of the chorismate precursor within the cytoplasm of the plant cell. However, other studies have suggested an important role for auxins in early feeding site development (see Section 9.5.6 and Box 9.3). An alternative role for chorismate mutase, supported by its presence in migratory nematodes and fungal pathogens, is the suppression of plant defence (Djamei *et al.*, 2011). One potential fate of chorismate is conversion, via a two-step reaction, to the plant defence signalling compound SA. Chorismate mutase could reduce the pool of chorismate available for conversion to SA, thus preventing normal activation of host defences. Indeed, expression of a chorismate mutase from *M. incognita* in *Nicotiana benthamiana* resulted in lower SA levels after pathogen infection (Wang *et al.*, 2018).

9.6.4. The effector IA7 from *Globodera pallida* influences the cell cycle in potato

Ever since it was known that nematodes influence the cell cycle during nematode feeding site formation, the key question has been how they are able to do so. We are currently still far from a complete understanding of this process, but the first insight has come from the study of a cyst nematode effector. GpIA7 from *G. pallida* associates with the potato epidermal growth factor receptor-binding protein and this results in altered expression of several key cell cycle components such as *cyclin D3;1* and *retinoblastoma related 1*. Expression of this effector in potato changes its growth and development, indicating a possible role in feeding site formation (Coke *et al.*, 2021).

9.6.5. Nematode peptides that mimic plant peptides

Studies on *A. thaliana* (see Box 9.2) have revealed that a transmembrane receptor-like kinase, CLAVATA1, regulates the balance between cell differentiation and cell division in the shoot apical meristem. CLAVATA1 and another protein, CLAVATA2, are thought to dimerize and to bind a small secreted peptide, CLAVATA3. Binding of CLAVATA3 to the extracellular side of this receptor complex may allow activation of the intracellular protein kinase domain and subsequent activation of signalling pathways. Bioinformatic analysis of a range of cyst nematodes has identified genes that encode short peptides similar to the CLAVATA3 ligand. The gene encoding the *H. glycines* CLAVATA3 peptide has been shown to complement *cle3* mutants of *A. thaliana*, confirming that the nematode CLE protein is a functional analogue of the plant peptide. The nematode peptide has been shown to bind to the CLE receptors in plants and to alter root development, providing compelling evidence that nematode CLE peptides play a role in manipulation of the host (Guo *et al.*, 2011).

Another CLE peptide effector found in *H. glycines* is almost identical to tracheary element differentiation inhibitory factor (TDIF) of *A. thaliana* (Guo *et al.*, 2017). TDIF controls maintenance of vascular stem cells, a pathway that may be of importance in NFS establishment. Indeed, plant mutants in this pathway revealed lower nematode infection and diminished syncytium size.

CLE peptides are not the only plant peptide mimics that nematodes use to assist plant parasitism. Others are related to CEP and IDA hormone peptides; see Gheysen and Mitchum (2019) for more details.

9.6.6. Nematodes also secrete non-protein effector molecules, e.g. cytokinins

Cytokinins are plant hormones that have a variety of roles in plant development, often acting in concert with auxins (Box 9.3). Secretions of root-knot and cyst nematodes contain auxins, but their biological significance is not known. Secretions of root-knot nematodes contain cytokinins at biologically significant levels. The major cytokinins present in root-knot nematode secretions are zeatins. These molecules are also present at lower levels in cyst nematodes, but in *H. schachtii* a cytokinin biosynthesis gene has been identified, strengthening the case for an important role for cytokinins in cyst nematodes (Siddique *et al.*, 2015). Thus, it is feasible that nematodes directly introduce cytokinins into plant cells and that these molecules influence plant developmental pathways.

9.7. Comparisons Between Cyst and Root-knot Nematodes

It is clear that there are intriguing parallels as well as clear differences between the invasion process and feeding sites of cyst and root-knot nematodes and that these parallels and differences are reflected in the substances secreted by the two groups of nematodes. The main features are summarized in Table 9.1. Phylogenetic analysis shows that cyst and root-knot nematodes are not directly related to each other, with each seeming to have evolved independently from different groups of migratory endoparasites (see Section 2.11.3). This implies that the ability to induce feeding sites has also evolved independently within each group. A comparison of the proteins secreted by each group is in agreement with this idea; the only effectors that are common between cyst and root-knot nematodes are also present in migratory endoparasites. These include cell wall degrading enzymes and chorismate mutase. Other effectors, including the large numbers of 'pioneer' sequences identified, are generally specific to cyst or root-knot nematodes.

9.8. Resistance and Avirulence Genes

Many plant resistance genes have been cloned and the encoded proteins fall into six classes, four of which (like the majority of plant resistance genes) consist of proteins with leucine-rich repeats (LRRs), structural motifs that have been shown to mediate protein–protein interactions (reviewed in Hammond-Kosack and Jones, 1997). Some encode transmembrane and mainly extracellular proteins, whilst some encode entirely cytoplasmically

Table 9.1. Comparison of some secreted proteins and feeding sites of cyst and root-knot nematodes.

	Cyst nematodes	Root-knot nematodes
Nematode secretions		
Subventral glands		
Cell wall degrading enzymes		
β1,4-endoglucanase	+	+
Pectate lyase	+	+
Polygalacturonase	N	+
Expansin	+	+
Xylanase	N	+
Cellulose-binding domain	+	+
Calreticulin	N	+
Chorismate mutase (also in dorsal gland in some species)	+	+
Dorsal gland		
SPRYSEC	+	N
Ubiquitin extension	+	N
CLAVATA3	+	N
14-3-3	N	+
Feeding sites		
Origin	Syncytium by fusion of protoplasts	Giant cells by acytokinetic mitosis
Cytology	Dense cytoplasm with many organelles	Dense cytoplasm with many organelles
Nuclei	Many large nuclei	Many large nuclei
Cell wall	Cell wall ingrowths close to xylem	Cell wall ingrowths close to xylem
	Cell wall degradation	–

N = not found; + = present; – = not upregulated.

located proteins. It is most likely that the localization of the *R*-gene product reflects the site of the cellular location of the interaction with the avirulence product. For example, a viral AVIR protein will be present inside the plant cell whilst a fungal AVIR protein may be localized in the intercellular plant space. Many R-proteins have a TIR or serine/threonine protein kinase domain, domains involved in intracellular signal transduction. Each *R*-gene is thought to have at least two functions: recognition of a specific AVIR-derived signal and activation of downstream signalling pathways to trigger the various defence responses, ultimately resulting in localized plant cell death.

Plant resistance genes to nematodes are present in several crop species and wild relatives (see Chapter 15) and their identification and incorporation into commercially viable cultivars are important factors in breeding programmes of tomato, potato and soybean, among others. Many of these resistance genes have been mapped at precise positions on plant chromosomes to facilitate the breeding process, and several of these genes have now been cloned (reviewed in Kandoth and Mitchum, 2013).

The best studied nematode resistance gene is the tomato gene *Mi-1.2*, which confers resistance to three species of root-knot nematodes. It is unusual compared to other resistance genes because it also confers resistance to three other very different organisms, the potato aphid *Macrosiphum euphorbiae*, the white fly *Bemisia tabaci* and the tomato psyllid

Bactericerca cockerelli. This broad activity may reflect different pathogens targeting a common host protein guarded by *Mi-1.2*. The *Mi* gene encodes a 1257 amino acid protein that is a member of the LZ-NBS-LRR family (Fig. 9.6). Swaps of protein domains between *Mi-1.2*, the functional gene, and *Mi-1.1*, a non-functional related protein, have shown that the LRR region has a role in the signalling process that leads to plant cell death (typical for the HR) and that the N-terminal part of the protein controls this cell death. This model implies that the N-terminus of the protein keeps the LRR in an inactive state, unless an avirulence signal from the pathogen releases this inhibition.

Two other cloned genes, *Gpa2* (a potato gene conferring resistance to *G. pallida*) and *Hero* (a tomato gene conferring resistance against *G. pallida* and *G. rostochiensis*), also belong to the LZ-NBS-LRR family (Fig. 9.6) but their overall sequence is not very similar to *Mi*. *Gro1-4* confers resistance to *G. rostochiensis* and belongs to the TIR-NBS-LRR class of resistance genes.

The first nematode resistance gene to be cloned was the *Hs1*$^{pro\text{-}1}$ gene (from wild beet), which confers resistance to *H. schachtii*. This gene is different from any other resistance gene cloned from plants and in backcrosses the presence of *Hs1*$^{pro\text{-}1}$ does not correlate with a resistant phenotype. Consequently, some controversy as to its precise role remains.

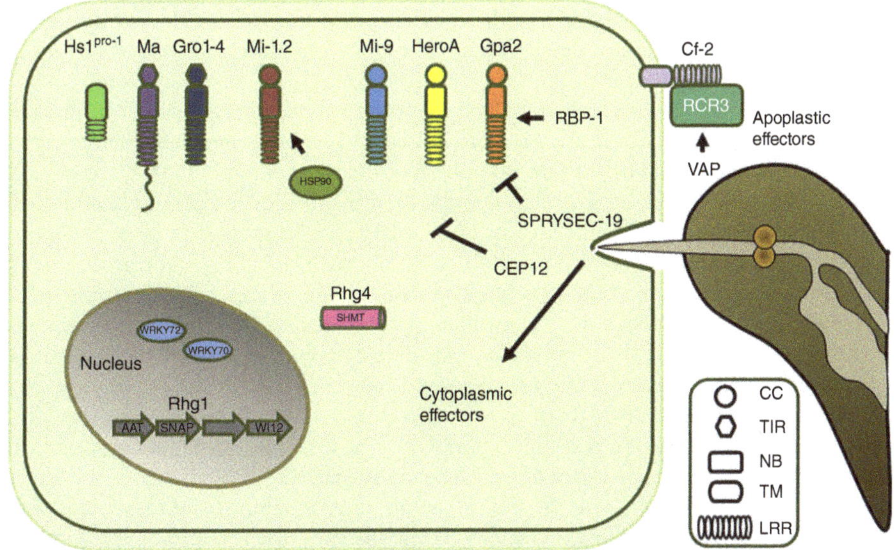

Fig. 9.6. Structure of plant resistance proteins active against plant-parasitic nematodes. The nematode genes that have been cloned to date fall into one of three classes: (i) TIR-NBS-LRR family to which *Gro1-4* and, among others, the virus resistance gene N and the bacterial resistance gene *RPP5* belong; (ii) LZ-NBS-LRR family to which three cloned nematode resistance genes and, among others, the bacterial resistance gene *RPM1* belong; and (iii) *Hs1*$^{pro\text{-}1}$, which is very different from all other resistance genes identified to date. Another class of plant resistance genes has an extracellular LRR domain; the fungal resistance gene *Cf-2*, which also gives resistance to *G. rostochiensis,* belongs to this family. Some genes that provide resistance *against H. glycines* (*rhg1* and *rhg4*) are not typical resistance genes. Abbreviations: CC, coiled-coil domain; LRR, leucine-rich repeat; NB, nucleotide binding site; TIR, Toll/interleukin-1/resistance domain; TM, transmembrane domain. (Figure modified from Kandoth and Mitchum, 2013.)

The gene encodes a relatively small protein (282 amino acids) that does not contain a leucine-rich repeat, the typical hallmark of most resistance genes (see Fig. 9.6).

Two *rhg* genes, which confer resistance against *H. glycines* in soybean, have been cloned and also show highly unusual properties compared with other resistance genes. Resistance conferred by the *rhg1* gene is due to the presence of multiple copies of a segment of DNA containing several different genes, none of which show a typical NB-LRR structure (Cook *et al.*, 2012). The α-SNAP protein encoded at the *rhg1* locus has differences in the amino acid sequence that cause a failure of its normal, essential function (Bayless *et al.*, 2016). This mutation is compensated for in resistant soybean plants by the presence of other normal α-SNAP proteins that are present due to ancient genome duplications in this species. However, the cytotoxic form of the protein accumulates specifically in the nematode feeding site, causing cell death and resulting in resistance (Bayless *et al.*, 2016).

The *rhg4* gene has been identified as a serine hydroxymethyl transferase (SHMT) (Liu *et al.*, 2012). Further functional studies suggested that the biochemical properties of the SHMT enzyme present in resistant plants were different to those from susceptible lines and that these differences may impact the biology of the developing feeding site and thus cause it to fail to provide nutrients to the developing nematode.

Studies on nematode genes that may encode the avirulence proteins recognized by resistance genes have not progressed as far as studies on similar proteins from other pathogens. At present three nematode effectors have been identified that may represent avirulence determinants. The root-knot nematode gene *Mi-Cg1* is required for *Mi-1.2* to confer resistance in tomato to *M. incognita* (Gleason *et al.*, 2008). Knocking-down expression of the *Mi-Cg1* gene by RNAi allows avirulent individuals to become virulent, suggesting that its product is recognized by *Mi-1.2*. However, the *Mi-Cg1* transcript does not seem to encode a secretory protein capable of interacting with an immune receptor. It is possible that this transcript is involved in regulation of an effector that is recognized by *Mi-1.2*. cDNA-amplified fragment length polymorphism (AFLP) analysis was used to compare near isogenic lines of *M. incognita* virulent and avirulent against *Mi-1.2* and a novel gene (*Map-1*) was identified encoding a secreted protein. Expression analysis suggested that variants of the protein containing fewer repeat sequences are expressed only in avirulent nematode species/lines. However, no experimental analysis showing that expression of *Map-1* in the presence of *Mi-1.2* leads to a HR has been published. A more complete example of a nematode avirulence gene is the *Gp-RBP-1* SPRYSEC gene from *G. pallida* (Sacco *et al.*, 2009). Transient co-expression of *Gp-RBP-1* and the nematode resistance gene *Gpa2* in leaf tissues induces a specific HR, making this effector the likely cause of avirulence in nematode populations.

In recent years it has become clear that the majority of plant resistance genes do not interact directly with pathogen avirulence factors. Instead, it is thought that the majority of resistance gene products monitor (or guard) a host protein and detect pathogen-induced changes in this host protein. This guard hypothesis neatly explains why some resistance genes (such as *Mi-1.2*) provide resistance against diverse pathogens; each of the pathogens targets the same host protein to promote infection and *Mi-1.2* detects these changes as it guards this common host target.

9.9. Concluding Remarks

The study of the molecular biology of plant–nematode interactions, like many other areas of biology, has been revolutionized in recent years by changes in sequencing technologies.

Good quality, well annotated genome sequences are available for many of the most economically important plant-parasitic nematodes and for many crop plants. Even where genomes have not been sequenced, detailed transcriptomic datasets are often available. New mapping techniques, based on enrichment sequencing targeted at resistance genes, are making it possible to identify candidate resistance genes against many pathogens, including nematodes (e.g. Armstrong *et al.*, 2019). Developments in genomic technologies are therefore underpinning increased understanding of both the fundamental processes underlying infection, as well as allowing new options for control to be developed.

Acknowledgements

Figure 9.1 contains images from NemaPix 1 (Mactode Publications), originally produced by J.D. Eisenbach. Figure 9.5 is reproduced from Gheysen and Fenoll (2002) with permission from Annual Reviews. The authors thank Bartel Vanholme for Fig. 9.2 and for comments on an earlier draft of this chapter.

10 Ecology

SOFIA R. COSTA*

CBMA – Centre of Molecular and Environmental Biology, University of Minho, Portugal

10.1. Introduction

Plant-parasitic nematodes feed on plants and have traditionally been studied for their impacts on crop performance and yield in agricultural conditions. For decades, nematology has been a discipline very much linked to plant pathology, and forms of plant protection against these disease-causing agents were sought. The vast majority of plant-parasitic nematodes dwell in the soil and feed in or on plant roots, which seems a very straightforward, simple pairwise interaction. For decades, nematologists have tried to develop tools to control plant-parasitic nematodes, to reduce their populations to levels that would not significantly affect yield beyond an economical threshold. Although relevant progress has

*sofia.costa@bio.uminho.pt

  Plant Nematology, 3rd edn (eds R.N. Perry, M. Moens and J.T. Jones)
DOI: 10.1079/9781800622456.0010

been made to mitigate crop losses due to nematode attack, a 'silver bullet' has never been found but we have progressively come up with better, smarter, more sustainable solutions. However, plant-parasitic nematodes are becoming more difficult to manage, as agricultural systems and edaphoclimatic conditions, as well as host plants themselves, have been changing. In this chapter, the ecology of root-parasitic nematodes is addressed, giving an overview of the interactions between them and other organisms and with their environment. Knowledge gained from the study of these interactions has provided – and continues to provide – the scientific bases for control methods for plant-parasitic nematodes (e.g. host resistance, biological control), whilst pushing ecological theory development in soil ecology in all types of terrestrial environments, and ultimately estimating how, and to what extent, these soil-dwelling plant parasites affect ecosystems.

10.2. The Soil Habitat

Soil is one of the most complex environments: a non-renewable resource formed over the course of millennia under the influence of diverse parent material (bedrock or sediment deposition), climate and biota. Soil – the resulting three-dimensional matrix of mineral and organic material, air, water and organisms – is the product of an almost infinite number of combinations of its formation factors and is extremely heterogeneous across spatial scales. This heterogeneity provides habitat to a breadth of organisms we have yet to understand fully, in what has been coined the third biotic frontier (Giller, 1996). Analysing the taxonomical and functional diversity of soil biota in their environment, discerning their interactions, and inferring on processes and functions in which they intervene, in an environment so unamenable to direct observation, has been the task of the area of soil ecology. Among the different horizons, each with distinct ratios of mineral to organic material, dominant physical–chemical processes and biological activity, the main focus of soil ecology has been on topsoil – the superficial layer down to ~20 cm depth, usually colonized by plant roots, and containing sufficient organic matter, air and water to sustain the soil biota that most directly impact the (above-ground) ecosystem.

10.2.1. Soil texture

Originating from the parent geological material, and continually altered by weathering processes, the mineral fraction is a major determinant of physical soil properties. Although mineral nutrients can be made available from geological material through chemical reactions that include enzymatic activity, their amounts are restricted by nutrient element constituents in the original minerals. Igneous, metamorphic or sedimentary rocks are widely diverse in elemental nutrient content, but also in the way they fragment into particles of smaller size. This is particularly important in soil, because the relative proportions of mineral particles of different sizes – texture – is a soil property that affects others, including soil water holding or drainage capacity, tillability, erodibility and aggregation. Soil texture classes (e.g. clay, sandy, silt–loam) are relatively easy to approximate in the field – without actually determining the percentage of mineral constituents sand, silt and clay – through the texture-by-feel method. Soil texture is a relatively stable property of soils; its easy determination and relationship with other soil properties (as above examples) makes it a key parameter in the standard soil analyses procured by farmers (Bot and Benites, 2005). The incidence and severity of attack by plant-parasitic nematodes have been related to soil texture classes, with higher densities and crop damage being associated mainly with sandy soils. A wide generalization has been made for root-knot nematode (*Meloidogyne* spp.) infestations being more associated with larger reproduction and greater crop losses in sandy rather than clay or heavy soils (Van Gundy, 1985). This is attributed to the size of soil pores and the water potential formed in these pores, both major physical determinants of nematode colonization ability (through their size exclusion limit). Although higher porosity and less drainage may occur in loam soils, the size of pores formed in sandy soils and their matric potential (the force of water binding to soil, dependent on surface tension and thickness of water films) may be more adequate for nematode movement and activity (see Chapter 8). The texture effect on densities of plant-parasitic nematodes is therefore associated with porosity, and hence not directly by texture itself, but to the arrangement of the mineral, organic, air, water and organisms: the soil structure.

10.2.2. Soil structure

Mineral materials and organic matter are not homogeneously distributed in the soil matrix, but rather form aggregates of various sizes and composition, contributing greatly to the heterogeneity of the soil environment and enabling a large complexity of microhabitats in soil pores. These pores form a vast network of corridors throughout the matrix, and are inhabited by a wide range of soil organisms that constantly interact with the remaining soil constituents. Nematodes can move through sandy soils with little structure, but in other types of soil they are most likely to inhabit macropores or interstitial spaces around soil macroaggregates (larger than 2 mm), whilst the smaller-sized microbiota colonize microaggregates themselves (Zhang *et al.*, 2013). As nematodes cannot exert sufficient force to break through soil aggregates, they can more easily move across longer distances through burrows formed by plant roots or macrofauna. Plant roots and associated mycorrhizal fungi are crucial for the formation and stability of soil aggregates, by binding and compressing soil particles. This is affected by root traits such as length, but also through exudation patterns that stimulate microbial activity and promote mycorrhizal fungi hyphal networks that bind soil particles together, also forming pathways for water flow (Gould *et al.*, 2016).

Porosity ultimately affects soil water retention capacity, determining the range between the extreme conditions of waterlogging (all pores filled with water and no air available) and the permanent wilting point (water only in micropores, unavailable to most organisms); both extremes severely constrain biological activity. According to the diameter of soil pores, they typically hold different proportions of air and water, both needed to sustain life. As nematodes are primarily aquatic organisms, they too are affected by soil water content, and especially by the matric potential. The relationship between plant-parasitic nematode movement and activity and the matric potential is covered in Chapter 8, Section 8.4.2 and Fig. 8.5.

It is important to keep in mind that neither water nor air in soil pores are 'clean'. Soil pores offer not just a physical route across soil for the establishment and movement of organisms, but also a means of short- and long-distance communication. In-pore water is in fact a complex solution of nutrients solubilized from the mineral fraction, the water-soluble fraction of organic matter and a range of metabolites originating from root exudates, as well as enzymes and compounds secreted or exuded by soil biota. Similarly, in-pore air is filled with not only atmospheric gases like molecular nitrogen, carbon dioxide and oxygen, but also volatile compounds that can travel longer distances more rapidly than water-soluble molecules (Reynolds *et al.*, 2010). Nematodes, as well as other soil organisms, use chemical cues and follow gradients to search for food sources (see Chapter 8). It is well established that plant-parasitic nematodes may respond to root exudates to stimulate hatching, locate suitable hosts or move towards their roots. Nematodes can communicate between themselves through ascarosides (see Chapter 7), glycolipid molecules that carry nematode behaviour- and physiology-altering messages according to different blends, concentrations or chemical composition. Interestingly, although ascarosides are constitutively and exclusively produced by nematodes, they can be perceived, transported and altered by plants and other soil organisms, thereby changing their 'message'. This is a critical feature of soil communication: as these molecules are released into soil pores, they are publicly broadcast. A wealth of different organisms evolved to eavesdrop and exploit communication among others, and can further hijack these messages and repurpose them (Yu *et al.*, 2021).

A key determinant of soil structure is soil organic matter content that, owing to its multiple roles in several functions, is also a routinely measured parameter in standard soil analyses. Organic matter can promote the formation and stability of soil aggregates whilst increasing porosity, absorbs water and adsorbs nutrients, ultimately providing habitats and resources for many soil organisms.

10.2.3. Soil organic matter

The brown colour of most soils is due to their organic matter content. This originates from living organisms, and can have numerous sources including exudations, secretions, waste and dead organisms, but a major part consists of plant materials, either from the shoot (leaf litter) or root tissues (rhizodeposition). Litter accumulates at the soil surface, preventing erosion and promoting water retention and infiltration. Organic matter in multiple stages of decomposition is incorporated into deeper soil layers by bioturbation, the intermixing of organic and mineral material by soil organisms, that further promotes soil aeration through the formation of macropores and tunnels through soil (Marhan and Scheu, 2006).

Through decomposition, weak organic acids (e.g. humic, fulvic) are formed, reducing but buffering soil pH that in turn increases nutrient availability to plants. As decomposition proceeds, nutrients in organic matter are mineralized and the decaying materials are left with progressively larger C:N ratios. The ensuing production of polysaccharides, combined with colonization of organic matter by decomposers, leads to stronger soil aggregation, and hence improved soil structure. Stable soil organic matter – humus – is made up of large amounts of complex, nutrient-poor and carbon-rich molecules such as cellulose and lignin, and consequently has a very low decomposition rate. This 'nutrient-spent' organic material is, nevertheless, of key importance to soil processes: it is highly hydrophilic and of weak negative charge. This means it can absorb water and adsorb cations (including nutrient ions Ca^{2+}, Mg^{+}, K^{+}) in forms that are then made available to plants (Bot and Benites, 2005).

The formation of humus is a prime foundation for carbon sequestration. In fact, soils are a relevant reservoir of carbon in the planet, and importantly so, as carbon in the atmosphere, e.g. CO_2 and methane, contributes to global climate change through the greenhouse effect (IPCC, 2019). Carbon in soil can be of an inert mineral origin (in the case where carbonate rocks form part of the parent material), but organic carbon originating from organic matter is a major governable component of soils. Recent strategies for **carbon farming** indeed aim to sequester and store carbon in soils and crops in agricultural production systems. Beyond the intended offsetting of greenhouse gas emissions by agricultural activities, carbon farming aims to serve multiple purposes for soil quality and health (MacDonald *et al.*, 2021). The importance of organic matter for soil biodiversity and function is addressed below for soil food webs (Section 10.2.4) and specifically for plant-parasitic nematode control (Section 10.7).

10.2.4. Soil food webs and energy channels

Through photosynthesis, plants combine mineral nutrients taken up from soil through their root systems with atmospheric carbon to form organic compounds. They are therefore producers, and form the basis of a food chain built on matter and energy from primary production. Primary production in the '**green**' food chain is consumed by herbivores – including plant-parasitic nematodes – that are themselves fed upon by higher-level consumers. Among root herbivores, plant-parasitic nematodes have a relevant role in returning plant-fixed carbon into the soil carbon pool when they are preyed upon or after their death (Gan and Wickings, 2020).

Primary production is plentiful in most ecosystems, with plants producing an estimated 100 gigatonnes (10^{14} kg) of biomass yearly across the planet. However, a considerable part of this plant biomass – up to 90% – is never consumed by herbivores, and inevitably ends up returning to soil as decaying organic matter (Gessner *et al.*, 2010). This organic matter forms the basis of a parallel food chain, that of decomposition, or the '**brown**' food chain. Organic matter is broken down by grazers and detritivores, transported and mixed through soil by ecosystem engineers, and consumed by bacteria and fungi, the decomposers. Neither bacteria nor fungi feed by ingesting material: digestion is extracellular. Both groups of organisms secrete enzymes that break down and solubilize complex molecules for subsequent uptake. In addition to their hyphal character that allows them to infiltrate, fragment and access more stable organic matter, fungi have a wider and more specialized arsenal of enzymes that can decompose all sorts of substrates, including nutrient-poor molecules with a large C:N ratio (Alison, 2006). The fungal decomposition channel is therefore associated with humus content, and more stable soil conditions. However, as digestion is extracellular, fungal-digested material can be taken up by 'cheating' bacteria, that are the most abundant and diverse organisms in soil. With their rapid turnover rates, bacteria rapidly colonize and exploit 'easy' resources (such as root exudates and excretions from other organisms) and can decompose nutrient-rich material. Therefore, the bacterial decomposition channel is associated with more labile and enriched conditions. Mineral nutrients – simple, small non-organic molecules originated by the decomposition process – are released, being available for absorption by plants. Importantly, with the often-negligible exception of algae or other autotrophs, no other organism in the soil food web can use mineral nutrients.

Decomposers themselves are food for their primary consumers, e.g. bacterial- and fungal-feeding nematodes, and their population densities reflect that of their prey. Free-living nematodes, as opposed to parasitic nematodes, do not need a host to complete their life cycle, and include bacterial and fungal feeders, omnivores and predators (Yeates *et al.*, 1993). Free-living nematodes should not be confused with the mobile, roaming stages of parasitic nematodes, as this term 'free-living' refers to their biology, not their movement. Bacterial- and fungal-feeding nematodes have an indirect, regulatory role in decomposition by controlling populations of decomposers (Ingham *et al.*, 1985).

Higher trophic level consumers include omnivorous and predatory nematodes that prey upon organisms feeding from the green or the brown energy channel, and bring together both channels (Fig. 10.1). By providing this overall loop in soil energetics, they provide both direct and indirect regulation of the food web, e.g. a generalist predator can increase and maintain its population by feeding on the most abundant organisms, irrespectively of whether they are in the brown or green energy channel (Moore *et al.*, 2004). Consequently, a large population of predators that increase by feeding on bacterial-feeding nematodes can become regulators of plant-parasitic nematodes.

Interestingly, nematodes are represented in all trophic levels of the soil food web (with the obvious exception of producers), and their community composition provides an overview of all soil energy channels and food web processes (Fig. 10.2). Together with their abundance in all biomes and environments, easy identification, and a wealth of supporting literature, this makes nematodes widely used bioindicators (e.g. Bongers and Bongers, 1998; Yeates and Bongers, 1999).

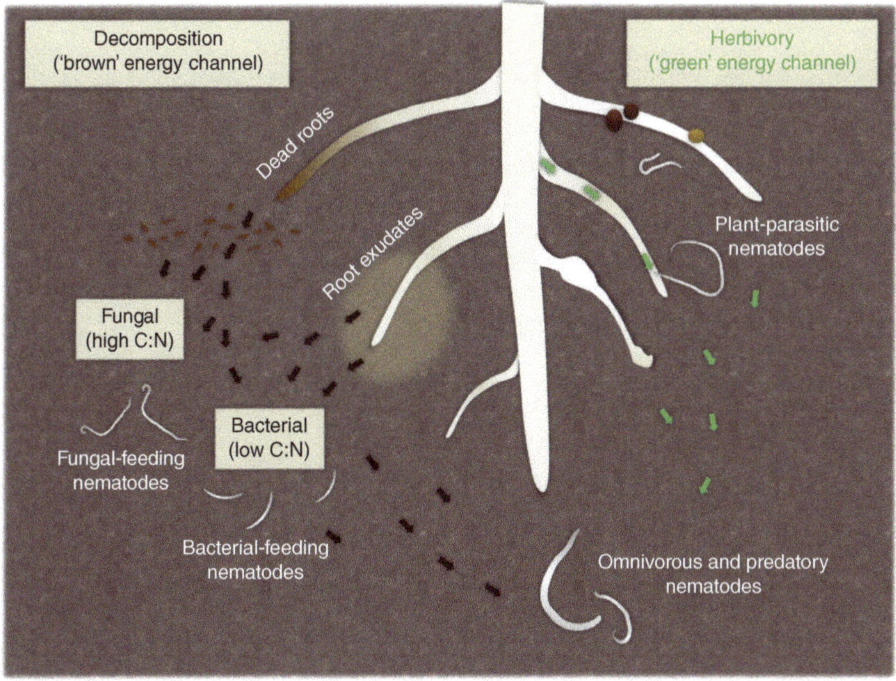

Fig. 10.1. Overview of nematode contributions and relationship with soil energy channels.

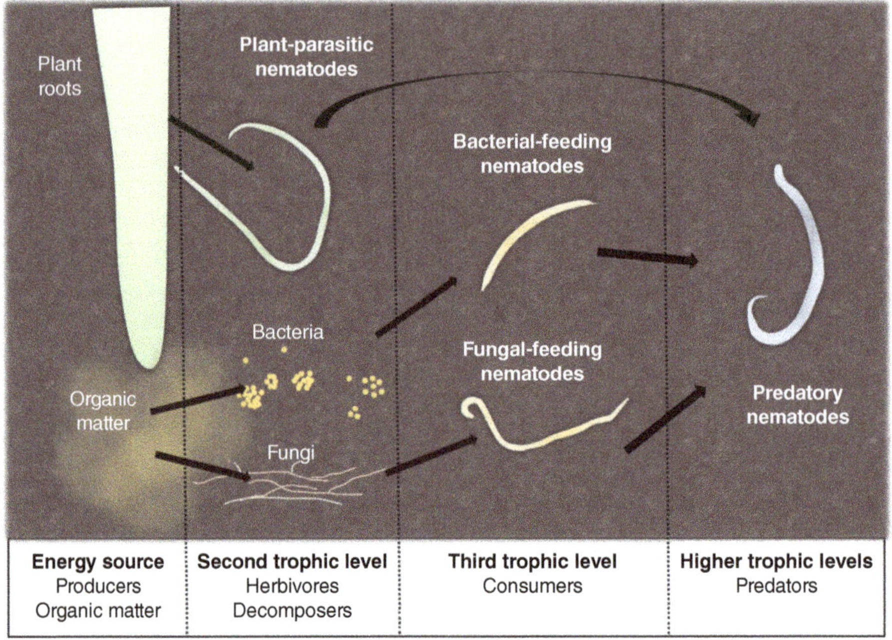

Fig. 10.2. Simplified food web depicting trophic levels and links between different functional groups of soil-dwelling nematodes.

Long, complex, self-regulating food webs (with numerous trophic levels including top predators) only establish under little disturbance, and enough energy (i.e. food) to support them, both from the green and the brown energy channels. Such conditions are rarely met in intensive production systems owing to a range of anthropogenic soil physical and chemical disturbances (Section 10.7). The simplified, shortened food webs in unsustainable agroecosystems lose balance and key functions, namely their regulatory role. In conditions where only the green channel is promoted by focusing on the needs of the crop plants, larger populations of plant-parasitic nematodes can develop, as they are left with fewer regulation hubs in the remaining soil food web. In such conditions where natural regulation is compromised, outbreaks of plant-parasitic nematodes are more common (van der Putten *et al.*, 2006), and their population sizes may effectively be limited only by host suitability and competition among them (but see Section 10.4). Both aspects depend heavily on plant-parasitic nematode diversity, identity and associated niche.

10.3. Functional Diversity of Plant-parasitic Nematodes

Plant-parasitic nematodes have historically been classified according to feeding location and host interactions as ecto-, semi-endo- or endoparasites (see Chapter 1, Section 1.4), and this reflects the strength and specialization in interactions with their hosts. Endoparasites, whether they are polyphagous or not, are considered specialized parasites that interact very closely with their hosts. Ectoparasites, however, are mostly considered generalists and, apart from the economic importance of some species that act as virus vectors (families Trichodoridae, Longidoridae) or other specific systems (e.g. stubby roots induced in grasses by trichodorids), their effect on crops is mostly disregarded. This does not mean their effects are negligible; they can have pronounced effects on plant performance and on interactions with other nematodes (Section 10.4.2).

10.3.1. Functional guilds and the Plant Parasite index

In ecological assessments, plant-parasitic nematodes are usually categorized in functional guilds by their life strategies, in a range of **colonizers to persisters**, an adaptation of ecological concepts of r-strategists and k-strategists, respectively. This **cp scale** reflects a spectrum of range-expanding, rapidly reproducing and short-lived nematodes (at cp 1) to large-bodied, disturbance-sensitive nematodes with high longevity and few offspring (cp 5). As all plant-parasitic nematodes are obligate parasites needing established plants to survive, none are considered pure r-strategists, and therefore none are classified as cp 1, a category exclusive to some free-living nematodes (Bongers, 1990). Many of the most scientifically and economically relevant plant-parasitic nematodes (according to Jones *et al.*, 2013) are classed as cp 3, and hence are expected to thrive in conditions of primary production availability and nutrient supply to plants. Theoretically, cp 2 nematodes are more prevalent in natural conditions under high plant–plant competition and low nutrient availability, whereas the higher cp-group trichodorids and longidorids, usually reported in low densities, may be more dependent on the nutritional quality of plant roots (Bongers *et al.*, 1997) (Box 10.1).

The **Sigma Maturity Index** (ΣMI) reflects the weighted average cp value of the entire nematode community, comprising both free-living and plant-parasitic nematodes (Yeates, 1994). Because diversity and abundance of plant-parasitic nematodes are very responsive to vegetation parameters and not always directly affected by soil quality, it is a seldom-used index. The maturity of assemblages of free-living and plant-parasitic nematodes is usually analysed separately as the Maturity Index, **MI** (for free-living nematodes; Bongers, 1990), and the Plant Parasite Index, **PPI** (for plant-parasitic nematodes, Bongers *et al.*, 1997). The MI is a successional structure index, with smaller values associated with early colonized or impacted habitats dominated by enrichment or general opportunists, and higher values indicating more stable, less disturbed habitats. The PPI was not proposed as an index of disturbance – and indeed may reveal opposite trends and trajectories to free-living nematode communities at the same site. The PPI tends to increase with fertilization and primary production, especially through nutrient allocation to roots (Freckman and Ettema, 1993).

Box 10.1. Colonizer–persister (cp) values of some common nematodes parasitic on plant roots.

cp 2 – *Cacopaurus*, *Gracilacus*, *Paratylenchus*
cp 3 – *Globodera*, *Helicotylenchus*, *Heterodera*, *Meloidogyne*, *Nacobbus*, *Pratylenchus*, *Radopholus*, *Rotylenchulus*, *Tylenchorhynchus*
cp 4 – *Nanidorus*, *Trichodorus*
cp 5 – *Longidorus*, *Xiphinema*
(information gathered from the Nemaplex Nematode Ecophysiological Parameters database)

Functional guild-based indices are highly informative and have allowed the diagnostics and understanding of the food web condition and trajectories facing numerous and diverse disturbances. However, most of these descriptors are based on free-living nematodes, not on plant-parasitic nematodes, and are therefore beyond the scope of this chapter. For a review on nematode-based indices and their applications, refer to Du Preez *et al.* (2022).

10.3.2. Metabolic footprints and herbivory pressure

A major advancement for nematode ecology has been the concept of nematode metabolic footprints that provide the much-needed quantitative detail and interrelationship with soil functions and ecosystem services (Box 10.2). Metabolic footprints reflect the carbon-based ecophysiology of nematodes, taking into account their biomass, activity and investment in reproduction, growth and respiration (Ferris, 2010). They can be calculated for individual nematode species, genera, family, trophic group, whole community or any grouping of nematodes of interest, and allow for the determination of the allocation and fate of carbon in food web components and processes. As nutrient availability increases, populations of free-living enrichment opportunists (mostly cp 1 bacterial feeders), increase their activity and abundance. **Enrichment** in the soil food web can contribute the required abundance of food to develop long food chains that can support higher trophic level omnivores and predators in complex food webs. This complexity, accompanied by a higher diversity and connectance within and among trophic levels in the soil food web, is described by its **Structure**. As nematodes are bioindicators of food web processes, the magnitude of nematode metabolic footprints indicates the activity of all food web organisms involved in a given process. For example, the **Enrichment Footprint** and the **Structure Footprint** are valuable indicators of the decomposition and nutrient cycling service, and

Box 10.2. Ecosystem Services.

Where functions performed by communities of organisms in their environment are considered beneficial from a human perspective, they are termed **Ecosystem Services**. These were first introduced by the **Millennium Ecosystem Assessment** (MEA, 2005) and classed into four interrelated categories:

Provisioning Services – products obtained from ecosystems, such as food, feed and fibre, and genetic resources. For example, plant-parasitic nematode resistance genes exploited in plant breeding.

Regulating Services – benefits obtained from the regulation of ecosystem processes, including pest and disease suppressiveness, and climate regulation. For example, predation of plant-parasitic nematodes by nematode omnivores and predators or other forms of food-web control in agricultural production systems; in natural ecosystems, plant-parasitic nematodes herbivory may regulate plant performance and distribution.

Cultural Services – non-material benefits such as knowledge systems, aesthetic and spiritual values. For example, ecotourism or historical significance of rural landscapes.

Supporting Services – underlying all the above, having an indirect or long-term benefit to people, such as soil formation and nutrient cycling. For example, the intervention of bacterial and fungal-feeding nematodes in decomposition.

of the natural regulation service, respectively. The **Herbivory Footprint** is calculated based on plant-parasitic nematode assemblages, yet describes herbivory pressure in a given system, i.e. how much carbon is being taken up from plants by herbivores. In agricultural conditions, herbivory is a relevant ecosystem disservice, i.e. from a human perspective, nematode feeding on crops is detrimental to the agroecosystem. Hence, the herbivory metabolic footprint is a highly valuable tool, providing estimates of how much carbon is being lost to pests and diseases (Sánchez-Moreno and Ferris, 2018).

As ecophysiological parameters, metabolic footprints are calculated based on nematode biomass and contributions to processes in the ecosystem, not solely on abundance of taxa. Fortunately, their calculation can be swiftly performed using the **Nematode Indicator Joint Analysis** (NINJA) app (Sieriebriennikov *et al.*, 2014), currently hosted at https://shiny.wur.nl/ninja/. However, results need to be interpreted under the NINJA assumptions of nematode parameters that use estimates of adult female body sizes. This can lead to very high Herbivory Footprints when, for example, large densities of infective second-stage juveniles (J2) of root-knot nematodes, which are non-feeding and a fraction of the size of adult females, are present in a given sample.

Even though plant-parasitic nematodes are classified into groups according to their taxonomy, life strategies or functions, they are not separated in soil or in roots. In fact, it is not unusual to find more than 10 species in a small field plot, different plant-parasitic nematode genera in a 100 ml soil sample, or even sharing the same host root – sometimes in very close proximity.

10.3.3. Plant-parasitic nematode niches

The niche is a well-established concept in ecology, albeit under several possible definitions and interpretations from its initial proposal by Hutchinson (1957). It describes the preferences of a given species in terms of their abiotic environment and food sources, and ultimately dictates what habitat they can colonize and how successfully they are expected to establish. Host sharing by different plant-parasitic nematodes, feeding from the same roots, may seem contradictory given that each species is bound to have a different niche. It is therefore important to distinguish between a fundamental and a realized niche. A **fundamental niche** is a multidimensional set of ideal conditions in which a given species performs best. A **realized niche** is an approximation of the fundamental niche in actual, real conditions, and under a set or limitations of varying strengths. When species niches have some overlap, plant-parasitic nematodes may find themselves having to share resources, and inevitably enter competition, which affects their performance and overall fitness – an assessment of how many offspring the population can produce. Common constraints to plant-parasitic nematodes in their realized niches include predator pressure, resource sharing, resistant hosts, suboptimal temperatures, among others, and are discussed in the following sections.

10.4. Regulation of Plant-parasitic Nematodes by the Soil Food Web

A variety of direct trophic and indirect interactions continually provide regulation of populations of plant-parasitic nematodes in natural ecosystems. Whereas some have been exploited for active plant-parasitic nematode control in agroecosystems, others may act intangibly or be difficult to implement as a curative tactic. In this complex network of organisms feeding on each other in soils, relevant loops in food webs are formed and contribute to their stability, but may need to be disentangled to be understood. Such are the sources for the often-unanticipated indirect effects that ripple through the food web.

Food web controls are classified according to the acting food level: (i) top-down control refers to the limitation of plant-parasitic nematode populations by higher trophic levels, i.e. exerted by predators and parasites; (ii) horizontal control relates to effects within the same trophic level of plant-parasitic nematodes, and concerns their competition; (iii) bottom-up control denotes the constraint of plant-parasitic nematode populations by the bottom trophic level in the primary production food chain: the plants; and finally (iv) indirect control consists of interactions resulting from trophic links among more than two populations, irrespective of their food web position (Fig. 10.3).

10.4.1. Top-down control

Plant-parasitic nematodes are themselves a food resource for several organisms in the soil food web. These include tardigrades, enchytraeids, predatory nematodes, protozoans, fungi and bacteria, with varying impacts on plant-parasitic nematode populations (Stirling, 1991). Some bacteria and fungi act upon plant-parasitic nematodes by producing nematotoxic compounds that can repel, paralyse or kill nematodes, indirectly affecting their populations by antibiosis. Such antagonists can then either feed on the nematodes or predispose them to attack by other organisms – an immobile nematode in soil is a 'sitting duck'. Natural enemies are those organisms that actively feed on plant-parasitic nematodes in a direct trophic interaction.

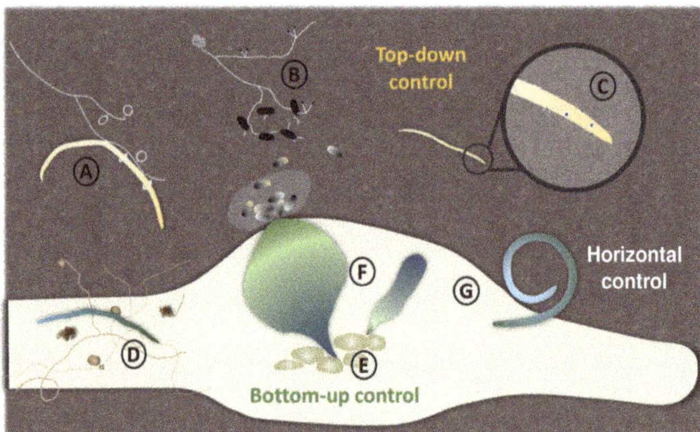

Fig. 10.3. Mechanisms exerting plant-parasitic nematode control *via* the food web. Top-down control: A: nematode-trapping fungus *Arthrobotrys dactyloides*; B: fungal parasite of sedentary stages *Pochonia chlamydosporia*; C: endospore-forming bacterial parasite *Pasteuria penetrans*. Bottom-up control: D: mycorrhizal fungi; E: plant root nutrients. Horizontal control: F: intraspecific; G: interspecific.

A wide diversity of natural enemies of plant-parasitic nematodes, with various degrees of specificity and strategies of capturing, infecting and feeding on nematodes, can suppress their populations. Predatory nematodes, for instance, hunt and feed on plant-parasitic nematodes and have been shown to be able to reduce their populations (Ferris *et al.*, 2012). But because they are among persister groups, in the cp 4 or cp 5 categories (see Section 10.3.1), they are frequently absent or present in very low densities in the agricultural conditions that are conducive to plant-parasitic nematodes. Long and complex food webs that can support a large abundance and diversity of higher trophic levels that regulate populations of plant-parasitic nematodes are seldom found in intensive agriculture.

Nevertheless, nematode-suppressive soils were encountered in agricultural conditions where even under continuous cereal monocropping, crop-specific nematodes, such as *Heterodera avenae*, could not increase their populations (see also Chapter 14). This was an exciting discovery back in the 1980s, followed by an intense research effort to characterize the underlying mechanisms of suppression (Kerry *et al.*, 1982). In these soils, a wealth of nematode microbial enemies was detected. These included bacteria and a range of fungal natural enemies, encompassing various strategies to nematode attack. Nematophagous fungi, those that consume nematodes, can be obligate nematode parasites (feeding only on nematodes) or facultative parasites, in which case they can additionally obtain their food saprophytically from soil organic matter (the brown energy channel). Facultative nematophagous fungi are widely considered generalists, which, bringing together the brown and the green energy channels, putatively gives them a stronger regulatory role in the food web. Among facultative parasites, most examples concern those that feed on sedentary stages of plant-parasitic nematodes: eggs, females of sedentary endoparasitic nematodes and cysts. But a large and diverse group of fungi involved in decomposition processes actively capture mobile nematode stages. These are the nematode-trapping fungi (e.g. *Arthrobotrys*, *Dactylella*) that, under the influence of numerous genetic and physiological factors, form traps to capture nematodes. Among these factors, the detection of nematode-produced ascarosides has been proposed as a trigger for trap production (Hsue *et al.*, 2013). It is still to be determined whether endoparasitic fungi (e.g. *Catenaria*, *Myzocytiopsis*) – mostly primitive fungi forming zoospores that follow nematode exudation gradients and infect them through body openings – actually use ascaroside signals as well.

Some nematophagous fungi have gained research interest for the development of plant-parasitic nematode biocontrol agents in agricultural production systems, due to their feeding preferences, amenability to multiplication and formulation in industrial systems, and ease of establishment in the field (see Chapter 14). Classical examples include *Pochonia chlamydosporia* and *Purpureocilum lilacinum*, both facultative generalists, albeit with a level of food preference, that parasitize sedentary plant-parasitic nematode forms, including cyst nematodes *Globodera*, and root-knot nematodes. As facultative parasites, their ability to feed on organic matter lends them to mass propagation in industrial conditions, and also means they can establish, increase and maintain their populations in agricultural conditions even when population densities of plant-parasitic nematodes are low (Manzanilla-Lopez *et al.*, 2013).

Among bacterial natural enemies of plant-parasitic nematodes, *Pasteuria penetrans*, a highly specific parasite of root-knot nematodes is the best-characterized example. These bacteria, closely related to *Bacillus*, can be specific at the subpopulation level, and form resistant endospores that are coincidentally their infective stage. Endospores can withstand chemical and physical disruption, maintaining viability even under extreme disturbance. This is a good ecological strategy to maintain its population in soil, but *P. penetrans* propagation and sporulation in laboratory conditions has been a challenge for its development as a biocontrol agent. As endospores are immobile structures, infection of plant-parasitic nematodes depends on nematode movement to enable contact. This dependence on nematode movement also implies that non-compacted, sandy soils offer better conditions for *P. penetrans* control of root-knot nematodes (Davies, 2009).

Due to legislative and industrial constraints, the application of microbial enemies as biocontrol agents to suppress populations of plant-parasitic nematodes in the field has often relied on the formulation of one or very few isolates. And yet, both in naturally suppressive agricultural soils and in natural ecosystems, a large diversity of microbial enemies, encompassing a range of ecological strategies and niches, has been found. This diversity is thought to be responsible for successful regulation of plant-parasitic nematodes, acting through various mechanisms. The highly nematode-specific *P. penetrans*, for example, can reduce root-knot nematode populations in coastal sand dunes and, in doing so, decrease their share of plant resources, indirectly facilitating lesion nematodes, *Pratylenchus* (Costa *et al.*, 2012). Depending on the specific or generalist nature of natural enemies, top-down control can therefore tip the balance among plant-parasitic nematode populations, where they are competitors for the same plant resource.

10.4.2. Horizontal and bottom-up control

Competition among plant-parasitic nematodes, whether it is inter- or intraspecific, has long been shown to reduce their fitness, i.e. reproductive capacity, on a given host. It must be noted, however, that plant-parasitic nematodes have never been shown actually to harm each other directly, exhibiting any form of aggression or territory protection against other nematodes (interference competition). Their horizontal control derives from exploitative competition – or resource limitation – and is therefore stronger where resources are limited, e.g. when their host is resistant and/or under abiotic or biotic stress. Even though it can be argued that plants can host much larger plant-parasitic nematode numbers than are usually detected (Norton, 1989), bottom-up (host effects) and horizontal control can be viewed as tightly interlinked mechanisms for the regulation of plant-parasitic nematodes, and should therefore be analysed in an integrative manner.

The larger the niche overlap of plant-parasitic nematodes, the less they can share their host by resource partitioning – taking up plant resources from different sites or origins – and the stronger horizontal control can become. This implies that stronger horizontal control can develop intra- rather than interspecifically, with nematodes of the same species exerting population control on themselves, as they have the same fundamental niche. The early Seinhorst models of population growth in a species–host combination indeed reflect the tapering curve of population increase when large densities are reached. Mechanisms to avoid intraspecific competition have arisen, namely the generation of male nematodes and larger male:female ratios in *Meloidogyne* when in high densities and/or in poor hosts (Davide and Triantaphyllou, 1968; Perpétuo *et al.*, 2021). Although they can form galls in roots, male nematodes in this genus do not feed, and hence the more males are produced, the more resources will be available for females. The population density determining the ratio of males to females depends on host suitability, and is expressed by its **carrying capacity**. This concept refers to the number of plant-parasitic nematodes that a plant host can support without losing its ability to grow and develop. As the carrying capacity is reached, whether or not more males are generated, the fitness of plant-parasitic nematodes decreases as they cannot obtain enough resources to sustain unrestricted reproduction.

In theoretical models of interspecific competition, a proportion of nematodes from a given species is considered as feeding equivalents to those of the competing species, and therefore take up a proportion of the carrying capacity of their shared host. Importantly, interspecific competition can be asymmetrical, whereby the different competitors capture unbalanced proportions of each other's host carrying capacity, making some species of plant-parasitic nematode stronger competitors than others.

The overarching effects of competition on populations of plant-parasitic nematodes are more complicated to assess when several interacting species are present. Although potentially having feeding preferences, generalist plant-parasitic nematodes can feed from different hosts, and benefit from plant diversity. With the capacity to move to the roots of another plant when their carrying capacity is reached, ectoparasites can putatively maintain their population densities, and in theory would suffer less pressure from, for example, specialist sedentary endoparasitic competitors. Ectoparasites have been shown to outcompete root-knot nematode endoparasites if they are feeding on poor hosts (Mateille *et al.*, 2020). Moreover, having established their feeding sites in susceptible host roots, root-knot nematodes cannot abandon roots and can become weaker competitors (Brinkman *et al.*, 2005).

Competition for host resources is not exclusive to plant-parasitic nematodes, and can occur among plant-parasitic nematodes and other pests and diseases, also being affected by above-ground herbivores. Further, it can occur between plant-parasitic nematodes and arbuscular mycorrhizal fungi that can outcompete root lesion nematodes in localized interactions in roots, possibly for colonization space (de la Peña *et al.*, 2006). A range of plant bacterial and fungal endophytes, including mutualistic rhizobia and mycorrhizal fungi, can promote plant-parasitic nematode control. Whether they directly antagonize plant-parasitic nematodes by the production of active compounds, increase plant health through better nutrition, or induce plant defences, these act in conjunction with the plant, and therefore indirectly act as bottom-up control.

10.4.3. Indirect effects

A range of indirect effects that contribute to regulation of plant-parasitic nematodes arise throughout the soil food web, with or without some form of mediation by the plants. Many of these are being estimated by manipulative micro- or mesocosm experiments, using treatment combinations of plants, plant-parasitic nematodes and other organisms, or by black-box approaches, through which entire functional or taxonomic groups are added or omitted to estimate overall effects. A large part of ecological theory and the basis for some of the knowledge on the above control methods was also developed from collecting field observations, in which nature itself manipulates the factors, and a wide range of organisms, some unknown to science, intervene. These usually involve sampling across different ecosystems and habitats, seasons, depths, vegetation types or chronosequences. As plant-parasitic nematodes have aggregate distributions due to their poor mobility and a series of biological filters to their establishment, composite samples are preferred to individual samples, and an approximately 100 ml subsample of the mixed composite sample processed for analysis. This, albeit a sound ecological method for hypothesis-testing at an appropriate scale, results in averaging of the assemblies of plant-parasitic nematodes and of their interacting organisms.

As technology and computer power progress, improvements in molecular techniques and statistical tools have allowed the unveiling and interpretation of the full assembly of living organisms in and around roots at a fine scale and often precise location. Although this methodology may have scalability issues, it has been yielding exciting results on fine rhizosphere assemblies and the mechanistics of their interactions. In fact, knowledge has been advancing by combining approaches and using the most appropriate set of techniques to address research questions. Molecular analysis in high-throughput conditions followed by established bioinformatics pipelines (see Geisen *et al.*, 2018) can be interpreted via co-occurrence network analysis and/or yield semiquantitative results with relative taxa proportions, and combined with factors determined by advanced sensing or fine analytical tools, by multivariate analyses and structural equation models.

The rhizosphere, the volume of soil colonized by roots and under their direct influence, has for decades been considered a hotspot for biological abundance, activity and diversity. Research on the microbiome associated with roots, comprising several bacteria, fungi and other microscopic organisms, including nematodes, has been uncovering several indirect mechanisms of plant-parasitic nematode control. As they move through the rhizosphere in search for suitable host roots, plant-parasitic nematodes encounter several microbiota that are mainly attracted to, and cultivated by, plants through rhizodeposition, and modulated to a certain extent by top-down control (Thakur and Geisen, 2019). Part of this microbiota is highly specific in adhering to the cuticle of the infective plant-parasitic nematodes and can antagonize nematodes directly or indirectly by inducing plant defence (Elhady *et al.*, 2017; Topalović *et al.*, 2022).

Natural soil suppressiveness to root-knot nematodes found in an organic agricultural production system has been linked to microbiome composition, and could be transferred to otherwise conducive conditions by inoculation of microbiome suspensions (Silva *et al.*, 2022). Whether microbiome-level suppressiveness results from plant recruitment of nematode antagonists has not been demonstrated. However, fine studies have revealed differences in the bacterial and fungal composition of the microbiome associated with plant roots depending on whether they are infected or non-infected with root-knot or cyst nematodes. This includes not only differential colonization of root segments by antagonistic bacteria, but also natural enemies such as *P. chlamydosporia* and *P. lilacinum*. Through their parasitic activity, nematodes change root exudation patterns and composition, but it is thus far unclear whether their antagonists and natural enemies are attracted to altered root exudates or to plant-parasitic nematodes infecting roots. If the perception of plant-parasitic nematodes by microbiota remains elusive, the small-scale variation in root-associated microbiota seems to be perceived by plant-parasitic nematodes themselves, which tend to prefer non-infected root segments (Yergaliyev *et al.*, 2020). This behaviour may reflect self-avoidance to reduce horizontal control, or avoidance of the top-down control by antagonists and natural enemies concentrated at sites infected by plant-parasitic nematodes.

A multitude of control mechanisms can be involved in the regulation of plant-parasitic nematode population density, and act continuously, albeit with different strengths depending on the particular interacting organisms and environmental conditions. Even so, plant-parasitic nematode outbreaks appear widespread in agricultural systems, frequently due to the exponential increase of a single species that may be selected, or released from natural regulation, in particular host-environment associations. These outbreak situations obviously deserve applied research interest aiming to attenuate damage to crops. To design ecologically sound strategies for plant-parasitic nematode control, insights can be gathered from the drivers of their diversity and distribution in natural ecosystems.

10.5. Plant-parasitic Nematodes in Nature

In natural ecosystems, plant-parasitic nematodes have co-evolved with their hosts and with the remaining soil community in a wide range of environmental conditions that span virtually all habitats colonized by plants. Due to their exclusive feeding on plants, plant-parasitic nematodes are directly involved in processes that maintain plant diversity and distribution. Numerous studies have demonstrated the ecosystem-level processes that regulate plant-parasitic nematode populations contribute to their population and community patterns, drivers and pressures. Here they are introduced at different scales, from local to regional and global patterns, but it is expected that the described effects act at multiple levels of spatial resolution.

10.5.1. Local effects: plant–soil feedbacks

As major contributors to both the green and the brown energy channels, plants shape their rhizosphere communities. Plant litter is a major input into the brown energy channel, and is promptly colonized by decomposers, whose communities develop to specialize in the decomposition (and humification) of this resource. '**Homefield advantage**' has been reported for forest trees, whose decomposer community more rapidly mineralizes nutrients from litter of tree species in their respective soil, indicating some degree of decomposer specialization (Ayres *et al.*, 2009). The rapid, efficient decomposition of plant litter makes nutrients trapped therein more readily available to the plant.

As roots grow through soil, root exudates are recognized by mutualists such as mycorrhizal fungi, rhizobial bacteria, growth promoters, etc. Mutualists are widely regarded as beneficial organisms to plants, promoting their health by obtaining limiting nutrients and improving their defence against herbivores. Most plants, with the notable exception of the Brassicaceae, form associations with mycorrhizal fungi in their roots, which are an energetically inexpensive way for the plant to extend its root prospecting zone. Benefits to the plant depend on the ability of mycorrhizal fungi to obtain water, mine nutrients and, importantly, solubilize phosphorous into plant-available nutrients. Mycorrhizal fungi that perform poorly in this respect still remain harboured into plant roots. The relationship is not always beneficial to the plant in terms of nutrient flows, but as discussed above, mycorrhizas can protect the plant from attack by plant-parasitic nematodes, which could reflect an important trade-off mechanism (Bell *et al.*, 2022). Although experimental evidence is still limited, the maintenance of plant associations with rhizobial mutualists that are poor nitrogen fixers are also suspected to involve trade-off mechanisms for protection against plant-parasitic nematode infection (Costa *et al.*, 2021).

In addition to attracting and establishing associations with mutualists, roots also attract antagonists – diseases and pests like plant-parasitic nematodes. Feeding by nematodes not only consumes plant resources, but it can also reduce their water and nutrient absorption, alter their exudation, and cause nutrient leaching, indirectly promoting neighbouring plants and changing interactions with other soil microbiota (Bardgett and Wardle, 2003). Ensuing root injuries, together with repressed plant defence, can form an entryway for secondary infection by other pests and diseases, and plant-parasitic nematodes have been associated with disease complexes in natural ecosystems.

Unable to move, all plants continually sit in the balance of the community of organisms they culture in their rhizosphere, and that in turn affects their performance and, consequently, their abundance and distribution. Plant–soil feedback theory has for more than 20 years been addressing how below-ground interactions are interlinked with the above-ground plant community patterns (van der Putten *et al.*, 2013). The relative strength of the interactions among plants, decomposers, mutualists and antagonists reflects their more negative or positive plant–soil feedback. Negative feedback is thought to be widespread in nature, and ultimately responsible for plant diversity. Plants that develop positive feedback with their soil community establish more successfully, becoming dominant and more prone to expand their habitat range.

As part of the triangle of plant–soil feedbacks, plant-parasitic nematodes have been shown to affect plant establishment and performance in natural ecosystems. In coastal sand dunes, plant-parasitic nematodes, together with a range of other soilborne pathogens, were responsible for the die-back of the dominant pioneer plant *Ammophila arenaria* (van der Putten *et al.*, 1993). Interestingly, along the sand dune succession, plant species were susceptible to plant-parasitic nematode communities of the subsequent plant in the pioneer-climax vegetation continuum, and therefore plant-parasitic nematodes reportedly have an established role in ecological succession (De Deyn *et al.*, 2003). By preferentially feeding and reproducing in roots of dominant earlier succession plants, plant-parasitic nematodes also accelerate succession in grasslands, and promote plant diversity. Plant diversity in turn offers heterogeneity and complementarity in the quality of resources available to the nematode communities, and hence promotes plant-parasitic nematode diversity, whilst reducing their abundance relative to root biomass (resource quantity). The varied resources also promote the free-living nematode community that may contribute to plant-parasitic nematode control, reducing their abundance. In plant species-poor communities, the overall diversity of plant-parasitic nematodes is reduced, and the abundance of different nematode species reflects their feeding preferences, which results in the dominance of specialist species of plant-parasitic nematodes (Dietrich *et al.*, 2021). Mature plant communities comprising a high diversity of plant species can have a dilution effect on specialist nematodes that cannot sustain their populations on sparse individuals of their preferred plant host. This specialist dilution effect is thought to be responsible for **overyielding**, a well-reported case of alleviation of the negative plant–soil feedback in plant species-rich communities responsible for increased overall primary production. Although overyielding has been attributed to the bottom-up control and top-down control of specialist plant-parasitic nematodes, a recent finding further sheds light on another potential mechanism: that of associational resistance. Nematode-resistant grassland species can alleviate the plant-parasitic nematode burden of their neighbouring plants, by granting them resistance to *Pratylenchus*, even if they are otherwise good hosts to the nematode (Liu *et al.*, 2022a). Interestingly, plant-parasitic nematode infection can be communicated to neighbouring plants that then anticipate nematode attack and react phenotypically to increase tolerance (Zhang *et al.*, 2021).

Taken together, the above findings suggest that, in agricultural conditions, monocultures are expected to perform worse due to the accumulation of specific plant-parasitic nematodes. Strong negative feedbacks on a given crop result in '**soil sickness**', which can be associated with an increased density and activity of plant-parasitic nematodes. Plant-soil feedback theory can also be applied in temporal scales: the cultivation of plant-parasitic nematodes as well as other antagonists, decomposers and mutualists by a given crop forms its soil legacy to the next plant. In agriculture, this '**soil memory**' is the foundation for crop rotation – especially with functionally and taxonomically distinct follow-ups that differ in plant-parasitic nematode host suitability, and with those that increase mutualisms, such as legumes. Considering the role of plant-parasitic nematodes in ecological succession, some form of contribution from different plant strata (i.e. annual vegetables, shrubs, trees) mimicking natural succession also deserves interest. Although not considering plant-parasitic nematodes explicitly in their rationale, this is the basis for agroforestry and syntropic types of agriculture. However, the success of their implementation on plant-parasitic nematode management has not been ascertained. The application of plant–soil feedback theory in agricultural systems has been reviewed by Mariotte *et al.* (2018).

10.5.2. Large-scale effects: vegetation and climate factors

One of the most impactful publications in the Ecology and Environment category was the mapping and quantification of soil nematode trophic group abundance and metabolism in global terms (van den Hoogen *et al.*, 2019). These data have been explored and combined with several other parameters to yield a thorough interpretation of major factors conditioning the distribution of plant-parasitic nematodes (and other trophic groups). With an estimated abundance of 1.25×10^{20} individuals in the top 10 cm of soil across the globe, plant-parasitic nematodes are the second most abundant trophic group of soil nematodes (second only to the numerous bacterial feeders) and second also in fresh biomass (after the larger omnivores). Their abundance is mostly associated with vegetation – and communities dominated by plant-parasitic nematodes related with the NDVI (normalized difference vegetation index), a measure of 'greenness' related to primary production. Plant-parasitic nematodes also follow the unexpected major latitudinal trend of higher densities in temperate broadleaf forests, boreal forests and tundra, rather than being more abundant in tropical biomes. Vegetation and climate are widely considered the main drivers of plant-parasitic nematode diversity and distribution in large spatial scales.

The analysis of distribution patterns of plant-parasitic nematodes has historically been done for single species, especially emerging and quarantine species, and yielded important inferences on their niches (preferred edaphoclimatic conditions, hosts, etc.), providing risk estimates on their expansion. More recently, abundance and community composition of plant-parasitic nematodes have been analysed to inform on macroecological patterns that allows for the detection and interpretation of major community-level drivers and pressures on organisms and ecosystems, that can be used to improve policies and management. On a regional scale and global scale, patterns of distribution have been sought according to land-use categories, but also to land use and management intensity. Data on the communities of plant-parasitic nematodes must then be combined with a collection of other local or remote observations and datasets. Spatially explicit sampling that is also accompanied with precise sampling date can further be used to obtain satellite (remote sensing) data. In this respect, the Copernicus programme through their Sentinel satellites now provides a range of relevant measurements at very fine spatial (down to 10 m resolution since 2018) and temporal scales, scanning any given site on Earth every 6 days (more information at: https://sentinels.copernicus.eu).

Although large-scale studies comparing communities of plant-parasitic nematodes across land-use types are still scarce, they have been yielding impactful results. In global patterns, human management of ecosystems leads to increased densities of plant-parasitic nematodes in soil relative to natural, primary ecosystems, irrespectively of land use type. Whether managed intensively or not, agricultural sites have on average higher plant-parasitic nematode densities than natural systems (Li *et al.*, 2022). Agricultural management and the conversion of natural to agricultural systems act as a filter of plant-parasitic nematode functional or trait diversity, leading to homogenization of their assemblages across different conditions. Expectedly, agricultural sites notoriously exclude plant-parasitic nematodes with larger body sizes and lower fecundity, overall reducing their functional diversity and the PPI (Archidona-Yuste *et al.*, 2021).

Across land use types assessed globally (natural, agricultural, pasture, urban), plant-parasitic nematode abundance is positively related to the NDVI and annual precipitation, but inversely related to mean annual temperature. In a regional scale with a narrower, temperate climatic range, the diversity of plant-parasitic nematodes in natural habitats increased with mean annual temperature, and functional diversity decreased with soil pH, suggesting edaphoclimatic factors drive assemblages in these ecosystems (Li *et al.*, 2020).

Precipitation conditions soil moisture, directly impacting motility, communication and – in extreme levels of drought or flooding – plant-parasitic nematode activity and life requirements. Indirectly, precipitation can also affect plant-parasitic nematodes via the response of vegetation. Water availability in soil influences nutrient allocation and biomass in plant roots, much affecting abundance of plant-parasitic nematodes, especially in natural and rainfed agricultural systems. But given that nematode soil moisture requirements match those of plants, a level of water scarcity can have little effect on plant-parasitic nematode diversity. In low soil moisture conditions, plant-parasitic nematodes show behavioural and physiological adaptations such as moving to deeper soil layers, coiling or entering anhydrobiosis. Spending their entire life cycle exposed in soil, ectoparasitic nematodes are regarded as more susceptible to drought than endoparasites, which can be protected from harsh conditions in plant roots. The colonizer strategy of enrichment opportunists in class cp 1 can form dauer larvae (dauers) that allows them to resist harsh environmental conditions. Although evidence for dauers in plant-parasitic nematodes is fragmentary (see Chapter 7), the J2 of *Meloidogyne*, for example, are able to survive in soil for months until finding a suitable host and can withstand (temporary) extreme conditions.

As they are poikilothermal organisms, nematode activity and life-cycle duration is vastly determined by temperature. At higher temperatures, densities of plant-parasitic nematodes are expected to increase due to faster life cycles, up to the nematode's temperature tolerance limit. Although soil temperatures do not fluctuate so widely as aboveground air temperatures, they can have considerable daily and seasonal variation, and hence cumulative temperature, expressed in day degrees (thermal time), is usually considered for ecological considerations (Trudgill *et al.*, 2005). The minimal (basal) and maximal temperature for nematode development depend on their ecological niche, with tropical species of plant-parasitic nematodes having a higher basal temperature than more temperate species. Within these extreme values, the duration of the life cycle in degree days is constant, assuming other environmental conditions are met. In fact, plant-parasitic nematode nutrition can also affect their life-cycle duration: both poor hosts and competition from other plant-parasitic nematodes can delay their reproduction (Wesemael and Moens, 2012).

The interactive effect of temperature and precipitation is evident in some temperate climates, where colder seasons are more rainy and warmer seasons are drier. Seasonality can affect community composition of plant-parasitic nematodes directly due to niche preferences in both temperature and soil moisture, but also indirectly due to the response of vegetation to environmental conditions. For example, in temperate climate, plant-parasitic nematodes in the cp 3 class can increase their populations in the autumn (increasing the PPI), following the trend in resumed vegetation development after the warmer (and drier) summer (Bakonyi and Nagy, 2000).

At a regional scale, and notwithstanding seasonality and vegetation effects, climatic factors such as mean annual temperature and annual precipitation have not been shown to produce significant differences in diversity of plant-parasitic nematodes in agricultural systems. Therefore, it is noteworthy that community-homogenizing agricultural impacts superimpose the climatic drivers on abundance and diversity of plant-parasitic nematodes (Li *et al.*, 2019).

10.6. The Changing Ecosystem

Although much can be learned from research on ecology of plant-parasitic nematodes in natural systems, it is important to keep in mind what – and how – knowledge can be applied to agricultural production systems. Since crops were first domesticated about 12,000 years ago, agricultural systems have drastically diverged from natural ecosystems. Through centuries of ploughing and fertilizing, topsoil in agricultural production conditions became deeper, finer, nutrient-enriched and more homogeneous, and thus more pliable to sowing and planting. The introduction of heavy machinery, a range of synthetic pesticides and mineral fertilizers, line-cropping and irrigation in the 1970s, have boosted yield, and led to the Green Revolution. This steep yield increase would not have been possible without major investments in crop improvement that saw the rise of hybrid cultivars, especially high-yielding varieties that, as implied, have record productivity, if managed under high-input intensive agriculture and stable environmental conditions. Climate change progressively alters abiotic and biotic patterns and their stability worldwide, posing a major challenge to agriculture.

10.6.1. Crop domestication

Natural selection is ultimately responsible for the range of mechanisms for regulation of plant-parasitic nematodes through the food web presented above (Section 10.4). In natural ecosystems, biotrophic pathogens such as plant-parasitic nematodes are expected to co-evolve with their hosts to become generalists or weaker parasites, under the threat that strong specialist parasites can have such deleterious effects on their hosts that they become (locally) extinct. Host–nematode interactions are rarely assessed in natural systems, but studies in coastal sand dunes seem to support weak negative effects of plant-parasitic nematodes on their hosts (van der Stoel *et al.*, 2006; Costa *et al.*, 2012). Such effects are not anticipated to occur in agricultural settings with high productivity, where plant-parasitic nematodes encounter domesticated crops. Crop domestication refers to the artificial selection of desired traits in wild plants and ancestral varieties. This artificial selection departs from natural selection through which plants co-evolve with their interacting organisms. Natural selection still occurs in domesticated crops as they are grown in agricultural conditions, resulting in unintended traits, such as reduced chemical defence (Soldan *et al.*, 2021).

Natural selection is guided by environmental challenges, and for plant-parasitic nematodes, and especially for endoparasitic nematodes, their environment *is* their host for a large part of their life cycle. The **Red Queen Hypothesis** (van Valen, 1976) proposes that biotic interactions form an essential part of natural selection and evolution. It considers that the net sum of fitness of all populations in a given community is constant, so if any of the populations gains fitness by, for example, having been bred for resistance to plant-parasitic nematodes, the fitness of all other populations is reduced. This acts as selective pressure on the affected populations, which then evolve strategies to improve their fitness by, for example, breaking resistance.

Crop domestication has led to crop improvements that, under appropriate management, guarantee plant survival and performance and overall improve its fitness, resulting in high yield. However, until the last few decades, major investments in crop domestication have seldom focused on promoting traits to increase plant defence or resistance against plant-parasitic nematodes or any other soil-dwelling antagonists. Instead, for centuries, crops have been selected and crossed for their high yield, organoleptic qualities, fast growth and rapid response to external inputs (fertilizers and water). Their continued cultivation in irrigated systems has decreased their investment in root biomass and water-prospective architecture so they have little water-use efficiency (Milla *et al.*, 2017). Under an acquisitive strategy, they take up and invest soil nutrients in N-rich above-ground plant parts, and consequently their leaf litter is highly labile, promoting the bacterial decomposition channel, and increasing bacterial feeder densities in soil (Palomino *et al.,* 2023). Being bred for, and grown in, high-nutrient input systems, domesticated crops do not depend so strongly as their wild progenitors on establishing efficient mutualisms (mycorrhiza, rhizobia) or on cultivating a beneficial soil microbiome, and hence are more prone to developing negative soil feedbacks (Carrillo *et al.*, 2018).

Although there is considerable variation among domesticated crops in the ability to resist, interact or communicate with other organisms in the rhizosphere, they generally have reduced investment in defence compared with their wild progenitors. Expectedly, domesticated genotypes suffer from a higher plant-parasitic nematode infection and lower mycorrhization rates than their wild progenitors or relatives. Their strong negative feedback leaves a soil legacy for ensuing crops that also have a larger plant-parasitic nematode burden, even though their performance is not always affected (Martin-Robles *et al.*, 2020).

Plants react to herbivory by releasing volatile organic compounds that attract natural enemies of their herbivores. The attraction of entomopathogenic (and, interestingly, of plant-parasitic nematodes) by insect-damaged maize roots has been lost in modern hybrids, demonstrating their poor ability to communicate with beneficial organisms (Rasmann *et al.*, 2005). The potential recruitment of natural enemies of plant-parasitic nematodes in the rhizosphere microbiome (Section 10.4.3) or other forms of communication involving plant-parasitic nematodes in domesticated crops has, to my knowledge, not been ascertained.

10.6.2. Global climate change

Agricultural paradigms have been changing as science and technology develop from putatively ancient slash and burn extensification to the conventional intensification of the 1970s, to precision agriculture and agroecology (see Section 10.7). But despite improvement, adaptation and transformation of agricultural management, other profound changes to agricultural production systems, and to ecosystems as a whole, are at work: those brought about by global climate change.

According to scenarios of increased global temperature, the changing climate is predicted to impact negatively on global food production, and high-confidence estimates reveal detrimental effects to agriculture in Africa, Australasia, parts of Asia, Europe and North America (IPCC, 2022). Climate change is expected to have numerous impacts on land (IPCC, 2019), partly due to the occurrence of extreme events, for which forecasting power is low. These extreme events, such as droughts, floods and heat waves, can lead to dramatic consequences for vegetation as well as exerting abiotic stress on plant-parasitic nematodes. Climate change is expected to exacerbate plant-parasitic nematode pressure on crops, but this is not always supported by the literature. In fact, depending on the magnitude and direction of climate change given the ambient (usual, unaltered) conditions of ecosystems, some effects may actually compromise activity and prevalence of plant-parasitic nematodes. The effects of climate change on plant-parasitic nematodes have been extensively reviewed (Dutta and Phani, 2023). Most studies on the effects of global change on species of plant-parasitic nematodes and communities have focused on three overarching and interacting parameters: increasing atmospheric CO_2 levels, changing precipitation amounts and patterns, and temperature increase.

Through a range of multiple effects on plants and various levels of the soil food web, atmosphere CO_2 enrichment can lead to net positive to neutral effects on plant-parasitic nematodes (Cesarz *et al.*, 2015; Zhou *et al.*, 2022). Higher CO_2 levels result in more efficient photosynthesis in dominant C3 plants, promoting plant growth and increasing carbon sequestration in soil, benefiting higher trophic groups and persister guilds. The increased root biomass available to plant-parasitic nematodes could increase their densities whilst rising plant tolerance. In addition, the plant growth investment at the expense of carbon may evidence N-limitation, compromising plant metabolism and investment in defence, further promoting plant-parasitic nematodes. However, N-limitation may also lower root nutritional quality, which, combined with a stronger regulatory role exerted by increased omnivore and predator populations, may restrict population development of plant-parasitic nematodes.

Temperature has direct effects on plant-parasitic nematode activity, life-cycle duration and, depending on optimal thermal requirements, their spatial distribution range. Indirectly, changes in plant host distribution and suitability can also affect nematode abundance. Host-specific plant-parasitic nematodes can be negatively affected by increasing temperatures if these alter the distribution of their hosts, but this effect may be more pronounced in natural ecosystems (Li *et al.*, 2020). In agricultural production systems, specific endoparasitic nematodes may expand their range (Jones *et al.*, 2017) if they are thermophilic or thermotolerant (*Globodera rostochiensis*, *Heterodera carotae*), or if they find a more suitable climate in previously overly cool conditions (longidorids), but otherwise their densities can be reduced (*G. pallida*).

Moderate warming increases plant-parasitic nematode activity and shortens life-cycle duration, allowing for additional generations to be completed within a cropping cycle. However, depending on warming intensity and duration, this can reduce the densities of plant-parasitic nematodes associated with crops (Liu *et al.*, 2022b). In cooler climates, moderate long-term warming or short-term severe warming both decrease nematode abundance and the PPI, whereas little effects are produced in temperate forests and croplands, except under drought conditions. In arid and semi-arid climates, where ambient temperature is high, further increases are expected to reduce abundances of plant-parasitic nematodes.

Increased temperatures may have more pronounced negative effects on nematodes exposed in soil such as ectoparasites, also depending on soil thermal conductivity as affected by mineral and organic matter content. Likewise, the increased activity of root-knot nematode J2 in soil can more rapidly deplete their energy reserves (see Chapter 7) and prevent them from infecting host roots, whereas a higher proportion of males can be formed in roots at higher soil temperatures, possibly due to plant stress, and especially under associated water scarcity. However, host suitability can become more favourable to plant-parasitic nematodes under moderate warming, as, for example, the reduction in expression of the root-knot nematode resistance gene *Mi* in tomato.

Although fewer studies have addressed the effects of altered precipitation on plant-parasitic nematodes, effects have been found to differ according to their functional group, the ambient precipitation of ecosystems and food web effects (Franco *et al.*, 2019; Ankrom *et al.*, 2020). Altered precipitation – both in quantity and in pattern – can affect plant-parasitic nematodes as they depend on soil moisture content for movement, communication and host finding. Increased precipitation can promote plant growth, providing a more favourable soil environment for microbiota and their primary consumers (including bacterial- and fungal-feeding nematodes) and contribute to food web complexity. Higher trophic groups such as omnivores and predators, but also nematophagous fungi, increase their population sizes and activity and regulatory potential on plant-parasitic nematodes. Accordingly, nematode abundance (of both ecto- and endoparasites) and herbivory footprint have been reportedly reduced under increased precipitation treatments in natural and agricultural systems. The correspondent decrease in the PPI in mesic grasslands under increased precipitation was attributed to losses in high cp-class longidorids and trichodorids. The reduction in precipitation overall causes the decline of all trophic groups, even though higher trophic groups are most affected, as well as of plant productivity.

Plant diversity, both in agricultural systems and in grasslands, helps mitigate the effects of climate change, even under extreme events of droughts and flooding. Combined effects of the above climate factors cause less pronounced nematode community structure changes in species-rich extensive agricultural than in intensive agricultural systems. Vulnerability of these intensive systems to warming and altered precipitation has been related to increased densities of plant-parasitic nematodes, lower diversity and complexity of the soil food web, making them more prone to yield losses. Intensive land use itself impairs functioning in agricultural systems more strongly than climate change effects (Siebert *et al.*, 2020).

If land-use intensity can cause such dramatic changes to the functioning of agricultural systems, it is then unequivocal that appropriate management can greatly enhance their resilience, from local to regional scales. However, the context-dependent, nearly unpredictable, trends in the response of various plant-parasitic nematode taxa and functional groups to changing environmental conditions makes the design of targeted strategies for their control very challenging. Their success can be further affected by plant hosts, climate factors and multitrophic interactions through the soil food web. The unsustainability of conventional agriculture is well recognized because it contributes to global climate change through water use and salinization, greenhouse gas emissions of fuel, chemical pesticide and fertilizer use, manufacturing and transport, and soil degradation. Rather than focusing solely on crop health, the sustainable, preventive management of plant-parasitic nematodes should be devised, taking into account their ecology, at the agroecosystem level.

10.7. Agroecology and Control of Plant-parasitic Nematodes

The promotion of agricultural system resilience to abiotic and biotic stresses, preserving and enhancing the ecosystem services provided by biodiversity, is the main objective of agroecology. To achieve this, agricultural production systems cannot be considered independent fractions of rural landscape mosaics, but part of an agroecosystem, that includes the surrounding natural and seminatural habitats, as well as human settlements, in a land-sharing perspective (FAO, 2014). This change in agricultural production paradigm makes use of agroecosystem-associated biodiversity to travel through and influence its three components, exploiting synergies among them.

The **natural, unmanaged component** provides physical protection against adverse weather, buffers the spread of pests and diseases, and overall serves as ecological infrastructure, acting as a source of beneficial organisms (plant mutualists, pest and disease antagonists) that may have been depleted in the **agricultural production component**. The **human component** has multiple facets that include actors in agricultural management, or distributors and consumers of, expectedly, fresh, nutritious and secure food. Agroecology arose in small family farming, low-input systems and became established as a fundamental and applied science, merging multidisciplinary topics on the ecology of agroecosystems. Its implementation also gained momentum as a social movement, and to date it cannot easily be dissociated from the cultural and political perspectives of farmer empowerment that have been promoting agroecological transition. Agroecology is not a mode of production such as, for example, organic or biodynamic farming, and so does not produce specifications for crop production. Rather, it is based on a set of principles that it advocates should be met to ensure sustainability of agroecosystems and their resilience to adverse conditions (Box 10.3).

The application of agroecological principles requires a solid body of research, so the discipline of agroecology was quickly adopted by research institutions worldwide. Nematology is a particularly well-positioned field to deliver established indicators of soil quality, processes and services to assess the success and guide the application of agroecological principles. Conversely, these principles can also guide fundamental and applied nematology research on control of plant-parasitic nematodes in agroecosystems. Taking into consideration the current knowledge of plant-parasitic nematode ecology as presented in the above sections, suggestions for their agroecologically sound management include: (i) increasing plant diversity; (ii) reducing tillage, synthetic pesticide and fertilizer application; and (iii) increasing soil organic matter content (Fig. 10.4). Although theoretically

Box 10.3. Principles of Agroecology (from Altieri and Nicholls, 2005).

- Ensure the recycling of biomass and optimize nutrient availability and flow through the system components;
- Secure favourable soil conditions for plant growth by managing organic matter and enhancing soil biological activity;
- Minimize losses due to flows of solar radiation, air and water by, for example, increased soil cover;
- Increase species and genetic diversification in time and space;
- Enhance beneficial biological interactions and synergisms among biodiversity components, promoting key ecological processes and services.

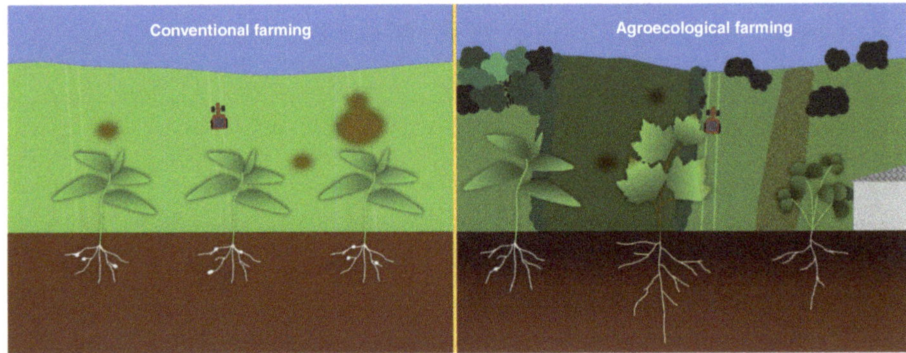

Fig. 10.4. Contrasting conventional and agroecological strategies to promote sustainable control of plant-parasitic nematodes associated with crops, and resulting below-ground effects on nematode communities. In **conventional** farming monocultures, crop plants have similar susceptibility to parasitism, so patches of poorly performing plants due to specific attack by plant-parasitic nematodes are more evident in the field. Domesticated crops with acquisitive strategies have poor investment in root biomass and poor defence, thereby suffering more damage due to plant-parasitic nematodes (galling by root-knot nematodes illustrated). In **agroecological** farming, land-sharing of agroecosystem components promotes associated biodiversity through investments in ecological infrastructure: natural vegetation, hedges and strip crops improve connectivity in the landscape mosaic. Increased plant functional and trait diversity provide resource heterogeneity and complementarity, promoting food web complexity and diluting effects of specialist plant-parasitic nematodes (and hence reducing damage and increasing plant biomass).

sound and not at all novel, these management options are still lacking research in an applied context, as the magnitude of 'increasing' or 'reducing' need to be estimated to be adopted successfully.

10.7.1. Increasing plant diversity

Plant diversity can be planned and increased via a range of practices but care must be taken in the selection of plants that are non-hosts for the dominant plant-parasitic nematode populations. This can be particularly challenging with, for example, root-knot nematodes and root lesion nematodes, that can have a wide range of hosts, including weeds, and calls for close collaboration between farmers and researchers (Schmidt *et al.*, 2017). However, given the differential host preferences of plant-parasitic nematodes, as well as indirect effects of plant diversity on the soil food web, increased plant diversity can act to reduce plant-parasitic nematode pressure on crops. Rather than focusing on a particular nematode species target as a curative strategy, increasing plant diversity is proposed to act on the nematode community, avoiding dominance and preventing outbreaks of single species. Agroecosystem-associated plant biodiversity in, for example, the field margins, hedges and other ecological infrastructure, can be harnessed to supply ecosystem services, particularly if slow-growing, late-successional, non-acquisitive plants are grown.

Intercropping increases the spatial diversity, arrangement and quality of food resources for plant-parasitic nematodes, decreasing the dominance and density of specialist species and thereby reducing plant–soil negative feedbacks to promote yield. If appropriate combinations of intercrops are used, nematode natural enemies can be promoted, and associational resistance effects can be exploited, further reducing the plant-parasitic nematode burden.

Crop rotation increases plant diversity temporally, reducing the negative legacy of accumulated plant-parasitic nematodes and scarcer mutualists of domesticated crop monocultures (Mariotte *et al.*, 2018). Admittedly, there are limited options for rotation in permanent crops, such as orchards or vineyards. In these systems, indigenous plants or seed mixtures can be grown as floor vegetation, managed by mowing and mulching, so as to maintain a permanent soil cover (Salomé *et al.*, 2016).

Cover crops can indeed have multiple functions in maintaining soil cover, thereby reducing erosion, promoting water infiltration and soil structure. The improved soil conditions harbour higher nematode trophic levels that then act in suppressing plant-parasitic nematodes (Ferris *et al.*, 2012). Cover crops maintain active green energy channels in soil, actively increasing its organic matter content and enabling resource complementarity to the soil food web (Zhang *et al.*, 2017), especially if functional crops such as legumes are used.

Lastly, the diversity of the main crop must be considered. Crop domestication, either through traditional crossings or by modern crop improvement techniques, has been an artificial process of granting crops specific traits at the expense of the ability of crops to interact with the remaining organisms. It is unlikely that highly selected crops will be able to perform under low-input agroecologically managed systems, or under changing climate conditions and extreme events. Strategies that make use of local knowledge for cash crop diversity include the recovery of ancestral varieties or locally adapted landraces for cropping and preservation of germplasm. Alternatively, breeding strategies should consider promoting trait plasticity rather than breeding for specific traits (Milla *et al.*, 2017). The impact of these suggested strategies for cash crop diversity on field populations of plant-parasitic nematodes is, however, difficult to predict, as crop identity effects appear to be idiosyncratic.

10.7.2. Reducing tillage, synthetic pesticide and fertilizer application

Tillage, chemical pesticides and mineral fertilizer were the basis of conventional intensification, but it is now well recognized that they all contribute to reduced soil structure and food web complexity, promoting plant-parasitic nematode populations in the medium to long term.

Conventional tillage by cultivation and soil inversion has been a standard practice to improve soil conditions for sowing and reducing weed pressure, but has not been shown to promote yield. The strength of immediate effects on the soil food web and plant-parasitic nematodes themselves appear to depend on tillage method (shallow or deep, inversion or non-inversion), soil texture and structure, climate, and plant-parasitic nematode identity (Schmidt *et al.*, 2017). As soil is mechanically impacted, larger body-sized nematodes (including omnivores and predators) are most affected, but the whole soil community is impacted by habitat loss through detrimental effects on soil aggregation. Particularly after inversion tillage, organic matter is exposed, rapidly decomposed and lost, as well as the nutrients contained therewith, further decreasing soil structure.

Synthetic pesticides can diffusely impact the soil food web through their overall high environmental toxicity, and have non-target, often unanticipated environmental effects, despite extensive testing prior to their release. Fungicides impact non-target organisms and have been shown to pose severe constraints to biological control agents such as *P. chlamydosporia* (Carvalho *et al.*, 2022), whereas nematicides (aldicarb, 1,3-dichloropropene) can impact other organisms, including free-living nematodes, as much as plant-parasitic species (Timper *et al.*, 2012). Thus, in spite of efficiently reducing plant-parasitic nematode populations, the lasting effects of nematicides on the regulatory ecosystem service of natural enemies of plant-parasitic nematodes reduces agroecosystem resilience. Thankfully, novel nematicides are being developed that explicitly test their effects on non-target soil organisms (see Chapter 17). Neutral (or positive) effects on the soil food web need to be clearly demonstrated to support their use.

Mineral fertilizers much increase the immediate availability of nutrients to plants, but synchrony and adequacy with plant demand is difficult to estimate, resulting in leaching to surrounding components of the agroecosystem. Nutrients in mineral fertilizers cannot be used by the brown energy channel of the soil food web, and thus promote the green energy channel – of plant-parasitic nematodes – even though improved plant nutrition can increase their tolerance. As a viable alternative, organic fertilization can increase soil organic matter content, improving its structure and promoting soil food web biomass and complexity, whilst reducing plant parasitic nematode populations (Schrama *et al.*, 2018).

10.7.3. Increasing soil organic matter content

The organic matter content is a major determinant of soil properties (Section 10.2) and an energy source for the brown channel, ultimately contributing to soil food web complexity and stability, and natural regulation of plant-parasitic nematodes. Organic matter is also a major source of carbon sequestration in soil, acting to mitigate climate change. It is therefore unsurprising that managing organic matter in soil is a key aspect in agricultural sustainability. The above suggestions on promoting plant diversity and reducing tillage, pesticides and fertilizers all contribute to maintaining or increasing organic matter content in soil. Organic matter addition in the form of organic fertilizers, particularly with a proportion of animal manure, is a useful practice to control plant-parasitic nematodes.

As organic fertilizers are added to soil, their decomposition process releases compounds toxic to plant-parasitic nematodes and stimulates the activity of decomposers that themselves can produce nematotoxic compounds (Oka, 2010). The decomposition of nutrient-rich organic fertilizers not only provides a balanced nutrient flow to plants, improving their performance, but also leads to increasing cp 1 nematodes – enrichment opportunists. These are a group of amplifiable prey for omnivores and predators that can further suppress plant-parasitic nematodes (Ferris *et al.*, 2012).

In summary, the provisioning service of agroecosystems can be aided sustainably by regulating and support ecosystem services that can be guaranteed by a proper management of biodiversity, under a sound ecological understanding of its processes (see Bommarco *et al.*, 2013). If biodiversity can be harnessed to supply crop nutritional balance and regulation of plant-parasitic nematodes, this results in lower dependence on external inputs and consequent externalities, such as diffuse pollution and greenhouse gas emissions. If a larger gap between attainable and actual yield can result from the conversion of conventional to agroecological management, this has not always been demonstrated, and has been shown to be offset within a few years. Importantly, it can lead to increased yield stability even under adverse weather fluctuations and extreme events. The better and more complete the understanding of the ecology of plant-parasitic nematodes in a range of natural to agricultural conditions, from fine roots to regional scales, the better their management can be tailored and improved in agroecosystems.

Part III Quantitative Nematology and Management

Preventing the spread of nematodes through effective quarantine legislation and prophylactic measures is essential to avoid previously uncontaminated areas from becoming infested. Once plant-parasitic nematodes are present in an area, effective sampling within cost and personnel constraints is necessary to define the extent of their spread. Limiting the damage caused by plant-parasitic nematodes depends on knowledge of nematode distribution patterns, population dynamics and their effects on plant growth. Sampling and damage assessment are the criteria for determining the economic rationale for management and control in relation to the economic benefits of increased crop returns. Integrated management strategies involve the use of several control components, including cropping schemes, biological control, the judicious use of chemicals and resistant cultivars. With the rapid advances in molecular information, development of resistant cultivars using a genomic approach that exploits improved understanding of crop genomics is being actively pursued. The seven chapters in this section reflect each of the quantitative assessments of nematode populations and distribution, and each component of the overall management strategies, including the burgeoning field of genetic engineering.

Plant Growth and Population Dynamics*

Corrie H. Schomaker**, Thomas H. Been and Misghina G. Teklu

Wageningen University and Research Centre, Plant Research International, The Netherlands and Biology Department, Ghent University, Belgium

11.1. Introduction

The main purpose of quantitative nematological research is to achieve an optimal economical protection of crops against plant-parasitic nematodes. To accomplish this, the costs of

*A revision of Schomaker, C.H. and Been, T.H. (2013) Plant growth and population dynamics. In: Perry, R.N. and Moens, M. (eds) *Plant Nematology*, 2nd edn. CAB International, Wallingford, UK.
**Corresponding author: schomakercorrie@gmail.com

DOI: 10.1079/9781800622456.0011

control measures must be adjusted to the costs of the expected yield reduction compared to the yield in a situation without the need for control. Such an adjustment requires quantitative knowledge of:

1. The relationship between a measure for the nematode activity (in practice mostly their population density at the time of planting) and plant response.
2. The population dynamics of nematodes in the presence of food sources (of different quality) and in the absence of food.
3. The effect of control measures on plant response and nematode population dynamics. The control measures may range from pesticide treatment, crop rotation and cultivation of crops that vary in suitability as a food source for nematodes.
4. Cost/benefit of the control measures.

11.2. Relationships of Nematodes with Plants

The majority of species of plant-parasitic nematodes live on or around plant roots (see Chapter 8). Nematode species can be divided into four types according to the plant parts they infest: (i) species that form galls in ovaries and other above-ground plant parts, e.g. *Anguina tritici* in wheat; (ii) leaf nematodes (*Aphelenchoides*) infecting leaf buds and causing malformations and necrosis in leaves of many ornamental plants and in strawberries; (iii) stem nematodes (*Ditylenchus dipsaci*) causing malformations, swellings, growth reduction and dry rot in above- and underground parts of plant stems such as onions, bulbs, rye, wheat, beet, potatoes and red clover; and (iv) root nematodes causing growth reduction in whole plants and malformations in underground plant parts (*Meloidogyne* spp., *Rotylenchus uniformis*), root necrosis and growth reduction (*Pratylenchus penetrans*, *Tylenchulus semipenetrans*) or growth reduction without any symptoms (*Globodera rostochiensis*, *G. pallida*, *Tylenchorhynchus dubius*).

In most cases of infestations by stem and root nematodes, the nematodes were already present in the soil at the time of planting. Damage in red clover and lucerne is often the result of seed infected with stem nematodes. Stem nematodes introduced into a field with infected onion seed are not known to reach densities that are high enough to cause immediate visible infestations. The introduction of nematodes on planting material such as bulbs and tubers and the spread of nematodes by machinery and other vectors is discussed in detail in Chapter 12. *Bursaphelenchus cocophilus* in coconut and oil palm and *B. xylophilus* in various pine species are unusual. Both are transmitted by a beetle. In fact, several species of beetles support the life cycle of *B. xylophilus*: pine sawyer beetles (*Monochamus* spp.) transport the nematodes to the pine trees where they feed on blue stain fungi. Bark beetles help to introduce the blue stain fungi into the trees and thus allow the nematodes to feed and multiply. *Bursaphelenchus xylophilus* (and possibly one or two other species) can feed on live trees as well as fungi.

11.3. Predictors of Yield Reduction

11.3.1. Symptoms

Figure 11.1 gives some examples of visible symptoms of nematode infestations. Some of them are very conspicuous but others are hardly visible. In some cases, visible symptoms

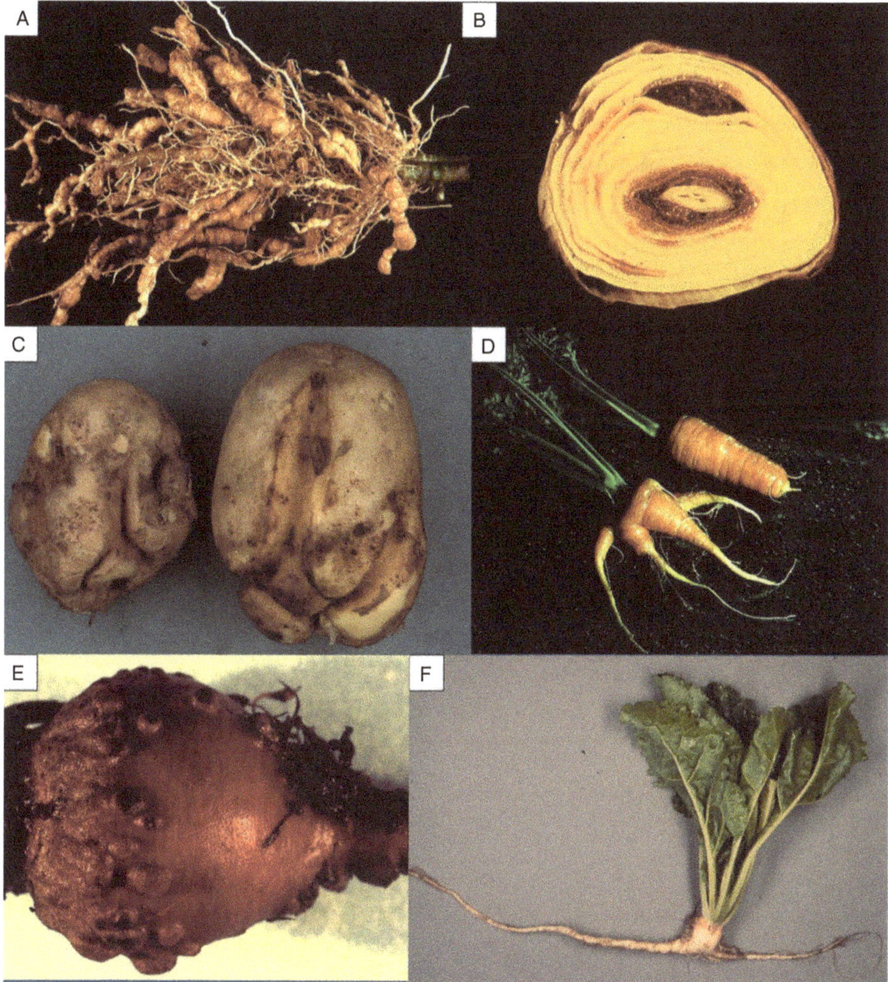

Fig. 11.1. Visual symptoms of nematode attack. A: *Meloidogyne incognita* on tomato roots. B: *Ditylenchus dipsaci* in onions. C: Deformations in potatoes caused by *Pratylenchus*. D: *Meloidogyne* in carrots. E: *Meloidogyne* in table beets. F: Deformation in sugar beet caused by trichodorids.

in plants can be used as a measure of yield reduction. For example, symptoms of nematode infestations in above- and underground parts of the stem are often easy to recognize, and yield reduction is closely related to the extent of the phenomena. Some root nematodes inflict conspicuous deterioration in roots or underground plant parts: *Meloidogyne* spp. cause root knots, and species of *Longidorus* and *Xiphinema* are responsible for bent, swollen root tips. If these plant parts are marketable products, the symptoms are closely related to yield reduction. If not, the relationship between symptoms and yield is much more complex. By contrast, other nematode species cause hardly any specific or recognizable symptoms and yet reduce yields severely. For example, *P. penetrans* causes root necrosis, and low to medium numbers of cyst nematodes, *T. dubius*, *R. uniformis*, *Helicotylenchus* spp., *P. crenatus* and *P. neglectus* cause no, or hardly any, symptoms, yet these species can cause considerable growth reduction in the plants they attack. *Longidorus elongatus* in some cases causes swollen root tips and a smaller root weight but does not affect the above-ground plant parts. Therefore, in general, visible symptoms of nematode infestation can seldom be used as a measure for growth and yield reduction.

11.3.2. Pre-plant density (P_i)

Only leaf and bud nematodes and *Bursaphelenchus* spp. can multiply fast enough to cause considerable damage very shortly after the first infection, even at very low densities. For these species, damage or yield reduction is almost independent of the numbers of nematodes at the time of planting (P_i). To control these nematodes, all sources of infection, such as infested plants, must be removed and breeding material must be free from nematodes. By contrast, the multiplication of most root nematodes is relatively slow, even on good hosts. Root nematodes only cause yield reduction when harmful densities are already present in the soil at the time of planting of a sensitive crop. On a small scale, root nematodes are distributed in the soil according to a negative binomial distribution with an aggregation coefficient (k) larger than 40 for most nematode species. This distribution is regular enough to assume that growing plant roots are continuously exposed to the attack of a nematode population with about the same density and, therefore, that the growth of annual plants (or the growth of perennial plants during the first year) is retarded at a constant rate.

11.3.3. Multiplication

Nematode damage to plants can have some influence on nematode multiplication but only when large nematode densities reduce the food source (often the root system) in sensitive plants. Conversely, multiplication of nematodes is of little importance to the growth and yield reduction they cause; the same amount of yield reduction may occur in resistant (non-host) and susceptible plants. Examples are *G. pallida* and *G. rostochiensis*, causing the same damage in resistant and susceptible potato cultivars (Fig. 11.2), *M. naasi* and *M. hapla* damaging beet during the first months after sowing, and damage by stem nematodes in flax, yellow lupin, maize and sun spurge. In these latter crops, marked yield reductions or growth aberrations have been found without any nematodes being detected in the plants. Therefore, the host status of a plant and its susceptibility to damage must be treated as independent qualities. The reason for this independence is that only a very small part of the damage caused by nematodes is caused by food withdrawal from host plants and the main part by the biochemical and mechanical disruptions that nematodes bring about in plants.

11.4. Different Response Variables of Nematodes

Crop returns are reduced by nematode attack as a result of reduction of crop weight per unit area, which is mostly equivalent to average weight of marketable product per plant,

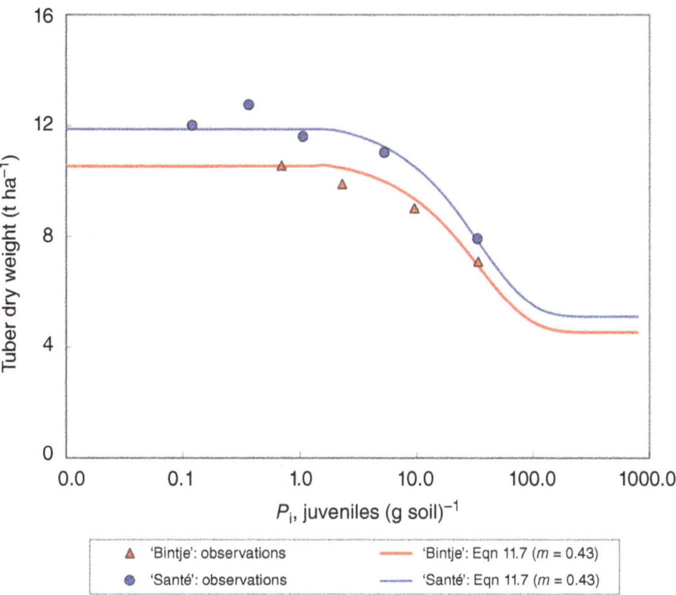

Fig. 11.2. The potato cultivars 'Bintje' (susceptible) and 'Santé' (resistant) were grown in a field infested with *Globodera pallida*. Although the yield potential of 'Santé' (11.9 t ha^{-1}) is greater than that of 'Bintje' (10.6 t ha^{-1}), the effect of the nematodes on tuber dry weight is the same: both cultivars have a relative minimum yield (m) of 0.43 and a tolerance limit T of 18 juveniles g^{-1} of soil. (L. Molendijk, unpublished data.)

and reduction of the value of the product per unit weight. For example, carrots attacked by root-knot nematodes (*Meloidogyne* spp.) may be worthless because of branching and deformation of the taproot (Fig. 11.1D), although they have the same weight per unit area as carrots without nematodes. Onions of normal weight but infected with a few stem nematodes (*D. dipsaci*) at harvest will, nevertheless, be lost in storage. Attack of potatoes by *G. rostochiensis* and *G. pallida* not only reduces potato tuber weight but also may reduce tuber size. However, small and medium densities of potato cyst nematodes attacking potatoes, and almost all root-infesting nematodes attacking crop plants of which the above-ground parts are harvested, hardly ever affect the value per unit weight of harvested product. Therefore, prediction of crop reduction by these nematodes can, in general, be based on models of the relation between nematode density at planting (P_i) and average weight of single plants (y) at harvest. In the following sections, the term 'yield' will be avoided. The yield in the agronomic sense must be derived from individual plant weights.

11.5. Stem Nematodes (*Ditylenchus dipsaci*)

To construct a model of the relationship between initial population density (immediately before planting), P_i, and the proportion, y, of uninfected plants (onions, flower bulbs), a theory is required concerning the mechanisms involved. The theory has to be translated into a mathematic model so that it can be tested. In fact, a mathematical analogue of the theory is formulated. To test or validate the mathematical model (and the theory), it is compared with mathematical patterns that are distinguishable in data derived from observations. At the same time, the values of system parameters are estimated under various experimental conditions. P_i and y are called system variables because they have different values in each experiment, in contrast to system parameters, which are constants. They have only one value in a certain experiment but they can vary between experiments because of changes in external conditions.

Seinhorst (1986) presented a competition model for the relationship between stem nematode densities (P_i) and the proportion of infected onion plants. As only nematode-free onions are marketable and the degree of infestation of single plants is irrelevant, only infected and non-infected onions were distinguished. To formulate the model, three assumptions were needed:

1. The average nematode is the same at all densities. This means that initial population density (P_i) does not affect the average body size or activity of the nematodes.
2. Nematodes do not affect each other's behaviour. They do not attract or repel each other directly or indirectly.
3. Nematodes are distributed randomly over the plants in a certain small area.

It is postulated that at $P_i = 1$ a proportion d of the onion plants is infected and that, therefore, a proportion $1 - d$ is left non-infected. Then, at density $P_i = 2$, a proportion d of already damaged plants is attacked (which has no additional effect as onions, once attacked, are worthless), plus a proportion d of the still non-infected proportion $(1 - d)$. So at $P_i = 2$ a proportion $d + d(1 - d)$ onions is attacked and $1 - d - d(1 - d) = (1 - d) - d(1 - d) = (1 - d)^2$ of the plants is left non-infected. At $P_i = 3$, again a proportion d of already damaged onions is damaged, which has no effect, and a proportion d of the non-infected plants $(1 - d)^2$ is attacked. Summing it all up, we see that at $P_i = 3$ the proportion of infected onions amounts to $d + d(1 - d)^2$ and that the proportion of non-infected onions is $1 - d - d(1 - d)^2 = (1 - d) - (1 - d)^2 = (1 - d)^3$. Schematically:

Population density, P_i =	Proportion of infected onions	Proportion of non-infected onions
1	d	$1 - d$
2	$d + d(1 - d)$	$1 - \{d + d(1 - d)\} = (1 - d)^2$
3	$d + d(1 - d) + d(1 - d)^2$	$1 - \{d + d(1 - d) + d(1 - d)^2\} = (1 - d)^3$
P		$(1 - d)^P$

In general: a nematode density $P_i = P$ leaves a proportion

$$y = (1-d)^P = z^P \tag{11.1}$$

of the onions non-infected. In Eqn 11.1, P is an integer, y is a variable (like P_i) and z is a parameter. The parameter z must be estimated. The expected value of z and its variance must be estimated in field experiments; the population density P_i can be estimated by taking soil samples with an appropriate sampling method (see Chapter 12). In Fig. 11.3, values of y $(= z^{P_i})$ are plotted for three different values of z. The values of y are not plotted against P_i, but against $\log P_i$. This log-transformation of P_i not only has the advantage that the shape of the curves is the same for all z, but also that, if P_i is estimated by counting nematodes from a soil sample, the variance of $\log P_i$ is constant (and independent of P_i), provided that P_i is not very small. The value of the parameter z is determined by conditions that influence the efficiency of nematodes in finding and penetrating plants. In patchy infestations of stem nematodes these conditions for attack appear to be more favourable in the centre of the patch than towards the borders. This results in an increase of z with increase of the distance from this centre and, thus, in persistency of the patchiness. The model also applies when nematodes spread from randomly distributed infested plants to neighbouring ones, leading to overlapping patches of infested plants.

In Box 11.1 we explain a numerical example of fitting Eqn 11.1 to experimental data. In this and further examples in Chapters 11 and 12 we use the software R. The R scripts, input data and results can be downloaded from the CABI website: https://www.cabidigitallibrary.org/doi/book/10.1079/9781800622456.0000.

Box 11.1. Fitting the stem nematode model. (The mathematical notation in this Box uses the R code of the "StemNematodes.R" script.)

When we want to fit Eqn 11.1 to experimental data, we can convert it into a linear model by applying two logarithmic rules:

```
log(a ^ b) = b * log(a)
log(a * b) = log(a) + log(b)                                  (11.2)
Then, log(Y) = log(z ^ Pi) = Pi * log(z)
```

Extending (11.2) with Ymax, the yield at Pi = 0, we get:

```
log(Y) = log(Ymax) + Pi * log(z)                              (11.3)
```

In our examples we use the software R, because of its widespread use and free availability. In the script "StemNematodes.R" we explain this analysis into more detail, step by step. We use the function lm(log(Y))~Pi to estimate the slope, log(z), and the intercept, log(Ymax). In the script, "StemNematodes.R", this is written as follows:

```
LM1 <-lm(dat$lY ~ dat$Pi)
```

Where lY is log(Y), the function summary(LM1) extracts detailed information about the regression analysis, including the adjusted (for the number of parameters) R^2. In the first column under "coefficients" we find our two parameters, the intercept log(Ymax) and slope log(z). The second column gives their standard errors. The call LM1$coefficients returns log(Ymax) and log(z) directly. Back transformation by taking the exponent of log(Ymax) and log(z) returns Ymax and z. **Note**: the standard errors in summary(LM1) are the standard errors of log(Ymax) and log(z). We can extract these standard errors from the variance/covariance matrix of LM1:

```
se(log(x)) = sqrt(diag(vcov(LM1)))
```

The diagonal of the matrix, diag(), the numbers from top left to bottom right of the matrix, represent the variances. Reading the matrix from bottom left to top right, returns the covariances. To obtain the standard errors of the back-transformed parameter estimates, we use the logarithmic rule:

```
var(log(x)) = log((CV(x)) ^ 2 + 1)
```

Then, taking the exponent on both sides, to get rid of the log, we get:

```
exp(var(log(x))) - 1 = (CV(x)) ^ 2
```

To be re-written as:

```
CV(x) = sqrt(exp(var(log(x))) - 1
```

Continued

Box 11.1. Continued.

Here, CV is the quotient of standard error and estimate of x. Multiplying the CV with the parameter estimate, gives the standard error. The results are summarized in Table 11.1.

For further information about linear regression in R see Dalgaard (2008). A good description of the lognormal distribution can be found in Wikipedia or in Mood *et al.* (1986).

Table 11.1. The results of the linear regression from the output file: "Output Stemnematodes.txt".

Field	log(Ymax)	log(z)	Ymax	z	CV.Ymax	CV.z	se.Ymax	se.z
Field1	4.6078	-0.05124	100.25	0.95	0.0064	0.00015	0.64	0.00014
Field2	4.6042	-0.03062	99.90	0.97	0.0066	0.00015	0.66	0.00015
Field3	4.6051	-0.00988	99.99	0.99	0.0061	0.00014	0.61	0.00014

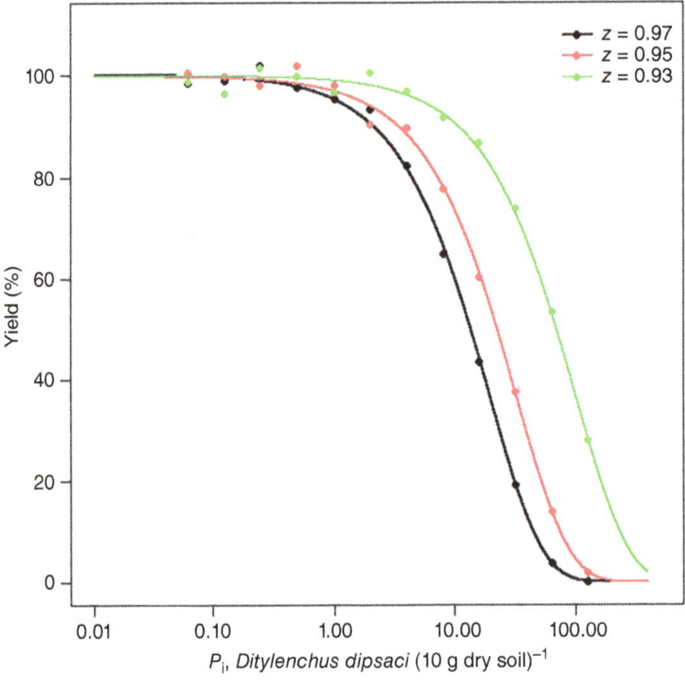

Fig. 11.3. Relation of stem nematodes and yield according to equation *y* = Ymax*z^Pi with three different values for *z* (*z* = 0.99, 0.97 and 0.95). The smaller *z*, the greater the activity of the nematodes and the smaller the yield

11.6. Root-invading Nematodes

Root-knot nematodes, *Pratylenchus* spp. and cyst nematodes are considered to be the most important root-invading tylenchid nematodes. Although some of these nematodes, especially root-knot nematodes, can also inflict qualitative damage in underground plant parts, they generally reduce crop yield in a less direct way than stem nematodes. Often there are no visible symptoms. Then, only the rate of growth and development of attacked plants is reduced, resulting in lower weight compared to plants without nematodes. To put it simply: in plants with nematodes the same thing happens later. In exceptional cases, nematode-infested plants reach the same or higher final weights than plants without nematodes, but at a later stage. In most cases, such a delay results in ripening of the crop being prevented by external conditions at the end of the plant growing season. This, economically most important phenomenon is defined as the first mechanism of growth reduction. Other mechanisms are explained in the next paragraph.

Seinhorst (1986), explaining the first mechanism of growth reduction, based a growth model on two simple concepts: (i) the nature of the plant (an element that increases in weight over time); and (ii) the nature of the plant-parasitic nematode (elements that reduce the rate of increase of plant weight and, in principle, the more so the larger the population density). To formulate the model for root nematodes three extra assumptions must be added to those made for stem nematodes (Section 11.5):

4. Root-infesting nematodes are distributed randomly in the soil.
5. Nematodes enter the roots of plants randomly in space and time. Therefore, the average number of nematodes entering per quantity of root and time is constant. This number is proportional to the nematode density P (number of nematodes per unit weight or volume of soil).
6. The growth rate of plants at a given time t after planting is the increase in total weight per unit time (dy/dt). Let this growth rate be r_0 for plants without nematodes and r_P for plants at nematode density P. According to Fig. 11.4, $r_0 = \tan(\alpha) = \Delta y/\Delta t_0$ and $r_P = \tan(\beta) = \Delta y/\Delta t_P$. Thus, for plants of the same total weight (and, therefore,

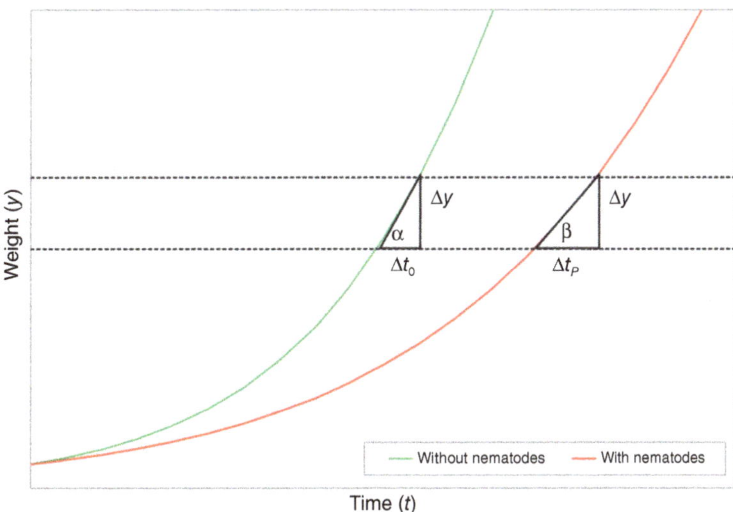

Fig. 11.4. Growth curves of plants without nematodes and with nematodes at density P; y = plant weight and t = time after planting. The variables t_0 and t_P are the times that plants without nematodes and at density P need to reach the same total weight y, respectively; $r_0 = \tan(\alpha) = \Delta y/\Delta t_0$ and $r_P = \tan(\beta) = \Delta y/\Delta t_P$ are the growth rates of plants without nematodes and at nematode density P, respectively.

of different age) with nematode density P and without nematodes, the ratio r_P/r_0 is constant during the growing period. Therefore,

$$\frac{r_P}{r_0} = \frac{t_0}{t_P} \qquad (11.4)$$

The relationship between population density of the nematodes and its total effect on the growth rate of the plants accords with Eqn 11.1. Eqn 11.1 is a continuous function for $0 \leq P \leq \infty$. There is one complication: all accurate observations on the relationship between the population density P of various nematode species and weight of various plant species indicate that there must be a maximum density, the tolerance limit T, below which the nematodes do not reduce plant weight. Therefore, Eqn 11.1 is adapted by replacing P by $P - T$ and we have to deal with a discontinuous function. In practice the transition between $P > T$ and $P \leq T$ will be smoother than it is in theory. Further, only in very few experiments were large nematode densities able to reduce plant weight to zero, whilst growth rates of attacked plants were never reduced to zero. Therefore, a second adaptation in Eqn 11.1 was the introduction of the minimum relative growth rate $k = r_P/r_0$ for $P \to \infty$. The equation constituting the growth model for plants attacked by root nematodes then becomes:

$$\frac{r_P}{r_0} = k + (1-k)z^{P-T} \quad \text{for}\, P > T \qquad (11.5)$$

and

$$\frac{r_P}{r_0} = 1 \quad \text{for}\, P \leq T$$

T, z and k are parameters (constants); r_P, r_0 and P are variables. The parameters z and k are constants smaller than 1. The value of the parameter k is independent of nematode density and time after planting but may vary between experiments. Growth curves of plants for different nematode densities can be derived from a growth curve of plants without nematodes with the help of Eqns 11.4 and 11.5. These curves may vary in shape but they must meet two conditions: (i) they must be continuous; and (ii) the growth rate must decrease continuously from shortly after planting. The frequently used logistic growth curve complies with these conditions. Fig. 11.5 gives an impression of the three-dimensional model with axes for total plant weight y, relative nematode density P/T and time t after planting.

From the model it can be deduced that nematodes reduce growth rates of plants by the production of a growth-reducing substance only during penetration in the roots but not when they have settled. This hypothesis is supported by Schans (1993), who observed stomatal closure in potato plants infected by *G. pallida* during the time of penetration and concluded that disturbance of cell development just behind the root tips interferes with the production of abscisic acid, which is known to act as a messenger for stomatal closure. Because of the constant number of nematodes penetrating per unit of root and time, the growth-reducing stimulus will then remain constant per unit weight of plant.

For nematodes that do not move once they have initiated a feeding site within the root, such as root-knot and cyst nematodes, z^P can be interpreted as the proportion of the food source that is left unoccupied by nematodes at density P. For species that are mobile during their whole life cycle, $1 - z^P$ is a measure of the ratio between the feeding

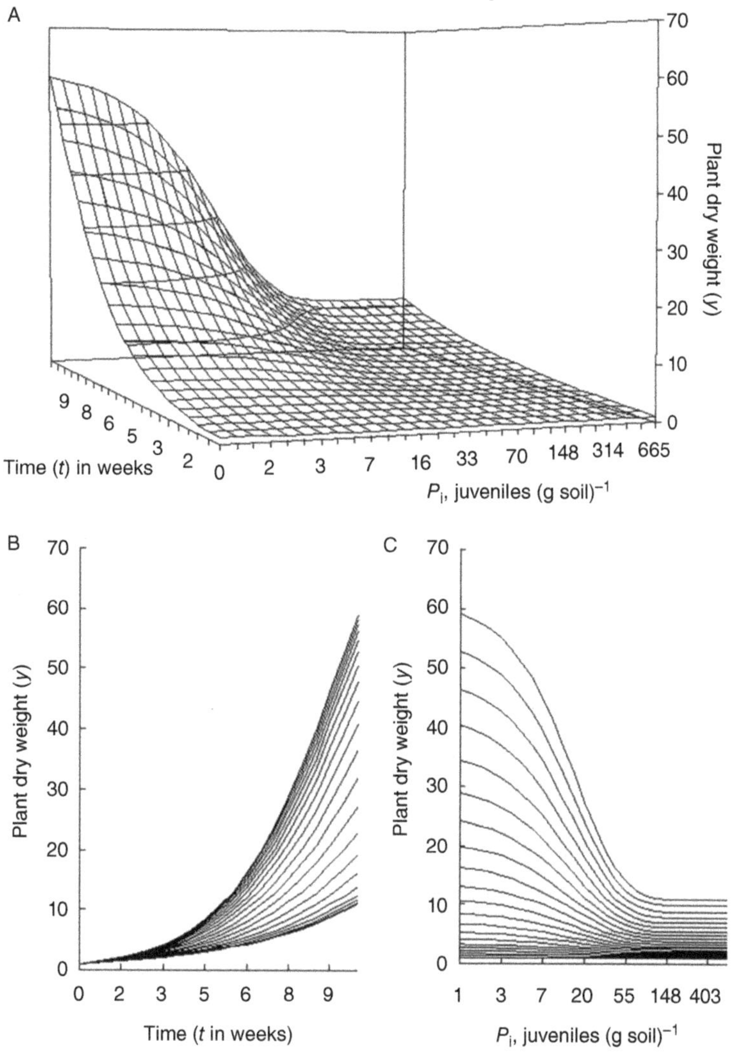

Fig. 11.5. Surface plots of the three-dimensional model (Eqns 11.4 and 11.5) representing the relation between weight, the relative nematode density P_i/T and time t after planting: A, at 230° rotation; B, at 0° rotation, showing the relation between plant weight and t at different nematode densities; C, at 270° rotation, showing the relation between plant weight and P_i.

times at density P and the maximum feeding times at $P \to \infty$. Therefore, $1 - z^P$ is proportional to a hypothetical growth-reducing substance that nematodes bring into the plants during penetration. The parameter k means that nematodes cannot stop plant growth completely and that, even at very large nematode densities, some growth remains: $r_P/r_0 = k$.

11.6.1. A simple model for nematode density and plant weight

The model described in Section 11.5 makes good biological sense but is not easy to use in everyday nematological practice. The primary results of experiments are almost always weights of plants attacked by known nematode densities at a given time after sowing or planting. To investigate whether these relationships are in accordance with the growth model, they must be compared with cross-sections orthogonal to the time axis (Fig. 11.5), through growth curves of plants for ranges of densities *P/T* and different values of *k*. If we describe these cross-sections mathematically, they appear to be in close accordance with Eqn 11.6:

$$\begin{aligned} y &= m + (1-m)z^{P_i - T} \text{ for } P_i > T \\ y &= 1 \qquad\qquad\qquad\quad \text{for } P_i \le T \end{aligned} \tag{11.6}$$

The parameter *m* is the minimum relative plant weight and usually slightly larger than *k*, the parameter *z* is a constant <1 with the same or a slightly smaller value than in Eqn 11.5 and the parameter *T* is the tolerance limit with the same value as in Eqn 11.5.

As in most experiments z^T deviates little from 0.95, Eqn 11.6 can be transformed into Eqn 11.7 for fitting the model to data:

$$y = m + (1-m)0.95^{\frac{(P_i - T)}{T}} \text{ for } P_i > T \tag{11.7}$$

Another mathematic formula to describe the relationship between population density and plant yield is the equation of Elston *et al.* (1991):

$$y = \frac{Y_{max}}{1 + \frac{P_i}{c}} \tag{11.8}$$

where *c* is the parameter for tolerance and Y_{max} the yield at $P_i = 0$. At smaller densities, and if the tolerance limit *T* is negligible, predictions by Eqns 11.7 and 11.8 do not deviate much. However, as Eqn 11.8 lacks a minimum yield, it overestimates yield reduction at medium and high nematode densities. The main shortcoming of Eqn 11.8 is that the starting values of its parameters cannot simply be estimated from the data and that it does not have a biological analogue. Therefore, Eqn. 11.8 does not contribute to theory building.

In Box 11.2 the results of a glasshouse experiment are presented and the main features of the regression analysis with Eqn 11.7 is discussed. Box 11.3 describes a more complicated regression analysis with Eqn 11.7 to compare the tolerance of two potato cultivars in a field experiment, followed by a Student's T-test as a function of the estimated parameters, their standard errors and the degrees of freedom. The R scripts, in and output data can be found on the CABI website at: https://www.cabidigitallibrary.org/doi/book/10.1079/9781800622456.0000.

11.6.2. Mechanisms of growth reduction

We can discriminate three kinds of growth reduction caused by nematodes: the 'first mechanism of growth reduction' operating at all population densities, and a 'second mechanism of growth reduction' additional to that of the first mechanism, with a noticeable effect only at medium to large nematode densities. 'Early senescence' of plants, attacked by high densities of nematodes, can be considered as a third mechanism.

Box 11.2. Fitting the Seinhorst model, Eqn 11.7, to experimental data.

Suppose the following data are the results from a glasshouse experiment that is performed to estimate tolerance of a cultivar for a nematode species:

P_i	0	0.125	0.25	0.5	1	2	4	8	16	32	64	128	256	512
y_i	10.4	10.2	10.3	10.6	10.5	10.3	10	9.5	9	7.7	7	6.2	6.1	6.2

P_i is the nematode density at the time of planting; y_i the average plant weight of five replicates at population density P_i. Pot size is chosen so that plants have the same space for their root system as under field conditions. For data analysis, the choice is basically between analysis of variance (ANOVA) and regression analysis. Regression is the preferred method if the predictor variable (P_i) represents a series and an increase or decrease in the response variable (y_i) is noticed. Moreover, ANOVA cannot estimate tolerance parameters. This dataset has an increasing series of P_i values and a decreasing series of y_i, and an estimate of T, the tolerance limit, and m, the relative minimum yield, is needed. Therefore, regression analysis is chosen. Several methods for regression analysis are available and the most simple and accurate is ordinary least squares (OLS) with trial and error, provided that two conditions are satisfied: (i) the response variable y is normally distributed and its variation is constant; and (ii) the distribution of the nematodes in the soil is random. When OLS is used, T and m are chosen so that the sum of squares (SS) of the deviations of y_i from the model (Eqn 11.7) is minimal. Expressed as a formula:

$$SS_i = \Sigma\left[y_i - y_{max}\left[m + (1-m)0.95^{\frac{(P_i - T)}{T}}\right]\right]^2 \quad \text{when } P_i > T \qquad (11.9)$$

$$SS_i = \Sigma\left[y_i - y_{max}\right]^2 \quad \text{when } P_i \le T$$

Y_{max} is the average plant dry weight at $P_i \le T$ and equals 10.38 in this particular experiment. We choose z^T to be 0.95 because that is the most probable value in relationships between P_i and plant weight (see Section 11.2.8). As the model is non-linear, the minimum value for SS and the best fit must be found by numerical trial and error. In the R script, y is plotted against log P_i to get an impression of a general pattern in the observations and to make the first estimates of the parameters Y_{max}, T and m.

Remembering that T is the largest nematode density P_i without yield reduction, an educated guess would be that T lies somewhere between 1 and 2. So $T = 1.5$ would provide a useful starting value. Now the starting value of Y_{max} can be estimated to be the average y at $P_i \le T$, being 10.4. The first estimate of m would be the average y at P_i values between 128 and 512, which is 6.16. The relative value, m, is then 6.16/10.4 = 0.59. Since there are three values to estimate m, this estimate is likely to be accurate.

When we have good starting values for the parameters, we can run our model. To simplify calculations, a Boolean operator, v, either 1 (TRUE) or 0 (FALSE), is used to distinguish between the two parts of Eqn 11.7. The model (fn1) is then defined as follows in the "*Glasshouse*.R" script:

```
fn1 <- function(p) {
  v <- (x < = p[2])
  model <- (p[1]+(1-p[1])*0.95^((x/p[2])-1))*p[3]*(1-v) + v*p[3]
  ssq <- sum((y-model)^2)
}
```

The parameters p[1], p[2] and p[3] represent m, T and Y_{max}, respectively. R will minimize the sum of squares, ssq, stepwise.

Continued

Box 11.2. Continued.

Table 11.2. Example of a result of nonlinear regression with R using ordinary least squares to estimate parameters *m*, *T* and Y_{max} in Eqn 11.7 in a glasshouse experiment. The sum of the minimum rest squares, *SQrest*, in R indicated as out$minimum, equals to 0.2125.

P_i	*Y*	Fitted model	*SQtotal*	*SQrest*
0.03	10.4	10.37	2.380	0.001
0.125	10.2	10.37	1.803	0.028
0.25	10.3	10.37	2.082	0.005
0.50	10.6	10.37	3.038	0.054
1.00	10.5	10.37	2.699	0.018
2.00	10.3	10.33	2.082	0.001
4.00	10.0	10.08	1.306	0.007
8.00	9.5	9.63	0.413	0.018
16.00	9.0	8.88	0.020	0.014
32.00	7.7	7.84	1.339	0.019
64.00	7.0	6.80	3.449	0.041
128.00	6.2	6.26	7.060	0.004
256.00	6.1	6.17	7.602	0.005
512.00	6.2	6.17	7.060	0.001
		sum *SQrest* = out$minimum = 0.2125		

In this analysis, R needs 16 iterations to reach the minimum, *SQrest* or out$minimum: 0.2126. We included the Hessian matrix, more precisely its inverse, to estimate the standard errors of the parameters, *sem*, *seT* and seY_{max}. The coefficient of determination, R^2, or *Rsq*, is defined as 1 – (*MSrest/MStotal*). *MStotal* is the variance of the *y*-variable, while *MSrest* equals *SSrest*, devided by *DF*, the degrees of freedom. *DF* are adjusted for the number of estimated parameters in the model. So, *DF* = *Nobs* (number of observations) – *Lp* (number of parameters). Table R1 in the script shows the results of the analysis, which is summarized in Table 11.3.

Table 11.3. Part of the R1 table produced by the "Glasshouse.R" script.

T	*m*	Y_{max}	seY_{max}	*seT*	*sem*	*MSrest*	*MStotal*	*Rsq*	*DF*
1.683	0.596	10.368	0.066	0.146	0.01006	0.0193	3.256	0.994	11

Note: in the script the tolerance limit, *T*, is indicated as *Te*. "T" has a special meaning in R, namely "TRUE". Therefore, we had to adjust it.

No matter the value of *Rsq*, always calculate the model values and make a graph to compare the observations with the model, as is done in Fig.11.6. At this point, it is important to realize that the statistical distribution of *T* is close to lognormal, while Y_{max} and *m* are normally distributed. The dimension of *T* is nematodes per unit soil, while *m* is a constant and dimensionless. Y_{max} is expressed as fresh or dry weight per pot.

The complete OLS results can be found using the "out" command, while the out$minimum command provides the sum of *SQrest*, the squares of the *y*-values minus the model values. In Table 11.2, *SQrest,* is compared to *SQtotal*, the squares of the *y* values minus the mean

Continued

Box 11.2. Continued.

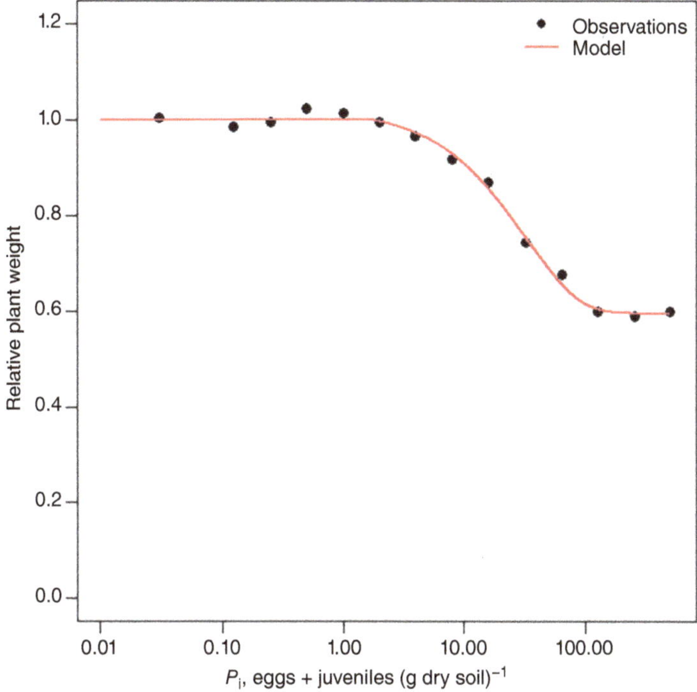

Fig. 11.6. Comparison of observations and the fitted model (Eqn 11.7) on tuber dry weights affected by potato cyst nematodes in a glasshouse experiment.

of *y*. The model follows the data accurately, while, considering *SQtot*, the mean of *y* is not a good representative of the data at all.

There are two pitfalls when using R^2. First, values of R^2 from different datasets cannot be used to compare goodness of fit. If we have $R^2 = 0.99$ in dataset 1 and $R^2 = 0.5$ in dataset 2, it cannot be concluded that the goodness of fit in dataset 1 is better than that of dataset 2. The reason is that R^2 also depends on the value of the parameters *T* and *m*. For example, in very tolerant cultivars, with *m* close to y_{max}, *MSrest* is not very small compared to *MStot*, so *Rsq* will never get close to 1. Nevertheless, the goodness of fit in both datasets may be perfect. A second pitfall is to use values of *Rsq* to compare mathematical equations to find the 'best' model. It is important to realize that *Rsq* is only a statistical navigation tool. The *Rsq* has only a minor function in biological theory building. A good model is a mathematical analogue of a consistent biological theory and can predict outcomes of future observations. The parameters of a good model have a clear biological meaning that enables 'translations' from mathematics into biology and vice versa.

In the first mechanism, Eqns 11.5 and 11.6 only apply to growth reduction that retards growth of plants at a constant rate and, occasionally, increases haulm length of plants. As long as only the first mechanism is active, water consumption during short periods is proportional to plant weight and, therefore, relative water consumption at

Box 11.3. Fitting the Seinhorst model, Eqn 11.7, to field data.

Comparing the tolerance of two cultivars.

To estimate the tolerance of crops or cultivars of crops against root nematodes, the parameters Y_{max}, T and m must be estimated. In the previous examples we showed how this was done for one crop at the time. In this example we compare the tolerance of two potato cultivars, 'Asterix' and 'VanGogh' for *Globodera pallida*. The data are made available by L.P.G. Molendijk. The two objects are numbered in the input file "Tolerance.txt" for the "*ToleranceTwoCultivars*.R" script: nr 1 is 'Asterix'; nr 2 is 'VanGogh'. The procedure of the non-linear regression analysis is as before, with two additions. First, we do calculations in a loop. Loops are used in programming to repeat a specific block of code. In this case, with two cultivars, we repeat the code for regression analysis. Second, a function is added to the script to do a T-test on the parameters. In this function we use the parameter estimates, their standard errors, and the degrees of freedom obtained from the regression analysis.

Before we start the analysis, we make an empty data structure (an array) that we are going to fill with the output data. The length *L* of this data structure equals 2. We start the loop with an accolade as follows: for (*i* in 1:*L*) { and will end the loop using }. Everything between the two accolades will be repeated twice. Then we define *x* and *y*, by referring to the numbering: nr = 1 or 2 using the *i* of the loop, follow the OLS procedure, now slightly different as all parameters are stored with an extension [*i*]:

e.g. m[i] <- out$estimate[1]

After 21 iterations R finds the minima of ssq, *SSrest*, to be 224.8 and 261.1, while *SStotal* amounts to 2255 and 2766 for 'Asterix' and 'VanGogh', respectively. The coefficients of determination, R^2, are 0.96 and 0.89. At the end of the regression procedure, the output in aggregated in table R2 and exported to the file: "Output Tolerance.txt", summarized in Table 11.4.

Table 11.4. The results of the regression analysis to estimate the tolerance parameters of the potato cultivars 'Asterix' and 'VanGogh'.

Cultivar	*m*	*T*	Y_{max}	se*m*	se*T*	seY_{max}	Rsq	DF	log(*T*)	se.log(*T*)
'Asterix'	0.426	0.553	45.93	0.05	0.134	0.96	0.96	29	–0.59	0.239
'VanGogh'	0.239	0.659	42.36	0.13	0.303	2.13	0.89	15	–0.42	0.438

Before we do the T-test, we must remember the dimensions and the statistical distributions of the parameters. Unlike the parameter *T*, both Y_{max} and *m* follow a normal distribution, so we can use the parameter estimates and their standard errors. For the parameter *T* we assume a lognormal distribution. Therefore, we do the T-test on log(*T*) and its standard error, estimated from the coefficient of variation. This procedure was explained in Box 11.1 about stem nematodes.

We end up with the additional data: *CV.Te*, *logTe*, *var.logTe* and *se.logTe* and three parameters, log(*T*), Y_{max} and *m* following a normal distribution.

The function for the T-test includes the following calculations:

- Diff the difference between the parameters
- Dfr the degrees of freedom of the difference
- SED the standard error of the difference
- t.score the difference divided by SED
- p.value the probability to accept the null hypothesis
- Kv the 0.975 quantile
- LSD the least significant difference, the confidence interval of the difference.

Continued

Box 11.3. Continued.

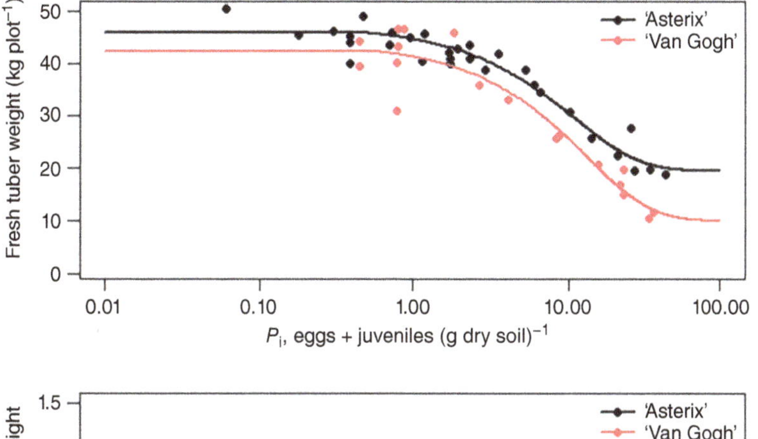

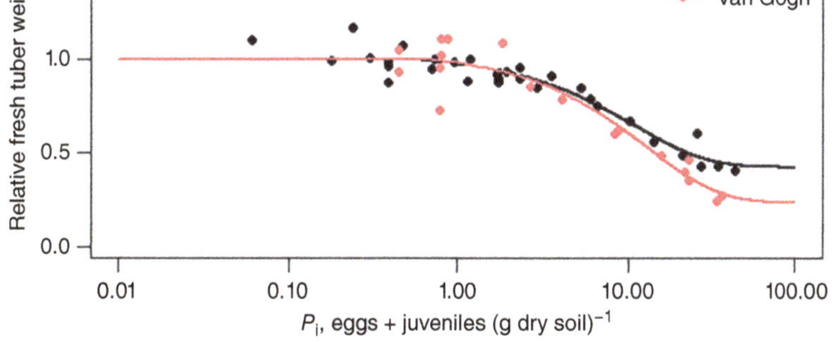

Fig. 11.7. A comparison of the observations with the fitted models. The top graph displays the actual yields while the y-axis of the bottom graph is relative, to better compare the parameters *m*. Remarkable is the shorter range of P_i values for 'Asterix', especially at the small densities.

For the difference, Diff, between two parameters to be statistically significant with α=0.05, it must be equal to or larger than the LSD. Then, zero is not part of the confidence interval of the difference.

From the results of the T-test we learn that the variance in the field experiments is not sufficiently small to discriminate between the tolerance of the two cultivars. The problem is that DF=15 for 'VanGogh' while DF=29 for 'Asterix', resulting in higher standard errors of all three parameter estimates for 'VanGogh'. This experiment illustrates one of the difficulties encountered with the estimation of tolerance in field trials.

different nematode densities and times after sowing or planting is a measure of relative plant weight. Actual plant weights are these relative weights times the actual weights of plants of the same age without nematodes, determined at the same time (Seinhorst, 1986).

In the second mechanism, water consumption per unit plant weight and the (active) uptake or excretion of K^+ and Na^+ are reduced, and the (passive) uptake of Ca^{2+} and the dry matter content of plants are increased (Seinhorst 1981; Been and Schomaker, 1986). It also tends to advance the development and ripening of seed to a certain limit. There is probably a negative correlation between age of the plant and the smallest P_i where the second mechanism manifests itself, i.e. the younger the plant, the smaller the P_i where the second mechanism can be noticed. For potato cyst nematodes, this density is rarely as small as $16T$ but more commonly, for other nematode species on other plant species, the density is $>32T$.

Early senescence was first observed in potato plants attacked by large numbers of *G. pallida*. Early senescence is a sudden ending of the increase of haulm length and weight and results in an early death of plants. There are indications that 'early senescence' coincides with initiation of tuber growth and, because of that, is negatively correlated with nematode density. The earliest occurrence was 9 weeks after planting in the early potato cultivar Ehud and the lowest nematode density of $25T$ (about 45 nematodes g^{-1} of soil) in the potato cultivar Darwina. Not all cultivars are equally sensitive. Early senescence was also observed in plants exposed to air pollution. So far, the causes are unknown.

11.6.3. *T* and *m* as measures of tolerance

Tolerance in plants can be quantified by expressing it in values of the tolerance limit T and the minimum yield m. The parameter T manifests itself at small nematode densities, m at larger ones. Yet we need a whole range of pre-plant nematodes densities (P_i) to estimate either one of the parameters.

The value of the tolerance limit, T, seems unaffected by differences in external conditions and can, therefore, be determined in pot experiments in both glasshouse and (more laborious and less accurate) field experiments. The only requirement of glasshouse tests is that large enough pots are used to guarantee about the same root density in the soil as in the field and to prevent the plants from becoming pot-bound. The latter affects the relationship between nematode density and plant weight and obscures the true value of T (Seinhorst and Kozlowska, 1977). The accuracy of the estimates is mainly a matter of uniformity of plant material and growth conditions (light, water content), carefully and uniformly filling and handling of the pots, including rotation of pots to avoid positional effects in the glasshouse.

The minimum yield, m, is more sensitive to external conditions than T. Values for potato cyst nematodes on potatoes varied between 0 on cultivar Bintje and 0.8 on cultivar Agria. Estimates of m for *Heterodera schachtii* on red beet varied between 0.3 and 0.65 and on Brussels sprouts between 0.32 and 0.65. More estimates of T and m are given in Table 11.5. Differences in tolerance in, for example, plant cultivars, should be established in one experiment under the same conditions and preferably conducted in a glasshouse. In addition, a sufficient number of values of m must be estimated in field experiments to establish a distribution function of m.

Estimation of T and m in field experiments is much more laborious than in pot experiments (Box 11.4). A full range of nematode densities, from 0 to at least 250 nematodes g^{-1} of soil, is needed and nematode density must be the only variable. For some species (e.g. potato cyst nematodes), ranges of nematode densities in infestation foci come closest to this requirement. Unfortunately, ranges of nematode densities cannot be created by applying different dosages of a nematicide as this has unpredictable effects on plant weight in addition to those caused by killing nematodes. The ranges of nematode densities needed to estimate the parameters in question must be determined in samples large enough to guarantee

Table 11.5. Data of 38 plant–nematode combinations (15 nematode species) out of more than 50 experiments on the relationship between P_i and total weight y, with $z^T = 0.95$.

Nematode species	Plant species	T[a]	m[b]
Globodera pallida and *G. rostochiensis*	*Solanum tuberosum* (susceptible and resistant)	1.8	0.2–0.6
Heterodera avenae	*Triticum sativum*	0.3	0
H. avenae	*Avena sativa*	1.4	0.27
H. avenae	*A. sativa*	0.85	0.05
H. avenae	*A. sativa*	0.35	0.6
H. avenae	*A. sativa*	1	0.6
H. carotae	*Daucus carota*	0.7	0
H. ciceri	*Cicer arietinum*	1.3	0.16
H. trifolii	*Trifolium repens*	0.8	0
H. trifolii	*T. repens*	0.7	0.12
H. goettingiana	*Pisum sativum*	3.4	0.28
Longidorus elongatus	*Fragaria vesca*	0.09	0.3
Meloidogyne artiellia	*Cicer arietinum*	0.13	0.1
M. incognita	*Beta vulgaris*	1.1	0.1
M. incognita	*Brassica oleracea*	0.5	0.05
M. incognita	*Capsicum annuum* (susceptible)	0.74	0.1
M. incognita	*C. annuum* (resistant)	0.74	0.4
M. incognita	*Coffea arabica*	1.4	0.4
M. incognita	*Cucumus melo*	0.19	0
M. incognita	*Helianthus annuus*	1.85	0.25
M. incognita	*Solanum lycopersicum* (resistant)	0.5	0.7
M. incognita	*S. lycopersicum* (susceptible)	4	0
M. incognita	*Nicotiana tabacum*	2	0
M. incognita	*Solanum melongena*	0.054	0.05
M. hapla	*Trifolium repens*	1.3	0.43
M. javanica	*Coffea arabica*	1.15	0.4
M. javanica	*Helianthus annuus*	0.25	0.43
M. javanica	*Oryza glaberrima*	2.2	0.035
M. javanica	*O. sativa*	0.18	0.04
Pratylenchus penetrans	*Daucus carota*	1.4	0.49
P. penetrans	*Digitalis purpurea*	5.6	0.16
P. penetrans	*Malus*	1.5	0.44
P. penetrans	*Vicia faba*	6.2	0.43
P. penetrans	*V. faba*	1.3	0.43
Tylenchorhynchus dubius	*Lolium perenne*	1.6	0.6
T. dubius	*L. perenne*	1.6	0.22
Xiphinema index	*Vitis vinifera*	0.15	0.05

[a]Nematodes g^{-1} of soil.
[b]Fraction of yield without nematodes.

a coefficient of variation (standard deviation/average) of egg counts less than 15%. At this coefficient of variation, density differences of 1:2 are just distinguishable. Soil samples of 4 kg plot^{-1} are then needed to estimate population densities of 1 egg g^{-1} of soil, given a coefficient of variation of the number of eggs per cyst of 16% and a negative binomial distribution of egg densities in samples from small plots with a coefficient of aggregation

Box 11.4. Experimental design to estimate *T* and *m*.

Pot or microplot experiments

Pot experiments can be conducted in glasshouses where external conditions, especially soil moisture and temperature, are controlled. Most nematodes die when temperatures rise above 30°C and both nematode multiplication and plant growth are reduced if soil moisture drops below 10% of the soil dry weight. If climate-controlled glasshouses are not available, microplot experiments in the open air with inoculated soil are the best alternative. In both types of experiments the volume of soil must be chosen so that the root system has the same space as in normal agricultural conditions. If the pots are too small, the plants will resume their normal growth rate (the same as without nematodes) as soon as the pot boundaries are reached, and the nematodes are depleted. To estimate *T* and *m*, nematode densities must be created according to a log series 2^x, x being whole negative and positive numbers. If x is chosen between –3 and 9 the density series: $2^{-3}, 2^{-2}, 2^{-1}, 2^0, 2^1, 2^2, 2^3, 2^4, 2^5, 2^6, 2^7, 2^8, 2^9$ is obtained which equals 0.125, 0.25, 0.5, 1, 2, 4, 8, 16, 32, 64, 128, 256, 512 nematodes g^{-1} of soil. The ratio between two succeeding nematode densities in this series is 2, which means that a ^{2}log series is attained. As mentioned, a log series is required to plot P_i against y. The number of nematodes that is needed in an experiment with a ^{2}log series of nematode densities is about twice the number of nematodes needed at the highest density. In an experiment with the above-mentioned series, using 1 l pots and five replicates 2 × 512 × 1000 × 5 = 5,120,000 nematodes are required. In theory, there is no objection against log nematode series with a base different from 2, as long as it is smaller. In practice it is not easy to create a dilution series with, for example, a factor 1.7 between succeeding nematode densities. During inoculation, care must be taken that the nematodes are distributed randomly through the soil, e.g. by placing them in well-distributed channels throughout the pot using syringes (Been and Schomaker, 1986). It is also important that the nematodes are present in the soil at the time of planting or sowing. If the soil is inoculated with nematodes days or weeks after planting, an artificial situation is created by introducing plants with an established, but non-infected, root system; this does not equate to the normal agricultural situation. Most root-infecting nematodes prefer penetration immediately behind a new root tip and, as the growing time of a root system is limited, mostly until the time of fruit setting, the number of penetrations will be equally limited. Then, the relative minimum yield will be increased.

Field experiments

In field experiments it is almost impossible to create densities, so pre-sampling is necessary to select plots with the required densities. Another requirement of a suitable experimental field is that the target nematode is the only variable that causes variation in yield. To estimate *T* and *m*, a field with both very small and large densities and some intermediate densities is needed. Fields with infestation foci are often very useful for experiments because of their gradually increasing population densities. In this case, the sampling errors can further be reduced by regression, making use of the focus model (linear regression on log nematode numbers and distance; see Chapter 12). The nematode densities according to the model are considered to be the 'true' densities. The deviations of the original data points from the model represent the variation due to sampling and laboratory procedures and can be removed. In fields where nematodes are distributed more uniformly, plant weights per plot often vary considerably between plots that have approximately the same nematode density. This makes curve fitting very difficult. This situation can be improved by dividing nematode densities into classes where ratios between minimum and maximum limits are less than 1.7. Average plant weight per density class is then plotted against average ^{10}log nematode density. Per class, average nematode density $P_i = P$ is estimated as the antilog of P, which is 10^P. The chosen base of the logarithm is arbitrary. If an ^{e}log (or natural logarithm) is chosen, the antilog is e^P (also written as $\exp(P)$). The sample sizes needed to estimate small, medium and large nematode densities accurately are discussed in Chapter 12.

of 50 for 1 kg soil. As the coefficient of variation per unit weight of soil is negatively correlated with nematode density, so is the required sample size. For example, to estimate densities of 0.5 or 0.25 eggs g^{-1} of soil with the same accuracy, eggs from soil samples of 10 and 20 kg, respectively, must be counted. Another requirement is that plots must on the one hand be small (e.g. 1 m^2) to reduce the effect of medium-scale density variation (see Chapter 12), whereas on the other hand a large enough area per small density interval must be available to guarantee a small variability of tuber weight per unit area. Again, there must be a sufficient number of plots at densities smaller than T to obtain the best estimate of the maximum yield with the smallest variance. It is most efficient to estimate T and m, as much as possible, from glasshouse experiments. Field trials are best used to confirm or reject these estimates for different combinations of pathotypes and cultivars under more natural environmental conditions. In Table 11.5 the estimates of T and m from data originating from field trials are presented.

11.6.4. Growth reduction in perennial plants

During the first year after planting, effects of nematode attack on perennial plants can be investigated in the same way as with annual plants. We cannot yet answer the question whether a reduction in growth and productivity should be expected in the second and subsequent years; this depends on nematode densities at planting, especially small densities. On citrus, *Radopholus similis* spreads from old roots to new roots at the periphery, thus rapidly increasing nematode numbers. The same tendency for numbers to increase rapidly and for migration is observed for stem nematodes in red clover and lucerne, in these cases via moist surfaces of plant leaves. Nematodes often cause no specific disease symptoms and annuals are more tolerant to second and later generations of nematodes than to a first generation present at planting (Seinhorst, 1995). Therefore, it is not certain that the presence of large nematode numbers in orchards with old trees, for example, will cause substantial reductions in productivity. Increase in productivity after treatment with contact nematicides or nematistats is no proof of nematode damage, as yield increases in treated fields even when damaging species of nematodes are absent. To investigate growth reduction of perennial plants by increasing nematode populations, patterns of weight of the total plant and its fruits must be studied at sufficiently wide ranges of nematode densities and at regular time intervals. To produce a clear pattern, a simple system with external conditions as constant as possible must be studied first. Later, more complex systems can be studied and their patterns compared with the patterns from the simple system.

11.7. Validation of the Model

Seinhorst (1998) collected all available data in the literature about the relationship between P_i of 14 tylench nematode species and the relative dry plant weight (y) of 27 plant species/cultivars several months after sowing or planting: in total 36 plant/nematode combinations out of 29 experiments. An overview is given in Table 11.5. As T and m varied it was necessary to standardize the variables P_i and y. To do this, for each separate experiment T and m were estimated and $y' = (y - m)/(1 - m)$ and P_i/T were calculated. The relationship between average y' and P_i/T appeared to be in close accordance with $y' = z^{P_i - T}$, where $z^T = 0.95$ for all nematode–plant combinations. Therefore, for most combinations of tylenchid nematodes and plant weight Eqn 11.6 can be rewritten as Eqn 11.7:

$$y = m + (1 - m)0.95^{\left(\frac{P_i}{T} - 1\right)} \quad \text{for } P_i > T$$

This relationship is obtained by log transformation of Eqn 11.6 and by substituting log(z) for (1/T)log(0.95). This common relationship for nematode–plant relationships implies, firstly, that nematodes that feed and multiply in very different ways still have the same effect on plant weight, despite external conditions, the host status of plants or the visible symptoms. A second implication is that host plants of any species of tylench nematodes are able to prevent growth reduction by these nematodes to the same degree, namely 0.95.

11.8. Effect of Nematicides

Nematicides can be divided into two groups: contact fumigants that kill nematodes directly and nematistats that, at the legally permitted concentrations, make nematodes immobile for a period of time (see Chapter 17). With respect to application, we can divide nematicides into fumigants and non-fumigants. All nematistats are non-fumigants, but not all non-fumigants are nematistats. Fumigants can be brought into the soil at a certain depth. Afterwards, they move actively upwards and downwards in the gaseous form.

11.8.1. Nematistats

The best-known nematistats are aldicarb, carbofuran and ethoprofos. The first two chemicals belong to the carbamates, the third to the organophosphates. Soil is treated immediately before planting and the chemicals make plants unattractive for nematodes. It has been observed that nematistats cause nematodes (*Globodera* spp. and *Ditylenchus dipsaci*) to leave plant roots but this is not an effective control measure. Good distribution through the soil is of utmost importance to the effectiveness of the nematicide as root systems are only protected when all roots are in close contact with the pesticide. Treated plants do not lose their ability to induce nematode hatch from eggs (e.g. some cyst nematodes and *Melodogyne* spp.) but juveniles become either immobilized or disoriented and cannot find their food source, the plant roots. If this effect lasts long enough, the nematodes will eventually starve (see Chapter 7). Therefore, mortality of the nematodes depends on the time that plants remain unattractive, and nematodes immobilized. Treatment of plants with nematistats delays nematode penetration into the roots and results in a certain fraction of the root system escaping nematode attack and thus remaining healthy. As a result, the minimum yield m is increased by the fraction of the root system untouched by nematodes. Nematode penetration is postponed until the chemical is no longer effective or when the roots grow into soil layers where the nematicide is not present. Usually, nematistats are distributed through the soil to a depth of 15–20 cm. Experiments on root-feeding nematodes demonstrate that nematistats increase m and hardly affect T. The parameter T is only affected (increases) when nematodes die because of a long-lasting effect of nematistats. The increase of m by nematistats may be negatively correlated with tolerance (L. Molendijk, personal communication). This seems logical: an intolerant cultivar with a relative minimum yield m of 0.1 leaves more room for improvement (1 – 0.1) than a tolerant cultivar with $m = 0.9$, where m can only increase by 0.1. Therefore, relative effectiveness, expressed as $m'/(1 - m)$, where m' is the difference in minimum yield between treated and untreated plants and m is the minimum yield of untreated plants, might be a stable variable to measure the effectiveness of nematistats.

11.8.2. Contact nematicides

Contact nematicides, which can be fumigant and non-fumigant (see Chapter 17), decrease the P_i. For cyst nematodes and *Meloidogyne*, hatching tests are often used to estimate percentage mortality. The higher the nematicide dose the longer these hatching tests must be continued as nematicides delay the hatching process (Schomaker and Been, 1999a). If the effect of nematistats and contact nematicides is compared in one field experiment where P_i is estimated immediately before application, we estimate the effect of the contact nematicide as an increase in T and the inviolability of the root system by the nematistats as an increase in m (Fig. 11.8). Increase of T is also observed in a glasshouse experiment if some of the nematodes die during the inoculation process or do not respond to root diffusate and do not invade the plant.

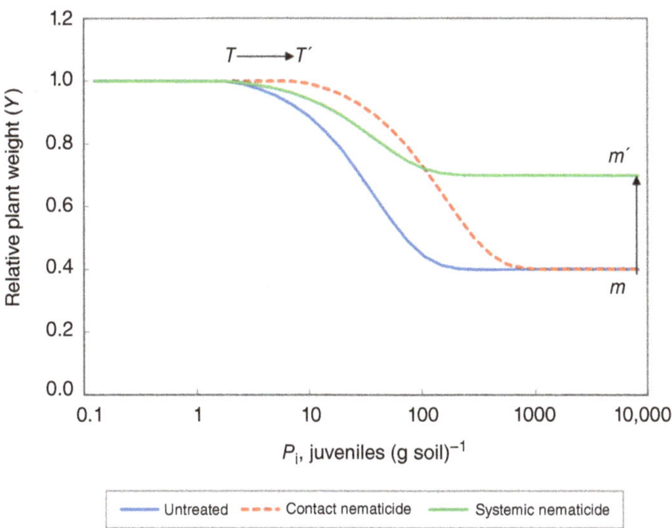

Fig. 11.8. Effect of nematicide treatments on plant yield. Treatment with a nematistat increases plant tolerance. The parameter m increases from 0.4 to 0.7. Nematistats can also cause mortality if their effectiveness lasts long enough. Mortality is seen as an increase in the tolerance limit T. Treatment with a contact nematicide causes a reduction of the initial population density. Compared to the original (pre-nematicide, pre-plant) population densities, this effect is observed as an increase in the tolerance limit T. In the graph a 75% mortality was presumed.

11.9. Population Dynamics

A population can be defined as a group of organisms that resemble each other genetically, morphologically and behaviourally, and are living in the same area or region. Population dynamics describes the general biological laws or processes that govern increase or decrease of organisms. As nematode densities are often good predictors of plant damage, the study of population dynamics is an important discipline in nematology. The increase and decrease of nematode numbers are relatively slow processes. Therefore, control measures to prevent nematode increase or to stimulate nematode decrease must be taken at an early stage. The time of planting a susceptible crop is often too late for control measures, even for most nematicide applications. In the population dynamics of plant-parasitic nematodes that live on crop plants two phases can be distinguished: (i) the growing period of plants on which nematodes can multiply; and (ii) the period when nematodes do not have access to plants, so no food source is available. Nematode populations can only increase during the first phase. For nematodes on annual plants there are several population dynamic models available that relate P_i to P_f. Here we have chosen a model that is suitable to predict future nematode densities. Such a model must meet three conditions: (i) the model must be a mathematical translation of a biological theory and its parameters must have a clear biological meaning; (ii) the model must be as simple as possible but as extensive as necessary; and (iii) the model must allow estimation of at least starting values of the parameters directly from datasets and not only by regression.

The models of Seinhorst comply with these conditions. In these models, at small nematode densities, nematode multiplication is restricted only by the amount of food that the nematodes can capture and utilize under the given conditions. As plants at small densities provide sufficient space for all nematodes, competition does not play an important role. At high densities, nematode multiplication is limited by competition and the total amount of food that the host can supply. If plants remain smaller as a result of nematode infestation, then the total amount of available food and P_i are negatively correlated. In the following sections, nematodes with one and nematodes with more than one generation per season are discussed separately.

11.9.1. Nematodes with one generation per season

For nematode species with one generation per year, which become sedentary after invasion, the population dynamic model is:

$$P_i = M\left(1 - e^{-a\frac{P_i}{M}}\right) \tag{11.10}$$

in which: P_i is initial nematode density (before planting) as juveniles g^{-1} of soil, P_f is final nematode density (after harvest) as juveniles g^{-1} of soil, a is the maximum rate of reproduction and M is maximum population density as juveniles g^{-1} of soil.

We assume that the offspring is proportional to the part of the root system that is exploited for food and that $P_f = aP_i$ if $P_i \rightarrow 0$ and $P_f = M$ if $P_i \rightarrow \infty$. This model is based on the same principles as Eqn 11.5. The occurrence of discrete, random events in space and/or time, such as the random encounters between nematodes and plant roots, are described by the Poisson distribution. The first term in the Poisson distribution, i.e. the likelihood for a plant root to escape nematode attack (zero encounters), is given by $\exp(-cP_i)$ where c is a constant. The probability of one or more encounters is then given by $1 - \exp(-cP_i)$. We can understand this by imagining plant roots as cylindrical surfaces divided into equal compartments that, per cross-section, are penetrated randomly by juveniles, often J2. As time and root growth go on, the cross-sections are moving up along the cylinder. If the juveniles can settle in more than one compartment at the same time, this would result in overlap in territories and a decrease in the number of eggs per female. If there is no overlap, only one juvenile per compartment can survive and number of eggs per female is not decreased. The size of the compartments depends on the place of the root in the root system and the growing conditions of the plant, but not on the density of the surviving juveniles. This simple model has some constraints. It makes no difference between the population dynamics in rooted and non-rooted soil and ignores reduction of plant roots by nematode infestation. Therefore, Eqn 11.11 applies only to small and medium densities where $P_i < M$.

As the values of a and M are determined not only by the qualities of plant roots as a food source, but also by external conditions, the final population density at one initial population density can vary markedly between fields and years. Therefore, it is impossible to predict the development of population densities in individual fields using only averages of a and M. To calculate the probability of all possible values of P_f, the frequency distributions of a and M are needed.

The population dynamics is not determined by the number of generations per season, but rather by the quality of the host as a food source for a particular nematode species.

11.9.2. Nematodes with more than one generation

For the population dynamics of migratory nematodes with more than one generation per year, the same basic principles apply as for sedentary nematodes with one generation that are discussed in Section 11.9.1. The only difference is that nematodes with more than one generation redistribute continuously while their numbers increase. The population increase per unit of time is proportional to the difference between P_i and the so-called equilibrium density. It is comparable with the 'law of diminishing returns'. This leads to the following relation between P_i and P_f:

$$P_f = \frac{aEP_i}{(a-1)P_i + E} \tag{11.11}$$

In which: E is equilibrium density (where $P_i = P_f$), a is the maximum multiplication rate and $M = aE/(a - 1)$ is the maximum population density.

For very small values of P_i the final density $P_f = aP_i$ and at very large values of P_i the final density $P_f = M$. At the equilibrium density, E, the amount of food supplied by the plant is just enough to maintain the population density at planting. Among other things, it enables us to compare and quantify the host suitability of different species or cultivars for certain nematodes. Figure 11.9 presents eight examples of the relationship between nematode populations at the beginning (P_i) and the end (P_f) of a growing season. Number 1 represents the population dynamics on a good host, 2 and 3 on a less good host and 4 and 5 on a poor host. These relationships can also be explained as the population dynamics of a good host under favourable (1), less favourable (2) and (3), or unfavourable (4) and (5) conditions. The relationships numbered 6, 7 and 8 represent the decrease in population density in the absence of hosts.

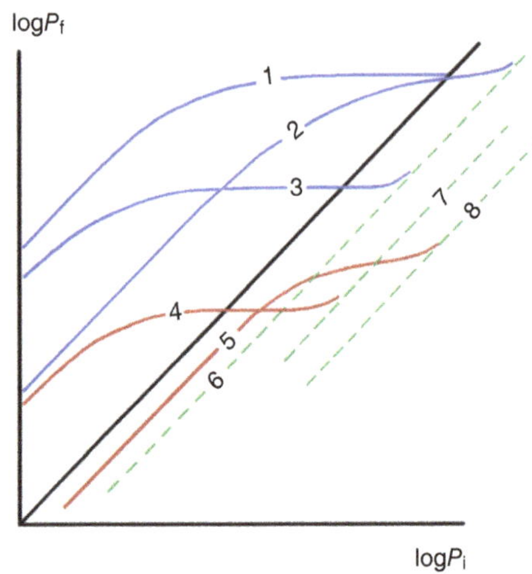

Fig. 11.9. Schematic presentation of the relation between initial and final population densities on a host. Solid lines: good (1) intermediate (2, 3) and poor (4, 5) hosts or good host grown under favourable (1), less favourable (2, 3) and unfavourable (4, 5) circumstances. Dotted lines (6, 7, 8) show the reduction of population densities in the absence of hosts dependent on the mortality rate of the nematode species. (From Seinhorst, 1981.)

Table 11.6 Average population dynamic parameters of four root nematode species, estimated in field experiments.

Nematode	*a*	*M* (juveniles (g soil)$^{-1}$)
Globodera rostochiensis	25	300
G. pallida	20	150
Pratylenchus penetrans	10	60
Meloidogyne chitwoodi	40–60	100

The parameters *a* and *M* are log normal distributed. The estimates for *a* and *M* for four nematode species are summarized in Table 11.6.

In Box 11.5 the results of a glasshouse experiment, where the host status for *P. penetrans* for two potato cultivars was compared, are discussed. Non-linear regression analysis was done with Eqn 11.12.

11.9.3. The effect of nematode damage and rooted area on population dynamics

Nematodes reduce plant growth and the size of plants. Stem nematodes in onions also decrease the number of plants. As a result, there is less food available for the nematodes when the P_i is larger. This is especially the case when host plants are intolerant. To account for this reduction in plant size we can expand Eqns 11.10 and 11.11 as follows for nematodes with one generation:

$$P_f = yM\left(1 - e^{\frac{-aP_i}{yM}}\right) \tag{11.12}$$

Box 11.5. Comparing the host status for *Pratylenchus penetrans*.

In a glasshouse experiment the host status of two potato cultivars, 'Festien' (A) and 'Seresta' (B), for *P. penetrans* was compared. There were 12 densities per cultivar, ranging from 0.1 to 242 nematodes g^{-1} of soil and four replicates per density. The procedure followed to analyse the data is described in the script "*PopulationDynamicsPP*".R. The overall procedure is the same as described in Box 11.3, but there are some extensions. The script can be divided in three sections:

1. The calculation of the average P_f values per P_i value, per cultivar in a loop. Before we can proceed, we must number the objects. The most robust method is using the paste function in R. We insert a new column, *Pop*, where we paste the *Cult* and P_i columns of the input file "*PopulationDynamicsPP.txt*" as follows: paste0(dat0$Cult, dat0$Pi). The zero behind paste() means that no space is allowed between the columns. The first five elements of this new column are: A0.1, A0.1, A0.1, A0.1, A0.22....
To turn the column into a vector we need c()

```
Vr <- c(dat0$Pop)
```

The rle() command is very useful to return each object and its length. However, rle() returns an encoded result. Therefore, to place these data properly in a data frame, we need the function unclass(). In the script, the data.frame, So, is then written as

```
So  <-data.frame(unclass(rle(Vr)))
```

For a better understanding of the procedure, we give a simple example of the rle command in the script. Within the scope of this book, it is impossible to explain all used R functions. Fortunately, there is a large amount of literature about R.

With information about each object and its length, we can easily number the data set. This numbering procedure is especially useful in large, irregular data sets.

The next step is to log transform the P_f values into lP_f and add them to the data set. Averages of log(P_f) values, *av.*lP_f, with the same number, are calculated in a loop. The back transformed average, *av.*P_f, is added with additional information about cultivar, P_i in a dataset. This table can be found in the output file "*AveragePf.txt*" of the script.

2. The non-linear regression analysis, with OLS, using Eqn 11.11. the model for migrating nematodes. The function fn is now (in R script):

```
fn <- function(p){
  MOD  <-(p[1]*x)/(x+(p[1]/p[2]))
  ssq   <-sum(((log(y))-(log(MOD)))^2)
}
```

Mind the brackets {}, delimiting the contents of the loop.

Both the model values and *y* (=*av.*P_f) are log transformed as neither is normally distributed. In the model, $x = P_i$. The starting values are calculated directly from the data: the starting value for *a*, the maximum multiplication rate, is *astart* <- max(*y*/*x*). The starting value for the parameter *M*, *Mstart*, is max(*y*). The results of the regression analysis are placed in a data.frame, R2. The table is exported as "*OutputPopulationDynamicstPP.txt*" and summarized in Table 11.7.

Graphs, shown in Fig. 11.10, were made to compare the observations with the models.

3. The last step is a T-test, using the parameter estimates and their standard errors. As neither *a* nor *M* are vectors, but two estimates, we cannot simply use the function

Continued

Box 11.5. Continued.

t-test() in R. As *a* and *M* are lognormal distributed, we log transform *a* and *M* into *loga* and *logM* and calculate their standard errors, *se.loga* and *se.logM*, from the coefficients of variation, the same way as we did in Box 11.3 for the parameter *T*.

The conclusion is that the maximum multiplication rate *a* on 'Seresta' (14.02) is larger than that on 'Festien' (7.1), but for the parameter *M* the situation is reversed: *M* on 'Festien' is larger (75.81) than *M* on 'Seresta' (47.13 nematodes g^{-1} of soil) The P_i~P_f lines for the cultivars are crossing at P_i=8.5 nematodes g^{-1} of soil (Fig. 11.10). This experiment illustrates again, that host status cannot be estimated at any P_i value.

Table 11.7. The regression output for comparison of host status.

Cultivar	*a*	*M*	*sea*	*seM*	*Rsq*	*DF*	*loga*	*logM*	*se.loga*	*se.logM*
'Festien'	7.14	75.81	0.65	7.95	0.98	9	1.97	4.33	0.09	0.10
'Seresta'	14.02	47.13	1.82	4.79	0.95	9	2.64	3.85	0.13	0.10

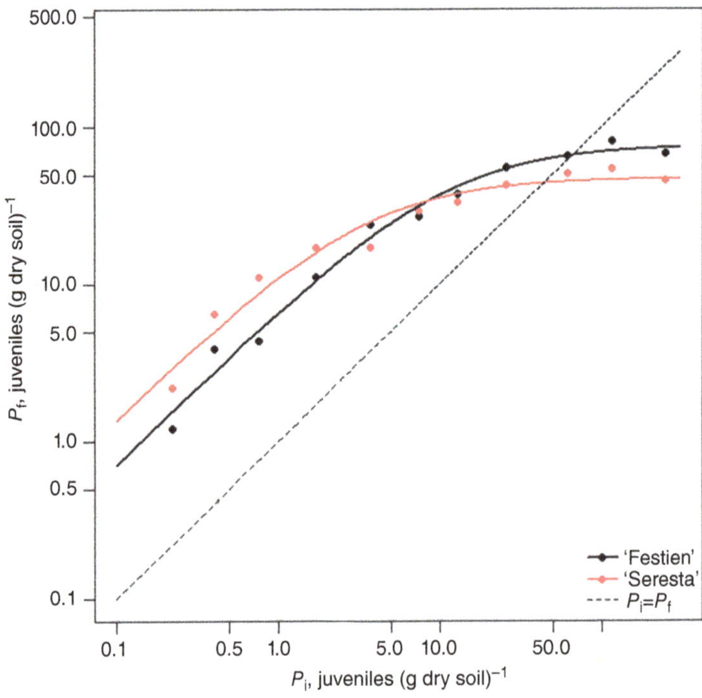

Fig 11.10. Comparison of the host status of two potato cultivars, 'Festien' and 'Seresta', for *Pratylenchus penetrans*. Observations are compared with the model according to Eqn. 11.11

and for nematodes with more than one generation:

$$P_f = y\,a\,E\frac{P_i}{(a-1)P_i + y\,E} \tag{11.13}$$

The variable *y*, described by Eqn 11.7, estimates the relative size of the root system affected by the nematodes. With root-invading nematode species, it can happen that part of the plant is not infested with nematodes if, for example, the food source is reduced and/or the nematodes are not in the vicinity of the roots. The larger the growth reduction of the plants, the smaller is the rooted area and the smaller the food source for the nematodes. Therefore, Eqns 11.2 and 11.3 only apply to the soil area containing roots. In the soil area without roots, the nematodes slowly decrease independently of P_i. To describe the population dynamics of the nematodes in the whole tillage (rooted and non-rooted) we can further expand Eqn 11.12:

$$P_f = r\,y\,M\left(1 - e^{-a\frac{P_i}{r}y\,M}\right) + (1 - r\,y)\alpha P_i \tag{11.14}$$

and Eqn (11.13):

$$P_f = r\,y\,a\,E\frac{P_i}{(a-1)P_i + r\,y\,E} + (1 - r\,y)\alpha\,P_i \tag{11.15}$$

where: *r* is the proportion of rooted soil at $P_i = 0$; *y* is the relative size of the root system, estimated from relative dry haulm weight; and α is the multiplication rate (≤1) of nematodes in the absence of hosts.

The proportion of rooted soil depends on plant anatomy and cropping systems. At large values of P_i in sensitive crops (*m* is small), the product *ry* comes close to zero and $P_f \rightarrow \alpha P_i$. The resulting population dynamic models are visualized in Fig. 11.11.

11.9.3.1. Resistance

Fewer females will mature on resistant cultivars compared to susceptible ones (for definitions see Section 15.2 in Chapter 15). The number maturing depends on the degree of resistance or its complement, susceptibility. Also, the number of offspring per female may be reduced, but this is not always the case. *Globodera pallida* females multiplying on cultivars resistant to *G. pallida* produce a smaller number of eggs per cyst than on susceptible cultivars but *G. rostochiensis* females have more eggs per cyst on resistant cultivars. In general, nematodes multiply less strongly on these cultivars than on susceptible ones and also sustain smaller maximum population densities.

If predictions about the population dynamics of nematodes on resistant cultivars are required, a stable measure of resistance is needed. The often-used P_f/P_i ratio is unsuitable because of its density dependence; remember that P_f/P_i is large at small densities and small at large ones. The parameters *a*, *M* (and *E*) vary strongly between fields and years under the influence of different external conditions, which makes them also unsuitable as stable measures for resistance. Therefore, the concept of '**relative susceptibility**' (RS) was introduced. Relative susceptibility was first described for *G. pallida* Pa3 on a selection of potato cultivars, some known as susceptible, others known as resistant against *Globodera*

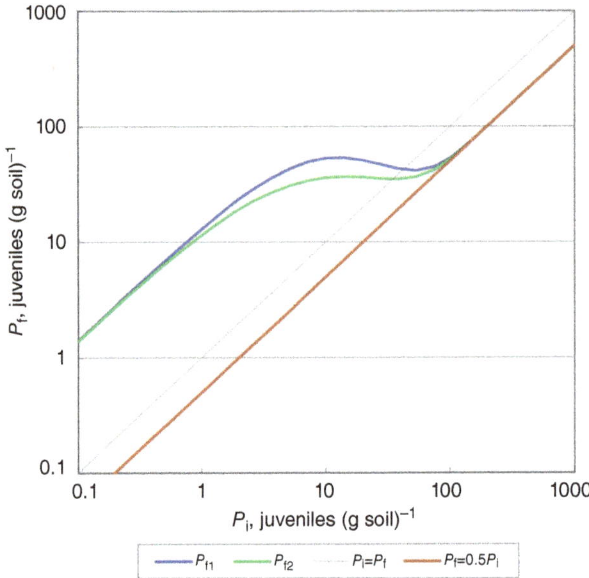

Fig. 11.11. Population dynamic models for nematodes with one generation, P_{f1}, and nematodes that multiply continuously, P_{f2}. Both models incorporate decrease of total mass of plant root by the nematodes, the rooted fraction of the soil and nematode mortality in the absence of food. In this figure, nematode mortality in the absence of food is presumed to be 50%; minimum yield (m) is set to 0.

rostochiensis Ro1 and or *G. pallida* Pa2. The concept of Relative Susceptibility also applies to other combinations of nematodes and their hosts, for example *Meloidogyne chitwoodi* on potato, fodder radish and sugar beet (Teklu *et al.*, 2014, 2016). It was defined as the ratio of the maximum multiplication rate a of a nematode population on the resistant cultivar and on the susceptible reference cultivar ($a_{resistant}/a_{susceptible}$) or the equivalent ratio of the maximum population density M on these cultivars ($M_{resistant}/M_{susceptible}$). These ratios present two equal measures of partial resistance or relative susceptibility, provided that the tested cultivar and the susceptible reference are grown under the same conditions in the same experiment. Figure 11.12 visualizes the relation between P_i and P_f of pathotype Pa3 of *G. pallida* on the partially resistant potato cultivar 'Darwina' and on the susceptible cultivar 'Irene' according to Eqn 11.14. The RS has been put into practice in The Netherlands and has proved to be independent of external conditions with one exception: during a hot summer the RS increased in two pot experiments when the temperature in the glasshouse exceeded 28°C. There are also reports that resistance for *Meloidogyne* spp. in tomato decreased (and susceptibility increased) under the influence of high temperatures. In temperate zones, this effect may be of little consequence, as temperatures in the soil are probably buffered sufficiently, but in tropical zones high temperatures may counteract resistance more frequently.

To make predictions about the population dynamics of a nematode population on resistant cultivars, two more tools are needed. First, estimates must be made of the expected values and the variances of a and M on the susceptible reference cultivar in

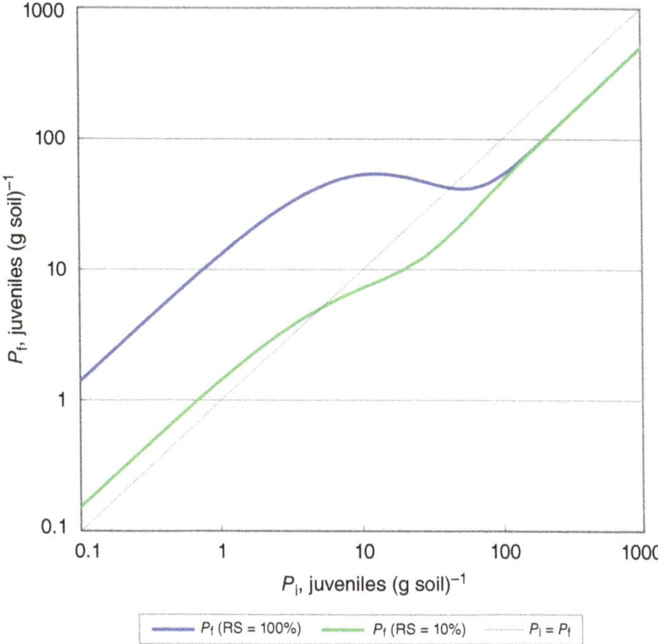

Fig. 11.12. Comparison of the population dynamics of potato cyst nematodes on a resistant and a susceptible variety. The relative susceptibility (RS) of the resistant variety is 10%; minimum yield (m) is set to 0. The tolerance of the susceptible and the resistant variety are the same. Note that tests at medium or high densities underestimate resistance.

different fields and different years. From this, frequency distributions for a and M can be made. Second, the RS of all cultivars resistant to a nematode species must be estimated under controlled conditions. For each nematode species, one or more carefully chosen populations are screened, depending on the variability in virulence. Some examples of partially resistant potato cultivars in The Netherlands are given in Table 11.8, together with the expected yield reductions if cultivars with these resistance qualities are grown in a 1:3 rotation.

Sufficient observations are available only for potato cyst nematodes to predict the population dynamics on resistant cultivars. Therefore, and because the same basic principles apply to all tylench nematode/plant relations, potato cyst nematodes are often used as model nematodes for other tylench species. Resistant cultivars give farmers an excellent tool to manage their nematode populations by keeping them at low densities that are not harmful. Farmers who have to deal with quarantine nematodes are often put in a difficult position, because governments demand that they should eradicate these nematodes. Fifty years of experience with some quarantine nematodes, such as potato cyst nematodes, has shown that these nematodes cannot be eradicated, not even by chemicals or fully resistant cultivars, on fields where hosts are grown frequently.

In general, nematodes inflict the same degree of damage in resistant and in susceptible cultivars but it is important to remember that both resistant and susceptible cultivars may vary in tolerance. For example, in various field trials with *G. rostochiensis*, the minimum

Table 11.8. A selection of 16 Dutch potato cultivars, their relative susceptibilities ($a_{resistant}/a_{susceptible}$) for two pathotypes (Pa2 and Pa3) of *Globodera pallida* and the percentage average yield reduction in a 1:3 rotation.

	Relative susceptibility (%)		Average yield reduction (%)	
Cultivar	Pa2	Pa3	Pa2	Pa3
'Irene'/'Bintje'	100	100	40.7	40.7
'Aveka'	0.14	0.4	0	0
'Aviala'	0.2	0.1	0	0
'Darwina'	0.3	12	0	0.6
'Kantara'	0.8	5	0	0
'Nomade'	1	4	0	0
'Producent'	6	56	0	25.6
'Seresta'	0.2	2	0	0
'Agria'	60	94	27.3	39.1
'Hommage'	9	53	0.1	24.1
'Innovator'	4.5	1	0	0
'Elles'	–	17	–	2.5
'Marijke'	45	51	20.1	23.1
'Maritiema'	2	46	0	20
'Santé'	5	30	0	11.8
'Sinora'	19	56.5	3.6	25.8
'Vechtster'	2.4	26	0	8.2

Note: the average relative minimum yield was taken to be 0.4. Relative susceptibilities are expressed as a percentage of the susceptible standards 'Irene' and 'Bintje'. For Pa3 the highly virulent 'Rookmaker' population was used.

yield *m* of the resistant cultivar 'Agria' was greater than *m* of the susceptible cultivar 'Bintje', which makes 'Agria' more tolerant than 'Bintje'. In most cases, tolerance and resistance to nematodes are independent characteristics in plants. That means that nematode multiplication is not related to plant damage. There are exceptions to this rule; tomato and potato cultivars resistant to *Meloidogyne* spp. are more tolerant than susceptible cultivars.

11.9.3.2. Population decline in the absence of hosts

Population decline in the absence of food is considered to be independent of nematode population density. Another theory is that the mortality rate of *G. rostochiensis* and *G. pallida* is greater in the first year after a potato crop than in subsequent years. This theory was supported by an extensive study by Andersson (1989) in north and south Sweden where a 60% and 80–90% decrease of *G. rostochiensis* was found in the first year after potatoes and much smaller reductions, on average 30% and 50%, respectively, in the following years. Findings on high organic soils in the Dutch starch potato area, with an average decline of 69% for *G. pallida* in the first year after potatoes and 20–30% in subsequent years, confirmed Andersson's observations. A Dutch project, where infestation foci of potato cyst nematodes were monitored from 2006 to 2012 on 21 fields on loamy

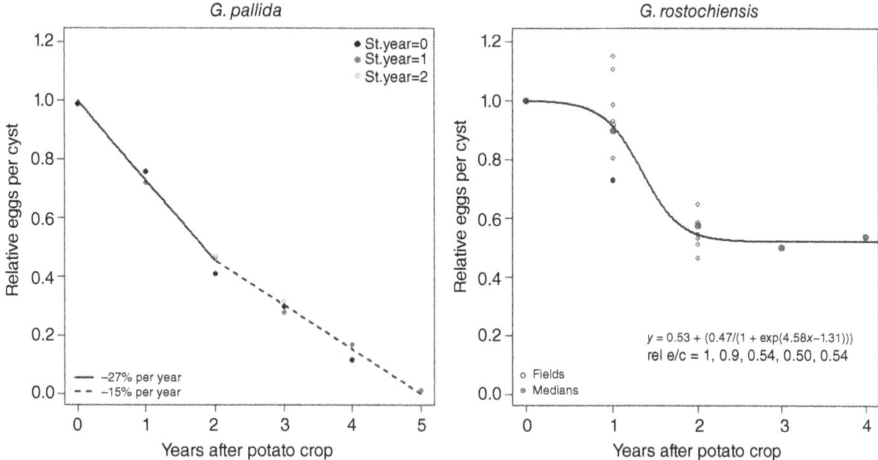

Fig. 11.13. Natural decline of *Globodera pallida* (left) and *G. rostochiensis* (right) on marine clay and loamy soils in The Netherlands. St., *Solanum tuberosum*; rel e/c, relative eggs per cyst.

soils, with 20 plots per field and 5 kg samples to minimize variation, shed a different light on this topic. As cyst numbers per plot changed because of their displacement by agricultural machines, relative eggs per cysts was chosen as a measure of decline. The relative eggs per cysts of *G. pallida* decreased according to a double linear model: 27% during the first 2 years; 15% during the last 3 years. In the same project, *G. rostochiensis* was followed on eight fields using the same methodology. The relative eggs per cysts for *G. rostochiensis* followed a logistic (S-shaped) pattern. No correlation was found between nematode decline and pH, silt, organic matter or previous crop. On most fields, the relative variation (standard deviation/mean) was smaller than 20%. For an overview, the decline rates of *G. pallida* and *G. rostochiensis* are compared in Fig.11.13.

Information about the mortality rate of *H. schachtii* is less exact but a decrease of 50% per year is probably the best estimate. At least 70% of J2 of *H. avenae* hatch the year after they developed, provided that they were exposed long enough to low temperatures. The same rate of decline applies for *M. naasi*. Juveniles of *Meloidogyne* species hatch in large numbers shortly after the moult to J2. Estimates of percentage hatch follow an exponential pattern and vary from 70% to 90%. These reductions are largely due to spontaneous hatching of an apparently fixed proportion of the eggs. Population densities of *Pratylenchus* spp., *Rotylenchus* spp. and *D. dipsaci* also decrease in the absence of food. Field observations are complicated as these nematodes have large host ranges, can maintain high densities on weed and even multiply on cut roots. Pudasaini *et al.* (2006a) found average mortality rates in *P. penetrans* populations of 64% with very little variance after maize, carrot and potato on sandy loam. Organic matter in these fields varied from 2.7% to 3.7%. The population decrease of *D. dipsaci* depends on soil type and can reach 90% on light, humus sandy soils; on light to heavy clay soils and loamy, sandy soils with poor humus, population densities of between 1 and 20 nematodes kg^{-1} are maintained, irrespective of whether hosts or non-hosts are grown.

Distribution Patterns and Sampling*

Thomas H. Been**, Corrie H. Schomaker and Misghina G. Teklu

Wageningen University and Research Centre, Plant Research International, The Netherlands and Biology Department, Ghent University, Belgium

12.1. Introduction

The spatial pattern of plant-parasitic nematode populations in an agricultural or natural ecosystem has two major components: (i) the **horizontal distribution**; and (ii) the **vertical**

*A revision of Been, T.H. and Schomaker, C.H. (2013) Distribution patterns and sampling. In: Perry, R.N. and Moens, M. (eds) *Plant Nematology*, 2nd edn. CAB International, Wallingford, UK.

**Corresponding author: thomas.been@wur.nl

DOI: 10.1079/9781800622456.0012

distribution of the organism throughout the soil or tillage. Both components will change in time because of different aspects of population dynamics, active and passive redistribution and spread.

The horizontal distribution can be divided, arbitrarily, into a micro-distributional component (within a field) and a macro-distributional component (growing regions, countries and parts of continents). The micro-distributional attributes of a nematode population are strongly linked to the population's life history, its feeding strategy and the availability of host plants. Sedentary endoparasitic nematodes deposit all their eggs at the same location, frequently in egg masses or in cysts, generating an initially highly aggregated spatial pattern. Ectoparasitic nematodes invest a proportion of their assimilated energy into movement and selection of feeding sites. As they deposit their eggs individually, a somewhat less aggregated pattern may result. Nematode micro-distribution is primarily mediated by the distribution of food sources. For plant-parasitic nematodes, spacing and morphology of the plant root system, the frequency of cropping hosts and redistribution by machinery are dominant determinants. The integral effect of biological and edaphic influences results in varying degrees of aggregation in the spatial pattern of nematode populations within fields. Macro-distribution is mediated by such factors as the length of time the nematode population has been present in the (agro) system. If the organism has been introduced from abroad, like *Globodera rostochiensis* and *G. pallida* in Europe, a gradual spread will occur from the initial infestation site(s) to fields in the same area, different growing areas and countries importing seed potatoes.

The vertical distribution of a nematode species is constrained by two main factors. First, the depth of the soil layer that, in theory, would be accessible to the roots of a host; this can be limited by bedrock or other impenetrable layers leaving a tilth of limited vertical dimensions. Second, the rooting pattern of the host, in particular the depth of the root system of the host, which will limit the depth a species can reach. As the morphology of the root differs between different plant species, e.g. compare carrots and maize, so will the nematodes' penetration of the tilth differ. Although the vertical distribution of plant-parasitic nematodes is largely dependent on the root distribution of host crops, some variation in the abundance of different nematode species with depth has been related to soil type and texture, temperature, moisture and biotic factors.

12.2. Practical Application

The spatial distribution of population densities is interesting from a scientific point of view but also has some practical implications. As will be demonstrated, spatial patterns of nematodes (and other pests, pathogens and diseases) vary not only among fields and regions, but also within fields and between plant and soil units. The variation within fields is of major importance in determining how samples have to be collected and what size of sample will be required to achieve a desired level of accuracy. It determines the methods to collect and process soil samples when surveys are carried out, or population densities in fields are estimated to advise farmers. How much soil has to be collected, how many cores are needed, which sampling pattern should be used and how much of the bulk sample has to be processed are questions that can only be answered when we formulate the purpose of the sampling method precisely and possess knowledge concerning the distribution patterns of the nematode under investigation.

Therefore, both the horizontal and vertical distribution patterns of nematode species will be discussed with emphasis on their origination and, most importantly, how this knowledge can be applied for practical use in the science of nematology and in the control of those plant-parasitic nematodes regarded as pests. Both components are of major importance for the following purposes.

- **To estimate population densities of the target nematode in small plots used in field experiments.** Usually, the initial population density needs to be established before the actual application of, for example, a certain crop or control method and the final population density after some time in order to obtain information on the effect of the treatment. The quality of both population estimations, expressed as the coefficient of variation (CV), has to be adapted to the required distinctive power of the experiment.
- **To determine the presence or absence of a certain nematode species.** This implies we want to detect a nematode, for example a quarantine organism, with a certain probability. We do not want to know the actual numbers present; it is just a question of detection: yes or no. As absolute certainty is impossible (we would have to dig up the whole field), we must define what has to be detected (e.g. the size of the infestation) and what degree of probability of detection will satisfy our needs.
- **To estimate population densities in a farmer's field of a certain size** (0.33 ha, 1 ha or 2 ha). We are now interested in the number of the target species per unit of soil, for example because we want to predict probable yield loss and have to decide whether or not a control measure has to be considered. The number of nematodes the sampling method yields should be as precise as required for this task, meaning that the variability of the estimate should be in an acceptable range in order to prevent gross over- or underestimation of the expected yield reduction caused by the nematode.

12.3. Horizontal Distribution

Within fields, plant-parasitic nematodes are usually clustered. Depending how deep one wants to venture, three to four 'scales of distribution' should be distinguished. Starting from very small to the largest, the following distribution patterns can generally be identified within a farmer's field.

12.3.1. Very small-scale distribution

The very small-scale distribution pattern of all plant-parasitic nematodes, but especially sedentary nematodes, is the result of the presence and distribution in time of roots throughout the tillage. Only where a root is present will nematodes aggregate. A so-called clumped or aggregated distribution develops, which can be considered as a population of subpopulations occupying small areas in the near vicinity of, or on, the root. It implies that each core taken will probe a different subpopulation and will show another population density, assuming that each subpopulation has an area larger than a single core but that cores are separated by distances larger than the diameter of each subpopulation. Although the resulting distribution is the origin of all other patterns that will emerge in the field, it is also the most difficult and laborious one to establish. One would need to collect systematically small volumes of soil and determine the presence of both roots and the target species until the selected volume is charted in both dimensions. Although this could be done for natural habitats, it will not be of any use in most agricultural systems where this pattern is destroyed when below-ground parts of agricultural crops are harvested. Apart from pure scientific interest, this distribution pattern has no practical importance when the aim is safeguarding agricultural produce. Far more interesting are the patterns emerging from this distribution. However, it tells us that a single core only samples a subpopulation and will not be a very useful density estimator for any area larger than that covered by the diameter of the auger used.

12.3.2. Small-scale distribution

The small-scale distribution describes the distribution pattern over small areas in the field. It is the result of growing the host plant in a grid pattern, defined by the distance in the row between individual plants and the distance between rows. In general, the area defined by the small-scale distribution is: (i) the largest area without a defined shape or gradient of population densities; or (ii) the largest area with an acceptable small variance of the population density estimator. In Fig. 12.1A and B, the small-scale distribution of *G. pallida* and *Paratrichodorus teres* is shown in a 1 m^2 plot with population densities presented per dm^2. The aggregated distribution of cysts in even such a small area is apparent and in fact applies to all plant-parasitic nematode species. As a result, the estimation of population densities with a limited margin of error is difficult even in such a small area. In order to estimate these errors, one first needs to describe the small-scale distribution of nematodes mathematically. This will enable the calculation of several interesting aspects for that area, e.g. the probability of finding 0, 1, 2 or more nematodes or cysts when taking one sample with an auger of a certain size. Similarly, one can calculate how much soil is required to detect a single nematode or cyst with a certain probability, or how much soil is required to get a reliable estimation of the population density in that area.

12.3.2.1. The negative binomial distribution

In the majority of the nematological literature, the spatial distribution of population densities is best described by the negative binomial distribution, irrespective of the area under investigation. The distribution also applies for the small-scale distribution and is as follows:

$$\Pr[x] = \binom{x+k+1}{k-1}\left(\frac{m}{m+k}\right)^{x}\left(\frac{k}{m+k}\right)^{k} \tag{12.1}$$

where: Pr[x] is the probability of finding a certain number of nematodes or cysts (x); m is the expected population density in the unit of soil collected; and k is the coefficient of aggregation.

(A)

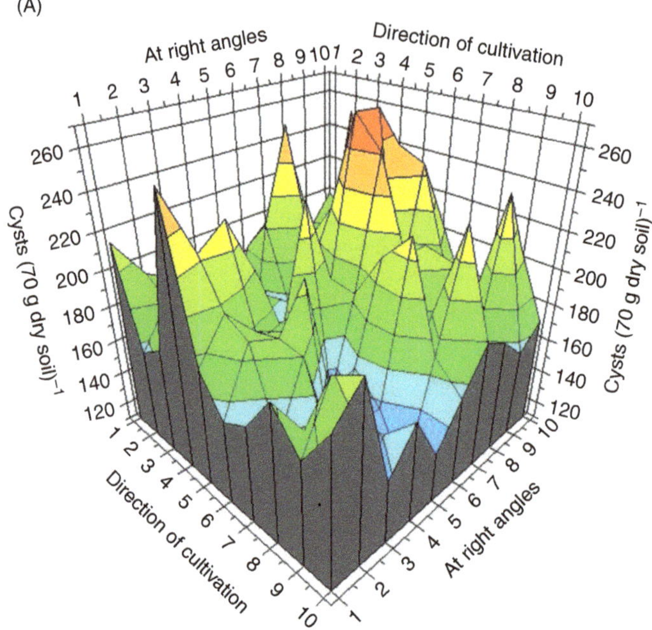

(B)

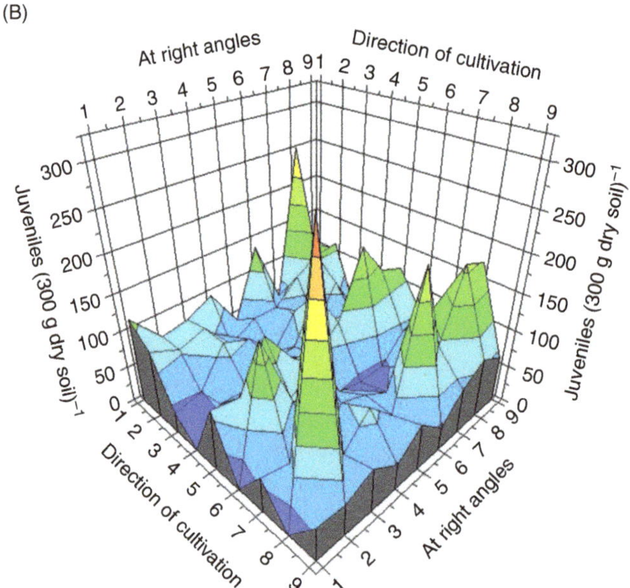

Fig. 12.1. Visualization of the mapped small-scale distribution in 1 m^2 of: (A) *Globodera pallida* representing the number of cysts per 70 g of dried soil; (B) *Paratrichodorus teres* representing the number of adults and juveniles per 300 g soil. Each data point represents an area of 1 dm^2.

Equation 12.1 can be simplified to the following equation when $x = 0$:

$$\Pr[0] = \left(\frac{k}{m+k}\right)^k \quad (12.2)$$

providing the probability of finding nothing and, therefore, used to calculate the detection probability (see Box 12.5).

The negative binomial distribution is one of a series of distributions describing an aggregated or clumped distribution of population densities. The aggregation factor, or coefficient of aggregation k, describes the degree of clumping of the population, with low numbers indicating high aggregation and high numbers less aggregation. When aggregation or clumping occurs in an area, the probability of finding another individual close to the one already located is greater when the distance between the two locations decreases. The distribution becomes identical to the Poisson distribution, showing a random distribution as k increases to infinity. Fractional k-values (<1.0) indicate that the distribution is approaching the logarithmic series which occurs when $k = 0$. As k-values differ from location to location, sometimes a 'common k' (Bliss and Owen, 1958) is used as an operational value for general use.

In Table 12.1, adapted from Seinhorst (1988) and updated with recent information, the aggregation factor k is presented for several relevant plant-parasitic nematode species. In some of these examples, e.g. for *Meloidogyne* spp., the aggregation factor listed is smaller (more aggregated) than it actually is in the field. Any error made in the estimation of the population density in the laboratory (subsampling of soil, efficiency of the extraction method, subsampling of suspension, counting error) is added to the variability found in the soil. Therefore, large laboratory errors will yield smaller values of k. Analysing subsamples from well mixed bulk samples, elutriated either by using the Oostenbrink elutriator or the zonal centrifuge, revealed a laboratory error of more than 50% for *Meloidogyne* spp., partly due to variation contributed by the organic fraction and probably caused by the presence or absence of egg masses in the root debris. For other free-living stages of plant-parasitic nematode species like *Pratylenchus* spp., laboratory error is lower; for cyst nematodes, because of sound methodology available, e.g. the upgraded Seinhorst cyst elutriator (Been *et al.*, 2007a), it is practically negligible. Therefore, the k-value in Table 12.1 is designated as k', indicating that it is an estimator of the real value.

Been and Schomaker (2000) used all data available for *Globodera* spp. to calculate a 'common k' and found a value of 70 for seed and consumption potatoes and 135 for starch potatoes for a 1.5 kg soil sample originating from 1 m^2. The latter are cropped in a 1:2 cropping frequency in completely infested fields, which obviously resulted in a lower aggregation on the small scale. Although the high k–value indicates that one could also use the Poisson distribution instead of the negative binomial distribution, this assumption is one of the many pitfalls of soil sampling (see Box 12.1).

12.3.2.2. How small is a small plot?

One of the problems in field experiments is the size of the plots. In field trials of resistant cultivars, pesticides or other nematode control measures, agronomists estimated crop yield at harvest by collecting the produce, e.g. tubers, from plots ranging from several

Table 12.1. Aggregation coefficient *k* of the negative binomial distribution for some nematode species as reported or derived from the literature and partly summarized by Seinhorst (1988).

Nematode species	Plot size (m^2)	Number of plots	Repeats per plot	k' (1.5 kg soil)
Globodera rostochiensis	1	4	10	72–88
G. rostochiensis	1	364	2	76
Pratylenchus crenatus	16	2	10	32–38
Rotylenchus uniformis	16	2	10	40–45
R. uniformis	1	58	2	60
Globodera pallida	30	29	2	361
G. pallida	10–20	78	2	55
G. pallida	20	1	40	81
G. pallida	30	1	40	166
G. pallida	45	1	40	12
Heterodera schachtii	30	30	2	70
H. schachtii	20	1	40	44
H. schachtii	30	1	40	64
H. schachtii	45	1	40	16
Meloidogyne arenaria	1	1	41	14–23
P. minyus	1	1	41	14–34
Paratrichodorus minor	1	1	41	4–5
Pratylenchus penetrans	1	3	75–80	37–46
H. schachtii	1	456	1	35–550
M. chitwoodi	1	20	10	15[a]
P. penetrans	1	36	10	80[a]
Trichodorids	1	15	10	50[a]
G. pallida (starch area)	1	41	10	135[a]
G. pallida (seed potatoes)	1	28	10	70[a]

[a]'Common *k*'.
Note: k' is an estimate of k, $k' = k$ if variation due to errors in laboratory procedures is negligible. At present, *k* for *Meloidogyne* spp. is seriously underestimated (more clustered) by the extraction methods in use.

square metres up to 100 m^2 or more per plot. If the same area is also used for collecting the corresponding soil sample to estimate the nematode population density of the plot, one in fact presumes that either the population density within the sampled area is uniform or that the variation of population densities within that area is small enough to still obtain a viable density estimator. If this is not so, this will result in erroneous correlations between nematode density and crop yield. Figure 12.4 shows the results of sampling a row of 1 m^2 plots in the direction of cultivation for *G. pallida*. Log nematode densities are plotted, and linear regression is applied to model the correlation. There is a clear trend of increasing population densities with increasing distance starting with a little more than 10 cysts and ending with >1200 cysts within 16 m distance. This trend is even stronger at right angles to the direction of cultivation. When a soil sample is taken from an area covering up to 100 m^2, an average over all encountered population densities in that area will be acquired. As the correlation between nematode density and yield loss is non-linear (see Chapter 11), the correlation will be biased when a bulk sample is collected from an area that contains a number of quite different population densities. In fact, Fig. 12.4 visualizes a part of the next distribution pattern encountered in the field: the medium-scale distribution. For most nematodes, the area of the small-scale distribution is confined to

Box 12.1. Pitfalls of soil sampling.

The aggregation factor k is dependent on the size of the sample collected. If k is expressed per 1.5 kg of soil as in Table 12.1, it can reach a high value. However, when a bulk sample is taken, a number of small cores will be collected using a certain sampling grid. For example, the old European and Mediterranean Plant Protection Organization (EPPO) sampling method required a bulk sample of 200 cm^3 of soil obtained by collecting 60 separate cores. Therefore, each core has a volume of approximately 3.33 cm^3 or 4–6 g of soil, dependent on the soil type. When one core is taken, the aggregation coefficient for that core will be proportional to the core size:

$$k = 135 \times \frac{4}{1500} \text{ up to } k = 135 \times \frac{6}{1500}$$

which yields a k-value between 0.36 and 0.53 for the core sample, assuming k = 135 for the 1.5 kg sample. This is a very small value indicating high aggregation and applicability of the negative binomial distribution. The method of establishing the aggregation factor k of the small-scale distribution is discussed in Box 12.2.

The negative binomial distribution fits to data of counts of any clumped biological entity. The small-scale distribution is well described by this frequency distribution. Even on the largest scale, the field, we can consider the distribution of hotspots as clumped and the negative binomial will apply or, better, will yield a value for the parameter k. In fact, it will almost always apply to nematode counts collected, irrespective of the area used to collect these data. However, every area will yield a different parameter value of k. Aggregation will, in most cases, increase with the size of the area sampled as more distribution patterns will be covered. Therefore, parameter values of k of areas of different size cannot be compared. Neither is there a mechanism to correct the parameter for a different area. Therefore, before starting out to estimate parameters for this, or any other, distribution, the size of area relevant for the purpose has to be determined carefully.

only a couple of square metres. For example, for potato cyst nematodes it was established that the optimum size for that distribution is 1 m^2 (1.33 × 0.75 m, keeping in mind the spacing of the rows and the between-row distance) and that an upper limit of 4 m^2 is acceptable if necessary. Soil samples taken from plots of 1 m^2 are therefore preferred in providing the best estimators of the population density and are used for scientific sampling in field experiments of all nematode species parasitizing arable crops. Haydock and Perry (1998) present a list of methods used for scientific research on potato cyst nematodes; only one complied with the requirements stated. In conclusion, we can state that the area covered by the small-scale distribution provides the area used to estimate population densities of plots in field experiments. When the area to be sampled and its corresponding k-value is known (see Box 12.2 on how to estimate k) we can use this knowledge to calculate the size of the soil sample required for plots in field experiments (Box 12.3).

12.3.2.3. How to sample a small plot

The aggregated distribution in even such a small area as 1 m^2 means that one cannot just take a core sample and use that to estimate the local population density. The best way is

Box 12.2. Estimating *k* and a 'common *k*' of *Pratylenchus penetrans*.

Let us presume that there is a need to develop a sampling method for scientific research, for example to sample plots in field experiments, for *P. penetrans* in order to obtain reliable estimations of the population density in these plots. The area used as plot size has to be chosen in such a way that within that area no measurable effect of the redistribution vectors of machines, tillage, etc., can be found, or that the effect is an acceptably small one. In the latter case, we choose the largest area that will yield an acceptable variance. If no prior information is available, one could use 1 m^2 plots, which have proved to be feasible in most research strategies. As most host crops will not cover the complete volume of soil, the population dynamics of the rooted and non-rooted part of the soil will differ. In the rooted part multiplication will occur, while in the non-rooted part densities will decrease. To avoid any bias, the length and width of the square metre has to be selected in such a way that a proportional part of the sample will be collected from the soil in the row and between the rows regardless of the situation of the plot in the field (and the visibility of the rows after cultivation).

The aggregation factor *k* of the negative binomial distribution can be estimated in several ways. (Actually, we will estimate *k*', an estimate of *k*; *k*' equals *k* if variation due to errors in laboratory procedures are negligible.) One possible way is to sample that area repeatedly, for example ten times. We now have ten estimations of the population density of this area. The easiest method is to use the moments (mean and variance) to estimate *k*. First one calculates the mean:

$$\bar{x} = \frac{1}{10}\sum_{j=1}^{10} x_j \qquad (12.3)$$

and the variance:

$$s^2 = \frac{1}{10-1}\sum_{j=1}^{10}\left(x_j - \bar{x}\right)^2 \qquad (12.4)$$

and uses the following equation to estimate k:

$$k = \frac{\bar{x}^2}{s^2 - \bar{x}} \qquad (12.5)$$

Although *k* is a parameter, we are dealing with a biological descriptive parameter and its value will vary between locations and in time as a result of different external conditions influencing the organism. Some researchers consider *k* as a function of the population mean and variance, which would indicate that it differs at different densities. However, this is for the most part the result of increasing laboratory and methodological error when nematode numbers are low. As a consequence, one wants to establish a *k* that is applicable anywhere. The above-described exercise has to be repeated on different plots in several fields and years and a so-called 'common *k*' (Bliss and Owen, 1958) can be calculated, which will also apply to the plots in fields that have to be sampled in the future. Another, elegant, way of estimating *k* is to calculate the coefficient of variation (CV) for all the plots sampled as described above. The CV is defined as:

Continued

Box 12.2. Continued.

$$CV = \frac{s}{\bar{x}} \tag{12.6}$$

By plotting the CV of each of the individual plots against their mean population density and fitting a regression line through these points, representing the CV according to the negative binomial distribution, an estimate will be obtained for the 'common *k*' value (Fig. 12.2). The formula to calculate the CV according to a negative binomial distribution is:

$$CV = \sqrt{\frac{1}{k} + \frac{1}{\bar{x}}} \tag{12.7}$$

Fitting this equation to the data points of *P. penetrans* resulted in a *k*-factor of 80 for 1.5 kg soil samples. We can also see in Fig. 12.2 that a certain number of nematodes have to be counted to obtain an acceptable CV. The amount of soil, from a large bulk sample, that needs to be processed to obtain a desired CV must be adapted to provide these numbers. Box 12.3 and Fig. 12.3 show how this can be used to determine the required sample size.

To calculate *k* for a single plot, use the "*k Single Plot.*R" script. How to calculate a 'common *k*' for a nematode species is demonstrated in the "*Common k.*R" script. The R scripts, input data and results can be downloaded from the CABI website at: https://www.cabidigitallibrary.org/doi/book/10.1079/9781800622456.0000.

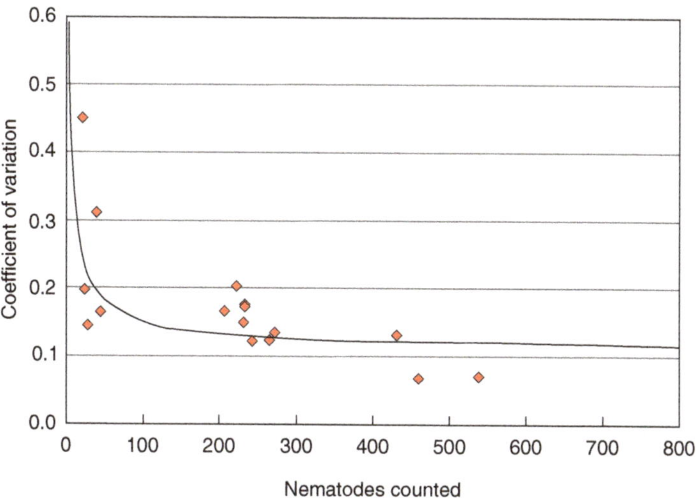

Fig. 12.2. The relationship between the coefficient of variation and the number of *Pratylenchus penetrans* counted (diamonds). Coefficient of variation according to the negative binomial distribution using Eqn 12.7 with *k* = 80 (solid line).

Box 12.3. Calculation of sample size for plots in field experiments.

The purpose of most quantitative nematological research is finding a relationship between population densities and a variable, e.g. final plant weight or effect of a control measure. Normally the *x*-values should be an independent value set by the researcher. However, in nematology, population density estimation is subject to experimental error, which increases with decrease in population density. Log or square root transformations often do not reduce covariance between variance and true population density enough to make statistical analysis feasible. A possible solution to this problem is adjusting the sample size of the bulk sample to the expected population density in order to achieve the desired variation, often expressed as coefficient of variation (CV). A possible way of estimating the required sample size is detailed below. (Meaning of abbreviations used: *e*, eggs or nematodes unit^{-1} of soil; *G*, sample size in units of soil; *c*, cysts unit^{-1} of soil; CV, coefficient of variation (standard deviation/mean); *ec*, eggs/cysts; *k*, coefficient of aggregation of the negative binomial distribution adapted to the unit soil; *m*, mean.)

For cyst nematodes:

The number of eggs is equal to the number of cysts times the eggs per cyst (*ec*):

$$e = c(ec) \tag{12.8}$$

Therefore:

$$\mathrm{CV}^2(e) = \mathrm{CV}^2(c(ec)) \tag{12.9}$$

(Applying the Taylor series) if *c* and *e*/*c* are independent:

$$\mathrm{CV}^2(e) = \mathrm{CV}^2(c) + \mathrm{CV}^2(ec) \tag{12.10}$$

If dependent:

$$\mathrm{CV}^2(e) = \mathrm{CV}^2(c) + \mathrm{CV}^2(ec) + 2\mathrm{cov}\left(\frac{c, ec}{c}\right) \tag{12.11}$$

Assuming that cysts are distributed according to the negative binomial distribution in restricted areas (small-scale distribution) and *c* and *ec* are independent:

Then: per 1 unit of soil (kg)

$$\mathrm{CV}^2(c) = \frac{1}{k} + \frac{1}{c} \tag{12.12}$$

If *G* units are taken

$$\mathrm{CV}^2(c) = \frac{1}{G}\left(\frac{1}{k} + \frac{1}{c}\right) \tag{12.13}$$

Eliminate CV(*c*): Eqns 12.10 and 12.13:

$$\mathrm{CV}^2(e) = \frac{1}{G}\left(\frac{1}{k} + \frac{1}{c} + \mathrm{CV}^2(ec)\right) \tag{12.14}$$

Continued

Box 12.3. Continued.

Solve G:

$$G = \frac{\left(\frac{1}{k} + \frac{1}{c} + \mathrm{CV}^2(ec)\right)}{\mathrm{CV}^2(e)} \tag{12.15}$$

Now we can choose the CV(e) that is required for the experiment and calculate G. The aggregation factor k should be known from previous studies (see Box 12.2). The CV of the number of eggs/cyst, CV(ec), as well as the expected number of cysts c should be known. The latter can be estimated by pre-sampling a couple of plots and interpolate densities in-between. Another way, and avoiding this problem, constitutes taking a bulk sample of such a size that extra subsamples are available if the standard elutriated sample is too small.

For k and CV(ec) a best estimate from the literature (Table 12.1) or one's own experiences with other field experiments can be used. For potato cyst nematodes, the best estimate of a 'common k' = 70 for a sample size of 1.5 kg soil (Been and Schomaker, 2000). For $\mathrm{CV}^2(ec)$ values ranging from 12% to 20% (mean 16%) are often found for 1.5 kg samples (Seinhorst, 1988), provided that they are estimated from a sufficient number of (small) plots. However, these values may be higher if pesticides are used frequently. As the unit of soil for the calculations is 1 kg, both the 'common k' and the $\mathrm{CV}^2(ec)$ have to be adapted to 1 kg units. The easiest way is to calculate a 'ke' which combines both variations:

$$\frac{1}{ke} = \frac{1}{k} + 0.16^2 \tag{12.16}$$

or

$$ke = \frac{1}{\frac{1}{k} + 16^2} \tag{12.17}$$

ke equals to 25 for 1.5 kg samples or 16.7 for 1 kg samples. The equation now can be written as

$$G = \frac{\left(\frac{1}{ke} + \frac{1}{c}\right)}{\mathrm{CV}^2(e)} \tag{12.18}$$

Graphs can be constructed giving the required sample size dependent on both the desired CV for eggs and juveniles and the actual or estimated population density in the plot. In Fig. 12.3 this relationship is presented to obtain egg densities of potato cyst nematodes for four levels of variation.

For free-living stages of plant-parasitic nematodes:

Equation 12.18 can be simplified to:

$$G = \frac{\left(\frac{1}{k} + \frac{1}{e}\right)}{\mathrm{CV}^2(e)} \tag{12.19}$$

'e' now stands for the number of nematodes. Again, a suitable estimate of k is required (Table 12.1) or one is chosen which is comparatively low (high aggregation = worst case approach).

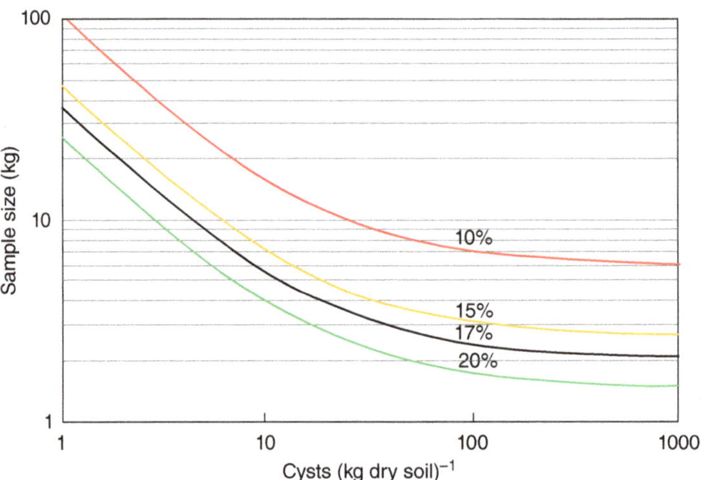

Fig. 12.3. Relationship between population density of cysts of *Globodera rostochiensis* and *G. pallida* and sample size required to obtain coefficients of variation of 10%, 15%, 17% and 20% of estimated egg densities, (Eqn 12.18) assuming a negative binomial distribution of cyst counts in samples with a value of k of 70 for 1.5 kg soil samples and a coefficient of variation of average numbers of eggs per cyst in 1.5 kg soil samples of 16% (k_{eggs} = 25). (From Seinhorst, 1988.)

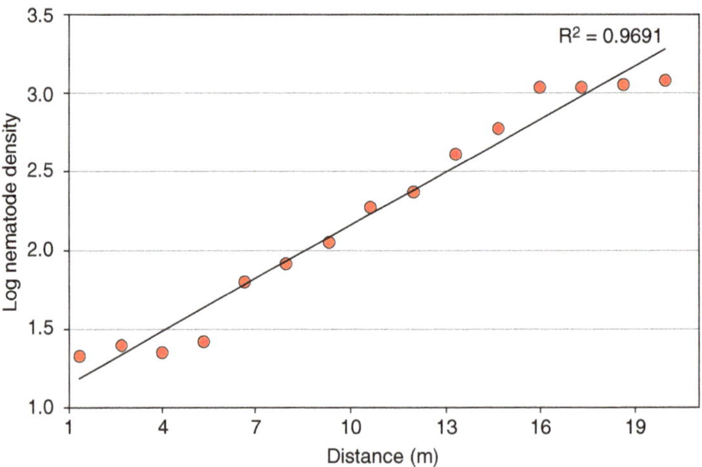

Fig. 12.4. Densities of *Globodera pallida* (log numbers per 1.5 kg soil) in a row of 1 m_2 plots along the direction of cultivation including linear regression fit.

to collect a number of core samples using a rectangular grid. This will guarantee that every area within the plot has the same probability of contributing individuals to the bulk sample and to the population density estimation. One of the questions that arises is how many samples have to be taken – large numbers of small sized cores or a few, larger cores. Two aspects are relevant:

1. The more, and smaller, cores taken to obtain a bulk sample of a certain size, the smaller the variation of the population density estimator. This is caused mainly by the fact that the area of the small-scale distribution will be divided into more parts, each contributing to the resulting density estimator. Table 12.2 presents a study of *Paratrichodorus teres* bulk samples (1.5 kg) collected with two augers of different diameter. Although no change in the average population density of *P. teres* was found, the smaller auger yielded a lower CV. Note that, as a result of the improved methodology, a higher aggregation factor *k* is obtained. To calculate the results yourself use the R script "*Auger Size*.R" with its accompanying datafile. The R scripts, input data and results can be downloaded from the CABI website at: https://www.cabidigitallibrary.org/doi/book/10.1079/9781800622456.0000.
2. Contrary to cyst nematodes, when sampling free-living stages of plant-parasitic nematodes, a too small auger diameter will cause mechanical damage to a large number of nematodes. Dead nematodes will not be recovered and underestimation of the true population density will occur. This is demonstrated abundantly by several authors and partly summarized by Seinhorst (1988) for *Rotylenchus uniformis*, *Pratylenchus crenatus*, *Tylenchorhynchus dubius* and *Paratrichodorus pachydermus*, with up to 1.6 times more individuals counted when larger diameter augers are used and compared to a 1 cm diameter auger.

For cyst nematodes the size of the auger is not a problem as cysts are not easily damaged. Also, the smaller the core, the easier and faster a core sample down to 25–30 cm is collected. Generally, using a 1.5 cm diameter auger and collecting 60–80 cores in a grid pattern, a soil sample of approximately 1.5 kg can be collected easily and quickly for almost all relevant nematode species. The 1.5 cm auger also proved to be superior to larger augers like the trichodorids auger and does not reduce nematode numbers by mechanical damage as can be derived from Table 12.2. It is not necessary to elutriate the whole bulk sample of 1.5 kg, provided enough individuals can be retrieved and counted from the first subsample processed and the required CV is achieved (Fig. 12.3; Box 12.3). If subsampling is employed the theory of subsampling has to be understood (Box 12.4).

Table 12.2. Results of a comparative study on the effect of auger size on the reliability of population density estimation of *Paratrichodorus teres*.

Statistics	Trichodorids auger (3.5 cm diameter)	Potato cyst nematode auger (1.5 cm diameter)
Number of cores	8	60
Bulk sample size (g)	1500	1500
Subsample size (g)	500	500
Mean (n = 10)	80.0	84.6
Standard deviation	25.4	13.8
Coefficient of variation	31.7	16.3
k-value	11.4	67.9

Note: There were no effects on the actual population density estimator averaged over ten replications. However, the reliability of individual replication is greatly improved by collecting a larger number of small cores when using the auger for potato cyst nematode sampling as expressed by the coefficient of variation. Note that as a result of the improved methodology, the *k*-value of the negative binomial distribution increased. There was no damage to the nematodes by the smaller auger as the mean value did not decrease.

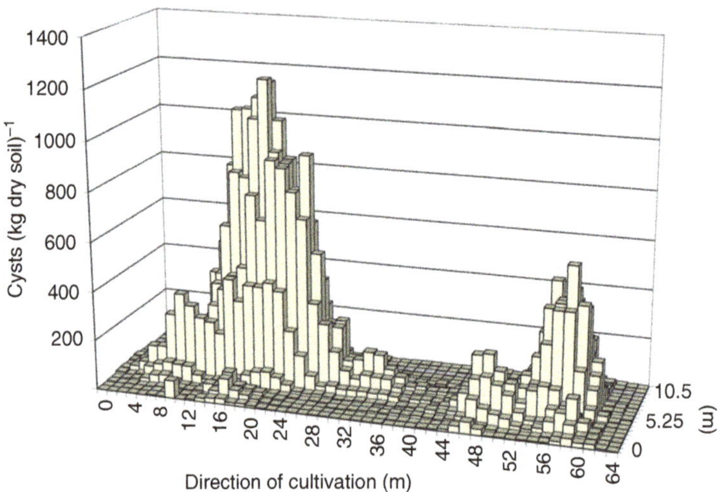

Fig. 12.5. Infestation of *Globodera pallida* on marine clay soil in The Netherlands. Each bar represents 1 m^2 with number of cysts per 1 kg of soil. Primary infestation is visible to the left and two secondary infestations to the right, located in the direction of cultivation.

Box 12.4. The effect of subsampling.

In many nematological field studies it is common practice to collect a bulk sample but to investigate only a part, taken from the bulk sample, after mixing. The size of such a subsample is mostly small, but never infinitely small, relative to the bulk sample. If a soil sample is perfectly mixed, the original negative binomial distribution of cysts or nematodes per unit weight is lost and the cysts or nematodes are randomly distributed in the soil (Seinhorst, 1988). If infinitely small subsamples were taken, the distribution of the nematodes between the subsamples would follow a Poisson distribution. However, in practice, subsamples are not so small in comparison with the bulk sample and, therefore, the numbers in the subsample and in the remainder of the bulk sample do not follow a Poisson distribution but rather a binomial distribution (McCullagh and Nelder, 1992). To visualize the mathematical problem, CVs of nematode counts in subsamples were calculated in a model study for bulk samples with 100 nematodes, assuming a binomial or a Poisson distribution of nematode counts between subsamples. The investigated proportion p of the bulk sample varied from 0.1 to 1. The two functions of the CVs (Eqns 12.20 and 12.21) are compared in Fig. 12.6. Two conclusions can be drawn. First, if the investigated proportion of the bulk sample is relatively small, the CVs calculated with the binomial and Poisson distributions are almost similar. Second, the assumption that nematode numbers between subsamples are Poisson distributed is basically wrong as this distribution misses a correction for finiteness. Application to subsampling of soil samples would imply a considerable CV even when the whole bulk sample is processed.

$$\text{Poisson distribution: CV} = \frac{\sqrt{x}}{\bar{x}} \qquad (12.20)$$

Continued

Box 12.4. Continued.

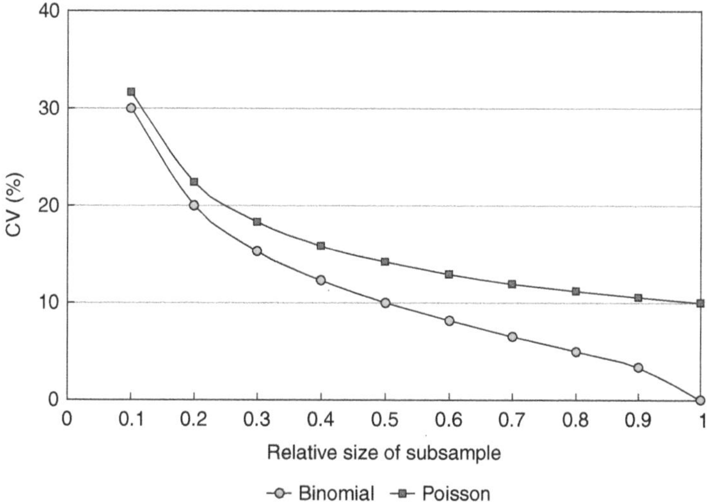

Fig. 12.6. Relation between relative size of one subsample from a bulk sample, and coefficient of variation (CV), in percent, according to a binomial or a Poisson distribution of cysts or juveniles. Number of cysts or juveniles in the bulk sample is set at 100. Multiply the CV of Eqn 12.7 and Eqn 12.21 by 100 to obtain the CV as a percentage.

$$\text{Binomial distribution: } CV = \sqrt{\frac{1-p}{px}} \qquad (12.21)$$

Incidentally, one can read in nematological literature that subsampling adds an extra factor of variation and one is advised to avoid subsampling if possible. The consequence is that only small samples, say 100 ml, are collected and processed. We now know that the extra error of subsampling is in accordance with a binomial distribution. As the error of the bulk sample from a small plot is in accordance with the negative binomial distribution, we can now explore how much error is added by subsampling and how the combined error of collecting the sample and subsampling in the lab can be reduced. Let us presume we have an elutriator that can process 100 cm^3 of soil, that k = 10 for a 1 kg soil sample and an expected population density of 100 nematodes/kg. We will collect 200 cm^3 of soil and elutriate 100 cm^3. Proportional with the size of the soil sample k will be 200/1000 × 10 = 2 for a 200 cm^3 sample and the expected value of recovered nematodes is 20 nematodes.

The field sampling error expressed as CV will be 74.2% according to Eqn 12.7. According to the binomial distribution an extra error of 22.4% will be added when half of the sample is processed. The CVs can be combined using the Taylor series; in this case the new CV equals:

$$CV_{new} = \sqrt{CV_{bulksample}{}^{2} + CV_{Subsample}{}^{2}} \qquad (12.22)$$

Therefore, the numeric solution is

$$CV_{new} = \sqrt{74.2^2 + 22.4^2} = 77.5$$

Continued

Box 12.4. Continued.

Subsampling added 3.3% to the CV we already suffered by collecting a bulk sample of 200 cm^3. Therefore, we can conclude that subsampling indeed adds an error, although a comparatively small one. However, let us consider the following alternative. We do not subsample and only collect and process 100 cm^3 of soil. The CV of the bulk sample according to Eqn 12.7 is now 104.9% and much larger than the previous one. If we decide to collect ten times as much soil (1000 cm^3), mix the soil thoroughly and extract and process a subsample of 100 cm^3 the combined error will be reduced to 38.7%. The amount of soil processed is still the same and an error is added by subsampling. However, the amount of soil collected from the field is ten times as much. As the aggregation factor k is directly correlated with the amount of soil collected, k for the bulk sample has increased from 1 (CV 104.9%) to 10 (CV 32.4%) while the subsample error has increased from 0 to 21.2%. However, the net result is a lower combined error. To explore this in more detail and step by step use the "*Sub Sampling.*R" script on the CABI website: https://www.cabidigitallibrary.org/doi/book/10.1079/9781800622456.0000.

Conclusion:

- Increasing the bulk sample will increase the k of the bulk sample and decrease the CV of the number of cysts or nematodes in the bulk sample.
- If we subsample after mixing, we maintain the large k of the bulk sample and the error we add only depends on the number of counted nematodes.
- The error of this processed sample (CV) can be smaller when the error reduction by a higher k-value is larger than the added error by subsampling.

Of course, optimization of this process depends on the k-value and the fraction of the bulk sample processed. If the size of the subsample is only a small fraction of the bulk sample, no extra gain in error reduction will be yielded. If nematode densities are too small, extra reduction of the estimation error is only possible by counting more nematodes, for example by elutriating more subsamples.

When we combine Eqn 12.7 with Eqn 12.21 the result equates to:

$$CV_{new} = \sqrt{\frac{1}{k} + \frac{1}{px}} \qquad (12.23)$$

where px is the number of nematodes that will be elutriated. So, in the end the only way to reduce the variance of the density estimator in the laboratory is to elutriate and actually count a sufficient number of nematodes. In the last stage, when a sample from the solution is taken to count the target nematode a Poisson error will be made. Enough nematodes have to be counted to stabilize that error. The golden rule is actually to count 200 nematodes to obtain a CV of 6.9%, which indicates that additional aliquots have to be counted when the first sample contains not enough numbers of the target nematode. Explore the "*Counting Error.*R" script to see how counting additional aliquots reduces variation.

12.3.3. Medium-scale distribution

The medium-scale distribution describes the change of population densities over larger areas than those covered by the small-scale distribution. It is the pattern that results when active and mechanical redistribution and spread acts on the small-scale distribution pattern.

Box 12.5. Detection of potato cyst nematodes in The Netherlands.

The following scheme was used to develop sampling methods for the detection of potato cyst nematodes with a predefined probability. Fields that, as a result of statutory soil sampling, were found to be infested with the potato cyst nematodes were provided by the Dutch Plant Protection Service. The statutory soil sampling method covers an area of 0.33 ha. First, the infestation was located within that area by collecting 1.5 kg soil samples from 1 m^2 plots using the scientific sampling method for density estimation. The plots were situated over the field in a grid pattern. Each 1 m^2 plot (1.33 × 0.75 m) was the centre point of, and representing, a larger plot of 8 × 3 m, with the longest distance in the direction of cultivation in which the greatest dispersion is to be expected. This sampling scheme provided a coarse map covering the infested area, as displayed in Fig. 12.7A. After location of the infestation, the infested area was sampled again, now by sampling each square metre. Again 1.5 kg of soil was collected and the number of cysts determined; the result of such an effort is displayed in Fig. 12.7B as a map and in Fig. 12.7C in three dimensions. According to this scheme several fields per growing area were mapped and finally more than 40 foci were available for analysis.

All foci had approximately the same shape, with population densities increasing exponentially towards the centre of the focus, the density of which is designated as the central population density. By calculating log densities, multiple regression analysis or generalized linear model analysis (McCullagh and Nelder, 1992) can be applied. The gradient parameters can now be calculated for all four directions with the central population density as starting point. A comprehensive mathematical model describing the shape and size of a focus, depending on the central population density, could be established (Schomaker and Been, 1999b).

However, for the development of detection methods a simplified model, which is symmetrical along both axes, was used with only two parameters (Eqn 12.24). Now, using the negative binomial distribution for the small-scale distribution, the possibility of detecting no cysts at all at a certain location within the focus, where a core sample of certain size will be collected, can be calculated using Eqn. 12.24. By calculating 1 minus the probability of finding no cysts in every core sample collected from within the focus, when a sampling grid of certain dimensions is superimposed on the infestation, the probability of detection can be computed. An evaluation of existing sampling methods using a specially designed software program called SAMPLE (Been and Schomaker, 2000) was carried out.

$$E[p(x,y)] = p(0,0)\, l^{x}\, w^{y} \qquad (12.24)$$

where: $p(0,0)$, population density in the centre of the focus; l, length gradient (direction of cultivation): average value = 0.83 (decline per metre); w, width gradient (at right angles to l): average value = 0.64 (decline per metre); x and y coordinates within focus defined as the distance in metres in relation to the focus centre, which has the x,y coordinates.

Using the simple model, a standard focus was postulated, which should be detected with 90% reliability. The central population density was set to 50 cysts kg^{-1} of soil for the following reasons:

- No growth reduction should occur when a potato crop was grown in order to obtain the maximum sanitary effect of a resistant cultivar; 50 cysts kg^{-1}, with an average cyst content of 250 eggs $cyst^{-1}$, would result in population densities of about 2 eggs g^{-1} of soil in a 1:3 rotation before the next host, which is close to the damage threshold for potato cyst nematodes.
- No visual damage should be noticeable in the potato crop at the time when visual inspections would be made.
- The degree of reliability was chosen as high as economically possible (considering the cost of sampling) to enable detection one or two crop cycles before the statutory soil sampling method of the government would detect the infestation. In that way, if potato cyst nematodes were present, the farmer would have ample time to implement control measures to prevent detection by the government in the future.

Continued

Box 12.5. Continued.

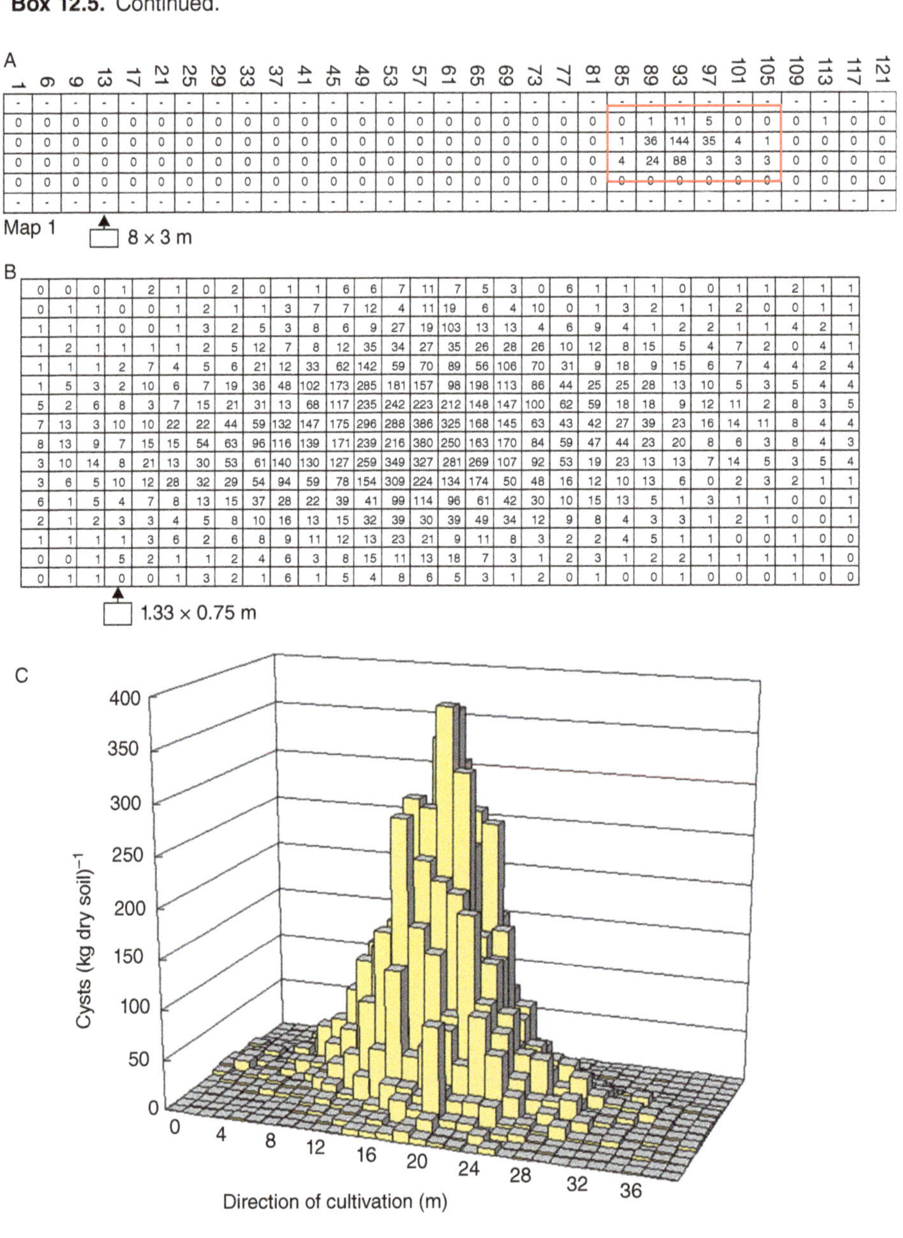

A

1	5	9	13	17	21	25	29	33	37	41	45	49	53	57	61	65	69	73	77	81	85	89	93	97	101	105	109	113	117	121
-	-	-	-	-	-	-	-	-	-	-	-	-	-	-	-	-	-	-	-	-	-	-	-	-	-	-	-	-	-	-
0	0	0	0	0	0	0	0	0	0	0	0	0	0	0	0	0	0	0	0	0	0	1	11	5	0	0	0	1	0	0
0	0	0	0	0	0	0	0	0	0	0	0	0	0	0	0	0	0	0	0	0	1	36	144	35	4	1	0	0	0	0
0	0	0	0	0	0	0	0	0	0	0	0	0	0	0	0	0	0	0	0	0	4	24	88	3	3	3	0	0	0	0
0	0	0	0	0	0	0	0	0	0	0	0	0	0	0	0	0	0	0	0	0	0	0	0	0	0	0	0	0	0	0
-	-	-	-	-	-	-	-	-	-	-	-	-	-	-	-	-	-	-	-	-	-	-	-	-	-	-	-	-	-	-

Map 1

8 × 3 m

B

0	0	0	1	2	1	0	2	0	1	1	6	6	7	11	7	5	3	0	6	1	1	1	0	0	1	1	2	1	1
0	1	1	0	0	1	2	1	1	3	7	7	12	4	11	19	6	4	10	0	1	3	2	1	1	2	0	0	1	1
1	1	1	0	0	1	3	2	5	3	8	6	9	27	19	103	13	13	4	6	9	4	1	2	2	1	1	4	2	1
1	2	1	1	1	1	2	5	12	7	8	12	35	34	27	35	26	28	26	10	12	8	15	5	4	7	2	0	4	1
1	1	1	2	7	4	5	6	21	12	33	62	142	59	70	89	56	106	70	31	9	18	9	15	6	7	4	4	2	4
1	5	3	2	10	6	7	19	36	48	102	173	285	181	157	98	198	113	86	44	25	25	28	13	10	5	3	5	4	4
5	2	6	8	3	7	15	21	31	13	68	117	235	242	223	212	148	147	100	62	59	18	18	9	12	11	2	8	3	5
7	13	3	10	10	22	22	44	59	132	147	175	296	288	386	325	168	145	63	43	42	27	39	23	16	14	11	8	4	4
8	13	9	7	15	15	54	63	96	116	139	171	239	216	380	250	163	170	84	59	47	44	23	20	8	6	3	8	4	3
3	10	14	8	21	13	30	53	61	140	130	127	259	349	327	281	269	107	92	53	19	23	13	13	7	14	5	3	5	4
3	6	5	10	12	28	32	29	54	94	59	78	154	309	224	134	174	50	48	16	12	10	13	6	0	2	3	2	1	1
6	1	5	4	7	8	13	15	37	28	22	39	41	99	114	96	61	42	30	10	15	13	5	1	3	1	1	0	0	1
2	1	2	3	3	4	5	8	10	16	13	15	32	39	30	39	49	34	12	9	8	4	3	3	1	2	1	0	0	1
1	1	1	1	3	6	2	6	8	9	11	12	13	23	21	9	11	8	3	2	2	4	5	1	1	0	0	1	0	0
0	0	1	5	2	1	1	2	4	6	3	8	15	11	13	18	7	3	1	2	3	1	2	2	1	1	1	1	1	0
0	1	1	0	0	1	3	2	1	6	1	5	4	8	6	5	3	1	2	0	1	0	0	1	0	0	0	1	0	0

1.33 × 0.75 m

Fig. 12.7. A. Locating an infection focus in a 0.33 ha (248 × 12 m) area previously designated as infected by a statutory soil sample. Cyst densities (cysts kg^{-1}) of *Globodera pallida* were estimated in the central square metre of each of 124 plots of 8 × 3 m (length × width; length in the direction of cultivation). B. Cyst densities (cysts kg^{-1}) within the rectangle outlined in Fig. 12.7A when every square metre is sampled. C. Three-dimensional representation of the cyst densities (cysts kg^{-1}) shown in Fig. 12.7B. Note the exponential increase of cyst densities towards the focus centre.

Continued

Box 12.5. Continued.

The statutory sampling method (EPPO), used before 2010, only detected the standard focus with an average detection probability of 12%. Several new methods have been developed (Been and Schomaker, 2000) that could detect this small infestation with 90% probability. Figure 12.8 visually displays the different foci that can be detected with 90% probability by the old statutory EPPO sampling method (600 ml ha^{-1}), which was applied until 2010, the new statutory sampling method of 1500 ml ha^{-1}, applied since 2010 (Seehofer, 2007), and the AMI:50 sampling method, targeted at seed potato production in a 1:3 cropping frequency. Foci detected by the AMI method have a 60 times smaller central population density and an 80 times lower number of cysts in the tilth of the infested area than those detected by the old EPPO method. In fact, foci with a central population density of 3000 cysts kg^{-1} of soil have never been found; the maximum found is 1500 cysts kg^{-1} in seed potatoes. At these densities all potatoes would suffer heavy damage and sampling would be unnecessary. The development and introduction in 2010 of the new 1500 ml ha^{-1} sampling method was a result of the research presented in this chapter.

In the "*Focus Detection*.R" script code is provided to calculate the map of any given infestation focus and calculate the average detection probability of any sampling method defined by sampling grid, core size, *k*-factor, etc., up to the subsample elutriated. See CABI website: https://www.cabidigitallibrary.org/doi/book/10.1079/9781800622456.0000.

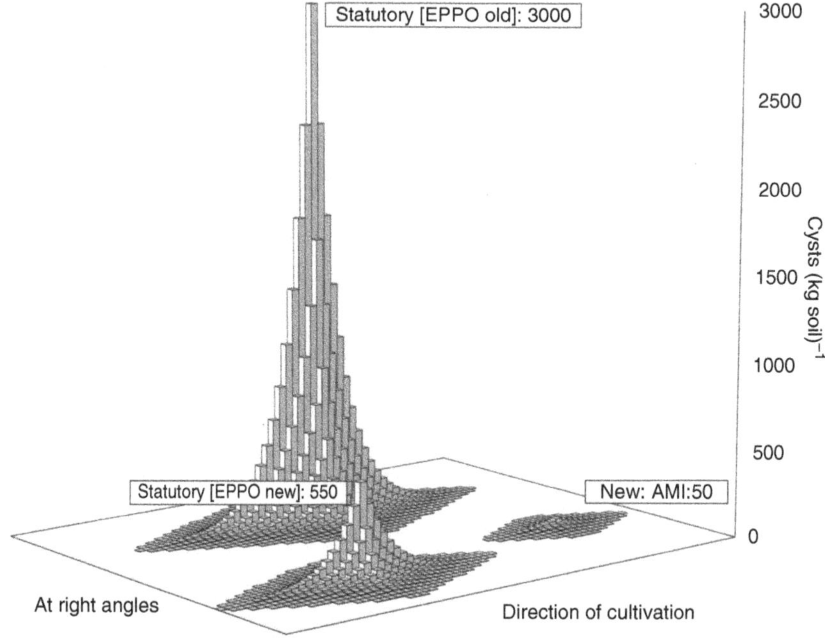

Fig. 12.8. Comparison of foci size detected with 90% average detection probability by the old statutory, the new statutory and the AMI:50 sampling method: 3000, 550 and 50 cysts kg^{-1} of soil in the centre of the infestation, respectively.

Soil will be lifted by machinery, mixed and displaced, either in the direction of cultivation (cultivation, ploughing and harvesting) or at right angles to it (ploughing, winter ploughing). This redistribution results in the horizontal growth of the primary point infestation over adjacent areas. Quite distinct shapes can result, such as the development of the so-called infestation focus or hotspot. Within the area covered by the medium-scale distribution there are different population densities at different locations that are related to each other. As most farmers in a growing area or country use the same kind of machinery, and sometimes even the same cropping frequency, the resulting infestation foci of a nematode species tend to be of the same shape at any given location. An oval-shaped spot (lens) in the field will appear where plant growth is retarded or, in extreme cases, completely inhibited. The area of the actual infestation is larger than that covered by the lens as densities at the border of the infestation are beneath the threshold for visible damage. The longest axis of the hotspot will lie in the direction of cultivation with slowly declining population densities, and its short axis, at right angles to the direction of cultivation, with sharply declining densities. Nematode densities will rise exponentially towards the centre (Fig. 12.5). In the case of extremely high population densities, host plants in the centre of the lens might actually die and the final population densities in the centre of the focus will be lower than those in the perimeter as a result of natural decline in the absence of the host. When the infestation focus is subject to intensive crop rotation with a host, population densities in the centre will reach a maximum (the carrying capacity of the nematode–host combination and its cropping frequency) over an increasing area. The method of calculating the detection probability (presence of one or more nematodes or cysts in a soil sample) if an infestation focus is present in the field is discussed in Box 12.5.

Although almost all nematodes newly introduced in a field will produce hotspots, there are considerable differences between nematode species in size, extent and rate of spread of the hotspot. Hotspot development of the stem nematode *Ditylenchus dipsaci* is caused by conditions that affect the efficiency of the nematodes in finding and penetrating plants. In patchy infestations of stem nematodes these conditions for attack appear to be more favourable in the centre of the patch than towards the borders, resulting in a decrease of nematode numbers with increase of the distance from the centre just like other hotspots, but with a persistence of the patchiness; the hotspots do not grow.

The dimensions of hotspots differ between nematode species. Whilst potato cyst nematodes, which are mainly distributed by mechanical redistribution of the cysts, cause the emergence of hotspots with a distinct and precisely defined shape of comparatively small dimensions (maximum 1400 m^2), species like *P. penetrans* and especially *Meloidogyne chitwoodi* and *M. fallax* – distribution of matrixes, but also eggs and juveniles – seem to cause quite large infestations in a short time (*M. chitwoodi*: >10,000 m^2). As a result, a potato cyst nematode hotspot can be found in fields that are otherwise free of the nematode, whilst when a *M. chitwoodi* hotspot is discovered, almost always a large area, and sometimes the complete field, is infested.

12.3.4. Large-scale distribution

Once the first hotspot is established, passive redistribution by machinery will not only cause the focus to grow, but also will, by way of clods of infested soil adhering to the machinery, result in secondary infestations when these clods are deposited further along the row. The secondary infestations will be found primarily in the direction of cultivation and, less frequently, at right angles to that direction. As long as susceptible hosts are grown, these infestations will grow, merge and finally result in completely infested fields, with maximum population densities defined by the carrying capacity of the host (see Chapter 11). The intermediate distribution pattern can be regarded as the combination of several medium scale patterns. Figure 12.5 presents this development stage for *G. pallida*. Been and Schomaker (1998) were unable to correlate the size of the primary focus, which can be regarded as being age dependent, with the number of secondary foci in the field. Several researchers used (infrared) aerial photography or crop scanning to visualize these agglomerates of foci in farmers' fields. Figure 12.9 displays a conglomerate pattern of the sting nematode, *Belonolaimus longicaudatus*, on a peanut farm.

Figure 12.10 presents the final stage of spread within a field of *G. pallida* after continuous growing of susceptible potato varieties in a 1:2 cropping rotation depicting the number of cysts per kg in the central square metre plot of each sampled 5 × 3 m block. No pattern can be distinguished. However, when investigated more closely by sampling every square metre of each 5 × 3 m block, nematode densities in consecutive square metres are not independent but are closely related to each other. This relationship can be used to describe the variability of population densities in these blocks and, if generalized, can be used to predict population densities between the grid points of other blocks. When a cross-section of mapped fields is available, sampling methods for full field infestations can be designed, provided this cross-section represents the majority of actual full field infestations. The developed sampling method will then also apply to the

Fig. 12.9. Aerial photograph showing a conglomerate of infestation foci of the sting nematode, *Belonolaimus longicaudatus*, on a peanut farm merging into a full field infestation. (Courtesy J.D. Eisenback, NemaPix 2, 1999.)

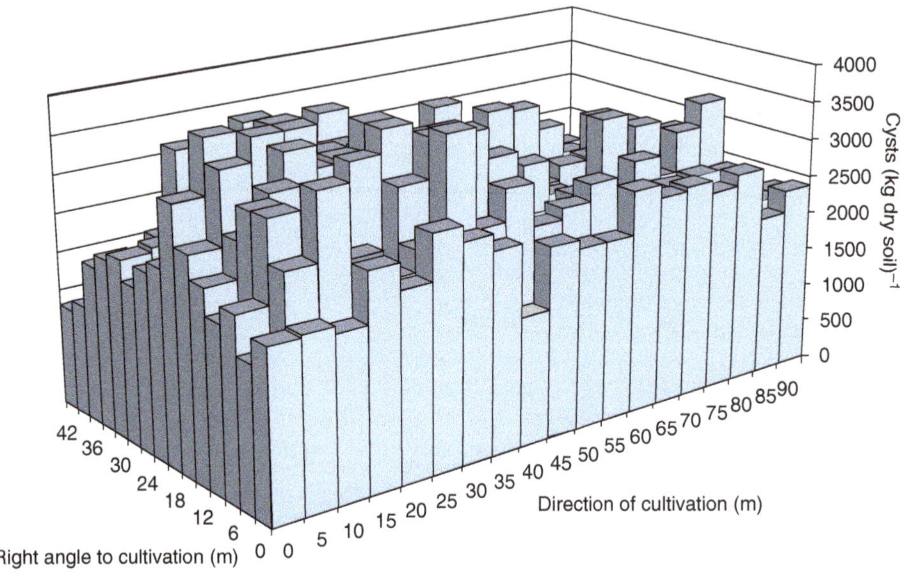

Fig. 12.10. Full field infestation (large-scale distribution at its climax) of *Globodera pallida* in the starch region of The Netherlands. Samples of 1.5 kg were collected from plots of 1 m^2 (1.33 × 0.75 m). Plots were spaced every 5 m in the direction of cultivation and every 3 m at right angles to that. The whole sample was processed.

majority of unknown (not mapped) infestations. Using this approach, sampling methods for full field infestations of the potato cyst nematode could be developed with a CV of 15% in The Netherlands. This uncertainty is acceptable to provide farmers with predictions about possible yield losses and population development in time.

12.3.5. Macro-distribution

The dispersion of a nematode species does not stop at the borders of a farmer's field. Any primary infestation in a growing area will ultimately spread to its neighbouring fields, which in turn will contribute to an accelerated spread throughout the area. Figure 12.11 illustrates the result of such a progression of the soybean cyst nematode, *Heterodera glycines*, in the USA. Some of the major factors involved in macro-distribution of nematodes are given in Box 12.6. Obviously, hygiene is of utmost importance to prevent the spread of species of plant-parasitic nematodes. However, in the majority of cases it will only be possible to slow down the spread; once a nematode population is present in an area, its gradual expansion throughout that area is inevitable even when strict phytosanitary precautions are taken. The most effective and successful strategy up to now has been the employment of the whole range of possible control and phytosanitary measures to maintain the pest at low population densities at which only small yield losses will occur, or to prevent the unlimited and excessive presence of quarantine pests in seed and export lots.

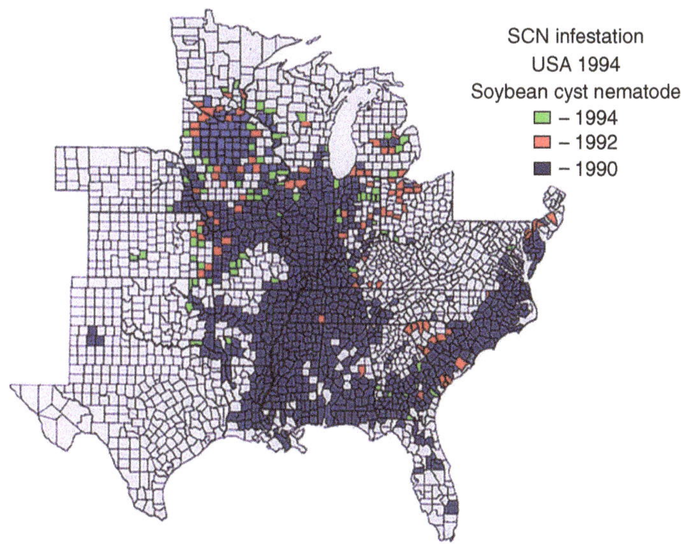

Fig. 12.11. Spread of soybean cyst nematode, *Heterodera glycines*, in the USA provided by R.A. Riggs of the University of Arkansas. (Courtesy Eisenback and Zunke, NemaPix 1, 2000.)

12.3.6. Other sampling approaches

There is a wide variety of approaches in the development of sampling methods for agriculture that have to yield population density estimates suitable for advisory services; some of those most commonly encountered in the literature will be summarized briefly here.

12.3.6.1. Mapping fields using core samples

One of the commonly used approaches is the mapping of a farmer's field by collecting single cores using a grid of predetermined dimensions and counting the nematodes in each of these cores. These data are then used to calculate sample statistics (mean, standard deviation, variation and CV) or are fitted to different distribution functions. The first error of this approach is that a single core will not give us a reliable local density estimate when the underlying spatial density distribution is not random – as we now know it never is. Therefore, a single core will only tell us something about the number of nematodes found in the volume of soil collected with that core. It gives us no information about the population densities outside the area covered by the core.

Using sample statistics, variance/mean (σ^2/μ) ratios are often examined to make approximate decisions about the frequency distribution of nematode counts. The ratio equals 1 for randomly distributed populations, is <1 for a regular distribution and >1 when the distribution is clumped. We now know that a uniform distribution is out of the question. A Poisson distribution would apply when active movement is easy and no attraction or repulsion of nematodes would occur, something hardly applying to nematodes with their limited movement and aggregation around hosts. A range of distributions describe an over-dispersed pattern, e.g. neyman type a, neyman type b, negative binomial. Again, the negative binomial

Box 12.6. Factors involved in macro-distribution.

1. Dispersion vectors

Propagation material. Nematodes can be retrieved from almost all material used for propagation. *Ditylenchus dipsaci* can be retrieved from seeds of different hosts, *Meloidogyne* spp. and *Globodera* spp. cysts can be found in and on potato tubers, *Pratylenchus* spp. and trichodorids (including tobacco rattle virus) in bulbs, and *P. penetrans* from rootstock of roses. To ensure that propagation material is free from nematodes, several methods, e.g. hot water treatment for bulbs, shallots, etc., are available (see Chapter 14). Legislation and statutory control measures are employed to prevent their presence in export lots (see Chapter 13).

Machinery. Clods of infested soil adhering to machinery will not only cause secondary foci in the same field but also initiate primary foci in previously non-infested fields tilled by the contaminated machinery. Soil fumigation equipment used to fumigate infested soils has been the primary vector dispersing potato cyst nematodes in The Netherlands.

Soil. Adhering soil, collected when produce is amassed for transport or packed for shipping, can import infestations when it is used to fill up low spots in a farmer's fields or to change the soil structure.

Wind. Wind can cause the spread of nematodes, especially cyst nematodes. For example, when part of the surface of the topsoil is frozen, winter storms lift up both soil and cysts, causing 'sandstorms' that cover large areas and result in the spread from one field to many others over large distances. This not only causes new infestations, but also enhances the dispersion of more virulent populations, emerging after the use of resistant hosts. To prevent sandstorms, green manure crops have been used successfully to contain the topsoil after harvest of the host. However, the green manure crop has to be carefully selected as almost all of these crops enhance the survival and even, being hosts, the multiplication of other plant-parasitic nematodes.

Water. Rain transports soil and plant-parasitic nematodes to trenches, which will spread the nematodes throughout an area. Using the water from these trenches for irrigation will further spread these nematodes. Drainage systems will finally deliver the nematodes to rivulets, streams and rivers.

2. Habitat suitability factors

Once a nematode species is established, variations in habitat suitability factors, such as soil type and texture, selection pressure from host distributions, cropping history, temperature and moisture will determine the extent of its spread over regions, countries and parts of continents. For example, both *G. rostochiensis* and *G. pallida* can be found in all temperate and some tropical (recently in east Africa) climatic regions and soil types where potatoes are cropped but these species are of minor importance, e.g. in Mediterranean countries as a result of unfavourable high soil temperatures. Soil type can be a barrier to nematode species; trichodorids or *Meloidogyne* spp. are not found in heavy clay soils.

distribution fits best to the majority of datasets available. However, we now deal with a completely different spatial distribution pattern: a large-scale distribution.

Fitting spatial sampling data to frequency distributions that are used as an offset for further calculations implies a reduction of information because all spatial information is omitted. In Fig. 12.12 a series of well-known spatial distribution patterns is displayed ranging from distinct patterns such as habitat borders, gradients and shapes, to random

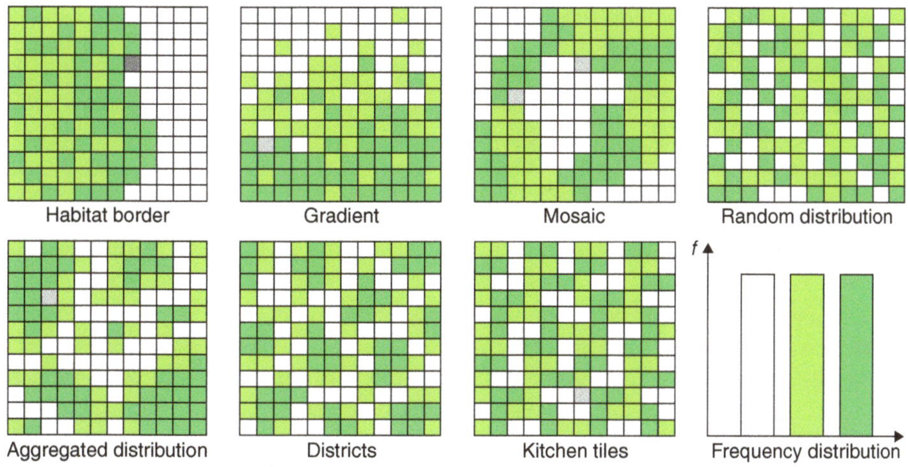

Fig. 12.12. Different spatial distribution patterns having the same frequency distribution (*f*). (After Ekschmitt, 1993.)

and aggregated patterns (Ekschmitt, 1993). The frequency distribution of the actual densities is uniform; each has the same probabilities. Sampling of these distributions in the field would result in bulk samples with nematode densities all having the same frequency distribution. For the development of sampling methods, especially for whole fields, this reduction of information is unfavourable.

12.3.6.2. Repeated collection of bulk samples from a field

Another approach for investigating the reliability of a population density estimate of a field by applying a particular sampling method is the repeated application of that method on the same field and calculating basic statistics. The mean, standard deviation and CV of the population densities of repeated bulk samples are calculated and used to evaluate the current sampling scheme.

This approach has some serious drawbacks. It will yield some information to optimize the current sampling method in order to obtain the desired accuracy of the density estimation. However, the resulting optimized method will only be reliable for the investigated field at that moment in time. It will not apply in the next year when population densities of the target nematode have changed or to another field in the same year where population densities are different from the start. It does not provide any insight into the underlying spatial distribution and, therefore, no information is available to calculate the effect of any change to the original method, e.g. a different sampling size or a different sampling pattern or grid, on the reliability of the population density estimator.

12.3.6.3. Sampling pattern/sampling grid

When a soil sample is collected from a field, usually separate core samples are collected from different locations within the field and combined into a bulk sample. The grid pattern used to collect this bulk sample has been an item for discussion and theorization in a number of investigations. Figure 12.13 shows some of the proposed and applied sampling patterns to traverse a square field, alongside which the separate cores have to be collected, starting with a diagonal pattern (A), a cross-diagonal pattern (B), a parallel-diagonal pattern (C), a W-shaped pattern (D), a zigzag pattern (E), a tilted square pattern (F), a perimeter pattern (G), an M-shaped pattern (H), a random pattern (I) and a rectangular grid pattern (J), as used in the case study detailed in Box 12.5. Of course, one wants to know the rationale behind these patterns. In general, the assumption was made that the distribution of nematodes throughout a field is random. A random pattern like pattern (I) then would be suitable but also inconvenient to execute as a standard procedure for statutory or advisory soil sampling. In fact, if randomness applies, it will not matter from which location in the field the separate cores samples are collected. However, as there were sufficient indications that the random distribution did not really apply and that differences in population densities throughout the field occur, most sampling patterns try to incorporate some of these different population densities by collecting cores while traversing the field in some way. The question arises as to whether there is a way of doing this that accords with current knowledge. Let us consider the infestation foci of the potato cyst nematode; its shape is described by Eqn 12.24, and especially by the length and width gradient, 0.83 and 0.64, respectively, describing the rate of decline of population densities from the centre of the focus in both

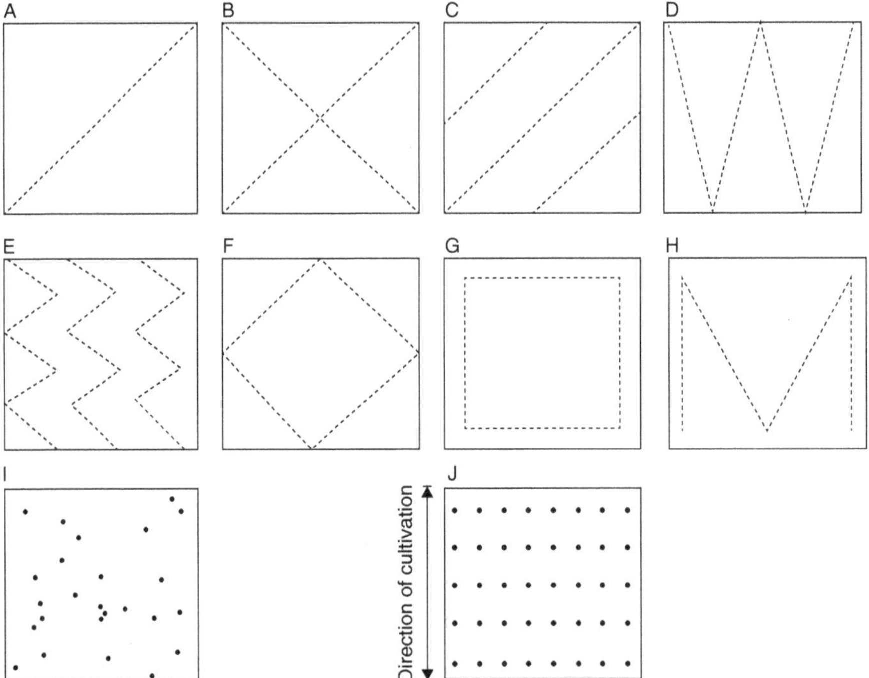

Fig. 12.13. Different patterns used to collect separate core samples from a field, which will form the bulk sample: A: diagonal pattern, B: cross-diagonal pattern, C: parallel-diagonal pattern, D: W-shaped pattern, E: zigzag pattern, F: tilted square pattern, G: perimeter pattern, F: M-shaped pattern, I: random pattern, J: rectangular grid pattern.

directions. An optimized grid would be adapted to the fact that the change in population densities in the direction of cultivation is smaller than the change at right angles to that direction, by taking more cores in the width and fewer in the length per unit distance (Been and Schomaker, 2000). The population density in the centre defines the size of the focus. The sampling grid should now be configured in such a way that at least one sample, but preferable more, will actually hit the infestation regardless of its location in the sampled field. This implies that the area of the field, divided by the area of the focus, will yield the minimum number of cores that have to be taken, each core representing an area as large as one infestation focus. This implies that all sampling patterns described in Fig. 12.13, except the zigzag pattern (E) and the stratified grid pattern (J), are disqualified. The use of stratified sampling using a rectangular grid adapted to the shape of the infestation as described by Eqn 12.20 and visualized schematically in Fig. 12.13 (J) presents the optimum solution. The zigzag pattern (E) can be considered as a small modification of the rectangular grid, with each second row of the grid shifted for half of the grid length, to facilitate sampling of two rows in one passage over the field. However, this approach yields only a marginal difference in walking distance and, in practical sampling schemes, is disregarded.

The defined sampling pattern, the rectangular grid, now yields one core sample hitting the infestation. This can be in the centre where densities are highest or in the border where densities decline to zero. Huge core sample sizes are required to obtain a high probability of detection when the latter occurs. Better results (smaller bulk sample; higher detection probability) will be obtained when smaller core samples are taken in a denser grid with the same length/width ratio. More core samples will actually hit the focus and the probability to hit higher densities in the central area will increase. As a result, detection probabilities will increase. To optimize this process of fine tuning and optimizing a sampling method for a desired detection probability a software program, SAMPLE V.5, is available.

The same reasoning applies when the whole field is infested. Although detection will be no problem, more, but smaller, cores will increase the reliability of the population density estimator (up to a certain level) as more areas of the field will contribute to the population density estimator.

12.4. Vertical Distribution

The possible extent of the vertical distribution of a plant-parasitic nematode species is important information in any sampling strategy. How deep has one to sample to make sure that the estimation of the population density over the sampled horizon will cover the actual population density the host will be exposed to during growth? Both scientists and extension workers need this information. Commercial sampling agencies, applying both statutory and advisory soil sampling methods for governments or for integrated pest management, respectively, want to know the necessary (in practice, the minimum) depth required to sample as the speed of collecting samples and, therefore, the costs involved depend on the depth at which cores have to be taken. Further, one wants to know whether the vertical extension of a nematode population will be reached by control measures because layers that are not accessed will serve as a source of new inoculum.

A standard depth of 25 cm is currently used in the majority of countries when collecting single cores to compose bulk samples for advisory purposes for cyst nematodes and the free-living stages of plant-parasitic nematodes. For trichodorids, occasionally samples are taken down to a depth of 50 cm. A standard depth of 25 cm might suffice for sedentary nematodes like cyst nematodes but can be insufficient for other nematode species. Free-living stages of plant-parasitic nematodes are reported to adapt to microclimatic changes in the soil layer and move to more favourable locations in the soil column (see Chapter 8). The most extreme behaviour reported is the active movement of trichodorids to deeper soil levels to avoid the effect of granular nematicides applied to the soil surface. More common is the effect of drought causing most non-cyst nematodes to be less abundant in the topsoil layers and more numerous in deeper soil layers. Depending on the aim of sampling and the biology of the target nematode, different depths of sampling might have to be chosen. Obviously, first of all, knowledge of the vertical distribution has to be acquired.

The literature available on the vertical distribution is limited. Extensive research to determine the adequate depth for soil sampling procedures has been carried out for potato cyst nematodes *M. chitwoodi*, *M. fallax* and *P. penetrans* in The Netherlands. Soil samples for potato cyst nematodes were taken up to a depth of 80 cm immediately after harvest and after the soil was substantially cultivated (e.g. winter ploughing) to reveal any changes in the vertical distribution. Although potato cyst nematodes were occasionally present at depths of 80 cm, and probably sometimes even deeper, cysts at this depth were only present in a minority of 32 fields investigated. Generally, 90% of the cyst population was found in the upper 35 cm of soil. It was concluded that, immediately after harvest of the potato crop, but also after cultivation, the vertical distribution of potato cyst nematodes in the upper 25 cm of soil was uniform (Table 12.3) and, therefore, sets no demands on the depths of sampling.

In Fig. 12.14, for five of the sampled fields the number of cysts per 500 g of dried soil from soil samples taken down to depths of 25 cm both in the ridges and between ridges are presented. Samples of the first 1 cm of the tilth were also taken and compared with the first 5 cm horizon but no accumulation of cysts after harvest could be detected in the upper centimetres. Whitehead (1977) investigated the vertical distribution of potato, beet and pea cyst nematodes in heavily infested soils (layers: 0–20, 20–40, 40–60 cm) and reported similar results for the number of eggs g^{-1} of soil. Cysts of *Heterodera goettingiana*, the pea cyst nematode, were rarely found below 20 cm.

Seinhorst (1988) monitored densities of *R. uniformis* and trichodorids per 3 cm layers down to a depth of 21 cm and showed that nematode numbers were significantly lower in the three upper layers compared to the others. This seems to be generally the case for all free-living stages of plant-parasitic nematodes. Pudasaini *et al.* (2006b) modelled the vertical distribution of *P. penetrans* under four different hosts throughout the year. The vertical distribution of the nematode was related to the presence of roots of the four different crops, black salsify, carrot, maize and potato, indicating that even for a single nematode species the vertical distribution can differ and is predominantly determined by the host (Table 12.4).

Clearly, the root system of the host is the most important factor influencing the vertical distribution of plant-parasitic nematodes. In temperate zones, drought and temperature only play a role in the upper few centimetres of the tilth. In tropical zones these two factors will have more impact on the vertical distribution.

Table 12.3. Average population densities of *Globodera* spp. per 500 g dry soil, per horizon of 5 cm, on sandy and marine clay soils down to a depth of 25 cm.

		Depth sampled (cm)					
		0–5	6–10	12–15	16–20	21–25	Mean
Sandy soils	Number of cysts	802	792	787	776	738	779
	% per horizon*	*100.0*	*99.8*	*98.4*	*98.9*	*89.8*	
	Cumulative average	802	797	794	789	779	
	% cumulative**	*103.0*	*102.3*	*102.0*	*101.3*	*100.0*	
Marine clay soils	Number of cysts	142	155	160	158	152	153
	% per horizon*	*100.0*	*107.8*	*107.9*	*108.4*	*105.7*	
	Cumulative average	142	149	152	154	153	
	% cumulative**	*92.8*	*97.4*	*99.3*	*100.7*	*100.0*	

*Each horizon relative to first horizon. **Cumulative average relative to tilth (25 cm).

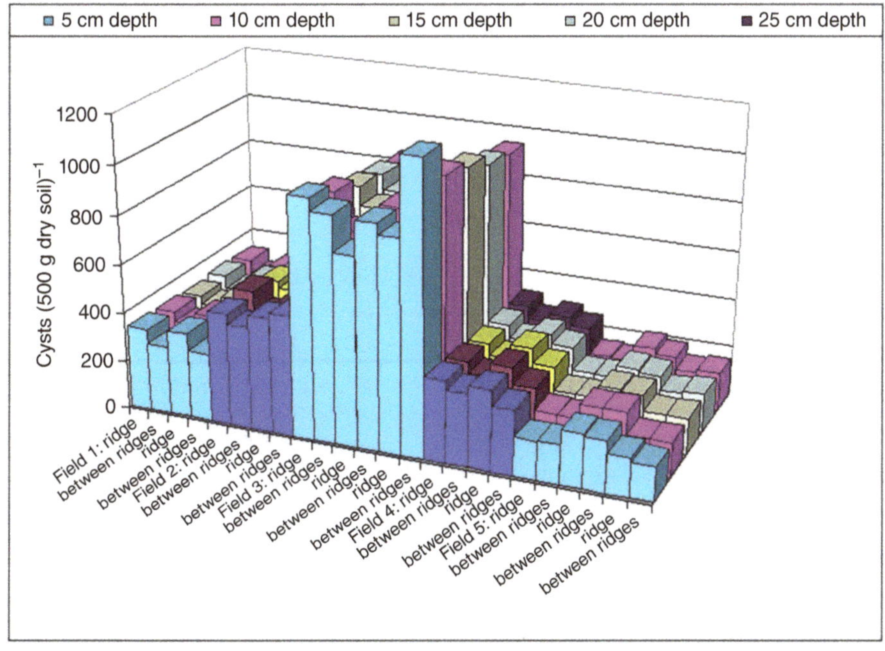

Fig. 12.14. The vertical distribution of *Globodera pallida* in five different fields at harvest and after soil cultivation (winter ploughing, harvesting) in five different layers down to 25 cm, in both the ridges and between ridges. First two entries per field: ridge and between ridges after harvest; second two entries per field: the same after winter ploughing; third two entries (if available): same after second harvest.

12.5. Setting up Field Experiments

The above information provides the basic methodology on how to set up and execute the sampling part of field experiments. Always we need to estimate the initial population density (P_i) and mostly also the final population density (P_f) in a sufficient number of

Table 12.4. The density of *Pratylenchus penetrans*, expressed as a percentage of the total number of *P. penetrans* recovered from 70 cm of soil column, in different soil layers in four crops.

	Percentage distribution of nematodes in different soil layers (in cm)						
Crops	0–10	12–20	21–30	31–40	41–50	51–60	61–70
Maize	**17**	**26**	**27**	**15**	**9**	**4**	**2**
Black salsify	**17**	**27**	**31**	**20**	**5**	0	0
Carrot	**27**	**37**	**27**	**7**	1	0	0
Potato	**23**	**34**	**27**	**14**	2	0	0

Values in bold indicate the presence of roots. (From Pudasaini *et al.*, 2006b.)

plots. Chapter 11 shows that the relation between the P_i and either the P_f or the yield losses caused is non-linear. To explore the relationship and to be able to estimate the parameters of the models used in Chapter 11 we need a range of population densities starting from zero – or, when not available, extremely low, if possible below the tolerance limit, T (see Chapter 11) – up to very high nematode numbers (reaching the maximum population density (M), the carrying capacity of the host). Sometimes even higher densities are available when a previous crop was a better host.

12.5.1. Plot size and sample collection

We know now that we have to collect a soil sample from a small area of 1 m^2 to a maximum 4 m^2 and we need to avoid certain errors. The relevant points are:

- Generally, 1 m^2 is chosen, the small-scale distribution, which provides us the most reliable population density estimation and therefore is used as the sampling unit for scientific research.
- To collect a representative sample from the aggregated distribution in 1 m^2, as shown in Figure 12.1A and B, one cannot just take one core sample or even only a few. To provide a reliable estimator of the population density, by minimizing its variation, we have to make sure that all possible densities in the plot have an equal probability to be represented in the collected bulk sample. The only way to achieve this is by collecting the bulk sample using a stratified pattern: core samples, taken in a regular grid pattern, each core of the same size and representing an identical area within the plot.
- A standard sample consists of 80 cores of approximately 20 g of soil, providing a bulk sample of approximately 1.6 kg. This is collected using a 25-cm-long auger, diameter 1.5 cm, which causes no physical damage to free-living nematodes.
- Adjust the square metre sampled to the pattern in which the crop is grown. As plants are grown in rows, sometimes on ridges (e.g. potatoes), which are not visible before planting or after harvest when the soil samples are collected, we have to adapt the square metre sampled to the planting grid: e.g. potatoes are grown on ridges 75 cm apart in the length of cultivation. In the ridges multiplication occurs; between the ridges, where no roots are present, populations decline. Applying the stratified sampling grid to a plot of 0.75 × 1.33 m, the latter in the direction of cultivation, will ensure that we

always sample one complete ridge and a complete area in-between two ridges, thus avoiding bias.

- The standard soil sample of 1.6 kg will not always suffice. When very low numbers of the target nematode are present the collected amount of soil has to be adapted to the nematode density present (Box 12.3). When numbers are low, one has to collect a larger soil sample; when numbers are high, only a part of the bulk sample has to be processed.

12.5.2. Dos and don'ts

- Do not go back to the plots in the field to collect extra soil when not enough nematodes have been collected. Too many errors can occur. It is recommended to collect larger soil samples than are actually required. Collecting larger soil samples is far cheaper than returning to the field to the exact same spot to collect an additional sample when needed. Moreover, a larger soil sample taken and mixed to extract a subsample has a smaller variation due to a proportionally increased aggregation factor *k*, which is not affected by subsampling (Box 12.4).
- Don't expect that your field is homogeneous. Numerous field experiments fail because of large variations over the field. These are sometimes explainable, e.g. by different previous crops grown on the experimental site in the last year, different soil types, different fertilization. Sometimes this is even caused when setting up the experiment, using fumigants or a range of crops with a different host suitability to obtain different nematode densities, but also causing differences in the soil structure, fertilization and, therefore, yield and multiplication. Sometimes they have to be attributed to unknown gradients. Spatial data analyses sometimes help to distinguish afterwards between two differently reacting areas in a field when enough data are collected. The primary lesson: do not tamper with the field in any way and make sure that the farmer did not treat or crop parts of the experimental site differently in the past.

12.5.3. Pre-sampling for P_i estimation

When preparing a field trial, the standard soil sample of 1.6 kg is used to get a first impression of the available population densities by mapping the selected (part of the) field. As we know, there is usually a linear relation between log nematode density and distance within and at right angles to rows – so between two adjacent plots (Fig. 12.4). Therefore, we can make educated guesses about these in-between densities from our pre-sampling effort and thus select locations of additional plots to obtain either intermediate, or repetitions of, population densities.

When fitting any of the models of Chapter 11 to data, a rule of thumb is to use a minimum of three densities per estimated parameters. In a glasshouse experiment, we would also use five replications per density to minimize the variation and most of the experiments would consist of 12 densities covering the whole range of possible densities, requiring 60 pots per cultivar or treatment. In a field test, there will be no exact replications per density and this is compensated for by using a large number of plots, e.g. 40 or more, which later can be combined in classes of densities for analysis, if needed.

Once the pre-sampling information is obtained, locations over the field can be collected to obtain the required number of plots and make sure that each treatment will have a range of required densities. The bulk sample size of each plot can be adjusted to obtain the required number of nematodes (Fig. 12.3). The golden rule: you should actually be able to count 200 nematodes of your target nematode from the solution obtained after elutriation in order to stabilize variation. If cyst nematodes are targeted, also try to collect a minimum of 200 cysts as these will come from different age classes and have different egg content. Again, then count at least 200 eggs from the suspension.

12.5.3.1 Using an infestation focus

The availability of an infestation focus of the target nematode is the easiest way to ensure that a complete range of required nematode densities is available. Farmers often know its location in the field by visual observation of symptoms during crop growth, which include retardation of growth (hot spots), the yellowing of leaves, etc. Generally, the centre of these spots indicates the highest population densities present. An example of how to determine the extent of an infestation focus of the potato cyst nematode is depicted in Fig. 12.15. A spot of retarded growth was observed at 240 m from the access path of the field in the direction of cultivation and a soil sample confirmed the presence of the potato cyst nematodes. A grid was put over the whole field of 3 × 8 m (the latter in the direction of cultivation) touching the probable centre of the infestation. From every third grid the bottom left square metre (0.75 × 1.33 m) was sampled using the scientific sampling method. This will now provide information on the extent of the infestation in the direction of cultivation. Also, around the centre at 240 m several rows of extra samples were collected at right angles to the direction of cultivation to the right and left, to determine the width of the infestation focus. It was then concluded that the focus was 100 m long and 36 m wide. By situating extra plots in between the original plots when the real P_is were collected, 192 samples could be collected ranging from 0.2 to 1394 cysts (kg soil)$^{-1}$ and 0.6 to 111.9 eggs (g soil)$^{-1}$, making it possible to obtain 48 plots per treatments. Samples for one treatment were taken from the edge of the focus up to the centre providing ranges of densities to feed the models. The volume of soil collected was adapted, based on the initial sampling, in such a way that at

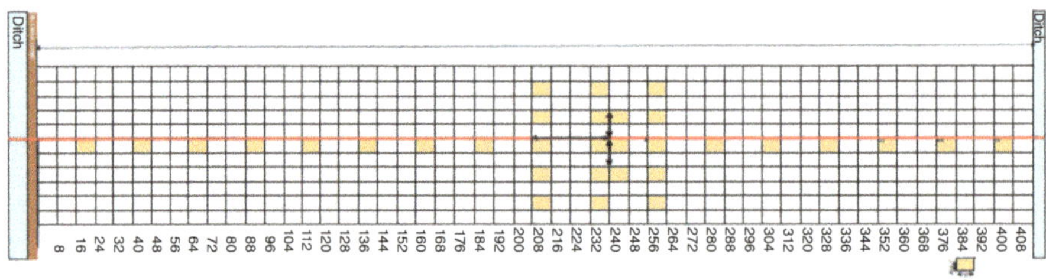

Fig. 12.15. The pre-sampling scheme of an infestation focus of the potato cyst nematode after location of growth reduction at 240 m from the access road of the field. Using a row of samples in the direction of cultivation the extent of the infestation was pinpointed, while a few samples to the left and right from the central line at 240 m provided information on the width of the infestation.

least 200 cysts could be elutriated and counted per plot. This resulted in varying sample sizes depending on the position of the plot within the infestation focus.

If no prior information is available over the exact location, but a statutory soil sample taken by a soil sampling agency provides evidence of the presence of an infestation focus (only a few cysts detected), a scheme as depicted in Fig. 12.7A can be used to locate its position in the farmer's field.

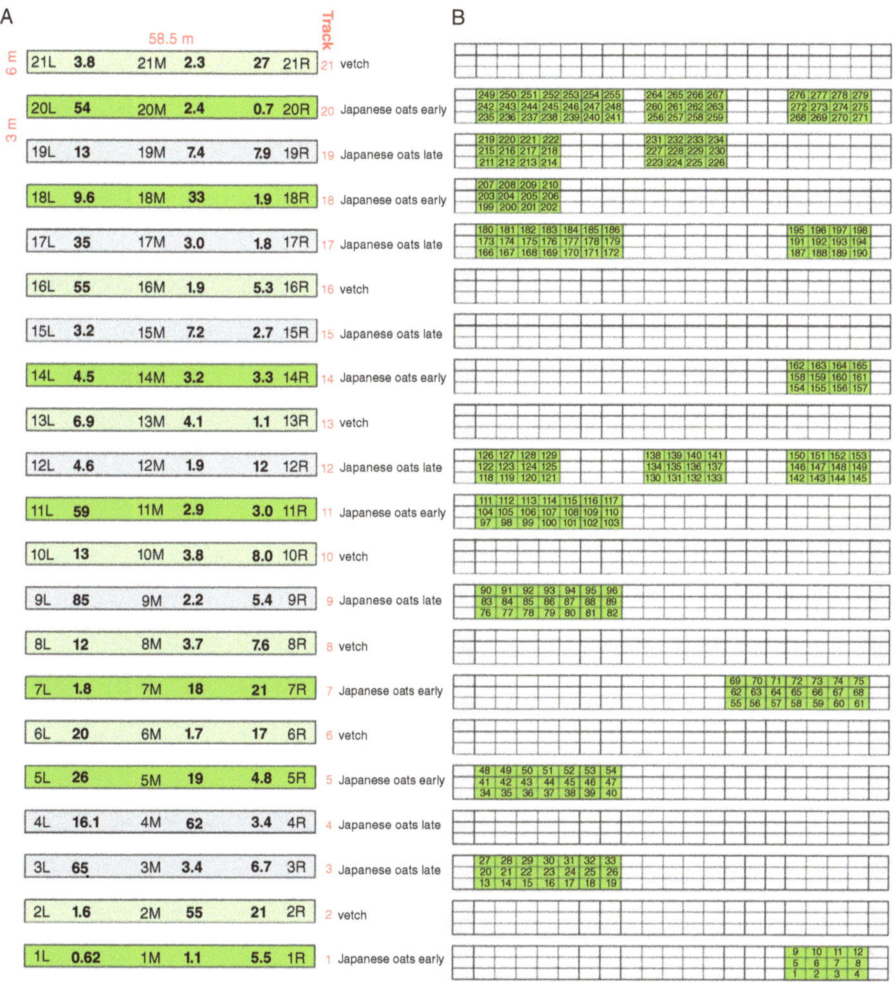

Fig. 12.16. A field, infested with *Meloidogyne chitwoodi*, was selected and an experimental area of 60 m wide and 200 m length established. (A) The area was divided into 21 tracks of 6 m wide and 58.5 m long. In each track three samples, using the scientific sampling method, were collected. Samples were located on the left (L), middle (M) or right (R) side of each track providing 63 samples. Numbers of juveniles (g soil)$^{-1}$ are provided in bold. (B) Based on the results, 299 plots were placed around those densities required to obtain a full range of densities for all cultivars tested.

12.5.3.2. *Using a full field infestation*

When only full field infestations are available and cyst nematodes are targeted, we can use a scheme as depicted in Fig. 12.7A. However, a larger area might be pre-sampled and more samples have to be collected to obtain the required range of population densities. When a nematode is involved with a more erratic large-scale distribution pattern, like *M. chitwoodi*, the search for the required nematode densities requires even more effort. We start using a scheme as depicted in Fig. 12.7A; however, the grid dimensions from which bulk samples of the scientific 1 m^2 sampling method are collected, might be larger as changes in nematode densities occur over larger distances. The length and width gradients of *M. chitwoodi* are much shallower than those of potato cyst nematode. Been *et al.* (2007b) found average length and width gradient of 0.85 and 0.73, respectively. Interpolation of densities will be the same as in the potato cyst nematode example but will not be as precise. After the pre-sampling effort, generally more plots are taken to collect P_i samples than will finally be used in the experiment. One starts to process those P_i samples expected to obtain the required nematode densities and chooses from the surplus of sampled plots when a shortage of certain densities arises, by processing plots from locations in which the desired densities are, most probably, available. In an experiment by Teklu *et al.* (personal communication) after pre-sampling 63 plots, as depicted in Fig. 12.16A, 299 plots were selected for the P_i measurement (Fig. 12.16B), of which 142 were actually elutriated and counted to provide 130 plots where both P_i and the P_f were estimated. Ultimately the required range of densities was obtained, but the process took more effort.

13 International Plant Health – Putting Legislation into Practice*

Sue Hockland[1], Thomas Prior[2]**, Jason D. Stanley[3], Renato N. Inserra[3] and Lisa M. Kohl[4]

[1]*Independent Plant Nematology Consultant, UK;* [2]*Fera Science Limited, UK;* [3]*Florida Department of Agriculture and Consumer Services, Division of Plant Industry, USA;* [4]*Animal and Plant Health Inspection Service, US Department of Agriculture, USA*

*A revision of Hockland, S., Inserra, R.N. and Kohl, L.M. (2013) International plant health – putting legislation into practice. In: Perry, R.N. and Moens, M. (eds) *Plant Nematology*, 2nd edn. CAB International, Wallingford, UK.

**Corresponding author: thomas.prior@fera.co.uk

 Plant Nematology, 3rd edn (eds R.N. Perry, M. Moens and J.T. Jones)
DOI: 10.1079/9781800622456.0013

13.1. Introduction and Terminology

For thousands of years, as people have migrated to new areas of the world, they have taken plants and plant products with them. However, in doing so people have also inadvertently moved associated pests. In addition to many biotic and abiotic factors (animal migration, water flow, winds, etc.), nematodes have been particularly dependent on international trade, and the migration of people between countries, for their present distribution. The difficulty associated with detecting nematodes and the lack of information about their biology and the damage they cause has contributed to the increased risk of them being moved unnoticed in trade on associated hosts or in soil residues. In the last 30 years or so international trade patterns have changed markedly, with an increase in the volume, frequency and speed of trade in ornamental plants, tubers and root crops. Such consignments inevitably increase the risk of plant-parasitic nematodes being introduced into new areas and affecting plant health.

As more has become known about their biology, plant-parasitic nematodes have increasingly been included in plant health legislation around the world and phytosanitary measures are being used to try to prevent their spread. Regulatory agencies implement phytosanitary measures to minimize the transport and spread of organisms harmful to plants, i.e. plant pests, by human activities. Plant pests comprise any species, strain or biotype of plant, animal or pathogenic agent injurious to plants and plant products, including plant-parasitic nematodes. Such regulatory systems are referred to by different terms in different countries, including 'plant health', 'plant quarantine' and 'plant protection' (Khan, 1989; Ebbels, 2003), and, increasingly, 'plant biosecurity', a term used in parallel with plant health worldwide, such as in New Zealand (https://www.mpi.govt.nz/biosecurity), in the UK (https://planthealthportal.defra.gov.uk/plant-biosecurity-strategy) and in the European Union (EU) (https://ec.europa.eu/food/plants/plant-health-and-biosecurity). In this chapter, we will retain the term 'plant health' as it is widely understood internationally, and will discuss this with an emphasis on key species of plant-parasitic nematodes.

The term 'pest' is defined by the *Glossary of phytosanitary terms* (ISPM 5), one of the International Standards for Phytosanitary Measures (ISPMs), as any species, strain or biotype of plant, animal or pathogenic agent injurious to plants and plant products, including, of course, plant-parasitic nematodes. Further ISPMs are used to describe not only the field of 'plant health' but also the phytosanitary measures for the exclusion, eradication, containment and suppression of plant pests, as well as surveys, risk assessments and closely related topics. They form part of the global programme for harmonization of policy and technical assistance administered by the Food and Agriculture Organization (FAO) of the United Nations (https://www.fao.org) and are available from https://www.ippc.int (a summary of abbreviations is given in Box 13.1). These standards, adopted until 1997 by the FAO Conference and now by the Commission on Phytosanitary Measures (CPM), the governing body of the IPPC, are recognized by the World Trade Organization (WTO) as global references for the harmonization of phytosanitary

Box 13.1. Regulatory abbreviations used in this chapter.

APHIS	US Department of Agriculture, Animal and Plant Health Inspection Service
CPM	Commission on Phytosanitary Measures
Defra	UK Department for Environment, Food and Rural Affairs
EFSA	European Food Safety Authority
EPPO	European and Mediterranean Plant Protection Organization
FAO	Food and Agriculture Organization
GATT	General Agreement on Tariffs and Trade
IPPC	International Plant Protection Convention
ISPM	International Standard for Phytosanitary Measures
NAFTA	North American Free Trade Agreement
NAPPO	North American Plant Protection Organization
NPPO	National Plant Protection Organization
PRA	Pest Risk Analysis
RNQP	Regulated Non-Quarantine Pest
RPPO	Regional Plant Protection Organization
WTO	World Trade Organization
WTO-SPS	World Trade Organization Agreement on the Application of Sanitary and Phytosanitary Measures

measures under the Agreement on the Application of Sanitary and Phytosanitary Measures (known as the WTO-SPS Agreement).

According to the IPPC, phytosanitary measures include any legislation, regulation or official procedure having the purpose of preventing the introduction and/or spread of plant pests and may normally only be applied to regulated pests, although the IPPC also provides for emergency measures to be taken against any pest that is a potential threat to the territory of a member country. The term 'quarantine pest' is defined by the ISPM *Glossary* as 'a pest of potential economic importance to the area endangered thereby and not yet present there, or present but not widely distributed and being officially controlled' (Box 13.2). Organisms are categorized as quarantine pests based on these defining criteria, which are evaluated through Pest Risk Analysis (PRA) (see Section 13.7). Examples of quarantine nematode pests for the EU are the pinewood nematode, *Bursaphelenchus xylophilus*, and the Columbian root-knot nematode, *Meloidogyne chitwoodi*. However, these nematode species are not quarantine pests at the national level for the continental USA, but maybe at a regional level, because both *B. xylophilus* and *M. chitwoodi* occur there and are not limited in distribution in the country.

Many plant-parasitic nematodes that may be of phytosanitary importance are intercepted in international trade by plant health inspectors at points of entry. Often these are unknown species that have the potential to become significant pests if allowed to enter and establish. Nematode pests that were previously unknown or are not well known may be subject to emergency quarantine actions to avoid possible introduction and spread before risks are better understood. The general lack of accurate country or regional lists can make it difficult to ensure that a given nematode species, especially a new species, is not already native and therefore should not be subject to quarantine regulations. The recognition of the existence and significance of *M. minor* in Europe is such an example. National Plant Protection Organizations (NPPOs) are the official services established by governments to discharge the functions specified by the IPPC,

Box 13.2. Definition of a quarantine pest (L. Kohl, personal communication).

When is an organism considered to be a quarantine pest?

If the pest is NOT present:	If the pest is present:
Must have the potential to cause economic impacts ▪ *Host plants are grown in the area* ▪ *Pest is adapted to the area's climate* ▪ *Pest is known to cause damage to host plants*	Must have the potential to cause economic impacts **AND** Is not widely distributed in the area **AND** Is under official control

and are almost universally provided with broad powers to take such actions in the light of new and unknown threats. However, emergency measures cannot continue indefinitely without evaluation to determine their appropriateness and technical justification. Measures should be modified as appropriate based on experience and the acquisition of new information. Cannon *et al.* (1999) describe how the UK adopted a systematic protocol to determine appropriate measures for nematode pests intercepted in the UK on Chinese penjing (dwarfed trees); other examples include certain species of root-knot nematodes (*Meloidogyne* species) intercepted by member states of the EU on imported rooted cuttings.

Against this background, the ISPM *Glossary* defines another specific category of regulated pests known as Regulated Non-Quarantine Pests (RNQPs). They may be distinguished from quarantine pests as they may be present (even widespread) in the importing country, and their impact is generally known; an established level of pest infestation may be tolerated if feasible and effective control measures are available. Their presence in plants for planting would affect the intended use of those plants with an economically unacceptable impact and which is therefore regulated within the territory of the importing contracting party with the aim of suppression, rather than the eradication or containment strategy applied to quarantine pests. According to the IPPC, regulatory restrictions for RNQPs can only be applied to specified plants for planting in cases where the presence of these organisms would severely impact the quality of propagative material. Actual RNQPs will differ from region to region, according to the relative importance of a particular pest, and whether it is native to the region or not. On occasion, a pest is a RNQP only until a PRA proves there is no case for regulation. Sometimes gathered evidence increases quarantine status; *M. enterolobii* was an example of an RNQP but is now currently recommended for regulation by the European and Mediterranean Plant Protection Organization (EPPO, https://www.eppo.int), as an EU Plant Health Annex II A pest, and has also been recommended for quarantine status in California and south-eastern states of the USA as shown in Section 13.9.4. More information on the concept and application of measures for RNQPs can be found in ISPM 16 on the FAO website (https://www.fao.org/3/y4223e/y4223e.pdf) or the IPPC website (https://www.ippc.int/en/core-activities/standards-setting/ispms/).

13.2. Historical Considerations

Nematodes that are moved on plants or plant parts are said to be following a pathway. A pathway, as defined by the IPPC, is any means that allows the entry or spread of a pest. Soil adhering to agricultural vehicles is an example of a pathway for nematodes. Indeed, historical records suggest that *Globodera rostochiensis* may have arrived in the USA in this way and interception records held by Fera Science Limited in the UK show that several nematode species, such as those in the genera *Heterodera* and *Pratylenchus*, can be spread by such means. Movement has also been facilitated by a lack of knowledge about the habits and crop losses induced by plant-parasitic nematodes. This problem continues even to the present day.

The exchange of crops and their nematode pests between continents increased after the European discovery of the Americas. Typically, nematode species were first described from a country that was not their centre of origin. *Tylenchulus semipenetrans* was introduced and first recorded in California on citrus and subsequently described from there and Florida by Cobb (1913), but its origins were in the Far East, from where it also spread to the Mediterranean region. Here a new biotype was recorded, which was able to infect and reproduce on olive (*Olea europaea*), a native tree of agricultural importance for the Mediterranean region. *Globodera pallida* and *G. rostochiensis*, known as the most common potato cyst nematodes (PCNs), were first classified as *Heterodera* species; *Heterodera rostochiensis* was first described in Germany (Wollenweber, 1923), and *Heterodera pallida* was described in the UK (Stone, 1972). The genus name *Heterodera* was changed to *Globodera* for all round-shape cyst nematodes, including the PCNs, by Behrens (1975).

Globodera species have evolved to their greatest extent in their centre of origin on native Solanaceae in the Andean regions of South America, where they were first recorded in 1952 (Wille and Bazán de Segura, 1952; Subbotin *et al.*, 2020). As international trade developed, so some populations of PCN were spread on seed potatoes and have become economically important pests of cultivated potatoes (*Solanum tuberosum*) worldwide. Most recently it has been shown that the formation of the Andean cordillera caused the emergence of cold-adapted solanaceous plants as well as the biodiversification of *Globodera* species that is still being recognized today (Grenier *et al.*, 2010; Subbotin *et al.*, 2010a, 2020). The spread of their greater genetic diversity poses a threat to the use of cultivars bred to be resistant to less genetically diverse populations occurring elsewhere in the world (Grenier *et al.*, 2010; Hockland *et al.*, 2012; Subbotin *et al.*, 2020). Other major crop and nematode movements included sugar beet (*Beta vulgaris*) with the beet cyst nematode (*H. schachtii*) and soybean (*Glycine max*) with the soybean cyst nematode (*H. glycines*), which were introduced into North America from Europe and Asia, respectively.

The involvement of particular plant-parasitic nematode species with crop loss was often not recognized until many years after they had been described. A plant-parasitic nematode of wheat, *Anguina tritici*, was first discovered by Needham in 1743 (Needham, 1743; see Box 5.2) but the role of nematode pests in suppressing plant yield was not realized until the mid-1800s. *Heterodera schachtii* was discovered in 1859, but it was not described and accepted as a causal agent of the crop decline until 1871. At this time, PCNs had not been distinguished from the beet cyst nematode, thus masking the importance of these species.

In recent decades trade in agricultural products worldwide has increased and changed considerably, thus increasing the risk of moving pests from one continent to another. For example, in 1998, the major suppliers of agricultural commodities to the USA were, in order of importance, Canada, Caribbean Basin, Mexico, EU, South America, Australia and New Zealand, and Far East countries (Brown, 1999). By 2004, South-east Asia had considerably increased the export of agricultural products to the USA. In Florida, the volume of maritime cargo has increased from less than 2 million t in 1990 to almost 5 million t in 2000 (Klassen *et al.*, 2002), reaching 16 million t in 2017 (U.S. Department of Transportation, 2017).

All horticultural imports (including trees, bulbs, cut flowers and other ornamentals, such as hardy nursery stock, cuttings and other young plants) have the potential to spread nematode pests, including newly identified strains or races of recognized species such as the stem and bulb nematode (*Ditylenchus dipsaci*), leaf and bulb nematodes (*Aphelenchoides* spp.) and root-knot nematodes (*Meloidogyne* spp.). For the EU, such imports via all routes have increased steadily over the last 20 years or so, from approximately 426,000 t in 2010 to approximately 588,000 t in 2021 (https://ec.europa.eu/eurostat/databrowser/view/ds-058213/legacyMultiFreq/table?lang=en&category=ext_go.ext_go_detail; and also: https://ec.europa.eu/eurostat/web/international-trade-in-goods/overview). UK imports of seed potatoes has varied in the last 20 years or so, peaking at over 34,000 t in 2013, and currently approximately 8500 t in 2020, but still posing a risk of importing soil and tubers containing plant-parasitic nematodes.

Despite the impact of an increasing volume of cargo around the world, there remain many serious pest species, or species that have the potential to become pests, that are not yet widespread in all areas where they could survive. In such cases, the prevention of further spread, control and eradication of such pests are major objectives for NPPOs around the world. Such objectives are closely supported by national and international schemes to produce healthy and vigorous planting material, and so plant health legislation has a place in the development of sustainable crop health systems by controlling the pests at their source. The challenges involved in putting such legislation into practice will be described here.

13.3. International Phytosanitary Frameworks

13.3.1. Development of legislative powers

The spread of plant pests in trade and their increasing association with crop losses in the mid-19th century stimulated international concern, leading to the first internationally agreed plant health measures in trade being taken against the American vine phylloxera, now known as *Viteus vitifoliae*, in 1878. This insect pest had spread rapidly throughout the vineyards of Europe, causing losses in France alone of approximately the equivalent of €72 million (US$96 million). The 'International Convention on Measures to be Taken against *Phylloxera vastatrix*' (the former name of the grape phylloxera *Dactulosphaira vitifoliae*) embodied many of the principles recognized today in international plant health. The most important of these, as listed by Ebbels (2003), are given in Box 13.3.

Use of this international convention against the American vine phylloxera highlighted various deficiencies, especially in relation to the lack of a clear definition of concepts and terms. Today, the CPM agrees to promote a common understanding of key phytosanitary

Box 13.3. Main principles of international plant health.

1. The responsibility to give an official written assurance on the pest-free provenance of host material being traded internationally.
2. The prohibition of international trade in certain kinds of material that might spread the pest.
3. The designation of official bureaux responsible for administering such trade.
4. Powers to inspect traded material and to take remedial action on items not complying with the requirements of international phytosanitary measures.
5. The prompt exchange of relevant information, particularly on new outbreaks.
6. That all these measures were to be embodied in national law by the participating countries.

concepts and terms, and to encourage the harmonization of phytosanitary measures under the WTO-SPS Agreement. This Agreement, one of the outcomes of the Uruguay Round of negotiations of the General Agreement on Tariffs and Trade (GATT) from 1986 to 1994, entered into force for most countries in 1995. It states that the purpose of an SPS measure should be limited to the protection of human, animal or plant life or health; it applies to all sanitary (food safety for humans; animal life and health) and phytosanitary (plant life and health) measures that can affect international trade. At the time the WTO-SPS Agreement was negotiated, countries were concerned that reduced tariffs and subsidies could lead to the use of non-tariff barriers to trade, including SPS measures, in order to protect domestic production. Thus, the WTO-SPS Agreement resulted out of a need to have rules in place to prevent, or at least reduce, the unjustifiable use of measures to block trade in agricultural commodities between countries. More details on this Agreement can be found on the website http://www.wto.org. Other free trade agreements exist, such as the North American Free Trade Agreement (NAFTA), which came into force in 1994 and is an agreement signed by the governments of Canada, Mexico and the USA, creating a trilateral trade bloc in North America.

Whilst the WTO-SPS Agreement recognizes the IPPC as the standard-setting body for developing international standards, guidelines and recommendations in respect of phytosanitary measures affecting trade, the IPPC has a longer history. Earlier efforts to expand international cooperation in the management of plant pest threats resulted in the creation of the IPPC in 1951. This multilateral treaty superseded all previous agreements and quickly became the premier international instrument upon which most national systems were based. The IPPC was revised in 1979 and again in 1997 and is now an international agreement on plant health with 177 member countries. It aims to protect cultivated and wild plants by preventing the introduction and spread of pests. The Secretariat of the IPPC is provided by the FAO of the United Nations, and the work programme is administered through the IPPC Secretariat and the CPM in cooperation with NPPOs and Regional Plant Protection Organizations (RPPOs) and NPPOs. The 1997 revision was a substantial one, setting out the phytosanitary concepts to be applied by contracting parties to be consistent with SPS obligations and also expanding and clarifying fundamental guiding principles associated with limiting the spread and impact of pests.

In line with the principles of the IPPC, national governments around the world develop their own plant health and quarantine regulations, which are usually enforced by an integral NPPO. A list of all NPPOs is provided by the IPPC (https://www.ippc.int/en/countries/

nppos/list-countries/). State or provincial governments within a country may also establish plant health agencies that function within these political subdivisions and often work in cooperation with the NPPO; usually, however, the national regulations for protecting the country from exotic pests take precedence over those at the state or provincial level (Khan, 1989). Additionally, for some countries, legislation passed by intergovernmental organizations is binding for member countries, such as in the EU (Khan, 1989; Ebbels, 2003).

Information on the latest EU plant health legislation can be found at: https://ec.europa.eu/food/plants/plant-health-and-biosecurity/legislation/plant-health-rules_en. This website also gives links to many other sites detailing work relating to plant health in Europe. The UK left the EU in January 2020 and plant health legislation here can be found at https://planthealthportal.defra.gov.uk. Similar information for the US federal government can be found at https://www.aphis.usda.gov/aphis/ourfocus/planthealth, but each US state may have its own legislation to deal with particular issues. A list of important websites relating to plant health is given in Box 13.4.

Box 13.4. International plant health websites. Legislation in plant health and related issues is constantly changing. Websites of relevant organizations are an important way of obtaining up-to-date information.

Website	Comments/website for:
https://www.aphis.usda.gov	US Department of Agriculture, Animal and Plant Health Inspection Service (APHIS). Contains information about APHIS's role in safeguarding US agriculture from exotic animal and plant pests.
https://www.eppo.int	European and Mediterranean Plant Protection Organization. Regional site for EPPO standards and information for European National Plant Protection Organizations.
https://www.eppo.int/RESOURCES/eppo_standards/pm7_diagnostics	EPPO Standards – PM 7 Diagnostics.
https://ec.europa.eu/eurostat	Home page of Eurostat, the source of EU statistics.
https://food.ec.europa.eu/plants/plant-health-and-biosecurity/legislation/plant-health-rules	EC legislation relating to plant health.
https://www.efsa.europa.eu/en/topics/topic/plant-health	Provides advice to the European Commission.
https://www.ippc.int	International Plant Protection Convention. Contains details of Regional Plant Protection Organizations and lists current ISPMs.
https://www.fao.org	Food and Agriculture Organization of the United Nations.
https://pra.eppo.int	EPPO Platform on Pest Risk Analysis.
https://planthealthportal.defra.gov.uk	Provides information specific to the UK.
https://www.wto.org	World Trade Organization.

13.3.2. Putting legislation into practice

Regional and interregional cooperation and harmonization in all areas of plant health is encouraged under Article IX of the IPPC and is performed by recognized RPPOs, intergovernmental organizations functioning as coordinating bodies for the NPPOs on a regional level. The European and Mediterranean Plant Protection Organization (EPPO) was established in 1951 as the first RPPO (https://www.eppo.int). The North American Plant Protection Organization (NAPPO) represents the USA, Canada and Mexico, and is listed with other such organizations at: https://www.ippc.int/en/ippc-community/.

RPPOs perform their role in various ways. They produce non-binding (i.e. without legal status) recommendations to member countries for harmonizing regulations among their members. Some hold working parties, panels and expert groups to address ongoing concerns such as phytosanitary regulations, PRA and diagnostics. They may also deal with specific pests and measures of concern to the region or other regions. One of the major roles of RPPOs is to raise awareness and provide guidance on newly identified and emerging pests that might pose a threat to agriculture or the natural environment in their region. This may be done through the production of lists of pests considered to be of quarantine concern, such as the EPPO A1 and A2 lists; these lists contain pests that are absent from the EPPO region and pests that are not widely distributed in the EPPO region, respectively (https://www.eppo.int/media/uploaded_images/ACTIVITIES/plant_quarantine/pm1-002-28-en.pdf; see also: https://www.eppo.int/media/uploaded_images/RESOURCES/eppo_standards/pm1/pm1-002-32-en_A1A2_2023.pdf). In addition, 'alert lists' raise awareness and provide an early warning to NPPOs regarding certain organisms moving in trade that may present a risk. Nematologists around the region may be consulted in the production of lists of pests that are of quarantine concern. Since the 1990s, such decision making has been facilitated by the development of PRA, which has become a major tool in plant health. Further information can be found on websites such as that for EPPO Platform on PRAs (https://pra.eppo.int), The European Food Safety Authority (EFSA) (https://www.efsa.europa.eu/en/topics/topic/plant-health), which provides scientific advice on pest risk assessments to the European Commission, and the UK Plant Health Information Portal, which also provides a wealth of information (https://planthealthportal.defra.gov.uk).

13.4. Early Legislation Enacted against Plant-parasitic Nematodes

Before 1900, quarantine actions initiated in several countries against exotic damaging organisms mainly concerned insects and fungi because damage by nematodes was not recognized. At that time, only sporadic information was available, such as for the cyst-forming nematodes and root-knot nematodes on crops such as cucumber, cereals, sainfoin and coffee.

Phytosanitary legislation specifically addressing plant-parasitic nematodes was introduced after 1900 and initially consisted mainly of isolated quarantine actions implemented against specific nematodes by a few countries concerned with preventing the spread of these nematodes to areas in which they had not been found. In the USA, the first regulatory action against any exotic nematode plant pest was implemented in 1909 by the US Department of Agriculture against a root-knot nematode infecting flowering cherry trees imported from Japan. The destruction of these trees almost caused a diplomatic crisis because the nematode-infected cherry trees were a gift from the Japanese government to the wife of the US President. This action, however, prompted approval of the first Plant Quarantine Act by the US Congress in 1912.

One species of PCN, *G. rostochiensis* (also known as the golden nematode), was detected in the USA for the first time in 1942 in Long Island, New York. It probably arrived with contaminated military equipment returning from Europe after the First World War. A golden nematode official control programme was enacted by the state of New York in 1944 to contain the spread of the nematode, action subsequently followed by the Golden Nematode Act in 1948, passed by the US Congress to protect the country's potato industry.

The first European legislation against nematodes was probably the Beet Eelworm Order in 1943 and following years, which was promulgated in Britain to manage the spread and increase in population levels of *H. schachtii* by adopting specific cultural practices. The Potato Root Eelworm Order, promulgated in Northern Ireland in 1945, was intended to suppress population levels of PCNs by enforcing the implementation of crop rotations and the use of certified seed potatoes.

13.5. International Phytosanitary Initiatives against Plant-parasitic Nematodes

There is a core of nematode species that are targeted by legislation around the world. Of course, not all species of plant-parasitic nematodes appear on every country's list of regulated pests because they may be endemic in some countries or regions and exotic in others. A list of plant-parasitic nematodes most frequently regulated in international trade over four decades is given in Table 13.1. Plant-parasitic nematodes, with their endoparasitic, sedentary and migratory habits, are a major target of phytosanitary measures because they can be easily transported unseen via roots, soil and above-ground parts of plants. The pinewood nematode *B. xylophilus,* a major forestry pest in some parts of the world that has also been introduced into two EU countries, Portugal (Mota *et al.*, 1999) and Spain (Robertson *et al.*, 2011), can be either mycetophagous or ectoparasitic and is spread by infected timber and timber products or by beetle vectors on a local scale. As this species has spread, it has been regulated by an increasing number of countries.

Reasons for the spread of other species over the years are a matter of speculation. In some instances, the significance of spread by uncertified potato tubers, bare-rooted plants for planting and excessive soil residues has been recognized, leading to additional regulation. However, the new appearance of *Longidorus diadecturus* in 2011 data, included in Table 13.1, appears to be a result of the expanding membership of the EU and other countries adopting the official EU list, rather than an increased risk in the spread or significance of this pest. At the time of the first EU list of quarantine nematodes, this species was thought to transmit viruses in its type country, USA, and was thus included. Recent data in Table 13.1 show that only in a minority of cases is there a significant decrease in importance of some species, which may be due to increased survey work showing the previously unknown distribution of some species, or establishment of some species in new areas; once a nematode species is detected and considered to be established in a region, a country has to decide on appropriate action. For example, *Xiphinema index*, which remains an important regulated pest in California and Florida, has been deregulated in other countries. *Heterodera schachtii*, widespread in Europe (which contributes almost 50% to world sugar beet production), has been the subject of intensive, rigorous control measures that include new resistant cultivars, but the species remains a regulated pest in South America and parts of Asia, where it does not yet occur.

Nematode pests of potatoes are amongst the most highly regulated because they are readily spread in infected tubers or associated soil residues and because potatoes destined

Table 13.1. Number of countries regulating species of plant-parasitic nematode in 1982, 2000, 2011 and 2022, in descending order as at 2022.[1]

Nematode species	1982	2000	2011	2022
Globodera rostochiensis	51	106	119	126
Globodera pallida	–*	55	80	89
Ditylenchus dipsaci	23	58	72	82
Bursaphelenchus xylophilus	–*	46	82	78
Ditylenchus destructor	12	53	65	75
Aphelenchoides besseyi	9†	70	54	70
Heterodera glycines	–*	52	55	64
Radopholus similis[2]	11	55	58	64
Meloidogyne fallax	–*	–*	39	52
Xiphinema americanum	–*	30	43	50
Meloidogyne chitwoodi	–*	–*	52	43
Xiphinema californicum	–*	–*	39	43
Meloidogyne enterolobii	–*	–*	–*	39
Longidorus diadecturus	–*	–*	33	35
Bursaphelenchus cocophilus	–*	21	13	30
Aphelenchoides fragariae	13	47	21	29
Anguina tritici	–*	24	21	25
Heterodera schachtii	16	22	14	20
Pratylenchus penetrans	–*	–*	25	17
Nacobbus aberrans	–*	38	52	10
Xiphinema index	–*	42	6†	10

[1]Data for species regulated in more than ten countries as recorded by Khan (1982), Lehman (2004), Hockland *et al.* (2006) and the APHIS/USDA Phytosanitary Certificate Issuance and Tracking System (PCIT) (2011 and 2022). Countries belonging to the EU were regarded as individual countries.

[2]2011 data for *Radopholus similis* includes countries regulating *R. citrophilus*, which has been proposed as a species in its own right but is now variously accepted as a subspecies or race of *R. similis*. In 2011, 11 countries listed only *R. similis* in their regulations, two countries listed only *R. citrophilus* and 45 countries listed both.

*Insufficient or no records at this time.

†Species regulated by fewer than ten countries but included here to show changes between 1982 and 2022. Since 2004, *L. diadecturus, M. chitwoodi* and *M. fallax* were regulated by more than ten countries. Since 2011, the combination of *B. cocophilus* and its weevil vector, *Rhynchophorus palmarum*, was regulated by more than 35 countries.

for consumption may instead be propagated. Two PCN species are the most common nematodes listed in Table 13.1. Other than these, *H. glycines* is the only cyst-forming nematode commonly regulated by countries. It is of major concern for several countries in North America (Canada), South America (e.g. Argentina, Brazil, Chile, Uruguay) and African countries such as South Africa. This nematode is of less concern in the Far East where it is native. The false root-knot nematode, *Nacobbus aberrans*, another major potato pest in Mexico and western South American countries, has not yet received the same level of concern as PCN species, although it has the potential to infect a wider range of crops including sugar beet (Manzanilla-López *et al.*, 2002).

In general, root-knot nematodes are not regulated as a group because the major economically important species are already widely distributed. However, several species that have emerged from Africa and the Americas in recent years have reached particular prominence, as can be seen in Table 13.1. *Meloidogyne chitwoodi* is a serious pest of potato and other economically important crops such as carrot. It is on the lists of prohibited pests in many countries. *Meloidogyne fallax*, another pest of potato, is also now recognized as an important pest in many countries. Other root-knot nematodes are being included on 'alert' lists worldwide: *M. citri*, a pest of citrus in China, which was thought to have a restricted host range but has now been shown to infect and reproduce on hosts such as tomato (Vovlas and Inserra, 2000); *M. ethiopica*, a very damaging pest of kiwi and grape in Brazil and Chile, respectively, recently introduced into Europe (Širca *et al.*, 2004); and *M. enterolobii*, which has been shown to overcome root-knot nematode resistant genes in some economically important crops (Brito *et al.*, 2007; Kiewnick *et al.*, 2009), is increasingly regulated as a quarantine pest and appears in Table 13.1 for the first time in this edition.

As endoparasitic nematodes are found within plant parts, they can easily be moved in infected plant material in trade. Migratory endoparasitic nematodes, such as the burrowing nematode *Radopholus similis*, are common in lists of quarantine pests. The occurrence of a *R. similis* race able to infect and damage citrus in Florida and the lack of reliable and rapid morphological and molecular analyses to identify this race has prompted a worldwide ban against the movement and spread of this economically important nematode to protect citrus-growing countries, and other countries producing *Musa* spp., ornamental plants and palms. The risk from introducing other damaging citrus nematode pests is believed to be sufficiently reduced by measures in place for *R. similis*, such that further restrictions are not deemed necessary by many countries. Specific restrictions against other endoparasitic migratory root feeders are imposed by Argentina, Brazil, Chile, Egypt, Mexico, Morocco, South Africa, Syria and Uruguay but the risk associated with these nematodes is considered less serious than that posed by *R. similis*. Listed root-lesion nematode pests include mainly tropical and subtropical species, such as *Pratylenchus coffeae*, *P. loosi*, *P. scribneri* and *P. zeae* and the temperate species *P. penetrans*.

Migratory endoparasitic species of foliar parts of plants are regulated by many countries; species belonging to the genera *Aphelenchoides* and *Ditylenchus* are most common in quarantine lists. Their frequent inclusion in the majority of the lists is due to the fact that these nematodes are easily transported inside plant tissues. These nematodes may cause symptoms that can be detected by visual inspection by well-trained inspectors, but at low levels such symptoms may not be apparent (Kohl *et al.*, 2010). Often the appearance of plants may also suggest nutritional disorders so laboratory confirmation is necessary. The insect-vectored species such as *B. xylophilus* and *B. cocophilus*, which also live inside foliar parts of plants, are also of major concern for many countries.

Ectoparasitic nematodes are found in the soil surrounding plant roots, rather than within plant material, so restrictions on the amount of soil associated with plants and plant products will reduce the risk of their introduction (nil tolerance or a few derogations internationally), so that these nematodes are less likely to follow trade pathways and hence they are rarely included in regulated pest lists. However, ectoparasitic nematodes that vector nepoviruses (dagger and needle nematodes) are of major concern for European and a few South American countries (e.g. Brazil and Uruguay). Nematodes in the *Xiphinema americanum sensu lato* species complex are among the most commonly regulated ectoparasitic species, as the overall risk of introduction of virus vectors in this group is considered higher than that of many other damaging ectoparasitic nematodes such as lance (*Hoplolaimus* spp.) and sting nematodes (*Belonolaimus* spp.), even though the pathway of introduction of all these species poses the same potential risk. Further details of nematodes listed by countries around the world can be obtained from the EPPO Plant Quarantine Data Retrieval (PQR) System, details of which can be found on the EPPO website (http://www.eppo.int), or the IPPC website (http://www.ippc.int).

13.6. Phytosanitary Problems Posed by Plant-parasitic Nematodes

Plant-parasitic nematodes pose particular problems to phytosanitary authorities because they rarely produce obvious symptoms on plants or plant products that can be detected by routine visual inspection. Likewise, nematological analysis of plant material or soil is often not feasible at points of entry because such analyses are time-consuming and require specific equipment, expertise and quarantine conditions for efficient nematode extraction. Types of nematodes that can be distinguished by symptoms they cause, such as the root-knot nematodes (*Meloidogyne* spp.), may not produce distinguishing symptoms at low levels of infestation. Thus, the most practical approach is to impose exclusion measures that prevent the importation of the substrate and any plant they inhabit (see Section 13.8).

13.7. Determining the Risk Posed by Plant-parasitic Nematodes Using Pest Risk Analysis

13.7.1. The development of formal procedures

Harmonization is a key principle of the WTO-SPS Agreement. This means that when NPPOs implement phytosanitary measures in trade to mitigate immediate or potential risk of pests to plant health within their jurisdictions, the measures must be based on available international standards (i.e. the ISPMs developed and adopted under the IPPC described earlier). If no international standard is available, or in cases where measures deviate from standards, measures need to be technically justified by a scientific risk assessment as described under the WTO-SPS Agreement. Under the IPPC, a scientific risk assessment is referred to by the term 'Pest Risk Analysis' (PRA). The IPPC definition of PRA includes evaluating biological or other scientific and economic evidence to determine whether a pest should be regulated (**pest risk assessment**) and the strength of any phytosanitary measures to be taken against it (**pest risk management**).

The IPPC has published a range of reference and concept standards and has begun developing pest and commodity-specific standards. Therefore, most current decisions on phytosanitary requirements, such as allowing or prohibiting the entry of a commodity or requiring a quarantine treatment of a commodity, must be justified by a PRA as described in ISPM 2 *Framework for pest risk analysis* and ISPM 11 *Pest risk analysis for quarantine pests including analysis of environmental risks and living modified organisms*. These two standards provide significant guidance for the international harmonization of PRA. All current ISPMs are published on the IPPC website, the International Phytosanitary Portal (IPP) at http://www.ippc.int.

Plant-parasitic nematodes being considered for quarantine status need to undergo a risk assessment (i.e. evaluation of the likelihood of pest entry, establishment and spread, and of the potential economic and environmental consequences should the pest become established) and subsequently a risk management plan (i.e. the identification and evaluation of pest risk management options to reduce the risk of introduction and spread of the regulated pest, followed by the selection of the most appropriate management option(s)).

As discussed earlier, an organism can only be considered a quarantine pest for a particular country if it is not present in that country or of limited distribution and under official control. Official control of regulated pests must be directed towards eradication or containment within the country and not merely a reduction or suppression of the population, and the official control programme must be implemented nationally. In the USA, for example, if a pest is present in more than one state but a state quarantine programme is implemented by only one of those states, then this would not meet the definition of 'official control' for the purpose of declaring a pest a regulated pest at the national level. Therefore, the quarantine status of a pest can change over time, not only as populations become established or eliminated, but also as regulatory action is implemented at various levels. Countries may also choose to deregulate widespread pests or pests that do not cause significant damage in order to reallocate resources to prioritize pests considered to be of higher risk; for example, post-EU exit, *M. fallax* in the UK is now designated a RNQP following its introduction and spread (see Section 13.9.5).

PRA can be initiated for a variety of reasons, such as if a particular pest is intercepted at points of entry, a new pest risk is identified by scientific research, or a pathway other than a commodity import (e.g. natural spread, international mail, garbage) is identified. A PRA can also be initiated because a country makes a request to the NPPO of a second country for authorization to allow the importation of an agricultural commodity (e.g. fruits and vegetables for consumption or plant products for propagation) into that second country. The importation of an agricultural commodity represents a pathway that pests can potentially follow to enter and establish in a new country.

In the preparation of a PRA initiated by a commodity import request, only quarantine pests likely to be carried on the particular commodity require complete risk analysis (i.e. risk assessment followed by risk management). The pathway of most fruits, herbs and vegetables for consumption is unlikely to carry plant-parasitic nematodes because these organisms usually only attack underground parts of plants. The exceptions to this include plant-parasitic nematodes that infect foliage (e.g. *Aphelenchoides* spp.) or endoparasitic nematodes that may be carried in edible roots and tubers such as beets and potatoes (e.g. *M. chitwoodi*, *N. aberrans* and *Pratylenchus* species). There is little evidence that nematode populations can become established if the host commodities are consumed after import as intended. Disposal of such commodities in landfill avoids dispersal onto susceptible hosts or their growing media and establishment in the importing country. Dispersal in compost, however, may pose risk of dissemination unless appropriate procedures are followed, requiring that the aerobically biodegraded raw material is composted at 55°C for 15 days (windrow composting method) or 5 days (in-vessel composting method) with frequent mixing to maintain uniform distribution of the temperature in the material to obtain a final product free of live plant-parasitic nematodes, other pests and weeds (Inserra *et al.*, 2006).

By contrast, without the implementation of specific phytosanitary measures, the pathway of plants for propagation, such as potato tubers for seed, bulbs and nursery stock, is much more likely to carry plant-parasitic nematodes (e.g. PCN, *D. dipsaci*) and result in the establishment of these nematodes in the importing country. Thus, compared to commodities for consumption, seeds and plants for planting pose a greater risk for the introduction and/or spread of nematode pests.

In addition to the type of plant commodity being imported, other factors typically used in assessing pest risk include, among others: (i) the biology of the pest (e.g. host range, feeding habits, life cycle, habitat, symptoms produced, overwintering/dormancy ability, dispersal ability, interaction with plant pathogens); (ii) the economic and environmental impacts of the pest in other parts of the world; and (iii) the host and environmental conditions in its geographical distribution, as well as those conditions in the importing country. The more knowledge of these factors one has, the more accurate the estimate of risk. Unfortunately, biological information is often lacking for many new pests intercepted in international trade, including most new nematodes. Furthermore, soil temperature data, so necessary for nematode assessments, is often non-existent and has to be extrapolated from air temperatures, a technique that has disadvantages, as discussed by Baker and Dickens (1993). Not only is soil temperature influenced by air temperature, but also by ground cover, soil texture, wetness, sun angle and day length. In addition, rainfall data often cannot be used because of the added complication of irrigation at monitoring sites.

During the final stage of the PRA, i.e. pest risk management, the estimated pest risk helps determine the most appropriate phytosanitary measures needed to reduce risks. There are several important WTO-SPS/IPPC principles that should be followed when deciding whether or not to allow importation of a commodity and, if so, what phytosanitary measures should be implemented to manage the risks associated with that commodity (many of these principles are embodied in the WTO-SPS Agreement).

The first of these principles is that of '**minimal impact**', also known as 'least restrictive measures', which states that measures must not restrict trade more than that required to achieve the appropriate level of protection for the importing country. In other words, the selected risk management measure(s) should be 'proportional to the risk identified in the pest risk assessment' (ISPM 2).

The principle of '**equivalence**' means that an importing country must recognize that different phytosanitary measures can potentially be used to achieve their appropriate level of protection. For example, if an irradiation or hot water treatment is equally effective in eradicating a particular plant-parasitic nematode pest on a particular commodity compared with a chemical management treatment, then exporting countries should be allowed to use one of the alternatives.

The WTO-SPS principle of '**non-discrimination**' states that importing countries should not discriminate between countries that have the same phytosanitary status; that is, countries that have the same pests for a certain commodity should not be treated differently in terms of what phytosanitary measures are required. For example, all countries with *G. rostochiensis* should be required to meet the same phytosanitary requirements for commodities (that may carry the nematode) to enter a given country.

Finally, the principle of '**transparency**' requires countries to provide information regarding their risk analysis procedures, including information justifying why certain phytosanitary measures were selected.

By following the key principles outlined above, and ensuring that decisions are based on scientific evidence through risk analysis, countries benefit from a more predictable, transparent, and fair-trading system while at the same time enhancing their ability to protect their own agriculture and natural environments.

13.7.2. Practical problems

The risks of *B. xylophilus* and *M. chitwoodi* presented by different trade pathways have been assessed (Tiilikkala *et al.*, 1995; Braasch *et al.*, 1996; Evans *et al.*, 1996). Initially, the trade patterns involving *M. chitwoodi* concentrated on the most important economic host, potato, but subsequent research has highlighted other pathways involving other root vegetables, such as carrots, and ornamentals including flower bulbs. The knowledge that *M. chitwoodi* can be transported within potato tubers forms the basis for most inspection procedures, but in the field it is often difficult to detect low populations that produce few symptoms on potato skins, and immature females, most often found in fresh tubers, are opaque and difficult to see during inspections, necessitating laboratory testing in suspect cases.

Many species, such as *M. chitwoodi*, are not host-specific and pose real problems for quarantine specialists developing phytosanitary control measures. Such measures work best where the species in question has a narrow host range, a slow rate of population increase and a rapid rate of decline under non-host crops. PCNs and *T. semipenetrans* are good examples of such species.

13.8. Phytosanitary Measures for Plant-parasitic Nematodes

As indicated in the previous section on PRA, selection of the most appropriate phytosanitary measures for managing plant-parasitic nematodes (or any pest) requires a detailed study of their biology and associated trade pathways, including the trade of crop hosts between countries and environmental factors favouring their establishment. Both direct and indirect measures are adopted to prevent the spread of nematodes and other organisms listed for exclusion (Box 13.5).

Disinfestation measures target nematodes in imported plants or plant products. For example, wood packing material (WPM) is a pathway for the pinewood nematode, *B. xylophilus*. ISPM 15 approves heat and methyl bromide treatments for WPM in international trade to control *B. xylophilus*. Usually, measures against the beetle vectors of this species are not practical. Nematodes are, however, posing increasing problems for phytosanitary inspectorates when enforcement of control is required. There are fewer chemical treatments that can be employed, which raises the importance of preventative or exclusion measures to reduce likely losses.

Exclusion measures include prohibiting the entry of a nematode's plant hosts and associated plant material that have a high risk of being contaminated, regardless of whether the nematode has been detected. However, the most important measure for nematode exclusion is the prohibition of the import of soil, growing media and packing

Box 13.5. Examples of typical phytosanitary measures used to prevent the spread of plant-parasitic nematodes include:

- a specific quarantine treatment of an import commodity, or
- prohibiting the entry of a host commodity likely to carry a particular nematode, or
- prohibiting the entry of untreated media, or
- that plants be imported with bare roots (i.e. free of media), or
- that plants should have been grown in approved growing media according to an officially agreed protocol, or
- that the commodities should originate from an area determined to be free of a particular nematode (i.e. a 'pest-free area').

materials that may harbour nematodes. Virtually every country bans the import of untreated soil because it may carry not only unknown populations of nematodes, but also pathogenic bacteria, actinomycetes, fungi, mites, certain life stages of insects and weeds. Consequently, countries generally require that plants be imported free of media, i.e. with bare roots. Unfortunately, even bare-rooted plant material is often contaminated with some soil particles; therefore there is a lowered, but still partial, pest risk.

The bare root requirement is also designed to allow easy inspection for galls or lesions on the roots. Any drying effects during transport may reduce adult nematode populations on plant surfaces but the extent to which this occurs cannot be relied on for quarantine purposes; indeed, nematologists in quarantine laboratories regularly find live nematodes in samples of soil residues from consignments. In particular, drying of roots does not kill or reduce the viability of nematode cysts, nematodes living in roots or eggs in gelatinous matrices from the root tissue. Many nematodes, such as *Pratylenchus* species, develop and reproduce inside the root tissues and are protected from any drying effects that may occur on root surfaces.

Requiring that plants be grown in approved growing media according to an official protocol is another form of exclusion measure. These growing media vary by country; furthermore, many countries require such media (e.g. peat, rock wool or perlite) to be sterile or previously unused, as nematodes are able to colonize artificial media if roots are present.

13.8.1. Certification and marketing schemes

These international schemes are administrative systems for phytosanitary and quality control of commercially produced plant material, including plants for propagation and plant products for consumption. They can be applied to production of both true seed and vegetative propagation, as described in detail by Ebbels (2003). Whilst some schemes are more concerned with genetic purity and related aspects such as germination potential, most include some aspect of freedom from pests such as plant-parasitic nematodes (e.g. *D. dipsaci*).

Official pre-cropping and growing season inspections are involved and soil, other growing media or suspect plant material may be sampled for laboratory analysis. Cost:benefit ratios of such practices have rarely been shown to be very favourable for diseases (Ebbels, 1988) but, although similar assessments have not been done for plant-parasitic nematodes, such schemes are credited with maintaining high-quality nematode-free stocks in several plants such as narcissus and strawberry.

13.8.2. Phytosanitary certificates

Countries wishing to export plants, plant products or other regulated articles that could present a risk to plant health have to satisfy the phytosanitary requirements of the importing country before it will accept such consignments. This is documented by providing a Phytosanitary Certificate, which states that the consignment meets the phytosanitary requirements of the importing country and may include additional declarations regarding specific pests or procedures. Inspections of the regulated plants or plant products can be done under the authority of inspectors, but the Phytosanitary Certificate must be issued before transit, by an inspector who is a government official.

13.9. Phytosanitary Measures and their Associated Cost versus Benefits

In spite of the impact of multilateral trade agreements and phytosanitary regulations, interceptions of regulated nematodes do occur, such as on infected container-grown plants or in potato tubers. On such occasions the consignment may be destroyed, returned to origin, diverted to a different end use (such as consumption) or diverted to another country to prevent the spread of pests, as invariably no effective treatment or other control exists. A few cultural methods may be employed where the nematode species have been identified as low risk. For example, root washing may be an option in the case of ectoparasitic species, but this may damage the plants.

It is difficult to eradicate a nematode species of particular economic importance when an outbreak has occurred on a crop or range of crops. Suppression (i.e. the application of phytosanitary measures to reduce the pest population) or containment (i.e. the application of phytosanitary measures in and around an infested area to prevent spread of a pest) may be more feasible and such campaigns may be cost-effective. In particular, they set an example to the agricultural and horticultural industries of the principles of pest control. On occasions, the benefits may be difficult to justify in monetary terms but there may be other environmental concerns or advantages to the trade in general. Cost:benefit analyses rely heavily on the availability and quality of data but the problem with most nematode species is that little is known about their economic effects; however, some economic data are provided in this section.

The implementation of phytosanitary measures following an outbreak protects the surrounding agricultural land from the dissemination of the introduced nematode pest species. However, these quarantine measures, including the ban of the cultivation of the host crop and implementation of official control measures, may need to be enacted for decades as long-term studies have demonstrated that the juveniles of some plant-parasitic nematodes, such as cyst-forming types, may remain viable in their cysts for more than 30 years the absence of the host, thus challenging attempts at deregulation (Holdago *et al.*, 2018). This has become a controversial issue due to pressure on the regulatory agencies from new trade agreements and political or agricultural interests to relax such phytosanitary measures.

It remains in national and international interests to restrict the spread of harmful organisms and so minimize their impact on trade, industries and the environment. The following examples explain how certain nematode species are being suppressed and contained, using five different phytosanitary programmes and their associated cost:benefits, if they are available.

13.9.1. Phytosanitary suppression programmes for *Radopholus similis* in the USA

The programme to suppress and prevent further spread of the citrus race of *R. similis* in Florida is very important because of the phytosanitary legislation involving this nematode in many countries worldwide (see Section 13.5). In the early 1950s, a serious decline of citrus orchards appeared, primarily in central Florida. In 1953, it was discovered that *R. similis* was the causal agent of the problem, known as spreading decline. Nematode-infected propagative material from Florida nurseries infected with both *R. similis* and *T. semipenetrans* had been disseminated into Florida's new and old citrus-growing areas where these nematodes were not native.

The management strategies implemented in the affected areas included surveys to delimit these infested areas, tree removal followed by soil chemical treatment (nicknamed 'push and treat') and isolation of orchards with chemical-treated buffer zones (barriers) around the infested areas. Concomitantly with the 'push and treat' programme, a citrus nursery certification programme was established, which required nursery stock to be grown under rigorous sanitation programmes to ensure that commercial citrus seedlings were free of *R. similis* and *T. semipenetrans.*

The combination of the initial suppression programme (push and treat), the barrier programme to isolate the infested orchards and the citrus nursery certification resulted in a reduction of nematode-infested hectares from 3798 ha in the 1950s to 1538 ha in 1984. The number of citrus nurseries infested by *R. similis* declined drastically from 278 in the 1950s to none from 1970 until now. Recent studies indicate that *R. similis* would have spread at least to an additional 18,000 ha without implementation of the phytosanitary programmes. The benefit of preventing the spread of *R. similis* to such an area of land potentially susceptible to the spreading decline was estimated at about US$1.4 billion (€1.05 billion) for a 35-year period. The cost of the suppression programmes was about US$100 million (€75.2 million) or 7% of the benefit value (Lehman, 2004).

The eradication approach of 'push and treat' was discontinued in the 1980s because of the serious environmental consequences resulting from the excessive use of fumigant nematicides. The citrus nursery nematode programme, which was initiated 50 years ago, has continued and the benefits growers received in 2000 was estimated to be US$17 million (€12.8 million) for *R. similis* and US$33 million (€24.8 million) for *T. semipenetrans.* The cost to the citrus industry of implementing the citrus certification programme was US$75,000 (€56,390) (Lehman, 2004).

The direct damage that *R. similis* causes to the citrus industry in Florida has had a rippling indirect and adverse effect on the ornamental industry of Florida; unfortunately, both the banana and citrus races of *R. similis* (see Chapter 5) are able to infect and reproduce on many ornamentals in many different families including plants in the Araceae, Laurantaceae, Marantaceae, Musaceae (which includes banana), Palmae and Strelitziceae, although the banana race does not attack citrus plants (Rutaceae). The absence of reliable morphological and molecular identification tools to distinguish between the races has resulted in a ban on all infected plant shipments from Florida by all the citrus-producing states in the USA and by many countries in order to protect their banana, citrus and ornamental interests. For the same reason this ban has also applied to other tropical states, such as Hawaii, and other countries (regardless of the race they have).

A financial disaster for the ornamental industry in Florida has been averted by the adoption of a nematode certification programme in the ornamental industry similar to that described earlier for citrus nurseries. Florida ornamental nurseries implement nematode eradication programmes using strict phytosanitary practices in order to export ornamentals to national and international markets. These programmes include the use of approved growing media such as sterile peat, clean sand, sawdust or wood shavings, or biologically inert fillers such as perlite or vermiculite. Clean propagative material is grown in these media and kept in clean containers not in contact with the ground, which may be inhabited by prohibited pests and pathogens. Other common sanitation requirements include clean irrigation water, weed control, appropriate sloping of the ground to avoid flooding of the nurseries, and construction of cement slabs or benches to protect the containers and growing media. Nurseries are periodically (usually annually) inspected and sampled by officers of the Florida Department of Agriculture and Consumer Services, and if prohibited nematodes are found the nursery is suspended from shipping plants until the production line is clean. Such phytosanitary programmes are expensive and are justified only by the high market values of ornamentals (more than US$1 billion (€0.75 billion) in Florida).

The synergistic effect of management programmes for other pests can occasionally lead to incidental control of nematodes. Such is the case for the recent implementation of a new citrus nursery certification programme in Florida to control the Asian citrus psyllid (ACP) (*Diaphorina citria*), transmitter of the bacterium *Candidatus* Liberibacter asiaticus, which causes Huanglongbing (Yellow Dragon Disease or Citrus Greening Disease), one of the most devastating citrus diseases in the world. A new citrus production system that requires citrus propagative material to be produced in hermetic glasshouses, impervious to the ACP, and where the citrus resets are grown on raised benches in sterile conditions, has resulted in a significant reduction in plant-parasitic nematodes.

13.9.2. Nematode phytosanitary exclusion programme for cotton and vegetables

As in some other states of the USA, plant health officials in California conduct inspections at the state's border stations to verify that agricultural commodities are free from its state-regulated pests, including nematode pests. Here the cost of excluding non-indigenous nematode pests has been estimated to be 3% of the potential crop loss value these nematodes would cause if accidentally introduced into the state. For the year 2000, this potential crop loss value was estimated to be US$600 million (€450 million).

In particular, the benefits of excluding the reniform nematode, *R. reniformis*, from the California cotton industry (valued at US$1 billion (€0.75 billion)) are estimated at US$7.2 million (€5.4 million). These benefits were calculated by assuming similar crop losses (7%) to those that have occurred in other US states. Similar benefits are achieved for the California melon and vegetable industry, valued at US$4 billion (€3 billion), because this pest causes 10% crop losses to cantaloupe and snap bean elsewhere. The annual cost of the certification programmes implemented in Florida to exclude California's state-regulated nematodes from California is US$100,000 (€75,190), which represents only about 1.3% of the losses that the California cotton industry would expect annually if *R. reniformis* became established in their cotton fields.

13.9.3. Eradication programmes for *Globodera rostochiensis* and *G. pallida* in the USA

The first regulatory measures against *G. rostochiensis* in New York were implemented at state level in 1944. Stringent phytosanitary programmes were established by the Golden

Nematode Act, promulgated in the USA in 1948; these have prevented the spread of *G. rostochiensis* within and outside the state of New York since that time. In 2012, APHIS removed some of the regulated areas in this state. The use of seed potatoes originating from states where *G. rostochiensis* has not been found has been critical in preventing the spread of this nematode in the USA. An outbreak of *G. pallida* occurred in 2006 in Idaho, where exclusionary measures were also enacted at a cost of US$58.8 million (€44.2 million) from 2006 to 2012 (Skantar *et al.*, 2007; Anon, 2011). Considering that PCN species can suppress potato yields by more than 10%, the benefits of excluding these nematodes from potato-growing areas in the USA are estimated to be US$300 million (€225 million) annually at 1995 values. These benefits are far greater than the US$445,000 costs (€334,595) (1996 data) required for preventing the spread of *G. rostochiensis* in and outside New York (Dwinell and Lehman, 2004).

13.9.4. Regulatory programmes to limit the spread of *Meloidogyne enterolobii* in the USA and Europe

Meloidogyne enterolobii, or the Pacara earpod tree root-knot nematode, has acquired major economic relevance in many countries due to its ability to overcome root-knot nematode resistant genes in some economically important crops (Brito *et al.*, 2007; Kiewnick *et al.*, 2009). In the USA, after its first detection in Florida, *M. enterolobii* was found in North and South Carolina where it became an important pest of sweet potato (*Ipomoea batatas* (L.) *Lam.*) causing plant stunting, galls and cracking of storage roots (Rutter *et al.*, 2019). Serious crop losses have prompted the implementation of a quarantine programme to prevent the dissemination of this nematode in many states including Arkansas, California, Louisiana, Mississippi, and North and South Carolina. As with many root-knot nematode species, the requirement for the presence of several life stages of the species to enable correct identification, and the wide host range of *M. enterolobii,* make implementation of the nematode certification programmes a challenge, but essential if the ornamental industry is to market ornamental stock.

Within Europe, *M. enterolobii* continues to be intercepted, associated with commodities originating from Asia, Africa, North and South America. This species has been placed on the EPPO A2 pest list and is a recent addition to the IIA EU Plant Health Annex list. Outbreaks under protected cultivation in France (Blok *et al.*, 2002) and Switzerland (Kiewnick *et al.*, 2008) demonstrate pathways of introduction of this species into the European region. For export of seed potatoes from the UK into Europe, evidence the crop has been grown on land designated a 'pest-free area' of *M. enterolobii* must be provided as part of EU export requirements.

13.9.5. The introduction and spread of *Meloidogyne fallax* in the UK

Meloidogyne fallax is a I/A2 EU annex listed nematode species, with an EPPO A2 designation. UK outbreaks of *M. fallax* in agricultural land were first reported from a field of organic leeks in 2013; in addition, the nematode population was detected in the nearest field boundaries, parasitizing red clover, creeping thistle and field pansy. Two further outbreaks were recorded from field-grown organic carrots in 2018 and 2020, with the latter in a rotation that included sports turf production. Potato volunteers in these fields were found to be infected with *M. fallax*. At all three outbreak sites, plant waste and associated soil residues resulting from the on-site processing of vegetable consignments produced from other EU member states were applied as an organic amendment to the land. Following a review by the UK Department for Environment, Food and Rural Affairs (Defra), management practices to reduce the risk of spread further were followed, including improved hygiene measures, a fallow period and growing of non-host crops (plants for planting, root vegetables, seed or ware potatoes were not permitted). Following this regime, the 2013 outbreak population had dropped to below detectable levels in 2017. From 2015, *M. fallax* has been confirmed infesting the pitch turf of Premier League football stadia and club training grounds; other non-quarantine plant-parasitic nematodes have also been recorded. It is possible the pest was introduced to these areas during re-laying of the pitches, where the sward is removed annually by machine 'planing' before a top layer of sand/compost mixture is re-applied, or during movement of EU machinery used to reinforce the natural root-system with artificial turf fibres. Following trace forward investigations, companies that produced the compost used during relaying of the turf had imported organic material from within the EU. The growing matrix removed from pitches had also been used to improve the soil on a small number of agricultural fields in the UK. Hygiene measures including boot and machine washing have been stipulated by Defra, in addition to control methods being trialled, such as a garlic extract and steam sterilization. In January 2021, following EU exit, *M. fallax* is now a RNQP for Great Britain under GB Plant Health Legislation and statutory action is taken only when *M. fallax* is found in association with seed potatoes. Following the new legislation, one UK finding has been recorded from amenity turf for export.

13.10. Future Challenges for the Control of Regulated Nematodes

The volume, frequency and speed of transport of plants and plant products around the world mean that it is impossible to inspect adequately every consignment that might contain nematodes. Therefore, one of the main challenges facing NPPOs is to target inspections so those consignments that pose the most risk to the agricultural industry in their countries are examined to ensure that commodities that cannot be adequately inspected are otherwise restricted. Accurate statistics help to provide the basis for good planning of phytosanitary services (Ebbels, 2003). Besides the volume and distribution of work, which will indicate the numbers of staff needed, statistics will provide information on the volume and fluctuation of various types of trade, the problems encountered and the association of problems with particular sources, areas or suppliers (https://ec.europa.eu/eurostat). This, in turn, helps to determine priorities for targeting inspections or monitoring and provides information for PRA.

The trend towards sustainable farming systems and the development of sophisticated integrated pest management programmes, combined with the loss of many chemical products used to control nematodes, means that another future challenge is to develop a better understanding of the biology of plant-parasitic nematodes so that as many cultural measures as possible can be used to suppress them at their source. Such measures might include, for example, increasing the interval between susceptible crops to reduce the rate of multiplication of pest species. One of the stipulations of seed potato growing regimes in Europe is specified intervals between cropping, so that the majority of PCN second-stage juveniles would have hatched by the time of planting of the next host crop. In addition, the growing desire to use plant waste for composting presents an additional risk unless appropriate composting procedures are taken to kill any nematodes therein. Many phytosanitary programmes serve as model programmes for crop health management in agriculture in general, but their implementation and success require the support and cooperation of the agriculture industry.

13.11. Challenges Facing Scientific Advisers and Researchers

Scientific service support is essential for the plant health services in any country or region, and for a phytosanitary programme to work well there needs to be good cooperation between the regulatory agencies, crop consultants, and farmers and growers, as illustrated in this chapter by the success of programmes aimed at citrus and ornamentals in Florida.

The ever-decreasing skills in identification and diagnosis are in demand as an increasing number of national and international standards for plant health services are established, not only for the production of clean propagation material, inspection and sampling procedures, but also to provide the basis for PRA and eradication and containment programmes. At the same time there are increasing demands to formalize quality procedures in laboratories, leading to the production of identification protocols that provide guidance for international agreement. This section discusses some of the challenges in putting science into practice to comply with phytosanitary legislation.

13.11.1. Morphological and molecular identification tools, as influenced by the development of Integrative Taxonomy

The identification of plant-parasitic nematodes is still largely dependent on recording morphological features and subsequent judgements by nematode identification specialists or taxonomists. Such judgements may, of course, differ at any one time, but are important in a sector where perhaps only one or a few specimens may be isolated from a sample and where an international consensus over the organism causing problems is vitally important. Original descriptions are important tools, as are authenticated reference slide collections. However, the international decline in taxonomic skills and the lack of resources for curation and conservation of collections have led to concerns over the whole basis of identification of regulated nematodes (Hockland, 2005). An example of the importance of such issues was the international controversy over the use of names *M. mayaguensis* and *M. enterolobii* for international phytosanitary measures. This has been resolved by morphometric and morphological comparisons by experts so that *M. mayaguensis* is now regarded as a junior synonym of *M. enterolobii* (Karssen *et al.*, 2012).

In Florida, the nematode certification programme that enables the ornamental industry to market ornamental stock in the states that regulate this, and other nematodes, is based on morphological analysis. Samples infected with *Meloidogyne* spp., including *M. enterolobii*, usually contain second-stage juveniles (J2), which are not identifiable morphologically without time-consuming morphometric analyses that ultimately may not be conclusive enough to justify regulatory decisions. Long-lasting bioassay tests to obtain mature females that may be identifiable by morphological or enzymatic analyses are impractical. These difficulties in the identification of *M. enterolobii* have been overcome by the adoption of appropriate molecular analyses (Tigano *et al.*, 2010), and hence enable the accurate detection of *M. enterolobii* and the certification of Florida operations to market their products in states where the nematode is regulated.

However, the EPPO diagnostic protocols for *M. enterolobii* (Anon, 2016) and other nematode species illustrate the continuing influence and inclusion of nematode reference collections and morphological identification procedures in developing identification standards. They play a vital and evolving role in the identification process, and can comprise live, dead, nucleic acids or '*in silico*' material (DNA sequences). These collections are fundamentally important for supporting correct identification of plant-parasitic nematodes and for taxonomic descriptions and revisions. Collections are also used for the production of reference material needed for the development and validation of diagnostic tests, inter-laboratory comparisons (test performance studies and proficiency tests), equipment calibration and the production of positive controls used in routine testing. Nematode collections are also important for training and quality assurance purposes (Anon, 2021).

Thus, plant health nematologists have, for many years, had to integrate information from different types of data and methodologies to confirm identification, and are in the forefront of **Integrative Taxonomy**, a formal name introduced in 2005 to reflect this multidisciplinary approach (Pante *et al.*, 2015). It has become essential for the accurate identification of plant-parasitic nematodes. It has provided evidence that important species on quarantine lists actually consist of morphologically and molecularly similar species having different host preference and hence regulatory significance. Whilst many questions in relation to such species remain, Palomares-Rius *et al.* (2014) have indicated that the proportion of these species is almost evenly distributed among major plant-parasitic nematode taxa, and suggest investigations into speciation according to host will be a major field of discovery of new species in the future as molecular techniques are increasingly incorporated into many diagnostic or taxonomic laboratories.

13.11.2. Molecular tools and their role in detection

An array of different technologies has been developed over the years, and is still evolving, to help specialists in morphological identification, especially in instances where identification is particularly difficult, or where only immature specimens (for which identification keys are rare) have been intercepted. They include polymerase chain reaction (PCR) and electrophoresis (see Chapter 2) and high-throughput sequencing (HTS). The application of HTS technologies requires the development of bioinformatics expertise to collect and manage sequence data, and the correct choice of tools for studies, e.g. detection versus intraspecific variation versus deep level phylogenomics. However, the new sequencing technologies are critical for studying recent species divergence as they utilize additional molecular markers to determine genetic differentiation, as it can proceed at different speeds throughout the genome (Palomares-Rius *et al.*, 2014). These technologies are now very reliable when used by specialists with expertise in both morphological and phylogenetic analyses. However, a preliminary, provisional identification by a morphological specialist is still essential to detect an unusual finding, for which few or no studies may have been made.

Analytical methods examining the genetic make-up of organisms are being continually refined and adapted to develop new phylogenetic models that are becoming an integral part of nematode systematics (De Ley and Blaxter, 2002). The associated technological equipment, though expensive, is becoming a familiar part of most diagnostic laboratories. Although advances in the use of molecular methods for the identification of diseases, especially viruses, has been significant, the pace of development of reliable, accredited diagnostic protocols in nematology has been relatively slow for species listed in legislation, although it is now advancing steadily. Recently, EPPO has taken a leading role in both development and accreditation practices, and collaborated with the European Co-operation for Accreditation (EA) to achieve greater cooperation in raising standards (https://www.eppo.int/RESOURCES/eppo_standards/pm7_diagnostics).

The emergence of a range of *Meloidogyne* species with the potential to cause economic damage (*M. enterolobii*, *M. ethiopica*, *M. floridensis*, *M. graminicola* and *M. minor*) has prompted more research in biochemical and molecular tools, but personnel in plant health services should have a full understanding of the limits of molecular technology; their real value for the future probably lies in the provision of screening tools, which would indicate any requirement to check identities further. This is because most current protocols for distinguishing regulated species may not include unregulated, native species of the genera that occur in the countries where interceptions or outbreaks occur, or new species that might be imported from elsewhere. Thus, plant health identification and detection services need to encompass a range of scientific skills, old and new, if unnecessary statutory action is to be avoided, e.g. for other unregulated species of *Bursaphelenchus*, *Globodera* or *Meloidogyne*. Furthermore, research into the use of molecular tools for detection of plant-parasitic nematodes in soil needs to continue to be developed with care and the implications fully understood, if the status of infestations is to be truly represented. Thus, the integrated role of experienced diagnosticians, taxonomists and molecular scientists in nematode identification and detection for phytosanitary services remains a vital one.

13.11.3. Science versus legislation

Science continues to provide new information about the bionomics of species, especially identification. This poses problems for phytosanitary legislation, which requires clarity and consistency to avoid misinterpretation, but it often takes time to publish amended legal documents. The names and identity of regulated plant-parasitic nematodes need to be established; however, as with other plant pests, this invariably poses a problem in taxonomy where the taxonomic details of some nematodes are frequently changing in line with new species concepts. Consequently, this demands an awareness that some species might be subject to many taxonomic changes and there may exist many synonyms in the legislation of some countries; this needs to be recognized if confusion is to be avoided and if correct phytosanitary action, including control, is to be taken to avoid unnecessary economic losses.

The controversy surrounding *R. citrophilus* and *R. similis*, which are both listed in European legislation, is a case in point. *Radopholus similis* was thought to consist of different pathotypes but Huettel *et al.* (1984) concluded that the banana race and the citrus race were distinct species; the name *R. similis* was restricted to the banana race and the citrus race was described as *R. citrophilus*. Subsequently, Kaplan *et al.* (1997) synonymized *R. citrophilus* with *R. similis*; Valette *et al.* (1998) proposed *R. citrophilus* as a junior synonym of *R. similis*, although in 2000 Siddiqi proposed it as a subspecies of *R. similis* and Elbadri *et al.* (2002), using molecular techniques, demonstrated marked intraspecific variation in various isolates of *R. similis*.

Similarly, *Aphelenchoides besseyi* is a well-known, often-listed pest of rice but on other crops it is likely to be confused with other species (https://www.ippc.int/static/media/files/publication/en/2016/11/DP_17_2016_En_2016-11-01_iaK6Hls.pdf). Recently morphologically similar specimens of *Aphelenchoides* species found on strawberry indicated an *A. besseyi* complex (Subbotin *et al.*, 2021b). Taxonomic uncertainty may cause more confusion for quarantine specialists involved in PRA work, as, for example, the host lists previously attributed to such species have to be used with care. Likewise, the controversy over the names *M. enterolobii* and *M. mayaguensis* illustrated that even though identification issues may have been resolved (see Section 13.11.1), studies of assessments of risk need to consider literature for both species. Such difficulties require the expertise of taxonomists, whose numbers are sadly in decline but who provide the essential framework for taxonomy and identification by developing species concepts and theories for the classification and identification of organisms, and hence determine correct names, set standards for descriptions, determine key morphological characters, develop identification keys and catalogue data such as those for distribution.

Thus, the integration of morphological and molecular advances in identification can also result in scenarios that test phytosanitary legislation; it needs to remain directed at the damaging genotypes (or, more strictly, the absent damaging genotypes), and if these are difficult to identify or in flux then perhaps it is best that legislation continues to remain cautious, with, for instance, nematodes included in the *X. americanum* group being designated as *Xiphinema americanum sensu lato* rather than as individual, nominal species.

In 2008, *Globodera* spp. populations able to parasitize potato, but genetically distant from *G. pallida* and *G. rostochiensis* (well known collectively as PCN), were reported in Oregon and Idaho, USA (Skantar *et al.*, 2011). They were described as a new species, *G. ellingtonae* (Handoo *et al.*, 2012), and populations subsequently detected in Argentina and Chile have been studied in recent years (Hesse *et al.*, 2021). Although this species is also becoming known as a PCN, no phytosanitary action against *G. ellingtonae* has been implemented internationally at the time of writing. The recent implementation of a new PCN Directive in Europe has raised the issue of the identity of PCN pathotypes, or rather populations, which exhibit different pathogenicity in various parts of the world, especially in their hub of diversity in South America. It is essential that the potato cultivars bred and used in Europe with resistance only to the PCN populations that exist there are not exposed to new types of biodiversity (Anon, 2012; Hockland *et al.*, 2012). Science has an important role to play in providing evidence to this effect.

In the USA, *P. coffeae* is a regulated species in California and Florida to protect the citrus and other agricultural industries. Phylogenetic studies of Florida populations, identified morphologically as *P. coffeae*, showed that they are cryptic species that are feeding on grasses rather than citrus and therefore do not meet the definition of a quarantine pest (De Luca *et al.*, 2010). Other *Pratylenchus* species in Florida that were incorrectly identified morphologically include a population of *P. scribneri* from *Amaryllis* and *P. penetrans* from a fern species that were shown to be, by using phylogenetical analyses, new species, *P. hippeastri* and *P. bolivianus*, respectively. Recently, the regulated

Aphelenchoides besseyi sensu lato has been found to be, by using integrative taxonomy, a species-complex consisting of *A. besseyi sensu stricto*, *A. orzyae* and a new species *A. pseudobesseyi*, having different host preference (Subbotin *et al.*, 2021b). Thus, the rapid progress that has been made in the identification of plant-parasitic nematodes using these integrative approaches complicates the promulgation of appropriate phytosanitary legislation and requires international agreement.

13.11.4. The future of diagnostics

Whilst the highest standards of delivery have always been the aim of diagnostic laboratories worldwide, the development of international standards has placed increasing demands on attaining prescribed levels of quality in the delivery of services and research. International standards for phytosanitary measures and also for diagnostics have become increasingly important, but their adaptation in some areas, such as the identification of species, which entails the use of judgment by experts rather than the output from machines, has proved a difficult philosophy for accreditation schemes to embrace. In addition, the variability of resources available in individual laboratories means a range of proven protocols has to be included. Nevertheless, selected protocols are slowly achieving international status, and the establishment of national and international reference laboratories illustrates continuing efforts to harmonize and coordinate standards in plant health diagnostics (Anon, 2017).

The combination of scarce scientific resource and the cost of providing prescribed levels and speed of delivery have led some countries to negotiate contracts for science services with those countries that still have the ability to deliver. Inevitably, this will lead to centres of expertise serving a community in a particular geographical location or region. Whilst this may have economic advantages, it should not discourage the broad development of essential identification and diagnostic expertise that is vital for the whole basis of phytosanitary work.

Phytosanitary services continue to embrace the full potential offered by advances in molecular science, computer technology and the internet. The demise of taxonomic expertise at a local level is stimulating the creation of databases and networks to take advantage of scarce skills at short notice and RPPOs, such as EPPO, have a role to play in facilitating this (https://www.eppo.int/RESOURCES/eppo_databases/global_database). The global plant health community is becoming ever more closely connected through the internet and especially social media. Thus, it is hoped that the shared experience and expertise in this sector will result in the widespread development and application of new quality standards, and hence result in improved understanding of plant-parasitic nematodes for phytosanitary services. Effective phytosanitary services cannot be provided without the close international cooperation between nematologists with expertise on nematode taxonomy and biology, and scientists with skills in nematode phylogeny and molecular science.

Acknowledgements

The authors would like to acknowledge the valuable help in writing this chapter received from numerous colleagues working for phytosanitary services worldwide. Thanks are also due to staff at Fera Science Limited and the Department for Environment, Food and Rural Affairs (Defra), UK, for supporting Thomas Prior in quarantine nematology, and in particular Alan MacLeod (Defra, UK) for extracting import data. We would like to pay particular thanks to Paul Lehman and Leah Millar, Jonathan M. Jones, National Program Manager, USDA-APHIS-PPQ, for data provided for Section 13.9.3, and to Bob Balaam, National Operation Manager, Exclusion and Imports, USDA, APHIS, PPQ and Eunett James-Mack, Florida Agriculture Liaison, Customs and Border Patrol, US Department of Homeland Security for providing Florida import data for this publication.

14 Biological and Cultural Management*

Nicole Viaene[1]**, Danny L. Coyne[2] and Keith G. Davies[3]

[1]*Plant Unit, Flanders Research Institute for Agriculture, Fisheries and Food (ILVO), Belgium and Department of Biology, Ghent University, Belgium;* [2]*International Institute of Tropical Agriculture (IITA), Nigeria and Department of Biology, Ghent University, Belgium;* [3]*School of Life and Medical Sciences, University of Hertfordshire, UK and Department of Biology, Ghent University, Belgium*

---

*A revision of Viaene, N., Coyne, D.L. and Davies, K.G. (2013) Biological and cultural management. In: Perry, R.N. and Moens, M. (eds) *Plant Nematology*, 2nd edn. CAB International, Wallingford, UK.
**Corresponding author: nicole.viaene@ilvo.vlaanderen.be

DOI: 10.1079/9781800622456.0014

14.1. Introduction

Biological control of nematodes principally concerns the exploitation of microbial agents. As cultural methods of crop protection, including cultivation and applications of organic matter (soil amendments), have profound effects on microbial abundance, diversity and activity in soil, it is appropriate that both approaches to nematode management are considered together. Several aspects of biological and cultural control have been reviewed (Stirling, 2011; Davies and Spiegel, 2011).

This chapter is directed towards the **management** of nematodes as opposed to their **control**. Control implies the use of a single measure to reduce or eliminate nematode pests, which in most cases is not possible, whilst management involves the manipulation of nematode densities to non-injurious or sub-economic threshold levels using several measures with consideration of the whole production system; maintenance of diversity is an objective of management but not of control (Brown and Kerry, 1987). Consequently, of major importance is the need to take into consideration the impact of the nematode management strategies on biodiversity and the ecological balance in the soil.

Biological control is defined here as the management of plant diseases and pests with the aid of living organisms. This definition includes predators and parasites of organisms that kill or damage their hosts and also microbes that indirectly influence the establishment, function and survival of pathogens and pests. The plant may have significant influences on these interactions but plant features, such as root diffusates, are not considered as a component of biological control. Neither is plant resistance discussed here as a cultural management tactic because it is the main topic of Chapter 15. It should be noted, however, that resistant rootstocks can influence the soil microbial populations around roots. Grafting of preferred, but nematode-susceptible, cultivars onto hardier, nematode-resistant rootstocks has long been an accepted practice for perennial crops, such as coffee, and for annual crops, such as fruit-bearing vegetables. More direct changes in the plant rhizosphere through the incorporation of organic materials, which influence microbial communities and help manage diseases and pests, are considered as **cultural control** in this chapter. The link between biological control and cultural methods becomes evident when the exploitation of natural enemies as biological control agents for the management of plant-parasitic nematodes is discussed. Both biological and cultural control methods are important for the biomanagement of pests through the combination of measures other than chemical pesticides to improve soil quality and plant health. The application of bioactive compounds, which are derivatives of plants or metabolites derived from fungi or bacteria used as bio-nematicides, are discussed (see Chapter 17) but are not considered as biological control. Although many organisms derive their nutrition from nematodes, the most studied natural enemies of nematodes are bacteria and fungi.

There is considerable public and legislative pressure to reduce the use of nematicides because of their potential health and environmental risks, but few organisms have been developed as practical control agents and none is in widespread use. Apart from microbially induced suppression of nematode pests, research shows that biological control agents generally provide too little control to be effective alone and their successful use in sustainable management strategies will depend on their integration with other control measures. These include the use of genetic resistance, quarantine measures and various cultural control practices. Numerous reviews on the use of cultural control practices for nematode management are available (e.g. Chen *et al.*, 2004; Perry *et al.*, 2009; Sikora *et al.*, 2018, 2021). A single management option rarely leads to the sustainable management of a nematode problem. A successful nematode management strategy ideally will involve the selection of a combination of complementary components, providing they are applicable, appropriate and economical. Furthermore, a successful strategy should be flexible to pest and disease changes, as cropping systems and knowledge evolve over time. Selection of a practice inevitably depends on a multitude of considerations, not least the scale of the cropping system; a smallholder farmer in Africa, for example, has wholly different criteria against which to assess a situation than does the large-scale intensive cereal farmer in North America or the highly intensive ornamental crop producer in Europe. Thus, biological and cultural nematode management options described here are not necessarily universally applicable and must be adapted to meet local or individual needs. Such considerations are especially true for exploiting cultural control methods, such as the use of bulky soil amendments that may be only locally available. Moreover, much of the information on biological and cultural control practices emphasizes the importance of an understanding of the target plant-parasitic nematode including its identification, hosts and environmental preferences. However, small-scale farmers, especially those in developing countries, who would benefit much from these management strategies, have little knowledge or access to such information. Therefore, they need technical guidance and assistance from well-informed extension people with adequate training in nematology, which, in many circumstances, is also sorely lacking.

14.2. Suppressive Soils

Soils are considered suppressive when nematode multiplication on susceptible crops is less than that normally observed on the same cultivar in another soil, in similar abiotic conditions. Many soils that are suppressive to nematodes are often not recognized as such, because nematode problems have not occurred for several years. However, observations on abnormal declines in nematode populations or their damage in fields where plant-parasitic nematodes used to be a problem have led to the discovery of natural biological control and the organisms involved. Methods have been developed to assess suppressive soils (Box 14.1).

Most organisms found to be involved in nematode suppression are nematophagous fungi (e.g. *Pochonia chlamydosporia* (syn. *Verticillium chlamydosporium*), *Hirsutella rhossiliensis*, *Dactylella oviparasitica* and *Trichoderma* spp.) and bacteria (e.g. *Pasteuria penetrans*) that parasitize their nematode hosts. Microbes that compete for nutrients, produce toxins or

Box 14.1. Methods to assess the extent and nature of suppressive soils.

- Comparison of nematode multiplication in untreated soil and soil treated with a chemical or physical biocidal treatment.
- Inducing suppression by the transfer of microbial agents from a suppressive to a conducive soil.
- Addition of different numbers of infective nematodes to a conducive and a suppressive soil and compare nematode reproduction.
- Correlation of population densities of specific organisms in soils with increased levels of suppressiveness using biological, culturing and molecular methods to measure their abundance and activity (Borneman *et al.*, 2004).

induce host resistance, such as some rhizosphere bacteria (e.g. *Pseudomonas* spp., *Agrobacterium radiobacter* and *Bacillus subtilis*) may reduce nematode damage but may not provide the long-term control of nematode populations associated with suppressive soils.

Studies of suppressive soils have led to the identification of several important potential biological control agents, which may be abundant and occur in many fields and orchards, and even in natural environments (Stirling, 2011). Natural biological control may be widespread but, in most cases, it is inadequate to keep nematode populations below their damage threshold densities for crops grown in most agricultural systems. The challenge is to determine the specific conditions that are conducive for biological control agents to work in agricultural systems so that we can manipulate the environment, or the biological control agents, to suppress population densities of plant-parasitic nematodes adequately.

14.3. Biological Control Agents

It is not possible to include here all the natural enemies of nematodes that have been described and studied. Instead, we give an overview of the major groups with potential as biological control agents and include details of some of the most studied organisms.

14.3.1. Predators

Predatory nematodes, mites, insects and other invertebrates, such as some tardigrades and collembolans, feed on nematodes. Predators are common in soil and can also feed on other organisms. Predatory nematodes, belonging to the Mononchida, Dorylaimida, Diplogasterida or Apehelenchida, vary greatly in their biology, feeding mechanisms, prey specificity and habitat. Several species, e.g. *Mononchoides aquaticus* and *Fictor composticola*, have been studied in detail (Kanwar *et al.*, 2021). They can destroy tens to hundreds of nematodes per day using their teeth, enzymes or toxins.

Micro-arthropods, such as mites and springtails, also have a role in the regulation of nematode populations, especially in natural ecosystems where they may be abundant. However, the lack of specificity for plant-parasitic nematodes of most predators makes them unsuitable for use in biological control programmes aimed at specific nematode pests. Even though many predatory nematodes have short life cycles and can be cultured *in vitro,* their mass production and delivery to the soil is considered a challenge as they need to be fed with prey that also needs rearing. The use of predators as biocontrol agents has not been fully explored and could offer some niche application, e.g. species feeding on eggs of plant-parasitic nematodes might be useful where *Meloidogyne*, and other damaging genera with high egg production, is a problem.

14.3.2. Nematophagous fungi

Some nematophagous fungi are obligate parasites (Box 14.2), which need nematodes to survive, others are facultative or opportunistic parasites (Box 14.3), which can survive saprophytically, and others have characteristics that are intermediate between these two categories. Nematophagous fungi may be most readily divided into those that have extensive hyphal growth outside their hosts, such as the nematode-trapping fungi, and the opportunistic parasites of nematode eggs and those fungi that are mainly endoparasitic.

Box 14.2. Obligate parasites.

Types of organism: *Nematophthora gynophila*
Mode of action: parasitism via adhesive spores.

Advantages:
- virulent against active and sedentary nematodes;
- resting spores with long shelf life;
- responsive to pest densities.

Limitations:
- difficult to produce *in vitro*;
- limited spread in soil;
- narrow host range.

Box 14.3. Facultative parasites.

Types of organism: Trapping fungi; parasites of nematode eggs.
Mode of action: Parasitism via traps produced on modified mycelium and/or penetrative hyphae.

Advantages:
- easily produced *in vitro*;
- some species rhizosphere competent;
- wide host range.

Limitations:
- may be difficult to regulate switch from saprophytic to parasitic activity;
- efficacy dependent on plant species, nematode host and other soil conditions that affect saprophytic growth;
- several species do not form resting structures and may be difficult to formulate.

The trapping fungi immobilize nematodes using non-adhesive traps or sticky structures usually produced on mycelia before they infect their host. Opportunistic fungi colonize the rhizosphere and attack the sedentary stages (females and eggs) of nematodes developing on plant roots. The endoparasitic fungi penetrate nematodes after germination of their adhesive spores, which attach to the nematode cuticle.

Nematode-trapping fungi ensnare active nematodes using one or more types of mycelial trap. For example, *Arthrobotrys dactyloides* uses constricting rings, *Dactylella candida* makes non-constricting rings and adhesive knobs, *Monacrosporium cionopagum* forms adhesive branches and a two-dimensional adhesive network, whilst *M. ellipsosporum* traps nematodes with adhesive knobs or adhesive branches that may form loops. Some nematode-trapping fungi are good saprophytic competitors but trap few nematodes, while others are efficient in capturing nematodes but do not establish well in soil. Little is known about their growth and development in soil, especially the triggers that cause the switch from the saprophytic to the parasitic phase. The trapping fungi apparently need a carbohydrate source to proliferate but other factors, such as those leading to mycostasis (inhibition of spore germination and/or fungal growth), also play a role in their abundance and trophic state in soil. Genomics research has begun to exploit the key molecular mechanisms involved in the interaction between trapping fungi and nematodes (Tunlid and Ahrén, 2011). The genome sequencing of *A. oligospora* (strain ATCC 24927) (Yang *et al.*, 2011) paved the way for understanding the genetic background and proteins involved in the formation and function of the specialized predaceous structure and virulence determinants of this microorganism and provided insights into the evolution and the nature of parasitism. Difficulties in the establishment of nematode-trapping fungi in the soil, their limited trapping activity and especially their non-specific capture of plant-parasitic nematodes reduce their potential in biological control. However, specifically for *A. oligospora*, hopes are still high to make this fungus a successful biocontrol agent by selecting, engineering and recombining the best performing gentotypes, based on the knowledge gained through the many genomic and proteomic investigations in the last decade (Wang *et al.*, 2023).

Arthrobotrys oligospora is the best known and most studied nematode-trapping fungus (Wang *et al.*, 2023). It makes a three-dimensional hyphal network to trap soil-dwelling nematodes. The network is coated with an adhesive that contains lectins that bind to specific carbohydrates present on nematode cuticles. The network attracts nematodes more than vegetative mycelium and the presence of nematodes induces trap formation. However, most studies were performed *in vitro* and it is not clear if these interactions are relevant in soil. Immobilization of the nematodes may be enhanced by a nematotoxin produced by the trap. Once the nematode is trapped, a hypha penetrates the cuticle within 1 h and forms an infection bulb. Penetration is thought to be mostly mechanical but collagenase might play a role. Assimilative hyphae develop and the fungus colonizes the nematode body. Later, conidiophores develop from the cadaver and bear conidia in a succession of clusters. *Arthrobotrys oligospora* is a good saprophytic competitor capable of using a wide range of carbohydrates. As with most trapping fungi, it captures all kinds of nematodes. The fungus is also able to colonize plant roots and cause cell wall modifications without affecting plant growth.

Many fungi infecting vermiform plant-parasitic nematodes form adhesive spores that adhere to the cuticle of the nematode when it passes the fungus. These fungi are mostly obligate parasites and poor saprophytic competitors in soil but generally have a broad nematode host range. For example, *H. rhossiliensis* (Fig. 14.1) was found to reduce nematode invasion and, consequently, populations of *Meloidogyne javanica*, *M. hapla*,

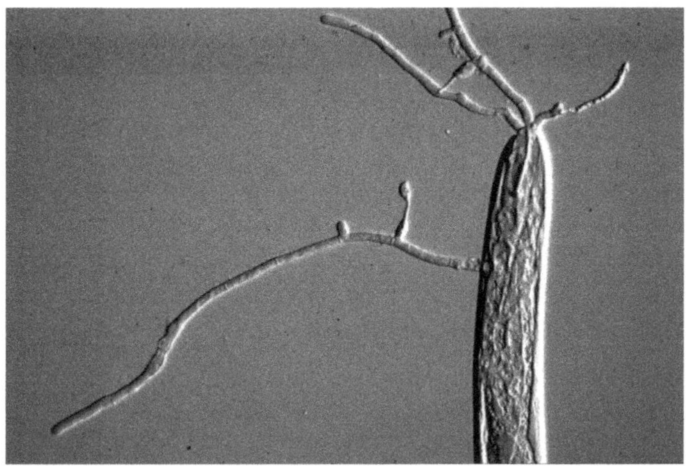

Fig. 14.1. Sporulating hyphae of *Hirsutella rhossiliensis* growing out of an infected juvenile of *Heterodera schachtii.* (Courtesy B. Jaffee, University of California, Davis.)

Heterodera glycines and *Criconema xenoplax*. Other examples of similar endoparasitic fungi that have been studied more closely are *Drechmeria coniospora*, *Nematoctonus* spp. and *Verticillium balanoides*. Not all endoparasitic fungi produce adhesive spores; some form zoospores that swim to the nematode cuticle, attach and encyst, often around the natural openings of the host. These encysted zoospores form a penetration tube that enters the nematode body, which the fungus colonizes by hyphal growth. The hyphae differentiate into sporangia where zoospores are produced. Some zoosporic fungi, such as *Catenaria anguillulae*, which attack vermiform nematodes, are believed to be opportunistic but others, such as *Nematophthora gynophila* and *Catenaria auxiliaris*, infect sedentary young female cyst nematodes and are obligate parasites. Zoospores require water-filled pores to be active and nematode infection is much affected by soil moisture levels. As many parasites of vermiform nematodes and the zoosporic fungi are difficult to culture *in vitro* and establish in soil, they are considered unsuitable for large-scale application as biological control agents.

By contrast, facultative fungal parasites of sedentary stages (eggs, developing juveniles and females) have attracted most interest because of their potential as biological control agents. Their target pest is immobile, and thus easier to infect, and these facultative parasites are able to survive saprophytically in the rhizosphere and most are relatively easy to mass-culture. They are not as specialized as the fungal parasites attacking soil-dwelling nematode stages and usually infect their host by simple hyphal penetration, sometimes with the formation of an appressorium. Many kinds of fungi have been isolated from sedentary stages but few have been studied in detail. *Cylindrocarpon destructans*, for example, is found regularly, as are species of *Fusarium*, *Gliocladium*, *Pochonia* and *Trichoderma*, all opportunistic fungi whose pathogenicity to nematodes may differ considerably between isolates. *Purpureocillium lilacinum* (syn. *Paecilomyces lilacinus*) is a well-studied parasite of a number of nematodes, including *Radopholus similis* and *Tylenchulus semipenetrans*, but most research has focused on the parasitism of *Meloidogyne* spp. and *Globodera rostochiensis* eggs. The fungus is abundant and active in subtropical and tropical regions. It is effective in reducing nematode damage to a range of crops in field trials and has been widely evaluated and developed by several small companies around the world. Soil application of the fungus has often resulted in variable levels of nematode control, although improvements through commercial development indicate that more consistent and promising results may be obtained. It is now marketed in various countries (e.g. BioAct®WP and Pl Plus®) for cyst and root-knot nematode management.

Pochonia chlamydosporia parasitizes females and eggs of cyst and root-knot nematodes, false root-knot nematode (*Nacobbus* spp.) and reniform nematodes (*Rotylenchulus reniformis*). Hyphae penetrate eggs after formation of an appressorium on the eggshell. A serine protease and chitinases, which degrade the eggshell, and a nematotoxin, phomalactone, produced by *P. chlamydosporia* may be involved in pathogenicity. Chlamydospores, which are resilient to environmental extremes, are produced as a survival structure and are used as an inoculum to establish the fungus in the soil and rhizosphere. Isolates of the fungus differ greatly in their ability to produce chlamydospores, colonize roots and infect nematodes. A range of biological (e.g. dilution plating on a selective medium) and molecular (e.g. real-time polymerase chain reaction (PCR)) methods have been developed to monitor the occurrence, abundance and activity of the fungus in the soil, rhizosphere and nematode egg masses. Rhizosphere colonization differs with plant species and is improved when the plant is infected by nematodes. The fungus also has host preferences: isolates obtained from root-knot nematodes are less able to infect cyst nematode eggs than isolates originating from cyst nematodes (Kerry and Hirsch, 2011). Single applications of the fungus at rates of 5000 chlamydospores g^{-1} of soil have provided control of root-knot nematodes on vegetable crops in tropical soils but results in Europe have been less satisfactory. The fungus is more effective when applied to low, pre-cropping densities of root-knot nematodes, preferably on poor hosts for the nematode, than when applied to large nematode infestations on highly susceptible crops. The factors that control the switch from saprophytic to parasitic activity are poorly understood. Even though *P. chlamydosporia* may be more abundant in soil with large amounts of organic matter, its parasitic activity may be no greater than in a mineral soil. Formulations based on fungal hyphae and conidia have been developed but chlamydospores are the most robust form of inoculum. Toxicological tests on chlamydospore-based inoculum have been successfully completed and the fungus is commercially available in some countries (e.g. Klamic® and Pochar®) (Manzanilla-López and Lopez-Llorca, 2017).

Trichoderma spp. commonly occur in soil and have received increasing attention for nematode management; isolates have been identified that successfully antagonize and control a wide range of plant pathogens, including nematodes (Sharon *et al.*, 2011). Research on various species and strains, e.g. *T. asperellum*-203, *T. asperellum* TR900 and *T. atroviride* IMI 206040, has shown control activity against nematodes, such as *R. similis* and, in particular, *Meloidogyne* species. Variability in the efficacy of different strains has been shown, with the application of strain combinations leading to increased levels of nematode control in some cases. Studies focusing on conidial attachment and parasitism reveal that a wide range of lytic enzymes (proteases and chitinases) are produced, the induction of which is likely to be environmentally mediated (Sharon *et al.*, 2011). Several mechanisms of action of *Trichoderma* have since been recognized, including mycoparasitism, antibiosis, enhanced plant tolerance against abiotic stresses and stimulation of plant-defences against pathogens.

When searching for strains suitable for the management of root-knot nematodes, studies indicate that isolates from egg masses may be more effective than those from soil.

Furthermore, assessing the genetic potential to produce chitinolytic enzymes was suggested as an additional selection criterion.

14.3.3. Endophytic fungi

Endophytic fungi (Box 14.4) grow within plant tissues without causing disease. Arbuscular mycorrhizal fungi (AMF), e.g. *Glomus* spp., are the best-known endophytes associated with plant roots; they form obligate symbiotic associations with plants. Their role in protecting the plant from nematode damage and in reducing nematode densities in the soil has been studied in different plant–nematode interactions, although most involve *Meloidogyne* spp. (Fig. 14.2) with relatively few studies looking at cyst and migratory nematodes (Schouteden *et al.*, 2015). AMF enhance plant growth by improving plant access to nutrients, especially P, and particularly under conditions of poor nutrient availability. AMF also aid access to and uptake of water by increasing root growth and branching. They can alleviate heavy metal toxicity and suppress pest and disease damage, including that of nematodes, by direct competition for nutrients and space, but also by induced resistance. They may also interfere with the production of root diffusates or produce nematotoxic compounds (Schouteden *et al.*, 2015). AMF are produced commercially as crop growth enhancers.

A range of fungi have been found to colonize plants endophytically and successfully reduce nematode infection and damage on various crops. Such fungi include *Neotyphodium* spp., *Trichoderma* spp., *P. chlamydosporia*, *P. lilacinum* and even non-virulent strains of *Fusarium oxysporum*. Their mode of action may rely on a toxic mechanism to reduce nematode infections in roots, induction of resistance or through competition for space. Increasing interest in endophytic pest and disease management has broadened the scope of research activity, with various fungal and bacterial isolates being assessed for potential use. The levels of nematode management can be acceptable, although isolates of the same species differ markedly in their efficacy against nematodes. Few, if any, commercial products

Box 14.4. Endophytes.

Types of organism: arbuscular mycorrhizal fungi (AMF) and other non-pathogenic root-colonizing fungi and bacteria.
Mode of action: competition for space and nutrients in roots and/or antagonism, increased (systemic) host defence/tolerance.

Advantages:
- active against wide range of nematodes including migratory endoparasites in roots;
- may suppress nematode colonization and multiplication;
- may promote plant growth and reduce nematode damage;
- easily produced *in vitro* and formulated;
- may be applied as a seed/seedling treatment.

Limitations:
- non-mycorrhizal fungi are closely related to plant pathogens and may be difficult to register for release;
- activity affected by crop cultivars.

based purely on endophytic activity are currently available, although isolates of *F. oxysporum* active against *R. similis* have undergone development and field testing on banana plantations in Central America and East Africa.

14.3.4. Bacteria

Most bacteria that interfere with nematode behaviour, feeding or reproduction do so indirectly by producing antibiotics, enzymes or toxins (Box 14.5). Many products, such as volatile fatty acids and nitrogenous substances, are formed by bacteria during

Fig. 14.2. Five-week-old soybean plants infected with 1000 *Meloidogyne incognita* per pot and treated with the nematicide carbofuran (left), with the control fungi *Glomus mosseae* and *Trichoderma pseudokingii* combined (right) and without treatment (centre). (Courtesy D. Coyne.)

Box 14.5. Antagonists.

Types of organism: rhizosphere bacteria.
Mode of action: toxins; modification of root diffusates and/or induced resistance.

Advantages:

- easily produced *in vitro*;
- may be applied as a seed/seedling treatment;
- active against a wide range of plant-parasitic nematodes;
- may promote plant growth and reduce nematode damage.

Limitations:

- may be effective for a relatively short time;
- no direct effect on nematode multiplication;
- activity affected by crop cultivar and soil environment.

decomposition of organic materials and may influence nematode populations in the soil and rhizosphere. Screening rhizobacteria or their metabolites (extracts of their cultures) in Petri dishes has led to the discovery of bacterial strains with strong antagonistic properties. However, the production and importance of such metabolites in the rhizosphere is not clear. *Burkholderia* spp., *Pseudomonas* spp., *Bacillus* spp. and *A. radiobacter* may reduce nematode invasion of roots through effects on nematode hatching and mobility or may induce plant resistance. Several of these bacteria are also plant-growth promoting bacteria (PGPB).

Of the few bacteria known to parasitize nematodes, *Pasteuria* spp. are endospore-forming bacteria and show greatest potential to be developed into biological control agents. *Pasteuria* endospores have been observed adhering to and parasitizing all of the economically important plant-parasitic nematodes. *Pasteuria* species and isolates show varying degrees of host-specificity; for example, a *Pasteuria* population isolated from cyst nematodes appeared to be more promiscuous than those isolated from root-knot nematodes (Mohan *et al.*, 2012). Molecular studies have led to the establishment of phylogenetic relationships between *Pasteuria* spp. and have provided tools to test if plants are infected with *Pasteuria*-infected nematodes (Rao *et al.*, 2012). The life cycle of *Pasteuria* spp. is initiated when dome-shaped endospores adhere to the cuticle as the nematodes move through soil (Fig. 14.3). It is thought that collagen-like fibres on the surface of the

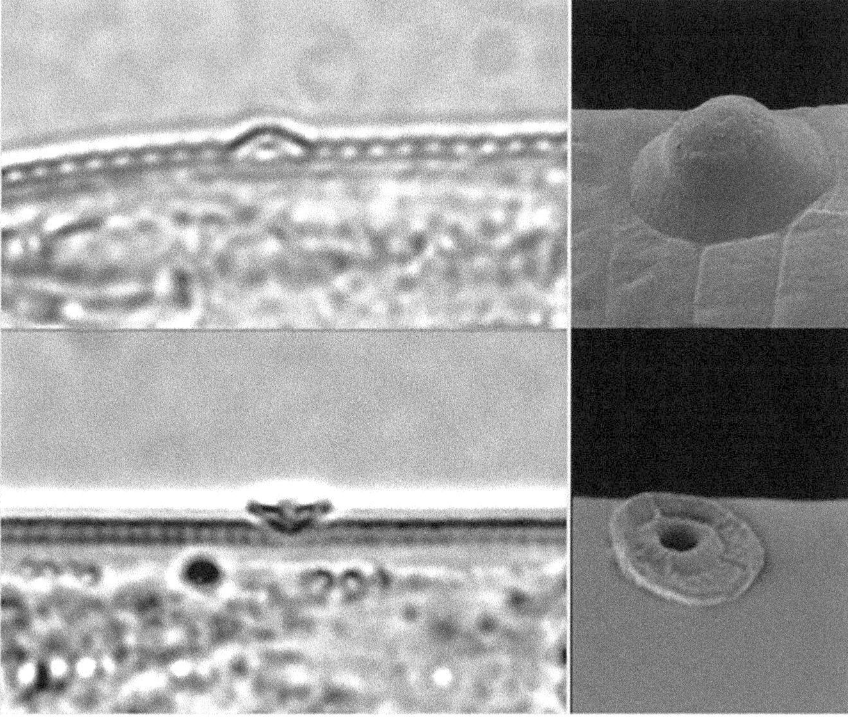

Fig. 14.3. Dome-shaped endospores obtained from *Heterodera cajani* on *H. cajani* (top, brightfield and scanning electron microscopy (SEM)) also attaching to *Globodera pallida* (bottom, brightfield and SEM). (Courtesy S. Mohan and K.G. Davies, Rothamsted Research, UK.)

endospore are involved in a Velcro-like attachment process with a cuticle receptor (Davies and Curtis, 2011). Depending on the nematode species, endospores germinate either immediately (e.g. in *Heterodera avenae*), or later when the nematode has entered the root and initiated a feeding site (e.g. *Meloidogyne* spp.) by producing an infection peg that penetrates the cuticle. Following germination, small rod-shaped bacteria develop exponentially to produce granular masses that eventually enter sporogenesis and form the next generation of spores (Davies *et al.*, 2011). In *Meloidogyne* spp. the infected female continues to develop, becoming infertile as up to 2 million endospores are produced by the bacterium destroying the reproductive system. These endospores are then released into the soil upon decomposition of roots and infected females. Spores can survive in air-dried soil for several years but successful infection of nematodes depends greatly on the distribution of the spores in the soil, which can be influenced by soil type, tillage practices, moisture and temperature. Juvenile nematodes in subsequent generations on the same plant only move short distances from the egg to nearby roots and are less infective than those in the first generation.

A natural decline of root-knot nematodes in fields of tobacco was shown to be caused by *Pasteuria* spp. but most studies on the efficacy of the bacteria as a biological control agent have been performed in pots because of difficulties in producing sufficient spores for large-scale trials. Declines in nematode numbers and root galling have been reported in small plots. Application is complicated because some isolates of the bacterium are very host specific and spore burden is not always correlated with virulence. The need for the thorough distribution of spores around the roots requires the inundative release of the bacteria at rates of $>10^4$ spores g^{-1} of soil. Previously, the main disadvantage of *Pasteuria* spp. for use in biological control was its obligate parasitism and hence difficulty for mass production but breakthroughs at Pasteuria Bioscience Inc. (USA) (now Syngenta) led to the *in vitro* culture of *Pasteuria* spp. The *in vitro* production of *Pasteuria* spp. for use against the sting nematode, *Belonolaimus longicaudatus*, on turf was marketed as Econem®. The species *Pasteuria nishizawae* is commercially available for use against *H. glycines* (soybean cyst nematode) as Clariva™ pn. The availability of mass production for niche markets paves the way for field-scale applications. Once more is known about the specificity of the populations towards different nematode species, *Pasteuria* spp. have increasing potential to be a most successful biological control agent. With the assembly of the genome of *P. penetrans* (Orr *et al.*, 2018) valuable information became available that could lead to the application of this bacterium in the field.

14.4. Interaction with Rhizosphere Microflora

The soil microflora in the rhizosphere is influenced not only by soil treatments, such as application of organic matter or soil disinfestation, but also by the presence of nematodes and by the plant species and cultivar. The population dynamics of the rhizosphere microflora are the result of direct and indirect effects of multitrophic interactions between the plant roots, the plant-parasitic nematodes and their natural enemies. These complex interactions between microorganisms and their environment are poorly understood. For example, root diffusates of nematode host plants were found to alter the ageing of the cuticle of second-stage juveniles, making them increasingly susceptible to hyperparasitic bacteria, whereas root diffusates from non-hosts did not (Mohan *et al.*, 2020). Research on these types of subtle interactions is currently in its infancy in relation to plant-parasitic nematodes, but in entomological situations such interactions have been developed into sustainable pest control strategies (Cook *et al.*, 2007). These multitrophic interactions between rhizosphere microorganisms, microbial biological control agents and plants are complex and may also involve competition for space and nutrients as well as antibiosis.

Molecular techniques, and more recently the use of metagenomic approaches, offer great possibilities to study changes in the microbial community of the root microbiome as species and isolates of microorganisms can be more easily monitored than with classic tools, such as selective media. For example, the bacterial microbiomes have been investigated in nematode conducive and suppressive soils (Elhady *et al.*, 2017) and it is expected that with these new techniques, increasing insight will be gained on the role of soil biodiversity and in the mechanisms of ecosystems resilience, i.e. the capacity of an ecosystem to respond to disturbance and recover quickly. This knowledge could possibly be used to improve mechanisms for the introduction or rather enhancement of biological control agents in agricultural soils, where plant-parasitic nematodes cause unacceptable crop losses.

14.5. Applying Biological Control Agents

For successful biological control, much depends on a thorough understanding of the population dynamics of the natural enemies and target pest, and of their interactions. For many organisms that have potential as biological control agents for nematodes such information is lacking. The pioneering work of Jaffee *et al.* (1992) in model systems revealed that parasitism by some microbial biological control agents is density-dependent, i.e. the probability of a specific host being parasitized increases with the density of the host. Although the activity of facultative parasites, including some nematode-trapping fungi, is density-dependent, their ability to survive saprophytically and their wide host range should enable them to be more effective at low pest densities than obligate parasites, such as *P. penetrans*, or those with poor saprophytic survival capabilities, such as *H. rhossiliensis*. The size of the inoculum reservoir required for significant infection rates and the relatively low transmission rates in soil tend to dampen fluctuations in the dynamics of the interactions between nematodes and their microbial parasites and pathogens. It is becoming increasingly clear that plant roots are adept at subtly manipulating their microbial communities to their own advantage (Cox *et al.*, 2019) and therefore a thorough understanding of these multitrophic interactions, at both the molecular and population level, needs to be placed within a theoretical framework if the development of rational biological control strategies is to be successful.

Currently, the application of biological control agents for the field management of plant-parasitic nematodes is notoriously difficult with most empirical studies providing inconsistent results. However, some agents may be practically exploited in nematode management and some, such as *P. lilacinum*, *Trichoderma* spp., *Pasteuria* spp. and *P. chlamydosporia*, have been successfully commercialized. The steps towards commercialization include selection of the most effective isolates, production of inoculum and formulation of the microbial agents. As synthetic nematicides continue to be prohibited for safety reasons it is important to understand that biological control cannot be a simple replacement for chemical control; application of microbiological control agents should not be considered as a silver bullet, but needs to be viewed within a framework of other management strategies, especially their interaction with cultural control methods (Stirling, 2011; Davies and Spiegel, 2011).

14.5.1. Selection of biological control agents and isolates

Due to the complexity of interactions that take place in a biodiverse rhizosphere, the choice of a biological control agent will depend on the nematode species and nematode stage to be controlled, as it is clear that some organisms are host-specific or attack only certain developmental stages and others feed on many kinds of nematodes. A combination of biological control agents with different modes of action or range of environmental tolerance limits may result in improved nematode control. However, no benefits and even reduced effects can occur, so careful selection of organisms for combination is required to avoid competition or antagonistic effects (Meyer and Roberts, 2002). In practice, combinations of agents may be too expensive to produce and register, so other approaches, such as soil amendments (see Section 14.6.3), also need to be considered.

Comparisons of different isolates of nematophagous fungi and bacteria have revealed that there exist enormous intraspecific differences in the performance of the microorganisms as potential biological control agents. Performance refers to the ability to parasitize the nematode and also to the survival capacity in soil under different environmental conditions (e.g. the production of survival structures, saprophytic growth), establishment in the rhizosphere (root-colonizing capacity) and soil (dependence on nematode hosts and production of resting structures). With improved knowledge on their biology, it is possible to focus selection for biological control agents. For example, when assessing *Trichoderma* spp. as biological control agents of *Meloidogyne* spp., those isolated from egg masses of *Meloidogyne* spp. appeared more effective than isolates recovered from soil (Affokpon *et al.*, 2011). In general, screening programmes to select isolates with biological control potential have resulted in less than 10% of the isolates being selected in simple bioassays and more being rejected in tests in soil. Hence, focus on selection of appropriate isolates is a key stage in the development of a biological control agent. However, care should be taken when interpreting results. For example, a recent study indicated that *P. chlamydosporia* was responsible for the suppression of *Meloidogyne* spp. in vegetable production systems in Spain (Ghahremani *et al.*, 2022), an interpretation similar to earlier studies of Kerry *et al.* (1982) where a fungal egg parasite was held responsible for the suppression of *H. avenae* in cereals. However, it was later revealed that in the same field a *Pasteuria* population was infecting more than 50% of the migratory infective juveniles and prohibiting them from reaching the root (Davies *et al.*, 1990), indicating that a mixture of microbes was responsible for the suppression of cereal cyst nematodes overall but each acting on different stages of the nematode's development. Therefore, a diversity of species and isolates of *P. penetrans* and *P. chlamydosporia* within a single site may enable the community to function in different environmental conditions, on different parts of the root system and against a variable host; it is unlikely a single isolate of a specific agent can be used as a single management strategy.

14.5.2. Inoculum production and formulation

Most fungi and bacteria with activity against nematodes are not obligate parasites and can be cultured on artificial media or grain products (e.g. wheat bran). Mass production of microbes on artificial media can undergo a process of attenuation, whereby the microbe loses its ability to parasitize its host; it is therefore important that the efficacy of the biological control agent is maintained, even after frequent sub-culturing. Although regular passage of the isolate through the nematode host to maintain its pathogenic capability has been recommended (Crump, 2004), it is not clear if this is important for facultative parasites that establish in soil through saprophytic growth.

Once inoculum has been produced, it needs to be formulated for storage. An ideal formulation has a long shelf life, allows for easy transportation and application, is readily distributed in soil and provides optimum conditions for growth and survival of the introduced agent. The formulation must also be compatible with other crop protection measures. For example, a whole science exists around the development of seed coatings, whereby individual seeds are coated with small dosages of fertilizer together with a combination of pesticides and biologically beneficial microbes that enhance growth of the seedling and additionally protect it from soilborne pests and diseases (Rocha *et al.*, 2019). It is therefore imperative that the various components of the seed coating are mutually compatible and not antagonistic. Alginate pellets have often been used to formulate spores, and mycelial fragments of several fungi have been tested and shown to be an efficient way to store, distribute and apply the fungi, while maintaining biological control activity. However, the pellets and their contents may be destroyed by collembola and mites, and at high concentrations alginates may be phytotoxic. Together with the biological control agents, amendments, such as chitin, can be added as a food source for the microflora. However, this practice does not always result in increased parasitic activity of the biological control agent.

14.5.3. Biodegradable organic carrier

Banana fibre was initially selected as a very suitable carrier material to deliver the nematicide abamectin to vegetative propagated seed materials, such as potato and yam. The banana fibre, when made into a paper-like matrix, deteriorates and decomposes gradually, releasing ultra-low but effective dosages of abamectin, resulting in an extended period of nematode knock-down. This has proven very effective against *G. rostochiensis* on potato in East Africa (Ochola *et al.*, 2022) and against *Scutellonema bradys* on yam in West Africa (Pirzada *et al.*, 2023). The technology has shown potential as a novel delivery mechanism for improving the efficacy of biological control agents, such as *T. asperellum* TR900 (Sustain®; RealIPM, Kenya), against root-knot nematodes. Furthermore, the unique properties of the fibre that enables it to bind strongly to abamectin also allow it to bind nematode host-location chemical signals (semiochemicals) from root diffusates. The technology therefore disrupts host-finding by nematodes, even without pesticide impregnation, and holds promise as an effective biologically based soilborne pest and disease management practice.

14.6. Integration of Biological Control with Other Control Measures

A grower's decision to include biological control methods as part of a nematode management strategy may be mediated by various reasons, including environmental concerns or strict regulatory restrictions on synthetic pesticide use/residues. Irrespectively, the integration of biological control needs to be considered with other control measures for both enhancing their impact and indeed avoiding the negative impact of pesticide applications, applied against other target pests, on the biological control products. Biological control strategies may involve the application of a biological control agent (inundative biological control) but can also include methods to increase the impact of the existing soil microflora on pest nematode population dynamics. The latter, also termed **conservation biological control**, aims at either a combination of adding an antagonist with the creation or maintenance of a suitable environment for optimum suppression of plant-parasitic nematodes, or can even indicate these cultural practices on their own, where no external antagonist is applied.

14.6.1 Crop rotation

As biological control agents alone rarely provide adequate control, they can be integrated with other methods such as crop rotation, resistant cultivars or antagonistic plants, either to reduce the nematode populations in soil or to promote the establishment of the biological control agent (Kerry and Hominick, 2002). Different crops produce different root diffusates and these affect microbial activity in the soil and rhizosphere. The rhizosphere-colonizing capacity of *P. chlamydosporia* differs greatly among host plants with nematode control greatest on roots that support extensive fungal growth but are poor hosts of the nematode. Hence, most nematode control is achieved by combining this fungus with a poor host for the plant-parasitic nematode to reduce soil infestation before the next susceptible host. The effects of crops on the establishment and functioning of other biocontrol agents has rarely been studied but is definitely a factor that plays a role in the field.

14.6.2. Soil disinfestation

Several control measures, such as soil disinfestation through soil solarization, biofumigation, steaming (see Section 14.9), creation of anaerobic conditions (see Section 14.9.5) or application of broad-spectrum biocides (see Chapter 17), facilitate the establishment of biological control agents introduced after these treatments. As the activity of the soil microbial community is reduced, the establishment and survival of biological control agents is enhanced, at least in the short term. Soil disinfestation also reduces the densities of plant-parasitic nematodes to levels that are manageable for biological control agents.

14.6.3. Soil amendments, green manures and biostimulants

In contrast to soil disinfestation, addition of organic matter to the soil through soil amendments, green manures and biostimulants results in an increase in density and diversity of microbial populations, including beneficial ones. Although there is some disagreement as to the specific description of a biostimulant (Yakhin *et al.*, 2016), all these approaches generally rely on the application of organic matter, without the external addition of a specific biological control agent (D'Addabbo *et al.*, 2019), and this often results in suppression of plant-parasitic nematode densities (see Section 14.11.1) and increase in saprophytic nematodes. The mechanism of this nematode suppression can be attributed to the release of nematicidal compounds by the organic material (e.g. glucosinolates, ammonium and organic acids), to the production of allelochemicals (e.g. antibiotics and chitinases) by the soil microflora that was able to increase in numbers because of the amendments, and to improved plant growth (e.g. by PGPB) (Widmer *et al.*, 2002; Oka, 2010; Thoden *et al.*, 2011). It is assumed that the incorporation of organic matter increases the abundance of microorganisms active against nematodes. However, addition of organic amendments may increase microbial competition and result in reduced nematode parasitism by poor competitors, such as *H. rhossiliensis*. Addition of organic amendments may affect the switch from the saprophytic to the parasitic state and some facultative parasites are less active in soils following application of soil amendments. The presence of adequate nutrients, supplied by the organic amendments, may suppress the production of enzymes that aid in nematode infection yet provide abundant materials and energy for population increase of the biological control agent. For example, the serine proteases from *P. chlamydosporia*, *P. lilacinum* and *P. suchlasporium* are repressed by the presence of glucose and easily metabolized nitrogen sources.

Defined amendments have also been used for nematode management. The addition of chitin causes an increase in chitinase-producing microbes in the soil and rhizosphere, which are thought to degrade the chitin-rich eggshells of nematodes. Although there may be an increase in the abundance of nematode antagonists after such applications, the major effect of chitin on nematodes, especially soon after application, is due to the release of ammonia. In general, large amounts of soil amendments are needed to reduce nematode infestations significantly and their effects on the soil microbial community are complex and difficult to interpret.

There is also interest in the use of chemical elements (e.g. silicon) that increases the general resilience of the plant to crop pests and diseases; for example, priming rice roots with silicon induced a defence response that protected plants against the rice root-knot nematode *Meloidogyne graminicola* (Zhan *et al.*, 2018). With the growing need for alternatives to synthetic chemical nematicides these approaches offer potential.

14.7. Nematode-free Planting Material

Problem avoidance should essentially form the basis of any pest management strategy. Use of nematode-free planting material provides an effective means of nematode management (or avoidance), often at a fraction of the cost fiscally and environmentally, to treating the cropping area, which may be impractical in any case. This nematode management tool requires that farmers understand the benefits of more costly, healthy planting material.

14.7.1. Production

Healthy planting material may be available through the use of certified seed and planting material, the production of seedlings/plantlets in sterile conditions and an effective quarantine system (see Chapter 13), which prevents the introduction of foreign nematode species. Cropping systems using nursery-grown plants for transplantation afford an excellent opportunity to provide nematode (and other soilborne disease) protection. Such strategies will limit nematode transmission to otherwise pest-free areas. In developing countries, subsistence farmers could prevent major losses from nematodes by using simple nurseries for nematode-free seedling production. Inert or sterilized potting media (e.g. sawdust, cocopeat, vermiculite and rice husks) can be used to obtain nematode-free stock. It is important to treat seedling containers or boxes regularly to maintain hygiene.

Where available through (international) trade at relatively low prices, healthy, certified seed or nursery material can reduce the need for farmers to produce their own pest-free material. Practices that are increasingly adopted are the use of tissue-culture propagation techniques, e.g. for ornamentals and bananas, and the floatation tray method as applied for tobacco and flower seedlings in Africa. Using vine or stem cuttings instead of tubers for the propagation, e.g. for sweet potato, potato and yam, avoids the propagation of nematode-infected materials. At the same time, such systems also provide the opportunity to deliver optimally seedling protectants, including beneficial microorganisms.

Where nematode-free propagation material is not guaranteed or available, disinfection of (potentially) infected material can provide a solution. Heat treatment and mechanical methods to separate infected from healthy seeds are widely practised methods to obtain nematode-free plants. However, if planting material is not treated or infested soil is used for seedling production before distribution, the practice can result in greater distribution of nematode (and other) pests, so caution is required.

14.7.2. Heat treatment

Hot-water treatment (therapy) is used to disinfest vegetatively propagated planting material such as tubers, bulbs, rhizomes, runners, woody rootstocks and also seeds. The practice relies on the application of sufficient heat over a sufficient length of time to prove lethal to nematodes without thermal injury to the planting material. The greater the difference in the thermal sensitivity between the plant host and the nematode pest, the greater is the potential success for decontamination. In Europe, the use of hot-water treatment for the management of *Ditylenchus dipsaci* led to the undoubted improvement in ornamental bulb health. In combination with infected root and corm tissue removal (paring), banana and plantain nematodes (*R. similis*, *Helicotylenchus multicinctus*, *Pratylenchus* spp. and *Meloidogyne* spp.) have been successfully managed by dipping rhizomes (suckers) in water at 53–55°C for 20 min. Other tuber, corm or bulb crops have been similarly successfully treated. Where fuel is scarce or costly, or temperature and timing regulation is problematic, such as in many smallholder farming systems, this approach may not be practical. However, a simplified technique using local materials (e.g. empty oil drums) has been adapted and uses boiling water in which pared banana suckers are immersed for short periods of just 30 s (Coyne *et al.*, 2010; Fig. 14.4). Hot-water treatment can also be further improved with pre-treatments, such as inducing cold-hardiness by storing at low temperature (e.g. roses), pre-warming (e.g. strawberry plants), pre-soaking (e.g. rice seeds) and/or immersing in cold water after heat exposure (e.g. grapevine rootings). Pre-soaking with solutions of hydrogen peroxide, sodium chloride or other salts can further improve the process. If conditions are unfavourable for efficient drying, the treatment of large volumes of seed may be inadvisable, as fungal contamination or premature germination may result. Spreading out dry seed under intense solar radiation can also help reduce nematode seed infections, e.g. *Aphelenchoides besseyi* in rice. Aerated steam-based thermotherapy is an adequate alternative to hot-water treatment for vulnerable plant materials, such as strawberry transplants, with leaves that might contain foliar nematodes and fragile roots with root-knot and lesion nematodes. When using hot-water treatment for nematode decontamination, it is necessary to test and modify the technique to suit individual circumstances. Differences exist between nematodes species, cultivars, and size and age of plant parts in their sensitivity to heat and exposure time.

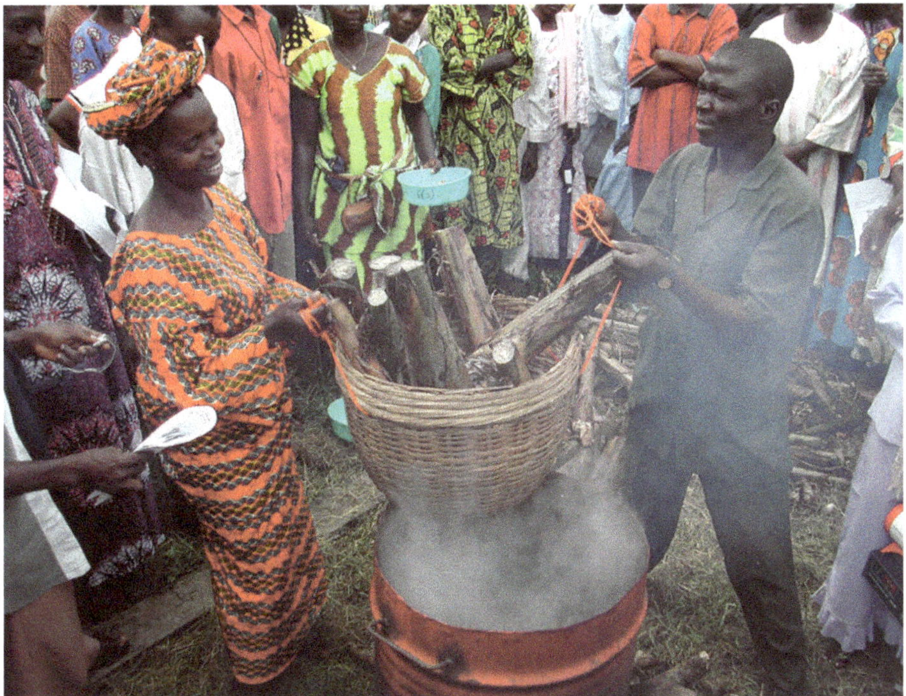

Fig. 14.4. Boiling water treatment of plantain suckers in a used oil drum to disinfest them of nematodes, in Nigeria. (D. Coyne.)

14.7.3. Mechanical methods

In the case of *Anguina tritici*, which infects wheat species (*Triticum* spp.), including emmer and spelt but also rye (*Secale cereale*), the infected seeds that have turned into galls (cockles) can be separated from healthy seed stocks by weight differential; as the galls are less dense, they can be winnowed or separated by floatation in salt solutions. They can also be separated using sieves, relying on size differences.

Seedborne nematodes that may infect other crops include *A. agrostis* and *A. funesta* on grasses, e.g. on bentgrass (*Agrostis* spp.) in the USA, as well as *A. arachidis* and *D. africanus* affecting groundnut in Africa. These seeds may also be cleaned using weight or size differentials of infected seed, or simply by visual selection of the infected distorted or discoloured seed (Brown and Kerry, 1987).

14.8. Sanitation

Approaches that limit the build-up or survival of nematodes should be practised wherever feasible. Simple sanitation measures to reduce nematode movement between sites via implements, machinery or through irrigation water or waste products are probably not a foremost consideration in most circumstances, but are of major importance as they contribute considerably to the spread and build-up of nematode populations. However, such practices can be time-consuming and people may be reluctant to implement them if the objective is not clearly understood.

Post-harvest sanitation and physical destruction of plant debris, such as straw and stubble burning, have traditionally played an important role in the management of some seed and stem nematodes such as *A. agrostis*, *A. besseyi*, *A. tritici* and *D. angustus*. Burning reduces the return of nematode inoculum to the soil after a susceptible crop. This is also the case for the traditional post-harvest burning of uprooted tobacco plants in Malawi and Zimbabwe as a means of managing *Meloidogyne* spp. However, environmental protection acts now restrict this practice in some parts of the world, limiting the use of this method for sanitation purposes.

In intensive potato production areas in Europe, outgraded potato tubers left in the field after harvest can sprout and grow among plants of the next crop in the rotation, thus enabling potato cyst nematodes (*Globodera* spp.) to reproduce despite the strict rotation scheme. Removal of such volunteer plants, most often by destruction using selective herbicides, prevents build-up of cyst populations. Weed control in general prevents survival and increase of nematode densities, especially of those species with a broad host range. In the intensive potato industry, tare soil and soil adhering to tubers are transported by trucks over long distances, thereby distributing potato cyst nematodes. Washing potato tubers free of soil has proven to be effective in the USA for many years. As this practice is not always possible, disinfestation of soil is receiving more attention, especially in view of international trade.

For *Bursaphelenchus xylophilus*, which infects *Pinus* trees, the treatment of wood and wood products is a key measure towards reducing its spread. *Bursaphelenchus xylophilus* should be controlled in timber and wood chips by proper heating before removal from infested sites, a practice required for the international trade of wood products as a preventive measure (ISPM 15; see Chapter 13.8). A principal method of containment and management, also the case for the red ring nematode *B. cocophilus* infecting palm trees, is the removal and destruction by burning of dead and dying trees.

14.9. Physical Soil Treatments

Managing nematode population densities in the soil to levels below the damage thresholds can be achieved by preventing nematode multiplication or survival, or by killing them directly. During previous decades, soil disinfestation has relied heavily on broad-spectrum biocides, at least in developed countries (see Chapter 17). Alternative methods for disinfesting or sterilizing soil rely on physical principles, mainly heat and oxygen deprivation.

14.9.1. Dry heat

The use of fire is not the most efficient way to apply heat for disinfesting soil. The fuel requirements to permit lethal heat levels to penetrate sufficiently deep are uneconomical in most circumstances. However, it may be practical and effective for small areas of land, such as in the treatment of seedling nurseries. The use of slow-burning rice husks in Asia has provided nematode-free soil for use in rice and vegetable nurseries. Pre-drenching the soil, before placing fuel and igniting, improves the depth of penetration of lethal heat and the general efficiency of the process. Dry heat can be used to treat bulk and container soil using various heating devices, e.g. metal plates heated by electricity placed at regular spacing in the soil. Here soil should also be moist to distribute the heat efficiently. Depending on the method and soil type, changes in the physical and chemical nature of treated soil can cause unsatisfactory plant growth.

The use of microwave radiation to heat soil and control nematodes has been demonstrated for *H. schachtii*, *R. reniformis* and *Meloidogyne* spp. but this method is not applied widely as it appears to be an impractical method for field soil. Research continues, however, for alternatives to synthetic pesticides that may prove suitable for small volumes of soil or for nursery situations.

14.9.2. Steam

Steam heat has long been applied in heated glasshouses as an effective, although costly, means of soil sterilization. Soil type and prior soil tillage affect heat penetration and efficacy, as does water absorption capability. Dry soil has a greater capacity to absorb condensed water and to a greater depth than wet soil. Steam can be applied by pumping through a network of underground, perforated pipes (about 60 cm depth). Efficient distribution of heat through the soil, especially the surface layers, requires covering the soil with plastic sheeting. Blowing steam under plastic sheeting (sheet steaming), which has been anchored at the edges, provides surface soil sterilization. Sheet steaming, combined with buried pipes can create 'negative pressure steaming', a more energy efficient and effective method that draws the steam through the soil using an extractor fan. Steam should be applied for several hours to reach lethal temperatures in the deeper soil layers. Treated soil should be set aside for some time to permit the soil to stabilize as the release of phytotoxic chemicals (nitrite, ammonia) and change in pH may affect plant growth. Steam can be effectively employed for the use of small volumes of soil, such as for use in nurseries. Containers, such as used oil drums, can be semi-filled with water, placed over a heat source and adapted to deliver steam beneath the soil via a pipe attached to the steam outlet on the drum. Plastic sheeting is highly recommended to contain the steam in the soil being treated.

14.9.3. Solar heat

Solar radiation has been employed effectively to disinfest soil of nematodes, primarily in hot climates and for relatively shallow depths of soil (Gaur and Perry, 1991). Moistened soil covered with polyethylene sheeting, with the edges anchored, will significantly reduce nematode populations (Fig. 14.5). Moisture is necessary as this promotes biological activity and conducts the heat, improving efficacy. This system has been variously termed plastic, polythene or polyethylene mulching or tarping, solar heating, solar pasteurization or soil solarization, the latter term being the more generally accepted. The effect reduces with depth, but solarization for at least 4–6 weeks will increase soil temperatures to 35–50°C to depths of up to about 30 cm, depending on soil type and prior tillage. Results can be optimized by using double layers of polyethylene, thin (25–30 μm), transparent as

Fig. 14.5. Use of plastic sheeting for solarization of a small area of land in Malawi. (D. Coyne.)

opposed to black sheeting, and using solarization during periods of highest solar intensity. Thinner sheeting tends to be more effective but less durable and more easily damaged. Larger areas are also more practical to treat, as soil heating is less effective near the edges of the covered area. The length of time required for effective solarization is a great limitation and the method is most suited to nursery beds and in glasshouses rather than large field areas. Also, in areas of intensive use, the disposal of the large quantities of plastic sheeting can be a problem. Excellent nematode control in soil for use in potting composts, to raise seedlings or rooting cuttings, can be achieved for small volumes of soil, which are moistened, contained in sealed plastic bags and placed on a suitable surface in direct sunshine for two weeks. An additional benefit of solarization is the control of soilborne pathogens and weeds.

14.9.4. Flooding

Areas of land following natural inundation are mostly free of plant-parasitic nematodes. Areas such as river flood plains, riverbanks and seasonally flooded lakes have long been exploited for crop production, primarily to take advantage of fresh sediment deposits. Extended periods of flooding provide almost nematode-free conditions. Notable exceptions are the rice root nematodes, *Hirschmanniella* spp., and the foliar nematodes, *A. besseyi* and *D. angustus*. Some species of *Tylenchorhynchus*, *Meloidogyne*, *Helicotylenchus*, *Heterodera*, *Rotylenchus*, *Pratylenchus*, *Paratylenchus*, *Hemicycliophora*, *Xiphinema* and Criconematidae are also known to survive periods of flooding. If small in area, flooded sites may be used for nursery purposes or, alternatively, the soil may be removed for use in a nursery sited elsewhere. Artificially flooded areas, such as rice paddies, can also provide nematode-free conditions for post-rice crops that are non-hosts for the rice nematodes. Rice–vegetable and rice–wheat production systems tend to be quite sustainable despite the presence of nematode pests, including species of *Meloidogyne*. This is partly a consequence of the nematode suppressiveness of soils flooded for the rice crop. A disadvantage of flooding is that the field cannot be used to grow crops, and so provide income, for an extended period of time. Fields also need to be non-sloping. As the use of nematicides becomes more restricted, the practice of flooding has gained more attention in high-value crops where quarantine regulations require zero levels of certain nematode species, e.g. flower bulbs and potato fields in The Netherlands are kept inundated for 17 weeks to reduce *D. dipsaci* or *Globodera* spp., respectively, to undetectable levels (Runia *et al.*, 2014).

14.9.5. Anaerobic soil disinfestation

A technique called anaerobic soil disinfestation (ASD) – also called biological soil disinfestation (BST) – has been developed as an alternative to the use of methyl bromide (Bello *et al.*, 2004). ASD consists of incorporating copious quantities of organic matter into the soil, followed by irrigation to obtain a moisture level above field capacity and carefully sealing the amended soil with impermeable polyethylene sheeting for several weeks (Fig. 14.6). The decomposing organic matter (e.g. rice husks, grass or crop by-products) and the wet conditions result in anaerobic conditions when several toxic gaseous compounds are produced. The method requires sufficiently high temperatures and careful sealing to trap the gases and establish the anaerobic conditions necessary for killing soil pathogens, including nematodes. ASD is applied on sandy soils for strawberry, asparagus, flower and vegetable

Fig. 14.6. Different steps in anaerobic soil disinfestation (ASD), from right: organic matter, which is a cover crop grown in the field, in this case grass; incorporation into the soil of the organic matter; soil compaction with roller; sprinkling of water over the compacted soil layer; and covering soil with plastic film (Courtesy Seelen, a company in Maasbree, The Netherlands.)

production in Spain, The Netherlands and California. When the incorporated material consists of plant parts containing substances, such as glucosinolates or dhurrin that release toxic substances (e.g. isothiocyanate, hydrogen cyanide) upon decomposition (see Section 14.10.4), the technique is called biofumigation. As well as creating the anaerobic conditions, the toxin contribute to the control of plant-parasitic nematodes and other pests and pathogens.

14.10. Biologically Based Practices

14.10.1. Crop rotation

Seasonal rotation on the same area of land with different crops remains one of the most important methods of nematode management. Rotation with crops, which have different nutritional demands on the soil and have different pest problems, has obvious benefits to the maintenance of the system. Supposedly, the Incas practised crop rotation in response to damage caused by the host-specific potato cyst nematodes. However, intensification of cropping systems and the success of pesticides has seen reduced reliance on crop rotation in modern agriculture. The basic premise of crop rotation is to distance the time between susceptible hosts to the same nematode species using resistant, poor or inhibitory hosts, in order that population densities do not increase to damaging levels or decline below damage thresholds before the next fully susceptible crop is grown. The number of cropping sequences or the period between susceptible hosts can depend on, in particular, the host status of the rotation crops (and cultivars), the nematode species and the nematode population density at the harvest of the last susceptible crop. A rotation of at least 7 years between potato crops was traditionally necessary to prevent losses due to *G. rostochiensis* and *G. pallida* in Europe but with the integrated use of nematicides and cultivars resistant to *G. rostochiensis*, this has since been shortened (Phillips and Trudgill, 1998). While crop rotation is widely practised and is environmentally appealing, it has important limitations. The occurrence of nematode communities containing multiple pests or polyphagous species with wide host ranges, such as some species of *Meloidogyne* or *Pratylenchus*, limits the potential of using acceptable non-host crops for the rotation. Rotation crops may also facilitate the increase of alternative nematode pest species. The degree of control is dependent on the level of resistance of the rotation crops and the length of the rotation cycle, which may be too long to be acceptable, especially for specialist producers in intensive production systems and for subsistence farmers with limited land. Further difficulties arise where the non-host crops have no local market or are of limited value. Correct nematode identification and knowledge of the host range and cultivar susceptibility, therefore, determine the introduction of successful rotations. *Meloidogyne* spp. as a group presents the most formidable challenge to the implementation of successful rotations because of their broad host range. Accurate identification of species present is essential but the development of 'resistance-breaking' populations, the emergence of virulent strains/races or pathotypes and communities of multiple species of *Meloidogyne*, all pose a challenge. Poor control of weeds, which host polyphagous nematode pests, will also reduce the effectiveness of crop rotations. Nevertheless, crop rotation as a nematode management strategy can be employed efficiently and can provide an essential measure in integrated pest management programmes.

Heterodera glycines, *G. rostochiensis*, *D. dipsaci* and even some species of *Meloidogyne* have limited host ranges amongst crop species or, alternatively, occur in situations where they tend to be the dominant nematode species. They provide examples where crop rotation has provided effective nematode management. *Heterodera glycines* has few alternative host crops and cultivation of non-hosts such as maize and groundnut for 2 years can be sufficient to provide a full yield of soybean. This can be further reduced, or the number of consecutive soybean crops increased, through the integration of resistant cultivars (see Chapter 15). In California, successful management of *M. naasi* on barley has been accomplished through barley–oat rotations. The development of cultivars with resistance to key nematode pests continues to improve the flexibility of nematode management programmes based on crop rotation.

14.10.2. Trap cropping

Certain situations may accommodate the implementation of trap cropping, where a highly susceptible host is planted and grown for sufficient time to permit nematode invasion and development, but not to complete its life cycle. This strategy can only be employed for endoparasitic nematode management. The trap crop must be physically removed or destroyed (with herbicide or ploughed in) before the nematodes reproduce. An alternative approach involves the use of resistant crops or cultivars that stimulate nematode activity and/or support nematode invasion but do not permit completion of the life cycle. Examples of resistant crops that do attract nematodes are sunn hemp (*Crotalaria juncea*) for *Meloidogyne* spp., *Sesbania rostrata* for *Hirschmanniella oryzae* in rice and *Solanum sisymbriifolium* and African nightshade (*S. scabrum*) for the control of *Globodera* spp. on potato (see Chapter 8).

The trap crop is preferably planted quite densely so that roots and root diffusates reach as many nematodes as possible. Trap crops are usually removed before the main crop, but some are used in mixed cultures (e.g with perennials). Success depends upon proper planting techniques, precise timing and total crop destruction in the case of susceptible crops. Trap cropping is therefore rather costly and inconvenient but can be a very effective tool for nematode management.

14.10.3. Antagonistic plants

Plants with nematode antagonistic properties, due to root diffusates, can be used in rotation or as intercrops with susceptible crops against certain nematode species. Species of marigold (*Tagetes erectus*, *T. patula* and *T. minuta*) are good examples, which successfully reduce populations of *Pratylenchus* spp. and *Meloidogyne* spp. as well as other species. The mode of action of *Tagetes* spp. is attributed to alpha-tertienyl, a very toxic compound contained in the root cells, but also simply to its non-host status for several nematode species, although some Trichodoridae are reported to increase on marigold.

The value of the crop and of the antagonistic plants determine the suitability of this approach, although antagonistic crops with a potential market, such as asparagus (*Asparagus officinalis*), will undoubtedly improve the acceptability. For example, use of *Tagetes* spp. for nematode management has increased with its increased use as a source of xanthophylls for food colourants.

14.10.4. Cover crops

Numerous examples of plants that are antagonistic, resistant, suppressive or detrimental to the population development of plant-parasitic nematodes are available (see Whitehead, 1998; Sikora and Roberts, 2018). However, many such plants often have no direct commercial value, limiting their appeal, compared with those that have an alternative use. Additional functions may include use as cover crops, for soil conservation during the off-season (winter or dry period), or as forage for livestock (e.g. joint vetch (*Aeschynomene* spp.), hairy indigo (*Indigofera hirsute*) and stylo (*Stylosanthes* spp.)). Leguminous crops contribute to soil nitrogen availability, whilst crops that produce extensive foliage may be ploughed in as a green manure. Cover crops with nematode suppressive qualities vary with geographical location and target nematode species. Noteworthy examples include *Aeschynomene* spp., horsebean (*Canavalia ensiformis*), butterfly pea (*Centrosema pubescens*), *Crotalaria* spp., kudzu (*Pueraria phaseoloides*), castor (*Ricinus communis*) and, particularly, velvetbean (*Mucuna pruriens* and *M. deeringiana*). In tropical regions, grasses and cereals are generally poor hosts of *Meloidogyne* spp. and are often successful in reducing *M. javanica* and *M. incognita*. Additionally, when ploughed in, some crops produce or release nematotoxic compounds upon decomposition. Brassica crops such as rapeseed (*Brassica napus*) and mustard (*B. campestris*) contain glucosinolates, which become hydrolysed to the volatile isothiocyanate and other products with broad biocidal activity. These act as biofumigants following incorporation of brassicaceous residues into the soil, reducing soilborne pathogens including plant-parasitic nematodes. Brassica crops and cultivars vary in their glucosinolate content and their effects on nematodes vary according to the physiological stage of the nematode and the environmental conditions. Their role in reducing pathogen densities is also attributed to merely adding organic matter to the soil, thereby stimulating diverse soil microbial activity, which in turn can have a positive effect on plant growth while reducing the share of plant pathogens.

Taking advantage of the off-season to grow a poor host as a cover crop has numerous merits for the farmer. However, if this period coincides with a period of natural nematode dormancy (see Chapter 8) or low activity, only limited nematode management may be achieved. Additionally, it should be noted that green manures can also be good hosts for some plant-parasitic nematodes, especially when they remain in the field for extended periods with sufficiently suitable temperatures to allow population build-up. Therefore, follow-up of this technique is recommended when used for nematode management.

14.10.5. Fallow

The term **black fallow** is used when no vegetation is allowed to grow on land, whereas **fallow** refers more to a period without agricultural cropping, but where weeds can still maintain or even increase nematode populations. Keeping land free of vegetation through frequent tillage or herbicide application can reduce nematode populations primarily through starvation, although desiccation and exposure to solar heat may contribute. In general, the practice is not particularly attractive due to the increased risk of soil erosion and loss of production during the period of black fallow. However, it is sometimes the only available option to manage economically important nematode species with broad host ranges, such as *M. chitwoodi*, a quarantine nematode in the EU.

In traditional subsistence cropping systems in developing countries, smallholders rely on natural fallow periods for several seasons following crop production, for restoration of soil fertility, maintenance of soil structure and suppression of pests. However, this practice can be maintained only where sufficient land is available. Intensification of cropping practices and rising human populations increasingly limit the extent of this practice.

14.11. Amendments

The application of fertilizers or organic matter to soil is a readily accepted practice for improving crop production, primarily with regard to improving soil fertility and structure, but it can also lead to reduced plant-parasitic nematode population densities.

14.11.1. Organic matter

Amending soil with various sources of organic matter (Table 14.1) has led to reduced nematode problems, either by reducing population levels or by increasing yields

Table 14.1. Plants most readily noted for field nematode management.

Plant	Method	Plant part used
American jointvetch (*Aeschynomene americana*)	Rotation	–
Brassica spp. (cabbage types)	Incorporated	Whole plant
Bahiagrass (*Paspalum notatum*)	Rotation	–
Butterfly pea (*Centrosema pubescens*)	Rotation	–
Castor (*Ricinus communis*)	Incorporated/rotation	Oil cake
Cotton (*Gossypium* spp.)	Incorporated/rotation	Oil cake
Sunn hemp (*Crotalaria* spp.)	Incorporated/rotation	Whole plant
Giant star grass (*Cynodon nlemfuensis*)	Rotation	–
Groundnut (*Arachis hypogaea*)	Incorporated/rotation	Oil cake
Hairy indigo (*Indigofera hirsuta*)	Rotation	–
Horse bean (*Canavalia ensiformis*)	Rotation	–
Lemon grass (*Cymbopogon flexuosus*)	Incorporated	Leaves
Marigold (*Tagetes* spp.)	Incorporated/rotation	Whole plant
Mexican sunflower (*Tithonia diversifolia*)	Incorporated/rotation	Leaves
Mustard (*Brassica campestris*)	Incorporated/rotation	Leaves
Neem [margosa] (*Azadirachta indica*)	Incorporated	Leaves, oil cake, seeds
Pangola grass (*Digitaria decumbens*)	Rotation	–
Kudzu (*Pueraria phaseoloides*)	Rotation	–
Rye (*Secale cereale*)	Incorporated	Leaves
Sesame (*Sesamum indicum*)	Incorporated/rotation	Oil cake
Sorghum (*Sorghum bicolor*)	Incorporated/rotation	Leaves
Sunflower (*Helianthus annus*)	Rotation	–
Velvet bean (*Mucuna* spp.)	Rotation	–

Note: Plants grown in rotation are often incorporated in due course, but are included here as their cultivation in rotation (in some cases intercropped) or as a cover crop provides nematode pest reduction. Incorporation refers to the fact that plant parts are transported to the site for the purpose of mulching and incorporated into the soil. Organic mulches/amendments in general, irrespective of plant origin, tend to result in nematode suppression.

without affecting populations (Widmer *et al.*, 2002; Thoden *et al.*, 2011). The complete picture of the mechanisms of control is not yet elucidated. Nematode suppression is certainly attributable to increased saprophytic and antagonistic soil biota activity following the application of organic matter. It is also recognized that crops are less stressed and more tolerant of nematode parasitism when grown with mulches than in less favourable conditions. The release and consequent build-up of organic acids, phenolic compounds, ammonia or other compounds to concentrations toxic to nematodes are also prime factors in the nematicidal activity of different soil amendments (Ebrahimi *et al.*, 2016). Numerous nematode genera are reported to be affected, although most studies deal with species of *Meloidogyne*. Application of oilseed waste (cakes), such as castor, neem (also known as margosa) (*Azadirachta indica*), cotton (*Gossypium* spp.), groundnut and mustard, amongst others, appear particularly effective at reducing nematode population levels. Waste crop by-products, such as sawdust, fruit pulp, coffee husk, oil palm debris and molasses, are also attractive amendments. Although such waste products tend to be inexpensive, they are often only locally available and require transportation to the field. Consequently, their practical use is of limited value for widespread field implementation. However, locally they can offer an effective means of nematode management and soil fertility improvement. Amendment with animal waste products, such as manures, bonemeal and crustacean shells can also lead to reduced nematode populations. The addition of crustacean waste products, such as composts or meals which contain high amounts of chitin and nitrogen, has proved highly effective, leading to commercial products sold as fertilizer or soil improvers. Frass waste, a chitin-based by-product from black soldier fly production, is also attracting attention as a soil amendment for improving soil fertility (sold as Frassor fertilizer in The Netherlands), as well as for soilborne pest and disease suppression, including nematodes. Chicken and pig manures also appear particularly effective and, as with chitin, their activity probably depends on the release and build-up of nematotoxic levels of ammonia. In some banana production systems mulching with a range of organic material has led to reduced damage by *R. similis*, caused by several factors, including suppressing temperature increases, which would otherwise be more optimal for nematode multiplication. Several types of compost have also been shown to contribute to the reduction of harmful nematode densities. Applying animal manure or cultivating a green manure crop is a regular agricultural practice. The additional benefit of nematode management may thus be an important and useful supplementary gain. However, although the mechanisms surrounding the reduction of nematode populations or damage appear complex, in general, large quantities of material are necessary to be effective. Therefore, the material must be locally available and inexpensive.

In addition to amending soil with plant material, aqueous extracts or leachates from plants with nematicidal properties have received substantial attention. Commonly termed botanical pesticides, nematicidal compounds from certain plants are applied as a pesticide following extraction. Many plants with nematicidal properties are also applied as organic amendments, in the form of leaves, bark or pounded seeds, for example to the base of the planting hole or as a mulch. Neem (*A. indica*) is perhaps the most studied, with some convincing results against various nematode species when applied either as a mulch, dip or drench. Other notable examples include *Tagetes* spp., *Crotalaria* spp. and lemon grass (*Cymbopogon flexuosus*), whilst some seaweeds are less well known but produce a range of bioactive compounds.

14.11.2. Non-organic amendments

Various levels of nematode control following application of mineral fertilizers have been observed. Applications of certain fertilizers may be toxic to nematodes or suppress their multiplication and damage through changes in host nutrition. The interactions between N, P and K availability and nematode populations and/or damage has probably received most attention, especially K, which is also understood to have a general balancing effect on N and P. However, results are often contradictory or highly variable in the levels of control obtained (Coyne *et al.*, 2004). Indeed, this may depend on several factors, such as the fertilizer type and rate of application, the minerals applied and their chemical formulation. However, differences in biotic and abiotic factors between sites, in climate and environmental factors, which affect nematode population dynamics, undoubtedly compound the influence of mineral availability on nematodes. Moreover, the magnitude and complexity of the probable interactions makes it difficult to provide general recommendations on the use of mineral fertilisers for nematode management.

14.12. Time of Planting

Planting to avoid periods of peak nematode activity can be exploited in certain circumstances to reduce nematode damage. Crops planted during periods when temperatures are sub-optimal for nematode development can enable seedlings to be sufficiently advanced to withstand nematode attack. Crops may still be affected but, due to delayed nematode maturation, peak activity may occur too late in the crop cycle to result in heavy yield losses. Autumn sown, as opposed to spring sown, cereals in Europe suffer smaller yield losses to *H. avenae*; carrots are sown late in California to avoid *M. incognita* damage. Care should be taken, however, not to cause a shift in nematode species that are better adapted to the time of planting. Early planting of potatoes in north-west Europe to avoid damage by *G. rostochiensis* contributed to the shift towards *G. pallida*, the potato cyst species that hatches at colder temperatures.

14.13. Conclusions

This chapter combines our knowledge on the role of microbes in the suppression of plant-parasitic nematodes and cultural strategies to manage and maintain them below economic thresholds. Plant pathologists studying above-ground fungal epidemics have remarked on the extent of the genetic diversity that can occur in large pathogen populations (McDonald *et al.*, 2022). Consequently, the benefits of deploying germplasm combinations of both resistant and susceptible cultivars to improve crop durability is advocated as a management strategy against a particular pathogen, including nematodes. In a stochastically variable field environment, however, outcomes are not always obvious, and careful monitoring is needed. Nematode-suppressive soils, for example, built up over several years of monocropping, give insights into how cultural approaches can result in plant-parasitic nematode decline. The applied ecologist or agronomist may view the outcome as the result of natural enemies but this may not always be so obvious. Biological and cultural nematode management in the field is therefore the result of combinations of multiple processes and one has to keep an open mind for unexpected, counter-intuitive observations.

Nematode Resistance in Crops*

ABIODUN O. CLAUDIUS-COLE

Department of Crop Protection and Environmental Biology, University of Ibadan, Nigeria

15.1. Introduction

There are several constraints associated with the management options available for many important crops that are damaged by nematodes. Such constraints may be the potential or actual harmful environmental side effects of pesticides, as well as practical difficulties in their application to soil for nematode control. Other limitations are imposed by the cropping system, including economic and social factors affecting costs of treatment or avoidance of damage through crop rotations. Crop damage can be economic or have a direct impact on household food security, depending on the nature of the crop and the production system employed. Growers and agricultural scientists increasingly look to the use of resistant cultivars as an environmentally benign, durable and cost-efficient way of avoiding nematode-induced crop loss. Pressure to provide alternative nematode management technology is increasing as a result of the withdrawal of highly effective synthetic nematicides, a massive movement to 'greener' cultivation technologies such as conservation, precision and/or organic production approaches, as well as greater awareness among particularly first-world growers of soil health restoration and preservation. Even so, the development and use of resistance requires time and planning if these potential benefits are to be attained.

Host ranges differ among nematode species, with some showing a high degree of specialization. Cyst nematode species, in particular, tend to have host ranges restricted to one or a few related plant families. Major pest root-knot nematode species, notably those

*A revision of Starr, J.L., Mc Donald, A.H. and Claudius-Cole, A.O. (2013) Nematode resistance in crops. In: Perry, R.N. and Moens, M. (eds) *Plant Nematology*, 2nd edn. CAB International, Wallingford, UK.
Corresponding author: bi_cole@yahoo.com

  Plant Nematology, 3rd edn (eds R.N. Perry, M. Moens and J.T. Jones)
DOI: 10.1079/9781800622456.0015

reproducing asexually, have wide host ranges that contribute to their economic importance (see Chapter 3). Many other non-sedentary nematode parasites also have wide host ranges with some showing preferences for particular plant taxa on which they reproduce most and may become significant pests of crops (see Chapters 5 and 6). These differences in host ranges can be exploited to manage nematode populations and reduce yield losses through rotation of crop species. It is within plant taxa that are generally hosts of a particular nematode species that resistant plants may be found and exploited as sources of resistance.

This chapter describes the impacts of resistant genotypes on host plant growth and nematode populations and provide specific examples to indicate some successes, opportunities and challenges.

15.2. Concepts of Resistance

Resistance is the term used to describe one aspect of the outcome of the interaction between a plant and a parasitic nematode. Resistance to nematodes can be defined as the ability of a plant to reduce nematode reproduction such that no nematode reproduction occurs in a highly resistant plant, a low level of reproduction occurs in a moderately resistant plant and unhindered nematode reproduction occurs in a susceptible plant (Roberts, 2002). In the context of resistant cultivars, it refers to plants that have a phenotype different from at least some other plants of the species, and usually different from the commonly grown crop cultivars. In many cases, resistance is towards one extreme of a host efficiency range shown by an array of plant genotypes. Figure 15.1 illustrates not only resistance but also other theoretical potential outcomes of interactions between plants and nematodes. Partial resistance is the condition in which the rate of reproduction of the nematode is low in relation to the degree of soil infestation. Generally, partial resistance is not specific to different populations of one nematode species, even though it may vary in effectiveness against the different populations. The term **relative susceptibility** has been introduced to describe the population dynamics on resistant cultivars (see Section 11.9.3.1. in Chapter 11).

The phenotypic continuum **susceptible–resistant** is a measure of **host efficiency**. This is measured by comparing nematode reproduction on a number of host plant genotypes. Those that support no or little nematode reproduction are **resistant**; those that allow substantial reproduction are **susceptible**. Note that this phenotypic classification is used as an adjective to describe the plant, but is in fact derived from the genetic interaction between plant and nematode. The important practical relevance of this range of phenotypes is that resistance can be used in crops to manage nematode populations.

The other aspect of the interaction is how a nematode affects its host, the **tolerant–intolerant** (or **sensitive**) continuum in the upper part of Fig. 15.1. This is a measure of **host sensitivity** and may be determined by assessing plant growth in the presence of the nematode, or in crop terms, measuring yield loss caused by nematodes. In practice, the sensitivity of a single plant usually not only depends upon its genotype, but also is related to how many nematodes attack it. At very great nematode population densities, plants suffer measurable growth reduction, whereas when only a few nematodes attack the roots most plants grow well. This is of course why we write of nematode management, or more strictly of nematode population management, rather than population control. The objective is to reduce nematode populations and then maintain them at densities less than that at which economic crop loss occurs.

Similarly, host efficiency is affected by nematode population density. Nematode population densities are usually expressed in relation to the volume or mass of soil but a better

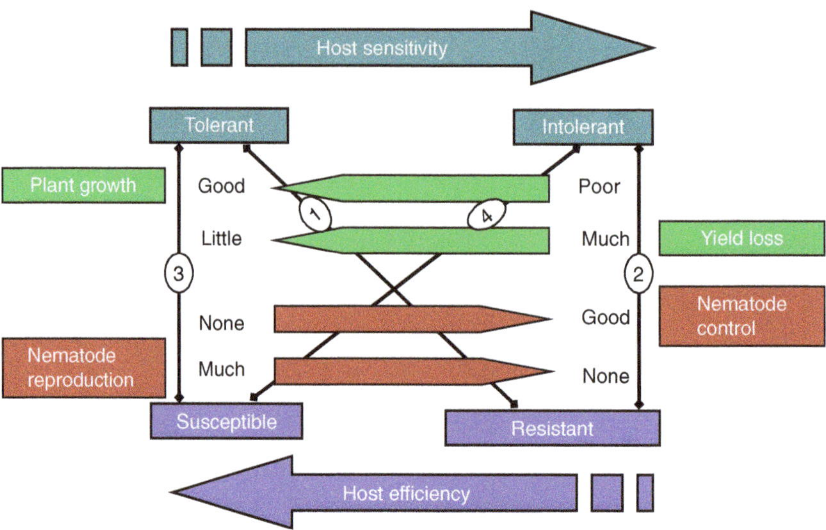

Fig. 15.1. The interrelationships between plants and nematodes defining resistance and susceptibility as extremes of host efficiency (measured by amount of nematode reproduction) and tolerance and intolerance as extremes of host sensitivity measured by plant growth (or by yield). The numbered lines represent potential extreme outcomes. (1) A resistant and tolerant cultivar – the ideal for nematode population management and crop production. (2) Resistant but intolerant, controlling nematode populations but liable to damage when sown in heavily infested fields. (3) Susceptible but tolerant, of limited value as nematode populations can increase to damaging densities. (4) Susceptible and intolerant, of no value where nematodes occur.

expression in relation to host efficiency would be nematode density per unit of available host tissue. As nematode density per unit of plant increases, a threshold is reached (governed in part by environment, host crop and nematode species) at which there is too little host resource to support maximum nematode reproduction. At populations greatly exceeding this threshold value or carrying capacity of the host, the final population density of nematodes per unit of soil is less than the initial density. This multiplication threshold is invariably greater than the damage threshold so the observation is of limited practical value but is important in quantifying plant–nematode relationships. Chapter 11 includes a full treatment of population dynamics, different types of threshold values and yield responses.

The outcome of the host sensitivity continuum is also subject to environmental influence through the often interacting effects of conditions on nematodes and plants. Thus, factors that favour nematode activity tend to increase plant sensitivity, whereas those that favour plant growth tend to increase tolerance. Often such influences may be favourable for both plant and nematode; soil moisture parameters, for example, may promote root growth and nematode movement and hence invasion. At other times, when nematodes have invaded and established feeding sites within roots, dry soils may adversely affect plant growth whilst not affecting the nematodes, thereby reducing the tolerance of the plants relative to growth in moister soils. Host efficiency is also subject to environmental influences although usually to a lesser degree, except that high temperatures may erode the effectiveness of some resistance.

Finally, in this section defining the concepts that are important to developing and using resistant cultivars, the variety of combinations of host efficiency and host sensitivity that exist in plants can be seen in Fig. 15.1, indicated by the numbered connecting lines as the following four extreme combinations:

1. A **tolerant and resistant plant** – the ideal for management of plant-parasitic nematode populations.
2. An **intolerant and resistant plant,** controlling nematode populations effectively but liable to damage when nematodes are numerous.
3. A **tolerant but susceptible plant,** of limited value where nematode reproduction may increase the population density above the damage threshold.
4. An **intolerant and susceptible plant,** of no value where nematodes occur.

Few cases of high levels of resistance coupled with high levels of intolerance have been documented (Fig.15.2). Tobacco carrying the *Rk* gene for resistance to *Meloidogyne*

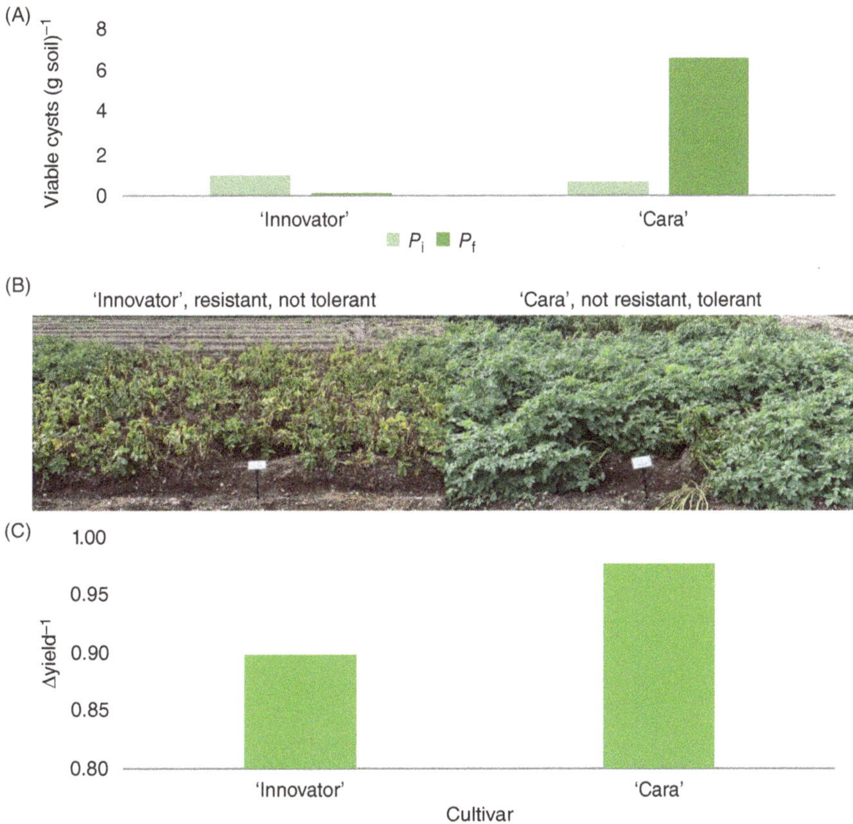

Fig. 15.2. Potato cyst nematode (PCN) resistance and tolerance under field conditions. 'Innovator' has strong resistance to *Globodera pallida*, whilst 'Cara' has no PCN resistance. This is reflected by the final populations (P_f) recorded in the soil following harvest and comparing it to the initial population (P_i) before planting (A). However, even though 'Innovator' is resistant to *G. pallida*, under a strong PCN infection pressure, it is yellow and wilted, whereas 'Cara' looks healthy (B). This reflects PCN tolerance. PCN tolerance can be measured through changes in yield where values for the inverse of Δyield that are further away from 1 have suffered more yield loss (C). Images courtesy of James Price. Data collected by Scottish Agronomy. (From Price *et al.*, 2024.)

incognita exhibits stunting when young plants are transplanted into soils infested with a large population of the nematode. Similarly, upland cotton genotypes (*Gossypium hirsutum*) with resistance to the reniform nematode, *Rotylenchulus reniformis*, introgressed from *G. longicalyx* are intolerant of the nematode. In practice, particularly in subtropical and tropical conditions where a nematode species can complete multiple cycles of reproduction on a single annual crop, effective resistance typically reduces nematode population densities per unit of plant such that, compared to the susceptible plant, resistant cultivars have significantly greater yield. In such situations, and when initial nematode densities are below the damage threshold, intolerant but resistant cultivars will yield more than susceptible ones.

15.3. Inheritance and Mechanisms of Resistance

Resistance to nematodes is expressed in many forms and is conditioned by a wide variety of genetic systems (Williamson and Kumar, 2006). Plant resistance is conditioned by a variety of genetic mechanisms, which may be mono-, oligo- or polygenic. The genes involved may be further classified by their effect on phenotypic expression, ranging from major genes (large effects) to minor genes (small effects). The apparent inheritance of resistance genes can be influenced by the genetic background in which they exist. In cases where resistance is conditioned by a single dominant gene with a major effect, typically a classic hypersensitive necrotic reaction by affected plant cells can occur. The *Mi* gene for resistance to *Meloidogyne* spp. in tomato is a well-characterized example of this type of resistance. Advances in genetics, genomics and bioinformatics have contributed to a better understanding of the molecular and genetic mechanisms of nematode resistance. These tools have enabled researchers to generate genomic resources and marker-trait associations on a large scale. Whole-genome resequencing, genotyping-by-sequencing, genome-wide association studies, haplotype analyses and DNA capture technology (enrichment sequencing) have been employed to map and dissect genomic locations for nematode resistance.

In some plants, resistance is conditioned by recessive genes. This has been observed in several cotton accessions that express moderate levels of resistance. Further, resistance in cotton is often transgressive, where progeny of crosses between two moderately resistant parents express a level of resistance that has greater than expected additive effects. In both groundnut and cotton, there are some resistant accessions where the best evidence suggests that resistance is conditioned by two genes. Further, it appears that one gene in each system is primarily responsible for suppression of nematode reproduction, whereas the other gene is primarily responsible for suppression of root galling. Resistance to root-knot nematodes in the two accessions of upland cotton, 'Clevewilt-6' (PI 65358) and 'Wild Mexico Jack Jones' (PI 593649), is conditioned by a single recessive gene. Mapping studies have confirmed that they are two distinct genes and progeny from a cross of these parents exhibit transgressive segregation. This transgressive segregation is due to recessive genes present in some susceptible genotypes (Wang *et al.*, 2008). However, after multiple backcrosses of progeny from this cross to susceptible parents, the resistance appears to be inherited as a dominant trait governed by two genes. Thus, that which initially appeared to be a recessive trait is expressed as a dominant trait after introgression into different genetic backgrounds.

Other forms of resistance may be conditioned by multiple genes and may also involve either a hypersensitive plant response or less dramatic responses. In some cases, the resistant plant fails to support well-developed giant cells (for root-knot nematodes; Figs 15.3 and 15.4) or syncytium complex (for cyst and reniform nematodes) and rates of nematode development and fecundity are reduced. In some soybean and groundnut accessions, resistance to root-knot nematodes is expressed as a reduction in the number of invading juveniles that establish feeding sites. Initial penetration of the roots is nearly equal between the susceptible and resistant plants but then there is a high rate of emigration from the roots of a resistant plant. Juveniles that do establish a feeding site in such resistant plants typically have a slower rate of development than do juveniles on susceptible plants. Roots of immune or resistant cassava cultivars form a callous tissue complex in response to nematode infection, thus making giant cell formation difficult.

The observed phenotypic resistance to *Ditylenchus africanus*, the groundnut pod nematode, in groundnut breeding lines point in a different direction (Steenkamp, 2008). Amplified fragment length polymorphism (AFLP) analyses of reciprocal two-way crosses between a susceptible (cultivar) and resistant (line) parent suggested that the resistance is quantitatively inherited (thus, is likely to be polygenic), although the phenotypic expression of resistance is relatively very strong. The linkage analysis indicates that three quantitative trait loci (QTLs) on two separate linkage groups are associated with the observed groundnut resistance to this nematode. Heritability of the trait and maintaining it in progeny is a challenge that would benefit from more in-depth knowledge of the genes or loci that are associated with the trait.

Understanding the physiological and molecular basis of resistance to most nematodes is a work in progress. Phytoalexins and other similar plant-produced compounds with antimicrobial activity have been implicated in some plant resistance responses. In lima bean (*Phaseolus lunatus*) and soybean (*Glycine max*) the accumulation of such compounds in necrotic lesions in response to *Pratylenchus scribneri* or *M. incognita* has been observed. These compounds appear to act as repellents to the nematodes. Various types

(A) Susceptible

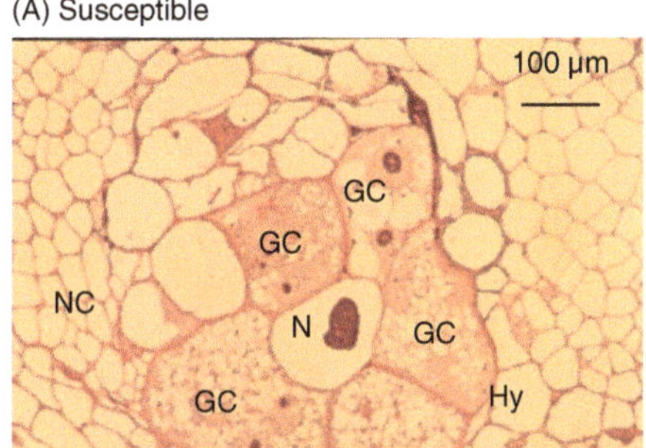

(B) Resistant

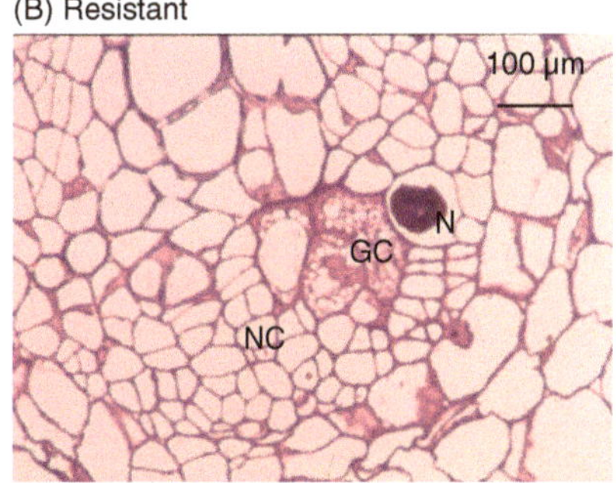

Fig. 15.3. Giant cell formation in susceptible ‘Prima2000’ (A) and resistant ‘LS5995’ (B) cultivars of soybean to *Meloidogyne incognita*. In the susceptible cultivar several large giant cells (GC) are evident, whereas in the resistant cultivar there are fewer and smaller giant cells. N, nematode; NC, normal cells; Hy, hypertrophy. (Courtesy H. Fourie.)

Fig. 15.4. Host plant responses of susceptible 'Prima2000' (A) and resistant 'LS5995' (B) cultivars of soybean to *Meloidogyne incognita.* (Courtesy H. Fourie.)

of phenolic compounds have been found to accumulate in sites of infection in roots of dicots and monocots that correlate with nematode resistance in a wide variety of plant–nematode combinations (Desmedt *et al.*, 2020).

The *Mi-1.2* gene from tomato is one of several genes for resistance to root-knot nematodes in the *Solanum peruvianum* germplasm and has been studied extensively (see Chapter 16). This gene was introgressed into cultivated tomato, *S. lycopersicum*, in the 1940s and is now widely used in commercial tomato production. The *Mi-1.2* resistance gene has several unique characteristics. It conditions resistance to three nematode species (*M. arenaria*, *M. incognita* and *M. javanica*, which may be due more to how closely related these species are to each other than to any unique attribute of the resistance gene), it is temperature sensitive and non-functional at temperatures exceeding 28°C. This gene also conditions resistance to potato aphids and whiteflies. *Mi-1.2* belongs to the NB-LRR class of plant resistance genes (containing a nucleotide binding site (NB) and a leucine-rich repeat (LRR)). The resistance is similar to resistance to other pathogens that is conditioned by this class of resistance genes. The interaction of the resistance gene with an effector molecule (avirulence gene product), either directly or indirectly, initiates a signal transduction pathway in the host cell. The end result of this pathway is cell death due to a complex of biochemical reactions. An important component of this signal transduction pathway is the protein Rme-1. This protein may act at the same step as the Mi-1.2 protein or upstream (Fig. 15.5). A key characteristic of such resistance responses is an oxidative burst with production of oxygen-free radicals, which then react with a host of other compounds. Affected host cells also respond with the production of various pathogenesis-related (PR) proteins and phytoalexins. Membrane integrity is compromised. The resulting host cell death inhibits further giant cell development and results in death of the invading nematode. In the case of the *Mi-1.2* gene, this host response is rapid, occurring within the first day or two after root penetration. In other resistant plants, a similar necrotic response may not be initiated for several days after invasion by the nematode. In such cases the initial development of the syncytia or giant cells appears normal before initiation of the necrotic response. The effector molecule from the nematode that initiates the signal transduction pathway has not been definitively identified but one candidate has been reported: the silencing of the *Cg-1* gene in a nematode renders them virulent on *Mi-1.2*-mediated resistance. The genes *Mi-2* through to *Mi-9*, which are also from accessions of *S. peruvianum*, have additional attributes (Williamson, 1998). Some of these genes are not temperature sensitive and others are active against nematode populations that are virulent on *Mi-1.2*. Unfortunately, none of these genes has yet been introgressed into cultivated tomato.

Not all major effect dominant resistance genes condition a necrotic hypersensitive host response. The resistance to *M. arenaria* in groundnut is conditioned by a single dominant gene that confers near immunity but no hypersensitive necrotic response is observed.

The identification of nematode resistance genes has been made difficult by the fact that they often reside in regions of the plant genome that have limited recombination. Further, these regions of the plant genome are often rich in resistance genes and resistance gene analogues. In several cases, the location of the resistance gene has been mapped to a particular region of a chromosome in addition to identifying the precise gene sequence responsible for nematode resistance. In the case of *Mi*, there are actually three closely

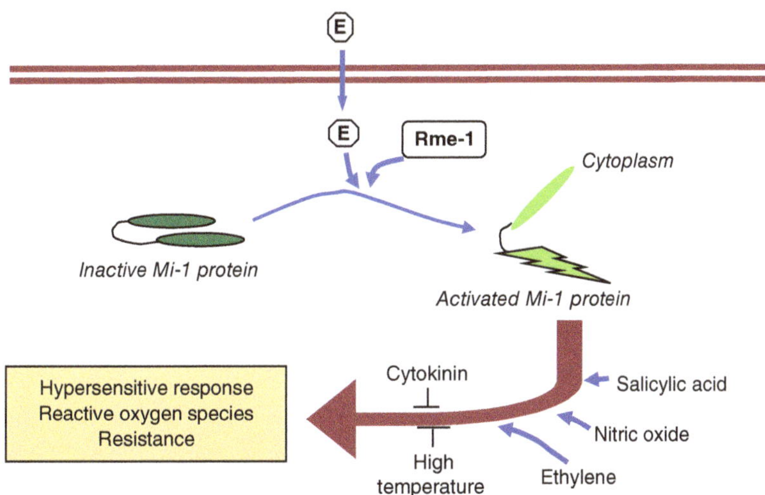

Fig. 15.5. Schematic representation of the interaction of the Mi-1.2 resistance gene protein in tomato with effector molecules from the nematode and other host components to initiate the resistant host response. The Rme-1 protein from the host may be involved in the activation of the resistance response. (From Williamson and Roberts, 2009.)

linked genes with a high degree of homology designated 1.1, 1.2 and 1.3, with only *Mi-1.2* being the active resistance gene. The genes *Gpa2* and *Gro1* for resistance to the potato cyst nematodes *Globodera pallida* and *G. rostochiensis*, respectively, have been sequenced and are also NB-LLR-type resistance genes. Diagnostic molecular markers for the *GpaIV* and *GpaV* major QTLs have been developed. The *H1* gene for resistance to *G. rostochiensis* has been found to be localized in chromosome V. *H1* resides in a large cluster of genes of the coiled-coil domain (CC)-NB-LRR type. Two genetic markers linked to the resistance gene *H1* have been used successfully to identify resistance in large populations of potatoes. Similarly, some diagnostic polymerase chain reaction (PCR) markers have been deployed for the *H3 (GpaIV)* source of resistance.

Resistance to the soybean cyst nematode, *H. glycines*, is complex with several QTLs contributing to the resistant phenotype, some of which are dominantly inherited (*Rhg4* and *Rhg5*) and some behaving as recessive genes (*rhg1–3*). The major QTLs controlling soybean resistance to soybean cyst nematode have been identified, localized and characterized (Yan and Baidoo, 2018). Resistance by the two loci with larger effects, *rhg1* and *Rhg4*, was found to be independent of LRR-kinase genes. Studies have provided evidence of two genes, those for an α-SNAP protein and a WI12 wound-inducible protein, being involved in the resistance response (Cook *et al.*, 2012; Lee *et al.*, 2015) (see Chapter 16). Interestingly, susceptible soybean has a single copy of these two linked genes, whereas resistant genotypes have several tandem repeats of these genes. Another recent report provides strong evidence for *Rhg4* being a serine hydroxymethyltransferase involved in serine ↔ glycine interconversions and essential for one-carbon metabolism in the plant. In addition, three different forms of α-SNAP proteins were observed among resistant lines with varying numbers of repeats, suggesting that the copy number and sequence of the α-SNAP protein in the *Rhg1* play active roles in soybean resistance to soybean cyst nematode. Two main types of resistance identified within the cultivated soybean are the Peking-type and PI 88788-type resistance. The Peking-type resistance is conferred by low-copy *SNAP* (soluble NSF attachment protein) gene at *rhg1-a* in combination with resistant-type *SHMT* (serine hydroxymethyltransferase) gene at *Rhg4* and *rhg*, and is resistant to soybean cyst nematode race 5. This resistance exhibits rapid syncytium degeneration. The PI 88788-type resistance is conferred by a high copy number of three genes, including *SNAP*, at *rhg1-b* with no presence of resistant-type *Rhg4* and is associated with slower syncytium degeneration (Liu *et al.,* 2017).

15.4. Virulence in Nematodes

The development of resistant cultivars has largely been based upon single, major genes (barley, groundnut, potato and tomato) or on a few genes (cotton, potato and soybean). During the course of breeding, it has become apparent that many plant–nematode interactions conform to predictions of the gene-for-gene hypothesis (Box 15.1). This postulates that for each plant resistance gene (*R*-gene) a pathogen has a corresponding avirulence gene (*Avr*-gene). The interactions between the products of these genes condition whether the plant is resistant and the nematode is avirulent, or conversely the plant is susceptible and the nematode is virulent. When resistance and avirulence are conditioned by the dominant allele, the four possible genotypic combinations give rise to the two phenotypes of each organism (Fig. 15.6). Such genetic interactions have been formally demonstrated for potato and soybean cyst nematodes. It is also likely that they apply to interactions

Box 15.1. Small grain cereals and cyst nematodes.

The use of resistance in cereals to *Heterodera* spp. is complicated by the diversity of the cyst nematodes. Resistant cultivars have been developed in Europe and India but have been most successful in Australia. It is likely that cereal cyst nematodes in Australia are introduced and are of limited genetic diversity. Australian populations appear to have a single virulence phenotype that is distinct from virulence phenotypes prevalent in Europe, although there are similarities with those in Spain and Germany.

The resistances used in wheat and barley in Australia are also effective against some, but not all, European nematodes. Oat that is fully resistant in Australia is not highly resistant in Europe. The use of resistant cereals allows the intensive culture of wheat in much of Australia, and with no emergence of virulence after several years. However, in fields in which cyst nematode populations have been controlled, species of migratory root-lesion nematodes (*Pratylenchus* spp.) increase and cause damage.

In Europe, notably in the UK and in Scandinavian countries, resistant cultivars of oat and barley have been bred and grown successfully for short periods. Virulent pathotypes, and also new species, have emerged on cultivars with major gene resistance. These virulent nematodes have not become a general problem because other factors control cereal cyst nematodes. These factors include soil suppressiveness by fungal parasites of nematode females and eggs, coupled with the greater tolerance of winter-sown crops and good growing conditions.

Rice has a notorious history of susceptibility to aggressive species of cyst, rice-root and root-knot nematodes. Little resistance has been found in Asian rice (*Oryza sativa*) but most African rice genotypes (*O. glaberrima*) are regarded as resistant to some of these nematode species. Resistance to *Heterodera sacchari* appears to be a qualitative trait in progeny of crosses between the rice species, whereas resistance to *Meloidogyne graminicola* appears to be quantitative.

Although there are numerous reports of resistance to various nematodes in maize, sorghum and pearl millets, little effort has been expended in the development of that resistance. In many instances, the nematodes are not viewed as pests of sufficient importance on these crops for resistance to nematodes to warrant attention from plant breeders.

with root-knot nematodes, many species of which reproduce asexually so that formal genetic proof is more difficult, requiring cloning of *Avr*-genes.

These genetic interactions and the phenotypes they control are the basis of the specificity that is found in some resistant cultivars. A key feature is that populations of both plants and nematodes differ in the numbers and frequencies of their *R*- and *Avr*-genes. Resistant cultivars of self-pollinated crops developed by modern plant breeding are pure lines and homozygous at their *R*-gene loci. Hybrid and out-crossing crops may be homozygous or heterozygous at the *R*-gene locus. Nematode populations may similarly be homozygous or heterozygous for their *Avr*-gene(s) or may have different alleles of more than one gene. A single *R*-gene will confer resistance only to avirulent individual nematodes, those with the matching *Avr*-gene. Where the crop and the pest have only a single gene pair interaction, the crop is resistant to the whole nematode population (perhaps in a field, or in all the fields of a region). Where the nematode population is more variable, there may be virulent individuals present that have the selective advantage of reproducing fully on that cultivar. If this occurs then the cultivar will appear to be less

	Avr/*Avr*	*Avr*/*avr*	*avr*/*Avr*	*avr*/*avr*
R/*R*	Ra	Ra	Ra	rV
R/*r*	Ra	Ra	Ra	rV
r/*R*	Ra	Ra	Ra	rV
r/*r*	rV	rV	rV	rV

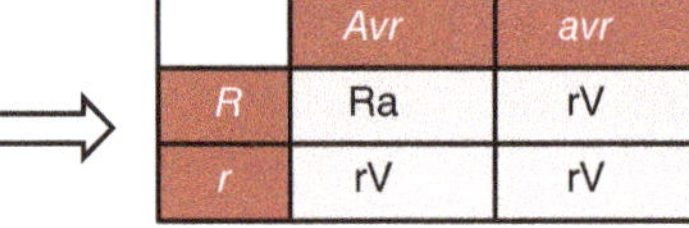

	Avr	*avr*
R	Ra	rV
r	rV	rV

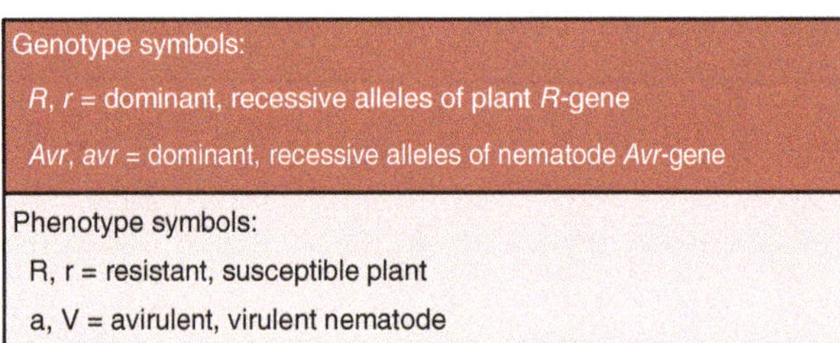

Genotype symbols:

R, *r* = dominant, recessive alleles of plant *R*-gene

Avr, *avr* = dominant, recessive alleles of nematode *Avr*-gene

Phenotype symbols:

R, r = resistant, susceptible plant

a, V = avirulent, virulent nematode

Fig. 15.6. How plant–nematode interactions conform to predictions of the gene-for-gene hypothesis: for each plant resistance gene (*R*-gene) a nematode has a corresponding avirulence gene (*Avr*-gene). The interactions between the products of these genes condition whether the plant is resistant and the nematode is avirulent or conversely the plant is susceptible and the nematode is virulent. When resistance and avirulence are conditioned by the dominant allele, the four possible genotypic combinations give rise to the two phenotypic combinations (resistant/avirulent and susceptible/virulent).

resistant. The degree and rate of erosion of the effectiveness of resistance depends upon the frequency of virulent nematodes in the population and rate of growth of the nematode population. For example, a study on *G. pallida* showed that continuous rearing of the selected field populations on partially resistant hosts produced significant changes in reproductive ability in many of the populations. A shift in the nematode population structure (frequency of virulence) will occur more rapidly in a species completing multiple generations per season than in species completing one generation per season.

Different terms have been used to describe this variation in virulence within nematode species. For cyst nematodes, populations distinguished by their virulence have been called pathotypes or races (see Chapter 4). In some cases, a field population may correspond precisely with a single pathotype (race). Other fields may have nematode populations with several virulence genes that have the potential to be selected by *R*-genes. The frequencies of these genes will determine the virulence phenotype of the population, exposure to different resistance genes will then cause changes in *Avr*-gene frequency and thus the phenotypic behaviour of the nematode population. Root-knot nematodes differing in responses to *R*-genes from more than one host plant species have been categorized using a concept of the biotype, which groups genetically similar individuals with a common phenotype (Table 15.1). Nematode pathotypes and races based on specific virulence traits should not be confused with 'host races' identified based on variation in host range. Variation in host range (plant species on which the nematode population will reproduce) is typically unrelated to the ability to reproduce on hosts carrying a specific *R*-gene.

Table 15.1. Hypothetical example of biotype designations for four isolates of a *Meloidogyne* spp. on tomato (*Solanum lycopersicum*; S. lyc) based on virulence on different *Mi* resistance genes. (From Roberts, 1993.)

Isolate	*mi*	*Mi1*	*Mi2*	*Mi3*	*Mi4*	*Mi5*	*Mi?*	Biotype
ABC	+	–	–	–	–	–	?	S.lyc 0/1.2.3.4.5
DEF	+	+	+	–	–	–	?	S.lyc 1.2/3.4.5
GHI	+	+	–	+	–	+	?	S.lyc 1.3.5/2.4
JKL	+	+	+	+	+	+	?	S.lyc 1.2.3.4.5/

mi, susceptible genotype; *Mi1* to *Mi5* are genes for resistance to *M. arenaria, M. incognita* and *M. javanica*; +, a compatible interaction –, an incompatible interaction.

This summary of the genetic basis of variation in resistance and virulence has been developed from the practical experiences outlined in the examples in this chapter and in the light of theory and experiments with plant interactions with other pathogens. This underpins the development of schemes to categorize plant host status and nematode virulence spectra. Pathotype or race identification schemes use selected plants with known resistance and at least one known susceptible as a control to test nematode populations of interest. The plants used are known as differentials. In practice, specific attributes of the individual crop and nematode have been major influences in shaping the schemes used to describe particular nematode population virulence phenotypes. Thus, schemes to categorize potato cyst nematode pathotypes include as differentials potato clones that are heterogeneous and nematode populations that were subsequently proven to be two species (Box 15.2). Schemes for cereal cyst nematode were developed using plants of several small grain cereal genera (barley, oat, rye and wheat) and applied to nematodes that subsequently have been allocated to several quite distinct species. *Heterodera glycines* (HG) Types are now the preferred designation for variation in virulence in phenotype in *H. glycines* to different sources of resistance in soybean (Box 15.3).

The function of pathotype classifications is to provide evidence as to the potential effectiveness of resistance against a range of populations of particular crop nematode pests. The information from these schemes has to be useful in plant breeding in directing the choice of resistance sources. Such a practical function means that the classifications do not always provide precise evidence of gene-for-gene interactions. Thus, many of the differential potatoes and cereals have 'cryptic' resistance genes, the existence of which was only revealed as the differentials were tested against more populations from wider areas. In the cereal cyst nematode scheme, it was even found that the susceptible control for the barley differentials was resistant to certain cereal cyst nematode pathotypes. Differential series for pathotype/race testing must be amenable to continuous revision in order to accommodate new sources of resistant plants and nematodes and new genetic information about them.

15.5. Origins and Functions

As agriculture has developed to present-day highly technological forms, there is increasing uniformity of cropping practices over quite large scales. At the same time, the degree of genetic diversity in crops has become increasingly restricted. Nematode pathotype and

Box 15.2. Potato and cyst nematodes.

Potato cyst nematodes (PCNs) are distributed worldwide but originate from South America, the centre of origin of the potato. *Globodera rostochiensis* has a more southerly native distribution than *G. pallida*. Both species were distributed around the world on tubers either directly from South America or as secondary spread from Europe.

Resistant cultivars with gene *H1* from *S. tuberosum* ssp. *andigena* CPC1673 have been widely grown in north-west Europe from about 1970. In The Netherlands 55% of ware and 99% of the starch potatoes are resistant to one or more PCN pathotypes. In the UK about 45% are resistant to pathotype Ro1. These cultivars still control *G. rostochiensis* Ro1 but growing them has led to the emergence, recognition and current dominance of *G. pallida*. Cultivars with partial resistance derived from *S. vernei* are grown with nematicides to control *G. pallida*. Pathotypes of *G. pallida* have not been clearly distinguished because of incomplete dominance of resistance genes and because nematode virulence is not fixed in field populations (Nijboer and Parlevliet, 1990). This is consistent with the observed selection for virulence in European populations of *G. pallida* on partially resistant potato cultivars (Turner *et al.*, 1983) and which threatens durable use of such cultivars. The majority of resistance genes or QTLs described for PCN are located within resistance gene clusters in the potato genome. Of the described QTLs from different potato wild species, 50% are mapped to the same region on chromosome V, where resistances for other potato pathogens are also described (Gartner *et al.*, 2021).

Resistance is clearly widespread in wild and cultivated *Solanum* species, with 52 of 63 species showing PCN resistance, and with 53% and 57% of 213 accessions resistant to *G. rostochiensis* and *G. pallida*, respectively, and 37% to both. Resistance sources to *G. rostochiensis* appeared to be centred in accessions from Bolivia and Argentina, whilst those resistant to *G. pallida* came from the whole geographic range, indicating selection not only by *G. pallida* but possibly also by the related cyst nematode species from the northern range, *G. mexicana* (Castelli *et al.*, 2003).

In Europe, potato and cyst nematode interactions vary with the PCN species involved. For *G. rostochiensis* the relationship is simplified by the genetic bottlenecks resulting from the introduction of limited genotypes of both organisms. However, with *G. pallida*, there is a more diverse and complex interaction with the potato host. In South America, genetic interactions are more complex such that it is more difficult to distinguish pathotypes and to exploit simple resistance. The evidence of co-evolution, represented by the widespread incidence of resistance in *Solanum* spp. matched by diversity in virulence in the nematode species, confirms that there is a multiplicity of resistance genes. Single resistance genes can be used effectively only where there is little variability in the nematode populations.

race schemes indicate that there is a diversity of resistance in crop plants and avirulence genes in their nematode pathogens. More *R*-genes are found in ancestral farmer-selected 'land races' of crops and in related wild species than in modern cultivars (see Section 15.6.1).

It seems certain that the resistance exploited in crop cultivars is derived from plant–nematode interactions, such as occur in natural ecosystems, but is utilized in crops in agricultural ecosystems that are very different in a number of significant ways. First, the uniformity of crop species in fields provides increased host density leading to high population densities of fewer nematode species than are encountered in natural ecosystems. Second, the narrow genetic base of the crop tends to select for genotypes of nematodes

Box 15.3. Soybean and cyst nematode: classification of virulence phenotypes.

Populations of *Heterodera glycines*, the soybean cyst nematode, exhibit considerable variation in virulence phenotype. The history of classification of virulence phenotypes illustrates the importance of the genetic interactions between plants and nematodes. Previously, 16 'races' were recognized based on female development differentials; however, the substantial variation obtained from different trials led to the development of more stable methods.

A classification scheme (Niblack *et al.*, 2002) defines 'HG Type' based on the development of females on a set of differential soybean accessions actually used as sources of resistance in formally released cultivars. The differentials are numbered sequentially based on the date of their release as a source of resistance. This system was developed with the clear understanding that it is not a genetic system but rather based on phenotypic behaviour of a specific population at a given point in time. It is intended to guide the selection of resistant cultivars carrying different combinations of resistance genes at the time at which the nematode population was characterized for HG Type.

Both the original race and the HG Type classification system use a Female Index (FI = number of females developing on the differential/number of females developing on the susceptible standard, expressed as a percentage) to measure virulence on a given differential host (see also Chapter 4).

The HG Type, by definition, is not a genetic tool. It can be expanded as new sources of resistance are introgressed into soybean cultivars. The system requires improvement for routine use to type large numbers of populations. However, used correctly, the HG Type system will reduce errors that commonly arose when the old race scheme was used to infer more from the race designation than was possible or appropriate. It has to be recognized that HG Type indicates population behaviour at one point in time. An avirulence rating does not mean an absence of the virulence phenotype in that population, rather it is present at a low frequency (<10%) at that point in time. The HG Type may change as the population evolves in response to environmental influences, in particular to the frequency at which different sources of resistance are deployed in a field.

that are well adapted to the crop and, therefore, reproduce freely on it. Where the crop has no *R*-genes, perhaps because cultivars have been bred by being grown in nematode-free soils where *R*-genes contribute no advantage and were lost during selection for yield, there may be no effect of the uniform crop on the variation in the nematode population. When single *R*-genes are introduced into such a system, it is therefore likely that it will select for increased frequencies of virulent nematode genotypes if virulence exists within the nematode population. When this happens, the resistance appears to have been eroded or 'broken down'. This is a practical shorthand term because the gene is still effective in controlling nematodes lacking the matching virulence traits. The resistance gene is not broken; rather, the nematode no longer produces a functional avirulence gene product that can interact with the resistance gene product to initiate the resistance response.

In nature, it is likely that resistance developed between co-evolving plants and nematodes. The incidence of extensive resistance sources among plant species that are ancestors of or related to crops suggests that the co-evolution is ancient. Comparisons of gene sequences thought to be involved in the parasitism of two cyst nematodes (*H. glycines* and *G. rostochiensis*) and a root-knot nematode (*M. incognita*) indicates that nematodes have diverged greatly from other organisms, also suggesting an ancient lineage (Baum *et al.*,

2004; Smant *et al.*, 2004). Molecular evidence suggests that two morphologically similar cyst nematodes (*H. glycines* and *H. schachtii*) diverged from each other between 7 and 4 million years ago (Radice *et al.*, 1988). This is additional evidence that indicates the very long timescale of plant–nematode co-evolution, compared to the domestication of most crops during the development of agriculture in the past 10,000 years.

Nematode crop pests have also been affected by the development of agriculture and some may be as different from nematodes in natural vegetation as are their crop hosts from wild plants. This may well be the case with the parthenogenetic root-knot nematodes (see Chapter 3; Trudgill and Blok, 2001). These have apparently undergone hybridization and polyploidization that has striking parallels with the evolution of several crop plant species. In some cases, the nematode pests have been introduced with their crop hosts into areas well beyond the apparent region or common centre of diversity where co-evolution occurred. This has certainly happened with potato cyst nematodes and potatoes from South America that had co-evolved close to the centres of diversity of both plant and nematode, which were introduced earlier into Europe and North America then more recently to parts of Asia, Oceania and Africa. Also, the cereal cyst nematode seems to have been introduced into Australia along with cereal cropping from Europe. Similarly, the clover cyst nematode appears to have been introduced from Europe into New Zealand, where it thrives on introduced white clover. Soybean and its cyst nematode have also been introduced from Asia into North and South America. These cyst nematodes have spread with their crop hosts in three continents, becoming significant pests within periods of less than 100 years. It is possible that the parthenogenetic root-knot nematodes have also spread in cropping systems throughout the world in relatively recent years.

15.6. Exploitation

15.6.1. Sources of resistance

There is a hierarchy of plant resources in which resistance to particular nematode species may be found. The development of resistant cultivars has been quicker and easier when sources of resistance are found in plants closely related to the susceptible cultivars. This is because there are no cytogenetic barriers to hybridization and fewer undesirable characteristics genetically linked to the resistance genes. When resistance is not available in closely related sources, more distant relatives need to be examined, including older varieties of the crop and also related, perhaps ancestral, species.

There are extensive bibliographies of nematode resistance and some germplasm collections have been thoroughly screened for resistance. Much of this information is widely available and provides a first screen to increase the likelihood of identifying sources effective against local populations. The existence of variation in *R*-genes and *Avr*-genes means that local testing is essential to validate reported resistance from other regions or countries. Various national and international germplasm collections also provide a wide range of material; some indication of what is freely available may be seen at the Germplasm Resources Information Network (GRIN) via http://www.ars-grin.gov/, which contains links to many germplasm repositories.

In wheat, a crop created during domestication of naturally occurring hybrids between wild species, resistance to cyst nematodes has been introduced by breeding from the ancestral wild species. Similarly, resistance to potato cyst nematode and to root-knot nematodes of tomato has been introduced to the crops from related but wild species. Such resistance is more difficult to use in breeding programmes and there may be many problems to be overcome before successful resistant cultivars can be produced.

There are usually cytogenetic barriers to hybridization between related species and, if these can be overcome, generally a number of undesirable features of the wild source are also transferred. The resulting increased breeding and selection efforts increases the costs and timescale of the breeding programme. Thus, Castelli *et al.* (2003) found cyst nematode resistance in 52 out of 63 wild and cultivated species of *Solanum*, the potato genus. Although not all wild species can be easily crossed with cultivated crops, some barriers can be overcome by a variety of procedures that in combination with resistance screening can transfer the gene(s) for resistance into plant breeding programmes (Box 15.4).

It is interesting that in wild species resistance often appears to be less specific than that of modern cultivars. Some of these sources are resistant to several species and some to more than one genus of nematodes. There is little evidence to indicate whether such sources have a single resistance mechanism effective against more than one nematode species. None the less, in some cases it seems possible that the same mechanism is involved; for example, sweet potato selected for resistance to *Pratylenchus coffeae* also proved to be resistant to *M. incognita*. It may be that in wild species such less specific resistance may be due to 'clusters' of resistance genes, which are disrupted by breeding and selection for a more specific resistance.

Resistance may lack durability because repeated use of single resistance genes often leads to a shift in the virulence characteristics of the nematode population, such that with

Box 15.4. Groundnut and *Meloidogyne arenaria*.

Resistance to *Meloidogyne arenaria* and *M. javanica* has been introgressed into groundnut (*Arachis hypogaea*) from a wild species. The cultivated groundnut is an allotetraploid ($2n = 40$) believed to contain an A and B genome, each with 20 chromosomes. Most wild *Arachis* species are diploids ($2n = 20$) with either the A or B genome and, therefore, are not directly compatible with the cultivated type. Two approaches have been used to introgress resistance to *M. arenaria* into cultivated groundnut from the diploid A-genome species *A. cardenasii*.

In a hexaploid pathway, the diploid species was crossed with the tetraploid, resulting in a sterile triploid. Treatment of embryos with colchicine resulted in the formation of a hexaploid, which had low to moderate fertility. Typically, several generations of backcrossing to a tetraploid recurrent parent would be necessary to achieve introgression of the resistance gene from the hexaploid into a tetraploid. In this case, however, a genetic 'error' resulted in the reversion of the hexaploid to a resistant tetraploid after one generation. This individual was fully compatible with cultivated groundnut and used as a parent in a traditional breeding scheme.

Another successful approach, the tetraploid pathway, was to cross a diploid B-genome individual (*A. batizocoi*) with an interspecific diploid A-genome hybrid individual (*A. diogoi* × *A. cardenasii*) to generate a sterile AB-genome diploid. Colchicine treatment of the sterile AB diploid resulted in a fertile AB-tetraploid, which was readily cross-compatible with cultivated groundnut and also carried the gene(s) for nematode resistance from the original A-genome parent *A. cardenasii*. The first groundnut cultivars with effective resistance to *M. arenaria* were subsequently developed from this tetraploid pathway by completing several backcross generations with a high-yielding cultivar as the recurrent parent.

time a specific resistance gene is no longer effective. This has been demonstrated with *Globodera* and *Heterodera* species on potato and soybean, respectively (Turner, 1990; Young and Hartwig, 1992), and for *M. incognita* and *M. javanica* with virulence to the *Mi* gene in tomato (Kaloshian *et al.*, 1996; Ornat *et al.*, 2001). However, if the nematode population in a given field or region lacks the appropriate diversity with respect to virulence, then there may be no selection for virulence with repeated use of a given resistance gene. This appears to be the case for the *H1* gene for resistance to *G. rostochiensis* in some regions. Similarly, repeated use of resistance may cause a shift in the species present in a field, with species against which the resistance is not effective becoming dominant. This has been documented for tobacco, where increased use of resistance to *M. incognita* led to an increase in the frequency of *M. javanica* against which the resistance was not effective, and in potato where use of resistance to *G. rostochiensis* led to an increased incidence of *G. pallida*. The point needs to be emphasized that variability with respect to virulence must exist within a population (either variation within a species or among species) for the use of resistance to select for a greater frequency of the virulence phenotype. Where the nematode is an introduced pest, the introduction process often represents a genetic bottleneck that reduces diversity relative to that which exists at the centre of origin. Thus, the number of potential virulence races in an introduced nematode population is typically fewer than the number that exists at the centre of origin for the nematode where it co-evolved with the host.

For most nematode pests there is no shortage of reports of resistance and many untapped sources are available. De Waele and Elsen (2002) and Peng and Moens (2003) list many sources of resistance to migratory parasitic nematodes. Peng and Moens reviewed 81 crop–nematode specific combinations and reported that there was resistance in 46 crops to 30 species of 12 nematode genera. Bred cultivars were said to be available for 25 of the 81 combinations, although not all are well adapted for production. Roberts (1982) and Sikora and Roberts (2018) have listed crops, including annuals, trees and other perennials, for which there are high-yielding cultivars available with resistance to species in eight genera of plant-parasitic nematodes (Table 15.2).

15.6.2. Selection criteria

The criteria for selecting plant germplasm as sources of resistance include the genetic relatedness of the source to the crop of interest. This is important as it affects the ways and ease with which resistance may be transferred, as indicated in the preceding and following sections. The aspects of selection criteria that are critical to the development of resistant cultivars derive from the definitions of resistance and tolerance and the factors that influence their expression (see Section 15.3). The application of selection criteria in breeding programmes (screening) has two main applications: the first is identification of sources, and the second is selection of resistant individuals from segregating populations during breeding. The principles that are relevant to the choice of method and of assessment criteria are broadly the same for both purposes. Done well, screening not only is accurate and ensures that resistance is effective against the appropriate range of nematode pathotypes or species, but also breaks genetic linkages with undesirable attributes. This provides germplasm suitable not only for cultivar development but also for the associated and relevant evaluations of nematode population control and crop loss reduction.

Table 15.2. Some food crops for which high-yielding cultivars with resistance to one or more nematode species are available (partly from Roberts, 1982; Sikora and Roberts, 2018).

Crop	Nematode	Crop	Nematode
Apricot	*Meloidogyne* spp.	Maize	*P. hexincisus*
Barley	*Heterodera avenae*	Peach	*M. incognita*
Bean, common	*M. incognita*, *M. javanica*, *Pratylenchus scribneri*	Potato	*Globodera pallida*, *G. rostochiensis*
Citrus	*Tylenchulus semipenetrans*	Oat	*D. dipsaci*, *H. avenae*
Clover	*Ditylenchus dipsaci*	Rice	*Aphelenchoides besseyi*, *D. angustus*
Cotton	*M. incognita*	Tobacco	*G. tabacum*, *M. arenaria*, *M. incognita*
Cowpea	*M. incognita*	Sweet potato	*M. arenaria*, *M. incognita*, *M. javanica*, *R. reniformis*
		Soybean	*M. incognita*, *M. javanica*, *H. gycines*
Grape	*Meloidogyne* spp., *Xiphinema index*	Tomato	*M. arenaria*, *M. incognita*, *M. javanica*
Groundnut	*M. arenaria*, *M. javanica*	Walnut	*Meloidogyne* spp.
Lucerne	*D. dipsaci*, *M. hapla*	Wheat	*H. avenae*, *P. neglectus*, *P. thornei*

Quality control is important: screening procedures should have a known degree of accuracy and must be reliably repeatable (Wesemael, 2021). The use of 'control' plants is essential: these are genotypes of known response, ideally including both a susceptible and a resistant control. The susceptible control assesses that the test has been successful by indicating the extent to which nematodes have invaded, developed and reproduced in the conditions of the test. This allows comparison among tests over time to ensure that nematode population densities and environmental factors are optimized to detect susceptibility. The resistant control, which obviously can only be introduced once a resistance source is identified, is the target by which other sources or breeding lines are identified. These two types of controls may include more than one genotype of each, and are also replicated throughout batches of 'unknown' plants to be screened. Replication is increased to improve accuracy or reduced to increase the amount of material that can be screened and is determined by the costs (and time) of growing and assessing each screening unit.

The nematodes used in screening may be introduced in a number of ways, as active invasive vermiform stages, eggs, egg masses or cysts, or by use of naturally infested soil. Generally, the order of this list corresponds to decreasing accuracy, but increasing robustness to environmental disturbance. The balance and choice of method depend upon technical resources available and the desired outcomes of the tests. There is also a parallel accuracy/robustness gradient related to the size and 'naturalness' of screening – a gradient with extremes represented by *in vitro* dixenic (only two organisms present) tissue culture methods to tests in fields or field plots. Eventually, and particularly for assessment of host sensitivity, relatively large-scale plot experiments are needed to confirm the value of the more refined tests.

It is also important to decide what diversity of nematodes should be included in tests. The alternatives are either to screen with a mixed inoculum or to use nematodes of a known but more narrowly defined genetic base: the mixtures may be of populations, for example with the major root-knot nematode asexual species, or of pathotypes of some cyst nematodes. In using mixtures, the expectation is of identifying resistance that is more broadly based. Inoculum with a narrow genetic base has the danger of identifying sources of limited effectiveness because it may be active only against a small portion of the larger nematode population. Theory strongly supports the use of narrowly based inoculum so as to maximize the numbers of resistance genes selected. In some cases, where broad-based resistance is available, it is sensible to maintain that by screening with a more broad-based inoculum. Other considerations affecting this decision include existing knowledge of variation in virulence or the likelihood of it occurring in the particular species. The choice is also affected by the goal of the breeding programme, for example, either to introgress a specific resistance gene into a crop or in initial screening of germplasm for new sources of resistance.

Assessments must take account of plant size to minimize false negatives by misclassifying as resistant such poorly growing plants that they are incapable of expressing their genetic susceptibility. This may occur when comparing nematode reproduction on wild species that lack the vigorous growth habit of the cultivated crop. A useful concept is one of effective nematode inoculum density to ensure sufficient root space for nematode invasion during the early stages of a screen. It is also important to consider the numbers of generations for which nematodes are allowed to reproduce as differences between tests (and even plants) may result from counting nematodes at one or more completed reproduction cycles. Generally, assessments of host efficiency are based upon a measure of reproduction by counting nematodes in the new generation. The nematodes counted may be the reproducing females or their progeny (in cysts, egg masses or even as next-generation invasive stages).

Sometimes a closely linked surrogate (marker) may be assessed, such as symptom expressions. These may include root symptoms such as galls, where these clearly indicate the establishment of reproducing root-knot nematodes, or knots on wheat roots characteristic of female cyst nematode feeding sites, or swollen buds typical of reproducing stem nematodes in lucerne. When such surrogates are assessed, their relationship to nematode reproduction needs to be well described if misclassifications are to be minimized. Although root-galling and reproduction of root-knot nematodes are typically positively correlated, cases of poor correlation where root-galling is not a good measure of resistance have been reported.

Identification of resistant individuals from segregating populations is increasingly reliant on marker-assisted selection (MAS) (Xu and Crouch, 2008). Identified DNA sequences derived either directly from the resistance gene or sequences closely linked to the resistance locus are used to develop various types of markers, such as the Kompetitive Allele Specific PCR (KASP) marker, utilized in molecular breeding for nematodes. Although the initial costs of development of such systems can be quite high, the potential for a highly specific, high-throughput system makes this technology very attractive. The rapidly increasing amount of genomic sequence data available for most crops is making the development of MAS more efficient. Successful breeding programmes screen thousands of individuals to identify those few that have the desired combination of yield potential, other desired agronomic or horticultural traits and resistance. Evaluation of such numbers of plants using phenotypic assays is much more costly and time-consuming than the use of high-throughput MAS systems. The favoured markers are those that are co-dominant with the resistance allele and thus allow one to select individuals that are homozygous for resistance, which is a highly desirable feature in a breeding programme. At present, most MAS systems are for resistance conditioned by single, major effect genes. Polygenic resistance requires the use of QTL markers. Unfortunately, a single QTL recovers only a portion of the available resistance, and when the original resistance is moderate, possibly only a 60–75% suppression of nematode reproduction, then a single QTL that recovers only 50% of the resistance phenotype results in a very low level of resistance in the selected progeny. The use of multiple QTLs is thus required to recover a substantial portion of the total available resistance. An important caveat to the use of MAS is that it does not eliminate the need for verification of the resistant phenotype in the final selections from a breeding programme by direct evaluation of the nematode–host interaction.

15.6.3. Breeding processes

The traditional plant breeding approaches to cultivar development are applicable to nematode resistance. In general terms, this involves four stages: (i) crossing to introduce resistance from the source into the acceptable crop background; (ii) screening the progeny to select those that are resistant; (iii) further crossing (either backcrossing to the recurrent susceptible parent or to other desirable yet susceptible parents); and (iv) repeated screening of each generation of hybrids or progeny to retain the resistance. Eventually, a range of resistant germplasm will be available for further selection accompanied by screening to produce useful resistant plants for multiplication and registration as a new cultivar or perhaps for further hybridization within the breeding programme.

The genetic structure of the crop plant and inheritance of resistance determine the appropriate breeding strategy and its integration with resistance screening. Boxes 15.1, 15.2, 15.4 and 15.5 detail some instances where resistant cultivars have involved transferring resistance from wild species, selection from among the heterogeneous plants of an existing outbreeding cultivar (lucerne) and by backcrossing from an unadapted genotype in pure line inbred crops (barley, soybean).

Transferring resistance from wild plants can take many generations to break linkages of the *R*-gene with genes for characters that are undesirable in crop plants. In tomato, the *Mi* gene from wild *S. peruvianum* confers resistance to root-knot nematodes. After crossing the wild and cultivated tomatoes, the developing embryo had to be excised and grown *in vitro* to obtain a viable hybrid (embryo rescue). Many generations of backcrossing were necessary before the resistance was available in genetic backgrounds suitable for use in breeding programmes to produce resistant cultivars.

Outcrossing crops are selected and released as cultivars that have a more or less uniform, recognizable collection of phenotypes. Even so, these may include a range of nematode resistance genotypes. In this case, it may be possible to select enough individual resistant plants for use as the parent plants to intercross and increase the proportion of resistant plants in a new variety that retains the desirable agronomic features of the original but more susceptible cultivar. A key feature of using outbreeding cultivars with less than 100% resistant plants is that it is essential that selection for resistance is made during early seed multiplication stages if the high levels of resistance are to be maintained (Box 15.5).

Box 15.5. Stem nematode resistant cultivars.

Resistance to the stem nematode *Ditylenchus dipsaci* has been exploited in a number of crops including inbred cereals (oat) and outbreeding herbage legumes (lucerne, clover). Stem nematode populations are characterized as host races, each associated with a few very susceptible host species or genera but usually able to reproduce to a lesser extent on a wider range of plants. None the less, major gene resistance has been very effective worldwide in lucerne and in the UK and Australia in oat.

In oat, resistance sources were identified among traditionally grown cultivars and transferred by backcrossing into adapted backgrounds. In the UK, the source was 'Old Grey Winter' and a single dominant gene is present in most successful winter oat cultivars grown there. Partly as a consequence of this resistance, but also because oat is grown less frequently than it once was, serious damage is rarely seen. The symptoms of resistance and susceptibility are so reliable that screening techniques are based upon the degree of swelling and necrosis. A similar approach is used with legumes to distinguish resistant and susceptible seedlings. In the field, there are big differences in growth and survival of resistant and susceptible plants on heavily infested sites.

In the USA, stem nematode resistance was identified in lucerne accessions from Turkestan, developed by mass selection as 'Lahontan' and subsequently introduced by crossing and selection. This resistance has been widely used and appears to be effective worldwide in modern cultivars. In Europe, resistant plants were identified in an older cultivar and mass selection produced parent plants for the first fully resistant cultivars. Mass selection has also been applied to red and white clover to provide cultivars with a higher proportion of resistant phenotypes. In all these outbreeding heterogeneous and polyploid crops, individual resistant plants are not heterozygous for the resistance gene. After outbreeding, progenies include some susceptible plants and it is important that selection for resistance is maintained during early generation seed multiplication.

In the field it may be difficult to recognize virulent strains, particularly on outbreeding crops, because after several generations of multiplication within a single plant genotype, nematode progeny disperse and themselves outbreed with nematodes selected on other genotypes. Reports of virulence in nematode populations in lucerne in the UK appear to be artefacts of inadequate control of the screening procedures. However, field populations virulent on resistant cultivars of field bean (*Vicia faba*) and white clover indicate that variation in virulence exists and is likely to be selected in the long term. There is also clear evidence for nematode isolate × plant genotype interactions in pea (*Pisum sativum*) (Plowright *et al.*, 2002).

In inbred crops, modern cultivars are homozygous and homogeneous; that is, they breed true for all characteristics. None the less, most inbred crops allow a degree of successful outcrossing. As a result, some of the older cultivars retain a small degree of heterozygosity in their resistance genotypes. These may be useful sources of resistance where the cultivar has been maintained by saving seed from enough plants (as saving seed from single plants leads rapidly to homozygosity in inbreds). Such mixtures may also contribute to observations that some cultivars have partial resistance when tested with heterogeneous nematode populations. In modern cultivars, the *R*-gene must be homozygous for the crop to breed true for resistance. Some crop cultivars, such as those of potato, are maintained as clones by vegetative propagation. In these, dominant resistance genes may be used in the heterozygous condition as there is no opportunity for segregation during propagation of seed tubers.

15.6.4. Use of resistant cultivars

The ideal resistant cultivar has all the advantages of susceptible cultivars, especially yield potential, together with the additional benefit of controlling nematode population increase. In practice, such resistant cultivars may be difficult to produce, particularly in crops where intensive breeding is leading to a rapidly changing portfolio of cultivars with many other improved attributes. In these cases, it is essential to use resistant cultivars in situations where nematodes are present and when control is advantageous to yields (Box 15.6).

To the grower, the most important benefit of a cultivar is the improved yield. In soil infested with nematode populations that exceed the damage threshold, yields of resistant

Box 15.6. Resistance in West Africa.

The root and tuber crops yam and cassava serve as major staples in West Africa. The major nematode pests associated with them are several root-knot nematodes and of special importance on yam is *Scutellonema bradys*. Resistance to both *Meloidogyne incognita* and *S. bradys* has been found in the yam species *Dioscorea dumetorum*, but not in the more than 300 cultivars of the most popularly consumed species of *D. alata* and *D. rotundata*. Breeding programmes are ongoing to develop nematode-resistant cultivars, with new accessions being developed and screened for nematode resistance. More success has been achieved in development of resistance in cassava for nematode management, with some cultivars rated as immune. These were improved cultivars that were developed for resistance to diseases.

The root-knot nematodes are the major pest of vegetables in West Africa. The warm humid climate allows for the development of several nematode generations within a season, resulting in an almost continuous disease cycle. Most of the tomatoes and peppers grown in the region are susceptible. Although the resistance in tomato is effective against three species (*M. arenaria*, *M. incognita* and *M. javanica*), the resistance in pepper is limited to *M. incognita*. Unfortunately, field populations often exist as mixed species, which limits the effectiveness of the available resistance in pepper. Resistant tomato cultivars were effective in suppressing nematode populations in Ghana where there was a mix of *M. javanica* and *M. incognita*. However, in Nigeria, where *M. enterobolii* was also present, the same resistant cultivars were not as effective. The molecular techniques used for identification of these *Meloidogyne* species also revealed the presence of populations that could not be identified and it is not known whether the currently available resistance will be effective against those populations. There has often been concern that the resistance in tomato is temperature sensitive and is not expressed at temperatures >28°C. This limitation can be overcome in the region by planting in the cooler months, when soil temperatures are below the critical threshold for at least the first several weeks of crop development and thus providing an opportunity for the crop to become well established before the onset of nematode parasitism.

Farmers in West Africa grow a variety of crops on small lots. Almost all the crops they grow are susceptible to one or more nematode species and the management options open to the farmers are limited. They have limited access to nematicides due to availability and/or cost. The effect of crop rotation is often not noticeable within the period they can afford to have a secondary crop growing. Further, biocontrol has so far not been successfully implemented in on-farm studies. Having resistance in some of the crops they grow has the potential to substantially increase farmers' productivity and income. Even the presence of partial resistance would increase their yields relative to susceptible cultivars.

cultivars are expected to be better than those of comparable susceptible cultivars. Suppression of nematode multiplication usually ensures that the crop grows better with fewer nematodes. Some intolerant resistant cultivars may be damaged by invasion even though they prevent nematode reproduction. In cases with annual crops and nematodes that have only a single generation per season, the yield benefits may not be great. However, in warm temperate to tropical crops, where nematodes complete multiple generations on annual crops, a yield increase through nematode population control will be achieved even in intolerant crops. Succeeding crops, whether resistant or susceptible, often benefit from nematode control provided by the resistant crop. Ogallo *et al.* (1997) showed that root-knot-susceptible lima beans yielded well after two successive crops of root-knot-resistant cotton but very poorly after susceptible cotton.

The benefits in yield are relative to those of cultivars with similar genetic yield potential. Where breeding has made rapid progress there may be susceptible cultivars with such potential that they yield more than a resistant cultivar in non-infested fields. The apparent negative effects of resistance on yield potential are probably mostly due to linkage drag, whereby genes with negative effects on yield potential are linked to resistance loci. No data are available that show a direct effect of resistance genes on reduced yield potentials. Indeed, as breeding programmes continue to work with resistance, the yield potential of the resistant genotypes usually increases. For example, the first groundnut cultivar with resistance to *M. arenaria* was selected from the fifth backcross generation in a breeding programme in which resistance derived from a wild species was introgressed into cultivated groundnut (Simpson and Starr, 2001). Yield of that first release was superior to the best susceptible cultivars in nematode-infested fields but yields of the resistant cultivar were not competitive in the absence of nematode parasitism (Church *et al.*, 2000). The second resistant cultivar, released after two additional backcross generations, had yield potentials nearly equal to that of the best susceptible cultivar and yields much better over a range of nematode population densities (Fig. 15.7).

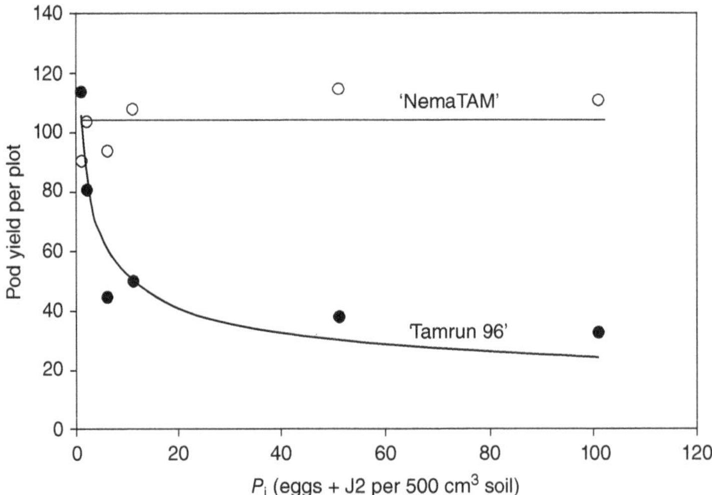

Fig. 15.7. Comparisons of the effects of *Meloidogyne arenaria* on the yields of susceptible ('Tamrun 96') and resistant ('NemaTAM') cultivars of groundnut (*Arachis hypogaea*) sown at a range of initial nematode population densities (P_i). J2, second-stage juvenile. (From Starr *et al.*, 2002.)

The development of soybean cultivars with resistance to *H. glycines* has been in progress for more than 40 years. Yield drag associated with nematode resistance has been reduced but not yet completely eliminated (Donald *et al.*, 2006).

15.6.5. Resistant cultivars and integrated nematode management

Resistant cultivars are rarely the single effective solution to managing nematode populations to avoid crop losses. Highly specific resistance effective against only one pathotype of a species or even a single species will not control other pathotypes or species in a polyspecific community. The durability of resistant cultivars may be enhanced by integration with other control measures. A current example of this is the combined use of soil treatments with nematicides and partially resistant potato cultivars to control the potato cyst nematode, *G. pallida*. In this case, the damage threshold approaches the detection limit and growers on infested soils choose to treat soils at planting with granular nematicides (see Chapter 17) to avoid damage, and use partially resistant cultivars to reduce the rate of population increase. In this way, growers can prevent yield losses and also minimize intervals between potato crops to increase their overall profitability. Durability of resistance to the soybean cyst nematode is enhanced by using crop rotation schemes to reduce the frequency with which it is necessary to grow a resistant cultivar.

Biological control may provide an environmentally benign approach that would benefit from being combined with resistant cultivars. In Western Europe, monocultures of cereal crops are now grown without damage from cereal cyst nematode. In recent history, an epidemic of these nematodes seriously threatened intensive cereal production through greatly reduced yields of third and fourth successive crops. However, under continuous cropping it was found that nematode numbers declined to less than the damage threshold and susceptible cultivars could be grown (see Chapter 14). The cereal cultivars grown in much of this area have been developed from the different land races that had been grown over several hundreds of years in different agro-ecological zones of Europe. Many of these land races, particularly of barley, oat and rye, had more than one gene for resistance to cereal cyst nematode. The effects of this resistance, often not fully effective today, in modern cultivars ensures that these have a degree of resistance in comparison to fully susceptible cultivars. So, in effect the natural decline of cereal cyst nematode associated with fungal parasites may be assisted by the unwitting use of partial resistance. Similar biological phenomena may be useful to control potato cyst nematodes (see Chapter 14).

In nature, root nematode populations are affected by a number of other soil organisms, including plant symbionts (mycorrhizal fungi). It is likely that the activity of these organisms, when coupled with moderate resistance, could be quite effective in the suppression of nematode population densities.

15.7. Successes and Opportunities

Resistance is currently available to several nematodes in a limited number of crops (Table 15.2), and there is a great need for the development of resistance to additional nematodes in numerous crops. It appears that available sources of resistance in crops are vastly underutilized, both in highly developed and in developing countries. Resistance to cyst nematodes is widely used in potato in Europe (particularly in The Netherlands and the UK) and in soybean in the USA, Brazil and Argentina. In both crops, diversity in virulence demands continued research and breeding to provide genes with sufficiently widespread effectiveness.

Resistance to *Meloidogyne* species in tomato is widely used commercially in California, but not in many other regions, especially in the tropics. Even though the *Mi-1.2* element of the *Mi* gene is not effective at temperatures above 28°C, it is effective during cooler months in many subtropical and tropical regions. Further, even if the resistance is effective only during the first few weeks of a growing season, before higher temperatures inactivate the system, this early season inhibition of nematode activity would probably be beneficial, especially when combined with other management tactics. There is increasing use of this gene in crops under protected cultivation in the Mediterranean regions of Europe but the emergence of virulence in Europe is a threat to the continued adoption of these cultivars. Recent screening of additional accessions of *S. peruvianum*, the source of the *Mi-1.2*, has revealed the presence of other nematode-resistance genes (designated *Mi-2* to *Mi-8*). These genes will be important to the more widespread use of resistance in tomato, and possibly other solanaceous crops, because some of these resistance genes are not heat sensitive and some are effective against *Meloidogyne* populations that are virulent on *Mi-1.2*.

Cereal cultivars (mainly wheat and barley) are used to control widespread cereal cyst nematode in Australia where there appears to be a single pathotype and emergence of virulence has not been a problem. In Europe, particularly in Scandinavian countries, major gene resistance to cereal cyst nematode is used in barley and oat cultivars and resistant wheat and barley have been introduced into northern India (Nicol *et al.*, 2004). In Europe, diversity in virulence within and among species of the cereal cyst nematode complex reduces the effectiveness of these cultivars.

Recently developed resistance in groundnut to *M. arenaria* and *M. javanica* should be useful in Africa, India and South-east Asia. Resistant rootstocks in perennial crops, such as peach and citrus, have been used successfully for several decades. More recently, the grafting of susceptible scions to resistant rootstocks has been used to manage root-knot nematodes on annual crops. This system allows one to overcome cytogenetic barriers to the creation of productive hybrids between more distantly related plants. This practice is being widely used on cucumber, melon, pepper and eggplant in South-east Asia, Taiwan, Morocco and Mediterranean regions of Europe.

Unfortunately, in many other cases the potential benefit of available resistance has not been realized. Cotton and common bean are examples where resistance to *M. incognita* has been identified and introgressed into modern crop genotypes but is still not widely available to growers. Considering the importance of cowpea and common bean as sources of dietary protein and their susceptibility and intolerance to *M. incognita*, one wonders what are the impediments to the greater utilization of resistance.

There are at least four possible explanations for why resistance is not more widely used. First, some reports of resistance may not have been accurate. Second, the currently available resistance is often linked to undesirable characteristics that made the growing of such cultivars unacceptable, thus requiring a substantial breeding effort to achieve a high-yielding cultivar. Third, the costs of developing resistant cultivars are not justified by the (perceived) importance of the nematode problem. The fourth, which in combination with the third is probably the chief explanation, is that naturally occurring genetic resistance tends to be too specific for use in intensive agriculture. Some of the examples (Boxes 15.1

to 15.5) illustrate the extent to which resistance has been compromised in nematode management programmes by the erosion of its effectiveness with the emergence of new virulent forms of the target or other species. It is, however, worth noting that resistance to nematodes may not be as valued to the grower as the yield of the crop. Therefore, where a grower does not experience the same level or better profit from using a resistant cultivar versus the cultivar they would usually plant, the grower is less likely to use the resistant cultivar.

One prospect is that the transfer of genes or gene combinations that provide widely effective resistance will be made easier and more predictable due primarily to MAS. This would also allow rotation of specific resistance genes to prolong durability. It is also essential to recognize that although resistant cultivars are described as cost free at the point of use, they do incur substantial costs in their development. Thus, accurate quantification of losses and benefits is essential if plant breeders are to develop and growers to exploit resistant cultivars. Their apparent environmental neutrality should be included in the balance sheet compared with chemical control approaches. We should also recognize that effective use of resistant cultivars requires substantial technological and educational inputs at the grower level.

16 Genetic Engineering for Resistance*

CATHERINE LILLEY AND PETER URWIN**

Centre for Plant Sciences, University of Leeds, UK

16.1. Genetic Engineering for Resistance: General Introduction

The use of natural nematode resistance for control can initially be highly effective. With developments in genomic and transcriptome sequencing coupled with cheap, high-throughput marker analyses, identification of quantitative trait loci (QTL) conferring nematode resistance is no longer the laborious undertaking it used to be. Next-generation sequencing platforms allow rapid single nucleotide polymorphism (SNP) discovery, which can be used to identify QTL. This technology has yet to be exploited fully in nematode resistance (*R*) gene discovery but upon identification resistance genes can then be introgressed into crop plants from wild or economically less desirable relatives using traditional breeding methods or by genetic manipulation. SNP markers are also the markers of choice for marker assisted selection (MAS) in breeding programmes due to their ease of use; SNP marker technologies such as KASPar (a homogeneous fluorescent resonance energy transfer (FRET)-based system) and various SNP chip formats enable inexpensive, rapid and robust results. New enrichment sequencing technologies show great promise for discovery of *R*-genes (Jupe *et al.*, 2013; Witek *et al.*, 2016) that act

*A revision of Cottage, A. and Urwin, P.E. (2006) Genetic engineering for resistance. In: Perry, R.N. and Moens, M. (eds) *Plant Nematology*, 2nd edn. CAB International, Wallingford, UK.
**Corresponding author: P.E.Urwin@leeds.ac.uk

  Plant Nematology, 3rd edn (eds R.N. Perry, M. Moens and J.T. Jones)
DOI: 10.1079/9781800622456.0016

against nematodes or other plant pathogens. Canonical *R*-genes encode nucleotide-binding, leucine-rich-repeat (NLR) proteins and NLR-specific enrichment sequencing (RenSeq) can be used to identify functional *R*-genes or to assess if a resistance phenotype is based on a known or uncharacterized gene. Mapping the H2 resistance effective against *Globodera pallida* in potato used two enrichment sequencing techniques in combination with bulked segregant analysis to identify not only SNPs linked to the resistance trait but also flanking markers that mapped the *H2* gene to the short arm of chromosome 5 (Strachan *et al.*, 2019).

Whilst advances in technologies now facilitate rapid resistance gene identification and the means to select for these in breeding programmes, they may have limited use. Resistances are often highly specific; for example, the tomato *Mi* gene confers resistance to *Meloidogyne incognita* but not to virulent populations of *M. javanica* (Tzortzakakis and Gowen, 1996) and current agricultural practice, such as the use of short rotations, can act as strong selectors for resistance breaking races or species. These can take the form of related nematode species, e.g. selecting *M. arenaria* instead of *M. incognita*, or virulent mutants of the same species; virulent *M. incognita* can develop within five generations on tomatoes with *Mi* conferred resistance (Noling, 2000) and one of the earliest uses of nematode resistance in soybean, derived from PI 88788, a germplasm accession from China, is now ineffective against most populations of *Heterodera glycines* (Mitchum, 2016; McCarville *et al.*, 2017). Genetic diversity amongst populations of the potato cyst nematode *G. pallida* can facilitate selection for increased virulence, although this can be mitigated by breeding plants with stacked resistance genes from two different sources. In this way, added protection was conferred against *G. pallida* populations that could overcome either single source of host resistance (Price *et al.*, 2024).

Despite technological advances, resistance QTL discovery and the resultant breeding programme require considerable time and financial commitments. Thus, the limited scope and durability of this solution to nematode parasitism may negate the investment for such a programme. An alternative solution is an engineered resistance, but what does this provide over the currently available methods? The immediate benefits of an engineered resistance are summarized in Box 16.1 and the main strategies are summarized in Table 16.1.

16.2. Genetic Engineering for Nematode Resistance: Use of Natural Resistances

Despite the identification of a number of canonical *R*-genes belonging to the NLR family and other genetic mechanisms conferring nematode resistance, there have been relatively few successes using these genes as part of an engineered resistance. In particular, many confer resistance to one species only and even to a specific pathotype. Transfer of *R*-genes

Box 16.1. Benefits of an engineered resistance.

- A knowledge-based, specific targeting of plant-parasitic nematodes.
- Ease and speed with which a specific desired trait can be introduced into a preferred plant variety.
- Resistance to a broader spectrum of plant-parasitic nematodes.
- Durability of resistance by targeting key steps in the infection cycle or by stacking resistance.

Table 16.1. Summary of targets of engineered nematode resistance, possible strategies and expected outcomes.

Resistance mechanism	Engineered resistance	Example	Outcome
Natural resistance			
Resistance (*R*) gene	Transfer of *R*-gene	*Mi* from resistant to susceptible cultivar	Hypersensitive response (HR) response
R-gene mediated signalling	Modification of signalling activator	*Hsp90* activation in response to nematode species	HR response
Avirulence factor (*Avr*) recognition	Modification of R-protein or Avr interacting protein	*Gpa2* responds to all RBP-1 variants	HR response
Targeting early nematode parasitism			
Hatching and attraction	Plants produce chemical deterrents	Over-production of *N*-formylloline in roots	Nematodes avoid plants
Penetration and migration	Expression of a sensory-disrupting peptide	Secretion from roots of a peptide that binds acetylcholine receptors	Attraction and migration disrupted, poor establishment
Targeting the nematode			
Intoxication	Expression of *Bacillus thuringiensis* (*Bt*) crystal (Cry) protein	Constitutive or local expression of *Bt* gene	Nematode death
Disruption of digestion	Expression of a protease inhibitor	Rice cystatin inhibits nematode cysteine proteases	Poor growth/reduced fecundity
Inhibition of protein production (RNA interference)	Expression of double-stranded RNA homologous to target transcripts	Targeted destruction of effector gene transcripts	Impaired parasitism

to a different host species has often resulted in loss of resistance. For example, transfer of *HeroA* (from tomato) into potato resulted in no significant resistance to *Globodera* spp. (Sobczak *et al.*, 2005); similarly, transfer of *Mi-1.2* (from tomato) to tobacco resulted in no significant resistance to *Meloidogyne* species (reviewed in Williamson, 1998). A further limiting factor of *Mi* resistance is that it is temperature-sensitive and breaks down over 28°C, limiting its use to temperate climes. The *R*-gene direct transfer approach has, however, proven successful for root-knot nematode species in some cases, and so this may be a fruitful line of future research. Resistance to *M. incognita* in susceptible tomato cultivars was conferred by expression of the *CaMi* gene from pepper (*Capsicum annuum*) (Chen *et al.*, 2007), whilst the tomato *Mi-1.2* gene expressed in transgenic lettuce (*Lactuca sativa*) provided resistance to the same nematode (Zhang *et al.*, 2010). In the latter case, the *R*-gene was transferred between distantly related species and yet maintained a level of efficacy. In a similarly unrelated transfer, the pepper resistance gene *Me3* reduced susceptibility to *M. incognita* by approximately 75% when expressed in transgenic lines of *Arabidopsis* (Liu *et al.*, 2023). Successful transfer of *R*-genes by a transgenic approach likely relies on conservation of downstream signalling components in the acceptor plant species and in some cases the presence or co-transfer of a required NLR partner.

Those novel resistance mechanisms that underlie the *Rhg1* and *Rhg4* resistance to *H. glycines* in soybean (see Chapter 15 for detail) might seem less amenable for transfer by genetic modification; however, they do offer insights into further engineered solutions for nematode resistance; overexpression of a combination of genes by either increasing copy number or altering transcript levels using promoters to produce elevated levels of transcripts and, hence, protein products can create an environment noxious to nematodes. Evidence suggests that the *Rhg1* locus interferes with mechanisms that are required for parasitism by all cyst nematodes, rather than relying on recognition of a specific nematode pathogen. This opens the way for engineering a broader spectrum resistance into a wider range of plant species. In a first step along this route, the three soybean genes underlying the *Rhg1* locus (encoding an amino acid transporter, an α-SNAP protein and a WI12 wound-inducible domain protein) were introduced into both potato (*Solanum tuberosum*) and *Arabidopsis*. The resulting plants displayed increased resistance to potato cyst and beet cyst nematode, respectively (Butler *et al.*, 2019), although efficacy did not reach the level observed in resistant soybean against avirulent populations of *H. glycines*. Further optimization and balancing of gene expression levels may improve resistance, whilst also addressing the negative effects observed on root growth and tuber production (Butler *et al.*, 2019).

16.2.1. Broadening *R*-gene resistance

There is evidence to suggest that the products of different sensor NLR *R*-genes share common downstream signalling cascades leading to an immune response, the intermediates of which could be engineered to respond to a wider range of nematode species. Research has identified some of these convergent signalling molecules; for example, RAR1 was identified as a common convergence point in signalling pathways initiated by fungi and viruses (Liu *et al.*, 2002), SGT1 was shown to interact with RAR1 (Azevedo *et al.*, 2002) and the heat shock protein HSP90 was identified as their molecular chaperone (Takahashi *et al.*, 2003). However, only HSP90 was shown to be essential for *Mi-1*-mediated resistance to aphids and nematodes (Bhattarai *et al.*, 2007). Another point of convergence in the downstream immune signalling response that could be targeted is the recently described helper NLR family (Jubic *et al.*, 2019). Each helper NLR functions as a downstream hub that can interact with multiple sensor NLRs. Transfer of *R*-genes between species may also be more successful if the corresponding helper NLR is co-expressed in the host plant. With an increase in the understanding of how resistance genes exert their function, in concert with an array of other components (reviewed by van Wersch *et al.*, 2020), it may be possible to broaden the specificity of a resistance or increase its durability by directed modification.

16.2.2. An alternative strategy of using natural resistances

In theory, determining *Avr* structure and utilizing ligand–receptor binding modelling may suggest engineering solutions that confer less receptor specificity, thus enabling *R*-gene mediated response to a wider range of nematode species. However, this approach may not be possible as a putative *Avr*-gene product (MAP-1) thought to interact with *Mi* was found exclusively in avirulent populations of *M. incognita* and other *Meloidogyne* species (Semblat *et al.*, 2001). The presence of this protein only in avirulent populations discounts R protein modification as no ligand would appear to exist in virulent populations. However, this may not be the situation for other *Avr*-gene products, which may differ between virulent and avirulent species by amino acid sequence, thus offering a means of receptor modification to achieve control. RBP-1 is a SPRY domain-containing protein secreted by *G. pallida*, which elicits a hypersensitive response (HR) response in potato plants expressing the *R*-gene *Gpa2*. RBP-1 is highly polymorphic and initiates a resistance response only if a proline residue is present at position 187 in the SPRY domain. However, whilst the leucine-rich repeat domain of *Gpa2* recognizes RBP-1 variants and initiates activation of the HR response, the interaction of *Gpa2* with RBP-1 is entirely dependent on a third protein RanGap2 (Sacco *et al.*, 2009).

The tacit assumption is that recognition of nematode Avr effector proteins triggers plant NLR protein-mediated cell death. However, in one case the opposite is true; the *Globodera rostochiensis* effector protein SPRYSEC-19 was shown to associate with the leucine-rich repeat domain of a SW5 resistance gene in tomato; this association did not result in programmed cell death and thus resistance, but conversely was shown to suppress it (Postma *et al.*, 2012). This was true for other (SW5B, Rx1, Gpa2 and RGH10) but not all NLR resistance proteins (where presence of SPRYSEC-19 did result in cell death). Similarly, Derevnina *et al.* (2021) showed that the SPRYSEC-15 effector directly binds to two helper NLRs, suppressing their activity. Thus plant pathogens, including nematodes, have evolved to counteract central nodes of the NLR immune signalling pathway through different mechanisms. Co-evolution with pathogen effectors may have driven NLR diversification into functionally redundant nodes in a massively expanded NLR network. These new insights into nematode effector protein function offers a further means of control as targeting nematode effector proteins that suppress cell death would enhance resistance. There is also the intriguing possibility of generating synthetic helper NLRs that could resurrect cryptic or defeated resistance proteins (Contreras *et al.*, 2023).

Currently very little is definitively established as regards *Avr*–*R*-gene product interactions; however, genomic sequences are now published for an ever-increasing number of plant-parasitic nematode species, providing new opportunities to identify *Avr* and other genes involved in parasitism.

16.3. Targets in the Nematode–Plant Interaction for Engineered Resistance

The first step in a knowledge-based approach to engineered resistance is to identify which steps in the interaction of nematodes with plants can be targeted to achieve an effective control. Figure 16.1 illustrates different stages of nematode plant parasitism that offer potential opportunities for control. Some of those stages that have been the focus of most research into engineered resistance will be discussed below.

16.3.1. Host-induced nematode hatching, attraction and root penetration

Previous chapters have described how various plant-parasitic nematodes locate host plants, penetrate and migrate through host tissues, feed and progress through their life

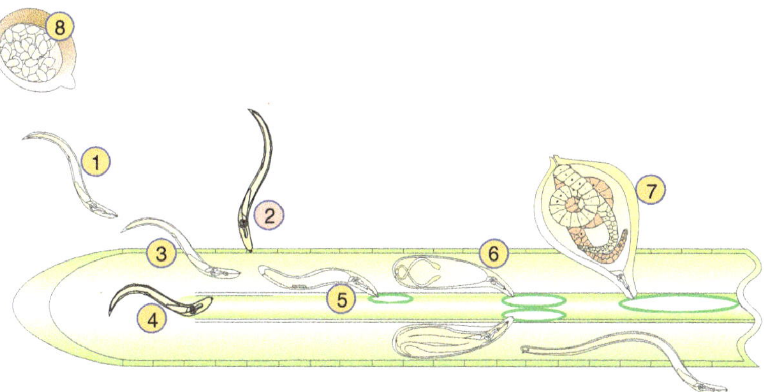

Fig. 16.1. Plant-parasitic nematode life cycle stages that are targets for genetically engineered resistance strategies. 1: Attraction of the nematode to the plant. 2: Grazing on the root (which is relevant for many ectoparasitic nematodes). 3: Penetration of the nematode into the plant. 4: Migration of the nematode through plant tissues. 5: Establishment of a specialized feeding site (as with endoparasitic nematodes). 6: Maturation and moulting with continued feeding from the plant. 7: Mating and egg production. 8: Maturation of eggs and hatching of juveniles (which begin the cycle of re-infection).

cycle. There are several mechanisms whereby nematodes are attracted to plants (see Chapter 8); these range from the detection of CO_2 gradients from plant respiration to the recognition of plant root diffusates, which in some species of nematodes can also stimulate hatching. In cereals, 2,4-dihydroxy-7-methoxy-1,4-benzoxazin-3-one (DIMBOA) is a naturally occurring hydroxamic acid that serves as a defence against a wide range of pests including insects, pathogenic fungi and bacteria, but acts as an attractant to the plant-parasitic nematode *Pratylenchus zeae* (Friebe *et al.*, 1998). *Pratylenchus scribneri* is attracted to the roots of tall fescue (*Festuca arundinacea*) by N-formylloline but only at concentrations of below 20 μg ml^{-1} of root extract as higher concentrations act as a repellent (Bacetty *et al.*, 2009). An olefinic compound β-myrcene is a potent attractant for the pinewood nematode, *Bursaphelenchus xylophilus*, and it is also an attractant for pinewood beetles of the genus *Monochamus* that vector *B. xylophilus*. Relatively few attractants have definitively been identified, but resistance could be achieved in plants that have been engineered to mask their attraction to nematodes or actively deter nematodes.

A novel defence that disrupts location and invasion of host roots by plant-parasitic nematodes has been developed to field efficacy. Nematodes must sense and respond appropriately to a range of chemical signals in order to achieve a successful parasitic interaction. Two distinct, synthetic peptides interfere with cyst nematode chemoreception by binding to either acetylcholinesterase or nicotinic acetylcholine receptors (nAChRs), both targets in the nematode cholinergic nervous system (Winter *et al.*, 2002). Transgenic plants were subsequently developed that secrete the peptides from their roots (Liu *et al.*, 2005; Green *et al.*, 2012). The acetylcholinesterase-inhibiting peptide suppressed the number of female *H. schachtii* that developed on *A. thaliana* by more than 80%, whilst expression in the root tips of potato plants resulted in almost 95% resistance to *G. pallida* (Lilley *et al.*, 2011). The nAChR-binding peptide is taken up from the environment by certain chemosensory sensilla within the anterior amphidial pouches and undergoes retrograde transport along some chemoreceptive neurons to their cell bodies and a limited number of interneurons. Chemoreception was only impaired when that transport had been completed (Wang *et al.*, 2011). Potato plants secreting this peptide from their root tips provided effective resistance (up to 77%) against potato cyst nematode in both containment glasshouse and field trials (Green *et al.*, 2012). This resistance strategy may effectively target other nematode species in addition to cyst nematodes as nAChRs are common to all nematodes studied (Sattelle, 2009). The infective stages of sedentary endoparasitic nematodes are vulnerable to sensory intervention prior to feeding cell initiation, whilst migratory endoparasites and ectoparasites remain motile and may be affected throughout their life cycle. This could potentially lead to greater resistance conferred by this strategy. Indeed, in a field trial all 10 transgenic plantain lines expressing the nAChR-binding peptide were significantly resistant at harvest to the migratory endoparasitic nematodes *Radopholus similis* and *Helicotylenchus multicinctus*. One of the peptide-expressing lines was >99% resistant and four others exceeded 90% resistance. Importantly, the yield of the best-performing line was about 186% that of non-transgenic plants based on larger bunches and less root damage leading to reduced plant toppling in storms (Tripathi *et al.*, 2015).

16.3.2. Nematode feeding and digestion

All plant-parasitic nematodes need to feed from their host plants in order to develop and reproduce. Ingestion of toxic or inhibitory proteins or other bioactive molecules from the plant is therefore a universal route for delivery of a resistance strategy. The ingested molecules may themselves interfere with nematode feeding and digestion, or they could target essential metabolic processes or reproduction.

16.3.2.1. Bacillus thuringiensis *(Bt) crystal (Cry) proteins*

Bacillus thuringiensis (*Bt*) crystal (Cry) proteins are best known for their insecticidal properties. The *Bt* toxin causes the death of insect larvae by binding to receptors in the epithelial cells of the larval gastrointestinal tract, which leads to pore formation and cell lysis (reviewed in Vachon *et al.*, 2012). Insect resistance conferred by the *Bt* gene has been used to target cotton bollworm, maize borers and potato beetles. Crops, predominantly maize and cotton, genetically engineered to express the *Bt* toxin cover over 100 million ha worldwide (ISAAA, 2018). The nematicidal potential of the *Bt* toxin was first demonstrated in *Caenorhabditis elegans*, where exposure to either Cry5B or Cry6A resulted in reduced fecundity and viability (Marroquin *et al.*, 2000). However, uptake of the *Bt* Cry protein as a possible control mechanism for plant-parasitic nematodes was not initially investigated as *H. schachtii* was shown to be unable to ingest proteins larger than 23 kDa (Böckenhoff and Grundler, 1994; Urwin *et al.*, 1997a) and *Bt* Cry6A is a 54 kDa protein, whilst Cry5B is even larger at 79 kDa. The feeding tubes of *Heterodera* and *Globodera* species appear to have very similar structures (Eves-van den Akker *et al.*, 2014b), so this size constraint may also exist for other cyst nematodes. However, this ingestion limit does not apply to all other plant-parasitic nematodes and expression of *Bt* Cry6A in tomato hairy root culture was shown to intoxicate *M. incognita*, as demonstrated by a fourfold reduction in progeny (Li *et al.*, 2007). By contrast, Cry5B-expressing roots supported significantly reduced numbers of galls. This was reflected in a reduced total egg production but there were no significant differences in the number of eggs per egg mass between transgenic and control lines (Li *et al.*, 2008). Therefore, Cry5B appears to exert its strongest effect on juvenile stages, whilst reproduction is most sensitive to Cry6A. It is likely that root-knot nematodes are able to ingest larger molecules than cyst nematodes as their feeding tubes differ significantly (Sobczak *et al.*, 1999; Eves-van den Akker *et al.*, 2014b, 2015).

The success of Bt Cry proteins against root-knot nematodes prompted research into smaller toxins or truncated Cry proteins that retained activity in assays against *C. elegans*. Cry proteins expressed in transgenic soybean plants were more effective against *H. glycines* when truncated at the C- and/or N-terminus. These smaller versions of Cry proteins typically ranged from 14 to 30 kDa, and so their improved bioactivity was presumed to be a function of more efficient ingestion (Bowen *et al.*, 2016). However, the large size of at least some Cry proteins may not preclude efficacy against cyst nematodes. Cry14Ab, with a molecular weight of 132 kDa, is resistant to proteolytic cleavage and yet the full-length protein expressed in transgenic soybeans is able to inhibit the establishment and development of *H. glycines* (Kahn *et al.*, 2021). Both glasshouse and field trials were conducted with the most resistant Cry14Ab lines. Some lines were categorized as highly resistant in glasshouse trials, and there was a significant reduction in nematode reproduction when the data from multiple transgenic lines in field trials conducted in replicate over two years were pooled (Kahn *et al.*, 2021). To date, Bt-expressing soybean plants are the only nematode-resistance technology to have received approval as genetically modified plants for planting and commercialization (ISAAA, 2023).

Thus, *Bt* has the potential to provide engineered resistance to a range of nematode species. Such resistance may have limited durability, however; the widespread and intensive use of *Bt* has resulted in the emergence of resistant pests in some crops in Australia, India and China, and resistance occurs frequently under artificial selection pressures in several species, including *C. elegans* (Barrows *et al.*, 2007).

16.3.2.2. Protease inhibitors

Transgenic expression of proteinase inhibitors (PIs) in plant roots is the most widely explored approach to engineered resistance to plant-parasitic nematodes. A range of different inhibitors, most of them naturally occurring plant proteins, have been shown to be detrimental to feeding nematodes, reducing their growth and fecundity. Inhibitors of all four main classes of proteinase (serine, cysteine, aspartic and metallo-proteinase) occur in plants and are often induced in response to wounding or herbivory. Correspondingly, proteinase genes and activity have been identified in plant-parasitic nematodes (Lilley *et al.*, 1996, 1997; Urwin *et al.*, 1997b; Neveu *et al.*, 2003; Fragoso *et al.*, 2009). A digestive role has been proposed for these enzymes, corroborated for some by expression in the intestine. With digestion of protein being a common requirement of nematodes, PI-based control could have efficacy against a wide range of species, irrespective of their parasitic strategy. This would have particular utility in those field situations where a number of different nematode pests occur concurrently.

Cysteine PIs, termed cystatins, have received the most attention. Initial experiments utilized the rice cystatin Oc-I, modifying its coding region to remove an amino acid and improve its inhibitory activity 13-fold over the native protein.

Expression of this engineered variant (Oc-IΔD86) in tomato hairy roots using the cauliflower mosaic virus (CaMV35S) promoter resulted in significantly smaller female *G. pallida* after 6 weeks when compared to control roots (Urwin *et al.*, 1995). Expression of Oc-IΔD86 in a second model system *Arabidopsis*, using the same promoter, allowed the cystatin to be tested against additional nematode species. The size of female *H. schachtii* and *M. incognita* was significantly reduced relative to controls with growth arrested before egg laying. This effect was correlated with detection of the cystatin in the feeding nematodes and reduced cysteine proteinase activity in the intestine of female *H. schachtii* recovered from plants (Urwin *et al.*, 1997b). The same *Arabidopsis* plants also suppressed growth and egg production of the reniform nematode, *Rotylenchulus reniformis*, with cystatin expression level influencing reproductive success (Urwin *et al.*, 2000). Although *A. thaliana* is not a favoured host for this nematode, the study is an example of a model system providing preliminary data to support later cystatin expression in crops of interest such as pineapple (Wang *et al.*, 2009b), where transformation is limited by a slow rate of regeneration. An alternative model host plant, lucerne (alfalfa), was used to demonstrate that the rice cystatins Oc-I and Oc-II expressed at a low level in lucerne under the control of a wound-inducible promoter conferred some resistance to the root-lesion nematode *Pratylenchus penetrans* (Samac and Smigocki, 2003).

A rather different approach was taken to inhibit cysteine proteinases of *H. glycines* parasitizing transgenic soybean hairy roots (Marra *et al.*, 2009). The propeptides that are cleaved from cysteine proteinase precursors can often act as inhibitors of their cognate enzymes (e.g. Silva *et al.*, 2004). The propeptide region of the *H. glycines* HGCP-I cathepsin L enzyme was expressed in roots and caused a reduction in the number and fecundity of female nematodes (Marra *et al.*, 2009). The prodomain inhibitor displays greater specificity for target enzymes than do typical plant PIs (Silva *et al.*, 2004) and whilst this may limit the utility of the approach to control a wide range of nematode species, it could have biosafety advantages for non-target organisms.

Although less widely studied, serine PIs have also demonstrated potential for nematode control. In another model system study, transgenic expression of the sweet potato serine PI, sporamin, inhibited growth and development of female *H. schachtii* parasitizing sugar beet hairy roots (Cai *et al.*, 2003). In this case, the severity of the effect was clearly correlated with the level of trypsin-inhibitory activity detected in the transformed root lines.

In terms of crop plants, engineered resistance based on PIs has been extensively tested in potato, primarily against *G. pallida*. The potential of plant PIs as anti-nematode effectors was first explored using the serine PI cowpea trypsin inhibitor (CpTI). CpTI expressed in transgenic potato influenced the sexual fate of newly established *G. pallida* (Hepher and Atkinson, 1992) and as a result the population comprised fewer cysts and was biased toward a predominance of the much smaller and less damaging males. Subsequent work focused on cystatins and culminated in successful field trials of transgenic potatoes. The best transgenic line of the fully susceptible potato 'Desiree', expressing chicken egg white cystatin from the constitutive CaMV35S promoter, displayed 70% resistance to potato cyst nematodes in the field (Urwin *et al.*, 2001). When the same construct was used to transform two potato cultivars, 'Sante' and 'Maria Huanca', that each display natural partial resistance to *G. pallida*, the best transgenic lines of each were enhanced to full resistance (Urwin *et al.*, 2003). Subsequent field trials demonstrated that both the modified rice cystatin (OcIΔD86) and a sunflower cystatin expressed in 'Desiree' afforded similar levels of protection to chicken egg white cystatin (Urwin *et al.*, 2003). Potato plants in which expression of the OcIΔD86 cystatin was limited mainly to the roots and, in particular, to the syncytia and giant cells induced by *G. pallida* and *M. incognita*, respectively, were shown to have similar resistance levels to those achieved with constitutive expression for both nematodes (Lilley *et al.*, 2004). Similar root-limited expression of the OcIΔD86 cystatin, using the *Arabidopsis* TUB-1 promoter, was able to suppress reproduction of *M. incognita* in eggplant (*Solanum melongena*) by up to 78% (Papolu *et al.*, 2016). In a further demonstration of the potential of PIs, a cystatin from the tropical root crop taro (*Colocasia esculenta*) was expressed constitutively in a root-knot nematode-susceptible tomato cultivar. There was a 50% reduction in the number of galls formed by *M. incognita* on the transgenic plants compared with wild-type plants and a larger reduction in the number of egg masses produced per plant (Chan *et al.*, 2010).

PI strategies have also been tested in banana and plantain. Cavendish dessert bananas that express the OcIΔD86 engineered variant of rice cystatin under the control of the maize ubiquitin promoter displayed 70 ± 10% resistance to *Radopholus similis* in a glasshouse trial (Atkinson *et al.*, 2004). The approach progressed to cooking varieties of *Musa;* the plantain 'Gonja' was transformed to express either a maize cystatin alone or both a cystatin and the nAChR-binding repellent peptide. The best level of resistance achieved for the cystatin lines in a glasshouse trial against the major pest species *R. similis* was 84% (Roderick *et al.*, 2012). A subsequent field trial of two of these maize cystatin lines in Uganda provided significant resistance of >90% at harvest for plants under combined challenge by *R. similis* and *H. multicinctus* (Tripathi *et al.*, 2015). There could be an additional advantage to cystatin-mediated nematode resistance in banana as cystatin impairs feeding and development of banana weevils (Kiggundu *et al.*, 2010).

To date, the only nematode resistance technology introduced into rice is the cystatin-based defence. Transgenic plants of four elite African rice varieties constitutively expressing the modified rice cystatin OcIΔD86 displayed 55% resistance to *M. incognita* (Vain *et al.*, 1998). Only a low level of cystatin expression was observed, possibly due to a suboptimal CaMV35S promoter or homology-dependent silencing of the transgene in combination with the endogenous *OcI* gene. In subsequent work, a maize cystatin has been expressed in the rice variety 'Nipponbare' under the control of a root promoter from *Arabidopsis thaliana* (TUB-1) that is known to be upregulated in the feeding cells of *M. incognita* parasitizing rice (Green *et al.*, 2002).

It is not only food crops that suffer significant losses to plant-parasitic nematodes and can therefore benefit from genetically engineered resistance. The same modified cystatin technology has been used to protect the cash crop Easter lily (*Lilium longiflorum*) from the root-lesion nematode *P. penetrans*. The total number of nematodes parasitising *in vitro* plants constitutively expressing the *OcIΔD86* transgene was significantly reduced by up to 75%. Importantly for growers, the bulbs of infected transgenic lily plants were healthier and grew more vigorously than those of wild-type plants (Vieira *et al.*, 2015a). This work provides another example of a promising line of research being progressed from *in vitro* studies to the field. The same transgenic lines planted in a field trial over two growing seasons still exhibited a degree of resistance to *P. penetrans* and displayed desirable growth and quality characteristics similar to non-transgenic lilies, although the level of resistance did not reach that observed *in vitro* (Westerdahl *et al.*, 2023).

16.4. RNA Interference (RNAi)

RNA interference (RNAi) is the process in which double-stranded RNA (dsRNA) triggers the silencing of specific target genes through mRNA degradation. In nematodes, ingested dsRNA is rapidly degraded into small interfering RNAs (siRNAs) that are incorporated into a protein complex called RISC (RNA-induced silencing complex). RISC will then bind to mRNAs that have sufficient sequence identity with the siRNAs ultimately resulting in their cleavage and subsequent degradation.

RNAi has been adopted as a tool for functional analysis of plant-parasitic nematode genes (Rosso *et al.*, 2009), and delivery of dsRNA from a host plant to bring about RNAi silencing of essential genes in the feeding nematode (sometimes described as host-induced gene silencing (HIGS)) has been proposed as a nematode resistance strategy (Lilley *et al.*, 2011). This would be an efficient approach for sustained introduction of dsRNA (or siRNAs) into the nematode over most of its life cycle and so should maximize the silencing of target nematode genes. To date, the approach has shown potential against several nematode species including both cyst and root-knot nematodes and screening methods have been developed to allow the evaluation of many gene targets. As with most of the engineered resistance strategies described here, easily transformable model systems are frequently used in the first instance to evaluate the potential of the approach for specific genes and nematode species. A so-called hairpin construct is created for transformation of the host plant. In this, a nematode target gene fragment present in both sense and antisense orientations and usually separated by an intron, is expressed under control of a plant promoter that is active in cells from which the nematode feeds. The transcribed RNA forms a hairpin structure with the intron sequence being removed by splicing to leave a linear dsRNA molecule. It is not clear if the target transcript suppression observed arises from ingestion of this dsRNA that is subsequently processed by the nematode of from uptake of plant-derived siRNAs.

The first successful demonstrations of HIGS against nematodes involved root-knot nematodes. A high level of resistance to root-knot nematode was achieved by targeting a parasitism gene expressed in the sub-ventral gland cells of *M. incognita* (Huang *et al.*, 2006). dsRNA complementary to the *16D10* gene was expressed in transgenic *Arabidopsis* and the resulting lines displayed a significant reduction (63–90%) in the number of galls and their size, with a corresponding reduction in total egg production. The high level of homology between the *16D10* sequences of different *Meloidogyne* species led to broad-range resistance against *M. incognita*, *M. javanica*, *M. arenaria* and *M. hapla*. Around the same time, almost complete resistance to *Meloidogyne* infection was reported in tobacco plants expressing dsRNA corresponding to splicing factor or integrase (Yadav *et al.*, 2006). Subsequently, *Arabidopsis* plants expressing dsRNA at least partially reduced transcript abundance of targeted parasitism genes in the pharyngeal gland cells of feeding *H. schachtii* (Patel *et al.*, 2008, 2010; Sindhu *et al.*, 2009). For six of eight genes tested this led to a significant reduction in female numbers of between 23% and 64%, with considerable variation between lines for some constructs (Sindhu *et al.*, 2009). Variable, non-significant effects were also observed when a fibrilin gene of *H. glycines* was targeted from chimeric soybean plants (Li *et al.*, 2010a). In the same study, RNAi of a coatomer subunit of this nematode resulted in a significant reduction in egg production. Soybean composite plants derived from hairy root cultures engineered to silence either of two ribosomal proteins, a spliceosomal protein or synaptobrevin, of *H. glycines* by RNAi resulted in 81–93% fewer females developing on the transgenic roots (Klink *et al.*, 2009), whilst a similarly high reduction in egg production was achieved by targeting mRNA splicing factor *prp-17* or an uncharacterized gene *cpn-1* (Li *et al.*, 2010b).

Examples of *in planta* RNAi to target nematode genes with effects on development and reproductive success continue to accumulate for a range of plant-parasitic nematodes (reviewed by Dutta *et al.*, 2015 and Banerjee *et al.*, 2017). The approach is clearly applicable to multiple plant-parasitic nematode species with different feeding habits, including those migratory nematodes that do not induce feeding sites. Reduced numbers of nematodes parasitizing dsRNA-expressing roots of soybean and walnut have been observed for *P. penetrans* (Vieira *et al.*, 2015b) and *P. vulnus* (Walawage *et al.*, 2013) respectively. Similarly, targeting either a cathepsin B (Li *et al.*, 2015) or cathepsin S (Li *et al.*, 2017) cysteine protease of *Radopholus similis* from transgenic tobacco plants increased resistance to the nematode. A potential weakness of RNAi, compared with some other engineered resistance strategies, is that the inherent sequence specificity could limit its utility to a single nematode species for each transgenic event. One study sought to overcome this and broaden the resistance spectrum by selecting sequence regions of orthologous genes that were highly conserved between different nematode species problematic on the banana crop, thus providing cross-genera control (Roderick *et al.*, 2018). Short dsRNAs corresponding to regions of either proteasomal alpha subunit 4 (*pas-4*) and actin-4 (*act-4*) highly conserved between *R. similis*, *P. coffeae* and *M. incognita* were expressed in carrot hairy roots. Targeting *pas-4* suppressed multiplication of *R. similis*, *P. coffeae* and *M. incognita* by at least 69% each for the best line, with *M. incognita* most severely affected. Targeting *act-4* had no impact on *M. incognita*, however, but reduced susceptibility to *R. similis* by >90%. This approach, combined with assessment of effects of such conserved sequences on non-target organisms, could be a promising way forward.

One strength of the RNAi approach is that the rapid recent increase in genomic and transcriptomic data for numerous and diverse plant-parasitic nematode species provides a large range of potential targets that can be screened *in vitro* to select those for plant transformation constructs. Research efforts involving RNAi tend to focus on either gland cell-expressed effector genes, presumed necessary for successful parasitism, or those genes that, based on predicted function, are essential for basic cellular processes and metabolism. Lethal phenotypes observed after RNAi of orthologous *C. elegans* genes are often used to inform the latter strategy. However, not every gene is amenable to successful RNAi silencing, and of those where a reduction in target transcript is detected, the effect can be variable. Reduction of transcript for a single gene may also not result in a clear detrimental phenotype, especially if there is genetic redundancy. Consequently, it is clearly proving a challenge to translate these results in model or *in vitro* systems to useful levels of resistance in the field as to date we are not aware of an RNAi-based resistance undergoing a field trial. It remains to be seen whether or not the efficacy of host-generated RNAi will work efficiently against all species of nematodes in the field.

16.5. Promoters for Optimal Expression

Genetically engineered solutions for nematode resistance require appropriate expression of the defence to ensure correct temporal and spatial delivery of the transgene product. To achieve the best results, and to confer the greatest level of specificity and biosafety, expression of the transgene should be as limited as possible to the nematode location. The rice cystatins Oc-I and Oc-II expressed at a low level in alfalfa under the control of a wound-inducible promoter conferred some resistance to *P. penetrans* (Samac and Smigocki, 2003). Invasion by cyst and root-knot nematodes occurs behind the root cap in the zone of elongation and disruption of chemoattraction and invasion requires promoter expression in outer root cells, so that peptides can be secreted into the rhizosphere. A promoter region (*AtMDK4-20*) of an *Arabidopsis* gene with homology to a maize gene known to express in the root cap was used to deliver a chemodisruptive peptide in both *Arabidopsis* and potato; in the case of the latter significantly improved resistance (approximately 60% greater) to *G. pallida* was achieved in comparison to expressing the peptide from the CaMV35S promoter (Lilley *et al.*, 2011).

Biotechnological control strategies of sedentary endoparasitic nematodes have often used the constitutive CaMV35S promoter, but more refined technology has utilized promoters with a more restricted pattern of expression. Promoters of TUB-1, a β-tubulin gene of *A. thaliana*, RPL16 that encodes an *Arabidopsis* ribosomal protein L16 and ARSK-1, a likely serine/threonine kinase, all direct expression of sufficient cystatin to provide partial resistance to *G. pallida* in the field and to *M. incognita* in containment. The ARSK promoter lines provide more resistance to *G. pallida* than *M. incognita*, whereas the other promoters were associated with less resistance against the cyst nematodes. All three promoters were active in the giant cells induced by *M. incognita* but only ARSK1 was also active in the syncytium of the cyst nematode (Lilley *et al.*, 2004). Molecular engineering can lead to promoters that are extremely specific. Deletion of the 5′-flanking region of a root-preferential promoter *Tob*RB7 resulted in a 300-bp promoter fragment just upstream of the coding region that remained active within the giant cells induced by *M. incognita* and silenced in root meristems (Opperman *et al.*, 1994). While such promoters may have specificity, the strength of expression must also be considered. When Fairbairn *et al.* (2007) targeted a transcription factor of *M. javanica* in an *in planta* RNAi biotechnological strategy (see Section 16.4) they tested the CaMV35S promoter and the Δ0.3*Tob*RB7 promoter. None of the lines containing the Δ0.3*Tob*RB7 promoter, in sharp contrast to those harbouring CaMV35S, showed any signs of silencing the targeted gene. When reporter plants made with GUS plants under the control of Δ0.3*Tob*RB7 were analysed only a small percentage of galls showed GUS activity and those that did revealed only weak activity (Fairbairn *et al.*, 2007).

The wealth of *Arabidopsis* resources makes this plant an attractive tool for identifying promoters that express in nematode feeding cells. These include primarily a fully sequenced genome and abundant expression data that have been used to identify syncytium and giant cell expressing transcripts (e.g. Jammes *et al.*, 2005; Fuller *et al.*, 2007; Szakasits *et al.*, 2009), and enhancer trap lines that have been used to identify feeding cell expressing promoter elements. Reporter tagged lines were instrumental in the identification of the *pyk20* gene, which encodes a transcription factor specific to Cruciferae (Puzio *et al.*, 1999) and is expressed in response to indoleacetic acid and kinetin. The tagged line showed reporter activity in early syncytia of *H. schachtii* (Puzio *et al.*, 1998). A syncytial-specific promoter from a tagged line that may represent this promoter has been used to achieve partial resistance to the potato cyst nematode (Ohl *et al.*, 1997). As the full genomic sequence of *Arabidopsis* is known, it is a relatively simple task to isolate putative promoter sequences from nematode-responsive genes using a polymerase chain reaction (PCR) approach. In *Arabidopsis*, 2 kbp of nucleotides 5′ to the initiation codon are generally sufficient to drive reporter gene expression. Clearly, this is not always the case as the promoter may exceed 2 kbp or proteins bound to sequences within introns, exons or 3′ untranslated sequences may interact with promoter elements and be necessary for a full expression profile. The promoters of myo-inositol oxygenase genes are predominantly expressed in syncytia induced by *H. schachtii* in *Arabidopsis* roots and may have a use in biotechnological strategies (Siddique *et al.*, 2009). A promoter of a defensin with strong expression in the syncytia of *H. schachtii* in *Arabidopsis* and limited expression elsewhere (in siliques) has been identified (Siddique *et al.*, 2011). The promoter was isolated from a defensin gene *Pdf2.1* which was previously identified as upregulated in syncytia from GeneChip array data (Szakasits *et al.*, 2009). The increasing number of genome and transcriptome resources for non-model plants should ensure that it becomes easier to identify and utilize appropriate promoters to drive resistance technologies in the crop of interest.

Many promoters are reported in the literature as expressing in the feeding cells of nematodes; however, caution should be exercised when utilizing these data in the context of engineering nematode resistance in crop plants. Frequently nematode infections are conducted in culture in media that may contain sucrose; when plants are grown in potting medium (soil, sand or compost), a different expression profile to that reported is often seen, particularly in root tissues.

16.6. Stacked Defences

Whilst there may be an immediate benefit in utilizing a successfully engineered resistance on its own, one should consider the longer-term implications and plan for a more sophisticated approach. Single resistances generate a strong selective pressure, especially in the predominant intensive farming/monoculture currently used in agriculture. For example, when resistance to *G. rostochiensis* was introduced (in the form of potato cultivars that are naturally resistant to *G. rostochiensis* as their genomes encode the *R*-gene *H1*), the initial success in the UK was soon negated by a shift in nematode populations. Currently *G. pallida* is prevalent, for which there is no commercial full resistance and 64% of the UK potato growing areas are now affected.

There are at least two approaches, which are not mutually exclusive, to try and ensure that the balance is shifted towards successful plant protection. The first approach is to stack several resistances to the pathogen in the plant, preferably targeting different aspects of the nematode's life cycle. Combining partial natural resistance with a protease inhibitor-based resistance has been shown to improve the overall resistance in comparison to that of single engineered resistance (Urwin *et al.*, 2003). However, a stacked approach does not always lead to improved resistance compared with a single transgene. Plantains expressing both a cystatin protease inhibitor and a chemodisruptive peptide displayed resistance similar to plants expressing either defence alone, although the resistance provided by the single transgenes was in most lines already >90% (Tripathi *et al.*, 2015). In this case the benefit should be durability rather than increased efficacy. The second approach is to utilize the engineered resistances as part of a panel of methods that can be used in integrated pest management that may include pesticides (if available), natural resistances, crop rotation, trap cropping and biocontrol measures.

16.7. The Research Approach to Engineering Nematode Resistance

When investigating a genetically engineered resistance, a number of factors need to be considered. First, the effective delivery of defined and quantified nematode inocula is important. In order to achieve this, nematodes should be relatively easy to culture and propagate *in vitro*, in containment and in field conditions. Nematode species and strain must be correctly identified and defined; for example, *M. javanica* rather than *M. incognita* or *G. pallida* pathotypes 2/3 rather than pathotype 1. Quantification of a viable inoculum may take the form of number of cysts/eggs per volume of soil, or a defined number of hatched juveniles applied in suspension. Second, the selection of a plant species or cultivar is not trivial; considerations are the ease of transformation and regeneration and the propagation time from parent to offspring. The obvious choice for proof-of-concept type research is the use of *Arabidopsis* with *H. schachtii* or *Meloidogne* spp. *Arabidopsis* has a short generation time of approximately 6 weeks, is easily transformed (by floral dip) and numerous plants can be grown in small areas *in vitro*. Similarly, crop plants of the family *Solanaceae*, e.g. potato and tobacco, have been used extensively as experimental models over the past two decades and can be readily transformed and regenerated. Before nematode resistance can be evaluated all putative transformants should be verified as containing the transgene (by PCR) and transgene copy number determined (using real-time quantitative PCR), and expression levels of resistance products assessed (using, e.g. real-time quantitative reverse transcription PCR or Western blotting). The transgene insertion site should also be determined to ensure that it has not interrupted a plant gene.

Symptoms of infection can vary considerably from plant to plant; additionally, the level of activity of an engineered resistance can vary considerably between plants. Therefore, it is necessary to try to standardize infection and cultivation as much as possible. The resistance trial design should include the test plants, susceptible controls and resistant controls (if available), populations of which should be infected and non-infected. All populations must have originated and been grown in the same way, i.e. in the first instance through tissue culture. The resistance is then measured as a population effect; instead of looking for one or two apparently highly resistant lines, an overall increase in resistance in the engineered population relative to the control population is sought.

Trial design enabling meaningful statistical evaluation necessitates large numbers of plants and scoring methods must either be simple enough to assess an experiment within a day or allow immediate fixing of plants and nematodes to facilitate assessment at a later date. Nematodes infecting plants can be counted on material grown and infected *in vitro* using standard techniques. Effects on nematode development due to the resistance mechanism can be assessed using image capture and measurement techniques (Atkinson *et al.*, 1996). As automation and machine-learning approaches improve, precise evaluation of nematode development over time and at large scale is becoming possible (Kranse *et al.*, 2022). Once resistance has been established in contained trials the next step is field trial evaluation, which evaluates not only the resistance in an agricultural setting but also enables the interaction of the transgenic crop with the environment to be assessed.

The ultimate aim of a genetically engineered resistance is to make that resistance commercially available, and freedom to operate should be determined as in many cases methodologies and target sequences are protected by intellectual property rights (reviewed by Rommens, 2010).

16.8. The Future

With the increase in understanding of plant–nematode interactions more sophisticated resistance mechanisms will become available. Recent advances in engineering resistance against other plant pathogens could pave the way for new technologies, one example being plant NLR (receptor)–nanobody fusions that trigger immune responses in the presence of a pathogen effector (Kourelis *et al.*, 2023). The major challenges will be to gain market acceptance, especially in a sceptical European environment. These challenges may be reduced if gene-editing technologies, with no residual presence of foreign genetic material, can be used to engineer nematode resistance. However, nematode resistance is probably one of the best mechanisms for demonstrating the benefits of a genetically modified technology if it is soundly based. In the long term, engineered resistances should just be seen for what they are – another tool in an armoury against persistent and damaging pests that are best used in an integrated pest management approach.

Acknowledgements

The authors would like to thank Dr Chris Thomas (Milton Contact Ltd, 3 Hall End, Milton, Cambridge CB4 6AQ, UK) and Dr Amanda Cottage for their contributions to this chapter.

Chemical Control of Nematodes*

Matthew A. Back[1]**, Martin C. Hare[1] and Tim C. Thoden[2]

[1]*Centre for Crop & Environmental Science, Harper Adams University, UK;*
[2]*Corteva Agriscience, München, Germany*

17.1. Background to Nematicides

17.1.1. Definitions and description of activity

The UK's Health and Safety Executive (HSE) defines pesticides, also known as plant protection products (PPPs), as 'substances used to control pests, weeds and disease'. Pesticides consist of one or more active substances co-formulated with other materials, while a nematicide is classed as a pesticide used to control unwanted nematodes. However, whilst some biological control agents for nematodes are classified for registration purposes as nematicides within the UK and EU, they will not be discussed here as they are considered separately in Chapter 14. Nematicides are chemical compounds that are lethal to nematodes,

*A revision of Haydock, P.P.J., Woods, S.R., Grove, I.G. and Hare, M.C. (2013) Chemical control of nematodes. In: Perry, R.N. and Moens, M. (eds) *Plant Nematology*, 2nd edn. CAB International, Wallingford, UK.
**Corresponding author: mback@harper-adams.ac.uk

  Plant Nematology, 3rd edn (eds R.N. Perry, M. Moens and J.T. Jones)
DOI: 10.1079/9781800622456.0017

whereas the term 'nematistat' (see Chapter 11) is frequently used for compounds that provide sub-lethal dosages that disrupt nematode behaviour. These latter chemicals work by paralysing the nematodes for a variable period of time, during which they may deplete their lipid reserves to such an extent that they are unable to invade the plant. Nematode recovery is possible from the sub-lethal effects of nematistats. The term nematicide is commonly used as an umbrella term encompassing both types of activity and will be used in this sense here.

Nematicides are applied primarily to reduce root/plant damage caused by nematodes and to increase productivity (Tobin *et al.*, 2008), which is achieved by reducing the numbers of nematodes feeding on (e.g. *Trichodorus* spp.) or in (e.g. *Meloidogyne* spp., *Globodera* spp.) plant tissues. Nematicides are also applied to prevent or reduce nematode reproduction and to limit the transmission of nematode-borne viruses to the plant (Dale *et al.*, 2004). The economic benefit to the crop is seen as a reduction in yield loss (and/or an increase in quality), and maintenance of future production and profitability. Nematicide application has been shown to reduce a wide number of plant-feeding nematodes significantly (Eisenhauer *et al.*, 2010); however, reductions in nematode population densities do not always occur, particularly where the nematode's reproductive rate is high or multiple generations of the nematode occur during the growth period of the crop, i.e. outside the period of nematicide activity. Consequently, nematicides may protect against yield loss but do not always have an effect on population density, meaning that other strategies such as cultivar choice and rotation need to be employed.

17.1.2. History

Nematicides have been in use since the late 19th century when the fumigant carbon disulphide was introduced. The development of further fumigants took place in the first half of the 20th century with the introduction of chloropicrin, 1,3-dichloropropene (1,3-D), methyl bromide, 1,2-dibromo-3-chloropropane (1,2-DBCP), 1,3-dichloropropene and 1,2-dichloropropane mixtures (DD), formaldehyde, metam sodium and dazomet. The remaining uses of methyl bromide were revoked for developed countries in 2005 under the Montreal Protocol for the reduction of gases contributing to global warming, although critical use exemptions still apply (Section 17.4.3). Similarly, 1,3-D is now banned in the EU (some critical use exemptions) but continues to be used in the USA and is registered for use in China (Qiao *et al.*, 2012). The second half of the 20th century saw the development of the organophosphates such as fenamiphos, ethoprophos and fosthiazate, together with the carbamates carbofuran, aldicarb and oxamyl (Table 17.1). In many countries around the world, these chemicals are still successfully used, but regulatory pressure has also increased due to their relatively high acute mammalian toxicity. Therefore, in the EU in 2023 only fosthiazate is still registered (https://ec.europa.eu/food/plant/pesticides/eu-pesticides-database/start/screen/active-substances).

Recent non-fumigant nematicide additions to the nematicide options include fluensulfone, a fluoroalkenyl (Oka *et al.*, 2009), fluopyram, a succinate dehydrogenase inhibitor (SDHI) fungicide (Faske and Hurd, 2015), fluzaindolizine, a imidazopyridine and a true nematicide (Lahm *et al.*, 2017), and cyclobutrifluram (pyridine carboxamide; SDHI). All these modern non-fumigant nematicides are characterized by much more favourable

Table 17.1 Globally important synthetic nematicides.

Active substance	Chemical group	Acute oral LD_{50} rat (mg kg^{-1})*	Year of discovery	Example trade name	State of formulation	Inventor/Main supplier
Abamectin	Avermectins	8.7[1]	1979	Avicta 500FS	Liquid	Syngenta
Aldicarb	Oxime carbamate	0.93	1965	Temik 15G	Microgranule	Bayer CropScience
Carbofuran	Carbamate	7.0	1965	Furadan 15G	Microgranule	FMC Corporation/Arysta
				Furadan 4F	Liquid	
Cadusafos	Organophosphorus	30.1	1982	Rugby 200 CS	Liquid	FMC Corporation
				Rugby 10G	Microgranule	
Cyclobutrifluram	Pyridine carboxamide	>5000[2]	2015	Tymirium	Liquid	Syngenta
Dazomet	Methyl isothiocyanate[2] liberator	415	1897	Basamid	Microgranule	BASF; AMVAC Chemical Corporation
1,3-dichloropropene	Halogenated hydrocarbon	150	1956	Telone II	Liquid	Dow Chemical (Corteva Agriscience)
				Telone EC	Liquid	
Ethoprophos	Organophosphorus	40	1966	Mocap 15G	Microgranule	AMVAC Chemical Corporation
				Mocap EC	Liquid	
Fenamiphos	Organophosphorus	>6	1967	Nemacur 15G	Microgranule	Bayer CropScience; AMVAC; Arysta
				Nemacur 3	Liquid	
Fluensulfone	Fluoroalkenyl	671	2004	Nimitz	Liquid	Adama Agricultural Solutions
Fluazaindolizine	Imidazopyridine	1187[3]	2011	Salibro (500SC)/ Reklemel Active	Liquid	Corteva Agriscience
Fluopyram	Pyridinyl-ethyl-benzamide	>2000	2013	Velum Prime/ Verango	Liquid	Bayer CropScience

Fosthiazate	Organophosphorus	57	1992	Nemathorin 10G Nemathorin 150EC	Microgranule Liquid	Syngenta
Metam sodium	Methyl isothiocyanate liberator	896	1951	Vapam HL	Liquid	Amvac Chemical Corporation
Oxamyl	Oxime carbamate	2.5	1974	Vydate 10G Vydate L	Microgranule Liquid	DuPont (Corteva Agriscience)
Spirotetramat	Tetramic acid derivative	>2000	2007	Movento	Liquid	Bayer CropScience

*Lethal dose (LD_{50}) values obtained from the Pesticide Properties DataBase (http://sitem.herts.ac.uk/aeru/ppdb/en/index.htm) except where stated.
[1]EU Annex I of Directive 98/8/EC Assessment Report 2011 (https://echa.europa.eu/documents/10162/f8b567f0-e610-2ec7-496b-3ec7720ff9e8).
[2]Notice of amendments to the Poisons Standard in relation to New Chemical Entities (NCEs) and Delegate-only decisions, Australian Government, 2022 (https://www.tga.gov.au/sites/default/files/2022-09/notification-amendments-poisons-standard-relation-delegate-only-final-decisions-new-chemical-entities-nces-september-2022.pdf).
[3]Fluazaindolizine: Comparison of Hazard Profile for Comparable Nematicides, USEPA, 2021 (https://www.regulations.gov/document/EPA-HQ-OPP-2020-0065-0011).

mammalian and ecotoxicity profiles (Desaeger *et al.*, 2020). Some of those actives have a global footprint while others – due to regulatory constraints – might only have regional footprints. The process of nematicide discovery and testing is outlined in Box 17.1. In the late 20th and early 21st century, there has been increased research and development to improve the efficiency of nematicide use, minimize their environmental impact and reduce their cost to the farmer. There has also been increased interest in and development of 'natural' nematicides, derived mainly from plant extracts and bacteria (e.g. avermectins from the soil bacterium *Streptomyces avermitilis* (Section 17.6)).

17.1.3. Global market size and usage

Nematicides remain an important component of nematode management programmes, whether used as part of an integrated management approach or as the sole control component (Hillocks, 2012; Qiao *et al.*, 2012; Sikora *et al.*, 2021).

The annual economic damage caused by nematodes has been reported to be as high as US$173 billion (Elling, 2013). Root-knot nematodes (*Meloidogyne* spp.), cyst nematodes (*Globodera* and *Heterodera* spp.) and root lesion nematodes (*Pratylenchus* spp.) are reported to be the four most important and widespread genera (Jones *et al.*, 2013). However, on a regional scale, major damage to various crops (including potato, carrot, sugar beet, etc.) has also been observed with stem nematodes (*Ditylenchus* spp.), (e.g. Northern Europe), stubby root nematodes (*Trichodorus* and *Paratrichodorus* spp.) (e.g. Northern Europe), the burrowing nematode in banana (*Radopholus similis*) (e.g. South America, Africa and Asia) or dagger nematodes in grapes (*Xiphinema* spp.) (e.g. the USA, Chile and the EU).

The global nematicide market was estimated to be worth around US$1 billion in 2018 and will reach around US$1.8 billion by 2030 (Fig. 17.1). Around 15% of this amount is spent on seed treatment nematicides – mainly in row crops such as maize, soybean and cotton. Most of this will be spent on soil-applied nematicides (both fumigants such as 1,3-D and metam sodium with a 40% market share and non-fumigant nematicides with a 50% market share) with key market segments in root vegetables (e.g. potatoes, carrots, sweet potatoes, ginger), fruiting vegetables (e.g. tomatoes, peppers, eggplants) and plantation crops (e.g. sugar cane, banana, coffee), but also cucurbits (e.g. cucumber, melon, pumpkin), berries (e.g. strawberry), trees (nuts and fruit) and turf and ornamentals. The market share of bionematicides only reached around 5% in 2018 but has recently started to increase significantly and is currently expected to be around 10%. In 2018 most nematicides were sold into potatoes (around US$170 million) followed by tomatoes (around US$95 million), sugar cane (around US$58 million), as well as strawberries (around US$40 million).

Key countries for nematicide sales are the USA, China, Japan, India and Brazil, but southern European countries around the Mediterranean also have a significant economic nematicide market. In Northern Europe, nematicides are widely used in the Benelux countries and in the UK (although this could decline significantly with the expected loss of oxamyl in 2025).

17.2. Nematicide Formulations, Modes of Action and Application

Nematicides can be classified according to their chemical group (e.g. carbamates, organophosphates), their mode of action (e.g. acetylcholinesterase inhibitor) and mode of application.

Box 17.1. Nematicide discovery and evaluation.

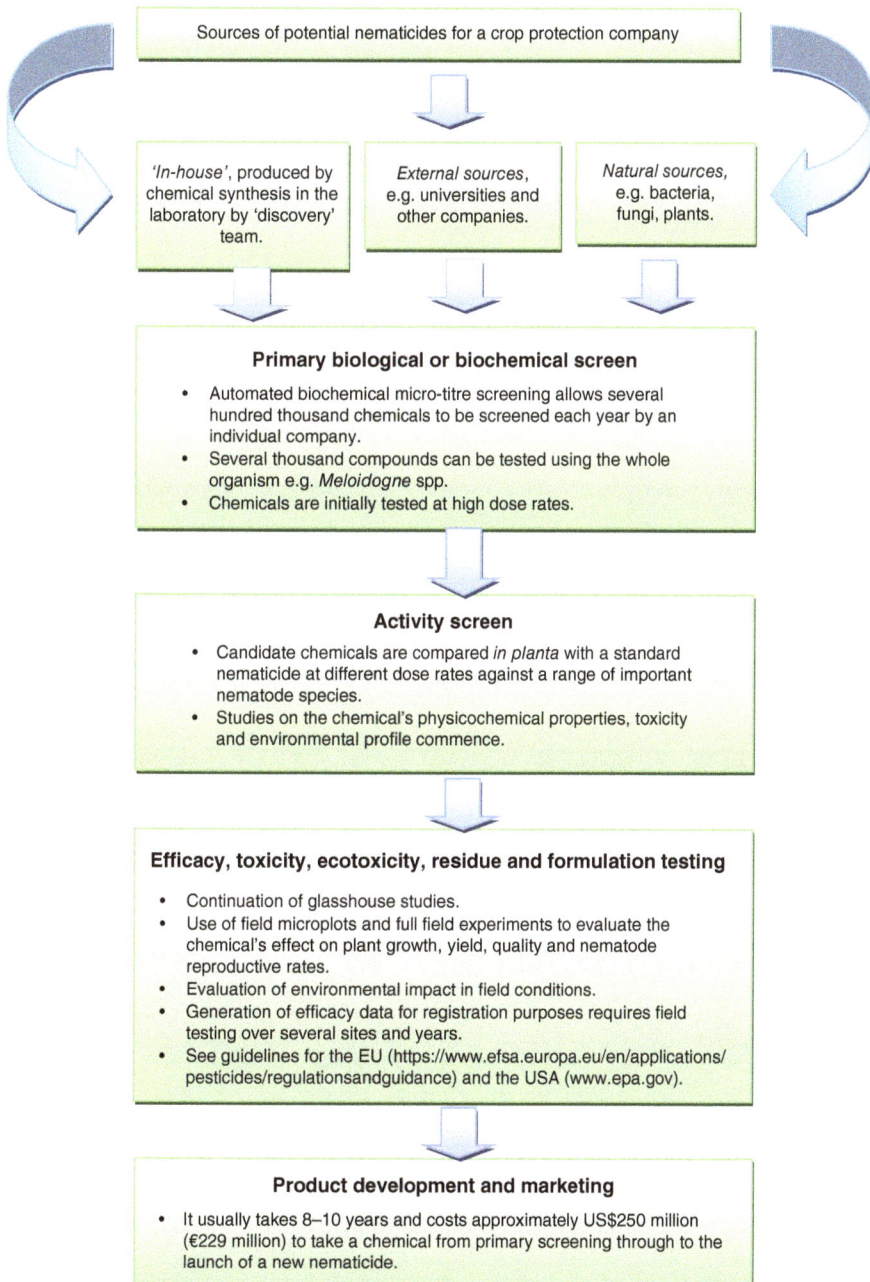

Sources of potential nematicides for a crop protection company

'In-house', produced by chemical synthesis in the laboratory by 'discovery' team.

External sources, e.g. universities and other companies.

Natural sources, e.g. bacteria, fungi, plants.

Primary biological or biochemical screen

- Automated biochemical micro-titre screening allows several hundred thousand chemicals to be screened each year by an individual company.
- Several thousand compounds can be tested using the whole organism e.g. *Meloidogne* spp.
- Chemicals are initially tested at high dose rates.

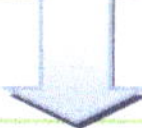

Activity screen

- Candidate chemicals are compared *in planta* with a standard nematicide at different dose rates against a range of important nematode species.
- Studies on the chemical's physicochemical properties, toxicity and environmental profile commence.

Efficacy, toxicity, ecotoxicity, residue and formulation testing

- Continuation of glasshouse studies.
- Use of field microplots and full field experiments to evaluate the chemical's effect on plant growth, yield, quality and nematode reproductive rates.
- Evaluation of environmental impact in field conditions.
- Generation of efficacy data for registration purposes requires field testing over several sites and years.
- See guidelines for the EU (https://www.efsa.europa.eu/en/applications/pesticides/regulationsandguidance) and the USA (www.epa.gov).

Product development and marketing

- It usually takes 8–10 years and costs approximately US$250 million (€229 million) to take a chemical from primary screening through to the launch of a new nematicide.

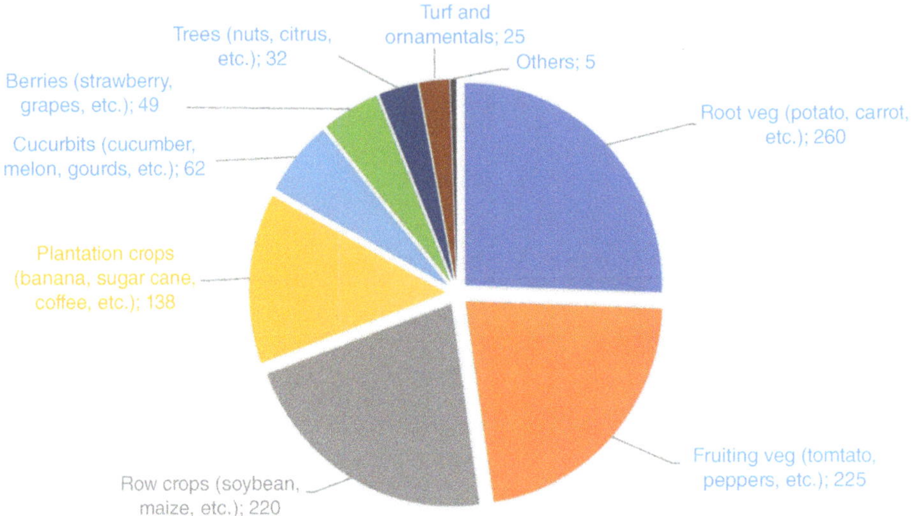

Fig. 17.1. The global market size of the nematicide market (fumigant and non-fumigant nematicides) in 2018 per crop group (in US$ million).

The Nematode Working Group within the Insect Resistance Action Committee (IRAC) have developed a nematicide classification scheme based upon mode of action (see https://irac-online.org/teams/nematodes/ for more detail). Currently there are nine groups, with carbamates and organophosphates forming subgroups of group N-1. In this chapter, nematicides will be grouped according to their mode of application and discussed by chemical group and mode of action. Whilst there is much data to support the efficacy of nematicides, there is a dearth of information to understand their precise activity in nematodes, most of the information being taken from their known effects in insects and mammals. Presently, nematicide products can be applied as seed treatments, as sprays applied in furrow, as granules and as soil fumigants. Active substances mostly act within the soil, external to the plant and target infective juveniles and adults. Some nematicides also inhibit juvenile hatching from egg masses or cysts. The mode of action of current active substances is shown in Table 17.2.

17.2.1. Seed treatments

Seed treatment is widely considered to be the most economically and environmentally sustainable method of pesticide application, requiring lower doses of active substances, while providing the option to combine a range of active substances for protection against soilborne fungi, insects and nematodes. Nematicide seed treatments (NSTs) may affect nematodes directly or indirectly by inducing plant resistance. The first NST for field-scale crops became available in 2006 (Beeman and Tylka, 2018), and since then, there has been a steady increase in the number of NSTs, particularly in the USA, South America and South Africa. Most of these products have been used to target important, large-scale field crops such as maize, cotton and soybean. Current chemical seed treatments include abamectin, cyclobutrifluram and fluopyram. Bionematicides are also available as seed

Table 17.2. Mode of action of currently available nematicides.

Active substance	IRAC group*	Mode of action
Carbamates: Aldicarb, carbofuran and oxamyl	N-1A	Acetylcholinesterase (AChE) inhibitor. AChE is responsible for breaking down acetylcholine at nerve synapses (neuromuscular junctions), terminating neurotransmission. When AChE is inhibited, acetylcholine builds up in excessive levels giving cholinergic hyperactivity, which causes paralysis in the nematode. Nematodes affected by AChE will eventually die due to depletion of lipid reserves.
Organophosphates: Caldusafos, ethoprophos, fenamiphos, fosthiazate and imicyafos	N1B	
Abamectin	N-2	Glutamate-gated chloride (GluCl) channel allosteric modulator. Abamectin amplifies glutamate in nerve synapses, preventing the transmission of electrical impulses in muscle.
Pyridinyl-ethyl-benzamide: Fluopyram	N-3	Succinate dehydrogenase inhibitor. Inhibits ATP production by binding to succinate dehydrogenase in the citric acid cycle of respiration (inner mitochondrial membrane).
Phenethyl pyridineamide: Cyclobutrifluram		
Cyclic keto-enol: Spirotetramat	N-4	Inhibitor of acetyl-CoA carboxylase involved in fatty acid synthesis. Affects fatty acid composition.
Fluoroalkenyl: Fluensulfone	N-UN (unknown)	Kearn *et al.* (2017) suggests that fluensulfone causes a metabolic impairment leading to reduced lipid synthesis.
Imidazopyridine: Fluzaindolizine		Unknown mode of action.
Halogenated hydrocarbon: 1,3-dichloropropene	N-UNX (presumed multisite inhibitors)	Specific mode of action unknown. Enters nematodes through the mouth and cuticle.
Volatile sulphur generator: Dimethyl disulfide (DMDS)		DMDS primarily affects mitochondrial respiration by blocking the activity of the enzyme cytochrome oxidase.
Methyl isothiocyanate (MITC) generators: Dazomet, metam sodium, allyl isothiocyanate (AITC)		MITC and AITC affect a variety of cellular enzymes inducing apoptosis (programmed cell death) and inhibiting cell cycle progression (cell division).
Phenyl oxadiazole: Tioxazafen		Disrupts ribosomal activity thereby affecting protein production.

*IRAC = Insect Resistance Action Committee.

treatments and are a key part of the market; *Bacillus firmus* (e.g. VOTiVO®), *B. amyloliquefaciens* (e.g. LUMIALZA®), and *B. subtilis* and *B. licheniformis* (ZIRONAR™) are bacteria that colonize plant roots, preventing nematodes from invading. Similarly, *Pasteuria nishizawae* has been used as seed treatment (CLARIVA™).

Owing to a lack of clear information on the movement of NSTs through the seed and root zone, Beeman and Tylka (2018) investigated the activity of fluopyram (ILeVO®) and *B. firmus* I-1582 (VOTiVO) on the soybean cyst nematode (*H. glycines*). *In vitro* experiments using exudates collected from fluopyram-treated seed reduced hatching and motility of *H. glycines* by <95%, whilst exudates collected from the radicles of the treated seed caused a 48% reduction in hatch. By contrast, VOTiVO seed did not affect any of the life stages of the nematodes. Fluopyram-treated seed also resulted in a clear reduction (>80%) in the infection of roots by *H. glycines* under glasshouse conditions.

Tioxazafen, marketed as Acceleron NemaStrike ST® (Bayer CropScience, formerly Monsanto Company), was withdrawn before being commercialized by the manufacturer in 2018 and 2020 due to health and safety concerns (skin irritation). Presently, it is unclear whether the product could be reintroduced in the future, perhaps with greater requirement for specific personal protective equipment (PPE) use. Acceleron® seed treatments are available in the USA, but with fluopyram (ILeVO). Tioxazafen has been shown to be effective against nematodes such as *M. nincognita* and the reniform nematode *Rotylenchulus reniformis* (Faske *et al.*, 2022), although the original supplier indicated suppression in a wider selection of nematode targets (Slomczynska *et al.*, 2015). The compound, made up of 1,2,4-oxadiazole and a thiophene ring, disrupts ribosomes, leading to nematode death (Wang *et al.*, 2022).

Cyclobutrifluram (Tymirium®), a succinate dehydrogenase inhibitor (inhibiting cellular respiration), was recently developed by Syngenta, and has broad spectrum activity against seed- and soilborne fungal pathogens and soilborne and foliar nematodes (e.g. cyst, root lesion and rook-knot nematodes). Victrato®, which incorporates Tymirium, is due to be released in Australia for managing root lesion nematodes (*Pratylenchus* spp.) and crown rot (*Fusarium pseudograminearum*) of wheat and barley. The product is currently registered in El Salvador and Argentina.

Recently, a novel approach was developed to tackle nematodes in East Africa. The 'Wrap and Plant' technique is not strictly speaking a seed treatment but uses a biodegradable lignocellulosic matrix (LCM) (banana paper) infused with a nematicide to cover (wrap) potato seed tubers. The technique was used in combination with a low dose of abamectin (1000 times lower than the label recommendation) to protect potato yields against potato cyst nematodes (*G. rostochiensis*) (Ochola *et al.*, 2022). In combination with the nematicidal activity of abamectin, the LCM absorbed root exudates, reducing potato cyst nematode hatching and host location by infective juveniles.

17.2.2. Soil treatments

17.2.2.1. Non-fumigant

Soil-applied nematicides that are non-fumigant, consist of liquids and granules. Nowadays, nematicides are more frequently formulated as a liquid, typically emulsifiable or soluble concentrates (EC or SL), which are applied as a broadcast or in-furrow spray or through drip irrigation. The current liquid formulations include the active substances fluensulfone (Nimitz®), fluopyram (Velum Prime®, Verango® and Indemnify®), fluazaindolizine (Salibro® or Reklemel®) and spirotetramat (Movento®). These are regarded as 'next-generation nematicides', having improved profiles with regard to environmental impact and human safety. Desaeger *et al.* (2020) draws attention to the fact that these nematicides have been subject to stringent regulatory criteria, which include assessments on their behaviour in soil, such as leaching, effect on beneficial organisms, soil persistence and degradation and metabolism pathways. These active substances control a wide range of plant-parasitic nematodes and are marketed in various formulations dependent on target organism, delivery system and crop. The range of crops protected by these products is as diverse as vegetables, root vegetables such as potatoes and carrots, bananas, cucurbits and tobacco. Reviews by Desaeger *et al.* (2020) and Oka (2020) provide excellent references on the performance of next-generation nematicides.

Generally, liquid nematicide formulations are soil applied. Methods of application include overall spray followed by incorporation to a specified depth, sprayed over the width of the furrow or row, shank injected followed by irrigation, or added to drip, trickle or overhead irrigation in a process known as chemigation (Fig. 17.2). Foliar application is also possible with some products such as Vydate L® (oxamyl) and Movento (spirotetramat). Spirotetramat is absorbed by the foliage and is translocated via the xylem and phloem; it is capable of moving in the apoplast by diffusion into intracellular spaces and xylem elements and the symplast by movement through intercellular pores known as plasmodesmata. The highly systemic activity of spirotetramat means that it provides protection against root-feeding nematodes. Vydate L can be applied through low-pressure sprinkler-type equipment including centre pivot, lateral move, solid-set, mini sprinklers and drip tape. Operators should always check with current recommendations from the

Fig. 17.2. The application of nematicide via trickle irrigation.

manufacturer to ensure that each formulation and application method is legal for each nematode problem, crop and country/state of use.

Where irrigation water is used as carrier for the nematicide it is of paramount importance that sufficient safety systems are used to prevent backflow into, and contamination of, the water source. As with all irrigation, but even more so with chemigation, it is essential to ensure that run-off and leaching do not occur as a result of either poor application technique or lack of knowledge of the soil moisture status at the time of irrigation. From an environmental standpoint it can be argued that trickle or drip irrigation offer the most environmentally acceptable forms of chemigation as these methods can reduce exposure of non-soil-dwelling organisms, such as birds, to nematicides.

Granules are almost dust-free, free-flowing granules of approximately 1 mm in diameter. The most widely available products are Nemathorin 10G® (fosthiazate) and Nimitz GR® (fluensulfone); these are formulated onto granules formed from materials such as sepiolite, brick dust and clay. Most of the carbamates (aldicarb, carbufuran and oxamyl) and organophosphates (cadusaphos, ethoprophos, fenamiphos and imicyafos) have been deregistered in many countries. Granular formulations can range in active substance content from 5% to 15%.

Granules must be metered onto the ground at the correct rate, and a number of methods have been developed to achieve this. The most reliable way of metering granules is by using a ground wheel driven method such as the Horstine Farmery Microband Applicator® (Fig. 17.3). This positive displacement machine eliminates the risk of over-treating when the tractor is stationary because dosing is generated by distance travelled and not the forward speed of the tractor. Cartridge inserts containing the correct metering unit for a particular product are available (Fig. 17.4). Modern electronic and hydraulic systems are also in use to achieve variable rate and precision application using a differential global positioning

Fig. 17.3. A Horstine Farmery Microband Applicator showing hoppers and 'fishtail' distribution units.

Fig. 17.4. A selection of granule metering cartridges. (Courtesy Horstine Farmery, Lincoln, UK.)

system (DGPS). The economics of sampling nematodes in the field seldom allows for the sampling intensity required to provide the necessary detailed field information. Nevertheless, electrical conductivity soil mapping could be a useful aid for identifying higher population densities of plant-parasitic nematodes to complement the accuracy of the application technology (Ortiz *et al.*, 2010; Hbirkou *et al.*, 2011). Therefore, issues regarding whether or not to treat low nematode populations and the potential for higher nematode population multiplication are a very real concern. Some product stewardship issues are apparent with respect to granule application and these are discussed in Section 17.4.4.

The application of granular nematicides for the control of plant-parasitic nematodes varies according to the nematode species and the crop to be protected. Factors such as root growth habit, i.e. tap rooted or fibrous rooting systems, nematode mobility and soil characteristics, will all affect the success of granular nematicide application. The high financial cost of nematicides means that they must be used efficiently and provide maximum benefit in terms of crop yield and nematode control. This is obtained by optimizing the distribution of the chemical in soil in the area of early root development (Fig. 17.5.).

At present, the recommendation given by nematicide manufacturers is to incorporate nematicide granules to a depth of 5–15 cm, depending on the crop, nematode problem and the nematicide concentration required in the soil water. In the field, different machines are used to achieve this recommendation and it is likely that some methods used by growers fail to achieve this objective. It is essential to know the incorporation characteristics of tillage machinery as shallow incorporation will not treat enough soil to give satisfactory returns in terms of crop yield and nematode control. Deep incorporation is also undesirable as the active substance of the nematicide is diluted in the greater volume of soil. Achieving the optimum depth of nematicide incorporation in relation to root growth is critical in order to maximize nematode control (Fig. 17.6).

Fig. 17.5. The incorporation of granules is best achieved by rotary cultivation. (Courtesy Horstine Farmery, Lincoln, UK.)

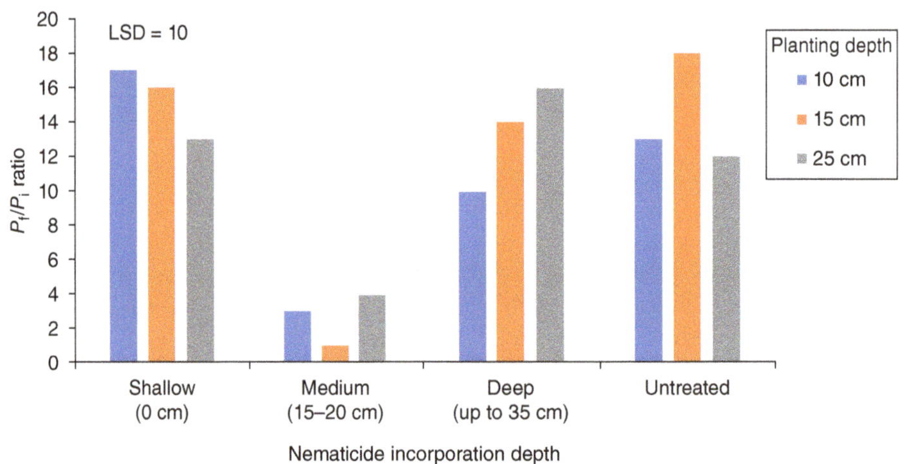

Fig. 17.6. The effect of depth of granular nematicide incorporation and depth of potato tuber planting on the multiplication of *Globodera pallida*. LSD, least significant difference. (Adapted from Woods and Haydock, 2000.)

Application timing is critical for granule application and is normally timed for 'as close to planting' as possible. This is to enable the product to provide effective doses for as long as possible during early phases of plant growth when growth retardation can be severe. Additionally, the 'harvest interval' (HI; days between application and the harvest of the crop) may also be critical for crops such as potatoes that are destined for 'early season' markets. The HI for fosthiazate when used for potatoes is 119 days.

17.2.2.2. Fumigant

Soil fumigation is a technique whereby chemical liquids, or solids, are incorporated into the soil where they volatilize, move through air spaces in the soil as a gas and provide control of a range of pests, weeds and disease. Fumigant nematicides consist of compounds based on halogenated hydrocarbons, volatile sulphur generators, allyl isothiocyanate, and those that release methyl isothiocyanate. In recent years, a significant proportion of fumigants have been deregistered due to their effects on non-target organisms. The majority of fumigant nematicides have been banned from use in the EU and in the UK, although registrations of metam sodium continue in countries such as Australia, Kenya, Zimbabwe and the USA.

The halogenated hydrocarbons include 1,3-D and methyl bromide, a compound that was phased out by The Montreal Protocol in 2005. These fumigants are thought directly to affect biochemical pathways in protein synthesis and respiration. Dimethyl disulphide has been considered as a chemical intermediate in refineries and the chemical industry for a long time, and its biological properties have been characterized and the compound was finally patented in December 2012 by Arkema (France) in Europe after decades without new soil fumigants. In soil, sodium *N*-methyldithiocarbamate (metam sodium) and dazomet degrade to release the active substance, methyl isothiocyanate. Methyl isothiocyanate and allyl v (e.g. Dominus®) cause cell death and inhibit cell division.

Fumigant nematicides are formulated as liquefied gases, volatile liquids or solids. The liquefied gases are held in liquid state under pressure but convert quickly back to gas when released into the soil or atmosphere. Volatile liquids are chemicals that remain as liquids while contained at normal atmospheric pressure but readily convert to gases in the soil environment, e.g. 1,3-D. Solids are normally granular formulations that ultimately degrade to release the active substance, e.g. methyl isothiocyanate, when placed into moist soil.

The effectiveness of soil fumigants can be influenced greatly by several soil factors: texture, structure, organic matter, moisture content and temperature at application. Coarse-textured soils can generally be fumigated more easily than fine-textured (clay) soils, as the coarse sandy soils contain more voids per unit volume. Unfortunately, this also makes them more reliant on good sealing. Clay and organic matter can adsorb large quantities of the applied chemical, either making it necessary to apply higher rates of chemical or unrealistic to use for control purposes in these soil types. Similarly, poor soil structure, e.g. compaction, can restrict gaseous movement and result in poor fumigant penetration and thus give poor nematode control. Soil moisture at the time of application can also be important for the efficacy of the fumigant. In general, it is useful to maintain soil moisture between 30% and 70% of field capacity, dependent on product and use (some need 40–60%). Consequently, there must be sufficient air spaces within the soil to allow for rapid dissemination of the gas but the soil must not be too dry as to allow rapid loss of the fumigant. There must also be sufficient moisture to interact with the fumigant and then act upon the organism in question. Suitable temperature is also of great importance for effective fumigation. Most fumigants require a range of 7–27°C to work well; below 7°C the fumigant may not volatilize or diffuse adequately and above 27°C volatilization may be too rapid and degradation or loss may occur. Figure 17.7 shows fumigant application within a field.

The application of the various fumigant formulations will vary according to the fumigant, pest and target zone, and most should be applied several weeks or months before planting to avoid crop damage due to their inherent phytotoxicity. Granular formulations can be broadcast on to the soil surface or injected, or incorporated by cultivation. There may be a requirement to then seal with sheeting, irrigation or surface consolidation. The application of the volatile liquids will greatly depend on the label recommendations and uses. They may be injected via chisel (shank) or subsoiler-type ploughs and commonly sealed with mechanical consolidation of the soil surface or suitable sheets. Liquefied gases can be injected into the soil with suitable equipment for field-scale fumigation or released into evaporation pans via tubes under sealing sheets for small-scale fumigation in glasshouses, etc. However, it is important to note that when the pests to be controlled exist in the upper soil horizons, up to 10 cm depth, the use of topsoil consolidation should be avoided as this consolidation effectively removes the pore spaces necessary for effective fumigant movement. In this case, the only option is to use a sealing sheet above the soil. Some dedicated product formulations also exist, such as 1,3-D as Telone EC®, which is a liquid formulation fumigant designed specifically for use in drip irrigation only.

Whether liquid or gas phase is considered, the dosage of the chemical is important. The dosage can be given as the amount of chemical required (concentration) and the time period over which it is required (time) to achieve kill of the nematode. This is given then as the concentration × time product (CTP).

Fig. 17.7. The application of 1,3-dichloropropene using contra-rotating drums to seal the soil surface.

17.3. Factors Affecting Nematicide Performance

The main objective of any nematicide application (fumigant and non-fumigant nematicides) is to interfere with the nematode life cycle either in the soil (contact nematicides) or in the plant (systemic compounds) and thereby reduce their potential for damage to the crop. In principle, this can be achieved by either: (i) a direct nematicidal effect (i.e. killing the target nematode); or (ii) any kind of mechanism that reduces the fitness of the nematode so that it no longer infects the plant (e.g. by reducing mobility, inhibiting host finding and orientation, reducing reproduction or hatching, etc.). It is important to understand that there is a wide range of parameters that potentially affect nematicide performance (see Fig. 17.8).

17.3.1. Intrinsic activity

The intrinsic activity of a nematicide (or its active substance) can be defined as the natural potency of a given molecule on a particular life trait of the target nematode. For example, if we look at the mortality of *Meloidogyne* juveniles within hours of exposure, we will see that nematicides such as cyclobutrifluram, fluopyram, fluensulfone, abamectin or some organophosphates inactivate many juveniles at concentrations as low as 1 to 5 ppm, whereas other nematicides, e.g. fluazaindolizine or some carbamates (e.g. oxamyl), require either longer exposure times or higher concentrations to achieve similar levels of mortality (Oka *et al.*, 2009; Faske and Hurd, 2015). However, as the latter materials impose high fitness costs on juveniles, this picture may change if, instead of juvenile mortality, juvenile root infectivity is used as a criterion of intrinsic activity (Thoden and Wiles, 2019).

Similarly, nematode species and life stages may differ in their susceptibility to a given nematicide at the same exposure time (e.g. Gourd *et al.*, 1993). Therefore, some nematicides, such as fluopyram or fluensulfone, can be considered broad spectrum, whilst others, such as fluazaindolizine, show remarkable variability in species sensitivity (Thoden *et al.*, 2020; Wram and Zasada, 2020). Similarly, differences in sensitivity between different populations of a given species have been observed (Thoden *et al.*, 2019).

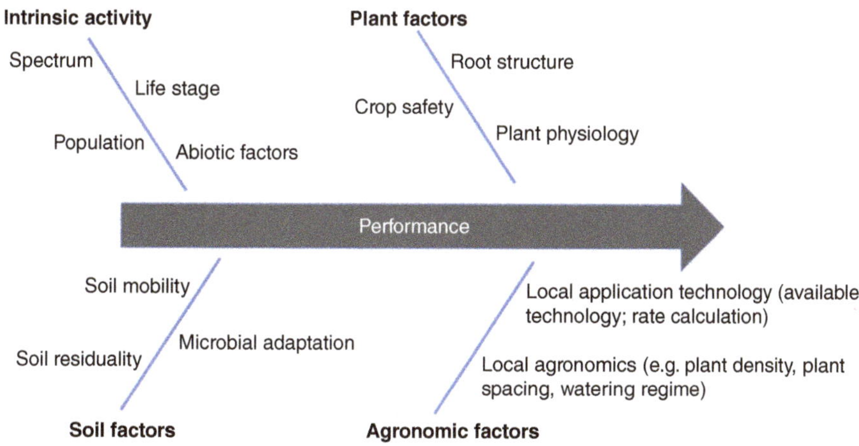

Fig. 17.8. Key factors affecting nematicide performance.

In addition to biotic factors that drive intrinsic activity, nematicide performance can also be influenced by abiotic factors such as temperature (e.g. lower efficacy at low temperatures; Oka, 2020) or pH (e.g. ionization state of the molecule and cuticular transmission).

17.3.2. Soil factors

Almost all nematicides are applied directly to the soil (either by soil injection, drip, surface, drench or furrow application) and their ultimate performance is determined by their ability to penetrate and remain in the root zone. Fumigant nematicides are mainly formulated as liquids that rapidly evaporate to a gas to move through open air spaces in the soil. Non-fumigant nematicides are now mainly formulated as liquids (often as a suspension concentrate (SC)), or with some granular formulations for certain niche markets, and are moved by downward flow of soil water. If a material is immobile, it will remain on the soil surface and not reach the targeted nematodes; if it is too mobile, it will leach below the required protection zone and, in the worst case, contaminate groundwater. Both the physicochemical properties of the active substance, e.g. water solubility, ionization state, K_{OC} (soil adsorption coefficient), and the physicochemical parameters of the soil, e.g. soil type, soil pH, organic matter content, etc., will have an effect on soil movement behaviour (e.g. Oka *et al.*, 2013). Compounds with high K_{OC} values are considered immobile (K_{OC} > 5000), whereas many of the modern nematicides such as fluopyram, fluensulfone or fluazaindolizine have average K_{OC} values in the range of 150–500, indicating moderate soil mobility (K_{OC} fluopyram > K_{OC} fluazaindolizine > K_{OC} fluensulfone). It is important to remember that the adsorption of a nematicide in the field will be strongly impacted by the organic matter content of a given soil.

In addition to soil mobility, the performance of a nematicide also depends on the residuality of the molecule or, in other words, the period during which it is active in the root zone. This is typically expressed in terms of its soil half-life (DT_{50}), which can vary widely for different nematicides (from a few days to a few months) and is highly dependent on the way in which a molecule is degraded in nature (e.g. chemical pathway by hydrolysis or microbial pathway) (Desaeger *et al.*, 2020). Similar to soil mobility, soil residuality is driven by site-specific factors (e.g. microbial activity) and can therefore vary greatly between sites. The adaptation of microbial populations after repeated applications should also be considered as a potential driver of performance and has been reported to be responsible for accelerated microbial degradation of some nematicides (Arbeli and Fuentes, 2007; Lagos *et al.*, 2019).

17.3.3. Plant factors and local agronomic practices

The success of a nematicide application also depends on certain crop-related factors. Points to consider are: (i) root morphology, which can vary widely between crops and therefore may require different patterns of product distribution; (ii) crop sensitivity to potential phytotoxic effects, which may require specific timing of nematicide application; and (iii) plant physiology, which may be related, for example, to certain root flushes associated with high nematode abundance in the soil (e.g. spring root flush in perennials).

Another factor that is often neglected but can have a significant impact on the performance of a nematicide is the local agronomic practice within a given crop or region.

Differences might be observed with respect to: (i) planting densities or row spacing; (ii) water regimes depending on soil type; and (iii) the way growers calculate their application rates (e.g. concentrated per hectare rates versus treated area only rates). The equipment available for application can also vary considerably between regions.

17.4. Legislation and Stewardship in Relation to Environment and Human Safety

17.4.1. Environmental legislation

Pesticides are amongst the most heavily regulated products used by humankind, and approval and registration processes are similar to those required for pharmaceuticals. In addition, pesticides must be rigorously tested for their environmental impacts and those with poor environmental profiles are no longer approved for use.

The use of pesticides was an issue debated at the Earth Summit in Rio de Janeiro in 1992. The outcome was a requirement to reduce the use of pesticides and many governments subsequently adopted policies to meet this objective. One of the most radical solutions had already been taken by the Danish government, who in 1986 through their Pesticide Action Plan had called for a 50% reduction by weight in the sales of pesticides within a ten-year period. The Danish government followed this up in 1991 by introducing a sustainable action plan and then a pesticide tax in 1996 based on pesticide load (Kudsk *et al.*, 2018). Pesticide taxation had already been introduced in 1984 in Sweden in order to reduce the impacts of pesticides on residues in food and the environment. Other countries took different approaches. The UK, for example, encouraged a reduction in pesticide use through the implementation of integrated crop management, a system that aimed to optimize all aspects of the production process resulting in an overall reduction in pesticide use. The UK's approach has developed into the promotion of integrated pest management, which fits with the EU thematic strategy on the sustainable use of pesticides and the Sustainable Use Directive (Directive 2009/128/EC). Approval processes for pesticides are complex and the data requirements for safety as well as efficacy are extensive. Before approval is likely to be given, regulators must be satisfied with the pesticide's profile in the following areas: chemistry of the active substance and product, mammalian toxicity, non-dietary human exposure, residues, fate and behaviour in the environment, ecotoxicology and efficacy, and that any risks posed by its use are acceptable. Individual countries may have their own legislation to deal with pesticide approval. However, in Europe issues specific to the active substance approval and subsequent plant protection product authorizations are dealt with on a coordinated basis in accordance with EU Regulation (EC) No. 1107/2009, which can be found on the official website of the EU (https://eur-lex.europa.eu). Active substances are approved for use within Europe as a whole. For use on non-protected crops, plant protection products are authorized initially for use in individual EU Member States on a zonal basis and then subsequently by mutual recognition within zones. The EU has been divided into three zones: the northern zone, the central zone and the southern zone. Initial product authorizations may be granted in multiple Member States within a zone following acceptance by a lead Member State acting on behalf of other Member States within the zone. This process reduces the bureaucratic requirements that existed under the now superseded EU Directive 91/414/EC, where the product approval process had to be performed in each Member State. Mutual recognition allows for an authorization of a product based on an existing authorization in a Member State within the same zone. Following its exit from the EU, the UK developed its own system for pesticide product authorization based on the EU system at the time of exit. The system now in use in the UK can be found on the Health and Safety Executive's website (https://www.hse.gov.uk/pesticides). A similar system of pesticide approval based on an evaluation of chemistry of the active substance and product, mammalian toxicity, etc., is operated in most other countries, e.g. the USA, and information on the US system can be found on the EPA website (http://www.epa.gov/pesticides). Unlike the EU, however, the USA, along with the majority of countries, has a regulatory system based on the risk associated with pesticide use rather than an initial assessment of the hazard of a pesticide. In addition, greater emphasis is placed on endangered species within the US system and less on product efficacy. The labelling and use direction claims of a product in the USA must be informed by efficacy data but these data are not routinely required to be submitted for registration purposes. Although the main areas for data requirement, e.g. ecotoxicology, are common to most countries, not all require the same data to be generated. Information on fish toxicity, for example, is required on *Cyprinus carpio* in Japan but Australia will accept data on *Oncorhynchus mykiss*, *Pimephales promelas*, *Danio rerio* or *Oryzias latipes*. However, both countries require data on the freshwater crustacean *Daphnia magna*. The Organisation for Economic Co-operation and Development (OECD) has guidance documents for pesticide registration that help facilitate global harmonization of pesticide registration systems (https://www.oecd.org) and these are used by many countries, e.g. South Africa.

17.4.2. Exposure during application

Nematicides have the potential to harm humans, and often old synthetic nematicides have high mammalian toxicities. Toxicity can be determined by experimentation, and a measure of toxicity such as an LD_{50} can be calculated and is used to derive a hazard classification for the product. An LD_{50} is a statistical value that represents the lethal dose required to kill 50% of a test population (often rats); the lower the LD_{50} the more toxic the nematicide. One of the first discovered nematicides, aldicarb, has an oral LD_{50} (rat) of 0.93 mg kg^{-1}, whereas the more recently discovered nematicide, cyclobutrifluram, has an LD_{50} of >5000 mg kg^{-1} (Table 17.1). There are three potential routes of contamination: inhalation, ingestion and absorption. Contamination through inhalation is uncommon under field conditions as particles small enough to be breathed into the lungs are generally produced in only small volumes during pesticide application.

Toxicity is not synonymous with hazard, and it is hazard that is more important when determining the risk to those using nematicides. Hazard is the potential of a substance to cause harm. A nematicide's hazard classification can be determined by its toxicity and its form, e.g. whether it is a gas, solid or a liquid, and how concentrated the formulation is. Therefore, the hazard of the nematicide-containing product must be determined. Hazard classifications are determined with reference to sets of trigger values, which result in a given warning symbol and risk phrase being placed on the product's label. For example, in the USA, a pesticide in toxicity category I on the basis of its oral, inhalation or dermal toxicity must have the word 'Poison' in red on the label and the skull and crossbones symbol must appear in immediate proximity to the word 'Poison'. Many countries, including the EU and the UK, have adopted the Globally Harmonized System of

Classification and Labelling of Chemicals (GSH) which is managed by the United Nations (UN). For packaged goods, i.e. a pesticide label, this system uses a suite of pictograms containing a black symbol, e.g. skull and cross bones, on a white background within a red diamond frame. Details of the GHS can be found on the UN website (https://unece.org).

The risk to the operator can be estimated from their exposure to the product. In the UK, the Control of Substances Hazardous to Health Regulations (COSHH) requires a risk assessment to be performed before using hazardous pesticides. Risk should be reduced by using the least hazardous product and then by reducing exposure. Exposure can be reduced in three principle ways: correct use of the product, engineering controls and the use of PPE. Correct use of the product can be achieved following adequate training and guidance in the use of the product. In addition, certificates can be obtained to demonstrate competence – these may be a legal requirement in some situations, e.g. within the EU when using professional pesticide products. Engineering controls can come in many forms but the main control measure would be an enclosed cab with an air filtration system on the sprayer/applicator. This provides a physical barrier to the spray/dust and reduces the reliance on PPE such as gloves, coverall and face shield. Another important engineering control measure is a closed transfer system. A closed transfer system allows the pesticide to be mixed into the sprayer or delivered to the applicator while reducing the possibility of splashing or accidental spillage (see Section 17.2.2.1 for granules). There are many different types of closed transfer system that have been designed to be used with either single or multitrip containers. The benefit of a multitrip container system is that the container does not need to be decontaminated on farm or disposed of as it is returned to the manufacturer for refilling. In addition, the use of such systems may reduce the requirement for operators to wear PPE. However, use of the appropriate PPE for a given operation is essential to provide an effective barrier against contamination. The topic of operator exposure, and the use of engineering controls and PPE, is covered in greater detail by Matthews (2015) and practical advice on the use of PPE is given in the UK's Code of Practice for Using Plant Protection Products (http://www.pesticides.gov.uk).

Reducing exposure to nematicides is very important as contamination can lead to poisoning. Symptoms of poisoning should be known before using a specific product, and these will be given on the product's label and materials safety data sheet (MSDS) together with first aid instructions. A section on poisoning by pesticides and first aid measures is included in Lainsbury (2023).

17.4.3. Residues in foodstuffs

Humans can also be exposed to pesticides through the consumption of contaminated food and water. Not all pesticide applications leave residues in the harvested crop, but some do. The issue of residues in animal feeds and foodstuffs is covered within pesticide approval processes. Potential residue levels should be toxicologically acceptable and maximum residue levels or limits (MRLs) are established to monitor the correct use of a pesticide. Individual countries can set MRLs in foods and feedstuffs, although in the EU, MRLs are set by the EU and are common to all Member States. International MRL standards (CXLs) are given in the Food and Agriculture Organization (FAO)/World Health Organization (WHO) Codex Alimentarius international food standards. The MRL is used as a trading standard. If the MRL is exceeded it does not necessarily mean that the residue will cause harm, but it does indicate that the pesticide has probably not been used correctly. If the pesticide has been applied to an approved crop in an approved way, i.e. correct dose rate and timing of application, the MRL should not be exceeded.

The toxicological acceptability of a residue is generally based on a 'no observed adverse effect level' (NOAEL), which is calculated from experimental observation of the most sensitive test species. From this value, the acceptable daily intake (ADI) (the amount of residue that could be consumed every day with the reasonable certainty that no adverse effect will occur) is calculated. The ADI is used to assess chronic exposure to a pesticide's residues. The NOAEL is divided by 10 to account for differences in sensitivity between humans and the test species, and 10 again to account for variation between individuals within the human population to give the ADI. Either of these two uncertainty factors could be greater than 10 if thought appropriate. To provide a worst-case scenario, the ADI is compared to the theoretical maximum daily intake (TMDI), which is calculated from the MRL and data on the 97.5th centile for consumption of the foodstuff. If the ADI is greater than the TMDI, i.e. the amount of pesticide residue eaten is less than that likely to cause harm, the residue is deemed to be toxicologically acceptable. A similar approach is taken with respect to acute exposure, and more realistic exposures to pesticide residues are also evaluated. The pesticide residue intake model (PRIMo) is the risk assessment tool used by the EU and it is available on the European Food Safety Authority (EFDA)'s website (https://www.efsa.europa.eu). Pesticides can end up in surface water and groundwater, both of which may be used for drinking water supply. Ultimately, if not degraded, pesticides will end up in the sea. The amount of pesticide permitted in drinking water in the EU is based on a maximum admissible concentration (MAC). The MAC in the EU is set at 0.1 $\mu g\ l^{-1}$ for an individual pesticide and 0.5 $\mu g\ l^{-1}$ for the total pesticide content. These values were set using the 'precautionary principle', whereas the WHO has issued guideline values for certain pesticides based on their health significance in drinking water. Most other non-EU countries also set allowable concentrations of pesticide in drinking water based on toxicology. In the USA, certain pesticides are classed as regulated chemical contaminants and the US Environmental Protection Agency (EPA) sets a Maximum Contaminant Level Goal (MCLG) for each pesticide. In addition, a Maximum Contaminant Level (MCL) is also set for each pesticide under the Chemical Contaminants Rules. Specific details of the US system can be found on the EPA's website (https://www.epa.gov/dwreginfo).

Nematicides must be kept from water by reducing contaminations from both point and diffuse sources. Following the guidance for use given on the product's label and any stewardship guidelines should minimize water contamination.

17.4.4. Product stewardship

As with all synthetic inputs into agriculture there is a growing need to justify the use of a nematicide and to improve the accuracy of its application. This has resulted in an increase in grower administration and increased complexity of crop protocols required by the largest retailers. Although it is permissible for produce to enter the food or feed chain with residues below the MRL, retailers are often requiring residues to be less than 90% of the MRL as a minimum. Growers, and manufacturers of nematicides, are working together to satisfy these new challenges set by the retailers. This partnership takes the form of product stewardship programmes. Such programmes are concerned with operator competence, correct identification of the nematode problem and the accurate and safe use of nematicides. Providing advice on operator training, certification in the use and application of pesticides and refresher training for latest developments are important, as educating the user is often the most effective way to improve the efficiency and safety of product use. Sampling soil to establish nematode population densities (see Chapter 12) is important before a decision on the appropriate treatment can be made. Decisions on nematicide usage should be made within the context of an integrated nematode management approach. Product manufacturers can provide calibration of nematicide applicators as a service to nematicide users. Detailed guidance on the correct application of nematicides will be found on the product label and in some cases best practice guidelines go beyond the regulatory requirements of the label to help protect wildlife and the environment. The analysis of treated produce for residues, ensuring that MRLs are not exceeded, is a useful end check on the quality of the nematicide application process. Likewise, testing of groundwater may be appropriate. Guidance on limiting environmental side effects from nematicides is increasingly important and should be part of any product stewardship scheme. For example, switching off the application equipment 1 m from the end of the row, before the applicator is lifted out of the soil, easily reduces the risk to birds from eating uncovered granules at the ends of rows. In addition, scouting for affected wildlife following nematicide application may be part of a stewardship scheme.

17.5. Semi-synthetic (Organic) Nematicides

Organic or semi-synthetic nematicides consist of plant and microbial extracts that have been refined and standardized to enable the supplier to provide approximate concentrations of the active substances. Such compounds are thought to be more acceptable to both legislators and the increasingly aware public. To be widely accepted, organic nematicides need to be reliable and have significant benefits over existing conventional nematicides, or find market niches where conventional products are unacceptable, e.g. in organic agriculture. Such products are available in formulations akin to their synthetic counterparts, i.e. granules and liquids, allowing them to be applied with conventional equipment. These commercialized products are an advancement of the raw extracts or soil amendments discussed in Chapter 14. Whilst the efficacy of many 'organic nematicides' is unproven, azadirachtin (the active substance contained in neem tree products) and garlic extracts have an increasing body of published data to confirm their efficacy. Organic nematicides may limit nematode damage by inducing paralysis, thus preventing plant invasion by juveniles. Alternatively, they have been shown to reduce hatch and reduce sensory perception. Table 17.3 provides a summary of some of the organic products that are available globally.

Garlic extracts are commonly used organic nematicides that are available for a range of crops including root crops, vegetables, ornamentals, peppers, tomatoes, cucurbits and melons. The activity of the garlic products comes from the conversion of alliin by alliinase to produce allicin, which is nematicidal. Furthermore, when allicin is incubated at 20°C for 20 h it decomposes into polysulphides, such as diallyl disulphide (Harunobu, 2006), which can induce oxidative stress in nematodes. Danquah *et al.* (2011) showed that a high dose of garlic extract was required to induce mortality of potato cyst nematode (*G. pallida*) juveniles (ED_{50} 983 μl l^{-1} for 24 h of exposure) when compared with other active substances such as oxamyl (ED_{50} 500 μg l^{-1}; Evans and Wright, 1982). The product

Table 17.3. Examples of organic/semi-synthetic products with nematicidal or nematostatic activities.

Name of product/ formulation/application	Manufacturer	Source of product	Proposed active substances and mode of action	Target nematodes
DiTera DF (liquid)	Valent BioSciences Corporation	Extracts of *Myrothecium verrucaria* (fungus) strain AARC-0255	Prevents sensory perception of host cues and reduces stylet thrusting. Inhibits nematode movement	Range of ecto- and endoparasitic nematodes
Majestene bionematicide (liquid applied to soil or foliage)	Pro Farm	Heat-killed *Burkholderia rinojensis* strain A396 (bacteria)	*Burkholderia* are known to produce a range of antibiotics siderophores, toxics and extracellular enzymes; manufacturer states that the product paralyses nematodes within 24 h of product application. Nematodes are also unable to complete their life cycle	Root lesion (*Pratylenchus* spp.), root-knot (*Meloidogyne* spp), dagger (*Xiphinema* spp.), stunt, reniform (*Tylenchorhynchus*) and soybean cyst (*Heterodera glycines*) nematodes
Oikos (liquid – emulsifiable concentrate or EC)	Sipcam Inagra	Neem seed (*Azadirachta indica*) extract	Azadirachtin and other neem-based triterpenoids	Range of ecto- and endoparasitic nematodes
Nemakill (liquid)	ExcelAg	Cinnamon (*Cinnamomum verum*) (32%), clove (*Syzygium aromaticum*) (8%) and thyme (*Thymus vulgaris*) (15%) essential oils	Proposed to have an inhibitory effect on acetylcholine esterase, leading to paralysis and death	Range of ecto- and endoparasitic nematodes
Cedroz	Eden Research	Terpenes (thymol and geraniol) extracted from essential oils	Geraniol and thymol prevent egg hatching and reduces movement.	Unspecified
Nemguard DE (granules applied via granule applicator) Nemguard SC (liquid spray applied through injection or drip irrigation) Nemguard PCN (granules applied via granule applicator) – potatoes only	Certis Belchim	Garlic (*Allium sativum*) extract	Diallyl polysulfides produced by breakdown of allicin by alliinase Mode of action – oxidation reactions that disrupt various cellular pathways and affect membrane integrity	Ectoparasites such as stubby root (*Trichodorus* spp.) and dagger nematodes (*Xiphinema* spp.) Endoparasites such as root-knot (*Meloidogyne* spp.), root lesion (*Pratylenchus* spp.) and cyst nematodes (e.g. *Globodera* spp.)

Nemguard DE® is a garlic extract based granular nematicide, supplied for the reduction of root-knot nematodes (*Meloidogyne* spp.) and ectoparasites such as *Trichodorus* spp. on root crops.

Azadirachtin has received considerable attention for treating a variety of nematodes in crops. This compound is classified as a tetranortriterpenoid (more specifically limonoid) and is found in abundance in the kernels of neem tree seeds (*Azadirachta indica*). The mode of action of azadirachtin is unclear but it has been shown to induce mortality of infective juveniles and inhibit hatching. For example, Nile *et al.* (2017) recorded a 100% reduction in hatching by southern root-knot nematodes (*M. incognita*) exposed to an undiluted neem extract.

The development of alternative 'organic' nematicides and research to increase efficiency and reduce environmental impact of existing nematicides both aim to get as close as possible to the 'ideal nematicide', the characteristics of which are indicated in Box 17.2.

17.6. Nematicides within Integrated Pest Management Programmes

The term Integrated Nematode Management (INM) was originally described by Bird (1981), who described the selection of two or more control measures, supported by research, development and technology. For a more in-depth overview of INM, refer to Sikora *et al.* (2021). Integrated Pest Management (IPM) is a systematic plan whereby control measures are strategically applied to limit environmental impact. The IPM measures are commonly represented in the form of a pyramid; Fig. 17.9 shows an adapted version of this schematic diagram for plant-parasitic nematode management. The base of the pyramid is broader highlighting the order of hierarchy of these measures. Ideally, nematicides are considered the final form of intervention, being applied when prior measures are unable to achieve the degree of management needed to produce food crops sustainably. There are certainly situations where a nematicide is needed to support other strategies being employed. For example, cultivars resistant to potato cyst nematodes

Box 17.2. Characteristics of the 'ideal nematicide'.

Cost: Inexpensive to develop, register, manufacture and market. Inexpensive to purchase. Economic to apply; gives reliable increase in economic gross margin for the grower.
Efficacy: Very effective at low rates of application.
Toxicity: Low acute and chronic (skin) toxicity to humans. Low toxicity to non-target organisms. Highly toxic to target nematode(s). Non-phytotoxic with no adverse effects on plant growth and vigour.
Persistence: Persistent enough to control fully the target nematode, then rapid degradation to harmless molecules.
Source: Naturally occurring or produced from living organisms.
Compatibility: Compatible with other control components, e.g. biocontrol agents.
Application: Chemically suited to range of formulations and application systems.
Mobility in plant: Good mobility to roots; can be applied to foliage and stems of perennials.
Solubility: Water soluble but not subject to leaching or run-off.
Residues: Leaves no detectable residues in harvested plant material.

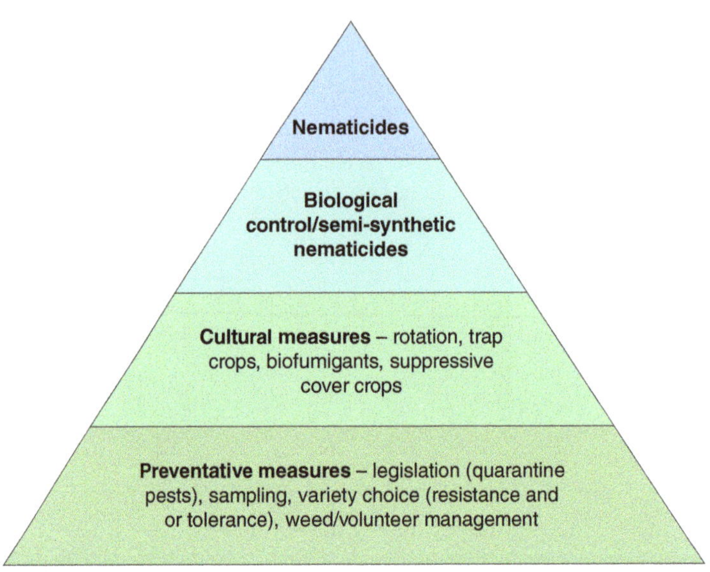

Fig. 17.9. The Integrated Pest Management pyramid adapted for the management of plant-parasitic nematodes.

(*Globodera* spp.) limit multiplication of females by causing necrosis around the site of feeding (syncytium), but unless tolerant, do not provide protection against yield loss. In this case, the combined use of cultivar and nematicide works well to protect the crop, while preventing the acceleration of population densities – the legacy effect.

Synergies between nematicides and other control measures have been shown elsewhere. The compatibility of arbuscular mycorrhizal fungi (AMF) with the nematicide aldicarb was shown in an outdoor pot experiment by Deliopoulos *et al.* (2008). AMF stimulates earlier and greater *G. pallida* hatch, which results in a higher proportion of juveniles being exposed to the nematicide leading to lower multiplication overall.

Finally, with increased focus on soil biology and its importance to productive soils, future nematicides need to demonstrate selectivity towards their nematode targets with minimal effects on non-target organisms. For example, Burns *et al.* (2023) report on a family of imidazothiazole nematicides, known as selectivins, that were shown to reduce root infection by the root-knot nematode *M. incognita* while having little effect on other phylogenetically diverse non-target organisms.

Note: The authors do not accept liability for any error or omission in the content, or for any loss, damage or other accident arising from the use of techniques, active substances or products mentioned in this chapter. Always consult the manufacturer and competent advisors before using any product.

References

Abad, P. and Opperman, C.H. (2009) The complete sequence of the genomes of *Meloidogyne incognita* and *Meloidogyne hapla*. In: Perry, R.N., Moens, M. and Starr, J.L. (eds) *Root-knot Nematodes*. CAB International, Wallingford, UK, pp. 363–376. DOI: 10.1079/9781845934927.0363

Abad, P., Gouzy, J., Aury, J.M., Castagnone-Sereno, P., Danchin, E.G. *et al.* (2008) Genome sequence of the metazoan plant-parasitic nematode *Meloidogyne incognita*. *Nature Biotechnology* 26, 909–915. DOI: 10.1038/nbt.1482

Aboul-Eid, H.Z. (1969) Histological anatomy of the excretory and reproductive systems of *Longidorus macrosoma*. *Nematologica* 15, 437–450. DOI: 10.1163/187529269X00777

Adlimoghaddam, A., Boeckstaens, M., Marini, A-M., Treberg, J.R., Brassinga, A-K.C. *et al.* (2015) Ammonia excretion in *Caenorhabditis elegans*: mechanism and evidence of ammonia transport of the Rhesus protein CeRh-1. *Journal of Experimental Biology* 218, 675–683. DOI: 10.1242/jeb.111856

Affokpon, A., Coyne, D.L., Htay, C.C., Lawouin, L. and Coosemans, J. (2011) Biocontrol potential of native *Trichoderma* isolates against root-knot nematodes in West African vegetable production systems. *Soil Biology and Biochemistry* 43, 600–608. DOI: 10.1016/j.soilbio.2010.11.029

Aguinaldo, A.M.A., Turbeville, J.M., Linford, L.S., Rivera, M.C., Garey, J.R. *et al.* (1997) Evidence for a clade of nematodes, arthropods and other moulting animals. *Nature* 387, 489–493. DOI: 10.1038/387489a0

Ahmed, M. and Holovachov, O. (2021) Twenty years after De Ley & Blaxter – how far did we progress in understanding the phylogeny of the phylum Nematoda? *Animals* 11, 3479. DOI: 10.3390/ani11123479

Ahmed, M., Roberts, N.G., Adediran, F., Smythe, A.B., Kocot, K.M. *et al.* (2022) Phylogenomic analysis of the Phylum Nematoda: conflicts and congruences with morphology, 18S rRNA, and mitogenomes. *Frontiers in Ecology Evolution* 9, 769565. DOI: 10.3389/fevo.2021.769565

Alexander, R.M. (2002) Locomotion. In: Lee, D.L. (ed.) *The Biology of Nematodes*. Taylor & Francis, London, UK, pp. 345–352. DOI: 10.1201/b12614

Ali, M.A., Azeem, F., Li, H. and Bohlmann, H. (2017) Smart parasitic nematodes use multifaceted strategies to parasitize plants. *Frontiers in Plant Science* 8, 1699. DOI: 10.3389/fpls.2017.01699

Alison, S.D. (2006) Brown ground: a soil carbon analogue for the green world hypothesis? *American Naturalist* 167, 619–627. DOI: 10.1086/503443

Altieri, M.A. and Nicholls, C.I. (2005) *Agroecology and the search for a truly sustainable agriculture*. United Nations Environment Programme – Environmental Training Network for Latin America and the Caribbean, Mexico DF, Mexico.

Álvarez-Ortega, S., Brito, J.A. and Subbotin, S.A. (2019) Multigene phylogeny of root-knot nematodes and molecular characterization of *Meloidogyne nataliei* Golden, Rose & Bird, 1981 (Nematoda: Tylenchida). *Scientific Report* 9, 11788. DOI 10.1038/s41598-019-48195-0

Amiri, S., Subbotin, S.A. and Moens, M. (2002) Identification of the beet cyst nematode *Heterodera schachtii* by PCR. *European Journal of Plant Pathology* 108, 497–506. DOI: 10.1023/A:1019974101225

Anderson, R.V. and Potter, J.W. (1991) Stunt nematodes: *Tylenchorhynchus, Merlinius*, and related genera. In: Nickle, W.R. (ed.) *Manual of Agricultural Nematology*. Marcel Dekker, New York, USA, pp. 529–586. DOI: 10.1201/9781003066576-12

Andersson, S., (1989) Annual population decline of *Globodera rostochiensis* in the absence of host plants. *Nematologica* 34, 254. [Abstract.]

Andrássy, I. (2007) Free-living nematodes of Hungary, II *(Nematoda errantia)*. In: *Pedozoologica Hungarica No.4* (Series Editors: Csuzdi, C. and Mahunka, S.). Hungarian Natural History Museum and Systematic Research Group of the Hungarian Academy of Sciences, Budapest, Hungary.

Andrássy, I. (2009) Free-living Nematodes of Hungary, III *(Nematoda errantia)*. In: *Pedozoologica Hungarica No. 5* (Series Editors: Csuzdi, C. and Mahunka, S.). Hungarian Natural History Museum and Systematic Research Group of the Hungarian Academy of Sciences, Budapest, Hungary.

Ankrom, K.E., Franco, A.L.C., Fonte, S.J, Gherardi, L.A., de Tomasel, C.M. *et al.* (2020) Ecto- and endoparasitic nematodes respond differently across sites to changes in precipitation. *Global Change Biology* 193, 761–771. DOI: 10.1007/s00442-020-04708-7

Anonymous (2011) Pale cyst nematode (*Globodera pallida*) eradication program – Idaho Falls, Idaho. USDA-APHIS, pp. 3. Available at: www.aphis.usda.gov/plant_health/plant_pest_info/potato/pcn.shtml (accessed 29 January 2024).

Anonymous (2012) Scientific opinion on the risks to plant health posed by European versus non-European populations of the potato cyst nematodes *Globodera pallida* and *Globodera rostochiensis*. *EFSA Journal* 10, 2644. DOI: 10.2903/j.efsa.2012.2644

Anonymous (2016) PM 7/103 (2) *Meloidogyne enterolobii. EPPO Bulletin*, 46, 190–201. DOI: 10.1111/epp.12293.

Anonymous (2017) EPPO PM 7/131 (1) Guidelines on the main tasks of Reference Laboratories for official plant pest diagnostics. *EPPO Bulletin* 47, 441–442.

Anonymous (2021) EPPO PM 7/148 (1) Guidelines for the management of nematode collections used for the production and maintenance of reference material. *EPPO Bulletin* 51, 507–548. DOI: 10.1111/epp.12798

Anthoine, G. and Mugniéry, D. (2005) Obligatory amphimixis and variation in chromosome number within and among South American populations of *Nacobbus aberrans* (Thorne, 1935) Thorne & Allen, 1944 (Nematoda: Pratylenchidae). *Nematology* 7, 783–787. DOI: 10.1163/156854105775143008

Aravind, R., Eapen, S.J., Kumar, A., Dinu, A. and Ramana, K.V. (2010) Screening of endophytic bacteria and evaluation of selected isolates for suppression of burrowing nematode (*Radopholus similis* Thorne) using three varieties of black pepper (*Piper nigrum* L.). *Crop Protection* 229, 318–324. DOI: 10.1016/j.cropro.2009.12.005

Arbeli, Z. and Fuentes, C. (2007) Accelerated biodegradation of pesticides: an overview of the phenomenon, its basis and possible solutions; and a discussion on the tropical dimension. *Crop Protection* 26, 1733–1746. DOI: 10.1016/j.cropro.2007.03.009

Archidona-Yuste, A., Navas-Cortés, J.A., Cantalapiedra-Navarrete, C., Palomares-Rius, J.E. and Castillo, P. (2016) Cryptic diversity and species delimitation in the *Xiphinema americanum* -group complex (Nematoda: Longidoridae) as inferred from morphometrics and molecular markers. *Zoological Journal of the Linnean Society* 176, 231–265. DOI: 10.1111/zoj.12316

Archidona-Yuste, A., Cantalapiedra-Navarrete, C., Castillo, P. and Palomares-Rius, J.E. (2019) Molecular phylogenetic analysis and comparative morphology reveals the diversity and distribution of needle nematodes of the genus *Longidorus* (Dorylaimida: Longidoridae) from Spain. *Contributions to Zoology* 88, 1–41. DOI: 10.1163/18759866-20191345

Archidona-Yuste, A., Wiegand, T., Eisenhauer, N., Cantalapiedra-Navarrete, C., Palomares-Rius, J.E. *et al.* (2021) Agriculture causes homogenization of plant-feeding nematode communities at the regional scale. *Journal of Applied Ecology* 58, 2881–2891. DOI: 10.1111/1365-2664.14025

Armstrong, M.R., Vossen, J., Lim, T.Y., Hutten, R.C.B., Xu, J. *et al.* (2019) Tracking disease resistance deployment in potato breeding by enrichment sequencing. *Plant Biotechnology Journal* 17, 540–549. DOI: 10.1111/pbi.12997

Ashton, F.T., Li, J. and Schad, G.A. (1999) Chemo- and thermosensory neurons: structure and function in animal parasitic nematodes. *Veterinary Parasitology* 84, 297–316. DOI: 10.1016/S0304-4017(99)00037-0

Askary, T.H. and Martinelli, P.R.P. (2015) *Biocontrol Agents of Phytonematodes*. CAB International, Wallingford, UK.

Assefa, A.D., Kim, S.-H., Mani, V., Ko, H.-R. and Hahn, B.-S. (2021) Metabolic analysis of the development of the plant-parasitic cyst nematodes *Heterodera schachtii* and *Heterodera trifolii* by capillary electrophoresis time-of-flight mass spectrometry. *International Journal of Molecular Sciences* 22, 10488. DOI: 10.3390/ijms221910488.

Atkinson, H.J. (1980) Respiration in nematodes. In: Zuckerman, B.M. (ed.) *Nematodes as Biological Models*, *Vol. 2*. Academic Press, London, UK, pp. 101–142.

Atkinson, H.J. and Taylor, J.D. (1983) A calcium binding sialoglycoprotein associated with an apparent egg shell membrane of *Globodera rostochiensis*. *Annals of Applied Biology* 102, 345–354. DOI: 10.1111/j.1744-7348.1983.tb02704.x

Atkinson, H.J., Urwin, P.E., Clarke M.C. and McPherson, M.J. (1996) Image analysis of the growth of *Globodera pallida* and *Meloidogyne incognita* on transgenic tomato roots expressing cystatins. *Journal of Nematology* 28, 209–215.

Atkinson, H.J., Grimwood, S., Johnston, K. and Green, J. (2004) Prototype demonstration of transgenic resistance to the nematode *Radopholus similis* conferred on banana by a cystatin. *Transgenic Research* 13, 135–142. DOI: 10.1023/b:trag.0000026070.15253.88

Ayres, E., Steltzer, H., Simmons, B.L., Simpson, R.T., Steiweg, J.M. *et al.* (2009) Home-field advantage accelerates leaf litter decomposition in forests. *Soil Biology and Biochemistry* 41, 606–610. DOI: 10.1016/j.soilbio.2008.12.022

Azevedo, C., Sadanandom, A., Kitagawa, K., Freialdenhoven, A., Shirasu, K. *et al.* (2002) The RAR1 interactor SGT1, an essential component of R gene-triggered disease resistance. *Science* 295, 2073–2076. DOI: 10.1126/science.1067554

Bacetty, A.A., Snook, M.E., Glenn, A.E., Noe, J.P., Nagabhyru, P. *et al.* (2009) Chemotaxis disruption in *Pratylenchus scribneri* by tall fescue root extracts and alkaloids. *Journal of Chemical Ecology* 35, 844–850. DOI: 10.1007/s10886-009-9657-x

Baker, R.H.A. and Dickens, J.S.W. (1993) Practical problems in pest risk assessment. In: Ebbels, D. (ed.) *Plant Health and the European Single Market. BCPC Monograph No. 54. Proceedings of a Symposium Organised by the British Crop Protection Council, the Association of Applied Biologists and the British Society of Plant Pathology, held at the University of Reading, 30 March–1 April 1993*, pp. 209–220.

Bakonyi, G. and Nagy, P. (2000) Temperature- and moisture-induced changes in the structure of the nematode fauna of a semiarid grassland – patterns and mechanisms. *Global Change Biology* 6, 697–707. DOI: 10.1046/j.1365-2486.2000.00354.x

Baldwin, J.G. and Hirschmann, H. (1973) Fine structure of cephalic sense organs in *Meloidogyne incognita* males. *Journal of Nematology* 5, 285–302.

Baldwin, J.G. and Mundo-Ocampo, M. (1991) Heteroderinae, cyst and non cyst-forming nematodes. In: Nickle, W.R. (ed.) *Manual of Agricultural Nematology*. Marcel Dekker, New York, USA, pp. 275–362.

Baldwin, J.G. and Perry, R.N. (2004) Nematode morphology, sensory structure and function. In: Chen, Z., Chen, S. and Dickson, D.W. (eds) *Nematology: Advances and Perspectives, Vol. 1: Nematode Morphology, Physiology and Ecology*. CAB International, Wallingford, UK, pp. 175–257.

Baldwin, J.G., Ragsdale, E. and Bumbarger, D. (2004) Revised hypotheses for phylogenetic homology of the stomatostylet in tylenchid nematodes. *Nematology* 6, 623–632. DOI: 10.1163/1568541042843559

Banerjee, N. and Hallem E.A. (2020) The role of carbon dioxide in nematode behaviour and physiology. *Parasitology* 147, 841–854. DOI: 10.1017/s0031182019001422

Banerjee, S., Banerjee, A., Gill, S.S., Gupta, O.P., Dahuja, A. *et al.* (2017) RNA interference: a novel source of resistance to combat plant parasitic nematodes. *Frontiers in Plant Sciences* 8, 834. DOI: 10.3389/fpls.2017.00834

Barcala, M., Garcia, A., Cabrera, J., Casson, S., Lindsey, K. *et al.* (2010) Early transcriptomic events in microdissected *Arabidopsis* nematode-induced giant cells. *Plant Journal* 6, 698–712. DOI: 10.1111/j.1365-313X.2009.04098.x

Bardgett, R.D. and Wardle, D.A. (2003) Herbivore-mediated linkages between aboveground and belowground communities. *Ecology* 84, 2258–2268. DOI: 10.1890/02-0274

Bargmann, C.I. (2006) Chemosensation in *C. elegans*. In: The *C. elegans* Research Community (ed.) *WormBook*, http://www.wormbook.org. DOI: 10.1895/wormbook.1.123.1

Bargmann, C.I. and Horvitz, H.R. (1991) Chemosensory neurons with overlapping functions direct chemotaxis to multiple chemicals in *C. elegans*. *Neuron* 7, 729–742. DOI: 10.1016/0896-6273(91)90276-6

Barker, K.R. and McGawley, E.C. (1998) Interrelations with other microorganisms and pests. In: Sharma, S.B. (ed.) *The Cyst Nematodes*. Kluwer Academic Publishers, Dordrecht, The Netherlands, pp. 266–292. DOI: 10.1007/978-94-015-9018-1_11

Barrett, J. (1991) Anhydrobiotic nematodes. *Agricultural Zoology Reviews* 4, 161–176.

Barrett, J. (2011) Biochemistry of survival. In: Perry, R.N. and Wharton, D.A. (eds) *Molecular and Physiological Basis of Nematode Survival*. CAB International, Wallingford, UK, pp. 282–310.

Barrett, J. and Wright, D.J. (1998) Intermediary metabolism. In: Perry, R.N. and Wright, D.J. (eds) *The Physiology and Biochemistry of Free-living and Plant-parasitic Nematodes*. CAB International, Wallingford, UK, pp. 331–353.

Barrios, A. (2014) Exploratory decisions of the *Caenorhabditis elegans* male: a conflict of two drives. *Seminars in Cell & Developmental Biology* 33, 10–17. DOI: 10.1016/j.semcdb.2014.06.003

Barrios, A., Nurrish, S. and Emmons, S.W. (2008) Sensory regulation of *C. elegans* male mate-searching behavior. *Current Biology* 18, 1865–1871. DOI: 10.1016/j.cub.2008.10.050

Barrows, B.D., Griffitts, J.S. and Aroian, R.V. (2007) Resistance is nonfutile: resistance to Cry5B in the nematode *Caenorhabditis elegans*. *Journal of Invertebrate Pathology* 95, 198–200. DOI: 10.1016/j.jip.2007.04.002

Baujard, P. and Martiny, B. (1995) Ecology and pathogenicity of the Hoplolaimidae (Nemata) from the Sahelian Zone of West-Africa. 7. *Helicotylenchus dihystera* (Cobb, 1893) Sher, 1961 and comparison with *Helicotylenchus* multicinctus (Cobb, 1893) Golden, 1956. *Fundamental and Applied Nematology* 18, 503–511.

Baum, T.J., Hiatt, A., Parrott, W.A., Pratt, L.H. and Hussey, R.S. (1996) Expression in tobacco of a functional monoclonal antibody specific to stylet secretions of the root-knot nematode. *Molecular Plant–Microbe Interactions* 9, 382–387.

Baum, T.J., Hussey, R.S. and Davis, E.L. (2004) Parasitism gene discovery in sedentary phytonematodes. In: Cook, R. and Hunt, D.J. (eds) *Proceedings of the Fourth International Congress of Nematology, 8–13 June 2002, Tenerife, Spain. Nematology Monographs and Perspectives 2*. Brill, Leiden, The Netherlands, pp. 581–588.

Bayless, A.M., Smith, J.M., Song, J., McMinn, P.H., Teillet, A. *et al.* (2016) Disease resistance by impairment of α-SNAP. *Proceedings of the National Academy of Sciences of the USA* 113, E7375–E7382. DOI: 10.1073/pnas.1610150113

Been, T.H. and Schomaker, C.H. (1986) Quantitative analyses of growth, mineral composition and ion balance of the potato cultivar Irene infected with *Globodera pallida* (Stone). *Nematologica* 32, 339–355. DOI: 10.1163/187529286X00426

Been, T.H. and Schomaker, C.H. (1998) Quantitative studies on the management of potato cyst nematodes (*Globodera* spp.) in The Netherlands. PhD Thesis, Agricultural University, Wageningen, The Netherlands.

Been, T.H. and Schomaker, C.H. (2000) Development and evaluation of sampling methods for fields with infestation foci of potato cyst nematodes (*Globodera rostochiensis* and *G. pallida*). *Phytopathology* 90, 647–656. DOI: 10.1094/phyto.2000.90.6.647

Been, T.H., van Bekkum, P.J., van Beers, T.G. and Beniers, A. (2007a) A scaled-up Seinhorst elutriator for extraction of cyst nematodes from soil. *Nematology* 9, 431–435. DOI: 10.1163/156854107781351990

Been, T.H., Korthals, G., Schomaker, C.H. and Zijlstra, C. (2007b) The MeloStop Project: sampling and detection of *Meloidogyne chitwoodi* and *M. fallax*. Report 138, Plant Research International B.V., Wageningen, The Netherlands.

Beeman, A.Q. and Tylka, G.L. (2018) Assessing the effects of ILeVO and VOTiVO seed treatments on reproduction, hatching, motility, and root penetration of the soybean cyst nematode, *Heterodera glycines*. *Plant Disease* 102, 107–113. DOI: 10.1094/PDIS-04-17-0585-RE

Behm, C.A. (1997) The role of trehalose in the physiology of nematodes. *International Journal for Parasitology* 27, 215–229. DOI: 10.1016/S0020-7519(96)00151-8

Behrens, E. (1975) *Globodera* Skarbilovich, 1959. Eine selbständige Gattung in der Unterfamilie Heteroderinae Skarbilovich, 1947 (Nematoda: Heteroderidae). *Vortragstagung zu Aktuellen Problemen der Phytonematologie*, 28 May 1975, Rostock, Germany, pp. 12–26.

Bell, C.A., Magkourilou, E., Urwin, P.E. and Field, K.J. (2022) Disruption of carbon for nutrient exchange between potato and arbuscular mycorrhizal fungi enhanced cyst nematode fitness and host pest tolerance. *New Phytologist* 234, 269–279. DOI: 10.1111/nph.17958

Bellafiore, S., Shen, Z., Rosso, M.-N., Abad, P., Shih, P. *et al.* (2008) Direct identification of the *Meloidogyne incognita* secretome reveals proteins with host cell reprogramming potential. *PLoS Pathogen* 4, e1000192. DOI: 10.1371/journal.ppat.1000192

Bello, A., López-Pérez, J.A., García-Álvarez, A., Sanz, R. and Lacasa, A. (2004) Biofumigation and nematode control in the Mediterranean region. In *Proceedings of the Fourth International Congress of Nematology, 8–13 June 2002, Tenerife, Spain*. Brill, Leiden, The Netherlands. DOI: 10.1163/9789004475236_016

Bert, W., Karssen, G., Van Driessche, R. and Geraert, E. (2003) The cellular structure of the female reproductive system within the Heteroderinae and Meloidogyninae (Nematoda). *Nematology* 4, 953–964. DOI: 10.1163/156854102321122575

Bhatla, N. and Horvitz, H.R. (2015) Light and hydrogen peroxide inhibit *C. elegans* feeding through gustatory receptor orthologs and pharyngeal neurons. *Neuron* 85 804–818. DOI: 10.1016/j.neuron.2014.12.061

Bhattarai, K.K., Li, Q., Liu, Y., Dinesh-Kumar, S.P. and Kaloshian, I. (2007) The *Mi-1*-mediated pest resistance requires *Hsp90* and *Sgt1*. *Plant Physiology* 144, 312–323. DOI: 10.1104/pp.107.097246

Biela, F., Dias-Arieira, C.R., Machado, A.C.Z., de Melo Santana-Gomes, S., Cardoso, M.R. *et al.* (2015) Genetic diversity of rice genotypes from Brazil based on their resistance to *Pratylenchus zeae*. *Tropical Plant Pathology* 40, 208–211. DOI: 10.1007/s40858-015-0031-y

Bird, G.W. (1981) Integrated nematode management for plant protection. *Plant Parasitic Nematodes* 3, 355–375.

Bird, A.F. and Bird, J. (1991) *The Structure of Nematodes*, 2nd edn. Academic Press, London, UK.

Bird, A.F. and McClure, M. (1976) The tylenchid (Nematoda) egg shell: structure, composition and permeability. *Parasitology* 72, 19–28. DOI: 10.1017/S0031182000043158

Blake, C.D. (1961) Importance of osmotic potential as a component of the total potential of the soil water on the movement of nematodes. *Nature* 191, 144–145. DOI: 10.1038/192144a0

Blanc-Mathieu, R., Perfus-Barbeoch, L., Aury, J.-M., Da Rocha, M., Gouzy, J. *et al.* (2017) Hybridization and polyploidy enable genomic plasticity without sex in the most devastating plant-parasitic nematodes. *PLoS Genetics* 13, e1006777. DOI: 10.1371/journal.pgen.1006777

Blaxter, M. (2011) Nematodes: the worm and its relatives. *PLoS Biology* 9, e1001050. DOI: 10.1371/journal.pbio.1001050

Blaxter, M. and Koutsovoulos, G. (2015) The evolution of parasitism in Nematoda. *Parasitology* 142(S1), S26–S39. DOI: 10.1017/S0031182014000791

Blaxter, M.L., De Ley, P., Garey, J.R., Liu, L.X., Scheldeman, P. *et al.* (1998) A molecular evolutionary framework for the phylum Nematoda. *Nature* 392, 71–75. DOI: 10.1038/32160

Blaxter, M., Koutsovoulos, G., Jones, M., Kumar, S. and Elsworth, B. (2016) Phylogenomics of Nematoda. In: Olson, P.D., Hughes, J. and Cotton, J.A. (eds) *Next Generation Phylogenetics*. Cambridge University Press, Cambridge, UK, pp. 62–83.

Bleve-Zacheo, T., Bongiovanni, M., Melillo, M.T. and Castagnone-Sereno, P. (1998) The pepper resistance genes *Me1* and *Me3* induce differential penetration rates and temporal sequances of root cell ultrastructural changes upon nematode infection. *Plant Science* 133, 79–90. DOI: 10.1016/S0168-9452(98)00021-1

Bliss, C.I. and Owen, A.R.G. (1958) Negative binomial distributions with a common *k*. *Biometrica* 45, 37–58. DOI: 10.1093/biomet/45.1-2.37

Blok, V.C. (2005) Achievements and future prospects of molecular diagnostics for plant-parasitic nematodes. *Canadian Journal of Plant Pathology* 27, 176–185. DOI: 10.1080/07060660509507214

Blok, V.C. and Powers, T.O. (2009) Biochemical and molecular identification. In: Perry, R.N., Moens, M. and Starr, J.L. (eds) *Root-knot Nematodes*. CAB International, Wallingford, UK, pp. 98–118.

Blok, V.C., Ehwaeti, E., Fargette, M., Kumar, A., Phillips, M.S. *et al.* (1997) Evolution of resistance and virulence in relation to the management of nematodes with different biology, origins and reproductive strategies. *Nematologica* 43, 1–13. DOI: 10.1163/004725997X00016

Blok, V.C., Wishart, J., Fargette, M., Berthier, K. and Phillips, M.S. (2002) Mitochondrial DNA differences distinguishing *Meloidogyne mayaguensis* from the major species of tropical root-knot nematodes. *Nematology* 4, 773–781. DOI: 10.1163/156854102760402559

Blok, V.C., Tylka, G.L., Smiley, R.W., de Jong, W.S. and Daub, M. (2018) Resistance breeding. In: Perry, R.N., Moens, M. and Jones, J.T. (eds) *Cyst Nematodes*. CAB International, Wallingford, UK, pp. 174–214.

Böckenhoff, A. and Grundler, F.M.W. (1994) Studies on the nutrient uptake by the beet cyst nematode *Heterodera schachtii* by *in situ* microinjection of fluorescent probes into the feeding structures in *Arabidopsis thaliana*. *Parasitology* 109, 249–254. DOI:10.1017/S003118200007637X

Bommarco, R., Kleijn, D. and Potts, G. (2013) Ecological intensification: harnessing ecosystem services for food security. *Trends in Ecology and Evolution* 4, 230–238. DOI: 10.1016/j.tree.2012.10.012

Bongers, T. (1990) The maturity index: an ecological measure of environmental disturbance based on nematode species composition. *Oecologia* 83, 14–19. DOI: 10.1007/BF00324627

Bongers. T. and Bongers, M. (1998) Functional diversity of nematodes. *Applied Soil Ecology* 10, 239–251. DOI: 10.1016/S0929-1393(98)00123-1

Bongers, T., Van der Meulen, H. and Korthals, G. (1997) Inverse relationship between the nematode maturity index and plant parasite index under enriched nutrient conditions. *Applied Soil Ecology* 6, 195–199. DOI: 10.1016/S0929-1393(96)00136-9

Borneman, J., Olatinwo, R., Yin, B. and Becker, J.O. (2004) An experimental approach for identifying microorganisms involved in specified functions: utilization for understanding a nematode-suppressive soil. *Australasian Plant Pathology* 33, 151–155. DOI: 10.1071/AP04007

Bot, A. and Benites, J. (2005) *The importance of soil organic matter: key to drought-resistant soil and sustained food production*. Food and Agriculture Organization of the United Nations – Soils Bulletin 80. Rome, Italy.

Bowen, D.J., Bunkers, G.J., Chay, C., Pitkin, J.W., Rydel, T.J. *et al.* (2016) *Pesticidal nucleic acids and proteins and uses thereof*. Patent Publication No. US9328356 B2.

Braasch, H. (2004) Morphology of *Bursaphelenchus xylophilus* compared with other *Bursaphelenchus* species. In: Mota, M. and Vieira, P. (eds) *The Pinewood Nematode* Bursaphelenchus xylophilus. *Nematology Monographs and Perspectives 3* (Series Editors: Hunt, D.J. and Perry, R.N.). Brill, Leiden, The Netherlands, pp. 127–143.

Braasch, H., Wittchen, U. and Unger, J.G. (1996) Establishment potential and damage probability of *Meloidogyne chitwoodi* in Germany. *EPPO Bulletin* 26, 495–509. DOI: 10.1111/j.1365-2338.1996.tb01492.x

Bretscher, A.J., Kodama-Namba, E., Busch, K.E., Murphy, R.J., Soltesz, Z. *et al.* (2011) Temperature, oxygen, and salt-sensing neurons in *C. elegans* are carbon dioxide sensors that control avoidance behavior. *Neuron* 69, 1099–1113. DOI: 10.1016/j.neuron.2011.02.023

Brinkman, E.P., Duyts, H. and van der Putten, W.H. (2005) Consequences of variation in species diversity in a community of root-feeding herbivores for nematode dynamics and host plant biomass. *Oikos* 110, 417–427. DOI: 10.1111/j.0030-1299.2005.13659.x

Brinkman, P. and Hallmann, J. (2021) Transporters of trouble: trichodorids and tobacco rattle virus in potatoes. In: Sikora R., Desaeger, J. and Molendijk, L.P.G. (eds) *Integrated Nematode Management: State-of-the-art and Visions for the Future*. CAB International, Wallingford, UK, pp. 327–332. DOI: 10.1079/9781789247541.0045

Brito, J.A., Stanley, J.D., Kaur, R., Cetintas, R., Di Vito, M. *et al.* (2007) Effects of the *Mi-1*, *N* and *tabasco* genes on infection and reproduction of *Meloidogyne mayaguensis* on tomato and pepper genotypes. *Journal of Nematology* 39, 327–332.

Brodie, B.B., Good, J.M. and Jaworski, C.A. (1970) Population dynamics of plant nematodes in cultivated soil: effect of summer cover crops in newly cleared land. *Journal of Nematology* 2, 217–222.

Brooker, R.W., Bennett, A.E., Cong, W.F., Daniell, T.J., George, T.S. *et al.* (2015) Improving intercropping: a synthesis of research in agronomy, plant physiology and ecology. *New Phytologist* 206, 107–117. DOI: 10.1111/nph.13132

Brooks, F.E. (2008) *Burrowing nematode disease. The Plant Health Instructor* (updated 2014). The American Phytopathological Society, St Paul, MN, USA. DOI: 10.1094/PHI-I-2008-1020-01

Brown, A.M.V. (2018) Endosymbionts of plant-parasitic nematodes. *Annual Review of Phytopathology* 56, 225–242. DOI: 10.1146/annurev-phyto-080417-045824

Brown, C.R., Mojtahedi, H., Zhang, L.-H. and Riga, E. (2009) Independent resistant reactions expressed in root and tuber of potato breeding lines with introgressed resistance to *Meloidogyne chitwoodi*. *Phytopathology* 99, 1085–1089. DOI: 10.1094/PHYTO-99-9-1085

Brown, D.J.F., Zheng, J. and Zhou, X. (2004) Virus vectors. In: Chen, Z.X., Chen, S.Y. and Dickson, D.W. (eds) *Nematology: Advances and Perspectives, Vol. 2: Nematode Management and Utilization*. CAB International, Wallingford, UK, pp. 717–770.

Brown, R.H. and Kerry, B.R. (1987) *Principles and Practice of Nematode Control in Crops*. Academic Press, Melbourne, Australia.

Brown, R.N., Jr (1999) Critical commodoties in the Caribbean basin: patterns of trade and market potential. In: Klassen, W. (ed.) *Mitigating the Effects of Exotic Pests on Trade and Agriculture, Part A. The Caribbean. Proceedings of T-STAR Workshop-X, 16–18 June 1999, Homestead, Florida*. Cooperative State Research, Education & Extension Service, USDA, Washington, DC, USA, pp. 27–44.

Bryant, A.S., Ruiz, F., Lee, J.H. and Hallem, E.A. (2022) The neural basis of heat seeking in a human-infective parasitic worm. *Current Biology* 32, 2206–2221. DOI: 10.1016/j.cub.2022.04.010

Bumbarger, D.J., Wijeratne, S., Carter, C., Crum, J., Ellisman, M.H. *et al.* (2009) Three-dimensional reconstruction of the amphid sensilla in the microbial feeding nematode, *Acrobeles complexus* (Nematoda: Rhabditida). *Journal of Comparative Neurology* 512, 271–281. DOI: 10.1002/cne.21882

Bumbarger, D.J., Riebesell, M., Rödelsperger, C. and Sommer, R.J. (2012) System-wide rewiring underlies behavioral differences in predatory and bacterial-feeding nematodes. *Cell* 152, 109–119. DOI: 10.1016/j.cell.2012.12.013

Burns, A.R., Baker, R.J., Kitner, M., Knox, J., Cooke, B. *et al.* (2023) Selective control of parasitic nematodes using bioactivated nematicides. *Nature* 618, 102–109. DOI: 10.1038/s41586-023-06105-5

Burr, A.H.J. (1984) Photomovement behaviour in somple invertebrates. In: Ali, M.A. (ed.) *Photoreception and Vision in Invertebrates*. Plenum Press, New York, USA, pp. 179–215.

Burr, A.H.J. and Robinson, A.F. (2004) Locomotion behaviour. In: Gaugler, R. and Bilgrami, A.L. (eds) *Nematode Behaviour*. CAB International, Wallingford, UK, pp. 25–62.

Busch, K.E., Laurent, P., Soltesz, Z., Murphy, R.J., Faivre, O. *et al.* (2012) Tonic signaling from O_2 sensors sets neural circuit activity and behavioral state. *Nature Neuroscience* 15, 581–591. DOI: 10.1038/nn.3061

Butcher, R.A. (2017) Decoding chemical communication in nematodes. *Natural Product Reports* 34, 472–477. DOI: 10.1039/C7NP00007C.

Butcher, R.A., Fujita, M., Schroeder, F.C. and Clardy, J. (2007) Small-molecule pheromones that control dauer development in *Caenorhabditis elegans*. *Nature Chemical Biology* 3, 420–422. DOI: 10.1038/nchembio.2007.3

Butler, K.J., Chen, S., Smith, J.M., Wang, X. and Bent, A.F. (2019) Soybean resistance locus *Rhg1* confers resistance to multiple cyst nematodes in diverse plant species. *Phytopathology* 109, 2107–2115. DOI: 10.1094/PHYTO-07-19-0225-R

Byrne, J., Twomey, U., Maher, N., Devine, K.J. and Jones, P.W. (1998) Detection of hatching inhibitors and hatching factor stimulants for golden potato cyst nematode, *Globodera rostochiensis*, in potato root leachate. *Annals of Applied Biology* 132, 463–472. DOI;10.1111/j.1744-7348.1998.tb05222.x

Cabrera, J., Barcala, M., Fenoll, C. and Escobar, C. (2016) The power of omics to identify plant susceptibility factors and to study resistance to root-knot nematodes. *Current Issues in Molecular Biology* 19, 53–72. DOI: 10.21775/9781910190357.07

Cai, D., Kleine, M., Kifle, S., Harloff, H.J., Sandal, N.N. *et al.* (1997) Positional cloning of a gene for nematode resistance in sugar beet. *Science* 275, 832–834. DOI: 10.1126/science.275.5301.832

Cai, D., Thurau, T., Tian, Y., Lange, T., Yeh, K-W. *et al.* (2003) Sporamin mediated resistance to beet cyst nematodes (*Heterodera schachtii* Schm.) is dependent on trypsin inhibitory activity in sugar beet (*Beta vulgaris* L.) hairy roots. *Plant Molecular Biology* 51, 839–849. DOI: 10.1126/science.275.5301.832

Cai, R., Archidona-Yuste, A., Cantalapiedra-Navarrete, C., Palomares-Rius, J.E. and Castillo, P. (2020) New evidence of cryptic speciation in the family Longidoridae (Nematoda: Dorylaimida). *Journal of Zoological Systematics and Evolutionary Research* 58, 869–899. DOI: 10.1111/jzs.12393

Cannon, R.J.C., Pemberton, A.W. and Bartlett, P.W. (1999) Appropriate measures for the eradication of unlisted pests. *EPPO Bulletin* 29, 29–36. DOI: 10.1111/j.1365-2338.1999.tb00789.x

Cantalapiedra-Navarrete, C., Liébnas, G., Archidona-Yuste, A., Palomares-Rius, J.E. and Castillo, P. (2012) Molecular and morphological characterisation of *Rotylenchus vitis* n. sp. (Nematoda: Hoplolaimidae) infecting grapevine in southern Spain. *Nematology* 14, 235–247. DOI: 10.1163/138855411X588175

Canto-Saenz, M. and de Scurrah, M.M. (1977) Races of the potato cyst nematode in the Andean region and a new system of classification. *Nematologica* 23, 340–349. DOI: 10.1163/187529277X00066

Carrillo, J., Ingwell, L.L., Li, X. and Kaplan, I. (2018) Domesticated tomatoes are more vulnerable to negative plant–soil feedbacks than their wild relatives. *Journal of Ecology* 107, 1753–1766. DOI: 10.1111/1365-2745.13157

Carrillo, M.A., Guillermin, M.L., Rengarajan, S., Okubo, R.P. and Hallem, E.A. (2013) O_2-sensing neurons control CO_2 response in *C. elegans*. *Journal of Neuroscience* 33, 9675–9683. DOI: 10.1523/JNEUROSCI.4541-12.2013

Carvalho, R.P., Vieira dos Santos, M.C., Almeida, M.T.M. and Costa, S.R. (2022) Effects of commercial pesticides on the nematode biological control agent *Pochonia chlamydosporia*. *Biocontrol Science and Technology* 32, 1220–1231. DOI: 10.1080/09583157.2022.2108759

Castagnone-Sereno, P., Danchin, E.G.J., Deleury, E., Guillemaud, T., Malausa, T. *et al.* (2010) Genome-wide survey and analysis of microsatellites in nematodes, with a focus on the plant-parasitic species *Meloidogyne incognita*. *BMC Genomics* 11, 598. DOI: 10.1186/1471-2164-11-598

Castelli, L., Ramsay, G., Bryan, G., Neilson, S.J. and Phillips, M.S. (2003) New sources of resistance to the potato cyst nematodes *Globodera pallida* and *G. rostochiensis* in the Commonwealth potato collection. *Euphytica* 129, 377–386.

Castillo, P. and Vovlas, N. (2005) *Bionomics and Identification of the Genus* Rotylenchus. *Nematology Monographs and Perspectives 3* (Series Editors: Hunt, D.J. and Perry, R.N.). Brill, Leiden, The Netherlands.

Castillo, P. and Vovlas, N. (2007) Pratylenchus *(Nematoda; Pratylenchidae): Diagnosis, Biology, Pathogenicity and Management. Nematology Monographs and Perspectives 6* (Series Editors: Hunt, D.J. and Perry, R.N.). Brill, Leiden, The Netherlands.

Čepulytė, R. and Būda, V. (2022) Toward chemical ecology of plant-parasitic nematodes: kairomones, pheromones, and other behaviorally active chemical compounds. *Journal of Agricultural and Food Chemistry* 70, 1367–1390. DOI: 10.1021/acs.jafc.1c04833

Cesarz, S., Reich, P.B., Scheu, S., Ruess, L., Schaefer, M. *et al.* (2015) Nematode functional guilds, not trophic groups, reflect shifts in soil food webs and processes in response to interacting global change factors. *Pedobiologia* 58, 23–32. DOI: 10.1016/j.pedobi.2015.01.001

Cha, D., Kim, D., Choi, W., Park, S. and Han, H. (2020) Point-of-care diagnostic (POCD) method for detecting *Bursaphelenchus xylophilus* in pinewood using recombinase polymerase amplification (RPA) with the portable optical isothermal device (POID). *PLoS One* 15, e0227476. DOI: 10.1371/journal.pone.0227476

Chan, Y.L., Yang, A.H., Chen, J.T., Yeh, K.W. and Chan, M.T. (2010) Heterologous expression of taro cystatin protects transgenic tomato against *Meloidogyne incognita* infection by means of interfering sex determination and suppressing gall formation. *Plant Cell Reports* 29, 231–238. DOI: 10.1007/s00299-009-0815-y

Channale, S., Kalavikatte, D., Thompson, J.P., Kudapa, H., Bajaj, P. *et al.* (2021) Transcriptome analysis reveals key genes associated with root-lesion nematode *Pratylenchus thornei* resistance in chickpea. *Scientific Reports* 11, 17491. DOI: 10.1038/s41598-021-96906-3

Chatzigeorgiou, M., Yoo, S., Watson, J.D., Lee, W.-H., Spencer, W.C. *et al.* (2010) Specific roles for DEG/ENaC and TRP channels in touch and thermosensation in *C. elegans* nociceptors. *Nature Neuroscience* 13, 861–868. DOI: 10.1038/nn

Chaves, N., Cervantes, E, Zabalgogeazcoa, I. and Araya, C.M. (2013) *Aphelenchoides besseyi* Christie (Nematoda: Aphelenchoididae), agente causal del amachamiento del frijol común. *Tropical Plant Pathology* 38, 243–252.

Chen, R.G., Li, H.X., Zhang, L.Y., Zhang, J.H., Xiao, J.H. *et al.* (2007) *CaMi*, a root-knot nematode resistance gene from hot pepper (*Capsicum annuum* L.) confers nematode resistance in tomato. *Plant Cell Reports* 26, 895–905. DOI: 10.1007/s00299-007-0304-0

Chen, Q.-W., Hooper, D., Loof, P.A.A. and Xu, J. (1997) A revised polytomous key for the identification of species of the genus *Longidorus* Micoletzky, 1922 (Nematoda: Dorylaimoidea). *Fundamental and Applied Nematology* 20, 15–28.

Chen, S. Chronis D. and Wang X. (2013) The novel GrCEP12 peptide from the plant-parasitic nematode *Globodera rostochiensis* suppresses flg22-mediated PTI. *Plant Signaling & Behavior* 8, e25359. DOI: 10.4161/psb.25359

Chen, Z.X., Chen, S.Y. and Dickson, D.W. (2004) (eds) *Nematology: Advances and Perspectives, Vol. 2: Nematode Management and Utilization*. CAB International, Wallingford, UK.

Chitamber, J.J. and Raski, D.J. (1984) Reactions of grape rootstocks to *Pratylenchus vulnus* and *Meloidogyne* spp. *Journal of Nematology* 16, 166–170.

Chitambar, J.J. and Subbotin, S.A. (2014) *Systematics of the Sheath Nematodes of the Superfamily* Hemicycliophoroidea. *Nematology Monographs and Perspectives 10* (Series Editors: Hunt, D.J. and Perry, R.N.). Brill, Leiden, The Netherlands.

Chitwood, B.G. (1949) Root-knot nematodes – Part 1. A revision of the genus *Meloidogyne* Goeldi, 1887. *Proceedings of the Helminthological Society of Washington* 16, 166–170.

Chitwood, D.J. (1998) Biosynthesis. In: Perry, R.N. and Wright, D.J. (eds) *The Physiology and Biochemistry of Free-living and Plant-parasitic Nematodes*. CAB International, Wallingford, UK, pp. 303–330.

Chitwood, D.J. and Masler, E.P. (2018) Biochemistry. In: Perry, R.N., Moens, M. and Jones, J.T. (eds) *Cyst Nematodes*. CAB International, Wallingford, UK, pp. 89–100. DOI: 10.1079/9781786390837.008.

Chitwood, D.J. and Perry, R.N. (2009) Reproduction, physiology and biochemistry. In: Perry, R.N., Moens, M. and Starr, J.L. (eds) *Root-knot Nematodes*. CAB International, Wallingford, UK, pp. 182–200.

Choe, A., von Reuss, S.H., Kogan, D., Gasser, R.B., Platzer, E.G. *et al.* (2012) Ascaroside signaling is widely conserved among nematodes. *Current Biology* 22, 772–780. DOI: 10.1016/j.cub.2012.03.024

Church, G.T., Simpson, C.E., Burow, M.D., Paterson, A.H. and Starr, J.L. (2000) Use of RFLP markers for identification of individuals homozygous for resistance to *Meloidogyne arenaria* in peanut. *Nematology* 2, 575–580. DOI: 10.1163/156854100509367

Cid Del Prado Vera, I. and Talavera, M. (2012) Criconematoidea. In: Manzanilla-López, R.H. and Marbán-Mendoza, N. (eds) *Practical Plant Nematology*. Bibliotheca Basica de Agricultura, Montecillo, Mexico, pp. 479–519.

Cid Del Prado Vera, I., Chizhov, V.N. and Subbotin, S.A. (2018) Molecular characterisation of gall-forming nematodes, *Mesoanguina amsinckiae* and *Anguina danthoniae* (Anguinidae: Tylenchida) from California, USA. *Russian Journal of Nematology* 26, 109–113.

Clarke, A.J. and Perry, R.N. (1985) Egg-shell calcium and the hatching of *Globodera rostochiensis*. *International Journal of Parasitology* 15, 511–516. DOI: 10.1016/0020-7519(85)90046-3

Clarke, A.J., Cox, P.M. and Shepherd, A.M. (1967) The chemical composition of the egg shells of the potato cyst-nematode, *Heterodera rostochiensis* Woll. *Biochem Journal* 104,1056–1060. DOI: 10.1042/bj1041056

Clarke, A.J., Perry, R.N. and Hennessy, J. (1978) Osmotic stress and the hatching of *Globodera rostochiensis*. *Nematologica* 24, 384–392. DOI: 10.1163/187529278X00506

Claudius-Cole, A. (2021) Importance and integrated nematode management of the yam nematode (*Scutellonema bradys*) in yam cropping systems of West Africa. In: Sikora R., Desaeger, J. and Molendijk, L.P.G. (eds) *Integrated Nematode Management: State-of-the-art and Visions for the Future*. CAB International, Wallingford, UK, pp.374–380. DOI: 10.1079/9781789247541.0052

Claverie, M., Dirlewanger, E., Bosselut, N., Van Ghelder, C., Voisin, R. *et al.* (2011) The *Ma* gene for complete-spectrum resistance to *Meloidogyne* species in *Prunus* is a TNL with a huge repeated C-terminal post-LRR region. *Plant Physiology* 156, 779–792. DOI: 10.1104/pp.111.176230

Clavero-Camacho, I., Cantalapiedra-Navarrete, C., Archidona-Yuste, A., Castillo, P. and Palomares-Rius, J.E. (2021) Remarkable cryptic diversity of *Paratylenchus* spp. (Nematoda: Tylenchulidae) in Spain. *Animals* 11, 1161. DOI: 10.3390/ani11041161

Cobb, N.A. (1913) New nematode genera found inhabiting fresh water and non-brackish soils. *Journal of the Washington Academy of Sciences* 3, 432–444. DOI: 10.5962/BHL.PART.20323

Coke, M.C., Mantelin, S., Thorpe, P., Lilley, C.J., Wright, K.M. *et al.* (2021) The GpIA7 effector from the potato cyst nematode *Globodera pallida* targets potato EBP1 and interferes with the plant cell cycle. *Journal of Experimental Botany* 72, 7301–7315. DOI: 10.1093/jxb/erab353

Contreras, M.P., Pai, H., Selvaraj, M., Toghani, A., Lawson, D.M. *et al.* (2023) Resurrection of plant disease resistance proteins via helper NLR bioengineering. *Sciences Advances* 9, eadg386. DOI: 10.1126/sciadv.adg3861

Cook, D.E., Lee, T.G., Guo, X., Melito, S., Wang, K. *et al.* (2012) Copy number variation of multiple genes at *Rhg1* mediates nematode resistance in soybean. *Science* 338, 1206–1209. DOI: 10.1126/science.1228746

Cook, R. and Noel, G.R. (2002) Cyst nematodes: *Globodera* and *Heterodera*species. In: Starr, J.L., Cook, R. and Bridge, J. (eds). *Plant Resistance to Parasitic Nematodes*. CAB International, Wallingford, UK, pp. 71–105.

Cook, R. and Rivoal, R. (1998) Genetics of resistance and parasitism. In: Sharma, S.B. (ed.) *The Cyst Nematodes*. Kluwer Academic Publishers, Dordrecht, The Netherlands, pp. 322–352.

Cook, S.M., Khan, Z.R. and Pickett, J.A. (2007) The use of push-pull strategies in integrated pest management. *Annual Review of Entomology* 52, 375–400. DOI: 10.1146/annurev.ento.52.110405.091407

Cook, S.J., Jarrell, T.A., Brittin, C.A., Wang, Y., Bloniarz, A.E. *et al.* (2019) Whole-animal connectomes of both *Caenorhabditis elegans* sexes. *Nature* 571, 63–71. DOI: 10.1038/s41586-019-1352-7

Coomans, A. (1962) Observations on the morphological structures in *Hoplolaimus pararobustus*. *Nematologica* 9, 241–254. DOI: 10.1163/187529263X00449

Coomans, A. (1979) The anterior sensilla of nematodes. *Revue de Nématologie* 2, 259–283.

Coomans, A. (1985) A phylogenetic approach to the classification of the Longidoridae (Nematoda: Dorylaimida). *Agriculture, Ecosystems & Environment* 12, 335–354. DOI: 10.1016/0167-8809(85)90006-4

Coomans, A. (1996) Phylogeny of the Longidoridae. *Russian Journal of Nematology* 4, 51–60.

Coomans, A. and Claeys, M. (2001) Structure of the female reproductive system of *Xiphinema americanum* (Nematoda: Longidoridae). *Fundamental and Applied Nematology* 21, 569–580.

Coomans, A. and Loof, P. (1986) Observations on the glands of the male reproductive system in dorylaims and its phylogenetic importance. *Revue de Nématologie* 9, 261–265.

Coomans, A., Huys, R., Heyns, J. and Luc, M. (2001) Character analysis, phylogeny, and biogeography of the genus *Xiphinema* Cobb, 1973 (Nematoda, Longidoridae). *Annales du Musée Royal de l'Afrique Centrale (Zoologie), Tervuren, Belgique* 287, 1–239.

Cordero, D.A. and Baldwin, J.G. (1990) Effect of age on body wall cuticle morphology of *Heterodera schachtii* Schmidt females. *Journal of Nematology* 22, 356–361.

Costa, M., Draper, B.W. and Priess, J.R. (1997) The role of actin filaments in patterning the *Caenorhabditis elegans* cuticle. *Development Biology* 184, 373–384. DOI: 10.1006/dbio.1997.8530

Costa, S.R., Kerry, B.R., Bardgett, R.D. and Davies, K.G. (2012) Interactions between nematodes and their microbial enemies in coastal sand dunes. *Oecologia* 170, 1053–1066. DOI: 10.1007/s00442-012-2359-z

Costa, S.R., Ng, J.L.P. and Mathesius, U. (2021) Interaction of symbiotic rhizobia and parasitic root-knot nematodes in legume roots: from molecular regulation to field application. *Molecular Plant–Microbe Interactions* 34, 470–490. DOI: 10.1094/MPMI-12-20-0350-FI

Cotton, J.A., Lilley, C.J., Jones, L.M., Kikuchi, T., Reid, A.J. *et al.* (2014) The genome and life-stage specific transcriptomes of *Globodera pallida* elucidate key aspects of plant parasitism by a cyst nematode. *Genome Biology* 15, R43. DOI: 10.1186/gb-2014-15-3-r43

Cox, D.E., Dyer, S., Weir, R., Cheseto, X., Sturrock, M. *et al.* (2019) ABC transporter genes *ABC-C6* and *ABC-G33* alter plant-microbe-parasite interactions in the rhizosphere. *Scientific Reports* 9, 19899. DOI: 10.1038/s41598-019-56493-w

Coyne, D. (2009) Pre-empting plant-parasitic nematode losses on banana in Africa: which species do we target? *Acta Horticulturae* 227–236. DOI: 10.17660/ActaHortic.2009.828.23

Coyne, D. and Affokpon, A. (2018) Nematode parasites of tropical root and tuber crops (excluding potatoes). In: Sikora, R.A., Coyne, D., Hallmann, J. and Timper, P. (eds) *Plant Parasitic Nematodes in Subtropical and Tropical Agriculture*, 3rd edn. CAB International, Wallingford, UK, pp. 252–289. DOI: 10.1079/9781786391247.0252

Coyne, D.L., Sahrawat, K.L. and Plowright, R.A. (2004) The influence of mineral fertiliser application and plant nutrition on plant parasitic nematodes in upland and lowland rice in Côte d'Ivoire and its implication in long-term experiments. *Experimental Agriculture* 40, 245–256. DOI: 10.1017/S0014479703001595

Coyne, D.L., Wasukira, A., Dusabe, J., Rotifa, I. and Dubois, T. (2010) Boiling water treatment: a simple, rapid and effective technique for producing healthy banana and plantain (*Musa* spp.) planting material. *Crop Protection* 29, 1478–1482. DOI: 10.1016/j.cropro.2010.08.008

Craig, J.P., Bekal, S., Niblack, T., Domier, L. and Lambert, K.N. (2009) Evidence for horizontally transferred genes involved in the biosynthesis of Vitamin B_1, B_5, and B_7 in *Heterodera glycines*. *Journal of Nematology* 41, 281–290.

Crisford, A., Calahorro, F., Ludlow, E., Marvin, J.M.C., Hibbard, J.K. *et al.* (2020) Identification and characterisation of serotonin signalling in the potato cyst nematode *Globodera pallida* reveals new targets for crop protection. *PLoS Pathology* 16, e1008884. DOI: 10.1371/journal.ppat.1008884

Croll, N. (1975) Behavioural analysis of nematode movement. *Advances in Parasitology* 13, 71–122. DOI: 10.1016/S0065-308X(08)60319-X

Crow, W.T. (2015) *Belonolaimus longicaudatus* Rau (Nematoda: Tylenchida: Belonolaimidae). Introduction, distribution, life cycle and biology, importance, symptoms, hosts, management and selected references. https://Entnemdept.Ufl.Edu/Creatures/Nematode/Sting_nematode.Htm.

Crow, W.T., Weingartner, D.P., Dickson, D.W. and McSorley, R. (2001) Effect of sorghum-sudangrass and velvetbean cover crops on plant-parasitic nematodes associated with potato production in Florida. *Journal of Nematology* 33, 285–288.

Crump, D. (2004) Biocontrol – a route to market. In: Cook, R. and Hunt, D.J. (eds) *Proceedings of the Fourth International Congress of Nematology, 8–13 June 2002, Tenerife, Spain. Nematology Monographs and Perspectives* 2. Brill, Leiden, The Netherlands, pp. 165–174.

Curtis, R.H.C. (2008) Plant–nematode interactions: environmental signals detected by the nematode's chemosensory organs control changes in the surface cuticle and behaviour. *Parasite* 15, 310–316. DOI: 10.1051/parasite/2008153310

Curtis, R.H.C., Robinson, A.F. and Perry, R.N. (2009) Hatch and host location. In: Perry, R.N., Moens, M. and Starr, J.L. (eds) *Root-knot Nematodes*. CAB International, Wallingford, UK, pp. 139–162.

Curtis, R.H.C., Jones, J.T., Davies, K.D., Sharon, E. and Spiegel, Y. (2011) Plant nematode surfaces. In: Davies, K. and Spiegel, Y. (eds) *Biological Control of Plant-parasitic Nematodes: Building Coherence between Microbial Ecology and Molecular Mechanisms*. Springer, Berlin, Germany, pp.115–144.

D'Addabbo, T., Laquale, S., Perniola, M. and Candido, V. (2019) Biostimulants for plant growth promotion and sustainable management of phytoparasitic nematodes in vegetable crops. *Agronomy* 9, 616. DOI: 10.3390/agronomy9100616

Da Rocha, M., Bournaud, C., Dazenière, J., Thorpe, P., Bailly-Bechet, M. *et al.* (2021) Genome expression dynamics reveal the parasitism regulatory landscape of the root-knot nematode *Meloidogyne incognita* and a promoter motif associated with effector genes. *Genes* 12, 771. DOI: 10.3390/genes12050771

Dale, M.F.B., Robinson, D.J. and Todd, D. (2004) Effects of systemic infections with *Tobacco rattle virus* on agronomic and quality traits of a range of potato cultivars. *Plant Pathology* 53, 788–793. DOI: 10.1111/j.1365-3059.2004.01093.x

Dalgaard, P. (2008) *Introductory Statistics with R*, 2nd edn. Springer, New York, USA.

Danchin, E.G.J, Thorpe, P., Perfus-Barbeoch, L., Rancurel, C., Da Rocha, M. *et al.* (2017) The transcriptomes of *Xiphinema index* and *Longidorus elongatus* reveal independent acquisition of parasitism genes by horizontal gene transfer in early-branching plant-parasitic nematodes. *Genes* 8, 287. DOI: 10.3390/genes8100287

Danquah, W.B., Back, M.A., Grove, I.G. and Haydock, P.P.J. (2011) *In vitro* nematicidal activity of a garlic extract and salicylaldehyde on the potato cyst nematode, *Globodera pallida*. *Nematology* 13, 869–885. DOI: 10.1163/138855411X560959

Davide, R.G. and Tryantaphyllou, A.C. (1968) Influence of the environment on development and sex differentiation of root-knot nematodes. *Nematologica* 14, 37–46. DOI: 10.1163/187529268X00624

Davies, K.G. (2009) Understanding the interaction between an obligate hyperparasitic bacterium, *Pasteuria penetrans* and its obligate plant-parasitic nematode host, *Meloidogyne* spp. *Advances in Parasitology* 68, 211–245. DOI: 10.1016/S0065-308X(08)00609-X

Davies, K.G. and Curtis, R.H.C. (2011) Cuticle surface coat of plant-parasitic nematodes. *Annual Review of Phytopathology* 49, 135–156. DOI: 10.1146/annurev-phyto-121310-111406

Davies, K.G. and Spiegel, Y. (eds) (2011) *Biological Control of Plant-parasitic Nematodes: Building Coherence between Microbial Ecology and Molecular Mechanisms*. Springer, Cham, Switzerland.

Davies, K.G., Flynn, C.A., Laird, V. and Kerry, B.R. (1990) The life-cycle, population dynamics and host specificity of a parasite of *Heterodera avenae,* similar to *Pasteuria penetrans*. *Revue de Nématologie* 13, 303–309.

Davies, K.G., Rowe, J.A. and Williamson, W.M. (2008) Inter and intra-specific cuticle variation between amphimictic and parthenogenetic species of root-knot nematode (*Meloidogyne* spp) as revealed by a bacterial parasite (*Pasteuria penetrans*). *International Journal of Parasitology* 38, 851–859. DOI: 10.1016/j.ijpara.2007.11.007

Davies, K.G., Rowe, J., Manzanella-López, R. and Opperman, C.H. (2011) Re-evaluation of the life-cycle of the nematode parasitic bacterium *Pasteuria penetrans* in root-knot nematodes, *Meloidogyne* spp. *Nematology* 13, 825–835. DOI: 10.1163/138855410X552670

Davies, K.G., Mohan, S. and Hallmann, J. (2018) Biological control of cyst nematodes through microbial pathogens, endophytes and antagonists. In: Perry, R.N., Moens, M. and Jones, J.T. (eds) *Cyst Nematodes*. CAB International, Wallingford, UK, pp. 237–270.

Davies, L.J. and Elling, A.A. (2015) Resistance genes against plant-parasitic nematodes: a durable control strategy? *Nematology* 17, 249–263. DOI: 10.1163/15685411-00002877

Davis, R.F., Galbieri, R. and Asmus, G.L. (2018) Nematode parasites of cotton and other tropical fibre crops. In: Sikora, R., Coyne, D., Hallmann, J. and Timper, P. (eds) *Plant Parasitic Nematodes in Subtropical and Tropical Agriculture*, 3rd edn. CAB International, Wallingford, UK, pp. 738–754. DOI: 10.1079/9781786391247.0738

de Almeida Engler, J., Van Poucke, K., Karimi, M., De Groodt, R., Gheysen, G. *et al.* (2004) Dynamic cytoskeleton rearrangements in giant cells and syncytia of nematode-infected roots. *Plant Journal* 38, 12–26. DOI: 10.1111/j.1365-313x.2004.02019.x

de Almeida Engler, J., Kyndt, T., Vieira, P., Van Cappelle, E., Bouldolf, V. *et al.* (2012) *CCS52* and *DEL1* genes are key components of the endocycle in nematode induced feeding sites. *Plant Journal* 72, 185–198. DOI: 10.1111/j.1365-313X.2012.05054.x

de Bono, M. and Bargmann, C.I. (1998) Natural variation in a neuropeptide Y receptor homolog modifies social behavior and food response in *C. elegans*. *Cell* 94, 679–689. DOI: 10.1016/s0092-8674(00)81609-8

de Bono, M. and Maricq, A.V. (2005) Neuronal substrates of complex behaviors in *C. elegans*. *Annual Review of Neuroscience* 28, 451–501. DOI: 10.1146/annurev.neuro.27.070203.144259

de Coninck, L. (1965) Classe des Nématodes – Généralités. In: Grassé, P. (ed.) *Traité de Zoologie* IV, 2, 1–217. Masson, Paris, France.

De Deyn, G.B., Raaijmakers, C.E., Zoomer, H.R., Berg, M.P., de Ruiter, P.C. *et al.* (2003) Soil invertebrate fauna enhances grassland succession and diversity. *Nature* 422, 711–714. DOI: 10.1038/nature01548

De Grisse, A.T. (1977) De ultrastruktuur van het zenuwstelsel in de kop van 22 soorten planten-parasitaire nematoden, behorende tot 19 genera (Nematoda: Tylenchida). DSc Dissertation, Ghent University, Ghent, Belgium.

de Jesus Rocha, A., dos Santos Ferreira, M., de Souza Rocha, L., Oliveira, S.A.S., Amorim, E.P. *et al.* (2020) Interaction between *Fusarium oxysporum* f. sp. *cubense* and *Radopholus similis* can lead to changes in the resistance of banana cultivars to Fusarium wilt. *European Journal of Plant Pathology* 158, 403–417. DOI: 10.1007/s10658-020-02081-y

de la Peña, E., Rodríguez-Echeverria, S., van der Putten, W. and Freitas, H. (2005) Mechanism of control of root-feeding nematodes by mycorrhizal fungi in the dune grass *Ammophila arenaria*. *New Phytologist* 169, 829–840. DOI: 10.1111/j.1469-8137.2005.01602.x

de la Peña, E., Moens, M., Van Aelst, A. and Karssen, G. (2006) Description of *Pratylenchus dunensis* sp. n. (Nematoda: Pratylenchidae), a root-lesion nematode associated with the dune grass *Ammophila arenaria* (L.) Link. *Nematology* 8, 79–88. DOI: 10.1163/156854106776179917

De Ley, I.T., De Ley, P., Vierstraete, A., Karssen, G., Moens, M. and Vanfleteren, J. (2002) Phylogenetic analyses of *Meloidogyne* small subunit rDNA. *Journal of Nematology* 34, 319–327.

De Ley, P. and Blaxter, M. (2002) Systematic position and phylogeny. In: Lee, D.L. (ed.) *The Biology of Nematodes*. Taylor & Francis, London, UK, pp. 1–30.

De Ley, P. and Blaxter, M.L. (2004) A new system for Nematoda: combining morphological characters with molecular trees, and translating clades into ranks and taxa. In: Cook, R. and Hunt, D.J. (eds) *Proceedings of the Fourth International Congress of Nematology, 8–13 June 2002, Tenerife, Spain. Nematology Monographs and Perspectives 2*. Brill, Leiden, The Netherlands, pp. 633–653.

De Ley, Van de Velde, M.C., Mounport, D., Baujard, P. and Coomans, A. (1995) Ultrastructure of the stoma in Cephalobidae, Panagrolaimidae and Rhabditidae, with a proposal for a revised stoma terminology in Rhabditida (Nematoda). *Nematologica* 41, 153–182. DOI: 10.1163/003925995X00143

De Luca, F., Troccoli, A., Duncan, L.W., Subbotin, S.A., Waeyenberge, L. *et al.* (2010) Characterisation of a population of *Pratylenchus hippeastri* from bromeliads and description of two related new species, *P. floridensis* n. sp. and *P. parafloridensis* n. sp. from grasses in Florida. *Nematology* 12, 847–868. DOI: 10.1163/138855410X495809

de Moura, R.M., de Enchandi, E. and Powell, N.T. (1975) Interactions of *Corynebacterium michiganense* and *Meloidogyne incognita* on tomato. *Phytopathology* 65, 1332–1335. DOI: 10.1094/Phyto-65-1332

De Waele, D. and Coomans, A. (1991) Occurrence of ecology of trichodorid nematodes in Belgium. *Revue de Nématologie* 14, 127–132.

De Waele, D. and Elsen, A. (2002) Migratory endoparasites: *Pratylenchus* and *Radopholus* species. In: Starr, J.L., Cook, R. and Bridge, J. (eds) *Plant Resistance to Parasitic Nematodes*. CAB International, Wallingford, UK, pp. 175–206.

Decraemer, W. (1991) Stubby root and virus vector nematodes *Trichodorus, Paratrichodorus, Allotrichodorus* and *Monotrichodorus*. In: Nickle, W.R. (ed.) *Manual of Agricultural Nematology*. Marcel Dekker, New York, USA, pp. 587–625.

Decraemer, W. (1995) *The Family Trichodoridae: Stubby Root and Virus Vector Nematodes*. Kluwer Academic Publishers, Dordrecht, The Netherlands.

Decraemer, W. (2012) Tokens of love: possible diagnostic value of mating plugs and refractive secretory uterine structures in *Trichodorus* (Diphtherophorina: Trichodoridae). *Nematology* 14, 151–158. DOI: 10.1163/138855411X581703

Decraemer, W. and Chaves, E. (2012) Longidoridae and Trichodoridae. In: Manzanilla-López, R.H. and Marbán-Mendoza, N. (eds) *Practical Plant Nematology*. Bibliotheca Basica de Agricultura, Montecillo, Mexico, pp. 579–617.

Decraemer, W. and Coomans, A. (2007) Revision of some species of the genus *Paralongidorus sensu* Siddiqi *et al.* (1993) with a discussion on the relationships within the family Longidoridae (Nematoda: Longidoridae). *Nematology* 9, 643–662. DOI: 10.1163/156854107782024776

Decraemer, W. and De Waele, D. (1981) Taxonomic value of the position of oesophageal gland nuclei of oesophageal gland overlap in the Trichodoridae (Diphtherophorina). *Nematologica* 27, 82–94. DOI: 10.1163/187529281X00089

Decraemer, W. and Geraert, E. (1992) *Criconema paradoxiger, Ogma civellae* and *O. paracivellae* sp. n. from Papua New Guinea (Nemata: Tylenchida). *Fundamental and Applied Nematology* 15, 355–366.

Decraemer, W., Baldwin, J., Eddlemans, C. and Geraert, E. (1996) *Criconema paradoxiger* (Orton Williams, 1982) Raski and Luc, 1985: cuticle ultrastructure and revalidation of the genus *Amphisbaenema*. *Nematologica* 42, 408–416. DOI: 10.1163/004525996X00028

Decraemer, W., Doucet, M.E. and Coomans, A. (1998) Longidoridae from Argentina with the description of *Paraxiphidorus brevistylus* sp. n. (Nematoda: Longidoridae). *Fundamental and Applied Nematology* 21, 371–388.

Decraemer, W., Karanastasi, E., Brown, D.J.F. and Backeljau, T. (2003) Review of the ultrastructure of the nematode body cuticle and its phylogenetic interpretation. *Biological Reviews* 78, 465–510. DOI: 10.1017/s1464793102006115

Decraemer, W., Cantalapiedra-Navarrete, C., Archidona-Yuste, A., Varela-Benavides, I., Gutiérrez-Gutiérrez, C. *et al.* (2019) Integrative taxonomy unravels cryptic diversity in the *Paratrichodorus hispanus*-group complex and resolves two new species of the genus and the molecular phylogeny of the family (Nematoda: Trichodoridae). *Zoological Journal of the Linnean Society* 185, 656–692. DOI: 10.1093/zoolinnean/zly059

Deliopoulos, T., Haydock, P.P.J. and Jones, P.W. (2008) Interaction between arbuscular mycorrhizal fungi and the nematicide aldicarb on hatch and development of the potato cyst nematode, *Globodera pallida*, and yield of potatoes. *Nematology* 10, 783–799. DOI: 10.1163/156854108786161427

Deng, M.-H., Zhong, L.-Y., Kamolnetr, O., Limpanont, Y. and Lv, Z.-Y. (2019) Detection of helminths by loop-mediated isothermal amplification assay: a review of updated technology and future outlook. *Infectious Diseases of Poverty* 8, 20. DOI: 10.1186/s40249-019-0530-z

Derevnina, L., Contreras, M.P., Adachi, H., Upson, J., Cruces, A.V. *et al.* (2021) Plant pathogens convergently evolved to counteract redundant nodes of an NLR immune receptor network. *PLoS Biology* 19, e3001136. DOI: 10.1371/journal.pbio.3001136

Desaeger J, Wram C and Zasada I. (2020) New reduced-risk agricultural nematicides – rationale and review. *Journal of Nematology* 52, e2020-91. DOI: 10.21307/jofnem-2020-091

Desaeger, J., Sikora, R.A. and Molendijk, L.P.G. (2021) Outlook: a vision of the future of integrated nematode management. In: Sikora, R.A., Desaeger, J. and Molendijk, L.P.G. (eds) *Integrated Nematode Management: State-of-the-art and Visions for the Future*. CAB International, Wallingford, UK, pp. 475–483. DOI: 10.1079/9781789247541.0065

Desmedt, W., Mangelinckx, S., Kyndt, T. and Vanholme, B. (2020) A phytochemical perspective on plant defense against nematodes. *Frontiers in Plant Science* 11, 602079. DOI: 10.3389/fpls.2020.602079

Devine, K.J. and Jones, P.W. (2002) Investigations into the chemoattraction of the potato cyst nematodes *Globodera rostochiensis* and *G. pallida* towards fractionated potato root leachate. *Nematology* 5, 65–75. DOI: 10.1163/156854102765216704

Devine, K.J., Byrne, J., Maher, N. and Jones, P.W. (1996) Resolution of natural hatching factors for the golden potato cyst nematode, *Globodera rostochiensis*. *Annals of Applied Biology* 129, 323–334. DOI: 10.1111/j.1744-7348.1996.tb05755.x

Devran, Z., Tülek A., Mıstanoğlu, I., Çiftçiğil, T.H. and Özalp, T. (2017) A rapid molecular detection method for *Aphelenchoides besseyi* from rice tissues. *Australasian Plant Pathology* 46, 43–48. DOI: 10.1007/s13313-016-0452-1

Dietrich, P., Cesarz, S., Liu, T., Roscher, C. and Eisenhauer, E. (2021) Effects of plant species diversity on nematode community composition and diversity in a long-term biodiversity experiment. *Oecologia* 197, 297–311. DOI: 10.1007/s00442-021-04956-1

Djamei, A., Schipper, K., Rabe, F., Ghosh, A., Vincon, V. *et al.* (2011) Metabolic priming by a secreted fungal effector. *Nature* 478, 395–400. DOI: 10.1038/nature10454

Djian-Caporalino, C., Fazari, A., Arguel, M.J., Vernie, T., VandeCasteele, C. *et al.* (2007) Root-knot nematode (*Meloidogyne* spp.) *Me* resistance genes in pepper (*Capsicum annuum* L.) are clustered on the P9 chromosome. *Theoretical and Applied Genetics* 114, 473–486. DOI: 10.1007/s00122-006-0447-3

Donald, P.A., Pierson, P.E., St Martin, S.K., Sellers, P.R., Noel, G.R. *et al.* (2006) Assessing *Heterodera glycines*-resistant and susceptible cultivar yield response. *Journal of Nematology* 38, 76–82.

Driscoll, M. and Kaplan, J. (1997) Mechanotransduction. In: Riddle, D.L., Blumenthal, T., Meyer, B.J. and Preiss, J.R. (eds) C. elegans II. Cold Spring Harbor Laboratory Press, Cold Spring Harbor, New York, USA, pp. 645–677.

Duarte, I.M., de Almeida, M.T., Brown, D.J.F., Marques, I., Neilson, R. *et al.* (2010) Phylogenetic relationships, based on SSU rDNA sequences, among the didelphic genera of the family Trichodoridae from Portugal. *Nematology* 12, 171–180. DOI: 10.1163/156854109X461721

Dubreuil, G., Magliano, M., Dubrana, M.P., Lozano, J., Lecomte, P. *et al.* (2009) Tobacco rattle virus mediates gene silencing in a plant parasitic root-knot nematode. *Journal of Experimental Botany* 60, 4041–4050. DOI: 10.1093/jxb/erp237

Duceppe, M.O., Lafond-Lapalme, J., Palomares-Rius, J.E., Sabeh, M., Blok, V. *et al.* (2017) Analysis of survival and hatching transcriptomes from potato cyst nematodes, *Globodera rostochiensis* and *G. pallida*. *Scientific Reports* 7, 3882. DOI: 10.1038/s41598-017-03871-x

Duncan, L.W., Inserra, R.N., Thomas, W.K., Dunn, D., Mustike, I. *et al.* (1999) Molecular and morphological analysis of isolates of *Pratylenchus coffeae* and closely related species. *Nematropica* 29, 61–80.

Dunn, C.W., Hejnol, A., Matus, D.Q., Pang, K., Browne, W.E. *et al.* (2008) Broad phylogenomic sampling improves resolution of the animal tree of life. *Nature* 452, 745–749. DOI: 10.1038/nature06614

Du Preez, G., Daneel, M., de Goede, R.G.M., Du Toit, M.J., Ferris, H. *et al.* (2022) Nematode-based indices in soil ecology: application, utility, and future directions. *Soil Biology and Biochemistry* 169, 108640. DOI: 10.1016/j.soilbio.2022.108640

Dusenbery, D.B. (1992) *Sensory Ecology*. W.H. Freeman, New York, USA.

Dusenbery, D.B. (1996) *Life at Small Scale: the Behaviour of Microbes*. Scientific American Library, New York, USA.

Dusenbery, D.B., Sheridan, R.E. and Russell, R.L. (1975) Chemotaxis-defective mutants of the nematode *Caenorhabditis elegans*. *Genetics* 80, 297–309. DOI: 10.1093/genetics/80.2.297

Dutta, T.K. and Phani, V. (2023) The pervasive impact of global climate change on plant-nematode interaction continuum. *Frontiers in Plant Science* 14, 1143889. DOI: 10.3389/fpls.2023.1143889

Dutta, T.K., Powers, S.J., Kerry, B.R., Gaur, H.S. and Curtis, R.H.C. (2011) Comparison of host recognition, invasion, development and reproduction of *Meloidogyne graminicola* and *M. incognita* on rice and tomato. *Nematology* 13, 509–520. DOI: 10.1163/138855410X528262

Dutta, T.K., Banakar, P. and Rao, U. (2015) The status of RNAi-based transgenic research in plant nematology *Frontiers in Microbiology* 5, 760.

Dwinell, L.D. and Lehman, P.S. (2004) Plant parasitic nematodes which are exotic pests in agriculture and forestry. In: Briton, K.O. (ed.) *Biological Pollution: An Emerging Global Menace*. American Phytopathological Society Press, St Paul, MN, USA, pp. 51–70.

Ebbels, D.L. (1988) The costs and benefits of seed potato classification. In: Clifford, B.C. and Lester, E. (eds) *Control of Plant Diseases: Costs and Benefits*. Blackwell Scientific Publications, Oxford, UK, pp. 115–122.

Ebbels, D.L. (2003) *Principles of Plant Health and Quarantine*. CAB International, Wallingford, UK.

Ebrahimi, N., Viaene, N. and Moens, M. (2015) Optimizing trehalose-based quantification of live eggs in potato cyst nematodes (*Globodera rostochiensis* and *G. pallida*). *Plant Disease* 99, 947–953. DOI: 10.1094/pdis-09-14-0940-re

Ebrahimi, N., Viaene, N., Vandecasteele, B., D'Hose, T., Debode, J. *et al.* (2016) Traditional and new soil amendments reduce survival and reproduction of potato cyst nematodes, except for biochar. *Applied Soil Ecology* 107, 191–204. DOI: 10.1016/j.apsoil.2016.06.006

Edwards, S.L., Charlie, N.K., Milfort, M.C., Brown, B.S., Gravlin, C.N. *et al.* (2008) A novel molecular solution for ultraviolet light detection in *Caenorhabditis elegans*. *PLoS Biology* 6, e198. DOI: 10.1371/journal.pbio.0060198

Eisenback, J.D. (1993) Interaction between nematodes in cohabitance. In: Khan, M.W. (ed.) *Nematode Interactions*. Chapman & Hall, London, UK, pp. 143–174.

Eisenback, J.D. and Hirschmann Triantaphyllou, H. (1991) Root-knot nematode: *Meloidogyne* spp. and races. In: Nickle, W.R. (ed.) *Manual of Agricultural Nematology*. Marcel Dekker, New York, USA, pp. 191–274.

Eisenhauer, N., Ackerman, M., Gass, S., Klier, M., Migunova, V.D. *et al.* (2010) Nematicide impacts on nematodes and feedbacks on plant productivity in a plant diversity gradient. *Acta Oecologica* 36, 477–483. DOI: 10.1016/j.actao.2010.06.004

Ekschmitt, K. (1993) Uber die Raumliche Verteilung von Bodentieren: Zur okologischen Interpretation der Aggregation und zur Probenstatistil. PhD Thesis, University of Bremen, Bremen, Germany.

Elbadri, G.A.A., De Ley, P., Waeyenberge, L., Vierstraete, A., Moens, M. *et al.* (2002) Intraspecific variation in *Radopholus similis* isolates assessed with restriction fragment length polymorphism and DNA sequencing of the internal transcribed spacer region of the ribosomal RNA cistron. *International Journal for Parasitology* 32, 199–205. DOI: 10.1016/S0020-7519(01)00319-8

Elhady, A., Gine, A., Topalovic, O., Jacquiod, S., Sorensen, S.J. *et al.* (2017) Microbiomes associated with infective stages of root-knot and lesion nematodes in soil. *PLoS One* 12, e0177145. DOI: 10.1371/journal.pone.0177145

Elling, A.A. (2013) Major emerging problems with minor *Meloidogyne* species. *Phytopathology*103, 1092–1102. DOI: 10.1094/PHYTO-01-13-0019-RVW

Elling, A.A., Mitreva, M., Recknor, J., Gai, X., Martin, J. *et al.* (2007) Divergent evolution of arrested development in the dauer stage of *Caenorhabditis elegans* and the infective stage of *Heterodera glycines*. *Genome Biology* 8, R211. DOI: 10.1186/gb-2007-8-10-r211

El-Sherif, M. and Mai, W.F. (1969) Thermotactic response of some plant parasitic nematodes. *Journal of Nematology* 1, 43–48.

Elston, D.A., Phillips, M.S. and Trudgill, D.L. (1991) The relationship between initial population density of potato cyst nematode *Globodera pallida* and the yield of partially resistant potatoes. *Revue de Nématologie* 14, 231–219.

Endo, B.Y. (1980) Ultrastructure of the anterior neurosensory organs of the larvae of the cyst nematode, *Heterodera glycines*. *Journal of Ultrastructure Research* 72, 349–366. DOI: 10.1016/S0022-5320(80)90070-2

Erkut, C., Penkov, S., Khesbak, H., Vorkel, D., Verbavatz, J.-M. *et al.* (2011) Trehalose renders the dauer larva of *Caenorhabditis elegans* resistant to extreme desiccation. *Current Biology* 21, 1331–1336. DOI: 10.1016/j.cub.2011.06.064

Esbenshade, P.R. and Triantaphyllou, A.C. (1985) Identification of major *Meloidogyne* species employing enzyme phenotypes as differentiating characters. In: Sasser, J.N. and Carter, C.C. (eds) *An Advanced Treatise on* Meloidogyne, *Volume I: Biology and Control*. North Carolina State University Graphics, Raleigh, NC, USA, pp. 135–140.

Esbenshade, P.R. and Triantaphyllou, A.C. (1987) Enzymatic and evolutionary relationships in the genus *Meloidogyne* (Nematoda: Tylenchida). *Journal of Nematology* 20, 8–18.

Escuer, M. and Arias, M. (1997) *Paralongidorus iberis* and *P. monegrensis* sp.n. from Spain with a polytomous key to the species of the genus *Paralongidorus* Siddiqi, Hooper & Khan, 1963 (Nematoda: Longidoridae). *Fundamental and Applied Nematology* 20, 135–148.

Esquibet, M., Grenier, E., Plantard, O., Andaloussi, F.A. and Caubel, G. (2003) DNA polymorphism in the stem nematode *Ditylenchus dipsaci*: development of diagnostic markers for normal and giant races. *Genome* 46, 1077–1083. DOI: 10.1139/g03-072

Estores, R.A. and Chen, T.A. (1970) Interaction of *Pratylenchus penetrans* and *Meloidogyne incognita acrita* as cohabitants on tomatoes. *Journal of Nematology* 4, 170–174.

Evans, A.A.F. (1987) Diapause in nematodes as a survival strategy. In: Veech, J.A. and Dickson, D.W. (eds) *Vistas on Nematology*. Society of Nematologists Inc., Hyattsville, MD, USA, pp. 180–187.

Evans, A.A.F. (1998) Reproductive mechanisms. In: Perry, R.N. and Wright, D.J. (eds) *The Physiology and Biochemistry of Free-living and Plant-parasitic Nematodes*. CAB International, Wallingford, UK, pp. 133–154.

Evans, H.F., McNamara, D.G., Braasch, H., Chadoeuf, J. and Magnusson, C. (1996) Pest Risk Analysis (PRA) for the territories of the European Union (as PRA area) on *Bursaphelenchus xylophilus* and its vectors in the genus *Monochamus*. *EPPO Bulletin* 26, 199–249. DOI: 10.1111/j.1365-2338.1996.tb00594.x

Evans, K. and Rowe, J. (1998) Distribution and economic importance. In: Sharma, S.B. (ed.) *The Cyst Nematodes*. Kluwer Academic Publishers, Dordrecht, The Netherlands, pp. 1–30.

Evans, S.G. and Wright, D.J. (1982) Effects of the nematicide oxamyl on life cycle stages of *Globodera rostochiensis*. *Annals of Applied Biology* 100, 511–519. DOI: 10.1111/j.1744-7348.1982.tb01417.x

Eves-van den Akker, S., Lilley, C.J., Jones, J.T. and Urwin, P.E. (2014a) Identification and characterisation of a hyper-variable apoplastic effector gene family of the potato cyst nematodes. *PLoS Pathogens* 10, e1004391. DOI: 10.1371/journal.ppat.1004391

Eves-van den Akker, S., Lilley C.J., Ault, J.R., Ashcroft, A.E., Jones, J.T. and Urwin, P.E. (2014b) The feeding tube of cyst nematodes: characterisation of protein exclusion. *PLoS One* 9, e87289. DOI: 10.1371/journal.pone.0087289

Eves-van den Akker, S., Lilley C.J., Jones, J.T. and Urwin, P.E. (2015) Plant-parasitic nematode feeding tubes and plugs: new perspectives on function. *Nematology* 17, 1–9. DOI: 10.1163/15685411-00002832

Eves-van den Akker, E.S., Stojiiković, B. and Gheysen, G. (2021) Recent applications of biotechnological approaches to elucidate the biology of plant-nematode interactions. *Current Opinion in Biotechnology* 70, 122–130. DOI: 10.1016/j.copbio.2021.03.008

Fairbairn, D.J., Cavallaro, A.S., Bernard, M., Mahalinga-Iyer, J., Graham, M.W. *et al.* (2007) Host-delivered RNAi: an effective strategy to silence genes in plant parasitic nematodes. *Planta* 226, 1525–1533. DOI: 10.1007/s00425-007-0588-x

Fallas, G.A., Sarah, J.-L. and Fargette, M. (1995) Reproductive fitness and pathogenicity of eight *Radopholus similis* isolates of banana plants (Musa AAA v. Poyo). *Nematropica* 25, 135–141.

FAO (2014) *Building a common vision for sustainable food and agriculture – principles and practices*. Food and Agriculture Organization of the United Nations, Rome, Italy.

Faske, T.R. and Hurd, K. (2015) Sensitivity of *Meloidogyne incognita* and *Rotylenchulus reniformis* to Fluopyram. *Journal of Nematology* 47, 316–321.

Faske, T.R., Brown, K. and Kelly, J. (2022) Toxicity of tioxazafen to *Meloidogyne incognita* and *Rotylenchulus reniformis*. *Journal of Nematology* 54, 20220007. DOI: 10.2478/jofnem-2022-0007

Feist, E., Kearn, J., Gaihre, Y., O'Connor, V. and Holden-Dye, L. (2020) The distinct profiles of the inhibitory effects of fluensulfone, abamectin, aldicarb and fluopyram on *Globodera pallida* hatching, *Pesticide Biochemistry and Physiology* 165, 104541. DOI: 10.1016/j.pestbp.2020.02.007

Ferris, H. (2010) Form and function: metabolic footprints of nematodes in the soil food web. *European Journal of Soil Biology* 46, 97–104. DOI: 10.1016/j.ejsobi.2010.01.003

Ferris, H., Venette, R.C., van der Meulen, H.R. and Lau, S.S. (1998) Nitrogen mineralization by bacterial-feeding nematodes: verification and measurement. *Plant and Soil* 203, 159–171. DOI: 10.1023/A:1004318318307

Ferris, H., Bongers, T. and de Goede, R.G.M. (2001) A framework for soil food web diagnostics: extension of the nematode faunal analysis concept. *Applied Soil Ecology* 18, 13–29. DOI: 10.1016/S0929-1393(01)00152-4

Ferris, H., Sánchez-Moreno, S. and Brennan, E.B. (2012) Structure, functions and interguild relationships of the soil nematode assemblage in organic vegetable production. *Applied Soil Ecology* 61, 16–25. DOI: 10.1016/j.apsoil.2012.04.006

Filipiak, A., Wieczorek, P. and Tomalak, M. (2019) A fast and sensitive multiplex real-time PCR assay for simultaneous identifcation of *Bursaphelenchus xylophilus*, *B. mucronatus* and *B. fraudulentus* – three closely related species from the *xylophilus* group. *European Journal of Plant Pathology* 155, 239–251. DOI: 10.1007/s10658-019-01767-2

Fine, A.E., Ashton, F.T., Bhopale, V.M. and Schad, G.A. (1998) Sensory neuroanatomy of a skin-penetrating nematode parasite *Strongyloides stercoralis*. II. labial and cephalic neurons. *Journal of Comparative Neurology* 389, 212–223. DOI: 10.1002/(SICI)1096-9861(19971215)389:2<212::AID-CNE2>3.0.CO;2-4

Foor, W.E. (1967) Ultrastructural aspects of oocyte development and shell formation in *Ascaris lumbricoides*. *Journal of Parasitology* 53, 1245–1261.

Foreman, J., Jackson, P., Aitken, K., Jingchuan, L., Liping, W. *et al.* (2007) Introduction and evaluation of clones derived from Chinese *Saccharum spontaneum* and *Erianthus* spp. *Proceedings of the Australian Society of Sugarcane Technologists* 29, 242–250.

Forghani, F. and Hajihassani, A. (2020) Recent advances in the development of environmentally benign treatments to control root-knot nematodes. *Frontiers in Plant Science* 11, 1125. DOI: 10.3389/fpls.2020.01125

Fortuner, R. (1987) A reappraisal of Tylenchina (Nemata). 8. The Family Hoplolaimidae Filipjev, 1934. *Revue de Nématologie* 10, 219–232.

Fortuner, R. (1991) The Hoplolaiminae. In: Nickle, W.R. (ed.) *Manual of Agricultural Nematology*. Marcel Dekker, New York, USA, pp. 669–720.

Fortuner, R. and Luc, M. (1987) A reappraisal of Tylenchina (Nemata). 6. The Family Belonolaimidae Whitehead, 1960. *Revue de Nématologie* 10, 183–202.

Fragoso, R.R., Lourenco, I.T., Batista, J.A.N., Oliveira-Neto, B., Silva, M.C.M. *et al.* (2009) *Meloidogyne incognita*: molecular cloning and characterization of a cDNA encoding a cathepsin-D like aspartic proteinase. *Experimental Parasitology* 121, 115–123. DOI: 10.1016/j.exppara.2008.09.017

Franco, A.L.C., Gherardi, L.A., de Tomasel, C.M., Andriuzzi, W.S., Ankrom, K.E. *et al.* (2019) Drought suppresses soil predators and promotes root herbivores in mesic, but not in xeric grasslands. *Proceedings of the National Academy of Sciences of the USA* 116, 12883–12888. DOI: 10.1073/pnas.1900572116

Franco, J.P. and Gonzalez, V.A. (2010) El nematode quiste de la papa *Globodera* sp. en Bolivia y Perú. In: *XXIV Congreso de la Asociación Latinoamericana de la Papa ALAP 2010*, pp. 291–292.

Franklin, M.T. (1951) *The Cyst-Forming Species of* Heterodera. Commonwealth Agricultural Bureaux, Farnham Royal, UK.

Franklin, M.T. (1969) *Heterodera latipons* n. sp., a cereal cyst nematode from the Mediterranean region. *Nematologica* 15, 535–542. DOI: 10.1163/187529269X00867

Franklin, M.T. (1972) *Heterodera schachtii*. In: *CIH Descriptions of Plant-Parasitic Nematodes*, Set 1, No. 1. Commonwealth Institute of Helminthology, St Albans, UK.

Franklin, M.T. and Siddiqi, M.R. (1992) *Aphelenchoides besseyi*. In: *CIH Descriptions of Plant-Parasitic Nematodes*. Set 1, No. 4. Commonwealth Agricultural Bureaux, Farnham Royal, UK.

Freckman, D.W. and Ettema, C.E. (1993) Assessing nematode communities in agroecosystems of varying human intervention. *Agriculture, Ecosystems and Environment* 45, 239–261. DOI: 10.1016/0167-8809(93)90074-Y

Friebe, A., Klever, W., Sikora, R. and Schnabl, H. (1998) Allelochemicals in root exudates of maize: effects on root lesion nematode *Pratylenchus zeae*. In: Romeo, J.T., Downum, K.R. and Verporte, R. (eds) *Phytochemical Signals and Plant–Microbe Interactions*. Plenum, New York, USA, pp. 71–93.

Frijters, R.J.J.M., Kalisvaart, J. and Verhage, A. (2021) Root-knot nematode resistance conferring gene. In: https://patentscope2.wipo.int/search/en/detail.jsf?docId=WO2021255272&_cid=JP1-KXLFAU-27935-15. The Netherlands.

Fuchs, M. (2017) Pyramiding resistance-conferring gene sequences in crops. *Current Opinion in Virology* 26, 36–42. DOI: 10.1016/j.coviro.2017.07.004

Fuller, V.L., Lilley, C.J., Atkinson, H.J. and Urwin, P.E. (2007) Differential gene expression in *Arabidopsis* following infection by plant-parasitic nematodes *Meloidogyne incognita* and *Heterodera schachtii*. *Molecular Plant Pathology* 8, 595–609. DOI: 10.1111/j.1364-3703.2007.00416.x

Futai, K. (2003) Role of asymptomatic carrier trees in epidemic spread of pine wilt disease. *Journal of Forest Research* 8, 253–260. DOI: 10.1007/s10310-003-0034-2

Gan, H. and Wickings, K. (2020) Root herbivory and soil carbon cycling: shedding "green" light onto a "brown" world. *Soil Biology and Biochemistry* 150, 107972. DOI: 10.1016/j.soilbio.2020.107972

Gartner, U., Hein, I., Brown, L.H., Chen, X., Mantelin, S. *et al.* (2021) Resisting potato cyst nematodes with resistance. *Frontiers in Plant Science* 12, 661194. DOI: 10.3389/fpls.2021.661194

Gaur, H.S. and Perry, R.N. (1991) The use of soil solarization for control of plant-parasitic nematodes. *Nematological Abstracts* 60, 153–167.

Gautier, C., Martinez, L., Fournet, S., Montarry, J., Yvin, J.-C. *et al.* (2020) Hatching of *Globodera pallida* induced by root exudates is not influenced by soil microbiota composition. *Frontiers in Microbiology* 11, 536932. DOI: 10.3389/fmicb.2020.536932

Geisen, S., Snoek, L.B., ten Hooven, F.C., Duyts, H., Kostenko, O. *et al.* (2018) Integrating quantitative morphological and qualitative molecular methods to analyse soil nematode community. *Methods in Ecology and Evolution* 9, 1366–1378. DOI: 10.1111/2041-210X.12999

Geraert, E. (1983) The use of the female reproductive system in nematode systematics. In: Stone, A.R., Platt, H.M. and Khalil, L.F. (eds) *Concepts in Nematode Systematics*. Academic Press, London, UK, pp. 73–84.

Geraert, E. (2006) *Functional and Detailed Morphology of the Tylenchida (Nematoda). Nematology Monographs and Perspectives 4* (Series Editors: Hunt, D.J. and Perry, R.N.). Brill, Leiden, The Netherlands.

Geraert, E. (2010) *The Criconematidae of the World. Identification of the Family Criconematidae (Nematoda).* Academia Press, Ghent, Belgium.

Geraert, E. (2011) *The Dolichodoridae of the World. Identification of the Family Dolichodoridae.* Academia Press, Ghent, Belgium.

Geraert, E. (2013) *The Pratylenchidae of the World. Identification of the Family Pratylenchidae (Nematoda: Tylenchida).* Academia Press, Ghent, Belgium.

Gerisch, B., Rottiers, V., Li, D., Motola, D.L., Cummins, C.L. *et al.* (2007) A bile acid-like steroid modulates *Caenorhabditis elegans* lifespan through nuclear receptor signaling. *Proceedings of the National Academy of Sciences of the USA* 104, 5014–5019. DOI 10.1073/pnas.0700847104

Gessner, M.O., Swan, C.M., Dang, C.K., McKie, B.G., Bardgett, R.D. *et al.* (2010) Diversity meets decomposition. *Trends in Ecology and Evolution* 25, 372–380. DOI: 10.1016/j.tree.2010.01.010

Ghahremani, Z., Escudero, N., Marín, I., Sanz, A., García, S. *et al.* (2022) *Pochonia chlamydosporia* is the most prevalent fungal species responsible for meloidogyne suppression in sustainable vegetable production systems. *Sustainability* 14, 16941. DOI: 10.3390/su142416941

Gheysen, G. and Fenoll, C. (2002) Gene expression in nematode feeding sites. *Annual Review of Phytopathology* 40, 191–219. DOI: 1146/annurev.phyto.40.121201.093719

Gheysen, G. and Fenoll, C. (2011) Arabidopsis as a tool for the study of plant-nematode interactions. In: Jones, J., Gheysen, G. and Fenoll, C. (eds) *Genomics and Molecular Genetics of Plant–Nematode Interactions*, Springer, Berlin, Germany, pp. 139–156.

Gheysen, G. and Mitchum, M.G. (2009) Molecular insights in the susceptible plant response to nematode infection. In: Berg, R.H. and Taylor, C.G. (eds) *Cell Biology of Plant Nematode Parasitism*. Springer, Berlin, Germany, pp. 45–81.

Gheysen, G. and Mitchum, M.G. (2019) Phytoparasitic nematode control of plant hormone pathways. *Plant Physiology* 179, 1212–1226. DOI: 10.1104/pp.18.01067

Giblin-Davis, R.M. (1993) Interactions of nematodes with insects. In: Khan, M.W. (ed.) *Nematode Interactions*. Chapman & Hall, London, UK, pp. 302–344.

Giblin-Davis, R.M., Davies, K.A., Williams, S. and Center, T.D. (2001) Cuticular changes in fergusobiid nematodes associated with parasitism of fergusonimid flies. *Comparative Parasitology* 68, 242–248.

Giblin-Davis, R.M., Davies, K.A., Morris, K. and Thomas, W.K. (2003) Evolution of parasitism in insect-transmitted plant nematodes. *Journal of Nematology* 35, 133–141.

Gibson, T., Farrugia, D., Barrett, J., Chitwood, D., Rowe, J. *et al.* (2011) The mitochondrial genome of the soybean cyst nematode, *Heterodera glycines*. *Genome* 54, 565–574. DOI: 10.1139/g11-024

Gilabert, A., Curran, D.M., Harvey, S.C. and Wasmuth, J.D. (2016) Expanding the view on the evolution of the nematode dauer signalling pathways: Refinement through gene gain and pathway co-option. *BMC Genomics* 17, 476. DOI: 10.1186/s12864-016-2770-7

Giller, P.S. (1996) The diversity of soil communities, the 'poor man's rainforest'. *Biodiversity and Conservation* 5, 135–168. DOI: 10.1007/BF00055827

Glazer, I., Orion, D. and Apelbaum, A. (1983) Interrelationships between ethylene production, gall formation, and root-knot nematode development in tomato plants infected with *Meloidogyne javanica*. *Journal of Nematology* 15, 539–544.

Gleason, C.A., Liu, Q.L. and Williamson, V.M. (2008) Silencing a candidate nematode effector gene corresponding to the tomato resistance gene *Mi-1* leads to acquisition of virulence. *Molecular Plant–Microbe Interactions* 21, 576–585. DOI: 10.1094/mpmi-21-5-0576

Gnanapragasam, N.C. and Mohotti, K.M. (2018) Nematode parasites of tea. In: Sikora, R., Coyne, D., Hallmann, J. and Timper, P. (eds) *Plant Parasitic Nematodes in Subtropical and Tropical Agriculture*, 3rd edn. CAB International, Wallingford, UK, pp. 584–616. DOI: 10.1079/9780851997278.0581

Gough, E.C., Owen, K.J., Zwart, R.S. and Thompson, J.P. (2020) A systematic review of the effects of arbuscular mycorrhizal fungi on root-lesion nematodes, *Pratylenchus* spp. *Frontiers in Plant Science* 11, 923. DOI: 10.3389/fpls.2020.00923

Gould, I.J., Quinton, J.N., Weigelt, A., De Deyn, G.B. and Bardgett, R.D. (2016) Plant diversity and root traits benefit physical properties key to soil function in grasslands. *Ecology Letters* 19, 1140–1149. DOI: 10.1111/ele.12652

Gourd, T.R., Schmitt, D.P. and Barker, K.R. (1993) Differential sensitivity of *Meloidogyne* spp. and *Heterodera glycines* to selected nematicides. *Journal of Nematology* (Supplement) 25, 746–751.

Goverse, A., Overmars, H., Engelbertink, J., Schots, A., Bakker, J. *et al.* (2000) Both induction and morphogenesis of cyst nematode feeding cells are mediated by auxin. *Molecular Plant–Microbe Interactions* 13, 1121–1129. DOI: 10.1094/MPMI.2000.13.10.1121

Grant, W. and Viney, M. (2011) The dauer phenomenon. In: Perry, R.N. and Wharton, D.A. (eds) *Molecular and Physiological Basis of Nematode Survival*. CAB International, Wallingford, UK, pp. 99–125.

Greco, N. and Di Vito, M. (2009) Population dynamics and damage levels. In: Perry, R.N., Moens, M. and Starr, J.L. (eds) *Root-knot Nematodes*. CAB International, Wallingford, UK, pp. 246–274.

Green, J., Vain, P., Fearnehough, M.T., Worland, B., Snape, J.W. and Atkinson, H.J. (2002) Analysis of the expression patterns of the *Arabidopsis thaliana* tubulin-1 and *Zea mays* ubiquitin-1 promoters in rice plants in association with nematode infection. *Physiological and Molecular Plant Pathology* 60, 197–205. DOI: 10.1006/pmpp.2002.0390

Green, J., Wang, D., Lilley, C.J., Urwin, P.E. and Atkinson, H.J. (2012) Transgenic potatoes for potato cyst nematode control can replace pesticide use without impact on soil quality. *PLoS One* 7, e30973. DOI: 10.1371/journal.pone.0030973

Grenier, E., Fournet, S., Petit, E. and Anthoine, G. (2010) A cyst nematode 'species factory' called the Andes. *Nematology* 12, 163–169. DOI: 10.1163/138855409X12573393054942

Grundler, F.M.W. and Hoffman, J. (2011) Water and nutrient transport in nematode feeding sites. In: Jones, J.T., Gheysen, G. and Fenoll, C. (eds) *Genomics and Molecular Genetics of Plant–Nematode Interactions*. Springer, Heidelberg, Germany, pp. 423–439.

Grundler, F.M.W., Sobczak, M. and Golinowski, W. (1998) Formation of wall openings in root cells of *Arabidopsis thaliana* following infection by the plant-parasitic nematode *Heterodera schachtii*. *European Journal of Plant Pathology* 104, 545–551. DOI: 10.1023/A:1008692022279

Grunewald, W., Karimi, M., Wieczorek, K., Van de Cappelle, E., Grundler, F. *et al.* (2008) A role for AtWRKY23 in feeding site establishment of plant-parasitic nematodes. *Plant Physiology* 148, 358–368. DOI: 10.1104/pp.108.119131

Grunewald, W., Cannoot, B., Friml, J. and Gheysen, G. (2009) Parasitic nematodes modulate PIN-mediated auxin transport to facilitate infection. *PLoS Pathogens* 5, e1000266. DOI: 10.1371/journal.ppat.1000266

Grunewald, W., De Smet, I., Lewis, D.R., Löfke, C., Jansen, L. *et al.* (2012) Transcription factor WRKY23 assists auxin distribution patterns during *Arabidopsis* root development through local control on flavonol biosynthesis. *Proceedings of the National Academy of Sciences of the USA* 109, 1554–1559. DOI: 10.1073/pnas.1121134109

Guo, F., Castillo, P., Li, C., Qing, X. and Li, H. (2022) Description of *Rotylenchus zhongshanensis* sp. nov. (Tylenchomorpha: Hoplolaimidae) and discovery of its endosymbiont *Cardinium*. *Journal of Helminthology* 96, e48. DOI: 10.1017/S0022149X22000384

Guo, X., Wang, J., Gardner, M., Fukuda, H., Kondo, Y. *et al.* (2017) Identification of cyst nematode B-type CLE peptides and modulation of the vascular stem cell pathway for feeding cell formation. *PLoS Pathogens* 13, e1006142. DOI: 10.1371/journal.ppat.1006142

Guo, Y., Ni, J., Denver, R., Wang, X. and Clark, S.E. (2011) Mechanisms of molecular mimicry of plant CLE peptide ligands by the parasitic nematode *Globodera rostochiensis*. *Plant Physiology* 157, 476–484. DOI: 10.1104/pp.111.180554

Gutbrod, P., Gutbrod, K., Nauen, R., Elashry, A., Siddique, S. *et al.* (2020) Inhibition of acetyl-CoA carboxylase by spirotetramat causes growth arrest and lipid depletion in nematodes. *Scientific Reports* 10, 12710. DOI: 10.1038/s41598-020-69624-5

Gutiérrez-Gutiérrez, C., Cantalapiedra-Navarrete, C., Montes-Borrego, M., Palomares-Rius, J.E. and Castillo, P. (2013) Molecular phylogeny of the nematode genus *Longidorus* (Nematoda: Longidoridae) with description of three new species. *Zoological Journal of the Linnean Society* 167, 473–500. DOI: 10.1111/zoj.12019

Gutiérrez-Gutiérrez, C., Mota, M., Castillo, P., Santos, M.T. and Palomares-Rius, J.E. (2017) Description and molecular phylogeny of one new and one known needle nematode of the genus *Paralongidorus* (Nematoda: Longidoridae) from grapevine in Portugal using integrative approach. *European Journal of Plant Pathology* 151, 155–172. DOI: 10.1007/s10658-017-1364-9

Habash, S.S., Radakovic, Z.S., Vankova, R., Siddique, S., Dobrev, P. *et al.* (2017) *Heterodera schachtii* Tyrosinase-like protein – a novel nematode effector modulating plant hormone homeostasis. *Scientific Reports* 7, 6874. DOI: 10.1038/s41598-017-07269-7

Haegeman, A., Vanholme, B., Jacob, J., Vandekerckhove, T.T.M., Claeys, M. *et al.* (2009a) An endosymbiotic bacterium in a plant-parasitic nematode: member of a new *Wolbachia* supergroup. *International Journal of Parasitology* 39, 1045–1054. DOI: 10.1016/j.ijpara.2009.01.006

Haegeman, A., Vanholme, B. and Gheysen, G. (2009b) Characterization of a putative endoxylanase in the migratory plant-parasitic nematode *Radopholus similis*. *Molecular Plant Pathology* 10, 389–401. DOI: 10.1111/j.1364-3703.2009.00539.x

Haegeman, A., Mantelin, S., Jones, J.T. and Gheysen, G. (2012) Functional roles of effectors of plant-parasitic nematodes. *Gene* 492, 19–31. DOI: 10.1016/j.gene.2011.10.040

Hahn, M.H., May De Mio, L.L., Kuhn, O.J. and Duarte, H.d.S.S. (2019) Nematophagous mushrooms can be an alternative to control *Meloidogyne javanica*. *Biological Control* 138, 104024. DOI: 10.1016/j.biocontrol.2019.104024

Halbrendt, J.M. (2021) A threat to stone fruit and grape production: tomato ringspot virus (ToRSV) transmission by *X. americanum* s.l. (*sensu lato*). In: Sikora R., Desaeger, J. and Molendijk, L.P.G. (eds) *Integrated Nematode Management: State-of-the-art and Visions for the Future*. CAB International, Wallingford, UK, pp. 207–214. DOI: 10.1079/9781789247541.0029

Hallmann, J. and Meressa, B.H. (2018) Nematode parasites of vegetables. In: Sikora, R., Coyne, D., Hallmann, J. and Timper, P. (eds) *Plant Parasitic Nematodes in Subtropical and Tropical Agriculture*, 3rd edn. CAB International, Wallingford, UK, pp. 346–410. DOI: 10.1079/9781786391247.0346

Hallmann, J. and Molendijk, L.P.G. (2021) Face to face: how *Paratylenchus bukowinensis* deals with vegetables. In: Sikora R., Desaeger, J. and Molendijk, L.P.G. (eds) *Integrated Nematode Management: State-of-the-art and Visions for the Future*. CAB International, Wallingford, UK, pp. 310–315. DOI: 10.1079/9781789247541.0043

Hamamouch, N., Li, C.Y., Seo, P.J., Park, C.M. and Davis, E.L. (2011) Expression of Arabidopsis pathogenesis-related genes during nematode infection. *Molecular Plant Pathology* 12, 355–364. DOI: 10.1111/j.1364-3703.2010.00675.x

Hammond-Kosack, K.E. and Jones, J.D. (1997) Plant disease resistance genes. *Annual Review of Plant Physiology and Plant Molecular Biology* 48, 575–607. DOI: 10.1146/annurev.arplant.48.1.575

Han, Z., Boas, S. and Schroeder, N.E. (2017) Serotonin regulates the feeding and reproductive behaviors of *Pratylenchus penetrans*. *Phytopathology* 107, 804–908. DOI: 10.1094/PHYTO-11-16-0397-R

Han, Z., Thapa, S., Reuter-Carlson, U., Reed, H., Gates, M. *et al.* (2018) Immobility in the sedentary plant-parasitic nematode *H. glycines* is associated with remodeling of neuromuscular tissue. *PLoS Pathogens* 14, e1007198. DOI: 10.1371/journal.ppat.1007198

Handoo, Z.A., Carta, L.K., Skantar, A.M. and Chitwood, D.J. (2012) Description of *Globodera ellingtonae* n. sp. (Nematoda: Heteroderidae) from Oregon. *Journal of Nematology* 44, 40–57.

Hannich J.T., Entchev, E.V., Mende, F., Boytchev, H., Martin, R. *et al.* (2009) Methylation of the sterol nucleus by STRM-1 regulates dauer larva formation in *Caenorhabditis elegans*. *Developmental Cell* 16, 833–843. DOI: 10.1016/j.devcel.2009.04.012

Hanounik, S.B. and Osborne, W.W. (1975) Influence of *Meloidogyne incognita* on the content of amino acids and nicotine in tobacco grown under gnotobiotic conditions. *Journal of Nematology* 7, 332–336.

Hartman, K.M. and Sasser, J.N. (1985) Identification of *Meloidogyne* species on the basis of differential host test and perineal pattern morphology. In: Barker, K.R., Carter, C.C. and Sasser, J.N. (eds) *An Advanced Treatise on* Meloidogyne, *Volume II: Methodology*. North Carolina State University Graphics, Raleigh, NC, USA, pp. 69–77.

Harunobu, A. (2006) Clarifying the real bioactive constituents of garlic. *The Journal of Nutrition* 136, 7165–7255. DOI: 10.1093/jn/136.3.716S

Haydock, P.P.J. and Perry, J.N. (1998) The principles and practice of sampling for the detection of potato cyst nematodes. In: Marks, R.J. and Brodie, B.B. (eds) *Potato Cyst Nematodes: Biology, Distribution and Control*. CAB International, Wallingford, UK, pp. 61–74.

Hbirkou, C., Welp, G., Rehbein, K., Hillnhütter, C., Daub, M. *et al.* (2011) The effect of soil heterogeneity on the spatial distribution of *Heterodera schachtii* within sugar beet fields. *Applied Soil Ecology* 51, 25–34. DOI: 10.1016/j.apsoil.2011.08.008

Hebert, P.D.N., Cywinska, A., Ball, S.L. and deWaard, J.R. (2003) Biological identifications through DNA barcodes. *Proceedings of the Royal Society of London B* 270, 313–321. DOI: 10.1098/rspb.2002.2218

Hedgecock, E.M. and Russell, R.L. (1975) Normal and mutant thermotaxis in the nematode *Caenorhabditis elegans*. *Proceedings of the National Academy of Sciences of the USA* 72, 4061–4065. DOI: 10.1073/pnas.72.10.4061

Helder, J. and Heuer, H. (2021) Let's be inclusive – the time of looking at individual plant parasitic nematodes is over, and new technologies allow for it. In: Sikora, R.A., Desaeger, J. and Molendijk, L.P.G. (eds) *Integrated Nematode Management: State-of-the-art and Visions for the Future*. CAB International, Wallingford, UK, pp. 403–407. DOI: 10.1079/9781789247541.0056

Hepher, A. and Atkinson, H.J. (1992) *Nematode control with protease inhibitors*. European Patent Publication Number 0 502 730 A1.

Hesse, C.N., Moreno, I., Acevedo Pardo O., Pacheco Fuentes, H., Grenier, E. *et al.* (2021) Characterization of *Globodera ellingtonae* populations from Chile utilizing whole genome sequencing. *Journal of Nematology* 53, e2021-88. DOI: 10.21307/jofnem-2021-088

Hewezi, T. (2020) Epigenetic mechanisms in nematode–plant interactions. *Annual Review of Phytopathology* 58 1, 119–138. DOI: 10.1146/annurev-phyto-010820-012805

Hewezi, T., Howe, P., Maier, T.R., Hussey, R.S., Mitchum, M.G. *et al.* (2008) Cellulose binding protein from the parasitic nematode *Heterodera schachtii* interacts with *Arabidopsis* pectin methylesterase: cooperative cell wall modification during parasitism. *Plant Cell* 20, 3080–3093. DOI: 10.1105/tpc.108.063065

Hewezi, T., Howe, P.J., Maier, T.R., Hussey, R.S., Mitchum, M.G. *et al.* (2010) Arabidopsis spermidine synthase is targeted by an effector protein of the cyst nematode *Heterodera schachtii*. *Plant Physiology* 152, 968–984. DOI: 10.1104/pp.109.150557

Hickerson, M.J., Carstens, B.C., Cavender-Bares, J., Crandall, K.A., Graham, C.H. *et al.* (2010) Phylogeography's past, present, and future: 10 years after Avise, 2000. *Molecular Phylogenetics and Evolution* 54, 291–301. DOI: 10.1016/j.ympev.2009.09.016

Hilliard, M.A., Bargmann, C. and Bazzicalupo, P. (2002) *C. elegans* responds to chemical repellents by integrating sensory inputs from the head and the tail. *Current Biology* 12, 730–734. DOI: 10.1016/s0960-9822(02)00813-8

Hilliard, M.A., Bergamasco, C., Arbucci, S., Plasterk, R.H.A. and Bazzicalupo, P. (2004) Worms taste bitter: ASH neurons, QUI-1, GPA-3 and ODR-3 mediate quinine avoidance in *Caenorhabditis elegans*. *The EMBO Journal* 23, 1101–1111. DOI: 10.1038/sj.emboj.7600107

Hillocks, R.J. (2012) Farming with fewer pesticides: EU pesticide review and resulting challenges for UK agriculture. *Crop Protection* 31, 85–93. DOI: 10.1016/j.cropro.2011.08.008

Hills, T., Brockie, P.J. and Maricq, A.V. (2004) Dopamine and glutamate control area-restricted search behavior in *Caenorhabditis elegans*. *Journal of Neuroscience* 24, 1217–1225. DOI: 10.1523/JNEUROSCI.1569-03.2004

Hockland, S. (2005) Role and development of invertebrate collections in quality of identification for quarantine pests. *EPPO Bulletin* 35, 165–169. DOI: 10.1111/j.1365-2338.2005.00796.x

Hockland, S., Inserra, R.N., Miller, L. and Lehman, P.S. (2006) International plant health – putting legislation into practice. In: Perry, R.N. and Moens, M. (eds) *Plant Nematology*, 1st edn. CAB International, Wallingford, UK, pp. 327–345.

Hockland, S., Niere, B., Grenier, E., Blok, V., Phillips, M. *et al.* (2012) An evaluation of the implications of virulence in non-European populations of *Globodera pallida* and *G. rostochiensis* for potato cultivation in Europe. *Nematology* 14, 1–13. DOI: 10.1163/138855411X587112

Holdago, R., Magnusson, C., Hammeraas, B., Rasmussen, I., Strandendes, K. *et al.* (2018) Potato cyst nematodes *Globodera* spp. occurrence, biology and management in Norway. *Nematropica 48, Proceedings of the 50th Annual Meeting of the Organisation of Nematologists of Tropical America*, Abstract 17.

Holden-Dye, L. and Walker, R.J. (2011) Neurobiology of plant parasitic nematodes. *Invertebrate Neuroscience* 11, 9–19. DOI: 10.1007/s10158-011-0117-2

Holterman, M., van der Wurff, A., van den Elsen, S., van Megen, H., Bongers, T. *et al.* (2006) Phylum-wide analysis of SSU rDNA reveals deep phylogenetic relationships among nematodes and accelerated evolution toward crown clades. *Molecular Biology and Evolution* 23, 1792–1800. DOI: 10.1093/molbev/msl044

Holterman, M., Schratzberger, M. and Helder, J. (2019) Nematodes as evolutionary commuters between marine, freshwater and terrestrial habitats. *Biological Journal of the Linnean Society* 128, 756–767. DOI: 10.1093/biolinnean/blz107

Hong, R.L., Riebesell, M., Bumbarger, D.J., Cook, S.J., Carstensen, H.R. *et al.* (2019) Evolution of neuronal anatomy and circuitry in two highly divergent nematode species. *eLife* 8. DOI: 10.7554/eLife.47155

Hooper, D.J. (1961) A redescription of *Longidorus elongatus* (de Man, 1876) Thorne & Swanger, 1936, (Nematoda, Dorylaimidae) and descriptions of five new species of *Longidorus* from Great Britain. *Nematologica* 6, 237–257. DOI: 10.1163/187529261X00072

Hooper, D.J. (1986) Drawing and measuring nematodes. In: Southey, J.F. (ed.) *Laboratory Methods for Work with Plant and Soil Nematodes*, Reference Book 402. HMSO, London, UK, pp. 87–94.

Hope, I.A. (2002) Embryology, developmental biology and the genome. In: Lee, D.L. (ed.) *The Biology of Nematodes*. Taylor & Francis, London, UK, pp. 121–145.

Hsue, Y.P., Mahanti, P., Schroeder, F.C. and Sternberg, P.W. (2013) Nematode-trapping fungi eavesdrop on nematode pheromones. *Current Biology* 23, 83–86. DOI: 10.1016/j.cub.2012.11.035

Hu, C., Kearn, J., Urwin, P., Lilley, C., O'Connor, V. *et al.* (2014) StyletChip: a microfluidic device for recording host invasion behaviour and feeding of plant parasitic nematodes. *Lab on a Chip* 14, 2447–2455. DOI: 10.1039/c4lc00292j

Huang, M. and Chalfie, M. (1994) Gene interactions affecting mechanosensory transduction in *Caenorhabditis elegans*. *Nature* 367, 467–470. DOI: 10.1038/367467a0

Huang, D., Yan, G., Gudmestad, N., Ye, W., Whitworth, J. *et al.* (2019) Developing a one-step multiplex PCR Assay for rapid detection of four stubby-root nematode species, *Paratrichodorus allius*, *P. minor*, *P. porosus*, and *Trichodorus obtusus*. *Plant Disease* 103, 404–410. DOI: 10.1094/PDIS-06-18-0983-RE

Huang, G.Z., Allen, R., Davis, E.L., Baum, T.J. and Hussey, R.S. (2006) Engineering broad root-knot resistance in transgenic plants by RNAi silencing of a conserved and essential root-knot nematode parasitism gene. *Proceedings of the National Academy of Sciences of the USA* 103, 14302–14306. DOI: 10.1073/pnas.0604698103

Huang, X., Xu, C.-L., Yang, S-H., Li, J.-Y., Wang, H.-L. *et al.* (2019) Life-stage specific transcriptomes of a migratory endoparasitic plant nematode, *Radopholus similis* elucidate a different parasitic and life strategy of plant parasitic nematodes. *Scientific Reports* 9, 6277. DOI: 10.1038/s41598-019-42724-7

Huettel, R.N., Dickson, D.W. and Kaplan, D.T. (1984) *Radopholus citrophilus* n.sp., a sibling species of *Radopholus similis*. *Proceedings of the Helminthological Society of Washington* 51, 32–35. DOI: 10.1111/j.1365-2338.2008.01248.x

Humphreys-Pereira, D.A. and Elling, A.A. (2013) Intraspecific variability and genetic structure in *Meloidogyne chitwoodi* from the USA. *Nematology* 15, 375–386. DOI: 10.1163/15685411-00002684

Hunt, D.J. (1993) *Aphelenchida, Longidoridae and Trichodoridae: Their Systematics and Bionomics*. CAB International, Wallingford, UK.

Hunt, D.J. (2008) A checklist of the Aphelenchoidea (Nematoda: Tylenchina). *Journal of Nematode Morphology and Systematics* 10, 99–135.

Hunt, D., Bert, W. and Siddiqi, M.R. (2012) Tylenchidae and Dolichodoridae. In: Manzanilla-López, R.H. and Marbán-Mendoza, N. (eds) *Practical Plant Nematology*. Bibliotheca Basica de Agricultura, Montecillo, Mexico, pp. 209–250.

Hutchinson, G.E. (1957) Concluding remarks. Population studies: animal ecology and demography. *Cold Spring Harbor Symposia on Quantitative Biology* 22, 415–427. DOI: 10.1101/SQB.1957.022.01.039

Ichinohe, M. (1952) On the soybean nematode, *Heterodera glycines* n. sp., from Japan. *Magazine of Applied Zoology* 17, 1–4.

Ichinohe, M. (1988) Current research on the major nematode problems in Japan. *Journal of Nematology* 20, 184–190.

Ichiishi, K., Ekino, T., Kanzaki, N. and Shinya, R. (2022) Thick cuticles as an anti predator defence in nematodes. *Nematology* 24, 11–20. DOI: 10.1163/15685411-bja10107

Iliff, A.J., Wang, C., Ronan, E.A., Hake, A.E., Guo, Y. *et al.* (2021) The nematode *C. elegans* senses airborne sound. *Neuron* 109, 3633–3646.e7. DOI: 10.1016/j.neuron.2021.08.035

Ingham, R.E., Trofymow, J.A., Ingham, E.R. and Coleman, D.C. (1985) Interactions of bacteria, fungi, and their nematode grazers: effects on nutrient cycling and plant growth. *Ecological Monographs* 55, 119–140. DOI: 10.2307/1942528

Inserra, R.N., Ozores-Hampton, M., Schubert, T.S., Stanley, J.D., Brodie, M.W. *et al.* (2006) Guidelines for compost sanitation. *Proceedings of the Soil and Crop Science Society of Florida* 65, 31–37.

Iordache, M.-D., Mantas, V., Baltazar, E., Pauly, K. and Lewyckyj, N. (2020) A machine learning approach to detecting pine wilt disease using airborne spectral imagery. *Remote Sensing* 12, 2280. DOI: 10.3390/rs12142280

IPCC (2019) *Climate Change and Land: an IPCC special report on climate change, desertification, land degradation, sustainable land management, food security, and greenhouse gas fluxes in terrestrial ecosystems*. Intergovernmental Panel on Climate Change. UN Environment Programme, Nairobi, Kenya.

IPCC (2022) *Climate Change 2022: Impacts, Adaptation and Vulnerability*. Contribution of Working Group II to the Sixth Assessment Report of the Intergovernmental Panel on Climate. Cambridge University Press, Cambridge, UK. DOI: 10.1017/9781009325844

ISAAA (International Service for the Acquisition of Agri-biotech Applications) (2018) Global status of commercialized biotech/GM crops in 2018: biotech crops continue to help meet the challenges of increased population and climate change. *ISAAA Brief No. 54*. ISAAA, Ithaca, NY, USA.

ISAAA (International Service for the Acquisition of Agri-biotech Applications) (2023) GM Events with Nematode Resistance. Available at: https://www.isaaa.org/gmapprovaldatabase/gmtrait/default.asp?TraitID=47&GMTrait=Nematode%20Resistance (accessed 1 February 2024).

Ito, T., Araki, M. and Komatsuzaki, M. (2015) No-tillage cultivation reduces rice cyst nematode (*Heterodera elachista*) in continuous upland rice (*Oryza sativa*) culture and after conversion to soybean (*Glycine max*) in Kanto, Japan. *Field Crops Research* 179, 44–51. DOI: 10.1016/j.fcr.2015.04.008

Jacob, J.E., Vanholme, B., Van Leeuwen, T. and Gheysen, G. (2009) A unique genetic code change in the mitochondrial genome of the parasitic nematode *Radopholus similis*. *BMC Research Notes* 2, 192. DOI: 10.1186/1756-0500-2-192

Jaffe, H., Huettel, R.N., Demilo, A.B., Hayes, D.K. and Rebois, R.V. (1989) Isolation and identification of a compound from soybean cyst nematode, *Heterodera glycines*, with sex pheromone activity. *Journal of Chemical Ecology* 15, 2031–2043. DOI: 10.1007/BF01207435

Jaffee, B., Phillips, R., Muldoon, A. and Mangel, M. (1992) Density-dependent host-pathogen dynamics in soil microcosms. *Ecology* 73, 495–506. DOI: 10.2307/1940755

Jammes, F., Lecomte, P., de Almeida-Engler, J., Bitton, F., Martin-Magninette, M.-L. *et al.* (2005) Genome-wide expression profiling of the host response to root-knot nematode infection in *Arabidopsis*. *Plant Journal* 44, 447–458. DOI: 10.1111/j.1365-313X.2005.02532.x

Janssen, T., Karssen, G., Verhaeven, M., Coyne, D. and Bert, W. (2016) Mitochondrial coding genome analysis of tropical root-knot nematodes (*Meloidogyne*) supports haplotype based diagnostics and reveals evidence of recent reticulate evolution. *Scientific Reports* 6, 22591. DOI: 10.1038/srep22591

Janssen, T., Karssen, G., Couvreur, M., Waeyenberge, L. and Bert, W. (2017a) The pitfalls of molecular species identification: a case study within the genus *Pratylenchus* (Nematoda: Pratylenchidae). *Nematology* 19, 1179–1199. DOI: 10.1163/15685411-00003117

Janssen, T., Karssen, G., Topalović, O., Coyne, D. and Bert, W. (2017b) Integrative taxonomy of root-knot nematodes reveals multiple independent origins of mitotic parthenogenesis. *PLoS One* 12, e0172190. DOI: 10.1371/journal.pone.0172190

Jaouannet, M., Magliano, M., Arguel, M.J., Gourges, M., Evangelisti, E. *et al.* (2013) The root-knot nematode calreticulin Mi-CRT is a key effector in plant defense suppression. *Molecular Plant–Microbe Interactions* 26, 97–105. DOI: 10.1094/mpmi-05-12-0130-r

Jarne, P. and Lagoda, P.J.L. (1996) Microsatellites, from molecules to populations and back. *Trends in Ecology and Evolution* 11, 424–429. DOI: 10.1016/0169-5347(96)10049-5

Jeger, M., Bragard, C., Caffier, D., Candresse, T., Chatzivassiliou, E. *et al.* (2018) Pest categorisation of *Xiphinema americanum* sensu lato. *EFSA Journal* 16, 5298. DOI: 10.2903/j.efsa.2018.5298

Jeong, P.-Y., Kwon, M.-S., Joo, H.-J. and Paik, Y.-K. (2009) Molecular time-course and the metabolic basis of entry into dauer in *Caenorhabditis elegans*. *PLoS One* 4, e4162. DOI: 10.1371/journal.pone.0004162

Jepson, S.B. (1987) *Identification of Root-knot Nematodes* (Meloidogyne *species*). CAB International, Wallingford, UK.

Ji, H.L., Gheysen, G., Denil, S., Lindsey, K., Topping, J.F. *et al.* (2013) Transcriptional analysis through RNA sequencing of giant cells induced by *Meloidogyne graminicola* in rice roots. *Journal of Experimental Botany* 64, 3885–3898. DOI: 10.1093/jxb/ert219

Johnson, P.W., Van Gundy, S.D. and Thomson, W.W. (1970) Cuticle formation in *Hemicycliophora arenaria, Aphelenchus avenae* and *Hirschmanniella gracilis*. *Journal of Nematology* 2, 59–79.

Johnston, W.L. and Dennis, J.W. (2012) The eggshell in the *C. elegans* oocyte-to-embryo transition. *Genesis* 50, 333–349. DOI: 10.1002/dvg.20823

Johnston, M.J., McVeigh, P., McMaster, S., Fleming, C.C. and Maule, A.G. (2010a) FMRFamide-like peptides in root knot nematodes and their potential role in nematode physiology. *Journal of Helminthology* 84, 253–265. DOI: 10.1017/S0022149X09990630

Johnston, W.L., Krizus, A. and Dennis, J.W. (2010b) Eggshell chitin and chitin-interacting proteins prevent polyspermy in *C. elegans*. *Current Biology* 20, 1932–1937. DOI: 10.1016/j.cub.2010.09.059

Jones, A.T., Brown, D.J.F., McGavin, W.J., Rudel, M. and Altmayer, B. (1994) Properties of an unusual isolate of raspberry ringspot virus from grapevine in Germany and evidence for its possible transmission by *Paralongidorus maximus*. *Annals of Applied Biology* 124, 283–300. DOI: 10.1111/j.1744-7348.1994.tb04134.x

Jones, J.D.G. and Dangl, J.L. (2006) The plant immune system. *Nature* 444, 323–329. DOI: 10.1038/nature05286

Jones, J.T. (2002) Nematode sense organs. In: Lee, D.L. (ed.) *The Biology of Nematodes*. Taylor & Francis, London, UK, pp. 353–368.

Jones, J.T., Perry, R.N. and Johnston, M.R.L. (1993) Changes in the ultrastructure of the cuticle of the potato cyst nematode, *Globodera rostochiensis*, during development and infection. *Fundamental and Applied Nematology* 19, 433–445.

Jones, J.T., Haegeman, A., Danchin, E.G.J., Gaur, H.S., Helder, J. *et al.* (2013) Top 10 plant-parasitic nematodes in molecular plant pathology. *Molecular Plant Pathology* 14, 946–961. DOI: 10.1111/mpp.12057

Jones, L.M., Koehler, A.K., Trnka, M., Balek, J., Challinor, A.J. *et al.* (2017) Climate change is predicted to alter the current pest status of *Globodera pallida* and *G. rostochiensis* in the United Kingdom. *Global Change Biology* 23, 4497–4507. DOI: 10.1111/gcb.13676

Jones, M.G.K. and Northcote, D.H. (1972) Nematode-induced syncytium – a multinucleate transfer cell. *Journal of Cell Science* 10, 789–809. DOI: 10.1242/jcs.10.3.789

Jones, M.G.K. and Payne, H.L. (1978) Early stages of nematode-induced giant cell formation in roots of *Impatiens balsamina*. *Journal of Nematology* 10, 70–84.

Jones, P.W., Tylka, G.L. and Perry, R.N. (1998) Hatching. In: Perry, R.N. and Wright, D.J. (eds) *The Physiology and Biochemistry of Free-living and Plant-parasitic Nematodes*. CAB International, Wallingford, UK, pp. 181–212.

Jorgensen, E.M. (ed.) (2023) Neurobiology and behaviour. The *C. elegans* Research Community (ed.) *WormBook*. Available at: www.wormbook.org/toc_neurobiobehavior.html (accessed 1 February 2024).

Jubic, L.M., Saile, S., Furzer, O.J., El Kasmi, F. and Dangl, J.L. (2019) Help wanted: helper NLRs and plant immune responses. *Current Opinion in Plant Biology* 50, 82–94. DOI: 10.1016/j.pbi.2019.03.013

Jupe, F., Witek, K., Verweij, W., Śliwka, J., Pritchard, L. *et al.* (2013) Resistance gene enrichment sequencing (RenSeq) enables reannotation of the NB-LRR gene family from sequenced plant genomes and rapid mapping of resistance loci in segregating populations. *Plant Journal* 76, 530–544. DOI: 10.1111/tpj.12307

Kahn, T.W., Duck, N.B., McCarville, M.T., Schouten, L.C., Schweri, K. *et al.* (2021) A *Bacillus thuringiensis* Cry protein controls soybean cyst nematode in transgenic soybean plants. *Nature Communications* 12, 3380. DOI: 10.1038/s41467-021-23743-3

Kaloshian, I., Williamson, V.M., Miyao, G., Lawn, D.A. and Westerdahl, B.B. (1996) 'Resistance-breaking' nematodes identified in California tomatoes. *California Agriculture* 50, 18–19.

Kandoth, P. and Mitchum, M.G. (2013) War of the worms: how plants fight underground attacks. *Current Opinion in Plant Biology* 16, 457–463. DOI: 10.1016/j.pbi.2013.07.001

Kanwar, R.S., Patil, J. and Yadav, S. (2021) Prospects of using predatory nematodes in biological control for plant parasitic nematodes – a review. *Biological Control* 160, 104668. DOI: 10.1016/j.biocontrol.2021.104668

Kanzaki, N. and Giblin-Davis, R.M. (2012) Aphelenchoidea. In: Manzanilla-López, R.H. and Marbán-Mendoza, N. (eds) *Practical Plant Nematology*. Bibliotheca Basica de Agricultura, Montecillo, Mexico, pp. 161–208.

Kaplan, J.M. and Horvitz, H.R. (1993) A dual mechanosensory and chemosensory neuron in *Caenorhabditis elegans*. *Proceedings of the National Academy of Sciences of the USA* 90, 2227–2231. DOI: 10.1073/pnas.90.6.2227

Kaplan, D.T., Vanderspool, M.C. and Opperman, C.H. (1997) Sequence tag site and host range assays demonstrate that *Radopholus similis* and *R. citrophilus* are not reproductively isolated. *Journal of Nematology* 29, 421–429.

Karanastasi, E. and Brown, D.J.F. (2004) Interspecific variation in the site of Tobravirus particle retention in selected virus-vector *Paratrichodorus* and *Trichodorus* species (Nematoda: Triplonchida). *Nematology* 6, 261–272. DOI: 10.1163/1568541041218022

Karanastasi, E., Decraemer, W., Zheng, J.W., de Almeida, M.T.M. and Brown, D.J.T. (2001) Interspecific differences in the fine structure of the body cuticle of Trichodoridae Thorne, 1935 (Nematoda: Diphtherophorina) and review of anchoring structure of the epidermis. *Nematology* 3, 525–533. DOI: 10.1163/156854101753389130

Karczmarek, A., Overmars, H., Helder, J. and Goverse, A. (2004) Feeding cell development by cyst and root-knot nematodes involves a similar early, local and transient activation of a specific auxin-inducible promoter element. *Molecular Plant Pathology* 5, 343–346. DOI: 10.1111/j.1364-3703.2004.00230.x

Karssen, G. (2002) *The Plant-Parasitic Nematode Genus* Meloidogyne *Göldi, 1892 (Tylenchida) in Europe*. Brill, Leiden, The Netherlands.

Karssen, G., Van Hoenselaar, T., Verkerk-Bakker, B. and Janssen, R. (1995) Species identification of cyst and root-knot nematodes from potato by electrophoresis of individual females. *Electrophoresis* 16, 105–109. DOI: 10.1002/elps.1150160119

Karssen, G., Liao, J., Kan, Z., van Heese, E.Y.V. and den Nijs, L.J.M.F. (2012) On the species status of the root-knot nematode *Meloidogyne mayaguensis* Rammah & Hirschmann, 1988. *ZooKeys* 181, 67–77. DOI: 10.3897/zookeys.181.2787

Kearn, J., Lilley, C., Urwin, P., O'Connor, V. and Holden-Dye, L. (2017) Progressive metabolic impairment underlies the novel nematicidal action of fluensulfone on the potato cyst nematode *Globodera pallida*. *Pesticide Biochemistry and Physiology* 142, 83–90. DOI: 10.1016/j.pestbp.2017.01.009

Kern, E.M.A., Kim, T. and Park, J.-K. (2020) The mitochondrial genome in nematode phylogenetics. *Frontiers in Ecology and Evolution* 8, 250. DOI: 10.3389/fevo.2020.00250

Kerry, B.R. and Hirsch, P.R. (2011) Ecology of *Pochonia chlamydosporia* in the rhizosphere at the population, whole organism and molecular scales. In: Davies, K.G. and Spiegel, Y. (eds) *Biological Control of Plant-parasitic Nematodes: Building Coherence Between Microbial Ecology and Molecular Mechanisms*. Springer, Berlin, Germany, pp. 171–182.

Kerry, B.R. and Hominick, W.N. (2002) Biological control. In: Lee, D.L. (ed.) *The Biology of Nematodes*. Taylor & Francis, London, UK, pp. 483–510.

Kerry, B.R., Crump, D.H. and Mullen, L.A. (1982) Studies of the cereal cyst nematode *Heterodera avenae* under continuous cereals, 1975–1978. II. Fungal parasitism of nematode females and eggs. *Annals of Applied Biology* 100, 489–499. DOI: 10.1111/j.1744-7348.1982.tb01415.x

Khan, R.P. (1982) The host as a vector: exclusion as control. In: Harris, K.F. and Marmorsch, K. (eds) *Pathogens, Vectors and Plant Diseases: Approaches to Control*. Academic Press, New York, USA, pp. 123–149.

Khan, R.P. (1989) *Plant Protection and Quarantine,* Volumes 1–3. CRC Press, Boca Raton, FL, USA.

Khan, M.W. (1993) Mechanisms of interactions between nematodes and other plant pathogens. In: Khan, M.W. (ed.) *Nematode Interactions*. Springer, Dordrecht, The Netherlands, pp. 55–78.

Khanam, S., Bauters, L., Singh, R.R., Verbeek, R., Haeck, A. *et al.* (2018) Mechanisms of resistance in the rice cultivar Manikpukha to the rice stem nematode *Ditylenchus angustus*. *Molecular Plant Pathology* 19, 1391–1402. DOI: 10.1111/mpp.12622

Kidane, S.A., Meressa, B.H., Haukeland, S., Hvoslef-Eide, A.K.(T.) and Coyne, D.L. (2021) The Ethiopian staple food crop enset (*Ensete ventricosum*) assessed for the first time for resistance against the root-lesion nematode *Pratylenchus goodeyi*. *Nematology* 23, 771–779. DOI: 10.1163/15685411-bja10075

Kiewnick, S., Karssen, G., Brito, J.A., Oggenfuss, M. and Frey, J.E. (2008) First report of root-knot nematode *Meloidogyne enterolobii* on tomato and cucumber in Switzerland. *Plant Disease* 92, 1370. DOI: 10.1094/PDIS-92-9-1370A

Kiewnick, S., Dessimoz, M. and Franck, L. (2009) Effects of the *Mi-1* and the *N* root-knot nematode-resistance gene on infection and reproduction of *Meloidogyne enterolobii* on tomato and pepper cultivars. *Journal of Nematology* 41, 134–139.

Kiggundu, A., Muchwezi, J., Van der Vyver, C., Viljoen, A., Vorster, J. *et al.* (2010) Deleterious effects of plant cystatins against the banana weevil *Cosmopolites sordidus*. *Archives of Insect Biochemistry and Physiology* 73, 87–105. DOI: 10.1002/arch.20342

Kikuchi, T., Aikawa, T., Kosaka, H., Pritchard, L., Ogura, N. and Jones, J.T. (2007) Expressed sequence tag (EST) analysis of the pine wood nematodes *Bursaphenelchus xylophilus* and *B. mucronatus*. *Molecular and Biochemical Parasitology* 155, 9–17. DOI: 10.1016/j.molbiopara.2007.05.002

Kikuchi, T., Aikawa, T., Oeda, Y., Karim, N. and Kanzaki, N. (2009) A rapid and precise diagnostic method for detecting the pinewood nematode *Bursaphelenchus xylophilus* by loop-mediated isothermal amplification. *Phytopathology* 99, 1365–1369. DOI: 10.1094/phyto-99-12-1365

Kikuchi, T., Cotton, J.A., Dalzell, J.J., Hasegawa, K., Kanzaki, N. *et al.* (2011) Genomic insights into the origin of parasitism in the emerging plant pathogen *Bursaphelenchus xylophilus*. *PLoS Pathogens* 7, e1002219. DOI: 10.1371/journal.ppat.1002219

Kikuchi, T., Eves-van den Akker, S. and Jones, J.T. (2017) Genome evolution of plant-parasitic nematodes. *Annual Review of Phytopathology* 55, 333–354. DOI: 10.1146/annurev-phyto-080516-035434

Kim, T., Lee, Y., Kil, H.-J. and Park, J.-K. (2020) The mitochondrial genome of *Acrobeloides varius* (Cephalobomorpha) confirms non-monophyly of Tylenchina (Nematoda). *PeerJ* 8, e9108. DOI: 10.7717/peerj.9108/table-2

Klassen, W., Brodel, C.F. and Fieselmann, D.A. (2002) Exotic pests of plants: current and future threats to horticultural production and trade in Florida and the Caribbean Basin. *Micronesica Supplement* 6, 5–27.

Klink, V.P., Kim, K.-H., Martins, V., MacDonald, M.H., Beard, H.S. *et al.* (2009) A correlation between host-mediated expression of parasite genes as tandem inverted repeats and abrogation of development of female *Heterodera glycines* cyst formation during infection of *Glycine max*. *Planta* 230, 53–71. DOI: 10.1007/s00425-009-0926-2

Koffi, M.C., Vos, C., Draye, X. and Declerck, S. (2013) Effects of *Rhizophagus irregularis* MUCL 41833 on the reproduction of *Radopholus similis* in banana plantlets grown under in vitro culture conditions. *Mycorrhiza* 23, 279–288. DOI: 10.1007/s00572-012-0467-6

Kohl, L.M., Warfield, C.Y. and Benson, D.M. (2010) Population dynamics and dispersal of *Aphelenchoides fragariae* in nursery-grown lantana. *Journal of Nematology* 42, 332–341.

Kolombia, Y.A., Karssen, G., Viaene, N., Kumar, P.L., Joos, L. *et al.* (2017) Morphological and molecular characterisation of *Scutellonema* species from yam (Dioscorea spp.) and a key to the species of the genus. *Nematology* 19, 751–787. DOI: 10.1163/15685411-00003084

Kort, J., Ross, H., Rupenhorst, H.J. and Stone, A.R. (1977) An international scheme for identifying and classifying pathotypes of potato cyst nematodes *Globodera rostochiensis* and *G. pallida*. *Nematologica* 23, 333–339. DOI: 10.1163/187529277X00057

Kourelis, J., Marchal, C., Posbeyikian, A., Harant, A. and Kamoun, S. (2023) NLR immune receptor–nanobody fusions confer plant disease resistance. *Science* 379, 934–939. DOI: 10.1126/science.abn4116

Kranse, O.P., Ko, I., Healey, R., Sonawala, U., Wei, S. *et al.* (2022) A low-cost and open-source solution to automate imaging and analysis of cyst nematode infection assays for *Arabidopsis thaliana*. *Plant Methods* 18, 134. DOI; 10.1186/s13007-022-00963-2

Krieg, M., Pidde, A. and Das, R. (2022) Mechanosensitive body-brain interactions in *Caenorhabditis elegans*. *Current Opinions in Neurobiology* 75, 102574. DOI: 10.1016/j.conb.2022.102574

Kudsk, P. Jørgensen, L.N. and Ørum, J.E. (2018) Pesticide Load – a new Danish pesticide risk indicator with multiple applications. *Land Use Policy* 40, 384–393. DOI: 10.1016/j.landusepol.2017.11.010

Kühn, J. (1877) Vorläufiger Bericht über die bisherigen Ergebnisse der seit dem Jahre 1875 im Auftrage des Vereins für Rübenzucker-Industrie ausgeführten Versuche zur Ermittlung der Ursache der Rübenmüdigkeit des Bodens und zur Erforschung der Natur der Nematoden. *Zeitschrift des Vereins für die Rübenzucker-Industrie im Zollverein* 27, 452–457.

Kumari, S., Decraemer, W., De Luca, F. and Tiefenbrunner, W. (2010) Cytochrome c oxidase subunit 1 analysis of *Xiphinema diversicaudatum, X. pachtaicum, X. simile* and *X. vuittenezi* (Nematoda, Dorylaimida). *European Journal of Plant Pathology* 127, 493–499. DOI: 10.1007/s10658-010-9614-0

Kyndt, T., Nahar, K., Haegeman, A., De Vleesschauwer, D., Höfte, M. and Gheysen, G. (2012) Comparing the systemic defence-related gene expression changes upon migratory and sedentary nematode attack in rice. *Plant Biology* 14, 73–82. DOI: 10.1111/j.1438-8677.2011.00524.x

Kyndt, T., Vieira, P., Gheysen, G. and de Almeida-Engler, J. (2013) Nematode feeding sites: unique organs in plant roots. *Planta* 238, 807–818. DOI: 10.1007/s00425-013-1923-z

Kyndt, T., Zemene, H.Y., Haeck, AA, Singh, R., De Vleesschauwer, D. *et al.* (2017). Below-ground attack by the root-knot nematode *Meloidogyne graminicola* predisposes rice to blast disease. *Molecular Plant–Microbe Interactactions* 30, 255–266. DOI: 10.1094/MPMI-11-16-0225-R

Lagos, S., Perruchon, C., Katsoula, A. and Karpouzas, D.G. (2019) Isolation and characterization of soil bacteria able to rapidly degrade the organophosphorus nematicide fosthiazate. *Letters in Applied Microbiology* 68, 149–155. DOI: 10.1111/lam.13098

Lahari, Z., Nkurunziza, R., Bauters, L. and Gheysen, L. (2020) Analysis of Asian rice (*Oryza sativa* L.) genotypes reveals a new source of resistance to the root-knot nematode *Meloidogyne javanica* and the root-lesion nematode *Pratylenchus zeae*. *Phytopathology* 110, 1572–1577. DOI: 10.1094/PHYTO-11-19-0433-R

Lahm, G., Desaeger, J., Smith, B.K., Pahutskia, T.F., River, M.A. *et al.* (2017) The discovery of fluazaindolizine: A new product for the control of plant parasitic nematodes. *Bioorganic & Medicinal Chemistry Letters* 27,1572–1575. DOI: 10.1016/j.bmcl.2017.02.029

Lainsbury, M.A. (2023) *The UK Pesticide Guide 2023 – 36th Edition*. British Crop Production Council, Cambridge, UK.

LaMacchia, J.C. and Roth, M.B. (2015) *Aquaporins-2* and *-4* regulate glycogen metabolism and survival during hypoosmotic-anoxic stress in *Caenorhabditis elegans*. *American Journal of Physiology, Cell Physiology* 309, C92–C96. DOI: 10.1152/ajpcell.00131.2015

Lamberti, F. and Ciancio, A. (1993) Diversity of *Xiphinema americanum*-group species and hierarchical cluster analysis of morphometrics. *Journal of Nematology* 25, 332–343.

Lamberti, F., Molinari, S., Moens, M. and Brown, D.J.F. (2000) The *Xiphinema americanum* group. I. Putative species, their geographical occurrence and distribution, and regional polytomous identification keys for the group. *Russian Journal of Nematology* 8, 65–84.

Lamberti, F., Hockland, S., Agostinelli, A., Moens, M. and Brown, D.J.F. (2004) The *Xiphinema americanum* group. III. Keys to species identification. *Nematologia Mediterranea* 32, 53–56.

Langeslag, M., Mugniéry, D. and Fayet, G. (1982) Développement embryonnaire de *Globodera rostochiensis* et *G. pallida* en fonction de la température, en conditions contrôlées et naturelles. *Revue de Nématologie* 5, 103–109.

Lee, C., Chronis, D., Kenning, C., Peret, B., Hewezi, T. *et al.* (2011) The novel cyst nematode effector protein 19C07 interacts with the Arabidopsis auxin influx transporter LAX3 to control feeding site development. *Plant Physiology* 155, 866–880. DOI: 10.1104/pp.110.167197

Lee, D.L. (2002) Behaviour. In: Lee, D.L. (ed.) *The Biology of Nematodes*. Taylor & Francis, London, UK, pp. 369–387.

Lee, T.G., Kumar, I., Diers, B.W. and Hudson, M. E. (2015) Evolution and selection of *Rhg1*, a copy-number variant nematode-resistance locus. *Molecular Ecology* 24, 1774–1791. DOI: 10.1111/mec.13138

Leelarasamee, N., Zhang, L. and Gleason, C. (2018) The root-knot nematode effector MiPFN3 disrupts plant actin filaments and promotes parasitism. *PLoS Pathogens* 14, e1006947. DOI: 10.1371/journal.ppat.1006947

Lehman, P.S. (2004) Cost-benefits of nematode management through regulatory programs. In: Chen, Z.X., Chen, S.Y. and Dickson, D.W. (eds) *Nematology Advances and Perspectives, Vol. 2: Nematode Management and Utilization*. CAB International, Wallingford, UK, pp. 1133–1177.

Lemieux, G.A., Cunningham, K.A., Lin, L., Mayer, F., Werb, Z. *et al.* (2015) Kynurenic acid is a nutritional cue that enables behavioral plasticity. *Cell* 160, 119–131. DOI: 10.1016/j.cell.2014.12.028

Le Saux, R. and Quénéhervé, P. (2002) Differential chemotactic responses of two plant-parasitic nematodes, *Meloidogyne incognita* and *Rotylenchulus reniformis*, to some inorganic ions. *Nematology* 4, 99–105. DOI: 10.1163/156854102760082258

Li, J., Todd, T.C. and Trick, H.N. (2010a) Rapid *in planta* evaluation of root expressed transgenes in chimeric soybean plants. *Plant Cell Reports* 29, 113–123. DOI: 10.1007/s00299-009-0803-2

Li, J., Todd, T.C., Oakley, T.R., Lee, J. and Trick, H.N. (2010b) Host-derived suppression of nematode reproductive and fitness genes decreases fecundity of *Heterodera glycines* Ichinohe. *Planta* 232, 775–785. DOI: 10.1007/s00425-010-1209-7

Li, T.-M., Chen, J., Li, X., Ding, X.J., Wu, Y. *et al.* (2013) Absolute quantification of a steroid hormone that regulates development in *Caenorhabditis elegans*. *Analytical Chemistry* 85, 9281–9287. DOI: 10.1021/ac402025c

Li, X., Zhu, H., Geisen, S., Bellard, C., Hu, F. *et al.* (2019) Agriculture erases climate constraints on soil nematode communities across large spatial scales. *Global Change Biology* 26, 919–930. DOI: 10.1111/gcb.14821

Li, X., Zhu, H., Geisen, S., Bellard, C., Hu, F. *et al.* (2020) Agriculture erases climate constraints on soil nematode communities across large spatial scales. *Global Change Biology* 26, 919–930. DOI: 10.1111/gcb.14821

Li, X., Liu, T, Li, H., Geisen, S., Hu, F. *et al.* (2022) Management effects on soil nematode abundance differ among functional groups and land-use types at a global scale. *Journal of Animal Ecology* 91, 1770–1780. DOI: 10.1111/1365-2656.13744

Li, X.-Q., Wei, J.-Z., Tan, A. and Aroian, R.V. (2007) Resistance to root-knot nematode in tomato roots expressing a nematicidal *Bacillus thuringiensis* crystal protein. *Plant Biotechnology Journal* 5, 455–464. DOI: 10.1111/j.1467-7652.2007.00257.x

Li, X.-Q., Tan, A., Voegtline, M., Bekele, S., Chen, C.-S. *et al.* (2008) Expression of Cry5B protein from *Bacillus thuringiensis* in plant roots confers resistance to root-knot nematode. *Biological Control* 47, 97–102. DOI: 10.1016/j.biocontrol.2008.06.007

Li, Y., Wang, K., Xie, H., Wang, D.W., Xu, C.L. *et al.* (2015) Cathepsin B cysteine proteinase is essential for the development and pathogenesis of the plant parasitic nematode *Radopholus similis*. *International Journal of Biological Science* 11, 1073–1087. DOI: 10.7150/ijbs.12065

Li, Y., Wang, K., Lu, Q., Du, J., Wang, Z. *et al.* (2017) Transgenic *Nicotiana benthamiana* plants expressing a hairpin RNAi construct of a nematode *Rs-cps* gene exhibit enhanced resistance to *Radopholus similis*. *Scientific Reports* 7(1):13126.

Lian, Y., Koch, G., Bo, D., Wang, J., Nguyen, H.T. *et al.* (2022) The spatial distribution and genetic diversity of the soybean cyst nematode, *Heterodera glycines*, in China: it is time to take measures to control soybean cyst nematode. *Frontiers in Plant Science* 13, 927773. DOI: 10.3389/fpls.2022.927773

Liébanas, G., Clavero-Camacho, I., Cantalapiedra-Navarrete, C., Guerrero, P., Palomares-Rius, J.E. *et al.* (2022) A new needle nematode, *Longidorus maginicus* n. sp. (Nematoda: Longidoridae) from southern Spain. *Journal of Helminthology* 96, e40. DOI: 10.1017/S0022149X22000311

Lilley, C.J., Urwin, P.E., McPherson, M.J. and Atkinson, H.J. (1996) Characterisation of intestinally active proteases of cyst-nematodes. *Parasitology* 113, 415–424. DOI: 10.1017/s0031182000066555

Lilley, C.J., Urwin, P.E., Atkinson, H.J. and McPherson, M.J. (1997) Characterization of cDNAs encoding serine proteases from the soybean cyst nematode *Heterodera glycines*. *Molecular and Biochemical Parasitology* 89, 195–207. DOI: 10.1016/S0166-6851(97)00116-3

Lilley, C.J., Urwin, P.E., Johnston, K.A. and Atkinson, H.J. (2004) Preferential expression of a plant cystatin at nematode feeding sites confers resistance to *Meloidogyne incognita* and *Globodera pallida*. *Plant Biotechnology Journal* 2, 3–12. DOI: 10.1046/j.1467-7652.2003.00037.x

Lilley, C.J., Wang, D., Atkinson, H.J. and Urwin, P.E. (2011) Effective delivery of a nematode-repellent peptide using a root-cap-specific promoter. *Plant Biotechnology Journal* 9, 151–161. DOI: 10.1111/j.1467-7652.2010.00542.x

Lin, B., Zhuo, K., Chen, S., Hu, L., Sun, L. *et al.* (2016) A novel nematode effector suppresses plant immunity by activating host reactive oxygen species-scavenging system. *New Phytologist* 209, 1159–1173. DOI: 10.1111/nph.13701

Linsell, K.J., Riley, I.T., Davies, K.A. and Oldach, K.H. (2014) Characterization of resistance to *Pratylenchus thornei* (Nematoda) in wheat (*Triticum aestivum*): attraction, penetration, motility, and reproduction. *Phytopathology* 104, 174–187. DOI: 10.1094/PHYTO-12-12-0345-R

Liu, B., Hibbard, J.K., Urwin, P.E. and Atkinson, H.J. (2005) The production of synthetic chemodisruptive peptides *in planta* disrupts the establishment of cyst nematodes. *Plant Biotechnology Journal* 3, 487–496. DOI: 10.1111/j.1467-7652.2005.00139.x

Liu, C., Hu, Z., Wang, X., Geng, Y., Ma, C. *et al.* (2019) Rapid detection of the *Bursaphelenchus xylophilus* by denaturation bubble-mediated strand exchange amplification. *Analytical Sciences* 35, 449–453. DOI: 10.2116/analsci.18P461

Liu, S., Kandoth, P.K., Warren, S.D., Yeckel, G., Heinz, R. *et al.* (2012) A soybean cyst nematode resistance gene points to a new mechanism of plant resistance to pathogens. *Nature* 492, 256–260. DOI: 10.1038/nature11651

Liu, S., Kandoth, P.K. and Lakhssassi, N., Kang, J., Colantonio, V. (2017) The soybean *GmSNAP18* gene underlies two types of resistance to soybean cyst nematode. *Nature Communications* 8, 1–11. DOI: 10.1038/ncomms14822

Liu, X., Raaijmakers, C., Vrieling, K., Lommen, S.T.E. and Bezemer, T.M. (2022a) Associational resistance to nematodes and its effects on interspecific interactions among grassland plants. *Plant and Soil* 471, 591–607. DOI: 10.1007/s11104-021-05238-8

Liu, Y., Wang, W., Liu, P., Zhou, H., Chen, Z. and Suonan, J. (2022b) Plant-soil mediated effects of long-term warming on soil nematodes of Alpine meadows on the Qinghai–Tibetan Plateau. *Biology* 11, 1596. DOI: 10.3390/biology11111596

Liu, Y., Cao, H., Ling, J., Yang, Y., Li, Y. *et al.* (2023) Molecular cloning and functional analysis of the pepper resistance gene *Me3* to root-knot nematode. *Horticultural Plant Journal* 9, 133–144. DOI: 10.1016/j.hpj.2022.12.003

Liu, Y.L., Schiff, M., Marathe, R. and Dinesh-Kumar, S.P. (2002) Tobacco *Rar1, EDS1* and *NPR1/NIM1* like genes are required for N-mediated resistance to tobacco mosaic virus. *Plant Journal* 30, 415–429. DOI: 10.1046/j.1365-313x.2002.01297.x

Loof, P.A.A. (1991) The family Pratylenchidae Thorne, 1949. In: Nickle, W.R. (ed.) *Manual of Agricultural Nematology*. Marcel Dekker, New York, USA, pp. 363–421.

Loof, P.A.A. and De Grisse, A.T. (1974) Taxonomic and nomenclatural observations on the genus *Criconemella* De Grisse & Loof, 1965 *sensu* Luc & Raski, 1981 (Criconematidae). *Mededelingen van de Faculteit Landbouwwetenschappen, Rijksuniversiteit Gent* 54, 53–74.

Loof, P.A.A. and Luc, M. (1990) A revised polytomous key for the identification of species of the genus *Xiphinema* Cobb, 1913 (Nematoda: Longidoridae) with exclusion of the *X. americanum* group. *Systematic Parasitology* 16, 35–66. DOI: 10.1007/BF00009600

Lopez-Nicora, H.R. and Niblack, T.L (2018) Interactions with other pathogens. In: Perry, R.N., Moens, M. and Jones, J.T. (eds) *Cyst Nematodes*. CAB International, Wallingford, UK, pp. 271–304.

Lourenço-Tessutti, I.T., Souza Junior, J.D.A., Martins-de-Sa, D., Viana, A.A.B., Carneiro, R.M.D.G. *et al.* (2015) Knock-down of heat-shock protein 90 and isocitrate lyase gene expression reduced root-knot nematode reproduction. *Phytopathology* 105, 628–637. DOI: 10.1094/PHYTO-09-14-0237-R

Lozano-Torres, J.L., Wilbers, R.H.P., Warmerdam, S., Finkers-Tomczak, A., Diaz-Granados, A. *et al.* (2014) Apoplastic venom allergen-like proteins of cyst nematodes modulate the activation of basal plant innate immunity by cell surface receptors. *PLoS Pathogens* 10, e1004569. DOI: 10.1371/journal.ppat.1004569

Lu, C.-J., Meng, Y., Wang, Y., Zhang, T., Yang, G.-F. *et al.* (2022) Survival and infectivity of second-stage root-knot nematode *Meloidogyne incognita* juveniles depend on lysosome-mediated lipolysis. *Journal of Biological Chemistry* 298, 101637. DOI: 10.1016/j.jbc.2022.101637

Luc, M. (1987) A reappraisal of Tylenchina (Nemata). 7. The Family Pratylenchidae Thorne, 1949. *Revue de Nématologie* 10, 203–218.

Luc, M. and Coomans, A. (1992) Les nematodes phytoparasites du genre *Xiphinema* (Longidoridae) en Guyane et en Martinique. *Belgian Journal of Zoology* 122, 147–183.

Luc, M. and Fortuner, R. (1987) A reappraisal of Tylenchina (Nemata). 5. The Family Dolichodoridae Chitwood, 1950. *Revue de Nématologie* 10, 177–181.

Luc, M., Coomans, A., Loof, P.A.A. and Baujard, P. (1998) The *Xiphinema americanum* group (Nematoda: Longidoridae). 2. Observations on *Xiphinema brevicollum* Lordello & da Costa, 1961 and comments on the group. *Fundamental and Applied Nematology* 21, 475–490.

Ludewig. A.H. and Schroeder, F.C. (2013) Ascaroside signaling in *C. elegans*. In: The *C. elegans* Research Community (ed.). *WormBook*, http://www.wormbook.org. DOI: 10.1895/wormbook.1.155.1

Lýsek, H., Malínský, J. and Janisch, R. (1985) Ultrastructure of eggs of *Ascaris lumbricoides* Linnaeus, 1758. I. Egg-shells. *Folia Parasitologica* 32, 381–384.

Maafi, Z.T. and Decraemer, W. (2002) Description of *Trichodorus gilanensis* n. sp. from a forest park in Iran and observations on *Paratrichodorus tunisiensis* (Siddiqi, 1963) Siddiqi, 1974 (Nematoda: Diphtherophorina). *Nematology* 4, 43–54. DOI: 10.1163/156854102760082195

MacDonald, H., Frelih-Larsen, A., Lorant, A., Duin, L., Andersen, S.P. *et al.* (2021) *Carbon farming – making agriculture fit for 2030*. Study for the committee on Environment, Public Health and Food Safety (ENVI), Policy Department for Economic, Scientific and Quality of Life Policies, European Parliament, Luxembourg.

Machado, R.A.R. and von Reuss, S.H. (2022) Chemical ecology of nematodes. *Chimia International Journal for Chemistry* 76, 945–953. DOI: 10.2533/chimia.2022.945

Maggenti, A.R. (1981) *General Nematology*. Springer, New York, USA.

Mani, V., Reddy, C.S., Lee, S.-K., Park, S., Ko, H.-R. *et al.* (2020) Chitin biosynthesis inhibition of *Meloidogyne incognita* by RNAi-mediated gene silencing increases resistance to transgenic tobacco plants. *International Journal of Molecular Sciences* 21, 6626. DOI: 10.3390/ijms21186626

Manohar, M., Tenjo-Castano, F., Chen, S., Zhang, Y.K., Kumari, A. *et al.* (2020) Plant metabolism of nematode pheromones mediates plant-nematode interactions. *Nature Communications* 11, 208. DOI: 10.1038/s41467-019-14104-2

Manosalva, P., Manohar, M., von Reuss, S., Chen, S., Koch, A. *et al.* (2015) Conserved nematode signalling molecules elicit plant defenses and pathogen resistance. *Nature Communications* 6, 7795. DOI: 10.1038/ncomms

Manzanilla-López, R.H. and Hunt, D. J. (2012) Taxonomy and systematics. In: Manzanilla-López, R.H. and Marbán-Mendoza, N. (eds) *Practical Plant Nematology*. Bibliotheca Basica de Agricultura, Montecillo, Mexico, pp. 65–87.

Manzanilla-López, R.H. and Lopez-Llorca, L.V. (eds) (2017) *Perspectives in Sustainable Nematode Management Through* Pochonia chlamydosporia *Applications for Root and Rhizosphere Health*. Springer, Cham, Switzerland. DOI: 10.1007/978-3-319-59224-4

Manzanilla-López, R., Costilla, M.A., Doucet, M., Franco, J., Inserra, R.N. *et al.* (2002) The genus *Nacobbus* Thorne and Allen, 1944 (Nematoda: Pratylenchidae): systematics, distribution, biology and management. *Nematropica* 32, 149–227.

Manzanilla-López, R.H., Esteves, I., Finetti-Sialer, M.M., Hirsch, P.R., Ward, E. *et al.* (2013) *Pochonia chlamydosporia*: advances and challenges to improve its performance as a biological control agent of sedentary endo-parasitic nematodes. *Journal of Nematology* 45, 1–7.

Maosa, J.O., Singh, P.R., Asumaila, A.-G., Couvreur, M. and Bert, W. (2024) Characterisation of *Paratylenchus cacao* n. sp. from Ghana, and *P. enigmaticus* (Nematoda: Paratylenchinae) from Belgium. *Nematology* 26, 415–432. DOI: 10.1163/15685411-bja10314

Marhan, S. and Scheu, S. (2006) Mixing of different mineral soil layers by endogeic earthworms affects carbon and nitrogen mineralization. *Biology and Fertility of Soils* 42, 308–314. DOI: 10.1007/s00374-005-0028-7

Marin, D.H., Sutton, T.B. and Barker, K.R. (1998) Dissemination of bananas in Latin America and the Caribbean and its relationship to the occurrence of *Radopholus similis*. *Plant Disease* 82, 964–974. DOI: 10.1094/pdis.1998.82.9.964

Mariotte, P., Mehrabi, Z., Bezemer, T.M., De Deyn, G.B., Kulmatiski, A. *et al.* (2018) Plant–soil feedback: bridging natural and agricultural sciences. *Trends in Ecology and Evolution* 33, 129–142. DOI: 10.1016/j.tree.2017.11.005

Marra, B.M., Souza, D.S., Aguiar, J.N., Firmino, A.A., Sarto, R.P. *et al.* (2009) Protective effects of a cysteine proteinase propeptide expressed in transgenic soybean roots. *Peptides* 30, 825–831. DOI: 10.1016/j.peptides.2009.01.022

Marroquin, L.D., Elyassnia, D., Griffitts, J.S., Feitelson, J.S. and Aroian, R.V. (2000) *Bacillus thuringiensis* (Bt) toxin susceptibility and isolation of resistance mutants in the nematode *Caenorhabditis elegans*. *Genetics* 155, 1693–1699. DOI: 10.1093/genetics/155.4.1693

Martin, R.J., Purcell, J., Robertson, A.P. and Valkanov, N.A. (2002) Neuromuscular organisation and control in nematodes. In: Lee, D.L. (ed.) *The Biology of Nematodes*. Taylor & Francis, London, UK, pp. 321–344.

Martin-Robles, N., Garcia-Palacios, P., Rodriguez, M., Rico, D., Vigo, R. *et al.* (2020) Crops and their wild progenitors recruit beneficial and detrimental soil biota in opposing ways. *Plant and Soil* 456, 159–173. DOI: 10.1007/s11104-020-04703-0

Masson, A.-S., Vermeire, M.-L., Leng, V., Simonin, M., Tivet, F. *et al.* (2022) Enrichment in biodiversity and maturation of the soil food web under conservation agriculture is associated with suppression of rice-parasitic nematodes. *Agriculture, Ecosystems & Environment* 331, 107913. DOI: 10.1016/j.agee.2022.107913

Mateille, T., Tavoillot, J., Goillon, C., Pares, L., Lefevre, A. (2020) Competitive interactions in plant-parasitic nematode communities affecting organic vegetable cropping systems. *Crop Protection* 135, 105206. DOI: 10.1016/j.cropro.2020.105206

Matthews, G. (2015) *Pesticides: Health, Safety and the Environment*, 2nd edn. John Wiley & Sons Ltd, Chichester, UK.

Mayer, E. (1942) *Systematics and the Origin of Species*. Columbia University Press, New York, USA.

McCarter, J.P., Mitreva, M.D., Martin, J., Dante, M., Wylie, T. *et al.* (2003) Analysis and functional classification of transcripts from the nematode *Meloidogyne incognita*. *Genome Biology* 4, R26. DOI: 10.1186/gb-2003-4-4-r26

McCarville, M.T., Marett, C.C., Mullaney, M.P., Gebhart, G.D. and Tylka, G.L. (2017) Increase in soybean cyst nematode virulence and reproduction on resistant soybean varieties in Iowa from 2001 to 2015 and the effects on soybean yields. *Plant Health Progress* 18, 146–155. DOI: 10.1094/PHP-RS-16-0062

McCullagh, P. and Nelder, J.A. (1992) *Generalized Linear Models*, 2nd edn. Chapman & Hall, London, UK.

McDonald, B.A. and Linde, C. (2002) Pathogen population genetics, evolutionary potential, and durable resistance. *Annual Review of Phytopathology* 40, 349–379. DOI: 10.1146/annurev.phyto.40.120501.101443

McDonald, B.A., Suffert, F., Bernasconi, A. and Mikaberidze, A. (2022) How large and diverse are field populations of fungal plant pathogens? The case of *Zymoseptoria tritici*. *Evolutionary Applications* 15, 1360–1373. DOI: 10.1111/eva.13434

McDonald, M.R., Ives, L., Adusei-Fosu, K. and Jordan, K.S. (2021) *Ditylenchus dipsaci* and *Fusarium oxysporum* on garlic: one plus one does not equal two. *Canadian Journal of Plant Pathology* 43, 749–759. DOI: 10.1080/07060661.2021.1910345

McKenry, M.V. (2021) *Mesocriconema xenoplax* predisposes *Prunus* spp. to bacterial canker. In: Sikora R., Desaeger, J. and Molendijk, L.P.G. (eds) *Integrated Nematode Management: State *of-the-art and Visions for the Future*. CAB International, Wallingford, UK, pp.199–206. DOI: 10.1079/9781789247541.0028.

MEA (2005) *A Report of the Millenium Ecosystem Assessment: Ecosystems and Human Well-being*. Island Press, Washington DC, USA.

Mei, Y., Thorpe, P., Guzha, A., Haegeman, A., Blok, V.C. *et al.* (2015) Only a small subset of the SPRY domain gene family in *Globodera pallida* is likely to encode effectors, two of which suppress host defences induced by the potato resistance gene *Gpa2*. *Nematology* 17, 409–424. DOI: 10.1163/15685411-00002875

Mejias, J., Truong, N.M., Abad, P., Favery, B. and Quentin, M. (2019) Plant proteins and processes targeted by parasitic nematode effectors. *Frontiers in Plant Science* 10, 970. DOI: 10.3389/fpls.2019.00970

Mendoza, A.R. and Sikora, R.A. (2009) Biological control of *Radopholus similis* in banana by combined application of the mutualistic endophyte *Fusarium oxysporum* strain 162, the egg pathogen *Paecilomyces lilacinus* strain 251 and the antagonistic bacteria *Bacillus firmus*. *Biocontrol* 54, 263–272. DOI: 10.1007/s10526-008-9181-x

Mendy, B., Wang'ombe, M.W., Radakovic, Z.S., Holbein, J., Ilyas, M. *et al.* (2017) Arabidopsis leucine-rich repeat receptor-like kinase NILR1 is required for induction of innate immunity to parasitic nematodes. *PLoS Pathogens* 13, e1006284. DOI: 10.1371/journal.ppat.1006284

Meyer, S.L.F. and Roberts, D.P. (2002) Combinations of biocontrol agents for management of plant-parasitic nematodes and soilborne plant-pathogenic fungi. *Journal of Nematology* 34, 1–8.

Milla, R., Garcia-Palacios, P. and Maresanz, S. (2017) Looking at past domestication to secure ecosystem services of future croplands. *Journal of Ecology* 105, 885–889. DOI: 10.1111/1365-2745.12790

Miller, L.I. and Gray, B.J. (1968) Horsenettle cyst nematode, *Heterodera virginiae* n. sp., a parasite of solanaceous plants. *Nematologica* 14, 535–543. DOI: 10.1163/187529268X00228

Miller, L.I. and Gray, B.J. (1972) *Heterodera solanacearum* n. sp., a parasite of solanaceaous plants. *Nematologica* 18, 404–413. DOI: 10.1163/187529272X00674

Millet, J.R.M., Romero, L.O., Lee, J., Bell, B. and Vásquez, V. (2022) *C. elegans* PEZO-1 is a mechanosensitive ion channel involved in food sensation. *Journal of General Physiology* 154, e202112960. DOI: 10.1085/jgp.202112960

Milligan, S.B., Bodeau, J., Yaghoobi, J., Kaloshian, I., Zabel, P. *et al.* (1998) The root knot nematode resistance gene *Mi* from tomato is a member of the leucine zipper, nucleotide binding, leucine-rich repeat family of plant genes. *Plant Cell* 10, 1307–1319. DOI: 10.1105/tpc.10.8.1307

Mitchum, M.G. (2016) Soybean resistance to the soybean cyst nematode *Heterodera glycines*: an update. *Phytopathology* 106, 1444–1450. DOI;10.1094/phyto-06-16-0227-rvw

Moens, T., Araya, M., Swennen, R. and De Waele, D. (2006) Reproduction and pathogenicity of *Helicotylenchus multicinctus, Meloidogyne incognita* and *Pratylenchus coffeae*, and their interactions with *Radopholus similis* on *Musa*. *Nematology* 8, 45–58. DOI: 10.1163/156854106776179999

Moens, M., Perry, R.N. and Starr, J.L. (2009) *Meloidogyne* species – a diverse group of novel and important plant parasites. In: Perry, R.N., Moens, M. and Starr, J.L. (eds) *Root-knot Nematodes*. CAB International, Wallingford, UK, pp. 1–17.

Moens, M., Perry, R.N. and Jones, J. (2018) Cyst nematodes – life cycle and economic importance. In: Perry, R.N., Moens, M. and Jones, J.T. (eds) *Cyst Nematodes*. CAB International, Wallingford, UK, pp. 1–26.

Mohan, S., Mauchline, T.H., Rowe, J., Hirsch, P.R. and Davies, K.G. (2012) *Pasteuria* endospores from *Heterodera cajani* (Nematoda: Heteroderidae) exhibit inverted attachment and altered germination in cross-infection studies with *Globodera pallida* (Nematoda: Heteroderidae). *FEMS Microbiology Ecology* 79, 675–684. DOI: 10.1111/j.1574-6941.2011.01249.x

Mohan, S., Kumar, K.K., Sutar, V., Saha, S., Rowe, J. *et al.* (2020) Plant root-exudates recruit hyperparasitic bacteria of phytonematodes by altered cuticle aging: implications for biological control strategies. *Frontiers in Plant Science* 11, 763. DOI: 10.3389/fpls.2020.00763

Mokrini, F., Viaene, N., Waeyenberge, L., Dababat, A.A. and Moens, M. (2018) Investigation of resistance to *Pratylenchus penetrans* and *P. thornei* in international wheat lines and its durability when inoculated together with the cereal cyst nematode *Heterodera avenae*, using qPCR for nematode quantification. *European Journal of Plant Pathology* 151, 875–889. DOI: 10.1007/s10658-018-1420-0

Mood, M.H., Graybill, F.A. and Boes, D.C. (1986) *Introduction to the Theory of Statistics*. McGraw Hill, New York, USA.

Moore, J.C., Berlow, E.L., Coleman, D.C., de Ruiter, P.C., Dong, Q. *et al.* (2004) Detritus, trophic dynamics and biodiversity. *Ecology Letters* 7, 584–600. DOI: 10.1111/j.1461-0248.2004.00606.x

Mor, M., Cohn, E. and Spiegel, Y. (1992) Phenology, pathogenicity and pathotypes of cereal cyst nematodes, *Heterodera avenae* and *Heterodera latipons* (Nematoda: Heteroderidae) in Israel. *Nematologica* 38, 494–501. DOI: 10.1163/187529292X00469

Mori, I. and Ohshima, Y. (1995) Neural regulation of thermotaxis in *Caenorhabditis elegans*. *Nature* 376, 344–348. DOI: 10.1038/376344a0

Mota, M.M. and Vieira, P. (eds) (2008) *Pine Wilt Disease: A Worldwide Threat to Forest Ecosystems*. Springer, Berlin, Germany.

Mota, M.M., Braasch, H., Bravo, M.A., Penas, A.C., Burgermeister, W. *et al.* (1999) First report of *Bursaphelenchus xylophilus* in Portugal and in Europe. *Nematology* 1, 727–734. DOI: 10.1163/156854199508757

Mueller, J. (2021) *Hoplolaimus columbus*: a prime candidate for site-specific management in cotton and soybean production. In: Sikora R., Desaeger, J. and Molendijk, L.P.G. (eds) *Integrated Nematode Management: State-of-the-art and Visions for the Future*. CAB International, Wallingford, UK, pp. 80–86. DOI: 10.1079/9781789247541.0012

Mundt, C.C. (2018) Pyramiding for resistance durability: theory and practice. *Phytopathology* 108, 792–802. DOI: 10.1094/phyto-12-17-0426-rvw

Mwamula, A.O., Decraemer, W., Kim, Y.H., Ko, H-R., Na, H. *et al.* (2020) Description of a new needle nematode *Paralongidorus koreanensis* n.sp., and two known *Xiphinema* spp. Cobb, 1913 from turfgrass in Korea. *European Journal of Plant Pathology* 15, 1–20. DOI: 10.1007/s10658-019-01846-4

Mwangi, J.M., Niere, B., Finckh, M.R., Krüssel, S. and Kiewnick, S. (2019) Reproduction and life history traits of a resistance breaking *Globodera pallida* population. *Journal of Nematology* 51, e2019-79.

Mwangi, J.M., Mwangi, G.N., Finckh, M.R. and Kiewnick, S. (2021) Biology, pathotype, and virulence of *Globodera rostochiensis* populations from Kenya. *Journal of Nematology* 53, e2021-03.

Myers, K.N, Conn, D. and Brown, A.M.V. (2021) Essential amino acid enrichment and positive selection highlight endosymbiont's role in a global virus-vectoring pest. *mSystems* 6, e01048-20. DOI: 10.1128/mSystems.01048-20

Nahar, K., Kyndt, T., De Vleesschauwer, D., Höfte, M. and Gheysen, G. (2011) The jasmonate pathway is a key player in systemically induced defense against root knot nematodes in rice. *Plant Physiology* 157, 305–316. DOI: 10.1104/pp.111.177576

Needham, J.T. (1743) A letter concerning certain chalky tubulous concretions, called malm; with some microscopical observations on the farina of the red lily, and of worms discovered in smutty corn. *Philosophical Transactions of the Royal Society of London* 42, 634–641. DOI: 10.1098/rstl.1742.0101

Nelson, F.K., Albert, P.S. and Riddle, D.L. (1983) Fine structure of the *Caenorhabditis elegans* secretory-excretory system. *Journal of Ultrastructure Research* 82, 156–171. DOI: 10.1016/S0022-5320(83)90050-3

Nematode Ecophysiological Parameters (2003) http://nemaplex.ucdavis.edu/Ecology/EcophysiologyParms/EcoParameterMenu.html; database updated 1 August 2023 (accessed 8 August 2023).

Neveu, C., Semblat, J.-P., Abad, P. and Castagnone-Sereno, P. (2002) Characterization of a cDNA related to K-Cl cotransporters in the root-knot nematode *Meloidogyne incognita*. *DNA Sequence* 13, 117–121. DOI: 10.1080/10425170290030024

Neveu, C., Abad, P. and Castagnone-Sereno, P. (2003) Molecular cloning and characterization of an intestinal cathepsin L protease from the plant-parasitic nematode *Meloidogyne incognita*. *Physiological and Molecular Plant Pathology* 63, 159–165. DOI: 10.1016/j.pmpp.2003.10.005

Ngala, B., Mariette, N., Ianszen, M., Dewaegeneire, P., Denis M-C. *et al.* (2021) Hatching induction of cyst nematodes in bare soils drenched with root exudates under controlled conditions. *Frontiers in Plant Science* 11, 602825. DOI: 10.3389/fpls.2020.602825

Nguyen, H.T., Trinh, Q.P., Couvreur, M., Nguyen, T.D. and Bert, W. (2021) Description of *Hemicycliophora cardamomi* sp. n. (Nematoda: Hemicycliophoridae) associated with A*momum longiligulare* T.L. Wu and a web-based key for the identification of *Hemicycliophora* spp. *Journal of Helminthology* 95, e2. DOI: 10.1017/S0022149X20000966

Nguyen, H.T., Nguyen, T.D., Le, T.M.L., Trinh, Q.P. and Bert, W. (2022) Remarks on phylogeny and molecular variations of criconematid species (Nematoda: Criconematidae) with case studies from Vietnam. *Scientific Reports* 12, 14832. DOI: 10.1038/s41598-022-18004-2

Nguyen, V.C., Villate, L., Gutierrez-Gutierrez, C., Castillo, P., Van Ghelder, C. *et al.* (2019) Phylogeography of the soil-borne vector nematode *Xiphinema index* highly suggests Eastern origin and dissemination with domesticated grapevine. *Scientific Reports* 9, 7313. DOI: 10.1038/s41598-019-43812-4

Niblack, T., Arelli, P.R., Noel, G.R., Opperman, C.H., Orf, J.H. *et al.* (2002) A revised classification scheme for genetically diverse populations of *Heterodera glycines*. *Journal of Nematology* 34, 279–288.

Nicol, J. and Rivoal, R. (2008) Global knowledge and its application for the integrated control and management of nematodes on wheat. In: Ciancio, A. and Mukerji, K.G. (eds) *Integrated Management and Biocontrol of Vegetable and Grain Crops Nematodes*. Springer, Berlin, Germany, pp. 251–294.

Nicol, J., Rivoal, R., Taylor, S. and Zaharieva, M. (2004) Global importance of cyst (*Heterodera* spp.) and lesion (*Pratylenchus* spp.) nematodes on cereals: distribution, yield loss, use of host resistance and integration of molecular tools. In: Cook, R. and Hunt, D.J. (eds) *Proceedings of the Fourth International Congress of Nematology, 8–13 June 2002, Tenerife, Spain*. *Nematology Monographs and Perspectives 2*. Brill, Leiden, The Netherlands, pp. 233–251. DOI: 10.1163/9789004475236_027

Nijboer, H. and Parlevliet, J.E. (1990) Pathotype specificity in potato cyst nematodes, a reconsideration. *Euphytica* 49, 39–47. DOI: 10.1007/BF00024129

Nile, A.S., Nile, S.H., Keum, Y.S., Kim, D.H., Venkidasamy, B. *et al.* (2017) Nematicidal potential and specific enzyme activity enhancement potential of neem (*Azadirachta indica* A. Juss.) aerial parts. *Environmental Science Pollution Research* 25, 4204–4213. DOI: 10.1007/s11356-017-0821-5

Niu, J.-h., Guo, Q.-x., Jian, H., Chen, C.-l., Yang, D. *et al.* (2011) Rapid detection of *Meloidogyne* spp. by LAMP assay in soil and roots. *Crop Protection* 30, 1063–1069. DOI: 10.1016/j.cropro.2011.03.028

Noel, G.R. and Atibalentja, N. (2006) '*Candidatus* Paenicardinium endonii', an endosymbiont of the plant-parasitic nematode *Heterodera glycines* (Nemata: Tylenchida), affiliated to the phylum Bacteroidetes. *International Journal of Systematic and Evolutionary Microbiology* 56, 1697–1702. DOI: 10.1099/ijs.0.64234-0

Noling, J.W. (2000) Effects of continuous culture of a resistant tomato cultivar on *Meloidogyne incognita* soil population density and pathogenicity. *Journal of Nematology* 32, 452.

Noling, J.W. (2021) Sting nematode management in Florida strawberry. In: Sikora R., Desaeger, J. and Molendijk, L.P.G. (eds) *Integrated Nematode Management: State-of-the-art and Visions for the Future*. CAB International, Wallingford, UK, pp. 182–191. DOI: 10.1079/9781789247541.0026

Norton, D.C. (1989) Abiotic soil factors and plant-parasitic nematode communities. *Journal of Nematology* 21, 299–307.

Nyczepir, A.P. and Thomas, S.H. (2009) Current and future management strategies in intensive crop production systems. In: Perry, R.N., Moens, M. and Starr, J.L. (eds) *Root-knot Nematodes*. CAB International, Wallingford, UK, pp. 412–443.

Ochola, J., Cortada, L., Mwaura, O., Tariku, M., Christensen, S.A. *et al.* (2022) Wrap-and-plant technology to manage sustainably potato cyst nematodes in East Africa. *Nature Sustainability* 5, 425–433. DOI: 10.1038/s41893-022-00852-5

Ogallo, J.L., Goodell, P.B., Eckert, J. and Roberts, P.A. (1997) Evaluation of NemX, a new cultivar of cotton with high resistance to *Meloidogyne incognita*. *Journal of Nematology* 29, 531–537.

Ohl, S.A., van der Lee, F.M. and Sijmons, P.C. (1997) Anti-feeding structure approaches to nematode resistance. In: Fenoll, C., Grundler, F. and Ohl, S. (eds) *Cellular and Molecular Aspects of Plant–Nematode Interactions*. Kluwer Academic Press, Dordrecht, The Netherlands, pp. 250–261. DOI: 10.1007/978-94-011-5596-0_19

Oka, Y. (2010) Mechanisms of nematode suppression by organic soil amendments – a review. *Applied Soil Ecology* 44, 101–115. DOI: 10.1016/j.apsoil.2009.11.003

Oka, Y. (2020) From old-generation to next-generation nematicides. *Agronomy* 10, 1387. DOI: 10.3390/agronomy10091387

Oka, Y., Shuker, S. and Tkachi, N. (2009) Nematicidal efficacy of MCW-2, a new nematicide of the fluoroalkenyl group, against the root-knot nematode *Meloidogyne javanica*. *Pest Management Science* 65, 1082–1089. DOI: 10.1002/ps.1796

Oka, Y., Shuker, S. and Tkachi, N. (2013). Influence of soil environments on nematicidal activity of fluensulfone against *Meloidogyne javanica*. *Pest Management Science* 69, 1225–1234. DOI: 10.1002/ps.3487

Oku, H., Shiraishi, S., Kurozumi, S. and Ohta, H. (1980) Pine wilt toxin, the metabolite of a bacterium associated with a nematode. *Naturwissenschaften* 67, 198–199. DOI: 10.1007/BF01086307

Oliveira, C.J., Subbotin, S.A., Álvarez-Ortega, S., Desaeger, J., Brito, J.A. *et al.* (2019) Morphological and molecular identification of two Florida populations of foliar nematodes (*Aphelenchoides* spp.) isolated from strawberry with the description of *Aphelenchoides pseudogoodeyi* sp. n. (Nematoda: Aphelenchoididae) and notes on their bionomics. *Plant Disease* 103, 2825–2842. DOI: 10.1094/PDIS-04-19-0752-RE

Olsen, S.K., Greenan, G., Desai, A., Müller-Reichert, T. and Oegema, K. (2012) Hierarchical assembly of the eggshell and permeability barrier in *C. elegans*. *Journal of Cell Biology* 198, 731–748. DOI: 10.1083/jcb.201206008

Oota, M., Tsai, A.Y.-L., Aoki, D., Matsushita, Y., Toyoda, S. *et al.* (2020) Identification of naturally occurring polyamines as root-knot nematode attractants *Molecular Plant* 13, 658–665. DOI: 10.1016/j.molp.2019.12.010

Opperman, C.H., Taylor, C.G. and Conkling, M.A. (1994) Root-knot nematode-directed expression of a plant root-specific gene. *Science* 263, 221–223. DOI: 10.1126/science.263.5144.221

Orlando, V., Chitambar, J.J., Dong, K, Chizhov, V.N., Mollov, D. *et al.* (2016) Molecular and morphological characterisation of *Xiphinema americanum*-group species (Nematoda: Dorylaimida) from California, USA, and other regions, and co-evolution of bacteria from the genus *Candidatus Xiphinematobacter* with nematodes. *Nematology* 18, 1015–1043. DOI: 10.1163/15685411-00003012

Ornat, C., Verdejo-Lucas, S. and Sorribas, F.J. (2001) Population of *Meloidogyne javanica* in Spain virulent to the *Mi* resistance gene in tomato. *Plant Disease* 85, 271–276. DOI: 10.1094/pdis.2001.85.3.271

Orr, J.N., Mauchline, T.H., Cock, P.J., Blok, V.C. and Davies, K.G. (2018) *De novo* assembly of the *Pasteuria penetrans* genome reveals high plasticity, host dependency, and BclA-like collagens. *bioRxiv* 485748. DOI: 10.1101/485748

Ortiz, B.V., Perry, C., Goovaerts, P., Vellidis, G. and Sullivan, D. (2010) Geostatistical modelling of the spatial variability and risk areas of southern root-knot nematodes in relation to soil properties. *Geoderma* 156, 243–252. DOI: 10.1016/j.geoderma.2010.02.024

Orton Williams, K.J. (1972) *Macroposthonia xenoplax*. In: *CIH Descriptions of Plant-Parasitic Nematodes*. Set 1, No. 12. Commonwealth Agricultural Bureaux, Farnham Royal, UK.

Orton Williams, K.J. (1974) *Dolichodorus heterocephalus*. In: *CIH Descriptions of Plant-Parasitic Nematodes*. Set 4, No. 56. Commonwealth Agricultural Bureaux, Farnham Royal, UK.

Page, R.D.M. and Charleston, M.A. (1998) Trees within trees: phylogeny and historical associations. *Trends in Ecology and Evolution* 13, 356–359. DOI: 10.1016/S0169-5347(98)01438-4

Palomares-Rius, J.E., Castillo, P., Liébanas, G., Vovlas, N., Landa, B.B. *et al.* (2010) Description of *Pratylenchus hispaniensis* n. sp. from Spain and considerations on the phylogenetic relationship among selected genera in the family Pratylenchidae. *Nematology* 12, 429–451. DOI: 10.1163/138855409X12559479585043

Palomares-Rius, J.E., Cantalapiedra-Navarrete, C. and Castillo, P. (2014) Cryptic species in plant-parasitic nematodes, *Nematology* 16, 1105–1118. DOI: 10.1163/15685411-00002831

Palomares-Rius, J.E., Hedley, P., Cock, P.J.A., Morris, J.A., Jones, J.T. *et al.* (2016) Gene expression changes in diapause or quiescent potato cyst nematode, *Globodera pallida*, eggs after hydration or exposure to tomato root diffusate. *PeerJ* 4, e1654. DOI: 10.7717/peerj.1654

Palomares-Rius, J.E., Cantalapiedra-Navarrete, C., Archidona-Yuste, A., Subbotin, S.A. *et al.* (2017a) The utility of mtDNA and rDNA for barcoding and phylogeny of plant-parasitic nematodes from Longidoridae (Nematoda, Enoplea). *Scientific Reports* 7, 10905. DOI: 10.1038/s41598-017-11085-4

Palomares-Rius, J.E., Escobar, C., Cabrera, J., Vovlas, A. and Castillo, P. (2017b) Anatomical alterations in plant tissues induced by plant-parasitic nematodes. *Frontiers in Plant Science* 8, 1987. DOI: 10.3389/fpls.2017.01987

Palomino, J., Garcia-Palacios, P., De Deyn, G.B., Martinez-Garcia, L.B., Sánchez-Moreno, S. *et al.* (2023) Impacts of plant domestication on soil microbial and nematode communities during litter decomposition. *Plant and Soil* 487, 419–436. DOI: 10.1007/s11104-023-05937-4

Pante, E., Schoelinck, C. and Puillandre, N. (2015) From integrative taxonomy to species description: one step beyond. *Systematic Biology* 64, 152–160. DOI: 10.1093/sysbio/syu083

Papolu, P.K., Dutta, T.K., Tyagi, N., Urwin, P.E., Lilley, C.J. *et al.* (2016) Expression of a cystatin transgene in eggplant provides resistance to root-knot nematode, *Meloidogyne incognita*. *Frontiers in Plant Science* 7, 1122. DOI: 10.3389/fpls.2016.01122

Papolu, P.K., Dutta, T.K., Hada, A., Singh, D. and Rao, U. (2020) The production of a synthetic chemodisruptive peptide in planta precludes *Meloidogyne incognita* multiplication in *Solanum melongena*. *Physiological and Molecular Plant Pathology* 112, 101542. DOI: 10.1016/j.pmpp.2020.101542

Patel, N., Hamamouch, N., Li, C.Y., Hussey, R.S., Mitchum, M. *et al.* (2008) Similarity and functional analyses of expressed parasitism genes in *Heterodera schachtii* and *Heterodera glycines*. *Journal of Nematology* 40, 299–310.

Patel, N., Hamamouch, N., Li, C.Y., Hewezi, T., Hussey, R.S. *et al.* (2010) A nematode effector protein similar to annexins in host plants. *Journal of Experimental Botany* 61, 235–248. DOI: 10.1093/jxb/erp293

Patil, J., Powers, S.J., Miller, A.J., Gaur, H.S. and Davies, K.G. (2013a) Effect of root nitrogen supply forms on attraction and repulsion of second-stage juveniles of *Meloidogyne graminicola*. *Nematology* 15, 469–482. DOI: 10.1163/15685411-00002697

Patil, J., Miller, A.J. and Gaur, H.S. (2013b) Effect of nitrogen supply form on the invasion of rice roots by the root-knot nematode, *Meloidogyne graminicola*. *Nematology* 15, 483–492. DOI: 10.1163/15685411-00002694

Peña-Santiago, R. (2021) *Morphology and Bionomics of Dorylaims (Nematoda: Dorylaimida). Nematology Monographs and Perspectives 13*. (Series Editors: Hunt, D.J. and Perry, R.N.). Brill, Leiden, The Netherlands.

Peng, D., Gaur, H.S. and Bridge, J. (2018) Nematode parasites of rice. In: Sikora, R., Coyne, D., Hallmann, J. and Timper, P. (eds). *Plant Parasitic Nematodes in Subtropical and Tropical Agriculture*, 3rd edn. CAB International, Wallingford, UK, pp. 120–162 DOI: 10.1079/9781786391247.0120

Peng, H., Peng, D., Hu, X., He, X., Wang, Q. *et al.* (2012) Loop-mediated isothermal amplification for rapid and precise detection of the burrowing nematode, *Radopholus similis*, directly from diseased plant tissues. *Nematology* 14, 977–986. DOI: 10.1163/156854112X638415

Peng, Y. and Moens, M. (2003) Host resistance and tolerance to migratory plant-parasitic nematodes. *Nematology* 5, 145–177. DOI: 10.1163/156854103767139653

Penkov, S., Mende, F., Zagoriy, V., Erkut, C., Martin, R. *et al.* (2010) Maradolipids: diacyltrehalose glycolipids specific to dauer larva in *Caenorhabditis elegans*. *Angewandte Chemie International Edition* 49, 9430–9435. DOI: 10.1002/anie.201004466

Peraza-Padilla, W., Cantalapiedra-Navarrete, C., Zamora-Araya, T., Palomares-Rius, J.E., Castillo, P. *et al.* (2018) A new dagger nematode, *Xiphinema tica* n. sp. (Nematoda: Longidoridae), from Costa Rica with updating of the polytomous key of Loof and Luc (1990). *European Journal of Plant Pathology* 150, 73–90. DOI: 10.1007/s10658-017-1253-2

Perpétuo, L.S., da Cunha, M.J.M., Batista, M.T. and Conceição, I.L. (2021) *Solanum linnaeanum* and *Solanum sisymbriifolium* as a sustainable strategy for the management of *Meloidogyne chitwoodi*. *Scientific Reports* 11, 3484. DOI: 10.1038/s41598-020-77905-2

Perry, R.N. (1999) Desiccation survival of parasitic nematodes. *Parasitology* 119, S19–S30. DOI: 10.1017/S0031182000084626

Perry, R.N. (2002) Hatching. In: Lee, D.L. (ed.) *The Biology of Nematodes*. Taylor & Francis, London, UK, pp. 147–169.

Perry, R.N. (2005) An evaluation of types of attractants enabling plant-parasitic nematodes to locate plant roots. *Russian Journal of Nematology* 13, 83–88.

Perry, R.N. (2011) Understanding the survival strategies of nematodes. *CAB Reviews: Perspectives in Agriculture, Veterinary Science, Nutrition and Natural Resources* 6, 1–5. DOI: 10.1079/PAVSNNR20116022

Perry, R.N. and Aumann, J. (1998) Behaviour and sensory responses. In: Perry, R.N. and Wright, D.J. (eds) *The Physiology and Biochemistry of Free-living and Plant-parasitic Nematodes*. CAB International, Wallingford, UK, pp. 75–102.

Perry, R.N. and Gaur, H.S. (1996) Host plant influences on the hatching of cyst nematodes. *Fundamental and Applied Nematology* 19, 505–510.

Perry, R.N. and Maule, A.G. (2004) Physiological and biochemical basis of nematode behaviour. In: Gaugler, R. and Bilgrami, A. (eds) *Nematode Behaviour*. CAB International, Wallingford, UK, pp. 197–238.

Perry, R.N. and Moens, M. (2011) Survival of parasitic nematodes outside the host. In: Perry, R.N. and Wharton, D.A. (eds) *Molecular and Physiological Basis of Nematode Survival*. CAB International, Wallingford, UK, pp. 1–27.

Perry, R.N. and Trett, M.W. (1986) Ultrastructure of the eggshell of *Heterodera schachtii* and *H. glycines* (Nematoda : Tylenchida). *Revue de Nématologie* 9, 399–403.

Perry, R.N., Clarke, A.J. and Beane, J. (1980) Hatching of *Heterodera goettingiana in vitro*. *Nematologica* 26, 493–495. DOI: 10.1163/187529280x00422

Perry, R.N., Rolfe, R.N., Sheridan, J.P. and Miller, A.J. (2004) Electrophysiological analysis of sensory responses and stylet activity of second-stage juveniles of *Globodera rostochiensis*, and plant reactions to nematode invasion. *Japanese Journal of Nematology* 34, 1–10. DOI: 10.3725/jjn1993.34.1_1

Perry, R.N., Subbotin, S. and Moens, M. (2007) Molecular diagnostics of plant-parasitic nematodes. In: Punja, Z.K., De Boer, S. and Sanfaçon, S. (eds) *Biotechnology and Plant Disease Management*. CAB International, Wallingford, UK, pp. 195–226.

Perry, R.N., Moens, M. and Starr, J.L. (eds) (2009) *Root-knot Nematodes*. CAB International, Wallingford, UK.

Phillips, M.S. and Trudgill, D.L. (1998) Population modelling and integrated control options for potato cyst nematodes. In: Marks, R.J. and Brodie, B.B. (eds) *Potato Cyst Nematodes: Biology, Distribution, and Control*. CAB International, Wallingford, UK, pp. 153–165.

Picard, D., Sempere, T. and Plantard, O. (2007) A northward colonisation of the Andes by the potato cyst nematode during geological times suggests multiple host-shifts from wild to cultivated potatoes. *Molecular Phylogenetics and Evolution* 42, 308–316. DOI: 10.1016/j.ympev.2006.06.018

Pirzada, T., Affokpon, A., Guenther, R.A., Mathew, R., Agate, S. *et al.* (2023) Plant-biomass-based hybrid seed wraps mitigate yield and post-harvest losses among smallholder farmers in sub-Saharan Africa. *Nature Food* 4, 148–159. DOI: 10.1038/s43016-023-00695-z

Plantard, O., Picard, D., Valette, S., Scurrah, M., Grenier, E. *et al.* (2008) Origin and genetic diversity of Western European populations of the potato cyst nematode *Globodera pallida* inferred from mitochondrial sequences and microsatellite loci. *Molecular Ecology* 17, 2208–2218. DOI: 10.1111/j.1365-294x.2008.03718.x

Platt, H.M. and Warwick, R.M. (1983) *Free-living Marine Nematodes. Part I. British Enoplids*. Synopses of the British Fauna, No. 28 (New Series Editors: Kermack, D.M. and Barnes, R.S.K.). The Linnean Society of London and The Estuarine and Brackish-water Sciences Association. Cambridge University Press, Cambridge, UK.

Pline, M. and Dusenbery, D.B. (1987) Responses of plant parasitic nematode *Meloidogyne incognita* to carbon dioxide determined by video camera-computer tracking. *Journal of Chemical Ecology* 13, 873–888. DOI: 10.1007/BF01020167

Plowright, R.A., Caubel, G. and Mitzen, K.A. (2002) *Ditylenchus* species. In: Starr, J.L., Cook, R. and Bridge, J. (eds) *Resistance to Parasitic Nematodes*. CAB International, Wallingford, UK, pp. 107–139.

Postma, W.J., Slootweg, E.J., Rehman, S., Finkers-Tomczak, A., Tytgat, T.O. *et al.* (2012) The effector SPRYSEC-19 of *Globodera rostochiensis* suppresses CC-NB-LRR-mediated disease resistance in plants. *Plant Physiology* 160, 944–954. DOI: 10.1104/pp.112.200188

Powers, T. (2004) Nematode molecular diagnostics: from bands to barcodes. *Annual Review of Phytopathology* 42, 367–383. DOI: 10.1146/annurev.phyto.42.040803.140348

Powers, T.O. and Harris, T.S. (1993) A polymerase chain reaction method for identification of five major *Meloidogyne* species. *Journal of Nematology* 25, 1–6.

Prahlad, V., Pilgrim, D. and Goodwin, E.B. (2003) Roles for mating and environment in *C. elegans* sex determination. *Science* 302, 1046–1049. DOI: 10.1126/science.1087946

Price, J.A., Ali, M.F., Major, L.L., Smith, T.K. and Jones, J.T. (2023) An eggshell-localised annexin plays a key role in the coordination of the life cycle of a plant-parasitic nematode with its host. *PLoS Pathogens* 19, E1011147. DOI: 10.1371/journal.ppat.1011147

Price, J., Preedy, K., Young, V., Todd, D. and Blok, V.C. (2024) Stacking host resistance genes to control *Globodera pallida* populations with different virulence. *European Journal of Plant Pathology* 168, 373–381. DOI: 10.1007/s10658-023-02761-5

Pudasaini, M.P., Viaene, N. and Moens, M. (2006a) Effect of marigold (*Tagetes patula*) on population dynamics of *Pratylenchus penetrans* in a field. *Nematology* 8, 477–484. DOI: 10.1163/156854106778613930

Pudasaini, M.P., Schomaker, C.H., Been, T.H. and Moens, M. (2006b) The vertical distribution of the plant-parasitic nematode, *Pratylenchus penetrans*, under four field crops. *Phytopathology* 96, 226–233. DOI: 10.1094/PHYTO-96-0226

Puzio, P.S., Cai, D., Ohl, S., Wyss, U. and Grundler, F.M.W. (1998) Isolation of regulatory DNA regions related to differentiation of nematode feeding structures in *Arabidopsis thaliana*. *Physiological and Molecular Plant Pathology* 53, 177–193. DOI: 10.1006/pmpp.1998.0173

Puzio, P.S., Lausen, J., Almeida-Engler, J., Cai, D., Gheysen, G. *et al.* (1999) Isolation of a gene from *Arabidopsis thaliana* related to nematode feeding structures. *Gene* 239, 163–172. DOI: 10.1016/s0378-1119(99)00353-4

Qiao, F., Kong, L.-A., Peng, H., Huang, W.-K., Wu, D.-Q. *et al.* (2019) Transcriptional profiling of wheat (*Triticum aestivum* L.) during a compatible interaction with the cereal cyst nematode *Heterodera avenae*. *Scientific Reports* 9, 2184. DOI: 10.1038/s41598-018-37824-9

Qiao, K., Dong, S., Wang, H., Xia, X., Ji, X. *et al.* (2012) Effectiveness of 1,3-dicloropropene as an alternative to methyl bromide in rotations of tomato (*Solanum lycopersicum*) and cucumber (*Cucumis sativus*) in China. *Crop Protection* 38, 30–34. DOI: 10.1016/j.cropro.2012.03.007

Qing, X., Wang, M., Karssen, G., Bucki, P., Bert, W. *et al.* (2019) PPNID: a reference database and molecular identification pipeline for plant-parasitic nematodes. *Bioinformatics* 15, 36 1052–1056. DOI: 10.1093/bioinformatics/btz707

Quist, C.W., Smant, G. and Helder, J. (2015) Evolution of plant parasitism in the phylum Nematoda. *Annual Review of Plant Pathology* 53, 289–310. DOI: 10.1146/annurev-phyto-080614-120057

Radice, A.D., Powers, T.O., Sandall, L.J. and Riggs, R.D. (1988) Comparisons of mitochondrial DNA from the sibling species *Heterodera glycines* and *H. schachtii*. *Journal of Nematology* 20, 443–450.

Radivojević, M. (1998) Observations on spear replacement in *Xiphinema dentatum* (Nematoda: Longidoridae). *Nematologica* 44, 137–151. DOI: 10.1163/005325998X00036

Rao, U., Mauchline, T.H. and Davies, K.G. (2012) The 16s rRNA gene of *Pasteuria penetrans* provides an early diagnostic of infection of root-knot nematodes (*Meloidogyne* spp.) *Nematology* 14, 799–804. DOI: 10.1163/156854112X627318

Raski, D.J. and Luc, M. (1987) A reappraisal of Tylenchina (Nemata). 10. The superfamily Criconematoidea Taylor, 1936. *Revue de Nématologie* 10, 409–444.

Rasmann, S., Kölner, T.G., Degenhardt, J., Hiltpold, I., Toepfer, S. *et al.* (2005) Recruitment of entomopathogenic nematodes by insect-damaged maize roots. *Nature* 434, 732–737. DOI: 10.1038/nature03451

Reinke, V., Krause, M. and Okkema, P. (2013) Transcriptional regulation of gene expression in *C. elegans*. In: The *C. elegans* Research Community (ed.) *WormBook*, http://www.wormbook.org, DOI: 10.1895/wormbook.1.45.2

Reynolds, A.M., Dutta, T.K., Curtis, R.H., Powers, S.J., Gaur, H.S. and Kerry, B.R. (2010) Chemotaxis can take plant-parasitic nematodes to the source of a chemo-attractant via the shortest possible routes. *Journal of the Royal Society Interface* 8, 568–577. DOI: 10.1098/rsif.2010.0417

Rhoads, R.E., Dinkova, T.D. and Korneeva, N.L. (2006) Mechanism and regulation of translation in *C. elegans*. In: The *C. elegans* Research Community (ed.) *WormBook*, http://www.wormbook.org, DOI: 10.1895/wormbook.1.63.1

Riggs, R.D. (1992) Host range. In: Riggs, R.D. and Wrather, J.A. (eds) *Biology and Management of the Soybean Cyst Nematode*. American Phytopathological Society, St Paul, MN, USA, pp. 107–114.

Rivoal, R. and Cook, R. (1993) Nematode pests of cereals. In: Evans, K., Trudgill, D.L. and Webster, J.M. (eds) *Plant Parasitic Nematodes in Temperate Agriculture*. CAB International, Wallingford, UK, pp. 259–303.

Roberts, P.A. (1982) Plant resistance in nematode pest management. *Journal of Nematology* 20, 392–395.

Roberts, P.A. (1993) The future of nematology: integration of new and improved management strategies. *Journal of Nematology* 25, 383–394.

Roberts, P.A. (2002) Concepts and consequences of resistance. In: Starr, J.L., Cook, R. and Bridge, J. (eds) *Plant Resistance to Parasitic Nematodes*. CAB International, Wallingford, UK, pp. 23–41.

Robertson, L., Cobacho Arcos, S., Escuer, M., Santiago Merino, R., Esparrago, G. *et al.* (2011) Incidence of the pine wood nematode, *Bursaphelenchus xylophilus* Steiner & Buhrer, 1934 (Nickle, 1970) in Spain. *Nematology* 13, 755–757. DOI: 10.1163/138855411X578888

Robertson, W. (1976) A possible gustatory organ associated with the odontophore in *Longidorus leptocephalus* and *Xiphinema diversicaudatum*. *Nematologica* 21, 443–448. DOI: 10.1163/187529275X00202

Robertson, W. (1979) Observations on the oesophageal nerve system of *Longidorus leptocephalus*. *Nematologica* 25, 245–254. DOI: 10.1163/187529279X00262

Robinson, A.F. (1989) Thermotactic adaptation in two foliar and two root parasitic nematodes *Revue de Nématologie* 26, 46–58.

Robinson, A.F. (1994) Movement of five nematode species through sand subjected to natural temperature gradient fluctuations. *Journal of Nematology* 26, 46–58.

Robinson, A.F. (1995) Optimal release rates for attracting *Meloidogyne incognita, Rotylenchulus reniformis* and other nematodes to carbon dioxide in sand. *Journal of Nematology* 27, 42–50.

Robinson, A.F. (2004) Nematode behaviour and migrations through soil and host tissue. In: Chen, Z., Chen, S. and Dickson, D.W. (eds) *Nematology: Advances and Perspectives, Vol. 1: Nematode Morphology, Physiology and Ecology*. CAB International, Wallingford, UK, pp. 330–405.

Robinson, C.P., Schwarz, E.M. and Sternberg, P.W. (2013) Identification of DVA interneuron regulatory sequences in *Caenorhabditis elegans*. *PLoS One* 8, e54971. DOI: 10.1371/journal.pone.0054971

Rocha, I., Ma, Y., Souza-Alonso, P., Vosátka, M., Freitas, H. and Oliveira, R.S. (2019) Seed coating: A tool for delivering beneficial microbes to agricultural crops. *Frontiers in Plant Science* 10,1357. DOI: 10.3389/fpls.2019.01357

Roderick, H., Tripathi, L., Babirye, A., Wang, D., Tripathi, J. *et al.* (2012) Generation of transgenic plantain (*Musa* spp.) with resistance to plant pathogenic nematodes. *Molecular Plant Pathology* 13, 841–851. DOI: 10.1111/j.1364-3703.2012.00792.x

Roderick, H., Urwin, P.E. and Atkinson, H.J. (2018) Rational design of biosafe crop resistance to a range of nematodes using RNA interference. *Plant Biotech Journal* 16, 520–529. DOI: 10.1111/pbi.12792

Roget, D.K., Neate, S.M. and Rovira, A.D. (1996) Effect of sowing point design and tillage practice on the incidence of rhizoctonia root rot, take-all and cereal cyst nematode in wheat and barley. *Australian Journal of Experimental Agriculture* 36, 683–693. DOI: 10.1071/EA9960683

Román, J. and Hirschmann, H. (1969) Morphology and morphometrics of six species of *Pratylenchus*. *Journal of Nematology* 1, 363–386.

Rommens, C.M. (2010) Barriers and paths to market for genetically engineered crops. *Plant Biotechnology Journal* 8, 101–111. DOI: 10.1111/j.1467-7652.2009.00464.x

Rosso, M.N., Jones, J.T. and Abad, P. (2009) RNAi and functional genomics in plant parasitic nematodes. *Annual Review of Phytopathology* 47, 207–232. DOI: 10.1146/annurev.phyto.112408.132605

Rostad, H.E., Reen, R.A., Mumford, M.H., Zwart, R.S. and Thompson, J.P. (2022) Resistance to root-lesion nematode *Pratylenchus neglectus* identified in a new collection of two wild chickpea species (*Cicer reticulatum* and *C. echinospermum*) from Turkey. *Plant Pathology* 71, 1205–1219. DOI: 10.1111/ppa.13544

Roubtsova, T.V. and Subbotin, S.A. (2022) Molecular diagnostics of the pigeon pea cyst nematode, *Heterodera cajani* Koshy, 1967, using real-time PCR. *Nematology* 24, 85–90. DOI: 10.1163/15685411-bja10113

Roubtsova, T.V., Burbridge, J. and Subbotin, S.A. (2020) Molecular characterisation and diagnostics of some gall-forming nematodes of the family Anguinidae Nicoll, 1935 (Nematoda: Tylenchida) using COI mtDNA. *Russian Journal of Nematology* 28, 123–130. DOI: 10.24411/0869-6918-2020-10013

Ruess, L., Häggblom, M.M., García Zapata, E.J. and Dighton, J. (2002) Fatty acids of fungi and nematodes – possible biomarkers in the soil food chain? *Soil Biology and Biochemistry* 34, 745–756. DOI: 10.1016/S0038-0717(01)00231-0

Runia, W.T., Molendijk, L.P.G., van den Berg, W., Stevens, L.H., Schilder, M.T. *et al.* (2014) Inundation as tool for management of *Globodera pallida* and *Verticillium dahlia*. *Acta Horticulturae* 1044, 195–201. DOI: 10.17660/ActaHortic.2014.1044.22

Rutter, W.B., Skantar, A.M., Handoo, Z.A., Mueller, J.D., Aultman, S.P. *et al.* (2019) *Meloidogyne enterolobii* found infecting root-knot nematode resistant sweet potato in South Carolina, United States. *Plant Disease* 103, 755. DOI: 10.1094/PDIS-08-18-1388-PDN

Rybarczyk-Mydłowska, K.D., Mooijman, P.J.W., van Megen, H.H.B., van den Elsen, S.J.J., Vervoort, M.T.W. *et al.* (2012) Small subunit ribosomal DNA-based phylogenetic analysis of foliar nematodes (*Aphelenchoides* spp.) and their quantitative detection in complex DNA backgrounds. *Phytopathology* 102, 1153–1160. DOI: 10.1094/phyto-05-12-0114-r

Ryss, A.Yu. and Subbotin, S.A. (2017) Coevolution of wood-inhabiting nematodes of the genus Bursaphelenchus Fuchs, 1937 with their insect vectors and plant hosts. *Zhurnal Obschei Biologii* 78, N 3, 13–42 (in Russian).

Ryss, A., Vieira, P., Mota, M. and Kulinich, O. (2005) A synopsis of the genus *Bursaphelenchus* Fuchs, 1937 (Aphelenchida: Parasitaphelenchidae) with keys to species. *Nematology* 7, 395–458. DOI: 10.1163/156854105774355581

Sacco, M.A., Koropacka, K., Grenier, E., Jaubert, M.J., Blanchard, A. *et al.* (2009) The cyst nematode SPRYSEC protein RBP-1 elicits Gpa2- and RanGAP2-dependent plant cell death. *PLoS Pathology* 5, e1000564. DOI: 10.1371/journal.ppat.1000564

Salomé, C., Coll, P., Lardo, E., Metay, A., Villenave, C., Marsden, C. *et al.* (2016) The soil quality concept as a framework to assess management practices in vulnerable agroecosystems: a case study in Mediterranean vineyards. *Ecological Indicators* 61, 456–465. DOI: 10.1016/j.ecolind.2015.09.047

Samac, D. and Smigocki, A. (2003) Expression of oryzacystatin I and II in alfalfa increases resistance to the root-lesion nematode. *Phytopathology* 93, 799–806. DOI: 10.1094/PHYTO.2003.93.7.799

Sánchez-Moreno, S.and Ferris, H. (2018) Nematode ecology and soil health. In: Sikora, R., Coyne, D., Hallmann, J. amd Timper, P. (eds), *Plant Parasitic Nematodes in Subtropical and Tropical Agriculture*, 3rd edn. CAB International, Wallingford, UK, pp. 62–86. DOI: 10.1079/9781786391247.0062

Sasaki-Crawley, A., Curtis, R.H.C., Birkett, M., Powers, S., Papadopoulos, A. *et al.* (2010) Signalling and behaviour of potato cyst nematode in the rhizosphere of the trap crop, *Solanum sisymbriifolium*. *Annals of Applied Biology* 103, 45–51.

Sattelle, D.B. (2009) Invertebrate nicotinic acetylcholine receptors–targets for chemicals and drugs important in agriculture, veterinary medicine and human health. *Journal of Pesticide Science* 34, 233–240. DOI: 10.1584/jpestics.R09-02

Sawin, E.R., Ranganthan, R. and Horvitz, H.R. (2000) *C. elegans* locomotory rate is modulated by the environment through a dopaminergic pathway and by experience through a serotonergic pathway. *Neuron* 26, 619–631. DOI: 10.1016/s0896-6273(00)81199-x

Schans, J. (1993) Population dynamics of potato cyst nematodes and associated damage to potato. PhD Thesis, Wageningen Agricultural University, Wageningen, The Netherlands.

Schmidt, J.H., Bergkvist, G., Campiglia, E., Radicetti, E., Wiittwer, R.A. *et al.* (2017) Effect of tillage, subsidiary crops and fertilisation on plant-parasitic nematodes in a range of agro-environmental conditions within Europe. *Annals of Applied Biology* 171, 477–489. DOI: 10.1111/aab.12389

Schmidt-Rhaesa, A. (1997) Phylogenetic relationships of the Nematomorpha – a discussion of current hypotheses. *Zoologischer Anzeiger* 236, 203–216.

Scholz, U. (2001) Biology, pathogenicity and control of the cereal cyst nematode *Heterodera latipons* Franklin on wheat and barley under semiarid conditions, and interactions with common root rot *Bipolaris sorokiniana* (Sacc.) Shoemaker (telemorph: Cochliobolus (Ito et Kurib.) Drechs. Ex. Dastur.). PhD Thesis, University of Bonn, Bonn, Germany.

Schomaker, C.H. and Been, T.H. (1999a) Compound models, describing the relationship between dosage of (Z)- or (E)-isomers of 1,3-dichloropropene and hatching behaviour of *Globodera rostochiensis*. *Nematology* 1, 19–29. DOI; 10.1163/156854199507947

Schomaker, C.H. and Been, T.H. (1999b) A model for infestation loci of potato cyst nematodes *Globodera rostochiensis* and *G. pallida*. *Phytopathology* 89, 583–590. DOI: 10.1094/PHYTO.1999.89.7.583

Schouteden, N., De Waele, D., Panis, B. and Vos, C.M. (2015) Arbuscular Mycorrhizal Fungi for the biocontrol of plant-parasitic nematodes: a review of the mechanisms involved. *Frontiers in Microbiology* 6, 1280. DOI: 10.3389/fmicb.2015.01280

Schrama, M., de Haan, J.J., Kroonen, M., Verstegen, H. and van der Putten, W.H. (2018) Crop yield gap and stability in organic and conventional farming systems. *Agriculture, Ecosystems and Environment* 256, 123–130. DOI: 10.1016/j.agee.2017.12.023

Schuske, K., Beg, A.A. and Jorgensen, E.M. (2004) The GABA nervous system in *C. elegans*. *Trends in Neuroscience* 27, 407–414. DOI: 10.1016/j.tins.2004.05.005

Scott, E., Hudson, A., Feist, E., Calahorro, F., Dillon, J. *et al.* (2017) An oxytocin-dependent social interaction between larvae and adult *C. elegans*. *Scientific Reports* 7, 10122. 10.1038/s41598-017-09350-7

Seehofer, H. (2007) Council directive 2007/33/EC on the control of potato cyst nematodes and repealing Directive 69/465/EEC. *Official Journal of the European Union* 156, 12–22.

Seinhorst, J.W. (1981) Water consumption of plant attacked by nematodes and mechanisms of growth reduction. *Nematologica* 27, 34–51.

Seinhorst, J.W. (1982) The relationship in field experiments between population density of *Globodera rostochiensis* before planting potatoes and yield of potato tubers. *Nematologica* 28, 277–284. DOI: 10.1163/187529282X00312

Seinhorst, J.W. (1986) Effects of nematode attack on the growth and yield of crop plants. In: Lamberti, F. and Taylor, C.E. (eds) *Cyst Nematodes*. Plenum Press, London, UK, pp. 191–209.

Seinhorst, J.W. (1988) The estimation of densities of nematode populations in soil and in plants. *Växtskyddsrapporter, Jordbruk 51*. Research Information Centre of the Swedish University of Agricultural Sciences, Uppsala, Sweden.

Seinhorst, J.W. (1995) The reduction of the growth and weight of plants by a second and later generations of the nematode. *Nematologica* 41, 592–602. DOI: 10.1163/003925995X00530

Seinhorst, J.W. (1998) The common relation between population density and plant weight in pot and microplot experiments with various nematode plant combinations. *Fundamental and Applied Nematology* 21, 459–468.

Seinhorst, J.W. and Kozlowska, J. (1977) Damage to carrots by *Rotylenchus reniformis*, with a discussion of the cause of increase of tolerance during the development of the plant. *Nematologica* 23, 1–23. DOI: 10.1163/187529277X00165

Sekora, N.S., Lawrence, K.S., Agudelo, P., van Santen, E. and McInroy, J.A. (2010) Differentiation of *Meloidogyne* species with FAME analysis. *Nematropica* 40, 163–175.

Semblat, J.P., Rosso, M.N., Hussey, R.S., Abad, P. and Castagnone-Sereno, P. (2001) Molecular cloning of a cDNA encoding an amphid-secreted putative avirulence protein from the root-knot nematode *Meloidogyne incognita*. *Molecular Plant–Microbe Interactions* 14, 72–79. DOI: 10.1094/MPMI.2001.14.1.72

Seo, Y. and Kim, Y.H. (2017) Pathological interrelations of soil-borne diseases in cucurbits caused by *Fusarium* species and *Meloidogyne incognita*. *The Plant Pathology Journal* 33, 410–423. DOI: 10.5423/PPJ.OA.04.2017.0088

Shanthi, A. and Rajendran, G. (2006) Biological control of lesion nematodes in banana. *Nematologia Mediterranea* 34, 69–75.

Sharon, E., Chet, I. and Spiegel, Y. (2011) *Trichoderma* as a biological control agent. In: Davies, K.G. and Spiegel, Y. (eds) *Biological Control of Plant-parasitic Nematodes: Building Coherence Between Microbial Ecology and Molecular Mechanisms*. Springer, Berlin, Germany, pp. 183–201.

Shepherd, A.M., Clark, S.A. and Hooper, D.J. (1980) Structure of the anterior alimentary tract of *Aphelenchoides blastophthorus* (Nematoda: Tylenchida, Aphelenchina). *Nematologica* 26, 313–357. DOI: 10.1163/187529280X00279

Sher, S.A. (1963) Revision of the Hoplolaiminae (Nematoda). II. *Hoplolaimus* Daday, 1905 and *Aorolaimus* n. gen. *Nematologica* 9, 267–295. DOI: 10.1163/187529263X00476

Shimizu, K., Akiyama, R., Okamura, Y., Ogawa, C., Masuda, Y. *et al.* (2023) Solanoeclepin B, a hatching factor for potato cyst nematode. *Science Advances* 9, 4166. DOI: 10.1126/sciadv.adf4166.

Shivakumara, T.N., Dutta, T.K., Chaudhary, S., von Reuss, S.H., Williamson, V.M. *et al.* (2019) Homologs of *Caenorhabditis elegans* chemosensory genes have roles in behavior and chemotaxis in the root-knot nematode *Meloidogyne incognita*. *Molecular Plant–Microbe Interactions* 32, 876–887. DOI: 10.1094/MPMI-08-18-0226-R.

Shokoohi, E. and Duncan, L.W. (2018) Nematode parasites of citrus. In: Sikora, R., Coyne, D., Hallmann, J. and Timper, P. (eds) *Plant Parasitic Nematodes in Subtropical and Tropical Agriculture*, 3rd edn. CABI, Wallingford, UK, pp. 446–476. DOI: 10.1079/9781786391247.0446

Shrestha, U., Augé, R.A. and Butler, D.M. (2016) A meta-analysis of the impact of anaerobic soil disinfestation on pest suppression and yield of horticultural crops. *Frontiers in Plant Science* 7, 1254. DOI: 10.3389/fpls.2016.01254

Siddiqi, M.R. (1972a) *Tylenchorhynchus cylindricus*. In: *CIH Descriptions of Plant-Parasitic Nematodes*. Set 1, No. 7. Commonwealth Agricultural Bureaux, Farnham Royal, UK.

Siddiqi, M.R. (1972b) *Helicotylenchus dihystera*. In: *CIH Descriptions of Plant-Parasitic Nematodes*. Set 1, No. 9. Commonwealth Agricultural Bureaux, Farnham Royal, UK.

Siddiqi, M.R. (1972c) *Rotylenchus robustus*. In: *CIH Descriptions of Plant-Parasitic Nematodes*. Set 1, No. 11. Commonwealth Agricultural Bureaux, Farnham Royal, UK.

Siddiqi, M.R. (1973) *Xiphinema americanum*. In: *CIH Descriptions of Plant-Parasitic Nematodes*. Set 2, No. 29. Commonwealth Agricultural Bureaux, Farnham Royal, UK.

Siddiqi, M.R. (1974a) *Pratylenchoides crenicauda*. *CIH Descriptions of Plant-Parasitic Nematodes*. Set 3, No. 38. Commonwealth Agriculture Bureaux, Farnham Royal, UK.

Siddiqi, M.R. (1974b) *Aphelenchoides ritzemabosi*. In: *CIH Descriptions of Plant-Parasitic Nematodes*. Set 3, No. 32. Commonwealth Agricultural Bureaux, Farnham Royal, UK.

Siddiqi, M.R. (1974c) *Xiphinema index*. In: *CIH Descriptions of Plant-Parasitic Nematodes*. Set 3, No. 45. Commonwealth Agricultural Bureaux, Farnham Royal, UK.

Siddiqi, M.R. (1974d) Systematics of the genus *Trichodorus* Cobb, 1913 (Nematoda: Dorylaimida), with descriptions of three new species. *Nematologica* 19, 259–278.

Siddiqi, M.R. (1975) *Aphelenchoides fragariae*. In: *CIH Descriptions of Plant-Parasitic Nematodes*. Set 5, No. 74. Commonwealth Agricultural Bureaux, Farnham Royal, UK.

Siddiqi, M.R. (1976) New plant nematode genera *Plesiodorus* (Dolichodorinae), *Meiodorus* (Meiodorinae subfam. n.), *Amplimerlinus* (Amplimerliniiae) and *Cracilancea* (Tylodoridae grad. n.). *Nematologica* 22, 390–416. DOI: 10.1163/187529276X00391

Siddiqi, M.R. (1986) *Tylenchida Parasites of Plants and Insects*. Commonwealth Agricultural Bureaux, Farnham Royal, UK.

Siddiqi, M.R. (2000) *Tylenchida Parasites of Plants and Insects*, 2nd edn. CAB International, Wallingford, UK.

Siddiqi, M.R. (2002) *Ecuadorus equatorius* gen. n., sp. n. and *Nanidorus mexicanus* sp. n. (Nematoda: Trichodoridae). *International Journal of Nematology* 12, 197–202.

Siddiqi, M.R., Baujard, P. and Mounport, D. (1993) Description of *Paratylenchus pernoxius* sp. n. and *Paralongidorus duncani* sp. n. from Senegal, and the synonymization of *Longidoroides* with *Paralongidorus*. *Afro-Asian Journal of Nematology* 3, 81–89.

Siddique, S., Endres, S., Atkins, J.M., Szakasits, D., Wieczorek, K. *et al.* (2009) Myo-inositol oxygenase genes are involved in the development of syncytia induced by *Heterodera schachtii* in Arabidopsis roots. *New Phytologist* 184, 457–472. DOI: 10.1111/j.1469-8137.2009.02981.x

Siddique, S., Wieczorek, K., Szakasits, D., Kreil, D.P. and Bohlmann, H. (2011) The promoter of a plant defensin gene directs specific expression in nematode-induced syncytia in *Arabidopsis* roots. *Plant Physiology and Biochemistry* 49, 1100–1107. DOI: 10.1016/j.plaphy.2011.07.005

Siddique, S., Radakovic, Z.S., De La Torre, C.M., Chronis, D., Novak, O. *et al.* (2015) A parasitic nematode releases cytokinin that controls cell division and orchestrates feeding site formation in host plants. *Proceedings of the National Academy of Science of the USA* 112, 12669–12674. DOI: 10.1073/pnas.1503657112

Siddique, S., Radakovic, Z.S., Hiltl, C., Pellegrin, C., Baum, T.J. *et al.* (2022) The genome and lifestage-specific transcriptomes of a plant-parasitic nematode and its host reveal susceptibility genes involved in trans-kingdom synthesis of vitamin B5. *Nature Communications* 13, 6190. DOI: 10.1038/s41467-022-33769-w

Siebert, J., Ciobanu, M., Schädler, M. and Eisenhauer, N. (2020) Climate change and land use induce functional shifts in soil nematode communities. *Oecologia* 192, 281–294. DOI: 10.1007/s00442-019-04560-4)

Sieriebriennikov, B., Ferris, H, and de Goede, R.G.M. (2014) NINJA: an automated calculation system for nematode-based biological monitoring. *European Journal of Soil Biology* 61, 90–93. DOI: 10.1016/j.ejsobi.2014.02.004

Sikora, R.A. and Roberts, P.A. (2018) Management practices: an overview of integrated nematode management technologies. In: Sikora, R.A., Coyne, D., Hallmann, J. and Timper, P. (eds) *Plant Parasitic Nematodes in Subtropical and Tropical Agriculture*, 3rd edn. CAB International, Wallingford, UK, pp. 795–838. DOI: 10.1079/9781786391247.0795

Sikora, R.A., Coyne, D. and Quénéhervé, P. (2018) Nematode parasites of bananas and plantains. In: Sikora, R.A., Coyne, D., Hallmann, J. and Timper, P. (eds). *Plant Parasitic Nematodes in Subtropical and Tropical Agriculture*, 3rd edn. CAB International, Wallingford, UK, pp. 617–657. DOI: 10.1079/9781786391247.0617.

Sikora, R.A., Desaeger, J. and Molendijk, L.P.G. (eds) (2021) *Integrated Nematode Management: State-of-the-art and Visions for the Future*. CAB International, Wallingford, UK.

Silva, F.B., Batista, J.A., Marra, B.M., Fragoso, R.R., Monteiro, A.C. *et al.* (2004) Prodomain peptide of HGCP-Iv cysteine proteinase inhibits nematode cysteine proteinases. *Genetics and Molecular Research* 3, 342–355.

Silva, J.C.P., Nunes, T.C.S., Guimarães, R.A., Pylro, V.S., Costa, L.S.A.S. *et al.* (2022) Organic practices intensify the microbiome assembly and suppress root-knot nematodes. *Journal of Pest Science* 95, 709–721. DOI: 10.1007/s10340-021-01417-9

Simpson, C.E. and Starr, J.L. (2001) Registration of ‘COAN’ peanut. *Crop Science* 41, 918. DOI: 10.2135/cropsci2001.413918x

Sindhu, A.S., Maier, T.R., Mitchum, M.G., Hussey, R.S., Davis, E.L. and Baum, T.J. (2009) Effective and specific in planta RNAi in cyst nematodes: expression interference of four parasitism genes reduces parasitic success. *Journal of Experimental Botany* 60, 15–324. DOI: 10.1093/jxb/ern289

Singh, P.R., Karssen, G., Couvreur, M., Subbotin, S.A. and Bert, W. (2021) Integrative taxonomy and molecular phylogeny of the plant-parasitic nematode genus *Paratylenchus* (Nematoda: Paratylenchinae): linking species with molecular barcodes. *Plants* 10, 408. DOI: 10.3390/plants10020408

Singh, P.R., van de Vossenberg, B.T.L.H., Rybarczyk-Mydłowska, K., Kowalewska-Groszkowska, M., Bert, W. *et al.* (2022) An integrated approach for synonymization of *Rotylenchus rhomboides* with *R. goodeyi* (Nematoda: Hoplolaimidae) reveals high intraspecific mitogenomic variation. *Phytopathology* 112, 1152–1164. DOI: 10.1094/PHYTO-08-21-0363-R

Singh, S., Awasthi, L.P., Jangre, A. and Nirmalkar, V.K. (2020) Transmission of plant viruses through soil-inhabiting nematode vectors. In: Awasthi, L.P. (ed.) *Applied Plant Virology*. Academic Press, Cambridge, MA, USA, pp. 291–300. DOI: 10.1016/B978-0-12-818654-1.00022-0

Širca, S., Urek, G. and Karssen, G. (2004) First report of the root-knot nematode *Meloidogyne ethiopica* on tomato in Slovenia. *Plant Disease* 88, 680. DOI: 10.1094/pdis.2004.88.6.680c

Skantar, A.M., Handoo, Z.A., Carta, L.K. and Chitwood, D.J. (2007) Morphological and molecular identification of *Globodera pallida* associated with potato in Idaho. *Journal of Nematology* 39, 327–332.

Skantar, A.M., Handoo, Z.A., Zasada, I.A., Ingham, R.E., Carta, L.K. *et al.* (2011) Morphological and molecular characterization of *Globodera* populations from Oregon and Idaho. *Phytopathology* 101, 480–491. DOI: 10.1094/phyto-01-10-0010

Slatko, B.E., Gardner, A.F. and Ausubel, F.M. (2018) Overview of next-generation sequencing technologies. *Current Protocols in Molecular Biology* 122, e59. DOI: 10.1002/cpmb.59

Slomczynska, U., South, M.S., Bunkers, J.G., Edgecomb, D., Wyse-Pester, D. *et al.* (2015) Tioxazafen: a new broad-spectrum seed treatment nematicide. In: Maienfisch, P. and Stevenson, T.M. (eds) *Discovery and Synthesis of Crop Protection Products. Symposium Series 1204*. American Chemical Society, Washington, DC, USA, pp. 129–147. DOI: 10.1021/bk-2015-1204.ch010

Smant, G., Davis, E.L., Hussey, R.S., Baum, T.J., Rosso, M.-N. *et al.* (2004) On the evolution of parasitism genes. In: Cook, R. and Hunt, D.J. (eds) *Proceedings of the Fourth International Congress of Nematology, 8–13 June 2002, Tenerife, Spain. Nematology Monographs and Perspectives* 2. Brill, Leiden, The Netherlands, pp. 573–579.

Smart, G.C. and Nguyen, K.B. (1991) Sting and awl nematodes: *Belonolaimus* spp. and *Dolichodorus* spp. In: Nickle, W.R. (ed.) *Manual of Agricultural Nematology*. Marcel Dekker, New York, USA, pp. 627–668.

Smiley, R.W. (2021) Root-lesion nematodes affecting dryland cereals in the semiarid Pacific Northwest U.S.A. *Plant Disease* 105, 3324–3343. DOI: 10.1094/pdis-04-21-0883-fe

Smiley, R.W., Dababat, A.A., Iqbal, S., Jones, M.G.K., Tanha Maafi, Z. *et al.* (2017) Cereal cyst nematodes: a complex and destructive group of *Heterodera* species. *Plant Disease* 101, 1692–1720. DOI: 10.1094/pdis-03-17-0355-fe

Smythe, A., Holovachov, O. and Kocot, K. (2019) Improved phylogenomic sampling of free-living nematodes enhances resolution of higher-level nematode phylogeny. *BMC Evolutionary Biology* 19, 121. DOI: 10.1186/s12862-019-1444-x

Sobczak, M., Golinowski, W.A. and Grundler, F.M.W. (1999) Ultrastructure of feeding plugs and feeding tubes formed by *Heterodera schachtii*. *Nematology* 1, 363–374. DOI: 10.1163/156854199508351

Sobczak, M., Avrova, A., Jupowicz, J., Phillips, M.S., Ernst, K. and Kumar, A. (2005) Characterization of susceptibility and resistance responses to potato cyst nematode (*Globodera* spp.) infection of tomato lines in the absence and presence of the broad-spectrum nematode resistance *Hero* gene. *Molecular Plant–Microbe Interactions* 18, 158–168. DOI: 10.1094/MPMI-18-0158

Soldan, R., Fusi, M., Cardinale, M., Daffonchio, D. and Preston, G.M. (2021) The effect of plant domestication on host control of the microbiota. *Communications Biology* 4, 936. DOI: 10.1038/s42003-021-02467-6

Southey, J.F. (1993) Nematode pests of ornamental and bulb crops. In: Evans, K., Trudgill, D.L. and Webster, J. (eds) *Plant-Parasitic Nematodes in Temperate Agriculture*. CAB International, Wallingford, UK, pp. 463–500.

Spence, K.O., Lewis, E.E. and Perry, R.N. (2009) Host-finding and invasion by entomopathogenic and plant-parasitic nematodes: evaluating the ability of laboratory bioassays to predict field results. *Journal of Nematology* 40, 93–98.

Srinivasan, J., Kaplan, F., Ajredini, R., Zachariah, C., Alborn, H.T. *et al.* (2008) A blend of small molecules regulates both mating and development in *Caenorhabditis elegans*. *Nature* 454, 1115–1118. DOI: 10.1038/nature07168

Stanton, J.M., McNicol, C.D. and Steele, V. (1998) Non-manual lysis of second-stage *Meloidogyne* juveniles for identification of pure and mixed samples based on the polymerase chain reaction. *Australian Plant Pathology* 27, 112–115. DOI: 10.1071/AP98014

Starr, J.L., Morgan, E. and Simpson, C.E. (2002) Management of the peanut root-knot nematode, *Meloidogyne arenaria*, with host resistance. *Plant Health Progress* 3, 1. DOI: 10.1094/PHP-20021121-01-HM.

Steenkamp, S. (2008) Host plant resistance as a management tool for *Ditylencus africanus* (Nematoda: Tylenchidae) on groundnut (*Arachis hypogaea*). PhD Thesis. North-West University, Potchefstroom, South Africa.

Stein, K.K. and Golden, A. (2018) The *C. elegans* eggshell. In: The *C. elegans* research community (ed.) *WormBook*, http://www.wormbook.org. DOI: 10.1895/wormbook.1.179.1

Stewart, G.R., Perry, R.N. and Wright, D.J. (1994) Immunocytochemical studies on the occurrence of gamma-aminobutyric acid (GABA) in the nervous system of the nematodes *Panagrellus redivivus, Meloidogyne incognita* and *Globodera rostochiensis*. *Fundamental and Applied Nematology* 17, 433–439.

Stewart, G.R., Perry, R.N. and Wright, D.J. (2001) Occurrence of dopamine in *Panagrellus redivivus* and *Meloidogyne incognita*. *Nematology* 3, 843–848. DOI: 10.1163/156854101753625335

Stirling, G.R. (1991) *Biological Control of Plant Parasitic Nematodes: Progress, Problems and Prospects*. CAB International, Wallingford, UK.

Stirling, G.R. (2011) Biological control of plant-parasitic nematodes, a review of progress and opportunities. In: Davies, K.G. and Spiegel, Y. (eds) *Biological Control of Plant-parasitic Nematodes: Building Coherence Between Microbial Ecology and Molecular Mechanisms*. Springer, Berlin, Germany, pp. 1–38.

Stirling, G.R. (2014) *Biological Control of Plant-parasitic Nematodes: Soil Ecosystem Management in Sustainable Agriculture*, 2nd edn. CAB International, Wallingford, UK.

Stone, A.R. (1973) *Heterodera pallida* n. sp. (Nematoda: Heteroderidae), a second species of potato cyst nematode. *Nematologica* 18, 591–606. DOI: 10.1163/187529272X00179

Strachan, S.M., Armstrong, M.R., Kaur, A., Wright, K.M., Lim, T.Y. *et al.* (2019) Mapping the *H2* resistance effective against *Globodera pallida* pathotype Pa1 in tetraploid potato. *Theoretical and Applied Genetics* 132, 1283–1294. DOI: 10.1007/s00122-019-03278-4

Sturhan, D. (1963) Beitrag zur Systematik der Gattung *Longidorus*. *Nematologica* 4, 875–882.

Subbotin, S.A., Franco, J., Knoetze, R., Roubtsova, T.V., Bostock, R.M. *et al.* (2020) DNA barcoding, phylogeny and phylogeography of the cyst nematode species from the genus *Globodera* (Tylenchida: Heteroderidae). *Nematology* 22, 269–297. DOI: 10.1163/15685411-00003305

Subbotin, S.A. (2021a) Molecular identification of nematodes using polymerase chain reaction (PCR). In: Perry, R.N., Hunt, D.J. and Subbotin, S.A. (eds) *Techniques for Work with Plant and Soil Nematodes*. CAB International, Wallingford, UK, pp. 218–239.

Subbotin, S.A. (2021b) Phylogenetic analysis of DNA sequence data. In: Perry, R.N., Hunt, D.J. and Subbotin, S.A. (eds) *Techniques for Work with Plant and Soil Nematodes*. CAB International, Wallingford, UK, pp. 265–282.

Subbotin, S.A. and Burbridge, J. (2021) Sensitive, accurate and rapid detection of the northern root-knot nematode, *Meloidogyne hapla,* using Recombinase Polymerase Amplification assays. *Plants* 10, 336. DOI: 10.3390/plants10020336

Subbotin, S.A. and Kim, D. (2021) Molecular characterisation and phylogenetic relationships of sedentary nematodes of the genus *Meloinema* Choi & Geraert, 1974 (Nematoda: Tylenchida). *Nematology* 23, 507–515. DOI 10.1163/15685411-bja10056

Subbotin, S.A. and Skantar, A.M. (2018) Molecular taxonomy and phylogeny. In: Perry, R.N., Moens, M. and Jones, J.T. (eds). *Cyst Nematodes*. CAB International, Wallingford, UK, pp. 398–417.

Subbotin, S.A., Rumpenhorst, H.J. and Sturhan, D. (1996) Morphological and electrophoretic studies on populations of *Heterodera avenae* complex from the former USSR. *Russian Journal of Nematology* 4, 29–38.

Subbotin, S.A., Waeyenberge, L. and Moens, M. (2000) Identification of cyst forming nematodes of the genus *Heterodera* (Nematoda: Heteroderidae) based on the ribosomal DNA-RFLPs. *Nematology* 2, 153–164. DOI: 10.1163/156854100509042

Subbotin, S.A., Vierstraete, A., De Ley, P., Rowe, J., Waeyenberge, L. *et al.* (2001) Phylogenetic relationships within the cyst-forming nematodes (Nematoda, Heteroderidae) based on analysis of sequences from the ITS regions of ribosomal DNA. *Molecular Phylogenetics and Evolution* 21, 1–16. DOI: 10.1006/mpev.2001.0998

Subbotin, S.A., Sturhan, D., Rumpenhorst, H.J. and Moens, M. (2003) Molecular and morphological characterisation of the *Heterodera avenae* species complex (Tylenchida: Heteroderidae). *Nematology* 5, 515–538. DOI: 10.1163/156854103322683247

Subbotin, S.A., Krall, E.L., Riley, I., Chizhov, V.N., Staelens, A. *et al.* (2004) Evolution of the gall-forming plant parasitic nematodes (Tylenchida: Anguinidae) and their relationships with hosts as inferred from internal transcribed spacer sequences of nuclear ribosomal DNA. *Molecular Phylogenetics and Evolution* 30, 226–235. DOI: 10.1016/S1055-7903(03)00188-X

Subbotin, S.A., Mundo-Ocampo, M. and Baldwin, J.G. (2010a) *Systematics of Cyst Nematodes (Nematoda: Heteroderinae). Nematology Monographs and Perspectives 8A* (Series Editors: Hunt, D.J. and Perry, R.N.). Brill, Leiden, The Netherlands.

Subbotin, S.A., Mundo-Ocampo, M. and Baldwin, J.G. (2010b) *Systematics of Cyst Nematodes (Nematoda: Heteroderinae). Nematology Monographs and Perspectives 8B* (Series Editors: Hunt, D.J. and Perry, R.N.). Brill, Leiden, The Netherlands.

Subbotin, S.A., Akanwari, J., Nguyen, C.N., Cid del Prado Vera, I., Chitambar, J.J. *et al.* (2017) Molecular characterization and phylogenetic relationships of cystoid nematodes of the family Heteroderidae (Nematoda: Tylenchida). *Nematology* 19, 1065–1081. DOI: 10.1163/15685411-00003107

Subbotin, S.A., Franco, J., Knoetze, R., Roubtsova, T. V., Bostock, R. M. *et al.* (2020) DNA barcoding, phylogeny and phylogeography of the cyst nematode species from the genus *Globodera* (Tylenchida: Heteroderidae). *Nematology* 22, 269–297. DOI: 10.1163/15685411-00003305

Subbotin, S.A., Palomares-Rius, J.E. and Castillo, P. (2021a) *Systematics of Root-knot Nematodes (Nematoda: Meloidogynidae). Nematology Monographs and Perspectives 14* (Series Editors: Hunt, D.J. and Perry, R.N.). Brill, Leiden, The Netherlands. DOI: 10.1163/9789004387584

Subbotin, S.A., Oliveira, C.J., Álvarez-Ortega, S., Desaeger, J.A., Crow, W. *et al.* (2021b) The taxonomic status of *Aphelenchoides besseyi* Christie, 1942 (Nematoda: Aphelenchoididae) populations from the southeastern USA, and description of *Aphelenchoides pseudobesseyi* sp. n., *Nematology* 23, 381–413. DOI: 10.1163/15685411-bja10048

Sukhdeo, M.V.K., Sukhdeo, S.C. and Bansemir, A.D. (2002) Interactions between intestinal nematodes and their hosts. In: Lewis, E.E., Campbell, J.F. and Sukhdeo, M.V.K. (eds) *The Behavioural Ecology of Parasites*. CAB International, Wallingford, UK, pp. 223–242.

Sundaram, M.V. and Buechner, M. (2016) The *Caenorhabditis elegans* excretory system: a model for tubulogenesis, cell fate specification and plasticity. *Genetics* 203, 35–63. DOI: 10.1534/genetics.116.189357

Suzuki, K. (2002) Pine wilt disease – a threat to pine forests in Europe. *Nematology Monographs and Perspectives 3* (Series Editors: Hunt, D.J. and Perry, R.N.). Brill, Leiden, The Netherlands, pp. 25–30.

Swart, A., Malan, A.P. and Heyns, J. (1996) *Paralongidorus maximus* (Bütschli) Siddiqi (Dorylaimida: Longidoridae) from South Africa. *African Plant Protection* 2, 31–34.

Syifa, M., Park, S.-J. and Lee, C.-W. (2020) Detection of the pine wilt disease tree candidates for drone remote sensing using artificial intelligence techniques. *Engineering* 6, 919–926. DOI: 10.1016/j.eng.2020.07.001.

Szakasits, D., Heinen, P., Wieczorek, K., Hofmann, J., Wagner, F. *et al.* (2009) The transcriptome of syncytia induced by the cyst nematode *Heterodera schachtii* in *Arabidopsis* roots. *Plant Journal* 57, 771–784. DOI: 10.1111/j.1365-313X.2008.03727.x

Takahashi, A., Casais, C., Ichimura, K. and Shirasu, K. (2003) HSP90 interacts with RAR1 and SGT1 and is essential for RPS2-mediated disease resistance in *Arabidopsis*. *Proceedings of the National Academy of Sciences of the USA* 100, 11777–11782. DOI: 10.1073/pnas.2033934100

Tanino, K., Takahashi, M., Tomata, Y., Tokura, H., Uehara, T. *et al.* (2011) Total synthesis of solanoeclepin A. *Nature Chemistry* 3, 484–488. DOI: 10.1038/nchem.1044

Taylor, C. and Brown, D. (1997) *Nematode Vectors of Plant Viruses*. CAB International, Wallingford, UK.

Taylor, S.R., Santpere, G., Weinreb, A., Barrett, A., Reilly, M.B. *et al.* (2021) Molecular topography of an entire nervous system. *Cell* 184, 4329–4347.e23. DOI: 10.1016/j.cell.2021.06.023

Teillet, A., Dybal, K., Kerry, B.R., Miller, A.J., Curtis, R.H.C. *et al.* (2013) Transcriptional changes of the root-knot nematode *Meloidogyne incognita* in response to *Arabidopsis thaliana* root signals. *PLoS One* 8, e61259. DOI: 10.1371/journal.pone.0061259

Teklu, M.G., Schomaker, C.H. and Been, T.H. (2014) Relative susceptibilities of five fodder radish varieties (*Raphanus sativus* var. *Oleiformis*) to *Meloidogyne chitwoodi*. *Nematology* 16, 577–590. DOI: 10.1163/15685411-00002789

Teklu, M.G., Schomaker, C.H., Been, T.H. and Molendijk, L.P.G. (2016) A routine test for the relative susceptibility of potato genotypes with resistance to *Meloidogyne chitwoodi*. *Nematology* 18, 1079–1094. DOI: 10.1163/15685411-00003016

Teklu, M.G., Schomaker, C.H. and Been, T.H. (2018) The effect of storage time and temperature on the population dynamics and vitality of *Meloidogyne chitwoodi* in potato tubers. *Nematology* 20, 373–385. DOI: 10.1163/15685411-00003145

Thakur, M.P. and Geisen, S. (2019) Trophic regulations of the soil microbiome. *Trends in Microbiology* 27, 771–778. DOI: 10.1016/j.tim.2019.04.008

Thoden T.C. and Wiles J.W. (2019) Biological attributes of Salibro™, a novel sulfonamide nematicide. Part 1: Impact on the fitness of *Meloidogyne incognita*, *M. hapla* and *Acrobeloides buetschlii*. *Nematology* 21, 625–639. DOI: 10.1163/15685411-00003240

Thoden, T.C., Korthals, G. and Termorshuizen, A. (2011) Organic amendments and their influences on plant-parasitic and free-living nematodes: a promising method for nematode management? *Nematology* 13, 133–153. DOI: 10.1163/138855410X541834

Thoden T.C., Pardavella I.V. and Tzortzakakis E.A. (2019) *In vitro* sensitivity of different populations of *Meloidogyne javanica* and *M. incognita* to the nematicides Salibro™ and Vydate®. *Nematology* 21, 889–893. DOI: 10.1163/15685411-00003282

Thoden T.C., Alkader M.A. and Wiles J.A. (2020) Biological attributes of Salibro™, a novel sulfonamide nematicide. Part 2: impact on the fitness of various non-target nematodes. *Nematology* 23, 287–303. DOI: 10.1163/15685411-bja10041

Thompson, D.P. and Geary, T.G. (2002) Excretion/secretion, ionic and osmotic regulation. In: Lee, D.L. (ed.) *The Biology of Nematodes*. Taylor & Francis, London, UK, pp. 291–320.

Thompson, J.P., Reen, R.A., Clewett, T.G., Sheedy, J.G., Kelly, A.M. *et al.* (2011) Hybridisation of Australian chickpea cultivars with wild *Cicer* spp. increases resistance to root-lesion nematodes (*Pratylenchus thornei* and *P. neglectus*). *Australasian Plant Pathology* 40, 601–611. DOI: 10.1007/s13313-011-0089-z

Tian, X.-J., Wu, X.-Q., Xiang, Y., Fang, X. and Ye, J.-R. (2015) The effect of endobacteria on the development and virulence of the pine wood nematode, *Bursaphelenchus xylophilus*. *Nematology* 17, 581–589. DOI: 10.1163/15685411-00002892

Tigano, M., de Siqueira, K., Castagnone-Sereno, P., Mulet, K., Queiroz, P. *et al.* (2010) Genetic diversity of the root-knot nematode *Meloidogyne enterolobii* and development of a SCAR marker for this guava-damaging species. *Plant Pathology* 59, 1054–1061. DOI: 10.1111/j.1365-3059.2010.02350.x

Tiilikkala, K., Carter, T., Heikinheimo, M. and Venäläinen, A. (1995) Pest risk analysis of *Meloidogyne chitwoodi* for Finland. *EPPO Bulletin* 25, 419–435. DOI: 10.1111/j.1365-2338.1995.tb00576.x

Timper, P., Davis, R., Jagdale, G. and Herbert, J. (2012) Resiliency of a nematode community and suppressive service to tillage and nematicide application. *Applied Soil Ecology* 59, 48–59. DOI: 10.1016/j.apsoil.2012.04.001

Tobin, J.D., Haydock, P.P.J, Hare, M.C., Wood, S.R. and Crump, D.H. (2008) Effect of the fungus *Pochonia chlamydosporia* and fosthiazate on the multiplication rate of potato cyst nematodes (*Globodera pallida* and *G. rostochiensis*) in potato crops grown under UK field conditions. *Biological Control* 46, 194–201. DOI: 10.1016/j.biocontrol.2008.03.014

Topalović, O., Santos, S.S., Heuer, H., Nesme, J., Kanfra, X. *et al.* (2022) Deciphering bacteria associated with a pre-parasitic stage of the root-knot nematode *Meloidogyne hapla* in nemato-suppressive and nemato-conducive soils. *Applied Soil Ecology* 172, 104344. DOI: 10.1016/j.apsoil.2021.104344

Triantaphyllou, A.C. (1985) Cytogenetics, cytotaxonomy and phylogeny of root-knot nematodes. In: Sasser, J.N. and Carter, C.C. (eds) *An Advanced Treatise on* Meloidogyne, *Volume I: Biology and Control*. North Carolina State University Graphics, Raleigh, NC, USA, pp. 113–126.

Triantaphyllou, A.C. (1993) Hermaphroditism in *Meloidogyne hapla*. *Journal of Nematology* 25, 15–26.

Trinh, P.Q., Nguyen, C.N., Waeyenberge, L., Subbotin, S.A., Karssen, G. *et al.* (2004) *Radopholus arabocoffeae* sp. n. (Nematoda: Pratylenchidae), a nematode pathogenic on

Coffea arabica in Vietnam and additional data on *R. duriophilus*. *Nematology* 6, 681–693. DOI: 10.1163/1568541042843577

Trinh, P.Q., Wesemael, W.M.L., Tran, H.A., Nguyen, C.N. and Moens, M. (2012) Resistance screening of *Coffea* spp. accessions for *Pratylenchus coffeae* and *Radopholus arabocoffeae* in Vietnam. *Euphytica* 185, 233–241. DOI: 10.1007/s10681-011-0558-z

Tripathi, L., Babirye, A., Roderick, H., Tripathi, J.N., Changa, C. *et al.* (2015) Field resistance of transgenic plantain to nematodes has potential for future African food security. *Scientific Reports* 5, 8127. DOI: 10.1038/srep08127

Troccoli, A., Fanelli, E., Castillo, P., Liébanas, G., Cotroneo, A. *et al.* (2021) *Pratylenchus vovlasi* sp. nov. (Nematoda: Pratylenchidae) on raspberries in North Italy with a morphometrical and molecular characterization. *Plants* 10, 1068. DOI: 10.3390/plants10061068

Trudgill, D. (1985) Potato cyst nematodes: a critical review of the current pathotype scheme. *European and Mediterranean Organisation Bulletin* 15, 273–279. DOI: 10.1111/j.1365-2338.1985.tb00228.x

Trudgill, D.L. and Blok, V.C. (2001) Apomictic, polyphagous root-knot nematodes: exceptionally successful and damaging biotrophic root pathogens. *Annual Review of Phytopathology* 39, 53–77. DOI: 10.1146/annurev.phyto.39.1.53

Trudgill, D.L., Honek, A., Li, D. and van Straalen, N.M. (2005) Thermal time – concepts and utility. *Annals of Applied Biology* 146, 1–14. DOI: 10.1111/j.1744-7348.2005.04088.x

Tsai, A.Y., Iwamoto,Y., Tsumuraya, Y., Oota, M., Konishi, T. *et al.* (2021) Root-knot nematode chemotaxis is positively regulated by l-galactose sidechains of mucilage carbohydrate rhamnogalacturonan-I. *Science Advances* 7, eabh4182. DOI: 10.1126/sciadv.abh4182

Tunlid, A. and Ahrén, D. (2011) Molecular mechanisms of the interaction between nematode-trapping fungi and nematodes: lessons from genomics. In: Davies, K.G. and Spiegel, Y. (eds) *Biological Control of Plant-parasitic Nematodes: Building Coherence Between Microbial Ecology and Molecular Mechanisms*. Springer, Berlin, Germany, pp. 145–169.

Turner, S.J. (1990) The identification and fitness of virulent potato cyst nematodes (*Globodera pallida*) selected on resistant *Solanum vernei* hybrids for up to eleven generations. *Annals of Applied Biology* 117, 385–397. DOI: 10.1111/j.1744-7348.1990.tb04225.x

Turner, S.J. and Evans, K. (1998) The origins, global distribution and biology of potato cyst nematodes (*Globodera rostochiensis* (Woll.) and *Globodera pallida* (Stone). In: Marks, R.J. and Brodie, B.B. (eds) *Potato Cyst Nematodes: Biology, Distribution and Control*. CAB International, Wallingford, UK, pp. 7–26.

Turner, S.J., Stone, A.R. and Perry, J.N. (1983) Selection of potato cyst nematodes on resistant *Solanum vernei* hybrids. *Euphytica* 32, 911–917. DOI: 10.1007/BF00042173

Tytgat, T., De Meutter, J., Gheysen, G. and Coomans, A. (2000) Sedentary endoparasitic nematodes as a model for other plant-parasitic nematodes. *Nematology* 2, 113–121. DOI: 10.1163/156854100508827

Tzortzakakis, E.A. and Gowen, S.R. (1996) Occurrence of a resistance breaking pathotype of *Meloidogyne javanica* on tomatoes in Crete, Greece. *Fundamental and Applied Nematology* 19, 283–288.

Uchida, J.Y., Sipes, B.S. and Kadooka, C.Y. (2003) Burrowing nematode on anthurium: recognizing symptoms, understanding the pathogen, and preventing disease. *Plant Disease* PD-24. College of Tropical Agriculture and Human Resources (CTAHR), University of Hawaii at Manoa, Honolulu, HI, USA.

Uehara, T., Mizukobo, T., Kushida, A. and Momota, Y. (1998) Identification of *Pratylenchus coffeae* and *P. loosi* using specific primers for PCR amplification of ribosomal DNA. *Nematologica* 44, 357–368. DOI: 10.1163/005525998X00034

Uehara, T., Kushida, A. and Momota, Y. (1999) Rapid and sensitive identification of *Pratylenchus* spp. using reverse dot blot hybridization. *Nematology* 1, 549–555. DOI: 10.1163/156854199508441

Urwin, P.E., Atkinson, H.J., Waller, D.A. and McPherson, M.J. (1995) Engineered oryzacystatin-I expressed in transgenic hairy roots confers resistance to *Globodera pallida*. *Plant Journal* 8, 121–131. DOI: 10.1046/j.1365-313x.1995.08010121.x

Urwin, P.E., Lilley, C.J., McPherson, M.J. and Atkinson, H.J. (1997a) Resistance to both cyst and root-knot nematodes conferred by transgenic *Arabidopsis* expressing a modified plant cystatin. *Plant Journal* 12, 455–461. DOI: 10.1046/j.1365-313X.1997.12020455.x

Urwin, P.E., Lilley, C.J., McPherson, M.J. and Atkinson, H.J. (1997b) Characterisation of two cDNAs encoding cysteine proteases from the soybean cyst nematode *Heterodera glycines*. *Parasitology* 114, 605–613. DOI: 10.1017/S0031182097008858

Urwin, P.E., Levesley, A., McPherson, M.J. and Atkinson, H.J. (2000) Transgenic resistance to the nematode *Rotylenchulus reniformis* conferred by *Arabidopsis thaliana* plants expressing proteinase inhibitors. *Molecular Breeding* 6, 257–264. DOI: 10.1023/A:1009669325944

Urwin, P.E., Troth, K.M., Zubko, E.I. and Atkinson, H.J. (2001) Effective transgenic resistance to *Globodera pallida* in potato field trials. *Molecular Breeding* 8, 95–101. DOI: 10.1023/A:1011942003994

Urwin, P.E., Green, J. and Atkinson, H.J. (2003) Expression of a plant cystatin confers partial resistance to *Globodera*, full resistance is achieved by pyramiding a cystatin with natural resistance. *Molecular Breeding* 12, 263–269. DOI: 10.1023/A:1026352620308

U.S. Department of Transportation (2017) U.S. Waterborne foreign container trade by U.S. customs ports 2000–2017. Available at: https://www.maritime.dot.gov/data-reports/data-statistics/us-waterborne-foreign-container-trade-us-customs-ports-2000-%E2%80%93-2017 (accessed 29 January 2024).

Vachon, V., Laprade, R. and Schwartz, J-L. (2012) Current models of the mode of action of *Bacillus thuringiensis* insecticidal crystal proteins: a critical review. *Journal of Invertebrate Patholology* 111, 1–12. DOI: 10.1016/j.jip.2012.05.001

Vain, P., Worland, B., Clarke, M.C., Richard, G., Beavis, M. *et al.* (1998) Expression of an engineered cysteine proteinase inhibitor (OC-IΔD86) for nematode resistance in transgenic rice plants. *Theoretical and Applied Genetics* 96, 266–271. DOI: 10.1007/s001220050735

Valette, C., Mounport, D., Nicole, M., Sarah, J.-L. and Baujard, P. (1998) Scanning electron microscope study of two African populations of *Radopholus similis* (Nematoda: Pratylenchidae) and proposal of *R. citrophilus* as a junior synonym of *R. similis*. *Fundamental and Applied Nematology* 21, 139–146.

Van De Velde, M.C. and Coomans, A. (1989) A putative new hydrostatic skeletal function for the epidermis in monhysterids (Nematoda). *Tissue and Cell* 21, 525–533. DOI: 10.1016/0040-8166(89)90005-0

van den Berg, E. and De Waele, D. (1989) Further observations on nematodes associated with rice in South Africa. *Phytophylactica* 21, 125–130.

van den Berg, E. and Quénéhervé, P. (2012) Hoplolaimidae Filipjev, 1934. In: Manzanilla-López, R.H. and Marbán-Mendoza, N. (eds) *Practical Plant Nematology*. Bibliotheca Basica de Agricultura, Montecillo, Mexico, pp. 251–298.

van den Elsen, S., Ave, M., Schoenmakers, N., Landeweert, R., Bakker, J. *et al.* (2012) A rapid, sensitive, and cost-effective assay to estimate viability of potato cyst nematodes. *Phytopathology* 102, 140–146. DOI: 10.1094/PHYTO-02-11-0051

van den Hoogen, J., Geisen, S., Routh, D., Ferris, H., Traunspurger, W. *et al.* (2019) Soil nematode abundance and functional group composition at a global scale. *Nature* 572, 194–198. DOI: 10.1038/s41586-019-1418-6

van der Putten, W.H., van Dijk, C. and Peters, B.A. (1993) Plant-specific soil-borne diseases contribute to succession in foredune vegetation. *Nature* 362, 53–56. DOI: 10.1038/362053a0

van der Putten, W.H., Cook, R., Costa, S., Davies, K.G., Fargette, M. *et al.* (2006) Nematode interactions in nature: models for sustainable control of nematode pests of crop plants? *Advances in Agronomy* 89, 227–260. DOI: 10.1016/S0065-2113(05)89005-4

van der Putten, W.H., Bardgett, R.D., Bever, J.D., Bezemer, T.M., Casper, B.B. *et al.* (2013) Plant–soil feedbacks: the past, the present and future challenges. *Journal of Ecology* 101, 265–276. DOI: 10.1111/1365-2745.12054 van der Stoel, C.D., Duyts, H., van der Putten, W.H. and Persson, L. (2006) Population dynamics of a host-specific root-feeding cyst nematode and resource quantity in the root zone of a clonal grass. *Oikos* 112, 651–659. DOI: 10.1111/j.0030-1299.2006.13547.x

Van Gundy, S.D. (1985) Ecology of *Meloidogyne* spp. – emphasis on environmental factors affecting survival and pathogenicity. In: Sasser, J.N. and Carter, C.C. (eds) *An Advanced Treatise on* Meloidogyne, *Volume I: Biology and Control.* North Carolina State University Graphics, Raleigh, NC, USA, pp 177–182.

van Megen, H., van den Elsen, S., Holterman, M., Karssen, G., Mooyman, P. *et al.* (2009) A phylogenetic tree of nematodes based on about 1200 full-length small subunit ribosomal DNA sequences. *Nematology* 11, 927–950. DOI: 10.1163/156854109X456862

Van Valen, L. (1976) The Red Queen. *The American Naturalist* 110, 809–810. DOI: 10.1086/283213

van Wersch, S., Tian, L., Hoy, R. and Li, X. (2020) Plant NLRs: The whistleblowers of plant immunity. *Plant Communications* 1, 100016. DOI: 10.1016/j.xplc.2019.100016

Veronico, P., Gray, L.J., Jones, J.T., Bazzicalupo, P., Arbucci, S. *et al.* (2001) Nematode chitin synthases: gene structure, expression and function in *Caenorhabditis elegans* and the plant parasitic nematode *Meloidogyne artiella*. *Molecular Genetics and Genomics* 266, 28–34. DOI: 10.1007/s004380100513

Vicente, C., Espada, M., Vieira, P. and Mota, M. (2012) Pine wilt disease: a threat to European forestry. *European Journal of Plant Pathology* 133, 89–99. DOI: 10.1007/s10658-011-9924-x

Vieira, P., Wantoch, S., Lilley, C.J., Chitwood, D.J., Atkinson, H.J. *et al.* (2015a) Expression of a cystatin transgene can confer resistance to root lesion nematodes in *Lilium longiflorum* cv. 'Nellie White'. *Transgenic Research* 24, 421–432. DOI: 10.1007/s11248-014-9848-2

Vieira, P., Eves-van den Akker, S., Verma, R., Wantoch, S., Eisenback, J.D. *et al.* (2015b) The *Pratylenchus penetrans* transcriptome as a source for the development of alternative control strategies: mining for putative genes involved in parasitism and evaluation of *in planta* RNAi. *PLoS One* 10, e0144674. DOI : 10.1371/journal.pone.0144674

Vieira, P., Mowery, J., Eisenback, J.D., Shao, J. and Nemchinov, L.G. (2019) Cellular and transcriptional responses of resistant and susceptible cultivars of alfalfa to the root lesion nematode, *Pratylenchus penetrans*. *Frontiers in Plant Science* 10, 971. DOI: 10.3389/fpls.2019.00971

Vinciguerra, M.T. and Clausi, M. (2000) *Afractinca andrassyi* gen. n., sp. n. (Nematoda: Actinolaimidae) from Ivory Coast. *Journal of Nematode Morphology and Systematics* 2, 109–112.

Vlaar, L.E., Bertran, A., Rahimi, M., Dong, L., Kammenga, J.E. *et al.* (2021) On the role of dauer in the adaptation of nematodes to a parasitic lifestyle. *Parasites Vectors* 14, 554. DOI: 10.1186/s13071-021-04953-6

Voronov, D.A. and Nezlin, L.P. (1994) Neurons containing catecholamine in juveniles of eight species of free-living nematodes. *Russian Journal of Nematology* 2, 33–40.

Vovlas, N. and Inserra, R.N. (2000) Root-knot nematodes as parasites of citrus. *Proceedings of the International Society of Citriculture, IX Congress*, Orlando, FL, USA, pp. 812–817.

Vovlas, N., Troccoli, A., Palomares-Rius, J.E., De Luca, F., Liébanas, G. *et al.* (2011) *Ditylenchus gigas* n. sp. parasitizing broad bean: a new stem nematode singled out from the *Ditylenchus dipsaci* species complex using a polyphasic approach with molecular phylogeny. *Plant Pathology* 60, 762–775. DOI: 10.1111/j.1365-3059.2011.02430.x

Walawage, S.L., Britton, M.T., Leslie, C.A., Uratsu, S.L. *et al.* (2013) Stacking resistance to crown gall and nematodes in walnut rootstocks. *BMC Genomics* 14, 668. DOI; 10.1186/1471-2164-14-668

Walker, J.T. and Tsui, R.K. (1968) Induction of ovoviviparity in *Rhabditis* by sulfur dioxide. *Nematologica* 14, 148–149. DOI: 10.1163/187529268X00769

Wallace, H. (1958) Movement of eelworms III. The relationship between eelworm length, activity and mobility. *Annals of Applied Biology* 46, 662–668. DOI: 10.1111/j.1744-7348.1958.tb02249.x

Wallace, H. (1968) The dynamics of nematode movement. *Annual Review of Phytopathology* 6, 91–114. DOI: 10.1146/annurev.py.06.090168.000515

Wang, C., Ulloa, M. and Roberts, P.A. (2008) A transgressive segregation factor (*RKN2*) in *Gossypium barbadense* for nematode resistance with the gene *rkn1*in *G. hirsutum*. *Molecular Genetics and Genomics* 279, 41–52. DOI: 10.1007/s00438-007-0292-3

Wang, C., Bruening, G. and Williamson, V.M. (2009a) Determination of preferred pH for root-knot nematode aggregation using pluronic F-127 gel. *Journal of Chemical Ecology* 35, 1242–1251. DOI: 10.1007/s10886-009-9703-8

Wang, D., Jones, L.M., Urwin, P.E. and Atkinson, H.J. (2011) A synthetic peptide shows retro- and anterograde neuronal transport before disrupting the chemosensation of plant-pathogenic nematodes. *PLoS One* 6, e17475. DOI: 10.1371/journal.pone.0017475

Wang, D., Ma, N., Rao, W. and Zhang, Y. (2023) Recent advances in life history transition with nematode-trapping fungus *Arthrobotrys oligospora* and its application in sustainable agriculture. *Pathogens* 12, 367. DOI: 10.3390/pathogens12030367

Wang, D.D., Cheng, X.Y., Wang, Y.S., Pan, H.Y. and Xie, B.Y. (2013) Characterization and expression of *daf-9* and *daf-12* genes in the pinewood nematode, *Bursaphelenchus xylophilus*. *Forest Pathology* 43, 144–152. DOI: 10.1111/efp.12011

Wang, M-L., Uruu, G., Xiong, L., He, X., Nagai, C. *et al.* (2009b) Production of transgenic pineapple (*Ananas comosus* (L.) Merr.) plants via adventitious bud regeneration. *In Vitro Cellular and Developmental Biology – Plant* 45, 112–121. DOI: 10.1007/s11627-009-9208-8

Wang, X., Xue, B., Dai, J., Qin, X., Liu, L. *et al.* (2018) A novel *Meloidogyne incognita* chorismate mutase effector suppresses plant immunity by manipulating the salicylic acid pathway and functions mainly during the early stages of nematode parasitism. *Plant Pathology* 67, 1436–1448. DOI: 10.1111/ppa.12841

Wang, Y., Luo, X., Chen, Y., Peng, J., Yi, C. and Chen, J. (2022) Recent research progress of heterocyclic nematicidal active compounds. *Journal of Heterocyclic Chemistry* 60, 1287–1300. DOI: 10.1002/jhet.4616

Wang, Z., Stoltzfus, J., You, Y-j., Ranjit, N., Tang, H. *et al.* (2015) The nuclear receptor DAF-12 regulates nutrient metabolism and reproductive growth in nematodes. *PLoS Genetics* 11, e1005027. DOI: 10.1371/journal.pgen.1005027

Wasala, S.K., Brown, A.M.V., Kang, J., Howe, D.K., Peetz, A.B. *et al.* (2019) Variable abundance and distribution of *Wolbachia* and *Cardinium* endosymbionts in plant-parasitic nematode field populations. *Frontiers in Microbiology* 10, 964. DOI: 10.3389/fmicb.2019.00964

Webster, J.M. (1985) Interaction of *Meloidogyne* with fungi on plants. In: Sasser, J.N. and Carter, C.C. (eds) *An Advanced Treatise on* Meloidogyne, *Volume I: Biology and Control*. North Carolina State University Graphics, Raleigh, NC, USA, pp. 183–192.

Wei, W., Yanli X., Li, S., Zhu, L. and Song, J. (2015) Developing suppressive soil for root diseases of soybean with continuous long-term cropping of soybean in black soil of Northeast China. *Acta Agriculturae Scandinavica, Section B, Soil & Plant Science* 65, 279–285. DOI: 10.1080/09064710.2014.992941

Wesemael, W.M.L. (2021) Screening plants for resistance/susceptibility to plant-parasitic nematodes. In: Perry, R.N., Hunt, D.J. and Subbotin, S.A. (eds). *Techniques for Work with Plant and Soil Nematodes*. CAB International, Wallingford, UK, pp. 60–70. DOI: 10.1079/9781786391759.0004

Wesemael, W.M.L. and Moens, M. (2012) Screening of common bean (*Phaseolus vulgaris*) for resistance against temperate root-knot nematodes (*Meloidogyne* spp.). *Pest Management Science* 68, 702–708. DOI: 10.1002/ps.2316

Wesemael, W.M.L., Perry, R.N. and Moens, M. (2006) The influence of root diffusate and host age on hatching of the root-knot nematodes, *Meloidogyne chitwoodi* and *M. fallax*. *Nematology* 8, 893–899. DOI: 10.1163/156854106779799204

Westerdahl, B., Riddle, L., Giraud, D. and Kamo, K. (2023) Field test of Easter lilies transformed with a rice cystatin gene for root lesion nematode resistance. *Frontiers in Plant Science* 14, 1134224. DOI: 10.3389/fpls.2023.1134224

Westphal, A., Buzo, T.R., Maung Zin Thu Zar, Mckenry, M., Leslie, C.A. *et al.* (2021) Strategies for breeding walnut (*Juglans* spp.) rootstocks with resistance and tolerance to plant-parasitic nematodes. *Acta Horticulturae* 1318, 33–38. DOI: 10.17660/ActaHortic.2021.1318.5

Wharton, D.A. and Perry, R.N. (2011) Osmotic and ionic regulation. In: Perry, R.N. and Wharton, D.A. (eds) *Molecular and Physiological Basis of Nematode Survival*. CAB International, Wallingford, UK, pp. 256–281.

Wharton, D.A., Perry, R.N. and Beane, J. (1993) The role of the eggshell in the cold tolerance mechanisms of the unhatched juveniles of *Globodera rostochiensis*. *Fundamental and Applied Nematology* 16, 425–431.

White, J. Southgate, E., Thomson, J.N. and Brenner, S. (1986) The structure of the nervous system of the nematode *Caenorhabditis elegans*. *Philosophical Transactions of the Royal Society of London B, Biological Science* 314, 1–340. DOI: 10.1098/rstb.1986.0056

Whitehead, A.G. (1968) Taxonomy of *Meloidogyne* (Nematodea: Heteroderidae) with descriptions of four new species. *Transactions of the Zoological Society, London* 31, 263–401. DOI: 10.1111/j.1096-3642.1968.tb00368.x

Whitehead, A.G. (1977) Vertical distribution of potato, beet and pea cyst nematodes in some heavily infested soils. *Plant Pathology* 26, 85–90. DOI: 10.1111/J.1365-3059.1977.TB01029.X

Whitehead, A.G. (1998) *Plant Nematode Control*. CAB International, Wallingford, UK.

Widmer, T.L., Mitkowski, N.A. and Abawi, G.S. (2002) Soil organic matter and management of plant-parasitic nematodes. *Journal of Nematology* 34, 289–295.

Wieczorek, K., Hofmann, J., Blöchl, A., Szakasits, D., Bohlmann, H. *et al.* (2008) Arabidopsis endo-1,4-β-glucanases are involved in the formation of root syncytia induced by *Heterodera schachtii*. *Plant Journal* 53, 336–351. DOI: 10.1111/j.1365-313x.2007.03340.x

Wille, J.E. and Bazán de Segura, C. (1952) La anguilula dorada, *Heterodera rostochiensis*, una plaga del cultivo de las papas, recien descubierta en el Peru. *Centro Nacional de Investigación y Experimentación Agricola de 'La Molina', Lima, Peru, Boletín*, 48, 17 pp.

Willett, D.S., Filgueiras, C.C., Benda, N.D., Zhang, J. and Kenworthy, K.E. (2020) Sting nematodes modify metabolomic profiles of host plants. *Scientific Reports* 10, 2212. DOI: 10.1038/s41598-020-59062-8

Williams, T.D. and Siddiqi, M.R. (1972) *Heterodera avenae*. In: *CIH Descriptions of Plant-Parasitic Nematodes*. Set 1, No. 2. Commonwealth Agricultural Bureaux, Farnham Royal, UK.

Williamson, V.M. (1998) Root-knot nematode resistance genes in tomato and their potential for future use. *Annual Review of Phytopathology* 36, 277–293. DOI: 10.1146/annurev.phyto.36.1.277

Williamson, V.M. and Čepulytė, R. (2017) Assessing attraction of nematodes to host roots using pluronic gel medium. *Methods in Molecular Biology* 1573, 261–268. DOI: 10.1007/978-1-4939-6854-1_19

Williamson, V.M. and Hussey, R.S. (1996) Nematode pathogenesis and resistance in plants. *Plant Cell* 8, 1735–1745. DOI: 10.1105/tpc.8.10.1735

Williamson, V.M. and Kumar, A. (2006) Nematode resistance in plants: the battle is underground. *Trends in Genetics* 22, 396–403. DOI: 10.1016/j.tig.2006.05.003

Williamson, V.M. and Roberts, P.A. (2009) Mechanisms and genetics of resistance. In: Perry, R.N., Moens, M. and Starr, J.L. (eds) *Root-knot Nematodes*. CAB International, Wallingford, UK, pp. 309–325.

Winter, M.D., McPherson, M.J. and Atkinson, H.J. (2002) Neuronal uptake of pesticides disrupts chemosensory cells of nematodes. *Parasitology* 125, 561–565. DOI: 10.1017/S0031182002002482

Witek, K., Jupe, F., Witek, A.I., Baker, D., Clark, M.D. and Jones, J.D. (2016) Accelerated cloning of a potato late blight-resistance gene using RenSeq and SMRT sequencing. *Nature Biotechnology* 34, 656–660. DOI: 10.1038/nbt.3540

Witvliet, D., Mulcahy, B., Mitchell, J.K., Meirovitch, Y., Berger, D.R. *et al.* (2021) Connectomes across development reveal principles of brain maturation. *Nature* 596, 257–261. DOI: 10.1038/s41586-021-03778-8

Woods, S.R. and Haydock, P.P.J. (2000) The effect of granular nematicide incorporation depth and potato planting depth on potatoes grown in land infested with the potato cyst nematodes *Globodera rostochiensis* and *G. pallida*. *Annals of Applied Biology* 136, 27–33. DOI: 10.1111/j.1744-7348.2000.tb00005.x

Wollenweber, H. (1923) Krankheiten und Beschädingung der Kartoffel. Arbeiten *Forschungs Institut für Kartoffel Berlin* 7, 1–56.

Wram, C.L. and Zasada, I.A. (2020) Differential response of *Meloidogyne*, *Pratylenchus, Globodera,* and *Xiphenema* species to the nematicide fluazaindolozine. *Phytopathology* 110, 2003–2009. DOI: 10.1094/PHYTO-05-20-0189-R

Wram, C.L., Hesse, C.N. and Zasada, I.A. (2022) Transcriptional changes of biochemical pathways in *Meloidogyne incognita* in response to non-fumigant nematicides. *Scientific Reports* 12, 9875. DOI: 10.1038/s41598-022-14091-3.

Wrather, J.A., Anderson, T.R., Arsyad, D.M., Tan, Y., Ploper, L.D. *et al.* (2001) Soybean disease loss estimates for the top ten soybean producing countries in 1998. *Canadian Journal of Plant Pathology* 23, 115–121. DOI: 10.1094/pdis.1997.81.1.107

Wright, D.J. (1981) Nematicides: mode of action and new approaches to chemical control. In: Zuckerman, B.M. and Rhode, R.A. (eds) *Plant Parasitic Nematodes, Vol. III*. Academic Press, New York, USA, pp. 421–449.

Wright, D.J. (1998) Respiratory physiology, nitrogen excretion and osmotic and ionic regulation. In: Perry, R.N. and Wright, D.J. (eds) *The Physiology and Biochemistry of Free-living and Plant-parasitic Nematodes*. CAB International, Wallingford, UK, pp. 103–131.

Wright, D.J. (2004) Osmoregulatory and excretory behaviour. In: Gaugler, R. and Bilgrami, A.L. (eds) *Nematode Behaviour*. CAB International, Wallingford, UK, pp. 177–196.

Wright, D.J. and Perry, R.N. (1998) Musculature and neurobiology. In: Perry, R.N. and Wright, D.J. (eds) *The Physiology and Biochemistry of Free-living and Plant-parasitic Nematodes*. CAB International, Wallingford, UK, pp. 49–73.

Wright, K.A. and Carter, R. (1980) Cephalic sense organs and body pores of *Xiphinema americanum* (Nematoda: Dorylaimoidea). *Canadian Journal of Zoology* 58, 1439–1451. DOI: 10.1139/z80-198

Wright, K.A. and Perry, R.N. (1991) Moulting of *Aphelenchoides hamatus*, with especial reference to the formation of the stomatostyle. *Revue de Nématologie* 14, 497–504.

Wu, Y., Wickham, J.D., Zhao, L. and Sun, J. (2019) CO_2 drives the pine wood nematode off its insect vector. *Current Biology* 29, R619–R620. DOI: 10.1016/j.cub.2019.05.033

Wyss, U. (1997) Root parasitic nematodes: an overview. In Fenoll, C., Grundler, F.M.W. and Ohl, S. (eds) *Cellular and Molecular Aspects of Plant – Nematode Interations*. Klumer Academic Press, Dordrecht, The Netherlands, pp. 5–22. DOI: 10.1007/978-94-011-5596-0_2

Wyss, U. (2002) Feeding behaviour of plant parasitic nematodes. In: Lee, D.L. (ed.) *The Biology of Nematodes*. Taylor & Francis, London, UK, pp. 233–260.

Xie, Y. and Zhang, L. (2022) Transcriptomic and proteomic analysis of marine nematode *Litoditis marina* acclimated to different salinities. *Genes* 13, 651. DOI: 10.3390/genes13040651

Xu, Y. and Crouch, J.H. (2008) Marker-assisted selection in plant breeding: from publications to practice. *Crop Science* 48, 391–407. DOI: 10.2135/cropsci2007.04.0191

Xue, Q., Xiang, Y., Wu, X.-Q. and Li, M.-J. (2019) Bacterial communities and virulence associated with pine wood nematode *Bursaphelenchus xylophilus* from different *Pinus* spp. *International Journal of Molecular Sciences* 20, 3342. DOI: 10.3390/ijms20133342

Yadav, B.C., Veluthambi, K. and Subramaniam, K. (2006) Host-generated double stranded RNA induces RNAi in plant-parasitic nematodes and protects the host from infection. *Molecular and Biochemical Parasitology* 148, 219–222. DOI: 10.1016/j.molbiopara.2006.03.013

Yakhin, O.I., Lubyanov, A.A., Yakhin, I.A. and Brown, P.H. (2016) Biostimulants in plant science: a global perspective. *Frontiers in Plant Science* 7, 2049. DOI: 10.3389/fpls.2016.02049

Yan, G. and Baidoo, R. (2018) Current research status of *Heterodera glycines* resistance and its implication on soybean breeding. *Engineering* 4, 534–541. DOI: 10.1016/j.eng.2018.07.009

Yang, D., Chen, C., Liu, Q. and Jian, H. (2017) Comparative analysis of pre- and post-parasitic transcriptomes and mining pioneer effectors of *Heterodera avenae*. *Cell and Bioscience* 7, 11. DOI: 10.1186/s13578-017-0138-6

Yang, G., Zhou, B., Zhang, X., Zhang, Z., Wu, Y. *et al.* (2016) Effects of tomato root exudates on *Meloidogyne incognita*. *PLoS One* 11, e0154675. DOI: 10.1371/journal.pone.0154675

Yang, J., Wang, L., Ji, X., Feng, Y., Li, X. *et al.* (2011) Genomic and proteomic analyses of the fungus *Arthrobotrys oligospora* provide insights into nematode-trap formation. *PLoS Pathogens* 7, e1002179. DOI: 10.1371/journal.ppat.1002179

Ye, W. and Hunt, D.J. (2021) Measuring nematodes and preparation of figures. In: Perry, R.N., Hunt, D.J. and Subbotin, S.A. (eds) *Techniques for Work with Plant and Soil Nematodes*. CAB International, Wallingford, UK, pp. 132–151. DOI: 10.1079/9781786391759.0007

Yeates, G.W. (1971) Feeding types and feeding groups in plant and soil nematodes. *Pedobiologia* 11, 173–179.

Yeates, G.W. (1994) Modification and qualification of the nematode maturity index. *Pedobiologia* 38, 97–101.

Yeates, G.W. (2004) Ecological and behavioural adaptations. In: Gaugler, R. and Bilgrami, A.L. (eds) *Nematode Behaviour*. CAB International, Wallingford, UK, pp. 1–24.

Yeates, G.W. and Bongers, T. (1999) Nematode diversity in agroecosystems. *Agriculture, Ecosystems & Environment* 74, 113–135. DOI: 10.1016/S0167-8809(99)00033-X

Yeates, G.W., Bongers, T., de Goede, R.G.M., Freckman, D.W. and Georgieva, S.S. (1993) Feeding habits in soil nematode families and genera – an outline for soil ecologists. *Journal of Nematology* 25, 315–331.

Yergaliyev, T.M., Alexander-Shani, R., Dimerets, H., Pivonia, S., Bird, D. McK. *et al.* (2020) Bacterial community structure dynamics in *Meloidogyne incognita*-infected roots and its role in worm-microbiome interactions. *mSphere* 5, e00306-20. DOI: 10.1128/mSphere.00306-20

Young, L.D. and Hartwig, E.E. (1992) Problems and strategies associated with long-term use of nematode resistant cultivars. *Journal of Nematology* 24, 228–233.

Yu, L.Z., Song, S.Y., Yu, C., Jiao, B.B., Tian, Y.M. *et al.* (2020) Rapid detection of *Anguina agrostis* by loop-mediated isothermal amplification. *European Journal of Plant Pathology* 156, 819–825. DOI: 10.1007/s10658-020-01932-y

Yu, R., Luo, Y., Zhou, Q., Zhang, X., Wu, D. and Ren, L. (2021a) Early detection of pine wilt disease using deep learning algorithms and UAV-based multispectral imagery. *Forest Ecology and Management* 497, 119493. DOI: 10.1016/j.foreco.2021.119493.

Yu, Y., Zhang, Y.K., Manohar, M., Artyukhni, A.B., Kumari, A. *et al.* (2021b) Nematode signalling molecules are extensively metabolized by animals, plants, and microorganisms. *ACS Chemical Biology* 16, 1050-1058. DOI: 10.1021/acschembio.1c00217

Yushin, V.V., Gliznutsa, L.A. and Ryss, A. (2022) Ultrastructural detection of intracellular bacterial symbionts in the wood-inhabiting nematode *Bursaphelenchus mucronatus* (Nematoda: Aphelenchoididae). *Nematology* 24, 1073–1083. DOI: 10.1163/15685411-bja10192

Zasada, I.A. and Ferris, H. (2004) Nematode suppression with brassicaceous amendments: application based upon glucosinolate profiles. *Soil Biology and Biochemistry* 36, 1017–1024. DOI: 10.1016/j.soilbio.2003.12.014

Zasada, I. and Forge, T. (2021) Ectoparasitic nematodes: emerging challenges to wine grape production in the Pacific Northwest of North America. In: Sikora, R.A., Desaeger, J. and Molendijk, L. (eds) *Integrated Nematode Management: State-of-the-art and Visions for the Future*. CAB International, Wallingford, UK, pp. 192–198. DOI: 10.1079/9781789247541.0027

Zhan, L.P., Peng, D.L., Wang, X.L., Kong, L.-A., Peng, H. *et al.* (2018) Priming effect of root-applied silicon on the enhancement of induced resistance to the root-knot nematode *Meloidogyne graminicola* in rice. *BMC Plant Biology* 18, 50. DOI: 10.1186/s12870-018-1266-9

Zhang, L.Y., Zhang, Y.Y., Chen, R.G., Zhang, J.H., Wang, T.T. *et al.* (2010) Ectopic expression of the tomato *Mi-1* gene confers resistance to root knot nematodes in lettuce (*Lactuca sativa*). *Plant Molecular Biology Reporter* 28, 204–211. DOI: 10.1007/s11105-009-0143-y.

Zhang, P., Bonte, D., De Deyn, G.B. and Vandegehuchte, M.L. (2021) Leachates from plants recently infected by root-feeding nematodes cause increased biomass allocation to roots in neighbouring plants. *Scientific Reports* 11, 2347. DOI: 10.1038/s41598-021-82022-9

Zhang, S., Li, Q., Lü, Y., Zhang, X. and Liang, W. (2013) Contributions of soil biota to C sequestration varied with aggregate fractions under different tillage systems. *Soil Biology and Biochemistry* 62, 147–156. DOI: 10.1016/j.soilbio.2013.03.023

Zhang, W., Li, Y., Liu, Z., Li, D., Wen, X. *et al.* (2022) Transcriptome analysis of dauer moulting of a plant parasitic nematode, *Bursaphelenchus xylophilus* promoted by pine volatile ß-pinene. *Agronomy* 12, 2114. DOI: 10.3390/agronomy12092114.

Zhang, X., Ferris, H., Mitchell, J. and Liang, W. (2017) Ecosystem services of the soil food web after long-term application of agricultural management practices. *Soil Biology and Biochemistry* 111, 36–43. DOI: 10.1016/j.soilbio.2017.03.017

Zhang, Y., Lu, H. and Bargmann, C. (2005) Pathogenic bacteria induce aversive olfactory learning in *Caenorhabditis elegans*. *Nature* 438, 179–184. DOI: 10.1038/nature04216

Zhao, B.G., Wang, H.L., Han, S.F. and Han, Z.M. (2003) Distribution and pathogenicity of bacteria species carried by *Bursaphelenchus xylophilus* in China. *Nematology* 5, 899–906. DOI: 10.1163/156854103773040817

Zhao, M., Wickham, J.D., Zhao, L. and Sun, J. (2020) Major ascaroside pheromone component asc-C5 influences reproductive plasticity among isolates of the invasive species pinewood nematode. *Integrative Zoology* 16, 893–907. DOI: 10.1111/1749-4877.12512

Zhou, J., Wu, J., Huang, J., Sheng, X., Dou, X. and Lu, M. (2022a) A synthesis of soil nematode responses to global change factors. *Soil Biology and Biochemistry* 165, 108538. DOI: 10.1016/j.soilbio.2021.108538

Zhou, Q., Liu, Y., Wang, Z., Wang, H., Zhang, X. and Lu, Q. (2022b) Rapid on-site detection of the *Bursaphelenchus xylophilus* using Recombinase Polymerase Amplification combined with lateral flow dipstick that eliminates interference from primer-dependent artifacts. *Frontiers in Plant Science* 13, 856109. DOI: 10.3389/fpls.2022.856109

Zijlstra, C., Donkers-Venne, D.T.H.M. and Fargette, M. (2000) Identification of *Meloidogyne incognita, M. javanica* and *M. arenaria* using sequence characterised amplified region (SCAR) based PCR assays. *Nematology* 2, 847–853. DOI: 10.1163/156854100750112798

Glossary

Acquisitive strategy: ecological investment in rapid nutrient uptake and growth, resulting in higher competitive ability under nutrient enrichment.
Agroecosystem: the production, human and natural environment components of systems devoted to the production of food, feed or fibre.
Agroforestry: a mode of agriculture based on replicating natural succession in agricultural fields, exploring plant functional diversity and interactions through the combination of annual crops with woody perennials (trees and shrubs).
Alae: thickened wings of cuticle located laterally or sublaterally on the body.
Alginate: a polysaccharide obtained from seaweed, frequently used to formulate microbial biological control agents.
Allelochemical: volatile or non-volatile chemical mediating interspecific responses (e.g. nematodes responding to host **root diffusate**).
Amixis: see **Parthenogenesis**.
Amphid: anterior sensillum; paired lateral structures, rarely shifted dorsad, located post-labially or on the lip region; main anterior chemosensilla.
Amphidelphic: reproductive system with two uteri extending in opposite directions.
Amphimixis: sexual reproduction where male gametes (sperm cells) and female gametes (oocytes) fuse to form the zygote.
Anastomoses: cross-connections: (i) common in criconematid taxa where the body annuli do not form continuous rings; the junction between the dorsal and ventral part of the annuli usually results in a lateral zigzag line and (ii) the points of fusion between compatible fungal hyphae.
Ancestral varieties: wild or barely domesticated relative or parent material of cropped cultivars, including landraces, retaining genetic diversity and local adaptation.
Anhydrobiosis: a state of **dormancy** induced by desiccation.
Annuli: deep transverse grooves involving at least one of the two deeper main zones of the cuticle (median and basal zone) (cf. **striae**).
Antibiosis: inhibiting life.
Apomixis: mitotic **parthenogenesis**; the most common method of asexual reproduction, the only division that occurs is mitosis and the oocytes retain the diploid chromosome number (cf. **automixis**).
Appressorium: swelling of fungal hypha or germ tube that helps in attachment to, and penetration of, the host.
Areolation: occurs at the level of the lateral field when the transverse grooves (**annuli**) or transverse **striae** extend into the lateral field.
Ascarosides: glycolipid molecules constitutively expressed and tailored by nematodes according to physiological and environmental conditions, composed of an ascarylose sugar and a fatty acid chain.

Automixis: meiotic **parthenogenesis**; there is a first meiotic division in the oocytes that permits some genetic reorganization, even though the diploid number is restored by self-fertilization (cf. **apomixis**).

Bacillary bands: modified epidermal gland cells opening through a complex cuticular pore.

Bar code: small sequence fragment of DNA sequence used as a unique identifier to separate organisms into species.

Basal spiral fibre layers: helically arranged fibre layers (angle 54°44′ or more); they play a role in maintaining the internal turgor pressure.

Basal zone: innermost zone of the cuticle.

Biocontrol agent: parasitic, pathogenic, antagonistic or competitive organism that kills or damages another organism (e.g. the nematode host) and interferes with the establishment, function, growth, multiplication or survival of the host.

Biocontrol: see **Biological control**.

Biodynamic farming: a certifiable mode of agriculture aiming to close inputs and nutrient cycles at farm level, or local sources. Following the work of Rudolf Steiner, this holistic farming approach combines spiritual/mystical and ecological elements.

Biofumigation: use of organic amendments that release volatile breakdown products in soil that are toxic to pests and diseases.

Biological control: the total or partial destruction of pathogen and pest populations with the aid of other living organisms, excluding plant hosts and antagonistic plants.

Biological filter: an impedance to population growth or dispersal of environmental origin

Biotroph: an organism that obtains its nutrients only from another living organism.

Bioturbation: the physical movement of soil by organisms, resulting in the mixing of mineral and organic particles across horizons.

Black-box approach: experiment in which exact organisms or species may be ignored, but larger-scale overall effects are assessed.

Buccal cavity: stoma *sensu lato*, usually referring to cheilostome and pharyngostome (pharyngostome = gymnostome + stegostome)

Bursa: also referred to as caudal alae; wing-like subventral thickened body cuticle in the tail region of males.

Carbon farming: exploitation of farming and silviculture management to promote carbon sequestration.

Carbon sequestration: the fixation of carbon (C) in complex organic molecules

Carrying capacity: maximum population size limit in a given habitat; for plant-parasitic nematodes, this relates to host suitability restrictions to population increase.

Cash crop: the main cultivated species or cultivar in a given field and growing season, the cropping focus.

Caudalids: anal–lumbar commissures that link the pre-anal and lumbar ganglia.

Cellulases: enzymes that break β1–4 linkages between glucan sugar residues in cellulose.

Cephalic framework: sclerotized framework within the head.

Cephalic sensilla: papilliform, setiform or embedded (in parasitic nematodes) sensilla that form the third circlet of anterior sensilla; submedian in position (laterodorsal or lateroventral); derived from somatic sensilla.

Cheilorhabdia: sclerotized plates/rods in the cuticular lining of the cheilostome.

Cheilostome: stoma *sensu stricto* or lip cavity, with hexaradiate or triradiate symmetry, lined with body cuticle.

Chemical control: management strategies that utilize the application of synthetic or naturally derived nematicides.

Chlamydospore: type of resting spore with thick cell walls usually formed within fungal hyphae.

Chord: longitudinal running dorsal, lateral and ventral thickenings of the epidermis/hypodermis, containing the nuclei of the epidermal cells.

Chronosequence: an environmental series of conditions of different and determined age, employed as an alternative to long-term studies (e.g. 20-, 10- and 5-year plantations)

Circomyarian muscle cell: muscle cell in which the sarcoplasm is completely surrounded by contractile elements.

Coelomyarian muscle cell: spindle-shaped muscle cell laterally flattened so that the contractile elements are arranged not only along the epidermis but also along the sides of the flattened spindle; coelomyarian muscle cells bulge into the pseudocoel.

Conducive soil: a soil where multiplication of nematodes or other pathogens on a susceptible crop is favoured (i.e. the opposite of a **suppressive soil**).

Conidiophore: specialized fungal hypha on which conidia are produced.

Conidium (plural: conidia): any asexual fungal spore, formed on a conidiophore.

Conventional intensification: farming mode that relies on monocultures and external inputs (including mechanization, irrigation, chemical pesticides and fertilizers) to increased yields.

Corpus: (i) anterior part of pharynx, (ii) main dorsal part of gubernaculum.

Cortical zone: zone of the cuticle beneath the epicuticle.

Cover cropping: cultivation of plants for multiple functional roles, including preventing soil erosion, reducing pest, disease and weed pressure, and promoting fertility by retaining or increasing soil organic matter, nutrient and water contents.

Crop domestication: artificial, human-centred selection of plants for desirable traits.

Crop rotation: cultivation of different plant species, often botanically distinct or having differing nutrient needs and investment, in a given area, varying through growing season.

Cultural control: management strategies based on cultivation and cropping methods that lead to the reduction of pests.

Damage threshold: nematode population density above which crop growth and yield are suppressed (see also **economic threshold**).

Dauer: an alternative nematode developmental stage for surviving unfavourable conditions for long periods.

Decomposition: breakdown of organic molecules into simpler, mineral compounds

Deirids: cervical papillae or somatic sense organs in the form of protuberances that lack an opening, located at mid-lateral field.

Denticles: minute teeth-like structures.

Diapause: a physiological state of arrested growth and development, usually in response to unfavourable conditions, whereby development does not continue until specific requirements have been satisfied, even if favourable conditions return.

Didelphic: reproductive system with two uteri.

Differential host plant: host plant with the ability to distinguish between nematode pathotypes based on differential rates of nematode reproduction.

Diorchic: reproductive system with two testes.

Disease complex: interactive (often synergistic) assembly of pathogens feeding on the same host.

Distal: furthest from a specified centre; for example, from anterior end or from the outside (vulva, anus). By contrast, in *Xiphinema* distal is considered away from the vulva.

Dormancy: a suspension of growth and development, usually in response to unfavourable conditions, associated with a lowered metabolic rate; dormancy is usually subdivided into **quiescence** and **diapause.**

Ductus ejaculatorius: posterior part of *vas deferens* when differentiated and connected to the cloaca.

Economic threshold: nematode population density above which the extra cost of nematode control is exceeded by the increase in crop returns (see also **damage threshold**).

Ecosystem service: direct and indirect beneficial contributions of natural processes to humans.

Edaphoclimatic conditions: a combination of environmental (e.g. temperature, precipitation) and soil (e.g. structure, pH) characteristics.

Endoparasitic fungus: a fungus that penetrates its host and feeds from within it, and growth outside the host is limited.

Endophyte: a fungus or bacterium that lives inside a plant without causing disease.

Enrichment: a measure of nutrient availability.

Epicuticle: trilaminar outermost part of the body cuticle, first layer to be laid down during moulting.

Epidermal glands: various epidermal glands opening through pores or sometimes through special setae on the cuticle; unicellular and often associated with somatic receptors; most important epidermal glands are the ventral and caudal glands.

Epidermis: a thin cellular or acellular layer beneath the cuticle arranged into four longitudinal chords (dorsal, ventral and two lateral); secretes the cuticle, contains proteins associated with nutrient and ion transport, participates in osmotic and volume regulation in nematodes and is functionally important during embryogenesis.

Expansins: enzymes that break non-covalent linkages between cellulose chains.

Exploitative competition: indirect negative interaction between two organisms, including sharing the same limited food resource.

Expressed sequence tags (ESTs): sequences obtained from clones randomly selected from cDNA libraries.

Feeding sites: enlarged, metabolically active plant cells on which nematodes feed (see **giant cell** and **syncytium**).

Fenestration: the arrangement and form of the translucent cuticular areas around the vulva in cyst nematodes (circumfenestrate, bifenestrate, ambifenestrate).

Fitness: adequacy of a population to its environment, usually assessed by their number of offspring

Functional guild: a combination of trophic level and life strategy (colonizer-persister value)

Gene flow: incorporation of genes into the gene pool of one population from one or more other populations.

Genetic distance: measures of the degree of genetic differences between populations or species based on differences in allele frequencies.

Genital papillae: sensory organs located on the male tail or ventrally shortly anterior to the cloacal opening in male.

Genotype: precise genetic constitution of an organism.

Giant cell: large multinucleate cell formed by cycles of mitosis uncoupled from cytokinesis induced by some nematodes, including root-knot nematodes, in plant roots.

Gonad: ovary (female), testis (male).
Gonoduct: oviduct and uterus (female); *vas deferens* (male).
Green manure: fresh plant material ploughed into soil to decrease nematode infestations, either directly or indirectly through stimulation of the activity of the soil microbial community.
Green Revolution: a period of greatly increased crop (mainly grain) yields in the 1970s, resulting from intensification and technological advances in agronomy, including irrigation, mechanization, chemical inputs and high-yielding varieties.
Gubernaculum: sclerotized accessory guiding device for the spicules.
Harvest interval: period between application of control agent and crop harvest.
Hemizonid: a commissure that runs between the nerve ring (lateral ganglion) and the ventral nerve; it is located near the secretory–excretory pore, i.e. the ventral side of the body.
Hemizonion: a small commissure situated posterior to the hemizonid.
Hermaphroditism: a method of reproduction where both egg and sperm are produced by the same gonad in the same individual.
High-yielding varieties: crop cultivars obtained through plant improvement, that have high productivity in response to fertilization and irrigation.
Homology: correspondence between structures that is attributable to their evolutionary descent from a common ancestor.
Horizontal gene transfer: acquisition of genetic material from a source other than the parents.
Humus: stable, carbon-rich organic matter in soil.
Hybrid cultivar: a crop variety resulting from intra or interspecific crossing of two or more varieties/species.
Intercropping: cultivation of different plant species or cultivars in the same field, often in alternate lines.
Interference competition: direct negative interaction between two organisms, such as involving physical aggression, or defending territories.
Intestine: mesenteron or mid-gut; may be subdivided into (i) an anterior ventricular area; (ii) the mid-intestinal region; and (iii) a pre-rectum posteriorly.
Isozyme (or **Isoenzyme**): one of several forms of an enzyme, produced by different loci in the genome of an individual organism.
Kinesis: nematode movement resulting from undirected responses to changes in stimulus intensity.
Labial sensilla: anterior sensilla comprising two circlets, each with six sensilla (six inner labial sensilla and six outer labial sensilla) arranged in a hexaradiate pattern; mechano- and/or chemoreceptors.
Lateral lines: the borders of longitudinal ridges at the level of the lateral chords.
Lectin: a plant protein that binds to specific carbohydrates.
Longitudinal ridges (or **Lamellae**): usually restricted to the lateral region but may be present all around the body; they result from thickening of the body cuticle (e.g. of the median zone).
Macroecology: a study of large-scale, regional patterns and processes resulting from organism interactions with their environment.
Median zone: zone of the cuticle, located immediately beneath the cortical zone. In some nematodes, the median zone may be absent.
Mesocosm: an experimental unit representing a contained environment approximating natural conditions, usually established outdoors and larger than 1 l volume.
Metabolic footprint: carbon-based activity of nematodes in feeding, reproduction and respiration.

Metacorpus: posterior part of corpus (anterior part of pharynx), may be strongly muscular and provided with a valve.

Micro-arrays: glass slides bearing short stretches of oligonucleotide probes or cDNA clones used to measure expression levels of many thousands of genes simultaneously.

Microbiome: a community of microbiota, including bacteria, fungi, nematodes and other microscopic organisms

Microcosm: an experimental unit under controlled conditions representing a contained environment usually smaller than 1 l volume.

Microsatellite: short sequences of 1–6 base nucleotide repeats with highly variable length distributed widely throughout the genome.

Migration: movement of an individual or population in a direction oriented with respect to a stimulus field (can be accomplished by **taxis** or **kinesis**).

Minimum yield: the yield obtained irrespective of nematode population density.

Molecular probe: a molecule that is labelled and used to detect another molecule (target) through its specific affinity.

Monodelphic: reproductive system with one uterus.

Monorchic: reproductive system with a single, usually anterior, testis.

Mulch: a protective covering of various substances, usually organic, but can be inorganic (e.g. plastic) placed on the ground around plants to retain moisture, prevent frost, retard weeds, suppress disease and provide nutrients (organic).

Multiple sequence alignment: a matrix of sequence data, in which the individual columns represent homologous characters.

Mutualism: beneficial pairwise interaction between two symbiotic organisms.

Mycorrhiza: fungi that live in mutualistic symbiosis within or on the roots of a plant.

Nematicide: a general term used to describe a synthetic chemical that is used to control plant-parasitic nematodes.

Nematistat (or **Nematistatic**): a non-fumigant nematicide, which paralyses rather than kills plant-parasitic nematodes in the soil.

Nematophagous fungus: fungus that is able to obtain nutrition from nematodes by parasitizing them.

Ocelli: eyespots or photoreceptors; pigmented areas located laterally or dorsolaterally along, or partly inside, the pharynx.

Odontophore: supporting structure of the odontostyle, formed by a modification of the anterior pharyngeal region.

Odontostyle: hollow protrusible anterior part of a dorylaimid spear/stylet with a dorsal opening and supported by the odontophore.

Onchiostyle: stylet of Trichodoridae; a stylettiform dorsal tooth with a solid tip, consists of an onchium or tooth and an onchiophore or pharyngeal supporting structure.

Opisthodelphic: reproductive system where the uterus or uteri are directed posteriad.

Organic farming: a certifiable mode of agriculture managing soil fertility and crop protection through the use of organic, rather than chemical or synthetic, inputs.

Orthologous gene: a gene derived from the same copy of a gene in the most recent common ancestor.

Ovejector: common region of the two uteri and connected to the vagina.

Oviduct: part of gonoduct next to the ovary, uniform or differentiated into a narrow distal and a wider proximal part that may act as spermatheca.

Papilla (plural: papillae): sensillum, 1 μm or smaller in size.

Paralogous gene: a gene derived from different copies of a gene in the most recent common ancestor.

Parasite: an organism living within or on another organism (the host) from which it obtains food or other requirements, usually to the detriment of the host.

Parthenogenesis: asexual reproduction, or amixis, where males are not involved.

Partial resistance: any resistance that is less than 100% inhibition of nematode reproduction in comparison to a chosen standard.

Pathogen: a disease-causing organism.

Pathogenicity: relative ability of a nematode taxon or population to damage a given plant.

Pathotype: a group of individual nematodes with common gene(s) for (a)virulence and differing from gene or gene combinations found in other groups, which can be distinguished from others of the species by their pathogenicity on a specific host.

Pectate lyases: enzymes that break galacturon sugar polymers.

Perineal pattern: the cuticular lines and associated structures surrounding the vulval region of root-knot and cyst nematodes.

Pest: in relation to plant nematology, any species, strain or biotype of pathogenic agent injurious to plants or plant products.

Pharyngeal gland cells: large cells in the pharynx of nematodes that produce secretions thought to be important in host–parasite interactions.

Pharyngostome: triradially symmetrical specialized anterior part of the pharynx lined with pharyngeal cuticle (i.e. lacking the median zone) divided into a gymnostome, part not surrounded by arcade cells and a stegostome surrrounded by myoepitheel cells and muscle cells of the pharynx and further subdivided in pro-, meso-, meta- and telostegostome; pharyngostome together with cheilostome forms the buccal cavity.

Pharynx: (also called oesophagus) a complex muscular pumping organ with a triradiate lumen lined with cuticle, and unicellular glands; main part of stomodeum; variable in shape from simple cylinder to complex with differentiation in corpus (with or without a further subdivision into pro- and metacorpus), isthmus and postcorpus.

Phasmid: somatic sensillum, located laterally, usually on the tail, but may also be found in a pre-anal position.

Phenotype: observed morphological, physiological, biochemical, behavioural and other properties of an organism.

Pheromone: volatile and non-volatile chemical(s) mediating intraspecific responses (e.g. sex pheromones).

Phylogenetic tree: diagrammatic representation of the relatedness of a group of taxa.

Phylogeny: the evolutionary history of a group of taxa.

Physical control: management strategies that utilize physical means, such as heat and drying, to control pests.

Phytoalexins: antimicrobial compounds produced by plants in response to pathogen attack

Plant–soil feedback: the overall biological effect of organisms cultivated by plant roots on plant performance and distribution.

Platymyarian muscle cell: the whole contractile part of the muscle cell that is flat and broad and borders the epidermis.

Population dynamics: (i) changes in population size over time and space; and (ii) the forces that regulate populations over time and space.

Population: a group of individuals of the same species at a given time and space.

Postcorpus: posterior bulb of pharynx, with or without valve, glandular and muscular.

Post-vulvar (post-vulval) uterine sac: also called post-uterine sac, a rudimentary part of the reduced genital branch.

Primer: single-stranded synthetic oligonucleotide designed to hybridize to a known sequence of DNA as a starting point for DNA amplification.

Probe: see **Molecular probe.**

Procorpus (or **Precorpus**): anterior part of corpus (anterior part of pharynx).

Proctodaeum: see **Rectum.**

Prodelphic: reproductive system where the uterus or uteri are directed forward.

Proximal: closest to a certain centre that has to be specified.

Pseudogene: a non-functional member of a gene family.

Quarantine: official confinement of regulated articles, such as plant-parasitic nematodes, for observation and research or for further inspection, testing and/or treatment.

Quarantine pest: a pest of potential economic importance to the area endangered thereby and not yet present there, or present but not yet widely distributed and being officially controlled.

Quiescence: a state of arrested development in response to unfavourable conditions, which is reversible when favourable conditions return.

Race: a subspecies group of nematodes that infects a given set of plant varieties.

Rachis: a central protoplasmic core to which developing oocytes are connected by cytoplasmic bridges.

Radial filaments: filaments of condensed material present in the cortical zone of the cuticle.

Radial striae: either in cortical or basal zone of the body cuticle, consisting of longitudinal and transverse circumferential interwoven laminae, at high magnification appearing as rods separated by electron-light material.

Recruitment: selection of organisms to establish or renew a population

Rectum (or **Proctodaeum**): posterior ectodermal part connected to the intestine through a valve surrounded by a sphincter muscle; may possess associated rectal glands.

Resistance: the ability of a plant to inhibit reproduction of a given nematode relative to a susceptible plant that supports high levels of reproduction.

Resource partitioning: differentiation in food preferences or exploitation strategies among organisms

Restriction enzyme: an enzyme that cuts double-stranded DNA at specific short nucleotide sequences.

Rhizosphere: the space around the root that is influenced by the root and its diffusates.

RNAi (RNA interference): degradation of an RNA due to the presence of a double-stranded RNA molecule of the same, or very similar, sequence. A naturally occurring phenomenon used to protect against viruses and transposable elements that replicate via a double-stranded RNA intermediate, and also to control expression of endogenous genes.

Root diffusate: (also called root exudate and root leachate) wide range of mucilages and metabolities released by the plant into the rhizosphere that sustains 50 to 100 times the microorganism population found in the bulk soil. Contains chemicals involved in nematode hatching and root attraction.

Rugae: a series of cuticular ridges forming part of cuticular ornamentation.

Sampling: the collection of soil or plant tissue for the purpose of estimating a given attribute (e.g. the density of nematodes or crop damage) by examining a defined portion of the collected material.

Saprophyte: organism that obtains its nutrition from dead organic material.

Scutellum: a phasmid with a large plug, probably secreted by the glandular sheath cell.

Secretory–excretory pore: ventrally located pore, usually in the pharyngeal region, through which the secretory–excretory system opens to the exterior.

Semiochemicals: volatile and non-volatile chemicals that cause interactions between organisms (see **Allelochemical** and **Pheromone**).

Sensillum (plural: sensilla): sense organ consisting of: (i) a socket cell; (ii) a sheath cell; and (iii) a neural part of bipolar neurons whose dendrites end in dendritic processes or receptors.

Seta (plural: setae): sensillum, 2 μm in size or longer.

Setiform papillae: sensilla with a length intermediate between a papilla and a seta, i.e. 1.5 μm.

Sex pheromone: see **Pheromone**.

Sibling species: species that are difficult or impossible to distinguish by morphological characters.

Soil food web: the network of trophic relationships among soil-dwelling organisms in multiple feeding levels

Soil legacy: persistence along time of a belowground community or abiotic parameters as influenced by, e.g. a given crop or management practice.

Soil solarization: using solar energy to increase the temperature of soil, usually covered with a polythene sheet, to levels that are lethal to nematodes and some soil microbes.

Soil structure: arrangement of organic, mineral, water, air and living organism in the soil matrix

Soil texture: mineral composition of soil considering the percentual contributions of different grain sizes: sand, silt and clay.

Somatic muscle cell: mainly with spindle-like form, but with a complex, irregular, shape in those cells with a special function, e.g. in area of vulva; consists of three parts: (i) a contractile portion of the cell towards the epidermis; (ii) a non-contractile portion towards the body cavity; and (iii) an arm or process extending from the non-contractile portion of the cell towards the dorsal or ventral nerve; muscle cells anterior to the nerve ring send processes directly into the nerve ring.

Somatic muscles: body wall muscles arranged in a single layer of obliquely orientated and longitudinally aligned somatic muscle cells, divided into four sectors by the four epidermal chords. Meromyarian musculature: few (up to 5–6) rows of muscle cells are present per quadrant; common in small species. Polymyarian musculature: more than six rows of muscle cells per quadrant; spindle-shaped muscle cells laterally flattened.

Somatic nerves: dorsal, lateral and ventral longitudinal running nerves, in between four submedian (laterodorsal, lateroventral) longitudinal nerves (in literature, the somatic nerves may be referred to as cords).

Spermatheca: an enlarged portion either of the oviduct or of the uterus of the female genital system, used for storage of sperm.

Sphincter muscle: present at different sites (e.g. around the vagina).

Spicular pouch: formed from the spicular primordium, specialized cells of the dorsal wall of the cloaca.

Spicule: male copulatory apparatus, either simple or differentiated into a proximal head (capitulum), a shaft (calomus) and a blade that may be with or without a velum (a ventral membrane-like structure); usually paired.

Stomatostyle: hollow, protrusible needle-like structure of tylenchs; consists of three parts: (i) a cone with a ventral aperture; (ii) a shaft; and (iii) a posterior part that swells to form three basal knobs.

Strain: a subspecies group of nematodes, bacteria and fungi distinguishable from the rest of the species by a heritable characteristic that the individuals in the group have in common (biotype).

Striae: shallow or superficial transverse grooves not penetrating to the deeper layers of the body cuticle.

Structure (food web): a measure of food web length and complexity.

Struts: column-like supporting elements in the median zone of the body cuticle.

Stylet: see **Stomatostyle**.

Supplements: mainly pre-cloacal but also post-cloacal genital sensilla of different shape (e.g. papilliform or tubuliform), usually mid-ventral in position.

Suppressive soil: soil where multiplication of plant-parasitic nematodes on a susceptible crop is less than that normally observed on the same cultivar in another soil, in similar abiotic conditions, due to the presence of natural enemies to the nematodes (i.e. the opposite of a **conducive soil**).

Surface coat: a stage-specific coating covering the cuticle.

Susceptibility: the counterpart of resistance; a measure of the ability of a plant to support reproduction of a nematode species.

Sustainability: the successful management of resources for agriculture to satisfy changing human needs, while maintaining or enhancing the natural resource base and avoiding environmental degradation.

Syncytium: large multinucleate cell produced by cell wall breakdown and fusion of protoplasts that are induced by nematodes, including cyst nematodes, in plants.

Syntropic farming: regenerative agriculture aiming to promote balance (reduce entropy) in agroecosystems through a holistic form of replicating plant assemblage structures of natural ecosystems. A farming mode related to agroforestry and popularized by Ernst Götsch.

Systematics: the scientific study of the diversity of organisms and their interrelationships (cf. **taxonomy**).

Systemically acquired resistance (SAR): increased resistance in a remote part of a plant due to detection of a pathogen in another region of the plant.

Taxis: locomotion resulting from orienting the body to the stimulus direction.

Taxon (plural: taxa): taxonomic category from subspecies to Kingdom.

Taxonomy: the theory and practice of identifying, naming and classifying organisms (cf. **systematics**).

Tillage: mechanical manipulation of soil, e.g. by ploughing or hoeing, with or without inversion of soil layers

Tolerance limit: nematode population density below which there is no reduction in plant growth and, therefore, yield.

Tolerance: relative ability of a plant to sustain growth and yield when parasitized by a nematode; not related to resistance

Topsoil: the uppermost and most biologically active layer of soil, combining the surface litter and organic matter-rich layer and the intermixed mineral and organic layer of horizon A.

Transcription factors: proteins that activate expression of genes through interaction with promoters.

Transformant: modified organism obtained by changing (transforming) a piece of the genetic code of an organism.

Transverse striae: shallow transverse grooves restricted to the cortical zone.

Trapping fungus: fungus that captures its nematode prey in traps from specialized hyphae and/or adhesive structures.

Trophic level: a grouping of organisms classified according to their feeding (consumers) or generation of own food (producers); also considers feeding proximity to energy source (e.g. primary, secondary or higher-level consumers)

Uterus: part of gonoduct connected to the vagina; either a simple tube but usually more complex and subdivided into a glandular part close to the oviduct, a muscular median part and a wider, or tubular, part towards the vagina.

Vagina: a canal lined with body cuticle that connects the gonoduct(s) with the vulva; may be divided into three parts: (i) *pars distalis* (distal section); (ii) *pars refringens* (intermediate section); and (iii) *pars proximalis* (proximal section).

Vas deferens: male gonoduct, uniform or subdivided.

Vector: an organism capable of carrying a plant-parasitic nematode from one host to another. Also, nematodes may act as carriers (vectors) for microorganisms (viruses, bacteria and fungi).

Vesicular seminalis: a part of each testis or the anterior part of the *vas deferens*.

Virulence: (i) a measure of the ability of a nematode to cause damage, with a highly virulent isolate causing more damage than a weakly virulent isolate and (ii) a measure of the ability of a nematode to reproduce on a plant, in particular when comparing populations of the same species for their ability to reproduce on resistant plants.

Zoospore: infective spore produced by fungal-like organisms (oomycetes) that bears flagella(e) and swims in water.

Index

Page numbers in **bold** refer to tables, figures and boxes.

www.ingramcontent.com/pod-product-compliance
Lightning Source LLC
Chambersburg PA
CBHW040136200326
41458CB00025B/6281
* 9 7 8 1 8 0 0 6 2 2 4 2 5 *